Methods in Enzymology

Volume 113

GLUTAMATE, GLUTAMINE, GLUTATHIONE, AND RELATED COMPOUNDS

METHODS IN ENZYMOLOGY

EDITORS-IN-CHIEF

Sidney P. Colowick   Nathan O. Kaplan

Methods in Enzymology

Volume 113

Glutamate, Glutamine, Glutathione, and Related Compounds

EDITED BY

Alton Meister

DEPARTMENT OF BIOCHEMISTRY
CORNELL UNIVERSITY MEDICAL COLLEGE
NEW YORK, NEW YORK

1985

ACADEMIC PRESS, INC.

(Harcourt Brace Jovanovich, Publishers)

Orlando San Diego New York London
Toronto Montreal Sydney Tokyo

QP601
C733
vol. 113

COPYRIGHT © 1985 BY ACADEMIC PRESS, INC.
ALL RIGHTS RESERVED.
NO PART OF THIS PUBLICATION MAY BE REPRODUCED OR
TRANSMITTED IN ANY FORM OR BY ANY MEANS, ELECTRONIC
OR MECHANICAL, INCLUDING PHOTOCOPY, RECORDING, OR
ANY INFORMATION STORAGE AND RETRIEVAL SYSTEM, WITHOUT
PERMISSION IN WRITING FROM THE PUBLISHER.

ACADEMIC PRESS, INC.
Orlando, Florida 32887

United Kingdom Edition published by
ACADEMIC PRESS INC. (LONDON) LTD.
24–28 Oval Road, London NW1 7DX

LIBRARY OF CONGRESS CATALOG CARD NUMBER: 54-9110

ISBN 0–12–182013–0

PRINTED IN THE UNITED STATES OF AMERICA

85 86 87 88 9 8 7 6 5 4 3 2 1

Table of Contents

Section I. Glutamate

JUN 20 1986

Section II. Glutamine

JUN 20 1986

Section III. Glutathione

Section IV. Aspartate and Asparagine

Section V. α-Aminoadipate

Contributors to Volume 113

Article numbers are in parentheses following the names of contributors.
Affiliations listed are current.

WILLIAM ABBOTT (73), *Department of Biochemistry, Cornell University Medical College, New York, New York 10021*

PER ÅLIN (63, 64), *Department of Biochemistry, Arrhenius Laboratory, University of Stockholm, S-106 91 Stockholm, Sweden*

R. DONALD ALLISON (51), *Department of Biochemistry, Cornell University Medical College, New York, New York 10021*

MARY E. ANDERSON (70, 71), *Department of Biochemistry, Cornell University Medical College, New York, New York 10021*

YASUHIKO ASADA (18), *Department of Food Science, Kagawa University, Miki-cho, Kagawa 761-07, Japan*

H. A. BARKER (23), *Department of Biochemistry, University of California, Berkeley, California 94720*

BRIAN BOETTCHER (55), *Chevron Chemical Co., Richmond, California 94804*

JOHN S. BRABSON (81), *Department of Physical Science, Mills College, Oakland, California 94613*

RICHARD J. BRIDGES (73), *Department of Psychobiology, University of California, Irvine, California 92717*

INGER CARLBERG (59), *Department of Biochemistry, Arrhenius Laboratory, University of Stockholm, S-106 91 Stockholm, Sweden*

P. BOON CHOCK (29), *Laboratory of Biochemistry, National Heart, Lung, and Blood Institute, National Institutes of Health, Bethesda, Maryland 20205*

SOO IL CHUNG (46), *National Institute of Dental Research, National Institutes of Health, Bethesda, Maryland 20205*

GEORGES N. COHEN (76, 77), *Unité de Biochimie Cellulaire, Département de Biochimie et Génétique Moléculaire, Institut Pasteur, 75724 Paris, Cedex 15, France*

ARTHUR J. L. COOPER (10, 11, 12, 13, 14, 15, 16, 43, 44, 45, 78), *Department of Neurology, Cornell University Medical College, New York, New York 10021*

LINDA D'ARI (26), *Department of Biochemistry, University of California, Berkeley, California 94720*

LARRY DENNER (1), *Department of Cell Biology, Baylor College of Medicine, Houston, Texas 77030*

JACK E. DIXON (69), *Department of Biochemistry, Purdue University, West Lafayette, Indiana 47907*

MICHAEL W. DUFFEL (66), *Division of Medicinal Chemistry, College of Pharmacy, University of Iowa, Iowa City, Iowa 52242*

THOMAS E. DUFFY (45), *Department of Neurology, Cornell University Medical College, New York, New York 10021*

VÉRONIQUE ERIKSSON (67), *Department of Biochemistry, Arrhenius Laboratory, University of Stockholm, S-106 91 Stockholm, Sweden*

HARVEY F. FISHER (3), *Department of Biochemistry, Kansas University Medical Center, Laboratory of Molecular Biochemistry, VA Medical Center, Kansas City, Missouri 64128*

J. E. FOLK (46), *National Institute of Dental Research, National Institutes of Health, Bethesda, Maryland 20205*

MARGARET L. FONDA (2), *Department of Biochemistry, University of Louisville School of Medicine, Louisville, Kentucky 40292*

PAUL M. GALLOP (24), *Department of Biological Chemistry, Harvard Medical School and Harvard School of Dental Medicine, Department of Oral and Pathophysiology, Harvard School of Dental Medicine, Laboratory of Human Biochemistry, Children's Hospital Medical Center, Boston, Massachusetts 02115*

Owen W. Griffith (53, 56), *Department of Biochemistry, Cornell University Medical College, New York, New York 10021*

Caren M. Gundberg (24), *Department of Oral Biology and Pathophysiology, Laboratory of Human Biochemistry, Children's Hospital Medical Center, Boston, Massachusetts 02115*

Claes Guthenberg (62, 64), *Department of Biochemistry, Arrhenius Laboratory, University of Stockholm, S-106 91 Stockholm, Sweden*

Richard A. Hartline (82, 83), *Department of Chemistry, Indiana University of Pennsylvania, Indiana, Pennsylvania 15705*

Peter V. Hauschka (24), *Department of Oral Biology and Pathophysiology, Laboratory of Human Biochemistry, Children's Hospital Medical Center, Boston, Massachusetts 02115*

Jerry G. Henslee (22), *Abbott Laboratories, Abbott Park, North Chicago, Illinois 60085*

Louis B. Hersh (5), *Department of Biochemistry, The University of Texas, Health Science Center at Dallas, Dallas, Texas 75235*

Patricia Hoben (8), *Department of Biochemistry and Biophysics, University of California, San Francisco, California 94143*

John S. Holcenberg (31), *Department of Pediatrics and Biochemistry, University of Southern California, Childrens Hospital of Los Angeles, Los Angeles, California 90054*

Arne Holmgren (68), *Department of Physiological Chemistry, Karolinska Institute, S-104 01, Stockholm 60, Sweden*

William B. Jakoby (61, 65, 66), *Laboratory of Biochemistry and Metabolism, NIADDK, National Institutes of Health, Bethesda, Maryland 20205*

Helgi Jensson (63), *Department of Biochemistry, Arrhenius Laboratory, University of Stockholm, S-106 91 Stockholm, Sweden*

Mary Ellen Jones (22), *Department of Biochemistry, University of North Carolina at Chapel Hill, Chapel Hill, North Carolina 27514*

Wanda M. Jones (21), *The Rockefeller University, New York, New York 10021*

Hiroyuki Kagamiyama (17), *Department of Medical Chemistry, Osaka Medical College, Takatsuki, Osaka 569, Japan*

Deborah S. Kaseman (41), *Department of Biochemistry, Cornell University Medical College, New York, New York 10021*

Daniel Kern (6), *Laboratoire de Biochimie, Institut de Biologie Moléculaire et Cellulaire du C.N.R.S., 67084 Strasbourg Cedex, France*

James J. Kramer (22), *Department of Biochemistry, University of North Carolina at Chapel Hill, Chapel Hill, North Carolina 27514*

Elling Kvamme (30), *Neurochemical Laboratory, Preclinical Medicine, University of Oslo, Oslo 3, Norway*

Jacques Lapointe (6, 7), *Département de Biochimie, Faculté des Sciences et de Génie, Université Laval, Québec G1K 7P4, Canada*

Kerstin Larson (67), *Department of Biochemistry, Arrhenius Laboratory, University of Stockholm, S-106 91 Stockholm, Sweden*

Sylvain Levasseur (6), *Département de Biochimie, Faculté des Sciences et de Génie, Université Laval, Québec G1K 7P4, Canada*

Luyuan Li (54), *Department of Biochemistry, Cornell University Medical College, New York, New York 10021*

Jane B. Lian (24), *Department of Biological Chemistry, Harvard Medical School and Laboratory of Skeletal Diseases, Children's Hospital Medical Center, Boston, Massachusetts 02115*

Chin-Tarng Lin (1), *Department of Physiology, The Milton S. Hershey Medical Center, Pennsylvania State University, Hershey, Pennsylvania 17033*

Bengt Mannervik (59, 60, 62, 63, 64, 67),

Department of Biochemistry, Arrhenius Laboratory, University of Stockholm, S-106 91 Stockholm, Sweden

JAMES M. MANNING (21), *The Rockefeller University, New York, New York 10021*

MICHAEL R. MAURIZI (81), *Laboratory of Molecular Biology, National Cancer Institute, National Institutes of Health, Bethesda, Maryland 20205*

ALTON MEISTER (9, 27, 41, 42, 43, 44, 45, 47, 48, 49, 50, 52, 53, 54, 55, 57, 71, 73, 74, 78), *Department of Biochemistry, Cornell University Medical College, New York, New York 10021*

JOHN E. MORIN (69), *Wyeth Laboratories, Philadelphia, Pennsylvania*

KOUSAKU MURATA (73), *The Research Institute for Food Science, Kyoto University, Kyoto 611, Japan*

MITSUHIRO NOZAKI (17), *Department of Biochemistry, Shiga University of Medical Science, Seta, Ohtsu, Shiga 520-21, Japan*

SUSAN G. POWERS-LEE (4), *Department of Biology, Northeastern University, Boston, Massachusetts 02115*

MARIO PROULX (7), *Département de Biochimie, Faculté des Sciences et de Génie, Université Laval, Québec G1K 7P4, Canada*

JESSE C. RABINOWITZ (26), *Department of Biochemistry, University of California, Berkeley, California 94720*

SUE GOO RHEE (29), *Laboratory of Biochemistry, National Heart, Lung, and Blood Institute, National Institutes of Health, Bethesda, Maryland 20205*

W. BRUCE ROWE (28), *Corporate Research, Travenol Laboratories, Inc., Round Lake, Illinois 60073*

ANDREW P. SEDDON (54), *Department of Biochemistry, Cornell University Medical College, New York, New York 10021*

GAIL FOURÉ SEELIG (47, 48), *Schering Corporation, Bloomfield, New Jersey 07003*

KENJI SODA (17, 18, 19, 20), *Laboratory of Microbial Biochemistry, Institute for Chemical Research, Kyoto University, Uji, Kyoto 611, Japan*

DIETER SÖLL (8), *Department of Molecular Biophysics and Biochemistry, Yale University, New Haven, Connecticut 06511*

GUOXING SONG (1), *Department of Biophysics, Institute of Basic Medical Sciences, Chinese Academy of Medical Sciences, Beijing, China*

THOMAS S. SOPER (21), *Oak Ridge National Laboratory, Oak Ridge, Tennessee 37830*

EARL R. STADTMAN (29), *Laboratory of Biochemistry, National Heart, Lung, and Blood Institute, National Institutes of Health, Bethesda, Maryland 20205*

JAMES L. STEVENS (65), *Molecular Pharmacology Laboratory, Center for Drugs and Biologics, Food and Drug Administration, Bethesda, Maryland 20205*

GERD SVENNEBY (30), *Neurochemical Laboratory, Preclinical Medicine, University of Oslo, Oslo 3, Norway*

ROBERT L. SWITZER (81), *Department of Biochemistry, University of Illinois, Urbana, Illinois 61801*

KATSUYUKI TANIZAWA (18, 19), *Laboratory of Microbial Biochemistry, Institute for Chemical Research, Kyoto University, Uji, Kyoto 611, Japan*

SURESH S. TATE (50, 58), *Department of Biochemistry, Cornell University Medical College, New York, New York 10021*

MASANOBU TOKUSHIGE (80), *Department of Chemistry, Faculty of Science, Kyoto University, Kyoto 606, Japan*

INGEBORG AA. TORGNER (30), *Neurochemical Laboratory, Preclinical Medicine, University of Oslo, Oslo 3, Norway*

SEIZEN TOYAMA (20), *Department of Agricultural Chemistry, University of the Ryukyus, Nishihara, Okinawa 903-01, Japan*

FREDERIC A. TROY, II (25), *Department of Biological Chemistry, University of California School of Medicine, Davis, California 95616*

HIROSHI UENO (21), *The Rockefeller University, New York, New York 10021*

CHRISTER VON BAHR (62), *Department of Biochemistry, Arrhenius Laboratory, University of Stockholm, S-106 91 Stockholm, Sweden*

YASUO WAKABAYASHI (22), *Department of Biochemistry, Kyoto Prefectural University of Medicine, Kamikyo-Ku, Kyoto 602, Japan*

MARGARETA WARHOLM (62), *Department of Biochemistry, Arrhenius Laboratory, University of Stockholm, S-106 91 Stockholm, Sweden*

DANIEL WELLNER (72), *Department of Biochemistry, Cornell University Medical College, New York, New York 10021*

JOANNE M. WILLIAMSON (53, 75), *Department of Fermentation Microbiology, Merck Sharp and Dohme Research Laboratories, Rahway, New Jersey 07065*

JOHN C. WRISTON, JR. (79), *Department of Chemistry, University of Delaware, Newark, Delaware 19711*

JANG-YEN WU (1), *Department of Physiology, The Milton S. Hershey Medical Center, Pennsylvania State University, Hershey, Pennsylvania 17033*

TING-SENG XU (9, 57), *Shanghai Institute of Biochemistry, Chinese Academy of Sciences, Shanghai, China*

TOSHIHARU YAGI (17), *Department of Agricultural Chemistry, Kochi University, Nankoku, Kochi 783, Japan*

KAZUO YONAHA (20), *Department of Agricultural Chemistry, University of the Ryukyus, Nishihara, Okinawa 903-01, Japan*

TOHRU YOSHIMURA (19), *Laboratory of Microbial Biochemistry, Institute for Chemical Research, Kyoto University, Uji, Kyoto 611, Japan*

HOWARD ZALKIN (32, 33, 34, 35, 36, 37, 38, 39, 40), *Department of Biochemistry, Purdue University, West Lafayette, Indiana 47907*

Preface

Although chapters on glutamate, glutamine, glutathione, aspartate, asparagine, and α-aminoadipate have appeared in earlier volumes of this series, it seems appropriate, at this time, to bring out a separate volume devoted to these structurally related compounds. Progress in this area of biochemistry has been remarkably rapid, and the metabolic and physiological importance of these compounds places them well beyond the traditional areas of enzymology and biochemistry. For example, glutamate is now known to be of major significance as a neurotransmitter, and glutathione has become a compound of importance to those interested in cellular protection, radiation research, toxicology, pharmacology, endocrinology, agriculture, and other fields. As has always been true in biochemistry, progress depends crucially on reliable methods of isolation and characterization Although the chapters collected in this volume reflect the present state of development of this area, they also indicate some of the deficiencies in our knowledge and some of the growing points of current research.

I am truly grateful to Sidney Colowick and Nathan Kaplan, who not only initiated the planning of this volume, but who also gave essential advice during its preparation. My colleagues and I are deeply saddened by the recent passing of Sidney Colowick, whose career and accomplishments are reviewed in this volume by Nathan Kaplan. Sidney Colowick scrutinized the plans for this volume in great detail and made many useful suggestions. He was indeed of inestimable value to me in its preparation. Sidney Colowick was not only a distinguished scientist, but he contributed freely and extensively of his own time and energy to the efforts of others. I wish to record my own deep sense of loss at the passing of a generous friend and colleague, one who will surely be missed by many.

I wish to thank the contributors to this volume, including those who submitted their manuscripts before the deadline and those who eventually did so, for taking time from their research to prepare this material for publication. It is a pleasure to thank the staff of Academic Press for excellent assistance during all stages of the publication process. I also wish to acknowledge the efforts of Mrs. Susan D. Truog and Mrs. Edith H. Perryman for invaluable assistance in the preparation and collection of the manuscripts.

Alton Meister

SIDNEY P. COLOWICK
January 12, 1916–January 9, 1985

Sidney P. Colowick

It is difficult to write about a dear colleague and friend of more than forty years. There is so much to say, but words cannot easily express one's feelings. As I write this memorial, I recognize that my objectivity is compromised by my emotions.

Sidney Colowick was an unusual individual. Not only was he a brilliant and creative scientist, but he was also a compassionate and warm human being. Sidney was born and raised in St. Louis; he attended Washington University and obtained a degree in chemical engineering in 1936 at the age of 20. Jobs were difficult to obtain in those depressed times. He took a Civil Service test and passed an examination which qualified him to be a shrimp inspector. He was offered a position as an inspector in New Orleans, but his family thought him too young to settle in such a far away city. He was told of a laboratory opening in a biochemistry laboratory at the Washington University School of Medicine. After an interview with Carl Cori, he was given the position on a four-week trial basis. He was then given the position permanently even though in his first week he dropped a desiccator containing valuable compounds just outside of Carl Cori's office. He was associated with Carl and Gerty Cori for approximately ten years. At the age of 21, he published his first paper as coauthor with the Coris on the classical studies of glucose 1-phosphate. I take the liberty of listing his first few papers to illustrate the great significance of his early work.[1–4]

With his characteristic remarkable wit Sidney used to refer to himself as the "meat" in the "Cori Sandwich." The potential "shrimp inspector," who had known nothing about biochemistry or physiology, had within a few years established himself as one of the most promising young people in the emerging field of biochemistry. The Coris had great confidence in Sidney, and in 1938, at the age of 22, he published, as sole author, a paper in the *Journal of Biological Chemistry* on the synthesis of mannose 1-phosphate and galactose 1-phosphate.

Sidney's ability was appreciated by the Coris, and he became Carl's first graduate student. Gerty and Carl believed that the way in which the

[1] Cori, C. F., Colowick, S. P., and Cori, G. T. (1937). The Isolation and Synthesis of Glucose-1-Phosphoric Acid. *J. Biol. Chem. 121,* 465.

[2] Cori, G. T., Colowick, S. P., and Cori, C. F. (1938). The Formation of Glucose-1-Phosphoric Acid in Extracts of Mammalian Tissues and of Yeast. *J. Biol. Chem. 123,* 375.

[3] Cori, G. T., Colowick, S. P., and Cori, C. F. (1938). The Action of Nucleotides in the Disruptive Phosphorylation of Glycogen. *J. Biol. Chem. 123,* 381.

[4] Cori, G. T., Colowick, S. P., and Cori, C. F. (1938). The Enzymatic Conversion of Glucose-1-Phosphoric Ester to 6-Ester in Tissue Extracts. *J. Biol. Chem. 124,* 543.

remarkable advances in biochemical technology could be introduced into their laboratories was to have talented individuals who had been exposed to recent developments spend time in St. Louis. Hence, eminent scientists such as Arda Green, Gerhard Schmidt, Severo Ochoa, Herman Kalckar, and Luis Leloir contributed to the growth and importance of the Cori laboratory. Sidney was a beneficiary of these unusual circumstances. He was the only graduate student in a laboratory filled with brilliant investigators. His education was greatly enriched by this exposure, and his association with these unusual individuals promoted both his skills and creativity. They became lifelong friends of Sidney and his wife Maryda.

Herman Kalckar and Sidney became interested in hexokinase. Meyerhof had been researching this enzyme, but there were some mysterious characteristics of the enzyme which Meyerhof had difficulty understanding. Sidney and Herman initiated an investigation of this enzyme, and during the course of these studies they discovered myokinase, which is now known as adenyl kinase. The discovery of this enzyme proved to be a key factor in the understanding of transphosphorylation reactions in both yeast and animal cells.

Earl Sutherland who was a medical student at Washington University became interested in basic research and joined the Cori laboratory. He and Sidney became close friends and worked together on several problems. They were principally interested in the conversion of glucose to polysaccharides, and published a classical paper in 1942 on the formation of glycogen from glucose utilizing purified enzymes. This work which Sutherland was doing with the Coris and Sidney led to Earl dissecting the phosphorylase system which eventually led to the discovery of cyclic AMP. Sidney, along with Milton Slein and Lou Berger, carried out pioneering studies on yeast hexokinase. These led to crystallization of the enzyme in 1961 by Sidney in collaboration with K. A. Trayser. The crystalline yeast hexokinase was studied in great detail both with respect to its catalytic and physical properties.

The years in the Cori laboratory were most fruitful and pioneering in the "new" biochemistry. After ten years in the Cori laboratory, Sidney joined the staff of the Public Health Research Institute of the City of New York, replacing Herman Kalckar who was returning to Denmark. He remained at the Institute from 1946 to 1948. This was an unhappy period for Sidney, although it demonstrated Sidney's unusual character. He spent the years attempting to repeat the work of a colleague whose results were questionable. Sidney was not remorseful about the inability to reproduce the findings, and showed no open hostility to the individual. I have never met anyone in my lifetime who almost destroyed his own career and health to show the best in another individual.

I first met Sidney in St. Louis in 1944, and from that time on our lives were intertwined. Sidney moved with Otto Bessey in 1948 to the Department of Biochemistry at the University of Illinois Medical School in Chicago. He persuaded Bessey to offer me a position. I arrived in Chicago six months after Sidney did. Giulio Cantoni was also a member of the Department. Sidney had already made arrangements for Earl Sutherland to join the faculty a few months after my arrival. When I arrived Sidney was in the hospital with an ulcer. He had been given the responsibility of preparing a new laboratory syllabus. He wrote the syllabus all by himself, keeping just one experiment ahead of the class of over 300 students. Bessey assigned me to the syllabus task after Sidney became ill. I must have worked twenty hours a day on it. In two weeks I was so exhausted that it was difficult for me to believe that Sidney was able to carry out this task for three months. Two weeks after my arrival at the University of Illinois, Sidney was released from the hospital, and we then worked together to complete the syllabus. It seemed to me that working with Sidney was a "piece of cake" compared to the experience that I had as the only author.

The time spent in Chicago was both unhappy and unproductive for Sidney and me. It became apparent shortly after my arrival that Sidney and I had to leave, and a year after my arrival we both left.

Sidney and I were indeed fortunate to obtain positions at Johns Hopkins in the new McCollum-Pratt Institute headed by Bill McElroy. Ironically we both had been approached separately to be Director of the Institute. The change from Chicago to Baltimore was delightful. McElroy was a very effective leader and constructed a climate which was favorable and stimulating for research. Sidney and I collaborated on problems of pyridine nucleotides and metabolism. Although we worked together on several problems, we did have projects which we carried out separately. Our graduate and postdoctoral students were very close colleagues. As a result, the period spent at Hopkins was productive, exciting, pleasant, and rewarding. There was a great deal of interaction at the McCollum-Pratt Institute with other scientists in Baltimore as well as with the group of young biochemists at the National Institutes of Health which included Arthur Kornberg, Bernie Horecker, Herb Tabor, Al Meister, Leon Heppel, Chris Anfinsen, and Earl and Terry Stadtman. The groups were close, and Sidney played an important role as an intellectual leader.

During the period at Hopkins, Sidney made many noticeable contributions. Sidney, in collaboration with Maynard Pullman and Tony San Pietro, established the correct structure of NADH—that the hydrogen was in the para position and not in the position that Warburg and Karrer had postulated almost twenty years previously. Sidney summarized a large part of the work at Hopkins in an article in a book which was dedicated to me.[5]

I left Hopkins in 1957. Sidney moved to Vanderbilt in 1959 as American Cancer Society Professor in the Department of Microbiology. At Vanderbilt, Sidney focused his work on hexokinase, and made many ingenious and notable contributions on hexose transport in cultured animal cells. His latest paper appeared in a March 1985 issue of *PNAS* in collaboration with his last student Wendelyn Inman.[6] This paper clearly showed that growth factors are involved in the uptake of glucose. Sidney's interest in glucose metabolism began with his research in Cori's laboratory and remained a lifelong interest. Sadly, Sidney died roughly two months after his mentor, Carl Cori.

It is my strong belief that Sidney was the foremost representative of American-born biochemists who made modern biochemistry an "American discipline." His leadership qualities were admired and respected by all who knew him.

Sidney played an unusual role at Vanderbilt. His presence attracted Earl Sutherland. He gave much of himself to help develop the many talents that were at Vanderbilt so that this institution is now one of the world's leading institutions in the field of regulatory biology. There is no doubt that Sidney and Rollo Park were the major forces that transformed Vanderbilt into a major center of excellence. What Sidney meant to Vanderbilt and to the academic world was best expressed by these elegant words of Alexander Heard, Chancellor Emeritus of Vanderbilt, at a memorial service for Sidney on January 15, 1985.

> A university's distinction is found first of all and most of all in the intellectual merit of the members of its faculty. In our century, in our country, the American university has become a many splendored instrument of our culture, called upon and reaching out to serve humankind through ancient and novel means, toward ancient and novel ends. The heart of the university in western civilization is its duty to inquire and discover, and to interpret and communicate a useful harvest. In these central missions of the most influential institution of the twentieth century, the university, Sidney Paul Colowick excelled. He was a person of science, of the intellect, of the university, of the eternal human search to know and understand. He helped create Vanderbilt by doing most what a university is created to do.

[5] Colowick, S. P. (1982). Origins of Pyridine Nucleotide Research at the McCollum-Pratt Institute in the 50's. *In* "Pyridine Nucleotide Coenzymes" (J. Everse and K.-S. You, eds.). Academic Press, New York

[6] Inman, W. H., and Colowick, S. P. (1985). *Proc. Natl. Acad. Sci. U.S.A. 82,* 1346–1349.

I know of no one who was as helpful and gave so freely of his time as Sidney. He would spend days rewriting someone's manuscript or document. I have known him to rewrite a whole chapter for which he was given no credit. Sidney did not worry about credits; he did these activities because he felt strongly about his university involvement. I believe that nearly every colleague at Hopkins and Vanderbilt had a paper reviewed by Sidney before submission for publication.

Sidney, although most helpful, could hold very firm to his views. An example of this was a story told to me by a distinguished professor who had submitted a paper to the *Archives of Biochemistry and Biophysics* which was sent to Sidney, who was on the Editorial Board, for review. Sidney felt that there was something intrinsically wrong with the results. Of course, the professor disagreed. After nine months of exchanges, the professor finally agreed that he was wrong and altered his views.

Sidney was an extraordinary human being. He rarely said anything derogatory about individuals, many of whom were deserving of being chastised. He could have scientific disagreements, but these never became personal. He certainly maintained close friendships with people with whom he had scientific disagreements.

Sidney received many honors and awards. He received the Eli Lilly Award in Biochemistry in 1947. He was elected to both the National Academy of Sciences and the American Academy of Arts and Sciences. In 1978, he was chosen as the Harvie Branscomb Distinguished Professor by the Vanderbilt Board of Trustees. He was also the Charles Hayden—American Cancer Society Professor of Microbiology. He was a member of many advisory committees, and served on the editorial boards of a large number of journals.

Herman Kalckar once said of Sidney and Maryda that their hospitality was legendary. Not only did they open their door to all, they showed compassion and understanding to individuals whose lives were besieged by tragedy and unhappiness. Their warmth gave new hope and encouragement to many.

Sidney loved music, and in recent years obtained much enjoyment in performing in chamber groups. But Sidney had a virtue unusual for scientists: not only was he involved in social events, he also spent much time working for a number of worthy humanitarian causes.

I would like to say a few words regarding *Methods in Enzymology*. We were approached in 1951 by the late Mr. Kurt Jacoby of Academic Press to do a series comparable to *Methodem der Ferment Forschung* which was edited by Baumann and Myrback, four volumes of which were published in German in 1940 and which was a pet project of Mr. Jacoby. Although we resisted taking on the enterprise, we were unable to withstand Mr. Jacoby's forceful and convincing arguments. The planning and publishing of the first two volumes were difficult and time consuming, but Mr. Jacoby used combined doses of encouragement and nagging. We somehow managed to finish the first two volumes. Actually, Mr. Jacoby took great pride in our accomplishments and praised us whenever possible. We did four more volumes, and then we thought that to proceed further would be difficult because of the rapid growth of biochemistry and that it was almost impossible to intimately know the many new fields that had come into existence. Again, Mr. Jacoby became a tremendous influence; he would not let us quit. He insisted that we had created a most useful and successful series, and that it was our duty to continue. From discussions with Mr. Jacoby and James Barsky evolved the new format for the volumes which has continued to the present. We wished to change the title of the series, but Mr. Jacoby insisted that *Methods in Enzymology* was already an established name and that its contents could cover nearly every field in the biochemical sciences. Sidney and I in later times agreed that Mr. Jacoby had a potent foresight which we ourselves did not realize.

Although we were separated physically, the *Methods* kept us together. We talked on the phone several times a week, and we had the opportunity to see each other a number of times each year. As time passed we became closer. These contacts certainly enriched my life and my perspective about many things. It was "fun" talking to Sidney.

Now I have lost my friend, colleague, and partner. Biochemistry has lost a creative and critical mind. The world has lost a humane individual whose compassion and warmth influenced the lives of many. To Maryda and his daughters, I extend my deepest sympathy. No wife was more devoted than Maryda was to Sidney. His daughters can be proud of their father, both for his accomplishments and as a human being.

I am indebted to all the editors and contributors to the volumes; they have made the *Methods* a successful venture. I would also like to thank James Barsky and Shirley Light of Academic Press for their support and cooperation, particularly during the past several months. I hope that the *Methods* will continue to be the timely, useful, and high quality series that Sidney had strived for, and that in the future we will be able to implement some new concepts to the volumes that Sidney and I had been discussing.

NATHAN O. KAPLAN

METHODS IN ENZYMOLOGY

EDITED BY

Sidney P. Colowick and Nathan O. Kaplan

VANDERBILT UNIVERSITY
SCHOOL OF MEDICINE
NASHVILLE, TENNESSEE

DEPARTMENT OF CHEMISTRY
UNIVERSITY OF CALIFORNIA
AT SAN DIEGO
LA JOLLA, CALIFORNIA

METHODS IN ENZYMOLOGY

EDITORS-IN-CHIEF

Sidney P. Colowick and Nathan O. Kaplan

VOLUME VIII. Complex Carbohydrates
Edited by ELIZABETH F. NEUFELD AND VICTOR GINSBURG

VOLUME IX. Carbohydrate Metabolism
Edited by WILLIS A. WOOD

VOLUME X. Oxidation and Phosphorylation
Edited by RONALD W. ESTABROOK AND MAYNARD E. PULLMAN

VOLUME XI. Enzyme Structure
Edited by C. H. W. HIRS

VOLUME XII. Nucleic Acids (Parts A and B)
Edited by LAWRENCE GROSSMAN AND KIVIE MOLDAVE

VOLUME XIII. Citric Acid Cycle
Edited by J. M. LOWENSTEIN

VOLUME XIV. Lipids
Edited by J. M. LOWENSTEIN

VOLUME XV. Steroids and Terpenoids
Edited by RAYMOND B. CLAYTON

VOLUME XVI. Fast Reactions
Edited by KENNETH KUSTIN

VOLUME XVII. Metabolism of Amino Acids and Amines (Parts A and B)
Edited by HERBERT TABOR AND CELIA WHITE TABOR

VOLUME XVIII. Vitamins and Coenzymes (Parts A, B, and C)
Edited by DONALD B. MCCORMICK AND LEMUEL D. WRIGHT

VOLUME XIX. Proteolytic Enzymes
Edited by GERTRUDE E. PERLMANN AND LASZLO LORAND

VOLUME XX. Nucleic Acids and Protein Synthesis (Part C)
Edited by KIVIE MOLDAVE AND LAWRENCE GROSSMAN

VOLUME XXI. Nucleic Acids (Part D)
Edited by LAWRENCE GROSSMAN AND KIVIE MOLDAVE

VOLUME XXII. Enzyme Purification and Related Techniques
Edited by WILLIAM B. JAKOBY

VOLUME XXIII. Photosynthesis (Part A)
Edited by ANTHONY SAN PIETRO

VOLUME XXIV. Photosynthesis and Nitrogen Fixation (Part B)
Edited by ANTHONY SAN PIETRO

VOLUME XXV. Enzyme Structure (Part B)
Edited by C. H. W. HIRS AND SERGE N. TIMASHEFF

VOLUME XXVI. Enzyme Structure (Part C)
Edited by C. H. W. HIRS AND SERGE N. TIMASHEFF

VOLUME XXVII. Enzyme Structure (Part D)
Edited by C. H. W. HIRS AND SERGE N. TIMASHEFF

VOLUME XXVIII. Complex Carbohydrates (Part B)
Edited by VICTOR GINSBURG

VOLUME XXIX. Nucleic Acids and Protein Synthesis (Part E)
Edited by LAWRENCE GROSSMAN AND KIVIE MOLDAVE

VOLUME XXX. Nucleic Acids and Protein Synthesis (Part F)
Edited by KIVIE MOLDAVE AND LAWRENCE GROSSMAN

VOLUME XXXI. Biomembranes (Part A)
Edited by SIDNEY FLEISCHER AND LESTER PACKER

VOLUME XLVI. Affinity Labeling
Edited by WILLIAM B. JAKOBY AND MEIR WILCHEK

VOLUME XLVII. Enzyme Structure (Part E)
Edited by C. H. W. HIRS AND SERGE N. TIMASHEFF

VOLUME XLVIII. Enzyme Structure (Part F)
Edited by C. H. W. HIRS AND SERGE N. TIMASHEFF

VOLUME XLIX. Enzyme Structure (Part G)
Edited by C. H. W. HIRS AND SERGE N. TIMASHEFF

VOLUME L. Complex Carbohydrates (Part C)
Edited by VICTOR GINSBURG

VOLUME LI. Purine and Pyrimidine Nucleotide Metabolism
Edited by PATRICIA A. HOFFEE AND MARY ELLEN JONES

VOLUME LII. Biomembranes (Part C: Biological Oxidations)
Edited by SIDNEY FLEISCHER AND LESTER PACKER

VOLUME LIII. Biomembranes (Part D: Biological Oxidations)
Edited by SIDNEY FLEISCHER AND LESTER PACKER

VOLUME LIV. Biomembranes (Part E: Biological Oxidations)
Edited by SIDNEY FLEISCHER AND LESTER PACKER

VOLUME LV. Biomembranes (Part F: Bioenergetics)
Edited by SIDNEY FLEISCHER AND LESTER PACKER

VOLUME LVI. Biomembranes (Part G: Bioenergetics)
Edited by SIDNEY FLEISCHER AND LESTER PACKER

VOLUME LVII. Bioluminescence and Chemiluminescence
Edited by MARLENE A. DELUCA

VOLUME LVIII. Cell Culture
Edited by WILLIAM B. JAKOBY AND IRA PASTAN

VOLUME LIX. Nucleic Acids and Protein Synthesis (Part G)
Edited by KIVIE MOLDAVE AND LAWRENCE GROSSMAN

VOLUME LX. Nucleic Acids and Protein Synthesis (Part H)
Edited by KIVIE MOLDAVE AND LAWRENCE GROSSMAN

VOLUME 61. Enzyme Structure (Part H)
Edited by C. H. W. HIRS AND SERGE N. TIMASHEFF

VOLUME 62. Vitamins and Coenzymes (Part D)
Edited by DONALD B. MCCORMICK AND LEMUEL D. WRIGHT

VOLUME 63. Enzyme Kinetics and Mechanism (Part A: Initial Rate and Inhibitor Methods)
Edited by DANIEL L. PURICH

VOLUME 64. Enzyme Kinetics and Mechanism (Part B: Isotopic Probes and Complex Enzyme Systems)
Edited by DANIEL L. PURICH

VOLUME 65. Nucleic Acids (Part I)
Edited by LAWRENCE GROSSMAN AND KIVIE MOLDAVE

VOLUME 66. Vitamins and Coenzymes (Part E)
Edited by DONALD B. MCCORMICK AND LEMUEL D. WRIGHT

VOLUME 67. Vitamins and Coenzymes (Part F)
Edited by DONALD B. MCCORMICK AND LEMUEL D. WRIGHT

VOLUME 68. Recombinant DNA
Edited by RAY WU

VOLUME 69. Photosynthesis and Nitrogen Fixation (Part C)
Edited by ANTHONY SAN PIETRO

VOLUME 70. Immunochemical Techniques (Part A)
Edited by HELEN VAN VUNAKIS AND JOHN J. LANGONE

VOLUME 71. Lipids (Part C)
Edited by JOHN M. LOWENSTEIN

VOLUME 72. Lipids (Part D)
Edited by JOHN M. LOWENSTEIN

VOLUME 97. Biomembranes [Part K: Membrane Biogenesis: Assembly and Targeting (Prokaryotes, Mitochondria, and Chloroplasts)]
Edited by SIDNEY FLEISCHER AND BECCA FLEISCHER

VOLUME 98. Biomembranes [Part L: Membrane Biogenesis (Processing and Recycling)]
Edited by SIDNEY FLEISCHER AND BECCA FLEISCHER

VOLUME 99. Hormone Action (Part F: Protein Kinases)
Edited by JACKIE D. CORBIN AND JOEL G. HARDMAN

VOLUME 100. Recombinant DNA (Part B)
Edited by RAY WU, LAWRENCE GROSSMAN, AND KIVIE MOLDAVE

VOLUME 101. Recombinant DNA (Part C)
Edited by RAY WU, LAWRENCE GROSSMAN, AND KIVIE MOLDAVE

VOLUME 102. Hormone Action (Part G: Calmodulin and Calcium-Binding Proteins)
Edited by ANTHONY R. MEANS AND BERT W. O'MALLEY

VOLUME 103. Hormone Action (Part H: Neuroendocrine Peptides)
Edited by P. MICHAEL CONN

VOLUME 104. Enzyme Purification and Related Techniques (Part C)
Edited by WILLIAM B. JAKOBY

VOLUME 105. Oxygen Radicals in Biological Systems
Edited by LESTER PACKER

VOLUME 106. Posttranslational Modifications (Part A)
Edited by FINN WOLD AND KIVIE MOLDAVE

VOLUME 107. Posttranslational Modifications (Part B)
Edited by FINN WOLD AND KIVIE MOLDAVE

VOLUME 108. Immunochemical Techniques (Part G: Separation and Characterization of Lymphoid Cells)
Edited by GIOVANNI DI SABATO, JOHN J. LANGONE, AND HELEN VAN VUNAKIS

VOLUME 109. Hormone Action (Part I: Peptide Hormones)
Edited by LUTZ BIRNBAUMER AND BERT W. O'MALLEY

VOLUME 110. Steroids and Isoprenoids (Part A)
Edited by JOHN H. LAW AND HANS C. RILLING

VOLUME 111. Steroids and Isoprenoids (Part B)
Edited by JOHN H. LAW AND HANS C. RILLING

VOLUME 112. Drug and Enzyme Targeting (Part A)
Edited by KENNETH J. WIDDER AND RALPH GREEN

VOLUME 113. Glutamate, Glutamine, Glutathione, and Related Compounds
Edited by ALTON MEISTER

VOLUME 114. Diffraction Methods for Biological Macromolecules (Part A) (in preparation)
Edited by HAROLD W. WYCKOFF, C. H. W. HIRS, AND SERGE N. TIMASHEFF

VOLUME 115. Diffraction Methods for Biological Macromolecules (Part B) (in preparation)
Edited by HAROLD W. WYCKOFF, C. H. W. HIRS, AND SERGE N. TIMASHEFF

VOLUME 116. Immunochemical Techniques (Part H: Effectors and Mediators of Lymphoid Cell Functions) (in preparation)
Edited by GIOVANNI DI SABATO, JOHN J. LANGONE, AND HELEN VAN VUNAKIS

VOLUME 117. Enzyme Structure (Part J) (in preparation)
Edited by C. H. W. HIRS AND SERGE N. TIMASHEFF

VOLUME 118. Plant Molecular Biology (in preparation)
Edited by ARTHUR WEISSBACH AND HERBERT WEISSBACH

VOLUME 119. Interferons (Part C) (in preparation)
Edited by SIDNEY PESTKA

VOLUME 120. Cumulative Subject Index Volumes 81–94, 96–101

VOLUME 121. Immunochemical Techniques (Part I: Hybridoma Technology and Monoclonal Antibodies) (in preparation)
Edited by JOHN J. LANGONE AND HELEN VAN VUNAKIS

Section I

Glutamate

[1] L-Glutamate Decarboxylase from Brain

By JANG-YEN WU, LARRY DENNER,
CHIN-TARNG LIN, and GUOXING SONG

γ-Aminobutyric acid (GABA) has emerged as one of the major neurotransmitters in the vertebrate central nervous system.[1,2] L-Glutamate decarboxylase (EC 4.1.1.15) (GAD), which catalyzes α-decarboxylation of L-glutamate to form GABA and CO_2, is believed to be the rate-limiting enzyme that normally determines the steady-state levels of GABA in vertebrate and invertebrate nervous systems.[3,4] Furthermore, it has been shown that there is good correlation between GABA levels and GAD activity in vertebrate nervous system.[5-7] Hence, GAD is a better marker for GABAergic neurons than GABA per se, which may redistribute or be metabolized during the preparation of the tissues.[7,8]

This chapter discusses the results obtained in the authors' laboratory, dealing with various purification procedures, assay methods, and criteria of purity of GAD preparations. In addition, immunochemical characterization of anti-GAD serum and immunocytochemical studies of GAD are also included.

Assay Methods

GAD was assayed by a radiometric method which is based on the formation of $^{14}CO_2$ from either L-[1-^{14}C]- or L-[U-^{14}C]glutamate.[9,10] In a

[1] D. R. Curtis, *in* "GABA-Neurotransmitters" (P. Krogsgaard-Larsen, J. Scheelkruger, and H. Kofod, eds.), p. 17. Academic Press, New York, 1979.

[2] K. Krnjevic, *Physiol. Rev.* **54,** 418 (1974).

[3] E. A. Kravitz, *in* "The Neurosciences" (G. C. Quarton, T. Melnechuk, and F. O. Schmitt, eds.), p. 433. Rockefeller Univ. Press, New York, 1967.

[4] E. Roberts and K. Kuriyama, *Brain Res.* **8,** 1 (1968).

[5] K. Kuriyama, B. Haber, T. Sisken, and E. Roberts, *Proc. Natl. Acad. Sci. U.S.A.* **55,** 846 (1966).

[6] K. Kuriyama, T. Sisken, B. Haber, and E. Roberts, *Brain Res.* **9,** 165 (1968).

[7] C. F. Baxter, *in* "Handbook of Neurochemistry" (A. Lajtha, ed.), p. 289. Plenum, New York, 1970.

[8] F. Fonnum, *in* "Metabolic Compartmentation and Neurotransmission" (S. Berl, D. D. Clarke, and D. Schneider, eds.), p. 99. Plenum, New York, 1975.

[9] J.-Y. Wu, *in* "GABA in Nervous System Function" (E. Roberts, T. Chase, and D. Tower, eds.), p. 7. Raven Press, New York, 1976.

[10] J.-Y. Wu, Y. Y. T. Su, D. M. K. Lam, A. Schousboe, and O. Chude, *Res. Methods Neurochem.* **5,** 129 (1981).

METHODS IN ENZYMOLOGY, VOL. 113
Copyright © 1985 by Academic Press, Inc.
All rights of reproduction in any form reserved.

typical assay, the incubation vessel contained 10 μl of 40 m*M* sodium glutamate (5 μCi/ml) in 0.1 *M* potassium phosphate buffer containing 0.2 m*M* pyridoxal phosphate (PLP), pH 7.2. The reaction was started by injecting 90 μl enzyme solution in 50 m*M* potassium phosphate buffer, pH 7.2, containing 0.2 m*M* PLP and 1 m*M* 2-aminoethylisothiouronium bromide (AET) (standard buffer) into the incubation vessel. The incubation was carried out for 30 min at 37° and was terminated by injecting 20 μl of 0.5 *N* H_2SO_4 into the reaction mixture. The vessels were incubated for another 60 min to ensure a complete release of CO_2 and absorption in the Hyamine base.

An ordinary disposable culture tube (15 × 85 mm) (Fisher and Scientific Co., Pittsburgh, PA) sealed with a serum tube rubber stopper can be used as an incubation vessel. A plastic center well (Kontes, Vineland, NJ) that contained Hyamine solution is inserted through the center of the rubber stopper to absorb the CO_2 formed in the reaction mixture.

The validity of the assay was established by the identification of the reaction product as GABA and a stoichiometric relation between GABA and CO_2 as previously reported.[10,11]

Purification Procedures

GAD has been purified to homogeneity from brain of a variety of species including mouse,[12] catfish,[13] bovine,[14] rat,[15,16] and human.[15] The purification procedures involved the initial extraction of GAD activity with isotonic sucrose or water protected with PLP and followed by ammonium sulfate fractionation, column chromatographies on Sephadex G-200, calcium phosphate gel, and DEAE-Sephadex and finally with preparative gel electrophoresis or isoelectric focusing.

The successive steps in the purification of GAD from 500 rat brains are summarized in the table. Approximately 5.2% of the total activity was recovered as a purified enzyme preparation representing 1230-fold purification over the original homogenate.[10] A typical preparation from 500 rat brains is detailed as follows.

Starting Material

Five hundred rats were killed by decapitation, the whole brain was removed rapidly, and a 10% homogenate was made in ice cold homogeni-

[11] J.-Y. Wu, O. Chude, J. Wein, and E. Roberts, *J. Neurochem.* **30,** 849 (1978).
[12] J.-Y. Wu, T. Matsuda, and E. Roberts, *J. Biol. Chem.* **248,** 3029 (1973).
[13] Y. Y. T. Su, J.-Y. Wu, and D. M. K. Lam, *J. Neurochem.* **33,** 169 (1979).
[14] J.-Y. Wu, *Proc. Natl. Acad. Sci. U.S.A.* **79,** 4270 (1982).
[15] J. M. Blindermann, M. Maitre, L. Ossola, and P. Mandel, *Eur. J. Biochem.* **86,** 143 (1978).
[16] L. Denner and J.-Y. Wu, *Fed Proc., Fed. Am. Soc. Exp. Biol.* **42**(7), 2008 (1983).

PURIFICATION OF GAD FROM RAT BRAIN[a]

Sample	Total activity (units)[b]	Specific activity (units/mg × 10^3)	Yield (%)
Homogenate	209	1.8	100
Supernatant	147	2.3	70
DEAE-Cellulose	106	11.6[c]	51
Hydroxylapatite	65	104[d]	31
Gel filtration	31	312[e]	15
Gel electrophoresis	11	2204	5.2

[a] Purification of GAD was made from 500 brains.
[b] One unit = 1 μmol of product formed per minute at 37° under standard conditions.
[c] Peak fraction specific activity of 15.1.
[d] Peak fraction specific activity of 128.
[e] Peak fraction specific activity of 349.

zation medium which was consisted of 1 m*M* AET, 0.2 m*M* PLP, 1 m*M* EDTA, 1 m*M* benzamidine hydrochloride, 0.1 m*M* phenylmethylsulfonyl fluoride, and 20 μg/ml soybean trypsin inhibitor in a motor-driven glass homogenizer with a Teflon pestle. The homogenate was centrifuged at 100,000 *g* for 60 min. All operations and centrifugations were carried out at 0 to 4° and all buffer solutions contained 0.2 m*M* PLP and 1 m*M* AET, unless otherwise mentioned. Supernatants were decanted and stored at 4°. Pellets were reextracted once with the homogenization medium and once in the same medium containing 25 m*M* potassium phosphate. The supernatants thus obtained were combined and concentrated on a DC2 Hollow Fiber Dialyzer/Concentrator (Amicon).

Chromatography on DEAE-Cellulose

DEAE-cellulose, DE-52, was equilibrated with 25 m*M* potassium phosphate standard buffer and packed into a column to give a bed volume, 5.0 × 40 cm. The GAD solution from the preceding step was applied at a flow rate of 100 ml/hr and the column was washed with four beds volume of the equilibration buffer. The column was further eluted with five beds volume of linear gradient of potassium phosphate made of 25 m*M*, pH 7.0 and 300 m*M*, pH 6.4. The enzyme activity started to appear at 100 m*M* and peaked at 180 m*M* phosphate buffer. Fractions with specific activities above 50% of the peak fraction were pooled and concentrated on the DC2 concentrator followed by dialysis against 1 m*M* potassium phosphate standard buffer.

Chromatography on Hydroxylapatite

Hydroxylapatite was packed in a column (5.0 × 50 cm) to give a bed volume of 5.0 × 20 cm. The enzyme solution from the preceding steps was applied to the column which had been equilibrated with 1 m*M* potassium phosphate buffer, pH 7.2. After the application of the solution, the column was washed with one bed volume of the equilibration buffer followed by 10 bed volumes of 10 m*M* potassium phosphate. A linear gradient made of two and one half beds volume each of 10 and 100 m*M* of potassium phosphate buffer was used to elute the column. GAD activity started to appear at 20 m*M* and peaked at about 35 m*M* of phosphate buffer. Fractions with specific activities above 60% of the peak fraction were pooled, concentrated by ultrafiltration in a stirred cell with PM 30 membranes (Amicon), and dialyzed against 25 m*M* potassium phosphate buffer.

Chromatography on Gel Filtration

Two gel filtration columns, 2.6 × 95 cm Sephadex G-150 and 2.6 × 80 cm Ultrogel AcA 44, were set up in series and equilibrated with 25 m*M* phosphate buffer. The concentrated enzyme solution from hydroxylapatite column which had a volume less than 1% of the bed volume of the series gel filtration columns was loaded and eluted at 30 ml/hr with the same buffer. Fractions with specific activities above 80% of the peak fraction were pooled and concentrated as described previously. The final buffer concentration was 10 m*M*.

Preparative Polyacrylamide Gel Electrophoresis

The final purification was achieved with preparative gel electrophoresis. Preparative polyacrylamide gel electrophoresis was performed as previously described.[14] Gels, 1.5 mm × 15 cm × 40 cm, were prepared with 5% acrylamide, 0.25% *N*,*N*′-methylene-bis-acrylamide, 0.0575% (v/v) *N*,*N*,*N*′,*N*′-tetramethylethylenediamine, and 1 mg/ml ammonium persulphate, in 187 m*M* Tris containing 65 m*M* HCl, pH 8.4. After polymerization at room temperature for 30 min, gels were prerun at 4° for 30 min at 20 mA. The electrophoresis running buffer contained 25 m*M* Tris, 192 m*M* glycine, 1 m*M* AET, 0.2 m*M* PLP, 0.5% (v/v) 2-mercaptoethanol, pH 8.4. The most purified GAD samples from gel filtration were used. About 700 μl of the concentrated GAD solution containing 10% glycerol was applied and electrophoresis was carried out at 4° for 14–18 hr at 20 mA. Two adjacent parallel gel strips with 0.5 cm width each were cut from the center of the gel. One was stained with 0.05% Coomassie Brilliant Blue G-250 for protein pattern and the other was cut in 0.5 cm length and assayed for GAD activity. Slices measuring 1.5 mm × 0.5 cm × 0.5 cm were chopped into small pieces and placed in disposable culture tubes containing 75 μl of gel assay buffer which contained 0.2 *M* potassium phosphate, 1 m*M* AET, 0.2 m*M* PLP, and 1 m*M* EDTA, pH 6.0. The reaction was started by adding 7.5 μl L-[1-^{14}C]glutamic acid (5 μCi/ml, 40 m*M*), and the enzyme activity was measured as described above. Gel slices with the highest enzyme activity were extracted by homogenization in 2 ml water plus 2 ml gel assay buffer per cm^3 of gel. Gel was removed by centrifugation at 105,000 *g* for 30 min. Pellets thus obtained were reextracted with 2 ml 25 m*M* potassium phosphate per cm^3 of gel. The two extracts were combined and concentrated by ultrafiltration. The final concentration of buffer was adjusted to 10 m*M*.

Criteria of Purity

The purity of GAD preparations was established based on the following criteria: (1) the purified enzyme migrated as a single protein band with the location of the enzyme activity corresponding to the location of the protein band on polyacrylamide gel electrophoresis[9,12,14]; (2) the purified mouse brain GAD preparation to be homogeneous in size as judged from the linear plot of the logarithm of concentrations against the squares of the distances from the center of rotation to points of interest in high-speed sedimentation equilibrium runs in both H_2O and D_2O solutions and in dissociating conditions, e.g., guanidine HCl and 2-mercaptoethanol.[9,12]

Immunochemical Characterizations

Production of Antibodies

Polyclonal Antibody. In the past, we were able to obtain high titer antibodies with microgram quantities of antigens. Briefly, 3–50 μg of antigen in 0.5 ml 0.9% NaCl was emulsified thoroughly with an equal volume of complete Freund's adjuvant and injected into two sites of subscapular muscles. Rabbits were given biweekly injection over a period of 10 weeks. Antibodies usually could be detected after the fifth injection.

This technique has been successfully used in our laboratory for the production of antibodies against various proteins purified from the nervous system. For instance, a total of 15, 50, 50, 7, and 75 μg of purified

GABA-transaminase[17] and GAD from mouse brain,[18–20] choline acetyltransferase from electric organ of *Torpedo,*[21] neurofilament protein from *Myxicola,*[22] and GAD and cysteinesulfinic acid decarboxylase from bovine brain,[14] respectively, were sufficient to produce specific antibodies in rabbits.

Monoclonal Antibody. In addition to the conventional method, hybridoma technique originally developed by Köhler and Milstein[23,24] and modified by Kennett[25] was also used for the production of monoclonal antibody against GAD.[26,27] Monoclonal anti-GAD IgG was obtained either from culture medium or from the ascites fluids of mice which had been injected with anti-GAD producing clone. The detailed procedures have been described previously.[26,27]

Characterization of Antibodies. Four different immunochemical methods, namely, immunodiffusion and immunoelectrophoresis, neutralization of enzyme activities by antibodies and microcomplement fixation tests and enzyme immunoassay were used to characterize polyclonal anti-GAD.[19,20] Monoclonal anti-GAD was characterized by ELISA test and by immunoblot analysis.

Localization of L-Glutamate Decarboxylase

Perhaps the most important and fruitful application of anti-GAD is to identify those neurons and neuronal projections that use GABA as their neurotransmitter by immunocytochemical visualization of GAD. The precise cellular and subcellular locations of GAD have been visualized in cerebellum,[28–31] retina,[32–35] habenula,[36,37] substantia nigra,[38,39] olfactory

[17] K. Saito, A. Schousboe, J.-Y. Wu, and E. Roberts, *Brain Res.* **65,** 287 (1974).

[18] T. Matsuda, J.-Y. Wu, and E. Roberts, *J. Neurochem.* **21,** 159 (1973).

[19] K. Saito, J.-Y. Wu, and E. Roberts, *Brain Res.* **65,** 277 (1974).

[20] J.-Y. Wu, *in* "Neuroimmunocytochemistry" (A. C. Cuello ed.), p. 159. Wiley, New York, 1983.

[21] C. Brandon and J.-Y. Wu, *Soc. Neurosci. Abstr.* **3,** 404 (1977).

[22] R. J. Lasek and J.-Y. Wu, *Soc. Neurosci. Abstr.* **2,** 40 (1976).

[23] G. Köhler and C. Milstein, *Nature (London)* **256,** 495 (1975).

[24] G. Köhler and C. Milstein, *Eur. J. Immunol.* **6,** 514 (1976).

[25] R. H. Kennett, this series, Vol. 58, p. 345.

[26] J.-Y. Wu, C. T. Lin, L. Denner, Y. Y. T. Su, and D. S. Chan, *Trans. Am. Soc. Neurochem.* **13,** 92 (1982).

[27] J.-Y. Wu, *in* "Problems in GABA Research" (Y. O. Okada and E. Roberts, eds.), p. 40. Excerpta Medica, Amsterdam, 1982.

[28] K. Saito, R. Barber, J.-Y. Wu, T. Matsuda, E. Roberts, and J. E. Vaughn, *Proc. Natl. Acad. Sci. U.S.A.* **71,** 269 (1974).

[29] B. J. McLaughlin, J. G. Wood, K. Saito, R. Barber, J. E. Vaughn, E. Roberts, and J.-Y. Wu, *Brain Res.* **76,** 377 (1974).

bulb,[40] spinal cord,[41,42] corpus striatum,[43] nucleus reticularis thalami,[44] striate cortex,[45] hypothalamus and pituitary gland,[46] dentate gyrus,[47] and hippocampus,[48,49] and in tissue cultures[50,51] by means of immunocytochemical techniques using antiserum against the purified GAD preparations. The details of immunocytochemical procedures have been described previously.[32,52]

Comments

The purification procedures described here are quite reproducible and have been applied successfully to the purification of other brain enzymes including cysteinesulfinic acid decarboxylase,[14] GABA-transaminase,[9,53]

[30] V. Chan-Palay, S. L. Palay, and J.-Y. Wu, *Anat. Embryol.* **157,** 1 (1979).
[31] V. Chan-Palay, G. Nilaver, S. L. Palay, M. C. Beinfeld, E. E. Zimmerman, and J.-Y. Wu, *Proc. Natl. Acad. Sci. U.S.A.* **78,** 7787 (1981).
[32] J.-Y. Wu, C. Brandon, Y. Y. T. Su, and D. M. K. Lam. *Mol. Cell. Biochem.* **39,** 229 (1981).
[33] C. Brandon, D. M. K. Lam, and J.-Y. Wu, *Proc. Natl. Acad. Sci. U.S.A.* **76,** 3557 (1979).
[34] C. Brandon, D. M. K. Lam, Y. Y. T. Su, and J.-Y. Wu, *Brain Res. Bull.* **5,** Suppl. 2, 21 (1980).
[35] D. M. K. Lam, Y. Y. T. Su, L. Swain, R. E. Marc, C. Brandon, and J.-Y. Wu, *Nature (London)* **278,** 565 (1979).
[36] Z. Gottesfeld, C. Brandon, D. M. Jacobwitz, and J.-Y. Wu, *Brain Res. Bull.* **5,** Suppl. 2, 1 (1980).
[37] Z. Gottesfeld, C. Brandon, and J.-Y. Wu, *Brain Res.* **208,** 181 (1981).
[38] C. E. Riback, J. E. Vaughn, K. Saito, R. Barber, and E. Roberts, *Brain Res.* **116,** 287 (1976).
[39] C. E. Ribak, J. E. Vaughn, and E. Roberts, *Brain Res.* **192,** 413 (1980).
[40] C. E. Ribak, J. E. Vaughn, K. Saito, R. Barber, and E. Roberts, *Brain Res.* **126,** 1 (1977).
[41] B. J. McLaughlin, R. Barber, K. Saito, E. Roberts, and J.-Y. Wu, *J. Comp. Neurol.* **164,** 305 (1975).
[42] S. P. Hunt, J. S. Kelly, P. C. Emson, J. R. Kimmel, R. J. Miller, and J.-Y. Wu, *Neuroscience* **6,** 1883 (1981).
[43] C. E. Ribak, J. E. Vaughn, and E. Roberts, *J. Comp. Neurol.* **187,** 261 (1979).
[44] C. Houser, J. E. Vaughn, R. P. Barber, and E. Roberts, *Brain Res.* **200,** 341 (1980).
[45] A. E. Hendrickson, S. Hunt, and J.-Y. Wu, *Nature (London)* **292,** 605 (1981).
[46] S. R. Vincent, T. Hökfelt, and J.-Y. Wu, *Neuroendocrinology* **34,** 117 (1982).
[47] D. Goldowitz, S. R. Vincent, J.-Y. Wu, and T. Hökfelt, *Brain Res.* **238,** 413 (1982).
[48] P. Somogyi, A. D. Smith, M. G. Nunzi, H. Takagi, and J.-Y. Wu, *J. Nerurosci.* **3,** 1450 (1983).
[49] C. E. Ribak, J. E. Vaughn, and K. Saito, *Brain Res.* **140,** 315 (1978).
[50] P. Panula, J.-Y. Wu, P. Emson, and L. Rechardt, *Neurosci. Lett.* **22,** 303 (1981).
[51] P. Panula, J.-Y. Wu, and P. Emson, *Brain Res.* **219,** 202 (1981).
[52] J.-Y. Wu, C.-T. Lin, C. Brandon, D.-S. Chan, H. Mohler, and J. G. Richards, *in* "Cytochemical Methods in Neuroanatomy" (S. Palay and V. Palay, eds.), p. 279. Alan R. Liss, Inc., New York, 1982.
[53] A. Schousboe, J.-Y. Wu, and E. Roberts, *Biochemistry* **12,** 2868 (1973).

and choline acetyltransferase.[54] The use of microgram instead of milligram quantities of protein as antigen, as has been used in the conventional method of immunizing animals for the production of polyclonal antibody, is essential because some brain proteins, e.g., GAD, can be obtained as homogeneous preparation only in minute quantity. Furthermore, the chance of producing antibodies against trace impurities, which might still be associated with the highly purified preparations and escape detection by sensitive physical and chemical techniques, is much less because of the small amount (microgram quantities) of antigen used in the immunization of animals. The use of hybridoma technique[23–25] for the production of monoclonal antibodies as in the case of GAD[26,27,52] will be proven to be of great importance to other enzymes since monoclonal antibodies have been shown to be invaluable for the characterization, quantitative analyses, and purification of macromolecular antigens.

The approaches that have been taken in the authors' laboratory for the elucidation of GAD-containing neurons or GABAergic projections in various parts of the vertebrate central nervous system starting with the purification of GAD to homogeneity, followed by production and characterization of anti-GAD serum and finally the visualization of GAD at cellular and subcellular levels by immunocytochemical techniques has been proven to be very fruitful. Similar approaches also have been applied to other enzyme systems, particularly to those involved in the biosynthesis of neurotransmitters, e.g., choline acetyltransferase,[55] tyrosine hydroxylase,[56,57] and cysteinesulfinic acid decarboxylase.[58,59]

[54] C. Brandon and J.-Y. Wu, *J. Neurochem.* **30,** 791 (1978).

[55] V. Chan-Palay, A. G. Engel, S. L. Palay, and J.-Y. Wu, *Proc. Natl. Acad. Sci. U.S.A.* **79,** 6717 (1982).

[56] D. J. Reis, R. H. Benno, L. W. Tucker, and T. H. Joh, *in* "Cytochemical Methods in Neuroanatomy" (S. Palay and V. Chan-Palay, eds.), p. 205. Alan R. Liss, Inc., New York, 1982.

[57] V. M. Pickel, *in* "Cytochemical Methods in Neuroanatomy" (S. Palay and V. Chan-Palay, eds.), p. 229. Alan R. Liss, Inc., New York, 1982.

[58] V. Chan-Palay, C.-T. Lin, S. L. Palay, M. Yamamoto, and J.-Y. Wu, *Proc. Natl. Acad. Sci. U.S.A.* **79,** 2695 (1982).

[59] V. Chan-Palay, S. L. Palay, and J.-Y. Wu, *Proc. Natl. Acad. Sci. U.S.A.* **79,** 4221 (1982).

[2] L-Glutamate Decarboxylase from Bacteria

By MARGARET L. FONDA

$$^{-}O_2CCH_2CH_2CH_2\overset{\displaystyle NH_3^+ \atop |}{C}HCO_2^- + H^+ \rightarrow CO_2 + {}^{-}O_2CCH_2CH_2CH_2NH_3^+$$

Bacterial L-glutamate decarboxylase (L-glutamate 1-carboxy-lyase, EC 4.1.1.15) has been purified to homogeneity from *Escherichia coli* and from *Clostridium perfringens* and has been studied extensively. When *E. coli* is grown in medium containing L-glutamate, the inducible enzyme is produced in large amounts as the medium becomes acidic.

Assay Method

Principle. Enzyme activity may be determined using a manometric assay of CO_2 evolution or a radiometric assay of the conversion of [^{14}C]glutamate to $^{14}CO_2$[1] or γ-[^{14}C]aminobutyrate. The manometric assay is used to measure enzyme activity during purification and in steady state kinetic studies. Chloride, an activator, and the coenzyme are added to the assay mixture to produce optimum conditions.

Reagents

Pyridine–HCl, 100 mM, containing approximately 120 mM NaCl to give a total chloride concentration of 200 mM, pH 4.6.

L-Glutamate, 20 mM, in pyridine–HCl, 100 mM, total chloride, 200 mM, pH 4.6.

Pyridoxal phosphate (pyridoxal-P), 2 mM, prepared fresh weekly, protected from light, stored frozen at −20°.

Procedure.[2] Three milliliters of the assay mixture containing 100 mM pyridine–HCl, 200 mM total chloride, 20 mM L-glutamate, and 67 μM pyridoxal-P at pH 4.6 is placed in the main chamber of a Warburg flask. Approximately 1 U of glutamate decarboxylase is placed in the side arm. After equilibration at 25° for 5 min, the contents of the flask are mixed, and the volume of gas evolved is recorded at 1-min intervals for 6 min. The data collected with a constant-pressure respirometer are corrected to actual volumes of CO_2 evolved at standard conditions as described by Gregory and Winter.[3]

[1] L. P. Hager, this series, Vol. 17A, p. 857.

[2] M. L. Fonda, *Biochemistry* **11,** 1304 (1972).

[3] K. F. Gregory and H. C. Winter, *Anal. Biochem.* **11,** 519 (1965).

Copyright © 1985 by Academic Press, Inc.
All rights of reproduction in any form reserved.

Definition of Unit and Specific Activity. One unit of enzyme is defined as the amount of enzyme required to catalyze the formation of 1 μmol of CO_2 per min at 25°. Specific activity is expressed as units per milligram of protein. Protein concentration during purification is determined by the procedure of Lowry *et al.*[4] The concentration of purified enzyme is determined from its absorbance at 420 nm using a molar absorptivity, ε, of 10,000 M^{-1} cm^{-1} at pH 4.6.[5]

Purification Procedure from *E. coli*

The procedure for the growth of the *E. coli* (ATCC 11246) and isolation of the glutamate decarboxylase has been described previously.[6,7] All steps except for the heat step are carried out at 4°. After the second ammonium sulfate precipitation the enzyme is protected from light. Typical results of purification of the enzyme from *E. coli* harvested from 90 liters of medium or approximately 500 g of cell paste are shown in the table.

The purified enzyme is stored at 4° in the dark at a concentration of approximately 10 mg/ml of 100 m*M* pyridine–HCl, pH 4.6, containing 0.1 m*M* dithiothreitol and 1 m*M* pyridoxal-P. It is stable under these conditions for at least 1 year.

Properties

Molecular Properties. The molecular weight of *E. coli* glutamate decarboxylase determined by sedimentation equilibrium at pH 4.5 is approximately 310,000.[8] At pH 4.5 the enzyme has a sedimentation coefficient $s^0_{20,w}$ of 12.7 S[8–10] and a diffusion coefficient $D^0_{20,w}$ of 4.9 cm^2 sec^{-1}.[8] The enzyme dissociates at low protein concentrations, at pH values of 6.0 and higher, and at low temperature.[8,9] The bacterial enzyme is a hexamer with a subunit molecular weight of 50,000.[8] Electron microscopic data are interpreted as indicating the hexamer has dihedral symmetry with the subunits in octahedral arrangement.[10,11] The subunits appear to be identi-

[4] O. H. Lowry, N. J. Rosebrough, A. L. Farr, and R. J. Randall, *J. Biol. Chem.* **193,** 265 (1951).

[5] M. L. Fonda, *J. Biol. Chem.* **246,** 2230 (1971).

[6] M. L. Fonda and R. F. DeGrella, *Biochem. Biophys. Res. Commun.* **56,** 451 (1974).

[7] B. I.-Y. Yang and D. E. Metzler, this series, Vol. 62, p. 528.

[8] P. H. Strausbauch and E. H. Fischer, *Biochemistry* **9,** 226 (1970).

[9] R. Shukuya and G. W. Schwert, *J. Biol. Chem.* **235,** 1649 (1960).

[10] B. S. Sukhareva and A. S. Tikhonenko, *Mol. Biol. (Moscow)* **6,** 851 (1972).

[11] C. M. To, *J. Mol. Biol.* **59,** 215 (1971).

PURIFICATION OF GLUTAMATE DECARBOXYLASE FROM *Escherichia coli*

Step	Volume (ml)	Total protein (mg)	Total activity (units)	Specific activity (units/mg protein)	Purification (fold)	Yield (%)
Cell-free extract	1800	29,700	68,800	2.32	1.0	100
Streptomycin sulfate	1750	21,100	49,600	2.35	1.01	72
Ammonium sulfate precipitation (26–70%)	570	17,800	46,900	2.63	1.14	68
Heat treatment	550	10,200	44,600	4.37	1.89	65
Ammonium sulfate precipitation (30–50%)	38	5,890	39,200	6.66	2.87	57
DEAE-Sephadex	60	674	36,400	54.0	23.5	53
Sephacryl S-200	59	414	28,100	67.9	29.5	41

cal and contain an amino-terminal methionyl residue and a carboxyl-terminal sequence of Lys-His-Thr-OH.[8]

Pyridoxal-P Content. Each subunit contains a pyridoxal-P molecule covalently bound to a lysyl residue. Glutamate decarboxylase has two spectrally distinct forms. At low pH, the enzyme absorbs maximally at 415–420 nm and the coenzyme is bound as a protonated imine.[12] This enzymatically active form has a molar absorptivity of 10,000 M^{-1} cm^{-1} at 420 nm in 100 mM pyridine–HCl, pH 4.6[5] and has a large positive CD peak. In a transition centered at pH 5.6 and involving a conformational change and the loss of three or more protons, the imine is converted to a form with an absorption peak at 340 nm[13,14] and very little CD. This 340-nm form is catalytically inactive and is suggested to be a substituted aldamine.[14] The sequence around the active site lysyl residue is Ser-Ile-Ser-Ala-Ser-Gly-His-Lys(PylP)-Phe.[15]

The apoenzyme of glutamate decarboxylase is readily formed by transaminating the pyridoxal-P of the holoenzyme with α-methylglutamate.[7] The molar absorptivity of the apoenzyme subunits is 86,100 M^{-1} cm^{-1} at 280 nm.[5] The binding of pyridoxal-P to the apoenzyme shows saturation kinetics.[16,17] It occurs in at least two steps, a rapid equilibrium

[12] J. A. Anderson and H. W. Chang, *Arch. Biochem. Biophys.* **110,** 346 (1965).
[13] R. Shukuya and G. W. Schwert, *J. Biol. Chem.* **235,** 1653 (1960).
[14] M. H. O'Leary and W. Brummund, *J. Biol. Chem.* **249,** 3737 (1974).
[15] P. H. Strausbauch and E. H. Fischer, *Biochemistry* **9,** 233 (1970).
[16] M. H. O'Leary and J. M. Malik, *J. Biol. Chem.* **247,** 7097 (1972).
[17] M. L. Fonda, *Arch. Biochem. Biophys.* **170,** 690 (1975).

binding followed by a slow conformation change.[16] The K_D for the initial equilibrium step is approximately 0.2 mM.[16,17] When coenzyme is present in excess with respect to apoenzyme, K_{CO} or the concentration of pyridoxal-P needed to restore the apoenzyme to one-half of its full activity is 0.1 μM.[5]

Although a number of pyridoxal-P analogs bind to the apoenzyme, few form an active enzyme. With respect to activity obtained with pyridoxal-P, pyridoxal-P N-oxide is 29% as active, 2-nor pyridoxal-P 20%, and 6-methyl pyridoxal-P, 2-nor,6-methyl pyridoxal-P, O-methyl pyridoxal-P, α^5-methyl pyridoxal-P, and the homolog of pyridoxal-P with a 5-$CH_2CH_2OPO_3^{2-}$ substituent are 5% or less as active.[5,18]

The rate of binding pyridoxal-P to the apodecarboxylase is enhanced by halides in the order fluoride > chloride > bromide.[17] On the other hand, the rate of reconstitution is inhibited by phosphate, sulfate, and carboxylate anions, with sulfate being one of the best inhibitors.[17,19] Apparently dianions interact strongly with the phosphate binding site of the enzyme, and pyridoxal-P binds best when its phosphate group has two negative charges.[17,19] The apoenzyme is inactivated by phenylglyoxal which is thought to modify an arginyl residue at or near the pyridoxal-P binding site of the enzyme.[20] The apoenzyme is also inactivated by bromopyruvate which modifies one cysteinyl residue per subunit near the pyridoxal-P binding site.[21]

pH Optimum. The V_{max} for the decarboxylation of L-glutamate is at optimal values between pH 4.0 and 4.5.[2,22] The pH dependence of V_{max}/K_m is bell-shaped with a maximum between pH 4.4 and 4.6.[2] It appears that glutamate binds best with both its carboxylates ionized.[2,22]

Substrate and Reaction Specificity. L-Glutamate is the only naturally occurring amino acid that is decarboxylated by glutamate decarboxylase at a significant rate. The *E. coli* enzyme slowly decarboxylates γ-methylene-DL-glutamic, *threo*-β-hydroxy-DL-glutamic, L-homocysteine sulfinic, and L-homocysteic acids, but not D-glutamic, L- or D-aspartic, DL-α-amino adipic, DL-α-aminopimelic, L-cysteine sulfinic and L-cysteic acids, and DL-O-phosphoserine, L-alanine, glycine, and L-homocysteine.[2,23,24]

[18] M. L. Mechanik, Yu. M. Torchinsky, V. L. Florentiev, and M. Ya. Karpeisky, *FEBS Lett.* **13,** 177 (1971).

[19] M. H. O'Leary and S. W. Koontz, *Biochemistry* **19,** 3400 (1980).

[20] S.-H. Cheung and M. L. Fonda, *Arch. Biochem. Biophys.* **198,** 541 (1979).

[21] M. L. Fonda, *J. Biol. Chem.* **251,** 229 (1976).

[22] M. H. O'Leary, D. T. Richards, and D. W. Hendrickson, *J. Am. Chem. Soc.* **92,** 4435 (1970).

[23] A. D. Homola and E. E. Dekker, *Biochemistry* **6,** 2626 (1967).

[24] B. Jolles-Bergeret and M. Charton, *Biochimie* **53,** 553 (1971).

DL-α-Methylglutamate is slowly decarboxylated by glutamate decarboxylase and undergoes a side reaction, decarboxylative transamination, to CO_2 and levulinic acid.[25] The pyridoxal-P is converted to pyridoxamine phosphate which dissociates from the enzyme. Decarboxylative transamination is reported also to occur very slowly with high concentrations of L-aspartate and L-glutamate.[26]

L-Glutamate and L-α-methylglutamate are decarboxylated with retention of configuration.[27,28] The stereochemistry of proton transfer to form pyridoxamine phosphate in the decarboxylative transamination catalyzed by *E. coli* glutamate decarboxylase is the same as in the transamination catalyzed by aspartate aminotransferase.[29]

Mechanism-dependent inactivators of *E. coli* glutamate decarboxylase have been found. α-(Fluoromethyl)-glutamic acid appears to undergo the normal decarboxylation to form a reactive intermediate which alkylates an amino acid residue.[30] However, glutamate decarboxylase apparently removes the proton from the α-carbon of *R*-(−)-4-aminohex-5-ynoic acid[31,31a] and L-serine-O-sulfate[32] as an initial step of inactivation by these compounds.

Kinetic Properties and Mechanism. The K_m value for L-glutamate is dependent on pH. At pH 4.6 the K_m is 0.5–1.0 m*M*. The reaction is activated by halides, and the order of effectiveness is $Cl > Br > I > F$.[9,33]

The mechanism of action of pyridoxal-P dependent amino acid α-decarboxylation reactions is well established. It consists of initial noncovalent interaction of the substrate followed by transaldimination to form the enzyme–substrate aldimine. Decarboxylation produces a quinoid intermediate which undergoes a second transimination to regenerate native enzyme and the amine product. O'Leary *et al.*[34] have demonstrated that the initial binding of glutamate to the enzyme is at equilibrium and involves desolvation of substrate, that a conformation change probably accompanies the transaldimination step, and that transaldimination and decarboxylation are both partially rate determining at optimum pH.

[25] T. E. Huntley and D. E. Metzler, *in* "Symposium on Pyridoxal Enzymes" (K. Yamada, N. Katunuma, and H. Wada, eds.), p. 81. Maruzen, Tokyo, 1968.
[26] B. S. Sukhareva and A. E. Braunstein, *Mol. Biol. (Moscow)* **5,** 302 (1971).
[27] H. Yamada and M. H. O'Leary, *Biochemistry* **17,** 669 (1978).
[28] A. R. Battersby, J. Staunton, and J. Tippett, *J. Chem. Soc., Perkin Trans. 1* p. 455 (1982).
[29] B. S. Sukhareva, H. C. Dunathan, and A. E. Braunstein, *FEBS Lett.* **15,** 241 (1971).
[30] D. Kuo and R. R. Rando, *Biochemistry* **20,** 506 (1981).
[31] M. J. Jung, B. W. Metcalf, B. Lippert, and P. Casara, *Biochemistry* **17,** 2628 (1978).
[31a] C. Danzin, N. Claveric, and M. J. Jung, *Biochem. Pharm.* **33,** 1241 (1984).
[32] J. J. Likos, H. Ueno, R. W. Feldhaus, and D. E. Metzler, *Biochemistry* **21,** 4377 (1982).
[33] J. T. Gerig and L. Kwock, *Biochem. Biophys. Res. Commun.* **54,** 937 (1973).
[34] M. H. O'Leary, H. Yamada, and C. J. Yapp, *Biochemistry* **20,** 1476 (1981).

Inhibitors. Although glutamate decarboxylase shows strict specificity in binding and decarboxylating L-glutamate, cycloglutamates[35] and substituted dicarboxylic acids[2] are effective substrate-competitive inhibitors. Many aliphatic mono-, di-, and tricarboxylic acids compete with the substrate.[2,36–38] The inhibition by dicarboxylic acids shows a chain-length effect with glutaric, pimelic, and adipic acids functioning most effectively.[2] A chain-length effect also occurs in the inhibition by aliphatic monocarboxylic acids with *n*-valeric acid functioning as the best inhibitor.[37] The holoenzyme is not inhibited by iodoacetate, N-alkylmaleimides, or *p*-chloromercuribenzoate.[21]

Clostridium perfringens Glutamate Decarboxylase. The enzyme purified from *Cl. perfringens* is very similar to the *E. coli* enzyme. It is a hexamer with a molecular weight of 290,000.[39] It has a pH optimum of 4.7 and a K_m of 0.6 mM for L-glutamate at 37° in 100 mM pyridine–HCl, pH 4.7.[40]

[35] L. P. Sashchenko, E. S. Severin, D. E. Metzler, and R. M. Khomutov, *Biochemistry* **10,** 4888 (1971).
[36] C. Ressler and T. Koga, *Biochim. Biophys. Acta* **242,** 473 (1971).
[37] M. L. Fonda, *Arch. Biochem. Biophys.* **153,** 763 (1972).
[38] J. T. Gerig and L. Kwock, *FEBS Lett.* **105,** 155 (1979).
[39] I. Cozzani and G. Bagnoli, *Ital. J. Biochem.* **22,** 36 (1973).
[40] I. Cozzani, *Anal. Biochem.* **33,** 125 (1970).

[3] L-Glutamate Dehydrogenase from Bovine Liver

By HARVEY F. FISHER

$$\text{L-Glutamate} + \text{NAD(P)} + H_2O \rightleftharpoons \alpha\text{-ketoglutarate} + \text{NAD(P)H} + NH_4^+ + H^+ \quad (1)$$

L-Glutamate dehydrogenase is a pyridine nucleotide enzyme which catalyzes the reversible oxidative deamination of L-glutamate to α-ketoglutarate and ammonia. Although the reaction as written above is customarily referred to as the "forward reaction" the equilibrium actually lies far in favor of glutamate formation (the "reverse reaction").

Values for the equilibrium constant,

$$K = \frac{[\text{NAD(P)H}][\alpha\text{-ketoglutarate}][NH_4^+][H^+]}{[\text{NAD(P)}][\text{L-glutamate}]} \quad (2)$$

vary widely with ionic strength.[1] At 25°, $\mu = 0.3$, adopting the convention that the standard state for water is 55.6 M, $K = 7 \times 10^{-14}\ M^2$. The

[1] P. C. Engel and K. Dalziel, *Biochem. J.* **105,** 691 (1967).

Copyright © 1985 by Academic Press, Inc.
All rights of reproduction in any form reserved.

thermodynamic parameters (determined calorimetrically[2]) are $\Delta G^\circ = 17.9$ kcal, $\Delta H^\circ = 15.4$ kcal, $\Delta S^\circ = -8.4$ cal K^{-1}, and $\Delta C_p^\circ = -136$ cal K^{-1}.

While this enzyme is very widely distributed in nature, the mammalian forms constitute a more or less discrete class, all, for example, sharing the ability to use either NADP(H) or NAD(H) as coenzymes, along with a number of other common properties. Therefore, while the following remarks and procedures are intended to apply specifically to the beef liver enzyme, in most cases they apply at least in a qualitative sense to the corresponding enzymes isolated from other organs and from other mammalian species.

L-Glutamate dehydrogenase has a unique role in mammalian metabolism. The "reverse reaction" which it catalyzes is the only pathway by which ammonia can become bound to the α-carbon atom of an α-carboxylic acid in any mammalian tissue and thus constitutes the only source of *de novo* amino acid synthesis in such species; other amino acids can be formed only by consequent transamination reactions. It may be for this reason that the enzyme occurs at levels greater than 2 mg/ml in beef liver. Most of the enzyme is located in soluble form in the mitochondrial matrix in apparent equilibrium with a small inactive fraction bound to the inner membrane.[3]

The bovine enzyme is characterized by three sets of properties: (1) a reversible concentration-dependent association to higher molecular weight forms (a feature not found in all mammalian forms); (2) the formation of tight ternary enzyme-reduced coenzyme–substrate (or product) ternary complexes whose rates of dissociation modulate the observed steady-state reaction rates; and (3) a wide variety of effects from the binding of any of a number of nucleotide modifiers. Since all three of these features interact with each other, the resulting behavior is much more complex than that of other pyridine nucleotide dehydrogenases. The properties of the enzyme have been reviewed from the standpoint of nucleotide regulation and protein association by Frieden,[4] from that of complex formation and transient kinetic behavior by Fisher,[5] and on the basis of comparative primary structure by Smith.[6] Additional aspects are covered in other reviews.[7-9]

[2] S. Subramanian, *Biophys. Chem.* **7,** 375 (1978).
[3] C. de Duve, R. Wattiaux, and P. Baudhuin, *Adv. Enzymol.* **24,** 291 (1962).
[4] B. R. Goldin and C. Frieden, *Curr. Top. Cell. Regul.* **4,** 77 (1971).
[5] H. F. Fisher, *Adv. Enzymol.* **39,** 369 (1973).
[6] E. L. Smith, B. M. Austen, K. M. Blumenthal, and J. F. Nyc, *in* "The Enzymes" (P. D. Boyer, ed.), 3rd ed., Vol. 11, p. 293. Academic Press, New York, 1975.
[7] C. Frieden, *in* "The Enzymes" (P. D. Boyer, ed.), 2nd ed., Vol. 7, p. 3. Academic Press, New York, 1963.

Here we will describe briefly only those properties of the enzyme which are most directly related to its purification, its assay, and to rate studies of the reaction it catalyzes.

Properties of the Enzyme

Protein Properties. The lowest molecular weight form of the bovine liver enzyme which is stable in aqueous solution consists of a hexamer of identical folded peptide chains each having a molecular weight of 55,390.[10–13] This hexamer (MW = 332,000) dissociates irreversibly into unfolded peptide chains at a pH below 2 or above 11,[14,15] by 2 *M* guanidine hydrochloride,[16] or by 6 *M* urea. The hexamer is also the lowest molecular weight form for which catalytic activity can be demonstrated. It contains six fully independent catalytically active sites. The enzyme has an unusually high degree of thermal stability relative to that of other dehydrogenases and to those of most other proteins found in the liver, a fact which we make use of in the enzyme preparation procedure described below. The free enzyme in solution can survive several minutes of treatment at 56°. Most of the binary complexes formed between the enzyme and its cofactors and modifiers, however, are not stable at such temperatures. The glutamate dehydrogenase–NADPH complex, for example, will not survive temperatures above 45°.

Solutions of the enzyme suffer a considerable loss of activity upon freezing, and lyophilization usually results in a preparation containing at least 15% inactive enzyme. The enzyme is somewhat more stable in phosphate buffer than in Tris, although the reaction rate is faster in the latter buffer. If it is necessary or advantageous to work in Tris solutions, however, a 0.1 *M* Tris buffer containing as little as 0.01 *M* phosphate will provide adequate stability. Inactive enzyme at concentrations of 0.01 mg/ml or higher usually aggregates into a very high molecular weight insoluble

---

[8] C. Frieden, *in* "The Mechanism and Action of Dehydrogenases" (G. W. Schwert and A. D. Winer, eds.), p. 197. University Press, Lexington, Kentucky, 1969.

[9] H. F. Fisher, *in* "The Mechanism and Action of Dehydrogenases" (G. W. Schwert and A. D. Winer, eds.), p. 221. University Press, Lexington, Kentucky, 1969.

[10] K. Moon and E. L. Smith, *J. Biol. Chem.* **248,** 3082 (1973).

[11] H. F. Fisher, D. G. Cross, and L. L. McGregor, *Nature (London)* **196,** 4857 (1962).

[12] J. A. Olsen and C. B. Anfinsen, *J. Biol. Chem.* **197,** 67 (1952).

[13] A. B. Hooper, J. Hansen, and R. Bell, *J. Biol. Chem.* **242,** 288 (1967).

[14] H. F. Fisher, L. L. McGregor, and U. Power, *Biochem. Biophys. Res. Commun.* **8,** 402 (1962).

[15] H. F. Fisher, L. L. McGregor, and D. G. Cross, *Biochim. Biophys. Acta* **65,** 175 (1962).

[16] M. Cassman and H. K. Schachman, *Biochemistry* **10,** 1015 (1971).

form which is easily removed by precipitation, again a fact which is useful in recycling "used" enzyme.

Solvent perturbation studies have shown that in the intact hexamer all of the phenylalanine groups are completely buried and inaccessible to water. Therefore, the appearance of the characteristic phenylalanine pattern of small sharp differential peaks at 253, 260, 263, and 267 nm,[17] when even a dilute solution of the enzyme is measured against the same enzyme in a solvent of slightly different refractive index, may be taken as evidence of denatured enzyme.

Protein Association. At protein concentrations below 0.1 mg/ml, the enzyme is completely in the form of the hexamer. As the protein concentration is raised, however, these hexamers associate until, at a concentration of 3 mg/ml, the weight-average molecular weight has about tripled. The process is not complete at this point, however. As the protein concentration is raised further, the molecular weight continues to increase gradually but without limit, as linear aggregates form and lengthen. These observations also apply quantitatively to the pig[18] and human liver[19] enzymes, and at least qualitatively to the chicken liver enzyme.[20] The enzymes from dogfish,[21] tuna,[22] and rat liver,[23] and that from rat kidney[24] do not undergo any such protein concentration-dependent association.

While there was a good deal of confusion on the point in the literature some years ago, it is now accepted as an established fact that the specific reaction rate of the enzyme is independent of protein concentration, and that the catalytically active sites themselves are quite unaffected by this higher association of hexamers. The binding of some of the allosteric modifier ligands, however, affects and is dependent on the state of association of the enzyme; in the presence of such modifiers, of course, the specific reaction rate does become dependent on protein concentration. These effects will be considered below under "Mutual Interactions between Protein Association and Nucleotide Modifier Binding."

Substrate Specificity. In addition to L-glutamate, the enzyme can catalyze the reversible oxidative deamination of L-α-amino monocarboxylic

[17] D. G. Cross and H. F. Fisher, *Biochemistry* **5,** 880 (1966).
[18] P. Dessen and D. Pantaloni, *Eur. J. Biochem.* **8,** 292 (1969).
[19] H. Kubo, M. Iwatsubo, H. Watari, and T. Soyama, *J. Biochem.* (*Tokyo*) **46,** 1171 (1959).
[20] C. Frieden, *Biochim. Biophys. Acta* **62,** 423 (1962).
[21] L. Corman, L. M. Prescott, and N. O. Kaplan, *J. Biol. Chem.* **242,** 1383 (1967).
[22] F. M. Veronese, private communication with E. L. Smith, *in* "The Enzymes" (P. D. Boyer, ed.), 3rd ed., Vol. 11A, p. 308. Academic Press, New York, 1975.
[23] K. S. King and C. Frieden, *J. Biol. Chem.* **245,** 4391 (1970).
[24] S. Seyama, T. Saeki, and N. Katunuma, *J. Biochem.* (*Tokyo*) **73,** 39 (1973).

acids.[25-27] The pH optima for the monocarboxylic substrate reactions are 1 to 1.5 pH units higher than that of L-glutamate and the substrate K_m values are much higher (from 30 to 100 mM). The V_{max} values for these substrates, however, measured at their own pH optima are as high (and in a few cases a bit higher) as that of L-glutamate measured at its optimum pH. Statements to the contrary in the literature are based on measurements made at fixed and arbitrary pHs which yield rates less than 2% of the V_{max} rate. A number of investigators have used L-alanine as a substrate for studying monocarboxylic acids, presumably because it was employed in the original proof that monocarboxylic substrates used the same active site involved in the L-glutamate reaction.[27] L-Norvaline or L-α-aminobutyrate is a much better choice for such studies.

Coenzyme Specificity. Glutamate dehydrogenase is one of the relatively few individual enzymes whose active site can utilize either NADP(H) or NAD(H) at reasonably similar rates. The NAD(H) reaction is the faster of the two but, while quite probably of considerable physiological significance, it is considerably more complex and correspondingly more difficult to study for reasons described in the next section. The enzyme is also fully reactive with 3-acetylpyridine adenine dinucleotide, with a number of analogs of NAD in which the adenine moiety has been modified, and with NMNH. (The K_m for the latter coenzyme is in the mM range.)

Modifiers. While the six active sites of the hexamer appear to act completely independently showing no evidence of positive or negative cooperativity among themselves, the enzyme is nevertheless subject to allosteric regulation by such naturally occurring nucleotides as ADP and GTP as well as by its own coenzymes, substrates, and reaction products. Since these effects must be carefully considered in assays and other rate measurements of the enzyme catalyzed reaction, it is useful at this point to consider briefly certain aspects of the reaction mechanism itself. The oxidative deamination of L-glutamate proceeds in four sequential phases which are more or less well separated in time.

Phase I: Enzyme + NADP + glutamate
$$\rightleftharpoons \text{(enzyme–NADP–glutamate)}$$

The formation of this ternary complex in which the two ligands are mutually bound much tighter than in their respective binary complexes is over in less than 1 msec.

[25] K. H. Bassler and C. H. Hammar, *Biochem. Z.* **330,** 446 (1958).
[26] J. Struck and I. W. Sizer, *Arch. Biochem. Biophys.* **86,** 260 (1960).
[27] H. F. Fisher, *J. Biol. Chem.* **236,** 791 (1961).

Phase II: (Enzyme–NADP–glutamate)
$\rightleftharpoons$ (enzyme–NADPH–α-iminoglutarate)
Enzyme–NADPH–α-iminoglutarate) + H_2O
$\rightleftharpoons$ (enzyme–NADPH–α-ketoglutarate) + NH_3

Since the first of these two reaction (hydride transfer) is somewhat slower than the second reaction (hydrolysis) they appear as a single phase lasting about 100 msec. Under forcing conditions (40 m*M* glutamate, 160 μM NADP) about 85% of the total enzyme is in the form of the enzyme–α-NADPH–α-ketoglutarate complex at the conclusion of the phase. It is this ternary complex whose properties dominate the kinetic behavior of both the forward and reverse reactions. In it α-ketoglutarate and NADPH are each mutually bound some 200-fold more tightly than in the absence of the other. The 340 nm band of NADPH is blue-shifted to 332 nm, distinguishing this complex from all other glutamate dehydrogenase–NADPH complexes, whose 340 nm bands are red-shifted to 348 nm.

Phase III: (Enzyme–NADPH–α-ketoglutarate) + glutamate
$\rightleftharpoons$ (enzyme–NADPH–glutamate) + α-ketoglutarate

This phase involves the interconversion of two ternary complexes, and is characterized experimentally by the shift of the reduced nicotinamide peak from 332 to 348 nm. Its rate is dominated by the off-rate of α-ketoglutarate from its ternary complex. This rate of dissociation is remarkably slow for an enzyme process; its half-life is almost 2 sec.

Phase IV: (Enzyme–NADPH–glutamate)
$\rightleftharpoons$ enzyme + NADPH + glutamate

In this phase, which proceeds several times faster than the previous one, free NADPH is finally released to the solution and free enzyme, ready to begin a new catalytic cycle, first becomes available. The rate of this process (or, at low glutamate concentrations, that of the preceding phase) constitutes the steady-state rate customarily used as a measure of enzymatic activity.

While there are a number of other intermediate steps now known to occur in the reaction, and while those intermediates are important in understanding the chemistry of the catalytic mechanism, the four gross observable phases described above will suffice for understanding the bewildering and often contradictory profusion of effects on the rate of this reaction that have been reported in the literature.

It has been known for many years, for example, that ADP activates the glutamate reaction but inhibits the reaction when a monocarboxylic amino acid is the substrate. GTP has completely opposite effects, inhibiting the glutamate reaction, while activating that of a monocarboxylic substrate.

Iwatsubo[28] carried out the first stopped-flow experiments demonstrating the "burst" phenomenon (a rapid increase in absorbance, Phases I and II) followed by a much slower steady increase (Phases III and IV). He also observed that ADP and GTP exerted their well-known steady-state effects on the latter slow phase of the glutamate reaction, but that each had quite the opposite effect on the early burst phase; ADP inhibited the burst slightly, while GTP activated it slightly.

With alanine as a substrate, on the other hand, no burst occurred. ADP inhibited and GTP activated the single linear phase. These experiments, understood in the context of the four-phase mechanism described above, provide a simple and quite sound basis for explaining most of the many modifier effects reported since that time. In the glutamate reaction, the rate of release of NADPH from its ternary complexes is the rate-limiting step of the steady state. It is now known that ADP, binding at a site nonidentical to, but overlapping with the NADPH binding site, weakens NADPH binding, thus speeding up the slowest step in the reaction.[29] ADP also weakens the binding of NADP in Phase I, but this is a very fast step and slowing it does not affect the steady-state rate. GTP is now known to bind cooperatively with the enzyme–NADPH binary complex, decreasing the dissociation of NADPH from any of its complexes and thus slowing the rate-limiting step. Again, the effect of GTP on increasing the binding of NADP in Phase I has no effect on the overall steady-state rate. Monocarboxylic amino acids cannot engage in the mutually cooperative binding interactions which produce the tight ternary complexes characteristic of the dicarboxylic substrate–coenzyme complexes. Their entire reaction sequence consists of Phases I and II. Thus, steady-state rate measurements of the oxidative deamination of monocarboxylic amino acids show the same effects of ADP and GTP observed in Phases I and II of the dicarboxylic reaction and these effects are opposite to those observed in the rate-limiting step of that reaction.

Thus, the application of these two simple ideas—that the reaction rate is limited by the rate of NADPH dissociation (for dicarboxylic substrates) and that NADPH dissociation itself is subject to being increased or decreased by other ligands in a noncompetitive manner—permits the understanding, and in many cases, the prediction of many modifier effects in this system.

The behavior described above applies only to reactions in which

[28] M. Iwatsubo and D. Pantaloni, *Bull. Soc. Chim. Biol.* **49,** 1563 (1967).

[29] A. H. Colen, D. G. Cross, and H. F. Fisher, *Biochemistry* **13,** 2341 (1974).

NADP(H) itself serves as the coenzyme; the use of NADH as a coenzyme introduces a host of additional complications. NADH does bind to the same catalytic active site as does NADPH and participates in the catalytic reaction in the same manner. In addition, however, a second molecule of NADH can bind to the ADP binding site of the enzyme using the adenosyl-ribose-diphosphate moiety which comprises half of its structure. Indeed, it has been shown that in both cases the adenine ring of the ligand is bound to the enzyme.[30] NADPH cannot bind to this site, presumably because of steric hindrance by the third phosphate group. Since NADH binds to the ADP regulatory site somewhat more tightly than to the catalytic site, and since each of these binding modes mutually weakens the binding of the other in a noncompetitive manner, the reaction rate of the NADH reaction is severalfold faster than that of the NADPH reaction. At high concentrations of NADH, a third effect may be observed: NADH inhibition of the reaction. Again, this effect does not occur with NADPH as the substrate. While the details of the multiple binding of NADH to glutamate dehydrogenase are still being sorted out, it may be noted that a good many years ago Frieden noted the complex curvature of double reciprocal plots using NADH and warned of the multiple pitfalls of interpreting such rate measurements.[31]

Mutual Interactions between Protein Association and Nucleotide Modifier Binding. In previous sections, we have described protein association in the absence of allosteric modifier binding, as well as nucleotide modifier binding as it occurs at protein concentrations below the point of protein association. We have also indicated that such association neither affects nor is affected by the binding of NADPH, NADH, ADP (by itself), or any of the substrates or products. That statement does not hold true, however, for at least two important modifiers—GTP and the regulatory and inhibiting binding modes of NADH. GTP requires two subsites for effective binding. One is on, or made available in the presence of, enzyme-bound NADPH. The second appears to be located on a portion of the enzyme surface which becomes buried as the hexamers associate at high protein concentration.[32] As a result, the K_D for GTP binding increases dramatically as protein concentration is raised, and, as a necessary concomitant, the extent of protein dissociation decreases in the presence of a mixture of GTP and NADPH.[33] This explains the sigmoidicity observed in the binding of GTP at high protein concentrations.

[30] D. Pantaloni and P. Dessen, *Eur. J. Biochem.* **11,** 510 (1969).
[31] C. Frieden, *J. Biol. Chem.* **234,** 809 (1959).
[32] D. G. Cross and H. F. Fisher, *J. Biol. Chem.* **245,** 2612 (1970).
[33] C. Frieden and R. F. Colman, *J. Biol. Chem.* **242,** 1705 (1967).

As we have stated, ADP binding to either free enzyme or to an enzyme–NADPH complex neither affects nor is affected by protein concentration. However, NADH when bound to this same ADP binding site, because of its greater length, appears to extend into the interhexamer region of the associating protein.[5] As a result, it shows effects similar to that of GTP at high protein concentrations adding one more level of complexity to the already complicated picture of NADH binding. Since both GTP and NADH interact with the binding of NADPH, ADP, L-glutamate, and α-ketoglutarate, the effects of protein concentration are transmitted by that linkage even to those ligands whose binding is not intrinsically involved with protein association. Finally, it should be noted that at the physiological concentration of glutamate dehydrogenase (>2 mg/ml) the enzyme is about 50% in the associated form and that the complex set of interactions described above must be presumed to occur. For this reason, the possible role of glutamate dehydrogenase as an important metabolic regulator must be given serious consideration.

Inhibitors. Given that coenzyme release is the rate limiting step in glutamate oxidation, and that coenzyme binding is itself subject to noncompetitive inhibition or activation, it follows that any substance capable of binding at any site overlapping with that of NADPH itself, or with that of any other ligand whose site overlaps the NADPH binding site, may be expected to affect the reaction rate. It is not surprising, then, that well over 300 compounds have been reported to have significant effects on the reaction.

Most of these inhibitory or activating effects have been of a noncompetitive nature (where the point has been tested) and many have been done with NADH, where as we have shown, the possibility of quantitatively large but mechanistically trivial effects is quite large. Indeed, it may be stated that 50 m*M* of almost anything added to a glutamate dehydrogenase reaction has a better than even chance of causing a significant change in rate. The enzyme appears to have at least one hydrophobic patch which serves as part of the binding site for the reduced nicotinamide ring, and the corresponding ADP and GTP binding sites probably also contain such patches. A large number of compounds containing aromatic or hydrophobic moieties have been reported to bind to, and to affect the rate of glutamate dehydrogenase at quite low concentrations. However, these effects, too, are typically noncompetitive and of uncertain significance.

Of more interest is the extensive list of inhibitors competitive with L-glutamate. Caughey, Smiley, and Hellerman[34] found a group of com-

[34] W. S. Caughey, J. D. Smiley, and L. Hellerman, *J. Biol. Chem.* **224,** 591 (1957).

pounds with structures as chemically diverse as isophthalate, *m*-iodobenzoate, and 5-bromofuroate as well as glutarate and D-glutamate to be in this class. They found that the only thing these effective inhibitory compounds had in common was a critical distance of 7.45 Å between a pair of strong negative charges. This generalization has stood up very well over the years. It may be noted that the inhibitors listed above are noncompetitive inhibitors of α-ketoglutarate in the reverse direction. Rife and Cleland have recently found oxalylglycine to be an effective and fully competitive inhibitor against α-ketoglutarate.[35] This is the only known compound with this property. Finally, it should be mentioned that in addition to the substrate inhibition by NADH mentioned above, α-ketoglutarate is inhibitory in the reverse direction at concentrations above 10 m*M*.

Assay Method

The enzyme is assayed spectrophotometrically by following the decrease in absorbance at 340 nm using the reverse reaction [Eq. (1)]. A solution containing final concentrations of NADPH = 100 μM, α-ketoglutarate = 5 m*M*, NH_4^+ = 50 m*M*, EDTA = 0.1 m*M*, bovine serum albumin = 1 mg/ml; in 0.1 *M* phosphate buffer pH 7.6 is incubated in a 1-cm path-length quartz cuvette at 25°. The reaction is started by adding enzyme (if possible in an "adder–mixer") sufficient to provide a final concentration of about 1.7×10^{-4} mg/ml. These conditions provide a rate of 0.036 A/min which remains linear for at least 1.5 min, and avoid substrate inhibition effects. It is essential that the measured rate not exceed 0.04 A/min to ensure that initial rates are being measured. If a recording spectrophotometer is used, the 0 to 0.1 A full scale setting should be used, and the chart speed should be at least 10 in./min. For assays (using the strict meaning of the word—determination of the amount of active enzyme in pure preparation) NADH may be used, as it is less expensive and somewhat faster. For rate measurements, on the other hand, where kinetic, pH, or inhibition studies are involved, NADPH should be used for reasons described earlier. Assays done on homogenates, extracts, or other crude preparations may be expected to be subject to large errors due to the presence of any of a variety of modifiers. These may be estimated by adding the same sized aliquot of a boiled preparation used in the assay to a standard assay system using pure enzyme. The ratio of the activities of the pure assay system in the presence and absence of the "crude" aliquot multiplied by the rate in the crude assay provides a more valid measure of the enzyme content of the crude system.

[35] J. E. Rife and W. W. Cleland, *Biochemistry* **19,** 2238 (1980).

Purification Procedure[36,37]

Extraction. Chopped liver (6 kg) is homogenized in 3 volumes of water for 3 min; 50 g of Na_2SO_4 per liter of homogenate is added. This suspension is heated rapidly to 56°, held at this temperature for 3 min, cooled to 45°, and centrifuged at 13,000 *g* for 30 to 60 min at 25°. This operation has three effects: extraction of the mitochondrial enzyme by thermal and osmotic shock, salt stabilization of active enzyme, and elimination of inactive protein.

Sodium Sulfate Fractionation. Twenty grams of Na_2SO_4 per 100 ml of solution is added to the supernatant solution obtained in the previous step. The suspension is centrifuged at 13,000 *g* for 20 min at 20°, and the precipitate containing the enzyme is dissolved in 0.1 *M* phosphate buffer (pH 6.4). The volume is adjusted to provide a final Na_2SO_4 concentration of 7.5%. The protein solution is subjected to a second thermal treatment, precipitated by 17.5% Na_2SO_4, and collected by centrifugation (34,000 *g* for 20 min at 25°).

Crystallization. The precipitate from the preceding step is dissolved in the minimum volume of pH 6.4 buffer. The Na_2SO_4 concentration must be at least 10%. After standing at 5° for several days, two kinds of crystals form. They are separated by differential centrifugation which results in a heavy layer of catalase crystals and an upper light layer of glutamate dehydrogenase. The upper layer is crystallized 5 times by the procedure described above. The yield is 1.5 to 2 g of enzyme. Protein concentration is determined spectrophotometrically using an absorbance coefficient of 0.973 $cm^2\ mg^{-1}$.

Most work reported in the literature currently is done with commercial preparations usually in the form of ammonium sulfate suspensions. The following procedure may be used to prepare such material for experimental use. The ammonium sulfate suspension is dialyzed against 22 volumes of 0.1 *M* potassium phosphate buffer at 4° for 24 hr. The step is repeated seven times, changing the buffer every 2 hr. The preparation is centrifuged at 5000 rpm for 5 min, and the precipitate discarded. The supernatant is filtered first through a Millipore "AP" prefilter, then through a Millipore "RA" 1.2-μm filter, and a third time through a Norite treated 1.2-μm filter prepared in the following manner. Norite A (4.5 g) is stirred with 400 ml of water, allowed to settle, and decanted. The process is repeated three times and the Norite is slurried with 300 ml of 0.1 *M*

[36] H. Kubo, I. Yamano, M. Iwatsubo, H. Watari, T. Soyama, and J. Shiraishi, *Bull. Soc. Chim. Biol.* **40,** 431 (1958).

[37] For an earlier preparation of the beef liver enzyme, see this series, Vol. 2 [27]. For preparation of this enzyme from other sources, see this series, Vol. 17 [117, 118, 119].

phosphate buffer pH 7.6. Of the stirred slurry 20 ml is poured onto the 1.2-μm filter, vacuum applied, and the dry filter washed three times with 20 ml of buffer. The enzyme solution is then filtered through this treated membrane and finally refiltered through an untreated 0.45-μm filter. The resulting enzyme solution should show a A_{280}/A_{260} ratio of 1.90 to 1.93. If the ratio is less than 1.90, the Norite filtration treatment should be repeated. The procedure described here may be used to reclaim concentrated batches of used enzyme. Since inactive enzyme tends to precipitate at high concentrations. Enzyme which has been exposed to ADP, however, cannot be reclaimed in this manner. The specific activity of the pure enzyme, using the assay conditions described above is about 25 μM min^{-1} mg^{-1}.

[4] *N*-Acetylglutamate Synthase

By SUSAN G. POWERS-LEE

N-Acetylglutamate synthase (acetyl-CoA : L-glutamate *N*-acetyltransferase, EC 2.3.1.1) catalyzes the reaction shown in Eq. (1). The enzyme has been studied in bacteria,[1-3] and in the mitochondrion of yeast,[4] mam-

$$\text{L-Glutamate} + \text{acetyl-CoA} \rightarrow N\text{-acetyl-L-glutamate} + \text{CoA} \quad (1)$$

malian liver,[5-10] and rat small intestinal mucosa.[11] In bacteria and yeast, *N*-acetylglutamate synthase catalyzes the first step in the biosynthesis of arginine; this enzyme is subject to feedback inhibition and repression by arginine. In the liver, and apparently also in the small intestinal mucosa, the only known role of *N*-acetylglutamate is as the essential allosteric

[1] D. K. Marvil and T. Leisinger, *J. Biol. Chem.* **252,** 3295 (1977).
[2] D. Haas, V. Kurer, and T. Leisinger, *Eur. J. Biochem.* **31,** 290 (1972).
[3] A. T. Abdelal and O. V. Nainan, *J. Bacteriol.* **137,** 1040 (1979).
[4] B. Wipf and T. Leisinger, *J. Bacteriol.* **140,** 874 (1979).
[5] K. Shigesada and M. Tatibana, *Eur. J. Biochem.* **84,** 285 (1978).
[6] T. Sonoda and M. Tatibana, *J. Biol. Chem.* **258,** 9839 (1983).
[7] S. Kawamoto, H. Ishida, M. Mori, and M. Tatibana, *Eur. J. Biochem.* **123,** 637 (1982).
[8] S. Kawamoto and M. Tatibana, *FEBS Lett.* **151,** 117 (1983).
[9] F. X. Coudé, G. Grimber, P. Parvy, and P. Kamoun, *Biochem. Biophys. Res. Commun.* **102,** 1016 (1981).
[10] C. Bachmann, S. Krähenbühl, and J.-P. Colombo, *Biochem. J.* **205,** 123 (1982).
[11] C. Uchiyama, M. Mori, and M. Tatibana, *J. Biochem.* (*Tokyo*) **89,** 1777 (1981).

Copyright © 1985 by Academic Press, Inc.
All rights of reproduction in any form reserved.

effector for carbamyl phosphate synthetase I,[12,13] the enzyme that catalyzes the entry and rate-limiting step of the urea cycle. The rate of production of *N*-acetylglutamate seems to be an important mechanism of regulation for carbamyl phosphate synthetase I since the usual *N*-acetylglutamate level (0.1 m*M*) is about equal to the K_m of carbamyl phosphate synthetase I for *N*-acetylglutamate. Furthermore, the liver *N*-acetylglutamate synthase is specifically activated by arginine, and it has been suggested that this activation might serve as a link between amino acid metabolism, as gauged by the level of free arginine, and elimination of excess ammonia via the urea cycle.[14] Although the *N*-acetylglutamate-dependent liver carbamyl phosphate synthetase I reaction can also function as the first step of arginine biosynthesis if arginase does not act to replenish ornithine and complete the urea cycle, this role is quantitatively minor relative to the flux through the complete urea cycle. The finding that arginine activates, rather than inhibits, the liver *N*-acetylglutamate synthase presumably also reflects the major metabolic function of carbamyl phosphate synthetase I.

Assay Method

Principle and Precautions. The most generally used assay is based on the acetyl-CoA-dependent incorporation of radioactivity from [^{14}C]glutamate into *N*-acetyl-[^{14}C]glutamate and the subsequent separation of the radioactive components by chromatography on Dowex 50.[2,5] An assay has also been developed in which the incorporation of radioactivity from [^{14}C]acetyl-CoA into *N*-[^{14}C]acetylglutamate is determined.[5]

The following precautions are necessary in the assay of the rat liver enzyme[6]: (1) initial detection of the activity is possible only in extracts of isolated mitochondria, presumably because the substrates are utilized by other enzymes in whole tissue extracts; (2) glassware used for the assay must be silicone coated; (3) at least 1 milliunit of enzyme is required in the assay in order to ensure linearity of product formation as a function of the amount of enzyme present; and (4) the reaction must be carried out at 25° since there is a discontinuity in the Arrhenius plot of the enzyme reaction at 30°, with decreased activity above this temperature.[15] The *Escherichia coli* enzyme also loses activity as a result of dilution and therefore the

[12] L. A. Fahien, J. M. Schooler, G. A. Gehred, and P. P. Cohen, *J. Biol. Chem.* **239,** 1935 (1964).

[13] C. M. Allen and M. E. Jones, *Arch. Biochem. Biophys.* **114,** 115 (1966).

[14] M. Tatibana and K. Shigesada, *in* "The Urea Cycle" (S. Grisolia, R. Baguena, and F. Mayor, eds.), p. 301. Wiley, New York, 1976.

[15] M. Tatibana, S. Kawamoto, T. Sonoda, and M. Mori, *Adv. Exp. Med. Biol.* **153,** 207 (1982).

enzyme is diluted in buffer containing 30 mg/ml bovine serum albumin before it is added to the assay.[1]

One unit of activity is defined as that quantity of enzyme which catalyzes the formation of 1 pmol of *N*-acetylglutamate per minute.

Reagents

10 m*M* L-[^{14}C(U)]Glutamic acid, 5.0 Ci/mol
5 m*M* Acetyl-CoA
100 m*M* Tris–HCl plus 2.0 m*M* EDTA, pH 8.2
1 *M* HCOOH plus 100 m*M* *N*-acetylglutamate
0.1 *M* HCOOH
Column (0.2 × 1.0 cm) of Dowex 50-X8 (H^+ form, 200–400 mesh)

Procedure.[5] The assay mixture consists of 1 m*M* L-[^{14}C]glutamic acid, 0.5 m*M* acetyl-CoA, 1 m*M* EDTA, 50 m*M* Tris–HCl, pH 8.2, in a final volume of 0.1 ml. Acetyl-CoA is omitted from the control assay mixture. The reaction is initiated with the addition of enzyme and allowed to incubate at 37° (25° for the mammalian liver enzyme, see above) for 5–30 min. The reaction is stopped by the addition of 0.05 ml of the formic acid plus *N*-acetylglutamate. An aliquot (0.1 ml) is applied to the Dowex 50 column, washed with 1 ml of 0.1 *M* HCOOH, and the entire eluate collected. For at least partially purified enzyme, the radioactivity in the eluate may be counted directly. For crude extracts, the acetyl[^{14}C]glutamate must be further purified by chromatography on Whatman 3MM paper in ethyl ether/benzene/formic acid/water (22/9/7/2).

Purification Procedure

Source and Stabilization. *N*-Acetylglutamate synthase has been purified 300-fold to apparent homogeneity from *E. coli*[1] and *Salmonella typhimurium*.[3] The final specific activity of these preparations is 133 units/mg. In contrast, a 24,000-fold purification is necessary to produce nearly homogeneous *N*-acetylglutamate synthase (specific activity, 14 units/mg) from rat liver.[6] Only the rat liver protocol will be described in detail in this chapter. Purification of the synthase from bacteria involves chromatography on DEAE-cellulose and hydroxylapatite and utilizes the large decrease in affinity of the enzyme for hydroxylapatite in the presence of *N*-acetylglutamate. *N*-Acetylglutamate synthase has also been partially purified from small intestinal mucosa[11] (specific gravity, 31.6 microunits/mg), human liver[10] (specific activity, 2.86 milliunits/mg), and *Pseudomonas aeruginosa*[2] (specific activity not reported).

In order to stabilize the rat liver enzyme, it is necessary to include 0.1% Triton X-100 in the buffers and to use silicone-treated glassware or plastic tubes (polyethylene, polycarbonate, or polypropylene). In order to

TABLE I
SUMMARY OF PURIFICATION PROCEDURE FOR *N*-ACETYLGLUTAMATE SYNTHASE FROM RAT LIVER[6]

Step	Total protein (mg)	Specific activity[a] (units/mg × 10^3)	Overall yield (%)	Purification (fold)
1. Mitochondrial extract	10,000	0.574	66[b]	—
2. 40% $(NH_4)_2SO_4$	1,830	4.83	100[b]	8.4
3. 37% $(NH_4)_2SO_4$	1,010	8.76	100	15.3
4. DEAE BioGel A (pH 7.8)	340	19.3	74	33.6
5. Sephacryl S-200	225	21.5	55	37.5
6. DEAE BioGel A (pH 8.5)	136	43.2	66	75.3
7. Phenyl-Sepharose	72.0	68.9	56	120
8. Aminooctyl BioGel A	25.4	152	44	265
9. Affi-Gel Blue	4.43	413	32	720
10. Acetylglutaminyl BioGel A	1.96	719	16	1,253
11. Sucrose gradient	0.270	1,460	4.5	2,544
12. Isoelectric focusing	0.024	14,000	3.8	24,390

[a] Arginine (1 m*M*) was included in the standard assay mixture.
[b] Since a greater amount of activity was present after Step 2, this amount was treated as 100%.

stabilize the bacterial enzyme, 15% (v/v) glycerol, 10 m*M* *N*-acetylglutamate and 2 m*M* phenylmethylsulfonyl fluoride are included in the extraction buffer.

The purification procedure for rat liver *N*-acetylglutamate synthase is as previously described[6] and is summarized in Table I. Unless specified otherwise, all operations are carried out at 0–4° and all buffers contain 1 m*M* dithiothreitol and 0.1 m*M* EDTA. All buffers from Step 4 on also contain 0.1% Triton X-100.

Step 1. Mitochondrial Extract. Rat liver mitoplasts are isolated from 985 g of liver by the procedure of Schnaitman and Greenawalt,[16] with the omission of bovine serum albumin from the isolation medium. The mitoplasts are suspended in 493 ml of 50 m*M* potassium phosphate, pH 7.5, and 25-ml aliquots are disrupted by sonication at 20 kHz for 6 min. The sonicated solution is centrifuged for 60 min at 105,000 *g* and 4° to remove the membrane fraction.

Step 2. 40% Ammonium Sulfate Precipitation. Ammonium sulfate (136 g) is added to the mitochondrial matrix fraction (558 ml) to 40% saturation. The precipitate is collected by centrifugation and dissolved in 295 ml of 50 m*M* potassium phosphate, pH 7.8.

[16] C. Schnaitman and J. W. Greenwalt, *J. Cell Biol.* **38,** 158 (1968).

Step 3. 37% Ammonium Sulfate Precipitation. Ammonium sulfate (65 g) is added to the solution to 37% saturation. The precipitate is collected by centrifugation and dissolved in 50 ml of 10 m*M* potassium phosphate plus 0.1% (w/v) Triton X-100.

Step 4. First DEAE BioGel A Chromatography. The ammonium sulfate fraction is desalted on a column of Sephadex G-25 (fine) into 10 m*M* potassium phosphate, pH 7.8, and applied to a DEAE BioGel A column (2 × 10 cm) that has been equilibrated with the same buffer. The DEAE column is washed with 30 ml of the equilibration buffer and developed (flow rate, 100 ml/hr) with a linear gradient (200 ml) of 0 to 0.26 *M* KCl in 10 m*M* potassium phosphate and 0.5 *M* urea, pH 7.8. The synthase elutes at 0.12 to 0.17 *M* KCl. Active fractions are concentrated in an Amicon ultrafiltration cell (XM-50 membrane) to a volume of 5 ml.

Step 5. Sephacryl S-200 Chromatography. The enzyme solution is applied to a Sephacryl S-200 column (2.4 × 33 cm) equilibrated with 50 m*M* potassium phosphate, pH 7.8, and eluted with the same buffer at a flow rate of 60 ml/hr. Active fractions are pooled and adjusted to pH 8.5 with 0.1 *M* Tris.

Step 6. Second DEAE BioGel A Chromatography. The enzyme solution is applied to a DEAE BioGel A column (1.5 × 8 cm) equilibrated with 20 m*M* Tris–HCl, pH 8.5. The column is washed with 15 ml of the equilibration buffer and developed (flow rate, 50 ml/hr) with a linear gradient (160 ml) of 0 to 0.26 *M* KCl in 20 m*M* Tris–HCl and 0.5 *M* urea, pH 8.5. Active fractions are pooled, solid KCl is added to 2 *M* and the pH adjusted to pH 7.8 with 0.1 *M* KOH.

Step 7. Phenyl-Sepharose Chromatography. The enzyme solution is applied to a Phenyl-Sepharose CL-4B column (1.5 × 8 cm) equilibrated with 2 *M* KCl in 10 m*M* potassium phosphate, pH 7.8. The column is washed with 40 ml of the equilibration buffer, then 40 ml of 50 m*M* potassium phosphate, pH 7.8, and developed (flow rate, 50 ml/hr) with a linear gradient (120 ml) of 0.1% Triton X-100 in 50 m*M* potassium phosphate, pH 7.8, to 2% Triton X-100 in 10 m*M* potassium phosphate, pH 7.8. The enzyme elutes at 0.5 to 1% Triton X-100. Active fractions are concentrated in an Amicon ultrafiltration cell (XM-50 membrane) to a volume of 5.2 ml.

Step 8. Aminooctyl BioGel A Chromatography. The enzyme solution is applied to an aminooctyl BioGel A-0.5m[17] column (1.5 × 4 cm) equili-

[17] Aminooctyl- and acetylglutaminyl-BioGel A were prepared by the protocol of Cuatrecasas [P. Cuatrecasas, *J. Biol. Chem.* **245,** 3059 (1970)]. BioGel A-0.5 m (20 g wet weight) was activated by 2 g of cyanogen bromide and divided into two aliquots. Octamethylene diamine (5 g in 0.1 *M* sodium borate, pH 9.5) was added to one aliquot of activated resin and acetylglutamine (2 g in 0.1 *M* sodium phosphate, pH 7) to the other aliquot.

brated with 10 mM potassium phosphate, pH 7.8. The column is washed with successive 18-ml aliquots of 0, 0.05, and 0.50 M KCl in 10 mM potassium phosphate, pH 7.8. The synthase elutes at 0.5 M KCl and is desalted on a column of Phenyl-Sepharose CL-4B (1.2 × 2.5 cm). The chromatography is performed as in Step 7 except that the enzyme is directly eluted with 1% Triton X-100 in 10 mM potassium phosphate, pH 7.8.

Step 9. Affi-Gel Blue Chromatography. The enzyme solution is applied to an Affi-Gel Blue column (1.2 × 2.5 cm, resin from Biorad) equilibrated with 10 mM potassium phosphate, pH 7.8. The column is washed with successive 8-ml aliquots of 0, 0.05, and 0.5 M KCl in 10 mM potassium phosphate, pH 7.8. The enzyme activity elutes at 0.5 M KCl and is desalted on Phenyl-Sepharose CL-4B (0.9 × 2.5 cm column) as described in Step 8.

Step 10. Acetylglutaminyl BioGel A Chromatography. The enzyme solution is applied to a column of acetylglutaminyl BioGel A-0.5m[17] (0.9 × 2.5 cm) equilibrated with 10 mM potassium phosphate, pH 7.8. The column is washed with successive 4.5-ml aliquots of 0.1, 0.3, and 0.5 M KCl in 10 mM potassium phosphate, pH 7.8. Active fractions are pooled and concentrated to 0.2 ml in a collodion ultrafiltration device.

Step 11. Sucrose Density Gradient Centrifugation. The enzyme solution is layered on a gradient (4.5 ml) of 5 to 20% sucrose in 20 mM potassium phosphate, pH 7.8, and centrifuged at 139,000 g for 16 hr. Fractions (0.2 ml) are collected from the bottom of the tube and active fractions are pooled.

Step 12. Isoelectric Focusing. A linear density gradient isoelectric focusing column (20 ml volume) is prepared with 0 to 50% (w/v) sucrose and 0.5 to 1.5% ampholine (pH range 4–8). The enzyme solution is made 1% in ampholine (from a 40% solution) and 32% in sucrose (from a 60% solution) and then added to the center of the gradient. The focusing is carried out for 40 hr at 600 V and fractions (0.5 ml) are collected. The peak of synthase activity is at pH 6.3. Active fractions are pooled, concentrated to 0.1 ml in a collodion ultrafiltration device, and dialyzed against 10 mM potassium phosphate, pH 7.8.

The purification protocol may be interrupted after Steps 3, 5, 7, and 10. The enzyme preparations are stored at −80°.

Properties

Homogeneity.[6] The enzyme preparation obtained from Step 12 of the purification procedure yields one major band and three minor bands of higher molecular weight on sodium dodecyl sulfate–polyacrylamide gel electrophoresis. One major and three minor bands are also observed after polyacrylamide gel electrophoresis in the absence of sodium dodecyl sulfate. Only the major band has synthase activity when eluted from the gel, and the eluted protein has the same mobility as the major species observed in sodium dodecyl sulfate–polyacrylamide gel electrophoresis.

Molecular Properties. The rat liver *N*-acetylglutamate synthase has a sedimentation coefficient ($s_{20,w}$) of 8.1 S and an estimated molecular weight of 160,000, based on comparison of gel filtration and sedimentation behavior with that of standard proteins.[6] Polyacrylamide gel electrophoresis in sodium dodecyl sulfate gives a subunit molecular weight of 57,000, suggesting that the synthase has a trimeric structure.[6] Human liver *N*-acetylglutamate synthase has a similar estimated molecular weight of 190,000,[10] based on gel filtration behavior. *N*-Acetylglutamate synthase from *E. coli* has a subunit molecular weight of 51,700 and can undergo self-association to form a hexamer.[1]

Catalytic Properties. The rat liver enzyme is active over a broad pH range, with an optimum at pH 8; in the presence of arginine, the active range is narrowed and the optimum is at pH 8.5.[5] The optimum pH of the human liver enzyme is pH 8.5, in the presence or absence of arginine.[10] The *E. coli* synthase has optimal activity at pH 10.0 but is most stable at pH 7.0.[1]

Maximal velocities of 14 and 133 μmol of *N*-acetylglutamate formed per minute per milligram of protein are obtained for the rat liver and *E. coli* enzymes, respectively.[1,6] For the rat liver enzyme, the K_m values are 3.7 and 0.76 m*M* for L-glutamate and acetyl-CoA, respectively.[6] For the partially purified human liver synthase, the K_m values are 4.7 and 8.1 m*M* for L-glutamate and acetyl-CoA, respectively.[10] For the *S. typhimurium* enzyme, the K_m for acetyl-CoA is 4.6 m*M*.[3]

N-Acetylglutamate synthase displays a high degree of specificity for acetyl-CoA and L-glutamate as substrates. Of many acyl CoA derivatives tested (Table II), only propionyl-CoA could substitute for acetyl-CoA, with rates 4.5 and 4.3% of those with acetyl-CoA for the *E. coli* and rat liver synthases, respectively.[1,6] Propionyl-CoA has been found to be a competitive inhibitor of formation of *N*-acetylglutamate by the rat liver enzyme. It seems that the hyperammonemia associated with propionic and methylmalonic aciduria may arise from this inhibition since the K_i (0.71 m*M*) for propionyl-CoA is within the range of propionate found in patients' serum and since a decreased level of *N*-acetylglutamate would lead to lower carbamyl phosphate synthetase I activity.[18] Of a large number of amino compounds (Table II), only the following are found to serve

[18] F. X. Coude, L. Sweetman, and W. L. Nyhan, *J. Clin. Invest.* **64,** 1544 (1979).

TABLE II
COMPOUNDS THAT DO NOT SERVE AS SUBSTRATES FOR *N*-ACETYLGLUTAMATE SYNTHETASE[1,6]

Butyryl-CoA	Protein amino acids (except glutamate, glycine, glutamine)
iso-Butyryl-CoA	D-Glutamate
Valeryl-CoA	D-Aspartate
iso-Valeryl-CoA	β-Alanine
Palmitoyl-CoA	L-Ornithine
Crotonyl-CoA	L-Norvaline
Tiglyl-CoA	L-Citrulline
Acetoacetyl-CoA	L-Glutamate γ-monoethylester
Malonyl-CoA	L-Glutamate diethylester
Methylmalonyl-CoA	L-Glutamate γ-hydroxamate
Succinyl-CoA	L-Glutamate γ-hydrazide
β-Hydroxy-β-methylglutaryl-CoA	DL-α-Aminobutyrate
Glutaryl-CoA	DL-β-Aminobutyrate
	γ-Aminobutyrate
	α-Amino-iso-butyrate
	L-Valerate
	L-2-Aminovalerate
	L-α-Hydroxyglutarate
	γ-Aminolevulinate

as acetyl group acceptors[1,6]: (1) L-glutamine, with relative rates of 6.8 and 5.0% for the *E. coli* and rat liver enzymes, respectively; (2) L-2-aminoadipate (1.3%, for the *E. coli* enzyme); (3) DL-2-aminoadipate (5.2%, for the rat liver enzyme); (4) DL-2-aminopimelate (4.0%, for the rat liver enzyme); and (5) glycine (2.9%, for the rat liver enzyme).

It should be especially noted that L-aspartate is not a substrate for *N*-acetylglutamate synthase.[1,6] *N*-Acetyl-L-aspartate is synthesized by a distinct L-aspartate *N*-acetyltransferase which is found only in nervous tissue,[19] has been partially purified from rat brain,[20] and does not utilize L-glutamate as a substrate.[21]

N-Acetylglutamate inhibits the rat liver enzyme strongly (88% inhibition at 2 m*M*),[5] but has only a weak effect on the *E. coli* enzyme (50% inhibition at 25 m*M*).[1] Coenzyme A inhibits the two enzymes in a similar manner; for the *E. coli* enzyme, there is 50% inhibition at 2.5 m*M* CoA[1] and for the rat liver enzyme, there is 20% inhibition at 1 m*M*.[5]

[19] H. H. Tallan, S. Moore, and W. H. Stein, *J. Biol. Chem.* **219,** 257 (1956).
[20] F. B. Goldstein, *J. Biol. Chem.* **244,** 4257 (1969).
[21] F. B. Goldstein, *J. Biol. Chem.* **234,** 2702 (1959).

N-Acetylglutamate synthase from both *E. coli*[1] and rat liver[5] is inhibited by divalent cations, by sulfhydryl-blocking agents and by high ionic strength.

Regulatory Properties. *N*-Acetylglutamate synthase from *E. coli*,[22] *Saccharomyces cerevisiae*,[4] *S. typhimurium*,[3] and *P. aeruginosa*[2] is subject to strong feedback inhibition by the endproduct of the pathway, arginine. In the first two cases, it has been shown that the arginine inhibition is enhanced by the addition of acetyl-CoA or *N*-acetylglutamate and also that polyamines do not inhibit the enzyme reaction. Although the synthesis of the synthase from *P. aeruginosa* is not repressed by arginine,[2] the synthesis of the enzyme from the other organisms is subject to arginine-specific repression. The repression ratios are 10, greater than 25, and greater than 250 for the enzymes from *S. cerevisiae*,[4] *S. typhimurium*,[3] and *E. coli*,[22] respectively.

In contrast to the other *N*-acetylglutamate synthases, the enzyme from rat liver is strongly and specifically activated by arginine.[14] Presumably this activation serves as a link between an increased amino acid pool and an increase in ammonia removal via the urea cycle. Carbamyl phosphate synthetase I, which is the rate-limiting step of the urea cycle under most conditions, requires *N*-acetylglutamate as an allosteric activator and the usual level of *N*-acetylglutamate is about equal to the K_m for carbamyl phosphate synthetase I. Thus, an arginine-induced increase in the amount of *N*-acetylglutamate synthase would lead to an increase in the activity of the urea cycle. For the purified enzyme, there is a 4.2-fold stimulation of activity by arginine; half-maximal and maximal levels of activation are produced by 5 and 20 μM arginine, respectively.[6] The sensitivity of the enzyme to activation by arginine undergoes large changes postprandially[7] and upon parenteral treatment with inhibitors of nucleic acid and protein synthesis.[8]

Rat liver *N*-acetylglutamate synthase is activated up to 4-fold by cationic polypeptides and this activation is additive to that caused by arginine.[6] Triton X-100 also stimulates the enzyme activity, with a 4-fold activation at 0.1% (w/v); this activating effect is not additive to that of arginine.[15]

Although the activities of the *E. coli* and rat liver *N*-acetylglutamate synthases are affected by arginine in an opposite manner, the specific activity of both enzymes increases with increasing protein concentration, and, in both cases, arginine stabilizes the larger form of the enzyme.[1,5]

[22] T. Leisinger and D. Haas, *J. Biol. Chem.* **250,** 1690 (1975).

[5] *N*-Methyl-L-glutamate Synthase

By LOUIS B. HERSH

L-Glutamate + methylamine ⇄ *N*-methyl-L-glutamate + ammonia

Introduction

N-Methyl-L-glutamate synthase is an inducible enzyme found in several species of *Pseudomonas* grown on methylated amines.[1-8] The enzyme functions in a cycle in which free methylamine first reacts with L-glutamate to yield *N*-methyl-L-glutamate and free ammonia. The *N*-methyl-L-glutamate is then oxidized to release formaldehyde and regenerate L-glutamate. This cycle provides both ammonia and an activated C-1 compound for growth of the microorganism.

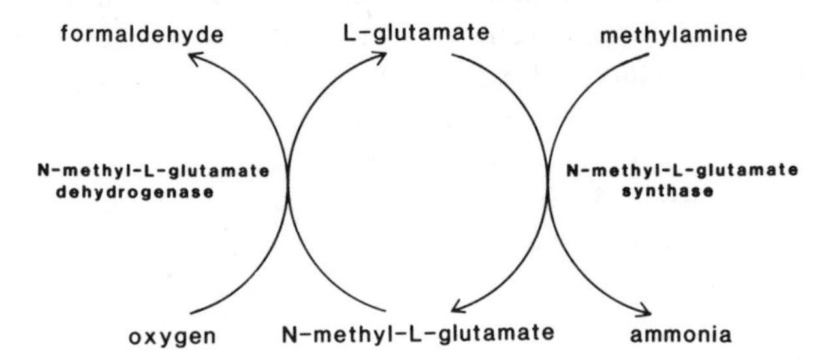

Assays Methods

Principle. The reaction between L-glutamate and methylamine is measured by following the incorporation of [^{14}C]methylamine into *N*-methyl-

[1] M. V. Shaw, L. Tsai, and E. R. Stadtman, *J. Biol. Chem.* **241,** 935 (1966).
[2] L. B. Hersh, J. A. Peterson, and A. Thompson, *Arch. Biochem. Biophys.* **145,** 115 (1971).
[3] H. F. Kung and C. Wagner, *J. Biol. Chem.* **244,** 4136 (1969).
[4] C. Watner and J. R. Quayle, *J. Gen. Microbiol.* **72,** 485 (1972).
[5] Y. Trotsenko and N. V. Loginova, *Mikrobiologia* **42,** 695 (1974).
[6] N. V. Logenova and Y. Trotsenko, *Mikrobiologia* **43,** 831 (1975).
[7] C. W. Bomforth and M. L. O'Connor, *J. Gen. Microbiol.* **110,** 143 (1979).
[8] J. Colby and L. J. Zatman, *Biochem. J.* **132,** 101 (1973).

Copyright © 1985 by Academic Press, Inc.
All rights of reproduction in any form reserved.

L-glutamate. Free methylamine is separated from *N*-methyl-L-glutamate by chromatography on Dowex 50-H^+.

Assay Mixture

100 m*M* potassium tricine buffer pH 8.3
100 m*M* [^{14}C]methylamine (SA 3×10^4 cpm/μmol)
20 m*M* L-glutamate
1 m*M* dithiothreitol
Total volume is 0.2 ml

Procedure

The reaction is initiated by the addition of an appropriate amount of enzyme and incubated at 37° for 15–30 min. Termination of the reaction is accomplished by the addition of 50 μl of 20% trichloroacetic acid, and the denatured protein removed by centrifugation. An aliquot (0.1 to 0.2 ml) is chromatographed on a 0.5 cm^3 column of Dowex-50-H^+ (200–400 mesh). The column is first washed with water (1–2 ml) to remove free methylamine, and then *N*-methyl-L-glutamate is eluted with 2 *N* NH_3 (2.5–3.0 ml). An aliquot of the ammonia effluent is neutralized with 2 *N* acetic acid and its radioactive content determined by liquid scintillation counting.

Definition of activity unit: 1 unit of activity is defined as the amount of enzyme producing 1 μmol of *N*-methyl-L-glutamate per min under the above assay conditions.

Alternative Assay Procedures

The reverse reaction can be measured by either of two assay methods. The first method involves measuring [^{14}C]methylamine released from [*methyl*-^{14}C]-*N*-methyl-L-glutamate while the second method involves measurement of methylamine release with an amine oxidase.

Reverse Reaction (N-Methyl-L-Glutamate + Ammonia → L-Glutamate + Methylamine. A reaction mixture containing 100 m*M* potassium tricine buffer, pH 8.4, 50 m*M* [*methyl*-^{14}C]-*N*-methyl-L-glutamate,[1] 100 m*M* ammonia, 1 m*M* dithiothreitol, and enzyme in a final volume of 0.2 ml is incubated in a diffusion bottle at 30°. The reaction is terminated by the addition of 50 μl of 40% sodium hydroxide, and the wick of the diffusion bottle is coated with 5 *M* H_2SO_4. The diffusion bottle is placed on a rotary wheel (Scientific Industries, Springfield, Mass.) and rotated through a 40° water bath. After 1.5 hr the [^{14}C]methylammonium sulfate is washed into scintillation fluid and counted.

Alternatively methylamine formation from nonradioactive *N*-methyl-

L-glutamate can be determined spectrophotometrically using the amine dehydrogenase reaction as described by Large *et al.*[9]

Preparation of N-Methyl-L-glutamate Synthase

N-Methyl-L-glutamate synthase is purified from cells of *Pseudomonas* MA (ATCC 23819) grown on methylamine as the sole carbon and nitrogen source.[1]

Preparation of Cell Extract. Frozen cells (75 g) are suspended in 2 volumes of 0.1 *M* Tris hydrochloride buffer, pH 8.5 containing 60 m*M* potassium chloride, 50 m*M* L-glutamate, 5 m*M* magnesium chloride, 5 m*M* 2-mercaptoethanol, 1 m*M* magnesium sodium EDTA, and 50 μM flavin mononucleotide. The cells are disrupted using either a French pressure cell at 9,000 to 16,000 psi or a "Bead Beater" for three 20 sec bursts. Deoxyribonuclease is added to a final concentration of 5 μg/ml and DNA digestion is allowed to proceed for 15 min at room temperature. The solution is centrifuged for 1 hr at 25,000 *g* and the pellet discarded.

First Ammonium Sulfate Fractionation. The enzyme is fractionated with ammonium sulfate by the dropwise addition of a saturated ammonium sulfate solution. The enzyme, which is precipitated between 30 and 60% ammonium sulfate, is collected by centrifugation and dissolved in a minimal volume of 40 m*M* potassium phosphate buffer, pH 7.4, containing 80 m*M* L-glutamate, 50 m*M* potassium chloride, 5 m*M* 2-mercaptoethanol, 5 m*M* magnesium chloride, 1 m*M* magnesium disodium EDTA, and 50 μM flavin mononucleotide (Buffer A).

Molecular Sieve Chromatography. The redissolved 30–60% ammonium sulfate fraction is chromatographed on a column of BioGel A 1.5m (5 × 93 cm) equilibrated with Buffer A. The enzyme elutes at approximately 1.6 column volumes.

Second Ammonium Sulfate Fractionation. To the pooled active fractions from molecular sieve chromatography is added solid ammonium sulfate to 40% saturation. The pH is maintained at 7 by the addition of 1 *N* sodium hydroxide. After stirring for 50 min, the precipitate is collected by centrifugation. The supernatant is adjusted to 50% ammonium sulfate by the addition of solid ammonium sulfate. The precipitate is collected by centrifugation and dissolved in a minimal volume of Buffer A in which the potassium phosphate buffer concentration is reduced to 10 m*M*, the L-glutamate concentration is reduced to 20 m*M*, and flavin mononucleotide is omitted (Buffer B). The enzyme is dialyzed against Buffer B overnight.

Ion Exchange Chromatography. The dialyzed enzyme is applied to a DEAE-cellulose column (175 ml) equilibrated with Buffer B. After wash-

[9] P. J. Large, R. R. Eady, and D. J. Murden, *Anal. Biochem.* **32,** 402 (1969).

PURIFICATION OF N-METHYL-L-GLUTAMATE SYNTHASE

Step	Total protein (mg)	Total activity (units)[a]	Specific activity (units/mg)	Recovery (%)
1. Extraction	6345	44.6	0.007	(100)
2. First ammonium sulfate fractionation	3298	44.2	0.013	99
3. Molecular sieve chromatography	752	33.8	0.045	76
4. Second ammonium sulfate fractionation	438	46.2	0.106	104
5. Ion-exchange chromatography	200	27.5	0.138	62

[a] A unit of activity corresponds to the formation of 1 μmol of *N*-methyl-L-glutamate per min.

ing with 0.5 column volumes of Buffer B, a linear salt gradient from 50 to 300 m*M* potassium chloride in Buffer B is applied to the column. The enzyme, which elutes at ~130 m*M* potassium chloride, is concentrated with an Amicon ultrafiltration cell using a PM 30 membrane and stored frozen.

A summary of the purification procedure is given in the table.

Physical Properties

The purified enzyme exhibits one major and two to three minor protein components when analyzed by PAGE.[10] In the presence of L-glutamate the enzyme sediments as a homogeneous protein with $s_{20,w}$ of 12. The enzyme exhibits a molecular weight of ~350,000–390,000 and is composed of subunits of molecular weight 30,000–35,000. The enzyme contains flavin mononucleotide as a prosthetic group.[10] The equilibrium constant for the reaction is ~1.0,[11] while the pH optima is 9.0.[10]

Kinetic Properties

The enzyme is not absolutely specific with respect to the amine substrate. In the forward direction ethylamine, propylamine, butylamine, and ethanolamine gave rates relative to methylamine of 16, 8, 5, and 18% respectively. With *N*-methyl-L-glutamate as substrate activities of various amines with respect to ammonia are methylamine 70%, ethylamine 32%, propylamine 18%, butylamine 7%, ethanolamine 27%, and cysteamine

[10] R. J. Pollock and L. B. Hersh, *J. Biol. Chem.* **246,** 4737 (1971).
[11] R. J. Pollock and L. B. Hersh, *J. Biol. Chem.* **248,** 6724 (1973).

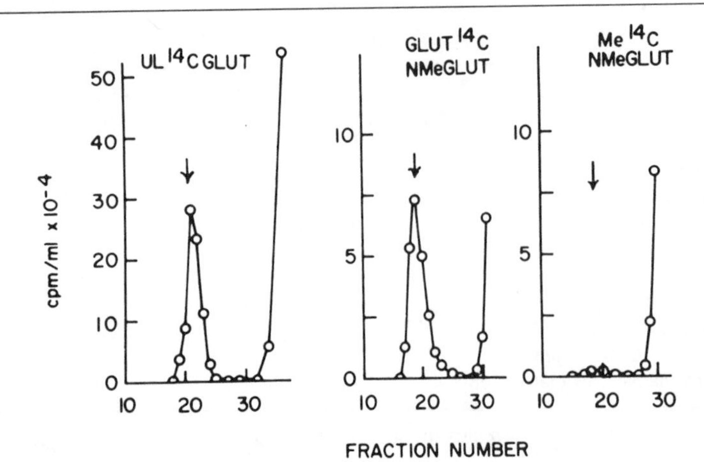

FIG. 1. Isolation of a glutaryl-enzyme species. *N*-Methyl-L-glutamate synthase was incubated for 10 min at 30° in 100 m*M* potassium tricine, pH 8.3, containing the indicated substrate at 1 m*M* concentration, then chromatographed at 4° on a Sephadex G-25 column. The labeled substrates employed were L-[U-^{14}C]glutamate (UL ^{14}C GLUT), *N*-[*glutaryl*-^{14}C]methyl-L-glutamate (GLUT ^{14}C NMeGLUT), and *N*-[*methyl*-^{14}C]methyl-L-glutamate [Me ^{14}C NMeGLUT]. The amount of labeling of the enzyme corresponds to 3.7, 1.6, and less than 0.01 mol of substrate per mole of active enzyme for L-[U-^{14}C]glutamate, *N*-[*glutaryl*-^{14}C]methyl-L-glutamate, and *N*-[*methyl*-^{14}C]methyl-L-glutamate, respectively. The arrow marks the position of maximum protein concentration, as determined by absorbance at 280 nm.

54%. Of a variety of amino acids tested only L-aspartate exhibited any activity (2.3%).

Initial velocity and product inhibition studies[11] showed that the *N*-methyl-L-glutamate synthase reaction follows a double-displacement kinetic mechanism. Incubation of the enzyme with either L-[^{14}C]glutamate or [*glutaryl*-^{14}C]-*N*-methyl-L-glutamate, but not [*methyl*-^{14}C]-*N*-methyl-L-glutamate resulted in the incorporation of radiolabel into the enzyme (Fig. 1) indicating the formation of a glutaryl-enzyme intermediate. Concurrent with the formation of the glutaryl-enzyme, the enzyme bound flavin mononucleotide is reduced (Fig. 2). The addition of an amine acceptor results in the reoxidation of the flavin mononucleotide at rates consistent with the glutaryl-enzyme serving as a true intermediate in the reaction.

The use of [α-^{3}H]glutamate as substrate showed that the α hydrogen although removed during formation of the glutaryl-enzyme intermediate does not exchange with solvent, but is found as the α hydrogen in the product *N*-methyl-L-glutamate. Substituting 5-deazaflavin mononucleotide for flavin mononucleotide permitted the demonstration of the transfer of the α hydrogen of glutamate to flavin.[12] Taken together these results are

[12] M. S. Jorns and L. B. Hersh, *J. Biol. Chem.* **250,** 3620 (1975).

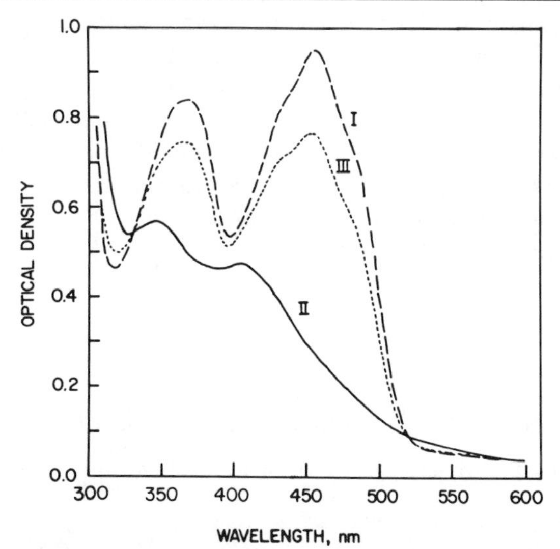

FIG. 2. The effect of substrates on the flavin oxidation state. The sample cuvette contained 100 mM potassium tricine, pH 8.3, 13.7 mg/ml of N-methyl-L-glutamate synthase (specific activity of 0.13 μmol/min/mg), 235 mM potassium chloride, 5 mM magnesium chloride, 4 mM dithioerythritol, 1 mM magnesium disodium EDTA, and the indicated concentrations of potassium L-glutamate, pH 8.3, and ammonium sulfate, pH 8.2, at 10° in a final volume of 1.2 ml. The reference cuvette contained the same components except that substrates were deleted and buffer was used in place of enzyme. After each addition of substrate, changes in the absorbance at 450 nm were monitored and the spectrum was recorded only after these changes had ceased. This equilibration period was about 30 min for each substrate addition. The spectra are I, native N-methyl-L-glutamate synthetase; II, sample I after addition of 1.75 mM L-glutamate; III, sample II after addition of 21 mM ammonium sulfate.

consistent with the following mechanism:

FIRST HALF-REACTION

$$\begin{array}{l} COO^- \\ | \\ H-C-NH_3^+ \\ | \\ CH_2 \\ | \\ CH_2 \\ | \\ COO^- \end{array} + \text{E-FMN} \rightleftharpoons \begin{array}{l} COO^- \\ | \\ C=NH_2^+ \cdot\cdot \text{E-FMNH}_2 \\ | \\ CH_2 \\ | \\ CH_2 \\ | \\ COO^- \end{array} \rightleftharpoons \begin{array}{l} COO^- \\ | \\ C=X-\text{E-FMNH}_2 \\ | \\ CH_2 \\ | \\ CH_2 \\ | \\ COO^- \end{array} + NH_4^+$$

SECOND HALF-REACTION

$$\begin{array}{l} COO^- \\ | \\ C=X-\text{E-FMNH}_2 \\ | \\ CH_2 \\ | \\ CH_2 \\ | \\ COO^- \end{array} + CH_3NH_3^+ \rightleftharpoons \begin{array}{l} COO^- \\ | \\ C=\overset{+}{N}H-CH_3 \cdot\cdot \text{E-FMNH}_2 \\ | \\ CH_2 \\ | \\ CH_2 \\ | \\ COO^- \end{array} \rightleftharpoons \begin{array}{l} COO^- \\ | \\ H-C-NH_2^+-CH_3 \\ | \\ CH_2 \\ | \\ CH_2 \\ | \\ COO^- \end{array} + \text{E-FMN}$$

This mechanism involves the use of enzyme bound flavin in oxidative catalysis. There is no net oxidation–reduction of the enzyme bound flavin. Glutamate oxidation in the first half-reaction occurs by flavin reduction, whereas reformation of the reduced carbon–nitrogen bond in *N*-methyl-L-glutamate occurs by flavin reoxidation in the second half-reaction. Although the bond between the glutaryl moiety and the enzyme has not been elucidated, it is presumably a Schiff's base formed with a lysine residue of the enzyme.

[6] Glutamyl-tRNA Synthetase from *Escherichia coli*

By JACQUES LAPOINTE, SYLVAIN LEVASSEUR, and DANIEL KERN

Introduction

In *Escherichia coli,* glutamic acid and glutamine are charged respectively on $tRNA^{Glu}$ and $tRNA^{Gln}$ in reactions catalyzed by distinct enzymes.[1,2] In general, each amino acid to be incorporated into proteins in *E. coli* or in other gram-negative bacteria is first activated by a specific aminoacyl-tRNA synthetase.[3] A different situation is observed in many gram-positive bacterial strains where no glutaminyl-tRNA synthetase (GlnRS) activity is detected and where the synthesis of glutaminyl-$tRNA^{Gln}$ proceeds via glutamyl-$tRNA^{Gln}$ [4–6]; the pure glutamyl-tRNA synthetase (GluRS) of *Bacillus subtilis*[7] catalyzes the aminoacylation of both $tRNA^{Glu}$ and $tRNA^{Gln}$ with glutamic acid.[8] The purification of this enzyme is described in the next chapter.

The structural gene for the monomeric GluRS of *E. coli* is at the locus *glt*X at min 50.[9,10] Using the thermosensitive mutant JP1449, altered in this locus and having a thermolabile GluRS, it was shown that the GluRS

[1] R. A. Lazzarini and A. H. Mehler, *Biochemistry* **3,** 1445 (1964).
[2] J. M. Ravel, S.-F. Wang, C. Heinemeyer, and W. Shive, *J. Biol. Chem.* **240,** 432 (1965).
[3] P. R. Schimmel and D. Söll, *Annu. Rev. Biochem.* **48,** 601 (1979).
[4] M. Wilcox and M. Nirenberg, *Proc. Natl. Acad. Sci. U.S.A.* **61,** 229 (1968).
[5] M. Wilcox, *Eur. J. Biochem.* **11,** 405 (1969).
[6] M. Wilcox, *Cold Spring Harbor Symp. Quant. Biol.* **34,** 521 (1969).
[7] M. Proulx, L. Duplain, L. Lacoste, M. Yaguchi, and J. Lapointe, *J. Biol. Chem.* **258,** 753 (1983).
[8] J. Lapointe, L. Duplain, and M. Proulx, submitted for publication.
[9] R. R. B. Russell and A. J. Pittard, *J. Bacteriol.* **108,** 790 (1971).
[10] J. Lapointe and G. Delcuve, *J. Bacteriol.* **122,** 352 (1975).

Copyright © 1985 by Academic Press, Inc.
All rights of reproduction in any form reserved.

or its main product, glutamyl-tRNA, is involved in the regulation of glutamine and/or glutamate biosynthesis in *E. coli*.[11] The structural gene of the GluRS on a 2.7 kb fragment of chromosomal DNA inserted into pBR322 was cloned in JP1449 by complementation of the ts mutation of this strain[12]: one of the transformant strains, HS7611, overproduces the wild-type GluRS about 25-fold compared to a wild-type strain of *E. coli*[12] (also compare Tables I and II).

This chapter describes two methods for the purification of this monomeric enzyme to homogeneity, and addresses the question of its association with a regulatory protein.

Purification of the Monomeric Glutamyl-tRNA Synthetase

Procedure

Homogeneously pure GluRS (25 mg) was obtained from 2 kg of *E. coli* MRE-600 wet cells by this procedure.[13] One molecule of pure enzyme catalyzes the formation of about 4 molecules of glutamyl-tRNA per second at 37°, at pH 8.6 which is optimal for this reaction.[13,14]

Reagents

Polyethylene glycol 6000 from H. T. Baker
Dextran T-500 and blue-Sepharose CL-6B from Pharmacia
Phenylmethylsulfonyl fluoride (PMSF), and 4-(2-hydroxyethyl)-1-piperazineethane-sulfonic acid (HEPES) from Sigma
Microgranular DEAE-cellulose (DE52), cellulose powder CF-11, and 3MM filter paper from Whatman
Hydroxylapatite (BioGel HTP) from Bio-Rad
Acrylamide, bisacrylamide, *N*,*N*,*N'*,*N'*-tetramethylethylenediamine (TEMED) and 2-mercaptoethanol from Eastman Chemicals
Omnifluor and uniformly labeled L-[^{14}C]glutamate from New England Nuclear

Biological Materials

Unfractionated tRNA from *E. coli* B (purchased from Schwarz/Mann) contains about 4% $tRNA^{Glu}$. It was protected against oxidation by 0.1 m*M* 2-mercaptoethanol in stock solutions (about 300 A_{260}/ml).

[11] J. Lapointe, G. Delcuve, and L. Duplain, *J. Bacteriol.* **123,** 843 (1975).
[12] H. Sanfaçon, S. Levasseur, P. H. Roy, and J. Lapointe, *Gene* **22,** 175 (1983).
[13] D. Kern, S. Potier, Y. Boulanger, and J. Lapointe, *J. Biol. Chem.* **254,** 518 (1979).
[14] D. Kern and J. Lapointe, *Eur. J. Biochem.* **106,** 137 (1980).

ATP was purchased from Sigma; it was neutralized with NaOH and the solution was stored at −20°.

E. coli MRE-600 was grown in a minimal medium containing per liter 7 g of $K_2HPO_4 \cdot 3H_2O$, 3 g of KH_2PO_4, 1 g of $(NH_4)_2SO_4$, 0.1 g of $MgSO_4 \cdot 7H_2O$, 0.5 g of Na_3 citrate · $2H_2O$, 1 mg of thiamine, and 1 μmol of $MnSO_4$ and of $Fe_2(SO_4)_3$. The growth was stopped during the exponential phase by rapidly cooling the culture (by addition of crushed ice to the culture), and the cells were harvested by centrifugation (Sharples centrifuge Type T-1P), frozen with liquid nitrogen and stored at −20°.

GluRs Assay. The formation of Glu-tRNA was followed in 0.1-ml reaction mixtures containing 50 m*M* sodium Hepes, pH 7.2, 16 m*M* $MgCl_2$, 2 m*M* ATP, 3 mg of tRNA/ml, 0.1 m*M* [^{14}C]glutamate (25 μCi/μmol) and the enzyme. When necessary, the enzyme was diluted in 10 m*M* sodium Hepes, pH 7.2, 20 m*M* 2-mercaptoethanol, and 1 mg of albumin/ml. After various incubation times at 37°, an aliquot was transferred on a disc of Whatman No. 3MM filter paper, washed successively during 15 min in each of three 5% trichloroacetic acid solutions at 0° (to remove the free [^{14}C]glutamate), two 95% ethanol solutions, and finally in diethyl ether. The dried filters (containing the acid-precipitated [^{14}C]Glu-tRNA) were placed in a solution of 4 g of Omnifluor/liter of toluene and counted for ^{14}C in a Beckman LS-355 scintillation counter.

Preparative Electrophoresis Procedure. The experiment was conducted in a Canalco gel electrophoresis apparatus using the procedure of Davis[15] modified as follows: the lower gel (40 ml) was formed at room temperature from a solution containing 7.5% acrylamide, 0.3% bisacrylamide, 0.365 *M* Tris–HCl, pH 8.9, 15 μl of *N,N,N′,N′*-tetramethylethylenediamine (TEMED), and 3 mg of ammonium persulfate. After the polymerization, unreacted persulfate ions were removed by a 30-min electrophoresis. The upper gel was omitted and the protein solution, previously dialyzed against 0.06 *M* Tris–HCl, pH 6.8, and 50% glycerol, was poured on the gel. During the electrophoresis, the external surface of the column was kept at 0 to 4° with a stream of water. The top of the upper gel and the bottom of the lower gel were connected respectively to the cathode and the anode by a buffer (pH 8.3) containing 6 g of Tris (base) and 28.8 g of glycine/liter.

Purification Steps. All the operations were performed between 0 and 4°. All the buffers contained 10% (v/v) glycerol, 20 m*M* 2-mercaptoethanol, and 0.1 m*M* PMSF as protective agents against proteases. In the buffer used for cell lysis, 10 m*M* PMSF was present. The centrifugations were made in a GSA rotor in a Sorvall RC2-B.

[15] B. J. Daivs, *Ann. N.Y. Acad. Sci.* **121,** 404 (1964).

Step 1: Cell Lysis. Wet cells (1 kg) were suspended in 2 liters of 10 m*M* potassium phosphate, pH 8.0, and broken by sonication during 10 min in a Raytheon sonic oscillator (model DF 101), by fractions of 75 ml. The lysate was centrifuged at 8000 rpm during 30 min to remove cell debris and intact cells, yielding 2350 ml of supernatant.

Step 2: Partition in a Polyethylene Glycol-Dextran Two-Phase System. Potassium phosphate, pH 8.0 (125 ml, 1 *M*) was added to the supernatant. Then, concentrated solutions of PEG-6000 and dextran T-500 were added to reach the final concentrations of 7 and 1.5%, respectively, in the supernatant. This suspension was mixed during 2 hr, and the two phases were separated by centrifugation at 5000 rpm during 20 min. The PEG-rich top phase contains most of the glutamyl-tRNA synthetase activity.

Step 3: Chromatography on DEAE-Cellulose. The top phase (3 liters) was diluted by addition of 2 liters of 10% glycerol, 20 m*M* 2-mercaptoethanol, to reduce the ionic strength. Half of the solution (2.5 liters) was adsorbed on a column (7 × 30 cm) of DEAE-cellulose (type DE52), which was then washed with 1 liter of 10 m*M* potassium phosphate, pH 7.5. The (macro)molecules left on the column were then eluted at about 400 ml/hr with a linear salt and pH gradient of 6 liters (20 m*M* potassium phosphate, pH 7.5 to 250 m*M* potassium phosphate, pH 6.5). The glutamyl-tRNA synthetase activity was eluted near the end of this gradient, and was already separated from most other aminoacyl-tRNA synthetases.[16] The most active fractions were pooled and dialyzed against 10 m*M* potassium phosphate, pH 6.8 (Fraction DEAE). The same purification step was conducted on the other half (2.5 liters) of the diluted top phase.

Step 4: Chromatography on Hydroxylapatite. The hydroxylapatite (BioGel HTP) was mixed with 10% (w/w) cellulose powder (Whatman CF-11) to increase the maximal flow rate through a column of this crystalline material. Fraction DEAE obtained from the chromatography of the 5 liters of diluted top phase was adsorbed on a hydroxylapatite column (6 × 12 cm) equilibrated against 10 m*M* potassium phosphate, pH 6.8. The column was washed at 300 ml/hr with 200 ml of the same buffer, then with 2 liters of a linear gradient from 20 to 200 m*M* potassium phosphate, pH 6.8. Two peaks of glutamyl-tRNA synthetase activity were eluted; the first, representing only a small percentage of the total activity, was eluted at a conductivity of 3.6 mmho (at 4°), whereas the second and major peak was eluted at 5.7 mmho. Only the most active fractions of this major peak were pooled (Fraction HA) and used in the following purification steps.

Step 5: Preparative Polyacrylamide Gel Electrophoresis: Last Purification Step. Fraction HA was concentrated 2- to 3-fold by dialysis against

[16] D. Kern and J. Lapointe, *Biochimie* **61,** 1257 (1979).

TABLE I
PURIFICATION OF THE *E. coli* GLUTAMYL-tRNA SYNTHETASE FROM 1 kg OF THE STRAIN MRE-600

Step	Total protein (mg)	Total activity (units)[a]	Specific activity (units/mg)	Recovery (%)
1. Cell extract	31,090	42,060	1.35	100
2. Liquid polymer extract	18,060	40,000	2.20	95
3. DEAE-cellulose	1,630	37,490	23.0	89
4. Hydroxylapatite	210	28,980	138.	69
5. Electrophoresis	25	19,750	790[b]	47

[a] One unit of enzyme catalyzes the formation of 1 nmol of Glu-tRNA/min at 37°.

[b] This specific activity corresponds to a turnover number of about 0.8 Glu-tRNA formed per second per molecule of GluRS at pH 7.2. A value of about 4/sec is obtained at the optimal pH (8.6).

30% polyethylene glycol and then against 0.01 *M* Tris, 0.077 *M* glycine (pH 8.3), 20 m*M* 2-mercaptoethanol, 50% glycerol. About 100 mg of proteins present in 8 ml of this concentrated Fraction HA was mixed with 0.1 ml of a saturated solution of bromphenol blue and layered on the top of a column (12×3.4 cm^2) of polyacrylamide gel whose preparation is described above. A constant current of 40 mA was passed through the gel, whose electric resistance gradually reached a constant value of about 10,000 Ω. The bottom surface of the gel was continuously washed with a 80 ml/hr stream of 0.01 *M* Tris, 0.077 *M* glycine (pH 8.3), 10% glycerol, and 20 m*M* 2-mercaptoethanol, which was collected in 8-ml fractions. Following the elution of bromphenol blue (4 hr of electrophoresis), the glutamyl-tRNA synthetase was the first protein eluted. This step had to be performed with three separate aliquots of 70 mg protein because of the low capacity of our column. The active fractions were pooled and concentrated (Fraction PAGE) by chromatography on a small column of DEAE-cellulose. The results of this purification are summarized in Table I.[13] Other purification procedures[17] (also second procedure of reference 13) led either to a less active or to a less pure fraction of GluRS. The pure enzyme is stable for more than a year when kept at −20° in 50 m*M* Na HEPES pH 8, 10 m*M* 2-mercaptoethanol, 0.2 m*M* dithiothreitol, and 50% glycerol. The enzyme lost no activity when kept frozen in the same buffer containing only 10% glycerol, and thawed a few times over a period of a year.

[17] G. Willick and C. Kay, *Biochemistry* **15,** 4347 (1976).

TABLE II

PURIFICATION OF THE *E. coli* GLUTAMYL-tRNA SYNTHETASE FROM 60 g OF THE OVERPRODUCING STRAIN HS7611

Step	Total protein (mg)	Total activity (units)[a]	Specific activity (units/mg)	Recovery (%)
1. Cell extract	2,630	97,560	37	100
2. 300,000 *g* supernatant	1,390	81,340	58	83
3. DEAE-cellulose	526	77,520	147	80
4. Hydroxylapatite	208	64,440	310	66
5. Blue-Sepharose CL-6B	40	52,260	1,310	54

[a] One unit of enzyme catalyzes the formation of 1 nmol of Glu-tRNA/min at 37°.

Procedure 2

For the purification of the GluRS from 60 g of the overproducing strain HS7611 (summarized in Table II) Procedure 1 was modified as follows: first, the two-phase polymer extraction was replaced by a 2 hr ultracentrifugation at 300,000 *g*; second, the low capacity preparative electrophoresis on polyacrylamide gel was replaced by affinity chromatography on blue-Sepharose CL-6B, as described below. All the other steps were the same as those of Procedure 1, with DEAE-cellulose and hydroxylapatite columns of size 4.4 × 20 cm and 6 × 12 cm, respectively.

Chromatography on Blue-Sepharose CL-6B. The active fractions eluted from the hydroxylapatite column were pooled and dialyzed against 10 m*M* KH_2PO_4 pH 7.0, 5 m*M* $MgCl_2$, 20 m*M* 2-mercaptoethanol, and 10% glycerol (which is the equilibrium buffer for the blue-Sepharose CL-6B column). These 210 mg of protein in 440 ml of equilibrium buffer were layered at a rate of 35 ml/hr onto a 1.7 × 30 cm column of blue-Sepharose CL-6B which was successively eluted with 250 ml of the same buffer and with a linear gradient (1.5 liter) of 0 to 0.4 *M* KCl in the same buffer at a rate of 35 ml/hr. The GluRS was eluted in a wide peak (Fig. 1) whose first half contained homogeneously pure enzyme as judged by analytical SDS–polyacrylamide gel electrophoresis.

Some Properties of the Glutamyl-tRNA Synthetase

The *E. coli* GluRS is a polypeptide of M_r about 56,000[13,17,18] whose shape is asymmetric.[17] Its extinction coefficient $\varepsilon_{280\text{ nm}}^{1\text{ mg/ml}}$ is 0.87.[13] It pos-

[18] D. M. Powers and A. Ginsburg, *Arch. Biochem. Biophys.* **191,** 673 (1978).

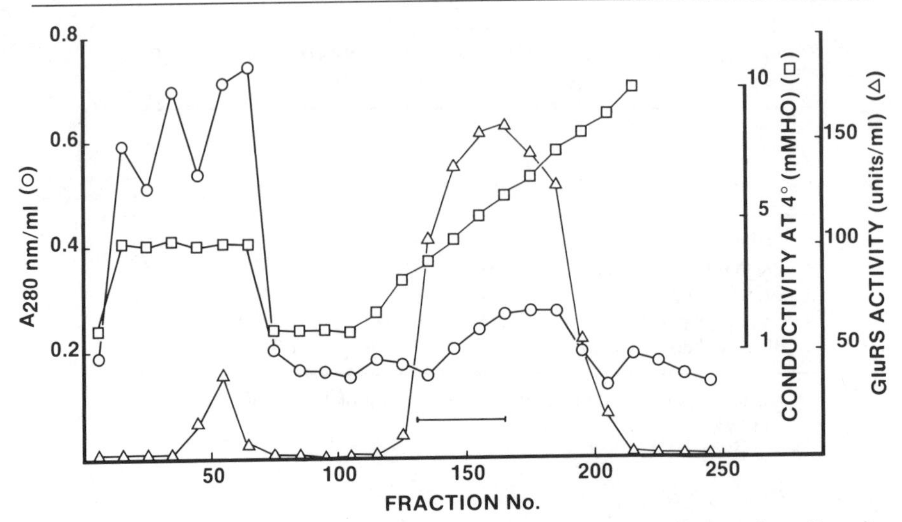

FIG. 1. Last step of the purification of the glutamyl-tRNA synthetase from *E. coli* HS7611, by chromatography on blue-Sepharose CL-6B. The fractions (6 ml) indicated by the horizontal line under the first part of the GluRS activity peak contained the pure enzyme.

sesses one binding site for each substrate (ATP-Mg, $K_D = 90\ \mu M$; L-glutamate, $K_D = 180\ \mu M$; $tRNA^{Glu}$, $K_D = 0.09\ \mu M$: values at pH 7.2), and is fully active, at least in vitro, in its monomeric form.[19,20]

GluRS is one of the three aminoacyl-tRNA synthetases of *E. coli* requiring the presence of a cognate tRNA to catalyze the activation of their amino acid substrate[3]; this requirement is related to the fact that glutamate binds to the GluRS only in the presence of $tRNA^{Glu}$.[19,20] A conformational change of the enzyme upon binding $tRNA^{Glu}$ has been detected by circular dichroism,[17] and is suggested by a kinetic investigation of the reaction mechanism of the GluRS.[21] Kinetic studies indicate that the formation of glutamyl-tRNA takes place via a two-step pathway; the intermediate complex, in which glutamate is activated, has been detected.[14,22,23]

The K_m values of the GluRS for its substrates have been determined under the standard aminoacylation conditions (except for the varying substrate), the nonvarying substrate concentrations being nearly saturat-

[19] D. Kern and J. Lapointe, *Biochemistry* **18,** 5809 (1979).
[20] D. Kern and J. Lapointe, *Nucleic Acids Res.* **7,** 501 (1979).
[21] D. Kern and J. Lapointe, *Eur. J. Biochem.* **115,** 29 (1981).
[22] D. Kern and J. Lapointe, *J. Biol. Chem.* **255,** 1956 (1980).
[23] D. Kern and J. Lapointe, *Biochemistry* **19,** 3060 (1980).

ing: K_m values of about 0.1 mM L-glutamate, 0.2 mM ATP, and 0.1 μM $tRNA^{Glu}$ have been obtained at pH 7.2.[13,17]

The optimal rate of the aminoacylation reaction catalyzed by the GluRS was observed at pH 8.6 at 47° in the presence of a $MgCl_2$/ATP ratio of 8 (with 2 mM ATP).[16,20] The ionic strength of the GluRS assay mixture described above is optimal for this reaction.[16] However, the rate of the aminoacylation reaction was measured at pH 7.2 at 37° because alkaline pH and high temperature destabilize the aminoacyl-tRNA ester bond.[24]

Copurification of a Regulatory Factor with the Monomeric Glutamyl-tRNA Synthetase

Using a procedure similar to the first one described above, Lapointe and Söll[25] observed the copurification of the GluRS with a polypeptide of M_r 46,000 able to protect the enzyme against heat inactivation and to increase its affinity for glutamate and ATP in the aminoacylation reaction.[26] Although this result was obtained several times, the same or other investigators also obtained the pure monomeric enzyme using the same procedure[17,18] (D. Kern and J. Lapointe, unpublished results). The reasons for this discrepancy are not known. The existence of such a factor in *E. coli* is also suggested by the existence in *E. coli* of mutations located far from the *glt*-X locus and resulting in a lower affinity of the GluRS for glutamate,[10,27] and by the copurification with the GluRS of *B. subtilis* of a polypeptide of M_r 46,000 with properties similar to those reported for the regulatory factor of the *E. coli* GluRS.[7] It is interesting to note that a polypeptide of the same size is copurified with the *E. coli* glutaminyl-tRNA synthetase[28] which is, as the *E. coli* GluRS, a small monomeric aminoacyl-tRNA synthetase containing no extensive sequence repeats.[13,29]

Acknowledgments

This work has been supported by Grant A9597 from the Natural Sciences and Engineering Research Council of Canada and by Grant EQ-1700 from the "Programme FCAC du Gouvernement du Québec."

[24] F. Schuber and M. Pinck, *Biochimie* **56,** 383 (1974).
[25] J. Lapointe and D. Söll, *J. Biol. Chem.* **247,** 4966 (1972).
[26] J. Lapointe and D. Söll, *J. Biol. Chem.* **247,** 4982 (1972).
[27] E. J. Murgola and E. A. Adelberg, *J. Bacteriol.* **103,** 178 (1970).
[28] P. Hoben and D. Söll, this volume [8].
[29] P. Hoben, N. Royal, A. Cheung, F. Yamao, K. Biemann, and D. Söll, *J. Biol. Chem.* **257,** 11644 (1982).

[7] Purification of Glutamyl-tRNA Synthetase from *Bacillus subtilis*

By MARIO PROULX and JACQUES LAPOINTE

Introduction

The glutamyl-tRNA synthetase (GluRS) of *Bacillus subtilis* is a monomer of M_r about 65,000 which aminoacylates both *B. subtilis* $tRNA^{Glu}$ and $tRNA^{Gln}$.[1,2] A specific amidotransferase catalyzes the formation of glutaminyl-$tRNA^{Gln}$ from the incorrect glutamyl-$tRNA^{Gln}$ in many gram-positive microorganisms including *B. subtilis*.[2-5] The monomeric GluRs "α" can be copurified with a polypeptide "β" of M_r 46,000 which increases the affinity of the enzyme about 10-fold for glutamate and for ATP, and stabilizes it against heat inactivation.[1]

We describe here procedures to purify the GluRS to homogeneity or as a complex with its regulatory factor β. The yield of pure GluRS is of the expected order of magnitude for an aminoacyl-tRNA synthetase (about 3 mg of enzyme from 250 g of wet cells) and its specific activity in a *B. subtilis* extract is of the same order of magnitude as that of the *Escherichia coli* GluRS.[6] However, the yield of this purification in terms of units of GluRS aminoacylation activity is very low (about 10%) and the turnover number of the pure enzyme is only about 0.07 Glu-tRNA per second per molecule of enzyme at 30° at pH 7.2 (it would be about 0.3/sec at the optimal pH of 8 at 37°). These results indicate that only a small percentage of the GluRS purified by this procedure is still active in the aminoacylation reaction. Because of the heat lability of this enzyme, the aminoacylation kinetics were conducted at 30°. A pure GluRS of higher specific activity is likely to result from a faster purification procedure using affinity chromatography or starting with a strain of *B. subtilis* overproducing this enzyme.

[1] M. Proulx, L. Duplain, L. Lacoste, M. Yaguchi, and J. Lapointe, *J. Biol. Chem.* **258,** 753 (1983).
[2] J. Lapointe, L. Duplain, and M. Proulx, submitted for publication.
[3] M. Wilcox and M. Nirenberg, *Proc. Natl. Acad. Sci. U.S.A.* **61,** 229 (1968).
[4] M. Wilcox, *Eur. J. Biochem.* **11,** 405 (1969).
[5] M. Wilcox, *Cold Spring Harbor Symp. Quant. Biol.* **34,** 521 (1969).
[6] J. Lapointe, S. Levasseur, and D. Kern, this volume [6].

Copyright © 1985 by Academic Press, Inc.
All rights of reproduction in any form reserved.

Purification Procedure

Reagents

Phenylmethylsulfonyl fluoride (PMSF) and 4-(2-hydroxyethyl)-1-piperazineethane sulfonic acid (HEPES) were obtained from Sigma
Microgranular DEAE-cellulose (DE52), cellulose powder CF-11, phosphocellulose (P11), and 3 MM filter paper from Whatman
Hydroxylapatite (BioGel HT) from Bio-Rad
Acrylamide, bisacrylamide, *N,N,N',N'*-tetramethylethylenediamine (TEMED) and 2-mercaptoethanol from Eastman Chemicals
Omnifluor and uniformly labeled L-[^{14}C]glutamate from New England Nuclear

Biological Materials

Bacillus subtilis 168 Tryp$^-$ is grown at 37° in APT medium[7] which efficiently prevents sporulation and the associated synthesis of large amounts of proteases and nucleases. Bacteria are harvested by centrifugation near the end of the logarithmic phase of growth, and frozen in liquid nitrogen. Alternatively, frozen cells of *B. subtilis* 168 Tryp$^-$ can be obtained from Grain Processing Corporation (Muscatine, Iowa).
Unfractionated *B. subtilis* tRNA prepared by the method of Vold[8] is generally used for GluRS assays. *E. coli* tRNA (commercially available, e.g., from Schwartz/Mann) can also be used since the *B. subtilis* GluRS aminoacylates one *E. coli* tRNA^{Gln} as efficiently as it does its natural tRNA substrates.[2] Stock solutions of tRNA (about 300 A_{260}/ml) contain 0.1 m*M* 2-mercaptoethanol to prevent oxidation of tRNA^{Glu}.
Hen egg white lysozyme, L-glutamate, and ATP are from Sigma. The stock solution of ATP (50 m*M*) is neutralized with NaOH and stored at −20°.
Uniformly labeled L-[^{14}C]glutamate (about 300 mCi/mmol) is from New England Nuclear; it is stored at 4°.

GluRS Assay

The formation of Glu-tRNA is conducted at 30° in 0.125 ml reaction mixtures containing 100 m*M* sodium Hepes, pH 7.2, 10 m*M* $MgCl_2$, 2 m*M* ATP, 80 A_{260} units/ml of unfractionated tRNA, 50 μM L-[^{14}C]glutamate (118 mCi/mmol), and the enzyme. The fractions to be assayed for GluRS

[7] J. B. Evans and C. F. Niven, *J. Bacteriol.* **62,** 599 (1951).
[8] B. Vold, this series, Vol. 29, p. 502.

activity are previously diluted in 100 m*M* sodium Hepes, pH 7.2, 20 m*M* 2-mercaptoethanol, 10% glycerol, and 1 mg bovine serum albumin per ml. The aminoacylation reaction is initiated by addition of enzyme, and stopped in various aliquots by transferring them on 2.5-cm discs of 3 MM Whatman filter paper and dropping them immediately after in a solution of 5% trichloroacetic acid at 0°. The discs are washed for 15 min at 0° in each of three solutions of 5% trichloroacetic acid to remove the free [^{14}C]glutamate, then in ethanol : ether (1 : 1), and finally in ether. The dried discs (containing the [^{14}C]Glu-tRNA) are placed in a solution of Omnifluor in toluene (4 g/liter) and counted for ^{14}C in a Beckman LS-355 liquid scintillation counter.

Buffers

A: 10 m*M* Tris–HCl, pH 8.0, 150 m*M* sodium chloride, 1 m*M* EDTA and 20% (W/W) sucrose
B: 10 m*M* Tris–HCl, pH 8.0, 10 m*M* magnesium chloride, 20 m*M* ammonium chloride, 20 m*M* 2-mercaptoethanol, 1 m*M* PMSF, and 10% glycerol
C: 10 m*M* Tris–HCl, pH 8.0, 1 m*M* magnesium chloride, 20 m*M* 2-mercaptoethanol, 0.1 m*M* PMSF, and 10% glycerol
D: 20 m*M* potassium phosphate, pH 6.8, 20 m*M* 2-mercaptoethanol, 0.1 m*M* PMSF, and 10% glycerol
E: 10 m*M* potassium phosphate, pH 7.0, 20 m*M* 2-mercaptoethanol, and 10% glycerol
F: 100 m*M* Tris–HCl, pH 6.7, 20 m*M* 2-mercaptoethanol, and 30% glycerol

Purification Steps

Unless otherwise indicated, all operations are conducted at 0 to 4°. To inactivate the serine proteases, phenylmethylsulfonyl fluoride is added to all buffers, from a 100 m*M* stock solution in 95% ethanol.[1,9]

Step 1. Protoplasts Preparation. Cells (250 g, wet weight) are washed once in a liter of Buffer A, spun, and resuspended in 600 ml of the same buffer in which 60 mg lysozyme are added. After 15 min of incubation at 37°, the protoplasts are centrifuged at 18,000 *g* for 30 min and resuspended in 400 ml of Buffer B.

Step 2. Cell Extract. The protoplasts are broken by passage through a French press under a pressure of 20,000 psi (140 MPa) at a rate of about 25 ml/min. Cell debris are removed by centrifugation at 18,000 *g* during 45 min. The supernatant is spun at 300,000 *g* during 2 hr to yield a ribosome-free cell extract.

[9] I. Schulze, J. Cratzith, and R. H. Gooding, this series, Vol. 9, p. 376.

Step 3. DEAE-Cellulose Column Chromatography. The cell extract is layered at a rate of 150 ml/hr on a 6 × 13 cm column of DEAE-cellulose (DE52) equilibrated against Buffer C. The column is washed with 1.5 liter of equilibration buffer, and the proteins are eluted with a linear gradient (8 liters) of 0 to 750 m*M* NH_4Cl in the same buffer at a rate of 250 ml/hr. The fractions containing GluRS activity are pooled and extensively dialyzed against Buffer D.

Step 4. Hydroxylapatite Column Chromatography. The DEAE-fraction is layered on a 3.5 × 8 cm column of hydroxylapatite Bio-Rad HT [mixed with 15% (v/v) of cellulose powder (Whatman CF11) to increase the flow rate] equilibrated with Buffer D. The column is washed successively with 150 ml of equilibration buffer and with a linear gradient (800 ml) of 20 to 200 m*M* K phosphate at a rate of 60 ml/hr. The active fractions are pooled and dialyzed against Buffer E.

Step 5. Phosphocellulose Column Chromatography. The hydroxylapatite fraction is layered on a 0.5 × 8 cm phosphocellulose column (P-11) equilibrated with Buffer E. The column is washed with 150 ml of equilibration buffer. Elution is performed with a linear gradient (1 liter) of 0 to 0.6 *M* KCl in the same buffer, at a flow rate of 40 ml/hr. The active fractions are pooled, dialyzed against Buffer C, and concentrated by adsorption on a small DEAE-cellulose column (1 × 2 cm) equilibrated with the same buffer. The enzyme is eluted with 5 ml of 0.5 *M* KCl in Buffer C and dialyzed against Buffer F.

Step 6. Polyacrylamide Gel Preparative Electrophoresis. The concentrated phosphocellulose fraction containing 0.0003% bromophenol blue is layered on the top of a 4 cm^2 × 10 cm column of polyacrylamide gel in a Canalco "Prep-Disc Electrophoresis" apparatus. The 10-cm-long separating gel and the buffers are prepared according to Davis.[10] Electrophoresis is performed under a constant potential difference of 400 V. When the bromophenol blue reaches the bottom of the gel (after about 4 hr), the elution is initiated (150 ml/hr) with a buffer containing 0.01 *M* Tris, 0.077 *M* glycine, pH 8.3, 20 m*M* 2-mercaptoethanol, and 10% glycerol. The active fractions are pooled, concentrated as in step 5, and dialyzed against Buffer C. A typical purification is summarized in the table.

Some Properties of the *B. subtilis* GluRS

The GluRS obtained by the procedure including the six steps described above (i.e., steps 1 to 4, 5B and 6B in the table) is a single polypeptide chain of M_r about 65,000 whose NH_2-terminal sequence has

[10] B. J. Davis, *Ann. N.Y. Acad. Sci.* **121,** 404 (1964).

PURIFICATION OF THE GLUTAMYL-tRNA SYNTHETASE FROM 250 g OF *Bacillus subtilis* 168 TRYP$^-$

Step	Total protein (mg)	Total activity (units)[a]	Specific activity (units/mg)	Recovery (%)
1. Cell extract	3400	1836	0.54	100
2. Ultracentrifugation	2330	1524	0.65	83
3. DEAE-cellulose, pH 8.0	978	876	0.90	48
4. Hydroxylapatite, pH 6.8	93	518	5.6	28
For $\alpha\beta$				
5A. Preparative electrophoresis	7.8	220	28	12
For α and β				
5B. Phosphocellulose	3.7	220	60	12
6B. Preparative electrophoresis	2.4	154	64	8.4

[a] One unit of enzyme catalyzes the formation of 1 nmol of glutamyl-tRNA/min at 30°.

been determined.[1] When stored at −20° in Buffer C containing 50% glycerol, this enzyme is stable for several months.

On the other hand, if the purification procedure does not include the phosphocellulose chromatography (i.e., steps 1 to 4 and 5A of the table) the GluRS fraction contains the monomeric enzyme of M_r 65,000 (α) and an approximately equimolar amount of a polypeptide of M_r 46,000 (β).

The K_m values of the pure monomeric GluRS (α) for $tRNA^{Glu}$, glutamate, and ATP are, respectively, 0.83, 280, and 170 μM, whereas those of the $\alpha\beta$ enzyme are 1.4, 24, and 21 μM. Other properties of the *B. subtilis* GluRS have been described previously.[1,2]

Acknowledgments

This work has been supported by Grant A9597 from the Natural Sciences and Engineering Research Council of Canada and by Grant EQ-1700 from the "Programme FCAC du Gouvernement du Québec."

[8] Glutaminyl-tRNA Synthetase of *Escherichia coli*

By PATRICIA HOBEN and DIETER SÖLL

Introduction

Each amino acid to be incorporated into proteins by mRNA directed protein synthesis is activated by a specific aminoacyl-tRNA synthetase. Bacteria have only one synthetase specific for each amino acid, and this enzyme recognizes all tRNA species (isoacceptors) for the cognate amino acid.[1] Glutamine and glutamate are activated by separate enzymes in *E. coli*. In contrast, gram-negative bacteria form Gln-$tRNA^{Gln}$ in a multistep process starting with the synthesis by glutamyl-tRNA synthetase (GluRS) of Glu-$tRNA^{Gln}$ and its subsequent transformation by an amidotransferase to Gln-$tRNA^{Gln}$.[2]

The structural gene for glutaminyl-tRNA synthetase (GlnRS), *glnS*, is located near min 14 on the *E. coli* chromosome.[3] Mutants of this synthetase giving rise to a thermolabile enzyme[3] or to a protein with altered tRNA recognition properties (mischarging)[4] have also been characterized. Studies with the temperature-sensitive *glnS* strain showed that the *in vivo* level of Gln-$tRNA^{Gln}$ did not affect the activity of GlnRS or of glutamine synthetase.[3] The cellular level of GlnRS is regulated both metabolically (A. Cheung, unpublished results) and by a gene unlinked to *glnS*.[5] The structural gene for GlnRS was isolated on a 6.5-kb DNA fragment and incorporated into pBR322 to form the plasmid pYY105.[6] This plasmid can rescue the temperature-sensitive phenotype of the *glnS* mutation. The transformed strain overproduces GlnRS about 12-fold. Introduction of the 6.5 kb *glnS* DNA fragment into the high expression plasmid PLc28[7] yields over 50-fold overproduction of the enzyme (M. Nichols, R. Swanson, and P. Hoben, unpublished results).

[1] P. Schimmel and D. Söll, *Annu. Rev. Biochem.* **48,** 601 (1979).
[2] M. Wilcox and M. Nirenberg, *Proc. Natl. Acad. Sci. U.S.A.* **61,** 229 (1968).
[3] A. Körner, B. B. Magee, B. Liska, K. B. Low, E. A. Adelberg, and D. Söll, *J. Bacteriol.* **120,** 151 (1974).
[4] H. Ozeki, H. Inokuchi, F. Yamao, M. Kodaira, H. Sakano, T. Ikemura, and Y. Shimura, *in* "Transfer RNA: Biological Aspects" (D. Söll, J. Abelson, and P. Schimmel, eds.), p. 341. Cold Spring Harbor Lab., Cold Spring Harbor, New York, 1980.
[5] A. Cheung, S. Morgan, K. B. Low, and D. Söll, *J. Bacteriol.* **139,** 175 (1979).
[6] F. Yamao, H. Inokuchi, A. Cheung, H. Ozeki, and D. Söll, *J. Biol. Chem.* **257,** 11639 (1982).
[7] E. Remaut, P. Stanssens, and W. Fiers, *Gene* **15,** 81 (1981).

Copyright © 1985 by Academic Press, Inc.
All rights of reproduction in any form reserved.

Two glutamine isoacceptors are found in *E. coli;* one tRNA is specific for the codon CAG, while the other recognizes CAA.[8] *E. coli* GlnRS catalyzes the esterification of one glutamine residue to the 2′-hydroxyl group of the 3′-terminal adenosine of both $tRNA^{Gln}$ species.[9,10] The mechanism of this reaction differs from the general mechanism of aminoacyl-tRNA formation, in which amino acid activation (aminoacyl adenylate formation) can be separated from the subsequent transfer of the activated amino acid to tRNA. In prokaryotes and eukaryotes, GlnRS, GluRS, and arginyl-tRNA synthetase (ArgRS) comprise a small class of monomeric, aminoacyl-tRNA synthetases which require the cognate tRNA for aminoacyl-adenylate formation.[1] In the cases of *E. coli* GlnRS (S. Berman and P. Hoben, unpublished results), and *E. coli*[11] and *B. subtilis*[12] GluRS, this unique activation mechanism appears to result from the inability of these enzymes to bind amino acid in the absence of their cognate tRNAs.

GlnRS was first purified from *E. coli* by Folk.[13] Here we describe a more rapid and higher yield purification scheme[14] based on the overproduction of GlnRS in an *E. coli* strain (JA221) carrying the cloned[6] *glnS* structural gene on a recombinant pBR322 plasmid (pYY105).

Assay Method

Principle. GlnRS activity is assayed by measuring the rate of formation of acid-precipitable [^{14}C]Gln-tRNA.

Reagents

Stock Solutions
- 1 *M* sodium cacodylate, pH 7.0
- 100 m*M* magnesium acetate
- 100 m*M* ATP (sodium salt) adjusted to pH 7.0
- 100 m*M* glutamate
- 100 A_{260} units/ml unfractionated tRNA
- L-[^{14}C]Glutamine (100–300 mCi/mmol)
- Enzyme dilution buffer: 50 m*M* sodium cacodylate, pH 7.0, 2 m*M* 2-mercaptoethanol, 1 mg/ml bovine serum albumin, 10% glycerol

[8] M. Yaniv and W. Folk, *J. Biol. Chem.* **250,** 3243 (1975).
[9] M. Sprinzl and F. Cramer, *Proc. Natl. Acad. Sci. U.S.A.* **72,** 3049 (1975).
[10] T. H. Fraser and A. Rich, *Proc. Natl. Acad. Sci. U.S.A.* **72,** 3044 (1975).
[11] D. Kern and J. Lapointe, *Biochemistry* **18,** 5809 (1979).
[12] J. Lapointe, S. Levasseur, and D. Kern, this volume [6].
[13] W. Folk, *Biochemistry* **10,** 1728 (1971).
[14] P. Hoben, N. Royal, A. Cheung, F. Yamao, K. Biemann, and D. Söll, *J. Biol. Chem.* **257,** 11644 (1982).

5% aqueous trichloroacetic acid
Whatman 3 MM filter disks (2.3 cm)

Procedures. The assay mixture (usually 100 μl) contains 50 mM sodium cacodylate, pH 7.0, 10 mM magnesium acetate, 2 mM ATP, 16 A_{260} units/ml of unfractionated tRNA, 1 mM glutamate, 4 μM L-[^{14}C]glutamine, and variable amounts (0.01–1 μg) of enzyme. The enzyme is always diluted in enzyme dilution buffer. When unfractionated extracts are assayed unlabeled glutamate is included in the reaction to prevent conversion of [^{14}C]Gln to [^{14}C]Glu by contaminating glutaminase; the labeled Glu would interfere with the measurement of Gln-tRNA formation. The aminoacylation reaction is initiated by the addition of diluted enzyme, and the reaction mixture is incubated at 37° for various times. Aliquots of this mixture are transferred to Whatman 3 MM paper filter disks, which are directly immersed into a solution of 5% trichloroacetic acid kept at 4°. Each 2.3-cm disk can hold up to 100 μl of solution. The filters are washed for 10 min each in 5% trichloroacetic acid (four times), then once in ethanol, and finally in ether. The dried filter disks are counted for ^{14}C radioactivity in a solution of Omnifluor in toluene (4 g/liter). One unit of GlnRS activity catalyzes the formation of one nanomole of Gln-tRNA in 10 min at 37°.

Purification Procedures

Reagents

Buffer A: 10 mM potassium phosphate, pH 8.0, 20 mM 2-mercaptoethanol, 10% (v/v) glycerol
Buffer B: 10 mM potassium phosphate, pH 7.2, 20 mM 2-mercaptoethanol, 10% (v/v) glycerol
Buffer C: 10 mM potassium phosphate, pH 6.5, 20 mM 2-mercaptoethanol, 10% (v/v) glycerol
Buffer D: 20 mM potassium phosphate, pH 7.5, 50 mM potassium chloride, 1 mM dithiothreitol, 50% (v/v) glycerol
KCl stock solution: 5 M
KH_2PO_4 stock solution: 0.5 M
K_2HPO_4 stock solution: 0.5 M

Bacterial Strain

JA221 (pYY105): *recA1, leuB6, trpE5, hsdR*$^-$, *hsdM*$^+$, *lacY,* C600 carrying *glnS* plasmid pYY105.[14]

All operations are carried out at 0–4°.

Step 1. S-100 Extract. Cells (10 g, wet weight) are washed once in 100 ml of Buffer A. The cells are centrifuged once at 17,000 g for 5 min and resuspended in the same buffer. The cell suspension (about 100 ml) is passed twice through a French pressure cell at 10,000 psi. The crude lysate is centrifuged for 30 min at 30,000 *g* to remove cell debris. The supernatant is centrifuged for 2 hr at 100,000 *g* to yield an S-100 fraction of approximately 50 ml.

Step 2. DEAE-Cellulose Column Chromatography. The S-100 fraction is dialyzed against Buffer B, then loaded onto a DEAE-cellulose column (Whatman DE-52, 2 × 15 cm) equilibrated with the same buffer. The column is washed with Buffer B until the absorbance (A_{280}) of the effluent is less than 0.1. Elution is carried out with a linear salt, reverse pH gradient (in a total volume of 300 ml) ranging from 20 m*M* potassium phosphate, pH 7.2, to 250 m*M* potassium phosphate, pH 6.5. GlnRS activity elutes at approximately 100 m*M* potassium phosphate, pH 6.9. The fractions containing GlnRS are pooled and extensively dialyzed against Buffer C.

Step 3. Phosphocellulose Column Chromatography. The DEAE-fraction is loaded onto a phosphocellulose column (Whatman P-11, 1.5 × 6 cm) equilibrated with Buffer C. The column is washed with equilibration buffer (about 3 column volumes) until the absorbance (A_{280}) of the effluent is less than 0.1. Elution is performed with a linear gradient (total volume, 100 ml) of potassium chloride (20–300 m*M*) in equilibration buffer, at a flow rate of about 10 ml/hr. One major protein peak containing GlnRS activity is eluted in the gradient at approximately 180 m*M* potassium chloride. The active fractions are pooled, and concentrated by ultrafiltration, followed by dialysis into Buffer D. A typical purification is summarized in the table.

Properties of the Enzyme

Storage and Stability. The pure enzyme is preferably stored in small aliquots in Buffer D at −20°. Under these conditions, only a marginal loss of specific activity is observed after 6 months.

PURIFICATION OF *E. coli* GlnRS[a]

Fraction	Volume (ml)	Total protein (mg)	Total activity (units)	Specific activity (units/mg)	Yield (%)
Cell extract (S-100)	50	855	27,360	32	100
DEAE-cellulose (DE-fraction)	40	54	27,000	500	99
Phosphocellulose (PC-fraction)	31	17	18,260	1,075	67

[a] Purification of Gln-tRNA synthetase from 10 g of JA221 (pYY105) cells.

Structural Properties. The enzyme obtained by the above procedure appears as a single Coomassie-blue stained band on 0.1% sodium dodecyl sulfate polyacrylamide gels. The complete primary structure of GlnRS has been obtained by a combination of DNA and protein sequencing techniques.[14,14a] GlnRS is a polypeptide of 553 amino acids with a molecular weight of 63,000 calculated from the DNA sequence.

However, when this purification procedure is carried out on extracts from *E. coli* strains *not* containing the *glnS* recombinant plasmid, we find that a single polypeptide (approximately 45,000 daltons) copurifies with GlnRS in about a 1 : 1 ratio through the phosphocellulose chromatography step. When this fraction is run on a BioGel P-150 column, most of the GlnRS and this polypeptide coelute in the void volume (P. Hoben, unpublished results). Therefore, GlnRS may be weakly associated with this as yet unidentified 45,000-dalton polypeptide. A polypeptide of similar size and properties has previously been identified in *E. coli*[15] and *B. subtilis*[12] GluRS preparations (see also the articles by Lapointe in this volume).

Substrate Specificity. The K_m values for ATP, glutamine, and $tRNA^{Glu}$ (in unfractionated tRNA) are 111, 110, and 0.1 μM, respectively.

[14a] P. Hoben, H. Uemura, F. Yamao, A. Cheung, R. Swanson, M. Sumner-Smith, and D. Söll, *Fed. Pr., Fed. Am. Soc. Exp. Biol.* **43,** 2972 (1984).

[15] J. Lapointe and D. Söll, *J. Biol. Chem.* **247,** 4982 (1972).

[9] Preparation of Optically Pure Isotopically Labeled L-Glutamate

By TINGSEN XU and ALTON MEISTER

Introduction

Commercially available preparations of radioactively labeled L-glutamate commonly contain small amounts of labeled D-glutamate. The presence of as little as a 0.1% impurity of D-glutamate in an L-glutamate preparation may lead to unreliable results, especially in studies on the metabolism of 5-oxo-L-proline.[1,2] A highly active enzyme (D-glutamate cyclase) is present in certain mammalian tissues (notably kidney and liver) that catalyzes rapid conversion of D-glutamate to 5-oxo-D-pro-

[1] R. Sekura, P. Van Der Werf, and A. Meister, *Biochem. Biophys. Res. Commun.* **71,** 11 (1976).

[2] W. L. Nyhan and H. Busch, *Cancer Res.* **18,** 385 (1958).

Copyright © 1985 by Academic Press, Inc.
All rights of reproduction in any form reserved.

line.[3–5] D-Glutamate may be formed in animals by the microbial flora. It also may enter through the diet. It is of interest that significant concentrations of 5-oxo-D-proline are found in human blood plasma[6] and that 5-oxo-D-proline is excreted in the urine of normal humans in substantial (50–400 μmol per day) amounts.[3]

Several procedures are described below which have been found to be useful in dealing with this problem. In one approach, commercially available radiolabeled L-glutamate is treated with partially purified D-glutamate cyclase[5] to convert the D-glutamate present to 5-oxo-D-proline; separation of labeled L-glutamate from labeled 5-oxo-D-proline is readily accomplished by chromatography on Dowex 50 (H^+). Another approach involves the synthesis of labeled L-glutamate by the action of glutamate dehydrogenase on labeled α-ketoglutarate.[7] Although this method seems straightforward, it may be complicated by the presence in some commercial samples of labeled α-ketoglutarate of a small amount of racemic labeled glutamate. The amount of D-glutamate contamination in an L-glutamate preparation can be determined by treating the sample with D-glutamate cyclase and measuring the 5-oxo-D-proline formed. Another procedure involves the use of L-glutamate decarboxylase[8]; in this method, the L-glutamate present is converted to labeled γ-aminobutyrate, which may be determined by amino acid analysis together with the residual D-glutamate.

Purification of L-Glutamate by Use of D-Glutamate Cyclase

D-Glutamate cyclase may be readily prepared in partially purified form from mouse kidney; a purification of 80- to 100-fold may be achieved.[5]

A solution (final volume, 4.0 ml) containing labeled L-glutamate (2–50 μmol), 40 mg of D-glutamate cyclase, and 0.05 *M* magnesium chloride is adjusted to pH 8.4 and shaken in a water bath at 37° for 6 hr. The protein is precipitated by adding 4 ml of 10% trichloroacetic acid, and after centrifugation at 16,000 *g* for 20 min, the clear supernatant solution is added to the top of a column (0.6 × 6 cm) of Dowex 50 (H^+), and the 5-oxo-D-proline is eluted with water. After it is established that no further radioactivity emerges from the column on washing with water, the column is treated with 8 ml of 3 *M* ammonium hydroxide. The ammonium hydroxide

[3] A. Meister, M. W. Buckenberger, and M. Strassberger, *Biochem. Z.* **338,** 217 (1963).
[4] A. Meister and M. W. Bukenberger, *Nature* (*London*) **194,** 557 (1962).
[5] A. Meister, this series, Vol. 17A, p. 860.
[6] A. G. Palekar, S. S. Tate, F. J. Sullivan, and A. Meister, *Biochem. Med.* **14,** 339 (1975).
[7] J. A. Olson and C. B. Anfinsen, *J. Biol. Chem.* **197,** 67 (1952).
[8] M. M. Camien, L. E. McClure, and M. S. Dunn, *Arch. Biochem.* **28,** 220 (1950).

effluent contains optically pure isotopically labeled L-glutamic acid. The solution is concentrated by flash evaporation at 30–40°. The residue is dissolved in a small volume of water (final pH 7–8) or 0.1 *M* HCl. Storage at values of pH between 2 and 5 promotes cyclization to 5-oxoproline. The recovery of labeled L-glutamate is usually about 90%.

Preparation of Isotopically Labeled L-Glutamate from Labeled α-Ketoglutarate

Labeled α-ketoglutaric acid is purified by passage through a small column of Dowex 50 (H^+). Thus, 2.5 ml of a solution containing 1–10 μmol of α-ketoglutaric acid is added to the top of a column (5 × 0.6 cm) of Dowex 50 (H^+), which is then washed with 6 ml of water. This procedure removes amino acids (such as labeled DL-glutamic acid, which may be present). The water effluent contains α-ketoglutaric acid; it may be lyophilized. The dry powder may be dissolved in an appropriate volume of water. Recoveries are usually about 90%.

A reaction mixture containing purified α-ketoglutarate (0.5–3 μmol), 300 μl of 1 *M* Tris–HCl buffer (pH 7.8), 60 μl of 2.5 *M* ammonium chloride, 40 μl of NADH (1–6 μmol), and 100 μl of L-glutamate dehydrogenase (100 units; added last) in a final volume of 2.5 ml is placed in a small vessel at 26–37°. The absorbance at 340 nm decreases to a stable value when the reaction is complete. The protein is precipitated by addition of sulfosalicylic acid (final concentration 5%). After centrifugation, the clear supernatant solution is added to the top of a column (5 × 0.6 cm) of Dowex 50 (H^+), which is washed with water until the effluent contains no radioactivity. Labeled L-glutamate is eluted from the column with 8 ml of 3 *M* ammonium hydroxide. The ammonium hydroxide effluent may be evaporated to dryness in a flash evaporator or lyophilized. The yield, based on α-ketoglutarate, is usually about 90%.

Analysis of Mixtures of Labeled D- and L-Glutamates Using L-Glutamate Decarboxylase

A reaction mixture (final volume, 1 ml) containing the labeled glutamates (0.5–5 μmol), 0.1 *M* sodium acetate buffer (pH 5.4), and L-glutamate decarboxylase (5–10 units) is shaken in a water bath at 37° for 1–5 hr. The protein is precipitated by adding sulfosalicylic acid (final concentration, 5%), and the mixture is centrifuged at 16,000 *g* for 20 min. An aliquot of the clear supernatant solution is analyzed on an amino acid analyzer for γ-aminobutyric acid and glutamic acid. Samples are also obtained from the analyzer for the determination of radioactivity. The γ-aminobutyric acid found is a measure of L-glutamate and the glutamate found in a measure of D-glutamate.

Comments on D-Glutamate and Its Enzymatic Cyclization to 5-oxo-D-Proline

D-Glutamate is a substrate of glutamine synthetase and of γ-glutamylcysteine synthetase and inhibits the action of these enzymes on L-glutamate.[1] The apparent K_i value for D-glutamate as an inhibitor of γ-glutamylcysteine synthetase is about 0.8 m*M*; the apparent K_m value for L-glutamate is about 1.8 m*M*. Glutathione synthetase is not significantly inhibited by D-glutamate. D-Glutamate also inhibits beef liver glutamate dehydrogenase competitively; the apparent K_i value is 0.17 m*M*. Administration of D-glutamate to mice has been found to decrease the levels of glutathione and (to a lesser extent) of glutamine in the kidney.[1]

Ratner administered DL-glutamate labeled with ^{15}N and deuterium to rats and observed that they excreted 5-oxo-D-proline; this compound had an unaltered ratio of isotopes. In these studies, all of the D-glutamate administered appeared as urinary 5-oxo-D-proline.[9] Later, observations of a similar nature were made.[10–12] Slices of rat liver and kidney have been found to catalyze the conversion of D-glutamate to 5-oxoproline,[12] and the enzyme that catalyzes the cyclization reaction was found in kidney and liver of several species including man.[3,4] The enzyme has been purified from mouse kidney.[4,5,13]

The enzymatic cyclization of D-glutamate to form 5-oxo-D-proline seems to constitute a special pathway for the elimination of D-glutamate. It may be of significance that D-amino acid oxidase, also highly concentrated in the kidney, exhibits little or no activity toward D-glutamate. The equilibrium of the reaction catalyzed by D-glutamate cyclase, like the nonenzymatic cyclization of glutamate, greatly favors cyclization. Thus, in contrast to the pathways for elimination of benzoate, phenylacetate, and various foreign compounds which are conjugated, hydroxylated, esterified, methylated, oxidized, or reduced, and which require energy and use a variety of intracellular compounds, the intramolecular acylation reaction of D-glutamate is relatively simple and efficient. Other aspects of the mechanism and significance of the mammalian pathway for elimina-

[9] S. Ratner, *J. Biol. Chem.* **152,** 559 (1944).
[10] F. Kogl, T. J. Barendregt, and A. J. Klein, *Nature (London)* **162,** 723 (1948).
[11] G. Hillmann, A. Hillmann-Elies, and F. Methfessel, *Nature (London)* **174,** 403 (1954).
[12] W. E. Wilson and R. E. Koeppe, *J. Biol. Chem.* **236,** 365 (1961).
[13] J. C. Unkeless and P. Goldman, *J. Biol. Chem.* **246,** 2354 (1971).

tion of D-glutamate, including certain speculations, have been considered.[1]

It is of interest that D-glutamate cyclase also acts on a variety of D-glutamate analogs, including α, β, and γ-substituted glutamates and α-aminotricarballylic acid; it does not act on D-glutamine, D-α-aminoadipate, or D-homoglutamine.[3,13]

[10] L-Glutamate-L-Amino Acid Transaminases

By ARTHUR J. L. COOPER

Introduction

Enzyme-catalyzed transamination was discovered in 1937.[1] It was initially thought that transamination reactions are limited to glutamate, aspartate, alanine, and the corresponding α-keto acids.[2] However, by the early 1950s it became apparent that many other transamination reactions occur. By 1955, at least 27 amino acid-α-ketoglutarate transamination reactions had been discovered in a wide variety of organisms.[3] By now, over one hundred amino acid-α-ketoglutarate transamination reactions have been described.[4] It would be a formidable task to list all the assay procedures that have been used. Instead, assay procedures for only those glutamate (α-ketoglutarate) transaminases of major metabolic significance and those that have been most thoroughly characterized are described here and in Chapters [11]–[16]. Another problem is that enzymes from different sources may require slightly different assay procedures. We have concentrated on assays developed for the adult rat tissue enzymes, but where appropriate have referenced assays to enzymes from other sources. Finally, workers have used different assay temperatures (usually 25, 30, or 37°), thereby further complicating comparisons.

Literally thousands of papers have been written on the properties, distribution and clinical importance of glutamate-aspartate and glutamate-alanine transaminases (see [11], [12]). In this chapter we will include

[1] A. E. Braunstein and M. G. Kritzmann, *Nature* (*London*) **140,** 503 (1937); *Enzymologia* **2,** 129 (1937).

[2] For example, see R. M. Herbst, *Adv. Enzymol.* **4,** 75 (1944).

[3] A. Meister, *Adv. Enzymol.* **16,** 185 (1955).

[4] For a review of the discovery of and metabolic importance of transaminases, see the review by A. J. L. Cooper and A. Meister, *in* "The Transaminases" (D. E. Metzler and P. Christen, eds.), pp. 534–563. Wiley, 1983.

Copyright © 1985 by Academic Press, Inc.
All rights of reproduction in any form reserved.

assays for other glutamate (α-ketoglutarate)-utilizing transaminases for comparative purposes.

Other L-Glutamate (α-Ketoglutarate) Transaminases

This section lists several α-ketoglutarate utilizing transaminases. Reference to purification and assay procedures are given, and where known the identity with other transaminases is noted.

6-N-Acetyl-β-lysine. This enzyme has been purified from *Pseudomonas* extracts and characterized.[5]

α-Aminoadipate. This enzyme is probably identical to halogenated aromatic amino acid transaminase (see the section on aromatic amino acid-α-ketoglutarate transaminases).

Cysteine. It has been shown that mitochondrial cysteine-glutamate transaminase is identical to mitochondrial glutamate-aspartate transaminase.[6] A cysteine-glutamate transaminase that is apparently not identical to glutamate-aspartate transaminase has been purified from rat liver and characterized by Ip *et al.*[7]

Cysteine Sulfinate. Cysteine sulfinate transaminase activity is widespread in both mitochondrial and soluble fractions of animal tissues. There has been some controversy as to whether or not the mitochondrial and soluble activities are due to the mitochondrial and soluble forms of glutamate-aspartate transaminase. However, the best evidence (at least for rat brain) now suggests that cysteine sulfinate transaminase is indeed identical to glutamate-aspartate transaminase.[8]

Histidine. This activity has not been shown to be catalyzed by a discrete enzyme in mammalian tissues. In bacteria the reaction is often catalyzed by a general aromatic amino acid transaminase.[9]

Histidinol Phosphate/Imidazole Acetol Phosphate. This enzyme catalyzes an important reaction in the biosynthesis of L-histidine. The assay system was originally described by Ames and co-workers[10] and has been modified by Chapman and Nester.[11]

[5] G. Bozler, J. M. Robertson, M. Ohsugi, C. Hensley, and H. A. Barker, *Arch. Biochem. Biophys.* **197,** 226 (1979).

[6] T. Ubuka, S. Umemura, Y. Ishimoto, and M. Shimomura, *Physiol. Chem. Phys.* **9,** 91 (1977).

[7] M. P. C. Ip, R. J. Thibert, and D. E. Schmidt, Jr., *Can. J. Biochem.* **55,** 958 (1977).

[8] M. Recasens, R. Benezra, P. Basset, and P. Mandel, *Biochemistry* **19,** 4583 (1981).

[9] C. G. Paris and B. Magasanik, *J. Bacteriol.* **145,** 257, 266 (1981).

[10] B. N. Ames and H. K. Mitchell, *J. Biol. Chem.* **212,** 687 (1955); B. N. Ames and B. L. Horecker, *ibid.* **220,** 113 (1956).

[11] L. F. Chapman and E. W. Nester, *J. Bacteriol.* **96,** 1658 (1968).

Kynurinine. See the section on halogenated aromatic amino acids.

ε-Lysine. This enzyme has been isolated from extracts of *Flavobacterium fuscum* and crystallized. It has a narrow specificity. The only other substrate (with α-ketoglutarate as acceptor) is ornithine, which is ~55% as active as lysine.[12] As with δ-ornithine transaminase, the product (Δ^1-piperideine-6-carboxylate) is assayed by reacting it with *o*-aminobenzaldehyde.

Methionine. This enzyme activity has not been well characterized but Ikeda *et al.* have isolated a leucine (methionine) transaminase from the soluble and mitochondrial fractions of rat liver.[13]

Taurine/Hypotaurine. A taurine transaminase has been purified from extracts of *Achromobacter superficialis* and crystallized.[14] The activity with various substrates (α-ketoglutarate as acceptor) was found to be in the order: hypotaurine > DL-β-aminoisobutyrate > β-alanine > taurine > β-aminobutyrate > 3-aminopropanesulfonate > DL-aminobutyrate. The product of the reaction with taurine (sulfatoacetaldehyde) is relatively stable, whereas the transamination product of hypotaurine (sulfinoacetaldehyde) is unstable and spontaneously decomposes to acetaldehyde and sulfite. Recently, a hypotaurine transaminase has been shown to be present in rat liver. It is inactive with taurine and activity with α-keto acid acceptors is in the order: pyruvate > α-ketoglutarate > oxaloacetate.[15]

Tryptophan. This activity has not been proved to be catalyzed by a discrete enzyme in mammalian tissues. In the brain, at least, the activity is due to soluble glutamate-aspartate transaminase.[16] Rat liver tyrosine transaminase exhibits some activity toward tryptophan.[17] In bacteria, the reaction is catalyzed by aromatic transaminases.[9]

[12] K. Soda, H. Misono, and T. Yamamoto, *Biochemistry* **7,** 4102 (1968); K. Soda and H. Misono, *ibid.* p. 4110.

[13] T. Ikeda, Y. Konishi, and A. Ichihara, *Biochim. Biophys. Acta* **445,** 622 (1976).

[14] S. Toyama, H. Misono, and K. Soda, *Biochim. Biophys. Acta* **523,** 75 (1978).

[15] J. H. Fellman, E. S. Roth, N. A. Avedovech, and K. D. McCarthy, *Life Sci.* **27,** 1999 (1980).

[16] It is now well established that a second cytosolic L-aromatic amino-L-glutamate transaminase and a mitochondrial L-aromatic-L-glutamate transaminase in pig heart preparations are, in fact, the soluble and mitochondrial forms of GOT [J. E. Miller and G. Litwack, *J. Biol. Chem.* **246,** 3234 (1971); E. Shrawder and M. Martinez-Carrion, *ibid.* **247,** 2486 (1972)]. It is also well established that the aromatic amino acid-L-glutamate transaminase activities of rat brain are due to the soluble and mitochondrial isozymes of GOT [see S. King and A. T. Phillips, *J. Neurochem.* **30,** 1399 (1978); T. Noguchi, M. Nakatani, Y. Minatogawa, E. Okuno, and R. Kido, *ibid.* **25,** 579 (1975)]. Apparently, L-tryptophan-(α-ketoglutarate)-oxaloacetate [transaminase from rat brain is also identical to sGOT; Y. Minatogawa, T. Noguchi, and R. Kido, *J. Neurochem.* **27,** 1097 (1976)]. mGOT is more active than sGOT in catalyzing transamination with the aromatic amino acids.

[17] G. A. Jacoby and B. N. La Du, *J. Biol. Chem.* **239,** 419 (1964).

[11] Glutamate-Aspartate Transaminase

By ARTHUR J. L. COOPER

Glutamate-aspartate transaminase is present in microorganisms, plants, and in all animal and human tissues thus far examined. Its ubiquity attests to its metabolic importance. In addition, in higher organisms, the enzyme exists in soluble (sGOT) and mitochondrial (mGOT) forms. Glutamate-aspartate transaminase catalyzes the following reaction [Eq. (1)][1]:

$$\text{L-Aspartate} + \alpha\text{-ketoglutarate} \rightleftharpoons \text{oxaloacetate} + \text{L-glutamate} \quad (1)$$

sGOT has been highly purified from pig heart,[2] pig brain,[3] beef heart,[4] beef brain,[5] chicken heart,[6] chicken liver,[7] human liver,[8] sheep liver,[9] rat brain,[10] rat liver,[11] and beef liver.[4,12] mGOT has also been highly purified from pig heart,[13] pig brain,[3] beef heart,[4] beef liver,[4] chicken heart,[7] chicken liver,[7] human liver,[8] sheep liver,[14] rat brain,[10] and rat liver.[11] A method for the simultaneous purification of gram quantities of sGOT and mGOT from pig heart has been described.[15] Both sGOT[2,5,7–11] and mGOT[7,8,10,11,13,14] exist in a number of electrophoretically distinct subforms. The amino acid content and immunological properties of the subforms of a given isozyme appear to be identical (see ref. 11 for a discus-

[1] For convenience soluble glutamate-aspartate transaminase (glutamate-oxaloacetate transaminase), mitochondrial glutamate-aspartate, and glutamate-alanine transaminase (glutamate-pyruvate transaminase) are referred to as sGOT, mGOT, and GPT, respectively.

[2] M. Martinez-Carrion, C. Turano, E. Chiancone, F. Bossa, A. Giartosio, F. Riva, and P. Fasella, *J. Biol. Chem.* **22,** 2397 (1967).

[3] H. George and S. Gabay, *Biochim. Biophys. Acta* **167,** 555 (1968).

[4] H. Wada and Y. Morino, *Vitam. Horm. (N.Y.)* **22,** 411 (1964); Y. Morino, H. Ito, and H. Wada, *Biochem. Biophys. Res. Commun.* **13,** 348 (1963).

[5] M. L. Krista and M. L. Fonda, *Biochim. Biophys. Acta* **309,** 83 (1973).

[6] L. H. Bertland and N. O. Kaplan, *Biochemistry* **7,** 134 (1968).

[7] E. J. Shrawder and M. Martinez-Carrion, *J. Biol. Chem.* **248,** 2140 (1973).

[8] H. Teranishi, H. Kagamiyama, K. Teranishi, H. Wada, T. Yamano, and Y. Morino, *J. Biol. Chem.* **253,** 8842 (1978).

[9] M. Campos-Cavieres and E. A. Munn, *Biochem. J.* **135,** 683 (1973).

[10] S. King and A. T. Phillips, *J. Neurochem.* **30,** 1399 (1978).

[11] Q. K. Huynh, R. Sakakibara, T. Watanabe, and H. Wada, *J. Biochem. (Tokyo)* **88,** 231 (1980).

[12] G. Marino, A. M. Greco, V. Scardi, and R. Zito, *Biochem. J.* **99,** 589 (1966).

[13] C. M. Michuda and M. Martinez-Carrion, *Biochemistry* **8,** 1095 (1969).

[14] A. Orlacchio, M. Campos-Cavieres, I. Pashev, and E. A. Munn, *Biochem. J.* **177,** 583 (1979).

[15] Y. Morino, S. Tanase, T. Watanabe, H. Kagamiyama, and H. Wada, *J. Biochem. (Tokyo)* **82,** 847 (1977).

Copyright © 1985 by Academic Press, Inc.
All rights of reproduction in any form reserved.

sion). The amino acid compositions of the mGOT and sGOT enzymes from rat liver, rat brain, pig heart, and human heart are known.[11] The complete amino acid sequence of pig heart sGOT,[16] mGOT,[17,18] and rat liver mGOT[19] is known.

The various assay procedures that have been used to measure GOT activity have been discussed in great detail by Bergmeyer and Bernt.[20] Tonhazy *et al.* described a procedure based on the differences in the absorbance profiles of basic solutions of the 2,4-dinitrophenylhydrazones of pyruvate (formed from decarboxylation of oxaloacetate with aniline citrate) and of α-ketoglutarate.[21] Reitman and Frankel used a similar procedure except that the aniline citrate step was omitted.[22] Cabuad *et al.*[23] devised a procedure in which the 2,4-dinitrophenylhydrazones of pyruvate is preferentially extracted into toluene. This method has the advantage that optimal concentrations of α-ketoglutarate can be employed.

Oxaloacetate produced in a fixed time according to Eq. (1) can be coupled to several diazonium salts. The resulting color is proportional to the amount of oxaloacetate formed and can be measured spectrophotometrically in the region 500–550 nm. Stable diazonium salts that have been employed include those of 4-amino-2,5-diethoxybenzanilide,[24] 6-benzamido-4-methoxy-*m*-toluidine,[24,25] and [1]*N*-butyl-4-methoxymetanilamide (Fast Red PDC dye).[26–29] Some of the procedures in which oxaloacetate is measured directly yield low values compared to the coupled

[16] Y. A. Ovchinnikov, C. A. Egorov, N. A. Aldanova, M. Y. Feigina, V. M. Lipkin, N. G. Abdulaev, E. V. Grishin, A. P. Kiselev, N. N. Modyanov, A. E. Braunstein, O. L. Polyanovsky, and V. V. Nosikov, *FEBS Lett.* **29,** 31 (1973).

[17] H. Kagamiyama, R. Sakakibara, H. Wada, S. Tanase, and Y. Morino, *J. Biochem. (Tokyo)* **82,** 291 (1977).

[18] D. Barra, F. Bossa, S. Doonan, H. M. A. Fahmy, G. J. Hughes, K. Y. Kakoz, F. Martini, and R. Petruzzelli, *FEBS Lett.* **83,** 241 (1977).

[19] Q. K. Huynh, R. Sakakibara, T. Watanabe, and H. Wada, *Biochem. Biophys. Res. Commun.* **97,** 474 (1980).

[20] H. U. Bergmeyer and E. Bernt, *in* "Methods of Enzymatic Analysis" (H. U. Bergmeyer, ed.), 2nd Engl. ed., pp. 727–751. Academic Press, New York, 1974.

[21] N. E. Tonhazy, N. G. White, and W. W. Umbreit, *Arch. Biochem. Biophys.* **28,** 36 (1950).

[22] S. Reitman and J. S. Frankel, *Am. J. Clin. Pathol.* **28,** 56 (1956).

[23] P. Cabaud, R. Leeper, and F. Wróblewski, *Am. J. Clin. Pathol.* **26,** 1101 (1956).

[24] A. L. Babson, P. O. Shapiro, P. A. R. Williams, and G. E. Phillips, *Clin. Chim. Acta* **7,** 199 (1962).

[25] M. Furuno and A. Sheena, *Clin. Chem. (Winston-Salem, N.C.)* **11,** 23 (1965).

[26] S. M. Sax and J. J. Moore, *Clin. Chem. (Winston-Salem, N.C.)* **13,** 175 (1967).

[27] S. Morgenstern, M. Oklander, J. Auerbach, J. Kaufman, and B. Klein, *Clin. Chem. (Winston-Salem, N.C.)* **12,** 95 (1966).

[28] M. K. Schwartz and O. Bodansky, *in* "Methods of Enzymatic Analysis" (H. U. Bergmeyer, ed.), 2nd Engl. ed., pp. 771–773. Academic Press, New York, 1974.

[29] J. J. Moore and S. M. Sax, *Clin. Chem. (Winston-Salem, N.C.)* **15,** 730 (1969).

assay (see below), possibly because of instability of oxaloacetate. Oxaloacetate readily decarboxylates at pH 7.0–8.0, 37°.

By far the most widely used assay of glutamate-aspartate transaminase involves coupling oxaloacetate produced in Eq. (1) with an excess of malate dehydrogenase [Eq. (2)].

$$\text{Oxaloacetate} + \text{NADH} + \text{H}^+ \rightleftharpoons \text{L-malate} + \text{NAD}^+ \quad (2)$$

The disappearance of absorbance at 345 nm due to NADH oxidation ($\varepsilon = 6.23 \times 10^3$) is continuously recorded spectrophotometrically.[30] This method was first described by Karmen,[31] and since this publication, the procedure has been modified by many authors. The optimal assay for measuring GOT in human serum has been reported to be as follows[32]: L-aspartate, 240 mM; Tris, 80 mM; pyridoxal 5′-phosphate, 100 μM; α-ketoglutarate, 12 mM; NADH, 180 μM; malate dehydrogenase, 7 μkat^{-1}; volume fraction of serum, 1/12, 30°, pH 7.8. The assay for sGOT and mGOT in rat tissues according to Magee and Phillips[33] is as follows: malate dehydrogenase, 7.6 μg; L-aspartate, 124 mM; potassium phosphate, 69 mM; NADH, 0.5 mM; potassium α-ketoglutarate, 6.9 mM; pH 7.4, 30°; final volume 0.29 ml. Bergmeyer and Bernt caution that for those tissues containing glutamate dehydrogenase and ammonia a blank in which aspartate is omitted must be included.[20]

The reaction can also be run in the reverse direction by measuring the rate of disappearance of oxaloacetate ($\varepsilon_{280\,nm} = 550$).[33] The reaction mixture (0.25 ml) contains monosodium glutamate, 60 mM; oxaloacetate, 1.7 mM; potassium phosphate, 69 mM, pH 7.4, 30°.

It is of interest to determine what the percentages of the two isozymes

[30] The disappearance of NADH fluorescence can also be used to measure GOT activity in the coupled assay [M. K. Schwartz and O. Bodansky, *in* "Methods of Enzymatic Analysis" (H. U. Bergmeyer, ed.), 2nd Engl. ed., pp. 768–771. Academic Press, New York, 1974].

[31] A. Karmen, *J. Clin. Invest.* **34,** 131 (1955). Unfortunately, the temperature at which the reaction was carried out was not specified.

[32] H. U. Bergmeyer, P. Scheibe, and W. Wahlefeld, *Clin. Chem.* (*Winston-Salem, N.C.*) **24,** 58 (1978). The assay procedure described for GOT is very similar to that recommended by the International Federation of Clinical Chemists (IFCC) for analysis of GOT in human sera. The major difference is that a prior incubation of the sample with pyridoxal 5′-phosphate is now recommended. See D. E. Bruns, J. Savory, A. C. Titheradge, R. E. Cross, and M. R. Wills [*Clin. Chem.* (*Winston-Salem, N.C.*) **27,** 156 (1981)] and C. C. Garber, D. H. Feldbruegge, and M. Hoessel [*Clin. Chem.* (*Winston-Salem, N.C.*) **27,** 614 (1981)] for automated procedures. Serum activity is generally regarded as due almost entirely to sGOT. See H. U. Bergmeyer and E. Bernt, *in* "Methods of Enzymatic Analysis" (H. U. Bergmeyer, ed.), 2nd Engl. ed., pp. 727–751. Academic Press, New York, 1974.

[33] S. C. Magee and A. T. Phillips, *Biochemistry* **10,** 3397 (1971).

are in various tissues. Older methods employed electrophoretic separation and activity staining (see review by Bergmeyer and Bernt[20]), or separation of cytosolic and mitochondrial fractions.[34] However, more recently immunochemical procedures have been devised. Using antibodies to the porcine mGOT and sGOT enzymes, Dooley *et al.* estimate that the total GOT activities of monkey, dog, rat, and mouse lives are 730, 790, 1260, 3800 U/g (temp. not specified) of which 87, 75, 64, and 59%, respectively, are due to the mitochondrial form.[35] The values for human liver are 150 U/g and 81%, respectively (30°).[36] The mitochondrial form also is of higher activity than the soluble form in rat brain[10,33] and in chicken liver.[37]

Rej[38] has recently described a continuous spectrophotometric assay in which glutamate is measured. The assay utilizes a glutamate dehydrogenase-diaphorase coupled reaction; the rate of formazan dye formation measured at 500 nm is directly related to enzyme activity. The assay can be adapted to measure any α-ketoglutarate-linked transaminase.

[34] Herzfeld and Greengard have determined the activity of GOT in soluble and particulate fractions of normal tissues of adult and fetal rat and in certain rat tumors. In the adult rats, the relative activity was found to decrease in the order heart > liver > kidney > skeletal muscle > brain > spleen > lung [A. Herzfeld and O. Greengard, *Biochim. Biophys. Acta* **237,** 88 (1971)].

[35] J. F. Dooley, K. M. Masullo, and P. H. Tse, *Clin. Chem.* (*Winston-Salem, N.C.*) **27,** 1038 (1981).

[36] R. Rej, *Clin. Chem.* (*Winston-Salem, N.C.*) **24,** 1971 (1978); R. Rej, C. R. Keese, and I. Giaever, *ibid.* **27,** 1597 (1981).

[37] R. Behra, P. Christen, and P. Sonderegger, *J. Biol. Chem.* **256,** 3381 (1981). sGOT/mGOT ratios for other chicken organs are also given.

[38] R. Rej, *Anal. Biochem.* **119,** 205 (1982).

[12] Glutamate-Alanine Transaminase

By ARTHUR J. L. COOPER

Glutamate-alanine transaminase is widely distributed. However, its activity in mammalian tissues tends to be somewhat lower than that of GOT. In the rat, highest activity is found in the liver[1] with small amounts in skeletal muscle,[1] heart,[1] brain,[2] and very little activity in kidney.[1] The enzyme has been obtained in a highly purified form from rat liver,[3] pig

[1] S. Hopper and H. L. Segal, *Arch. Biochem. Biophys.* **105,** 501 (1964).

[2] M. Benuck, F. Stern, and A. Lajtha, *J. Neurochem.* **18,** 1555 (1971).

[3] P. W. Gatehouse, S. Hopper, L. Schatz, and H. L. Segal, *J. Biol. Chem.* **242,** 2319 (1967); T. Matsuzawa and H. L. Segal, *ibid.* **243,** 5929 (1968).

Copyright © 1985 by Academic Press, Inc.
All rights of reproduction in any form reserved.

heart,[4] and beef liver,[5] and the mechanism of the reaction has been extensively studied. The pig enzyme exhibits a single band on starch gel electrophoresis[4] whereas the rat liver enzyme exhibits 2 bands on polyacrylamide gels.[3] The rat liver bands appear to be subforms.[6] GPT catalyzes the following reaction [Eq. (1)]:

$$\text{L-Alanine} + \alpha\text{-ketoglutarate} \rightleftharpoons \text{pyruvate} + \text{L-glutamate} \quad (1)$$

The reaction can be assayed in either direction. The reaction can be quantitated in the reverse direction by coupling α-ketoglutarate to the glutamate dehydrogenase reaction [Eq. (2)] and noting the rate of decrease of absorbance at 340 nm of NADH.[4]

$$\alpha\text{-Ketoglutarate} + NH_3 + \text{NADH} + H^+ \rightarrow \text{L-glutamate} + NAD^+ \quad (2)$$

Appropriate blanks are necessary since pyruvate is a substrate (albeit a poor one) of glutamate dehydrogenase. Most workers assay GPT in the direction of pyruvate formation either by colorimetric procedures or by coupling pyruvate formation to the lactate dehydrogenase reaction [Eq. (3)] and noting the rate of decrease of absorbance at 340 nm of NADH.

$$\text{Pyruvate} + \text{NADH} + H^+ \rightarrow \text{lactate} + NAD^+ \quad (3)$$

The various assay techniques have been reviewed by Bergmeyer and Bernt.[7] As for the GOT reaction, pyruvate formation can be measured by the 2,4-dinitrophenylhydrazone method of Reitman and Frankel.[8] The pyruvate-2,4-dinitrophenylhydrazone can be selectively extracted into toluene.[9] This modification has the advantage that the assay can be carried out with optimum α-ketoglutarate concentrations.

Saier and Jenkins describe a procedure in which pyruvate is treated with alkaline solutions of salicylaldehyde to form a red color.[4,10] The

[4] M. H. Saier, Jr. and W. T. Jenkins, *J. Biol. Chem.* **242,** 91, 101 (1967).

[5] B. Bulos and P. Handler, *J. Biol. Chem.* **240,** 3283 (1964).

[6] There has been some controversy as to whether or not GPT exists in soluble and mitochondrial isozymic forms. R. W. Swick, P. L. Barnstein, and J. L. Stange [*J. Biol. Chem.* **240,** 3334 (1965)] claimed that ~90% of the activity in rat liver is cytosolic whereas ~9% is mitochondrial and that the latter was a distinct enzyme. Saier and Jenkins[4] showed that about 10% of the activity in pig heart is particulate. A homogeneous preparation has been obtained from pig kidney mitochondria [G. De Rosa, T. L. Burk, and R. W. Swick, *Biochim. Biophys. Acta* **567,** 116 (1979)]. GPT activity in rat liver is markedly elevated in corticoid-treated rats [e.g., H. L. Segal, D. S. Beattie, and S. Hopper, *J. Biol. Chem.* **237,** 1914 (1962)].

[7] H. U. Bergmeyer and E. Bernt, *in* "Methods of Enzymatic Analysis" (H. U. Bergmeyer, ed.), 2nd Engl. ed., pp. 752–767. Academic Press, New York, 1974.

[8] S. Reitman and J. S. Frankel, *Am. J. Clin. Pathol.* **28,** 56 (1956).

[9] N. E. Tonhazy, N. G. White, and W. W. Umbreit, *Arch. Biochem. Biophys.* **28,** 36 (1950).

[10] W. T. Jenkins and M. Saier, Jr., this series, Vol. 17A, p. 159.

assay mix (1 ml) contains L-alanine (100 m*M*), α-ketoglutarate (10 m*M*), Tris (40 m*M*); final pH 8.1 at 25°. After incubation at 37° for 10 min, transamination is terminated by addition of 1 ml of 60% (w/v) KOH. Salicylaldehyde (0.5 ml, 2% in 95% ethanol) is added and the orange color is permitted to develop for 10 min at 37°. The color is stabilized by addition of 2.5 ml of ice water. One micromole of pyruvate gives an absorbance of ~0.35.[10] The sensitivity of the procedure is somewhat low and is therefore not suitable for tissues with low GPT activity. On the other hand the method has the advantage that a large number of assays can be conveniently determined in a short period of time. The method was used to obtain highly purified pig heart GPT.[4]

Bergmeyer *et al.* have suggested an optimal method for the measurement of GPT in human serum.[11] The assay mixture (1 ml) contains L-alanine (500 m*M*), α-ketoglutarate (15 m*M*), Tris (100 m*M*), pyridoxal 5′-phosphate (100 μ*M*), NADH (180 μ*M*), LDH (20 μkat^{-1}), serum (diluted 1/12), pH 7.5, 30°. The disappearance of absorbance of NADH at 340 nm is continuously recorded. The method employed by Segal and co-workers[1,6,12] for determining GPT activity in rat liver is similar to that of Bergmeyer *et al.* The assay mix (3 ml) contains L-alanine (17 m*M*), α-ketoglutarate (4.1 m*M*), NADH (170 μ*M*), potassium phosphate buffer (pH 7.3), and lactate dehydrogenase (5 units), 37°.

[11] H. U. Bergmeyer, P. Scheibe, and W. Wahlefeld, *Clin. Chem.* (*Winston-Salem, N.C.*) **24,** 58 (1978).

[12] H. L. Segal and T. Matsuzawa, this series, Vol. 17A, p. 153.

[13] Glutamate-Branched-Chain Amino Acid Transaminase

By ARTHUR J. L. COOPER

Branched-Chain Amino Acid-α-Ketoglutarate Transaminases

Branched-chain amino acid transaminase activity is widespread.[1] In mammalian tissues there appear to be several distinct enzymes. Taylor and Jenkins obtained an almost pure preparation of branched-chain amino

[1] For example, rat organ homogenates catalyze transamination between α-ketoglutarate and valine (leucine, isoleucine) and the order of activity is kidney > heart > skeletal muscle > liver [A. Ichihara and E. Koyama, *J. Biochem.* (*Tokyo*) **59,** 160 (1966)].

Copyright © 1985 by Academic Press, Inc.
All rights of reproduction in any form reserved.

acid transaminase from the soluble fraction of pig heart.[2] A short time later, Ichihara and colleagues purified a branched-chain amino acid transaminase from the soluble and mitochondrial fractions of pig heart.[3,4] Ichihara has given the name "type I" to these heart enzymes. The supernatant pig heart enzyme is active with at least eleven amino acids but it is most active with the branched-chain amino acids. The activity with various amino acids is in the order isoleucine > valine > DL-alloisoleucine > leucine.[2] Aki *et al.* established that type I activity is equally distributed between the soluble and mitochondrial fractions of most rat organs.[3,5] A "type II" enzyme was shown to be present in the soluble fraction of rat liver; this enzyme was at first thought to be specific for leucine but was later found to also utilize methionine as a substrate.[6] A distinct mitochondrial leucine (methionine) transaminase was subsequently purified from rat liver by Ichihara's group.[6] This leucine (methionine) transaminase appears to be lacking in pig tissue. Finally, a type III enzyme was isolated from hog brain and shown to be catalytically and immunochemically distinct from pig heart type I enzyme.[7] Type III activity was also found in the particulate fraction. In the rat, type III activity is present in brain, ovary and placenta.[6] The type III enzyme utilizes valine, leucine, and isoleucine almost equally well but has little activity with other amino acids.

The assay procedures for branched-chain amino acid transaminases involve selective extraction of the branched-chain α-keto acid 2,4-dinitrophenylhydrazone into a suitable organic solvent. In the procedure of Taylor and Jenkins[2,8] the reaction mixture contains 20 μmol of L-leucine, 20 μmol of α-ketoglutarate, 200 μmol of Tris (pH 8.6), and enzyme in a final volume of 3.0 ml, 37°. After incubation for 10 min the reaction is terminated by addition of 0.3% 2,4-dinitrophenylhydrazine in 2 *M* HCl. After 10 min at room temperature the solution is extracted with 5 ml of cyclohexane and the sealed tube is shaken vigorously for 20 sec. The phases are separated by brief centrifugation in a clinical centrifuge. The upper phase (4 ml) is removed and α-ketoisocaproate-2,4-dinitrophenylhydrazone is back extracted into 1.5 ml of 10% w/v sodium carbonate. To 1 ml of the sodium carbonate extract is added 2 ml of 1 *M* NaOH and the absorbance at 440 nm is determined and compared to absorbance values of standard amounts of α-ketoisocaproate carried through the pro-

[2] R. T. Taylor and W. T. Jenkins, *J. Biol. Chem.* **241,** 4391, 4396, 4406 (1966).
[3] K. Aki, K. Ogawa, A. Shirai, and A. Ichihara, *J. Biochem.* (*Tokyo*) **62,** 610 (1967).
[4] K. Aki and A. Ichihara, this series, Vol. 17A, pp. 807, 811, 815.
[5] K. Aki, K. Ogawa, and A. Ichihara, *Biochim. Biophys. Acta* **159,** 276 (1968).
[6] T. Ikeda, Y. Konishi, and A. Ichihara, *Biochim. Biophys. Acta* **445,** 622 (1976).
[7] K. Aki, A. Yokojima, and A. Ichihara, *J. Biochem.* (*Tokyo*) **65,** 539 (1969).
[8] W. T. Jenkins and R. T. Taylor, this series, Vol. 17A, p. 802.

cedure. One micromole gives an absorbance of ~1.4. The method used by Ichihara and colleagues is very similar except that pyridoxal 5′-phosphate and 2-mercaptoethanol are included in the assay mixture and the α-ketoisocaproate-2,4-dinitrophenylhydrazone is extracted into toluene.[4,6] Toluene is used because all the branched-chain α-keto acid-2,4-dinitrophenylhydrazones are equally well extracted.[4,6]

[14] Glutamate-Aromatic Amino Acid Transaminase

By ARTHUR J. L. COOPER

It has long been known that bacteria possess aromatic amino acid-α-ketoglutarate transaminase activity. Feldman and Gunsalus showed that *Escherichia coli, Pseudomonas fluorescens,* and *Bacillus subtilis* possess considerable aromatic (phenylalanine, tyrosine, tryptophan) transaminase activity.[1] Later Rudman and Meister showed that *E. coli* possesses at least two major transaminase systems of wide specificity. Transaminase A (α-ketoglutarate as amino acceptor) is most active with aspartate but also exhibits considerable activity with tryptophan, phenylalanine, and tyrosine. Transaminase B is most active with the aliphatic amino acids but it also exhibits some activity toward phenylalanine and tyrosine.[2] A highly purified preparation of an aromatic amino acid transaminase was obtained by Fujioka *et al.* from *Achromobacter eurydice* grown on L-phenylalanine as sole carbon source.[3] The specificity is somewhat broad and oxaloacetate and phenylpyruvate, in addition to α-ketoglutarate, act as amino group acceptors. In addition to L-phenylalanine, L-tyrosine, and L-tryptophan, the enzyme shows some activity toward L-aspartate and L-kynurenine.[3] Paris and Magasanik have shown that *Klebsiella aerogenes* possesses at least four distinct aromatic amino acid-α-ketoglutarate transaminases (I–IV).[4] Enzyme I has been highly purified and enzyme IV has been partially purified. Enzyme I has a much higher affinity for the aromatic amino acids than does enzyme IV; enzyme IV is highly active toward oxaloacete.

In the rat, L-aromatic amino acid-L-glutamate transaminase activity is widespread. However, activity with L-tyrosine is generally much greater than with L-phenylalanine and the major aromatic amino acid trans-

[1] L. I. Feldman and I. C. Gunsalus, *J. Biol. Chem.* **187,** 821 (1950).
[2] D. Rudman and A. Meister, *J. Biol. Chem.* **200,** 591 (1953).
[3] M. Fujioka, Y. Morino, and H. Wada, this series Vol. 17A, p. 585.
[4] C. G. Paris and B. Magasanik, *J. Bacteriol.* **145,** 257, 266 (1981).

Copyright © 1985 by Academic Press, Inc.
All rights of reproduction in any form reserved.

aminase of rat tissues is referred to as tyrosine transaminase.[5–7] The highest activity is found in the cytosolic fraction of rat liver homogenates. Jacoby and La Du found with a 500-fold purified enzyme from rat liver, that activity toward L-aromatic amino acids is in the order tyrosine = 3-iodotyrosine > 3,4-dihydroxyphenylalanine > phenylalanine > tryptophan.[8] Recently the rat liver tyrosine transaminase has been purified to homogeneity and shown to exist in three subforms (I–III).[6]

Nakano purified a halogenated tyrosine-α-ketoglutarate transaminase 30-fold from sonicated rat kidney mitochondria and confirmed its identity with thyroid hormone transaminase.[9] Tyrosine analogs with halogen (iodine, bromine, or chlorine) substituents in positions 3 and 5 are the most active substrates; 3-iodotyrosine is 70% as effective as diiodotyrosine.[9] Tyrosine, phenylalanine, and tryptophan are only slightly active with this enzyme. Later Tobes and Mason purified an α-aminoadipate-α-ketoglutarate transaminase from the soluble fraction of rat kidneys and showed that the enzyme also catalyzes L-kynurenine-α-ketoglutarate transamination.[10] The authors noted the similarity of this enzyme to that purified by Nakano and indeed showed that the rat kidney possesses an α-aminoadipate (kynurenine, halogenated tyrosine)-α-ketoglutarate transaminase in both soluble and mitochondrial fractions.[11] The soluble enzyme was purified almost to homogeneity.[10]

The assay system of Paris and Magasanik for *K. aerogenes* L-aromatic transaminase is as follows. The reaction mixture (1.5 ml) contains Tris (80

[5] Much of the work on the tyrosine transaminase has been carried out with the rat liver enzyme because the level of activity is quite high, and can be induced up to 30-fold by treatment with hydrocortisone analogs (see, for example, ref. 6). This inducibility was first discovered by E. C. C. Lin and W. E. Knox, *Biochim. Biophys. Acta* **26,** 85 (1957).

[6] J. L. Hargrove and D. K. Granner, *Anal. Biochem.* **104,** 231 (1980); J. L. Hargrove, M. Diesterhaft, T. Noguchi, and D. K. Granner, *J. Biol. Chem.* **255,** 71 (1980).

[7] It is now well established that a second cytosolic L-aromatic amino acid-L-glutamate transaminase and a mitochondrial L-aromatic-L-glutamate transaminase in pig heart preparations are, in fact, the soluble and mitochondrial forms of GOT [J. E. Miller and G. Litwack, *J. Biol. Chem.* **246,** 3234 (1971); E. Shrawder and M. Martinez-Carrion, *ibid.* **247,** 2486 (1972)]. It is also well established that the aromatic amino acid-L-glutamate transaminase activities of rat brain are due to the soluble and mitochondrial isozymes of GOT [see S. King and A. T. Phillips, *J. Neurochem.* **30,** 1399 (1978) and T. Noguchi, M. Nakatani, Y. Minatogawa, E. Okuno, and R. Kido, *ibid.* **25,** 579 (1975)]. Apparently, L-tryptophan-(α-ketoglutarate)-oxaloacetate transaminase from rat brain is also identical to sGOT [Y. Minatogawa, T. Noguchi, and R. Kido, *J. Neurochem.* **27,** 1097 (1976)]. mGOT is more active than sGOT in catalyzing transamination with the aromatic amino acids.

[8] G. A. Jacoby and B. N. La Du, *J. Biol. Chem.* **239,** 419 (1964).

[9] M. Nakano, *J. Biol. Chem.* **242,** 73 (1967); this series, Vol. 17A, p. 660.

[10] M. C. Tobes and M. Mason, *J. Biol. Chem.* **252,** 4591 (1977). See Hartline, this volume [83], for additional information.

[11] M. C. Tobes and M. Mason, *Life Sci.* **22,** 793 (1978).

m*M*, pH 8.0) (or potassium phosphate, pH 8.0), α-ketoglutarate (20 m*M*), L-aromatic amino acid, pyridoxal 5′-phosphate (7 μM), 37°. The reaction is stopped by addition of 0.5 ml of 20% metaphosphoric acid and 0.5 ml of the deproteinized sample is added to either 3 ml of 0.1 *M* arsenate (pH 6.5) (keto sample) or to 3 ml of 0.1 *M* arsenate–0.5 *M* borate (enol sample). The difference in absorbance at a given wavelength (see below) is determined; the absorbance characteristics of the various enol-borate complexes are indolylpyruvate ($A_{338\ nm} = 9.3 \times 10^3$); phenylpyruvate ($A_{300\ nm} = 5.9 \times 10^3$); imidazolylpyruvate ($A_{293\ nm} = 1.1 \times 10^4$). A similar enol-borate method was used by Jacoby and La Du to purify tyrosine transaminase from rat liver.[8] These authors also list the absorbance maximum and the extinction coefficients of 25 aromatic α-keto acid enol-borate complexes. Granner and colleagues used a somewhat more sensitive assay for measuring L-tyrosine-α-ketoglutarate transaminase activity,[6,12] which is based on a previous assay described by Diamondstone.[13] The *p*-hydroxyphenylpyruvate produced via transamination [Eq. (1)] is converted to *p*-hydroxybenzaldehyde in strong base [Eq. (2)].

$$\text{L-Tyrosine} + \alpha\text{-ketoglutarate} \rightleftharpoons p\text{-hydroxyphenylpyruvate} + \text{L-glutamate} \quad (1)$$

$$p\text{-Hydroxyphenylpyruvate} + O_2 \xrightarrow{OH^-} p\text{-hydroxybenzaldehyde} + \text{oxalate} \quad (2)$$

In 0.7 *M* NaOH the A_{max} is 331 nm and $\varepsilon = 19{,}000$.[13] The reaction mixture (0.93 ml) contains L-tyrosine (6 m*M*), potassium phosphate (100 m*M*), α-ketoglutarate (10.8 m*M*), and pyridoxal 5′-phosphate (57 μM). After incubation at 37° from 5 to 30 min, 0.7 ml of 10 *M* KOH is added and the solution is *immediately* shaken. After 30 min at room temperature the absorbance at 330 nm is determined.

The L-kynurenine-α-ketoglutarate transaminase assay according to Tobes and Mason[14] is given by Hartline, this volume [83].

[12] D. K. Granner and G. M. Tomkins, this series, Vol. 17A, 633.

[13] T. I. Diamondstone, *Anal. Biochem.* **16,** 395 (1966). For assaying fresh liver extracts that contain *p*-hydroxyphenylpyruvate oxidase, Diamondstone recommends addition of 3.5 m*M* diethyldithiocarbamate to inhibit this enzyme.

[14] M. C. Tobes and M. Mason, *Biochem. Biophys. Res. Commun.* **62,** 390 (1975).

[15] Glutamate-Ornithine Transaminases

By ARTHUR J. L. COOPER

δ-Ornithine-α-Ketoglutarate Transaminase

In 1951, Quastel and Witty demonstrated that rat liver preparations catalyze a transamination reaction between ornithine and pyruvate as noted by the disappearance of pyruvate and appearance of alanine.[1] It was not stated which nitrogen of ornithine is removed. Later Fincham showed that *Neurospora* extracts catalyze a δ-amino transfer [Eq. (1)].[2] The product, L-glutamate semialdehyde, spontaneously cyclizes [Eq. (2)].

$$\text{L-Ornithine} + \alpha\text{-ketoglutarate} \rightleftharpoons \text{L-glutamate semialdehyde} + \text{L-glutamate} \quad (1)$$

$$\text{L-Glutamate semialdehyde} \rightarrow \Delta^1\text{-pyrroline-5-carboxylate} \quad (2)$$

Meister showed that rat liver extracts also catalyze a δ-ornithine transamination reaction and that the α-keto acid specificity of the transaminase is quite broad.[3] Other workers showed the activity in rat tissues varies in the order: kidney > liver > heart ≅ spleen ≅ brain > skeletal muscle; in the liver the activity was shown to be almost entirely mitochondrial.[4] The rat liver enzyme has been purified to homogeneity and crystallized.[5] Jenkins and Tsai have also crystallized the pig kidney enzyme and shown that the activity with α-ketoacid acceptors is in the order: α-ketoglutarate > glyoxylate > pyruvate > oxaloacetate.[6]

It is interesting that several cases of an imbalance of ornithine metabolism were described in a Finnish population in the early 1970s.[7] The disease, which is autosomal recessive, is associated with hyperornithinemia and gyrate atrophy of the choroid and retina. It was later established by Trijbels *et al.*, from work with cultured fibroblasts, that the disease is associated with a deficiency of δ-ornithine transaminase.[8] A more conve-

[1] J. H. Quastel and R. Witty, *Nature (London)* **167,** 556 (1951).

[2] J. R. S. Fincham, *Biochem. J.* **53,** 313 (1953).

[3] A. Meister, *J. Biol. Chem.* **206,** 587 (1954). The reaction with glyoxylate is unusual in that both the reactant and the product are aldehydes.

[4] C. Peraino and H. C. Pitot, *Biochim. Biophys. Acta* **73,** 222 (1963).

[5] T. Matsuzawa, T. Katsunuma, and N. Katunuma, *Biochem. Biophys. Res. Commun.* **32,** 161 (1968); C. Peraino, L. G. Bunville, and T. N. Tahmisian, *J. Biol. Chem.* **244,** 2241 (1969).

[6] W. T. Jenkins and H. Tsai, this series, Vol. 17A, p. 281.

[7] O. Simell and K. Takki, *Lancet* **1,** 1031 (1973); K. Takki and O. Simell, *Br. J. Ophthalmol.* **58,** 907 (1974).

[8] J. M. F. Trijbels, R. C. A. Sengers, J. A. J. M. Bakkeren, A. F. M. De Kort, and A. F. Deutman, *Clin Chim. Acta* **79,** 371 (1977).

Copyright © 1985 by Academic Press, Inc.
All rights of reproduction in any form reserved.

nient clinical assay based on measurement of δ-ornithine transaminase activity in hair roots has been devised (see below).

The simplest assay of δ-ornithine transaminase is based on the formation of a dihydroquinazolinium between Δ^1-pyrroline-5-carboxylate and *o*-aminobenzaldehyde.[9] The assay of Matsuzawa *et al.*[5] is as follows: The reaction mixture (1 ml) contains 20 m*M* α-ketoglutarate, 20 m*M* L-ornithine, enzyme, and 100 m*M* potassium phosphate buffer, pH 8.0, 37°. After incubation for 30 min, the reaction is terminated by addition of 1 ml of 1% *o*-aminobenzaldehyde in 1 *M* HCl. The reaction mixture is boiled for 5 min, centrifuged, and the absorbance at 440 nm is determined, $\varepsilon = 2.59 \times 10^3$. The *o*-aminobenzaldehyde method has been adapted for use in high-performance liquid chromatography (HPLC).[10] This method is useful for analysis of small samples.

Several radiochemical assays have been devised. Phang *et al.* devised a method in which Δ^1-[U-^{14}C]pyrroline-5-carboxylate formed from L-[U-^{14}C]ornithine is separated on small Dowex 50 H^+ columns.[11] The method is more sensitive than the *o*-aminobenzaldehyde method. Trijbels *et al.* used DL-[U-^{14}C]ornithine and measured the product on an amino acid analyzer in which the effluent is measured by liquid scintillation counting.[8] Janssen *et al.* separated ^{14}C-labeled Δ^1-pyrroline-5-carboxylate as the *o*-aminobenzaldehyde complex on Dowex 50 H^+ columns.[12] The assay is sensitive enough to measure δ-ornithine transaminase in hair roots. By comparing the activity against glutamate dehydrogenase (also a mitochondrial enzyme) Janssen *et al.* confirmed the previous findings that patients with gyrate atrophy of the choroid and retina have a decreased level of δ-ornithine transaminase compared to controls.[12]

Phang *et al.*[11] and Trijbels *et al.*[8] were careful to check that labeled glutamate was not produced from L-[U-^{14}C]ornithine. Nevertheless, the assay technique using labeled ornithine has been criticized by Wong *et al.* because of the possibility of formation of labeled proline or labeled glutamate (from reduction or oxidation of Δ^1-[U-^{14}C]pyrroline-5-carboxylate, respectively) which will result in an underestimate of activity.[13] Some authors have attempted to measure δ-ornithine transaminase activity by assuming that conversion of L-[U-^{14}C]ornithine to L-[U-^{14}C]glutamate is not rate limiting and that the labeled glutamate produced from labeled

[9] See, for example, H. J. Strecker, *J. Biol. Chem.* **235,** 2045 (1960).

[10] J. J. O'Donnell, R. P. Sandman, and S. R. Martin, *Anal. Biochem.* **90,** 41 (1978).

[11] J. M. Phang, S. J. Downing, and D. Valle, *Anal. Biochem.* **55,** 272 (1973).

[12] A. J. M. Jannsen, T. Plakké, F. J. M. Trijbels, R. C. A. Sengers, and L. A. H. Monnens, *Clin Chim. Acta* **113,** 213 (1981).

[13] P. T.-H. Wong, E. G. McGeer, and P. L. McGeer, *J. Neurochem.* **36,** 501 (1981).

glutamate semialdehyde can be quantitated by decarboxylation of labeled glutamate followed by trapping of the labeled CO_2.[14] This method has also been criticized by Wong *et al.* who showed that this technique underestimates activity in rat brain preparations due to incomplete conversion of L-[U-^{14}C]glutamate semialdehyde to L-[U-^{14}C]glutamate.[13] Wong *et al.* devised a radiochemical method based on transamination of labeled α-ketoglutarate with L-ornithine to yield to labeled glutamate.[13] The procedure gives somewhat higher rates of activity than the previously utilized methods. The authors showed that, in the brain, δ-ornithine transaminase is largely mitochondrial and that its activity is particularly high in synaptosomes. The authors speculated that the enzyme may be important for the synthesis of glutamate in brain.[13] The assay system of Wong *et al.* is as follows: The reaction mixture (60 μl) consists of 0.05 μCi α-ketoglutarate (2 m*M*), L-ornithine (20 m*M*), and enzyme; final pH 7.1–7.2. The incubation tube (1 $\times$ 3.5 cm) is placed in a liquid scintillation vial together with a gelatin capsule containing 0.1 ml of hyamine hydroxide on filter paper. The vial is sealed with a rubber serum stopper and incubated at 37° for 1 hr. The reaction is stopped and decarboxylation is initiated by injection of 6.25 μg of bacterial glutamate decarboxylase (*E. coli,* ICN pharmaceuticals) in 0.25 ml of 0.1 *M* pyridine–0.1 *M* NaCl buffer, pH 4.0. After incubation at 37° for 30 min, 0.2 ml of 1 *M* H_2SO_4 is injected. After 4 hr the incubation tube is removed and the radioactivity in the capsule is determined by addition of a suitable scintillation fluid followed by quantitation in a liquid scintillation counter.

Recently Kato *et al.*[15] have devised a highly sensitive sandwich enzyme immunoassay of several protein antigens, including human δ-ornithine transaminase, using the FAB′ of rabbit antibody coupled to β-D-galactosidase together with rabbit antibody IgG loaded onto silicone pieces. The technique is extremely sensitive; 30 attomol (1 attomol = 10^{-18} mol or 6.02 $\times$ 10^5 molecules) of δ-ornithine transaminase can be detected.

α-Ornithine-α-Ketoglutarate Transaminase

Many texts state that the conversion of L-ornithine to L-proline occurs via the following sequence: L-ornithine $\rightarrow$ L-glutamate semialdehyde $\rightarrow$ Δ^1-pyrroline-5-carboxylate $\rightarrow$ L-proline. An alternative pathway is only occasionally considered, i.e., L-ornithine $\rightarrow$ α-keto-δ-aminovalerate $\rightarrow$

[14] M. J. Jung and N. Seiler, *J. Biol. Chem.* **253,** 7431 (1978).

[15] K. Kato, Y. Hamaguchi, S. Okawa, E. Ishikawa, K. Kobayashi, and N. Katunuma, *J. Biochem. (Tokyo)* **81,** 1557 (1977).

Δ^1-pyrroline-2-carboxylate → L-proline. There is ample evidence that α-keto-δ-aminovalerate (and its cyclic form) can be readily reduced to L-proline in a variety of organisms (see references quoted by Mestichelli *et al.*[16]). Mestichelli *et al.* showed by tracer experiments with DL-[5-^{14}C, 2-^{3}H]- and DL-[5-^{14}C, (*RS*)-5-^{3}H]ornithine, that the metabolic conversion of ornithine to proline in three plant species *(Nicotiana tabacum, Datura stramonium,* and *Lupinus angustifolius)* takes place with maintenance of the δ-hydrogen atoms but with loss of the α-hydrogen atom. This finding indicates that α-keto-δ-aminovalerate (and not glutamate semialdehyde) is the intermediate in the route from ornithine to proline in these plants.[17] This hypothesis is strengthened by the previous findings of Hasse *et al.*[18] These authors found that *Lupinus angustifolius* and *Phaseolus aureus* ROXB (mung bean) possess a transaminase that catalyzes transfer of the α-amino group of ornithine (and lysine) to an α-keto acid acceptor whereas *Pseudomonas fluorescens* and *Flavobacterium flavescens* possess a transaminase that catalyzes transfer of the terminal amino group of lyine (to yield Δ^1-piperideine-6-carboxylate). The various possible products [Δ^1-pyrroline-5(and 2)-carboxylate, Δ^1-piperideine-6(and 2)-carboxylate] were identified as their addition products with *o*-aminobenzaldehyde followed by paper chromatography or paper electrophoresis.[18] Mestichelli *et al.* state that "the time would be ripe for a reexamination of proline biosynthesis, and of the metabolic interrelation between ornithine, proline, and glutamic acid."[16] Since Δ^1-pyrroline-5-carboxylate and Δ^1-pyrroline-2-carboxylate reductase activities are distinct enzymes that are widely distributed in nature (see, for example, Meister *et al.*[19]), it would seem that the potential exists for generating L-proline from L-ornithine via δ-amino transfer or by α-amino transfer. It now seems that the preferred route in several plants is via α-amino transamination, but to what extent this route applies to for other organisms remains to be determined.

[16] L. J. J. Mestichelli, R. N. Gupta, and I. D. Spenser, *J. Biol. Chem.* **254,** 640 (1979).

[17] It is unfortunate that a DL mixture was used since at least some of the product may have arisen by a normally unimportant route, i.e., via a D-amino acid oxidase reaction.

[18] K. Hasse, O. T. Ratych, and J. Salnikow, *Hoppe-Seyler's Z. Physiol. Chem.* **348,** 843 (1967).

[19] A. Meister, A. N. Radhakrishnan, and S. D. Buckley, *J. Biol. Chem.* **229,** 789 (1957).

[16] Glutamate-γ-Aminobutyrate Transaminase

By ARTHUR J. L. COOPER

γ-Aminobutyrate (GABA) was first recognized in the central nervous system of mammals more than 30 years ago.[1] It is now well established that γ-aminobutyrate is a major inhibitory transmitter of the central nervous system of vertebrates.[2] GABA arises via decarboxylation of L-glutamate [Eq. (1)] and is subsequently metabolized via transamination to succinic semialdehyde [Eq. (2)]; succinic semialdehyde is then oxidized to succinate [Eq. (3)].

$$\text{L-Glutamate} \rightarrow \text{GABA} + CO_2 \quad (1)$$
$$\text{GABA} + \alpha\text{-ketoglutarate} \rightleftharpoons \text{succinic semialdehyde} + \text{L-glutamate} \quad (2)$$
$$\text{Succinic semialdehyde} + NAD^+ + H_2O \rightarrow \text{succinate} + NADH + H^+ \quad (3)$$
$$\text{SUM: } \alpha\text{-Ketoglutarate} + NAD^+ + H_2O \rightarrow \text{succinate} + CO_2 + NADH + H^+ \quad (4)$$

The enzymes catalyzing reactions (1)–(3) have been called the GABA shunt (or bypath). The shunt allows four of the five carbons lost from the tricarboxylic acid cycle at the level of α-ketoglutarate to be reincorporated into the cycle at the level of succinate [Eq. (4)].[3] Not surprisingly, much of the work on GABA metabolism has been elucidated from studies with brain enzymes and it is generally thought that nonneural tissues are not important for GABA metabolism.[4] However, GABA and enzymes involved in GABA metabolism also occur in nonneural tissues (see references quoted by Martin del Rio[5] and by Buzenet *et al.*[6]). It is also well

[1] E. Roberts and S. Frankel, *Fed. Proc., Fed. Am. Soc. Exp. Biol.* **9,** 219 (1950).

[2] See, for example, appropriate chapters, *in* "GABA in Nervous System Function" (E. Roberts, T. N. Chase, and D. B. Tower, eds.). Raven Press, New York, 1976.

[3] Estimates of the flux through the GABA shunt vary but best estimates (at least for guinea pig brain) suggest 8% of the carbon of the tricarboxylic acid cycle is directed through the GABA shunt [R. Balázs, Y. Machiyama, B. J. Hammond, T. Julian, and D. Richter, *Biochem. J.* **116,** 445 (1970)].

[4] For details of the regional cerebral distribution of GABA transaminase, see N. M. Van Gelder [*J. Neurochem.* **12,** 231 (1965)], J. C. Hyde and N. Robinson [*Histochemistry* **49,** 51 (1976)], and S. R. Vincent, H. Kimura, and E. G. McGeer [*Neurosci. Lett.* **16,** 345 (1980)].

[5] In fact, it has been shown that the rat oviduct contains a higher concentration of GABA than does the rat brain; glutamate decarboxylase, GABA-α-ketoglutarate transaminase, and succinic semialdehyde-NAD$^+$ oxidoreductase enzymes are also well represented in rate oviduct [R. Martin del Rio, *J. Biol. Chem.* **256,** 9816 (1981)].

[6] A. M. Buzenet, C. Fages, M. Bloch-Tardy, and P. Gonnard, *Biochim. Biophys. Acta* **522,** 400 (1978).

Copyright © 1985 by Academic Press, Inc.
All rights of reproduction in any form reserved.

known that GABA and the GABA shunt enzymes are present in a number of microorganisms.[7,8]

GABA-α-ketoglutarate transaminase has been highly purified from mouse brain,[9] rabbit brain,[10] rat brain,[11] human brain,[12] and pig liver.[6] The enzyme appears to be specific for L-glutamate as an α-amino acid substrate. However, the enzyme is active with a number of ω-amino monocarboxylic acids in addition to GABA. Thus, β-alanine, DL-β-aminoisobutyric acid, DL-β-hydroxy GABA, δ-aminovaleric acid, and ε-aminocaproic acid have been shown to be substrates of the mouse brain and pig liver enzymes.

Many methods for the assay of GABA-α-ketoglutarate transaminase have been devised. Sytinsky and Vasilijev described a spectrophotometric assay in which succinic semialdehyde is reacted with 3-methyl-2-benzthiazolone-2-hydrazone to form a colored product.[13] Alternatively, succinic semialdehyde formation can be measured fluorometrically following condensation with 3,5-diaminobenzoic acid[14] or spectrophotometrically following condensation with *o*-aminobenzaldehyde.[8] Several workers have used α-keto[U-^{14}C]glutarate as substrate; L-[U-^{14}C]glutamate is separated from labeled α-ketoglutarate on a Dowex 50 H^+ and it is then eluted with 2 *M* NH_4OH.[12] α-Keto [5-^{14}C]glutarate[15] or [1-^{14}C]glutarate[6,16] has also been used as substrate.[6,16] L-[1-^{14}C]Glutamate generated via transamination is decarboxylated with *E. coli* glutamate decarboxylase; labeled CO_2 is trapped in base and quantitated in a liquid scintillation

[7] For example, the three GABA shunt enzymes are present in the mitochondrial matrix of a streptomycin-bleached mutant of *Euglena gracilis* strain Z [M. Tokunaga, Y. Nakano, and S. Kitaoka, *J. Protozool.* **26,** 471 (1979)]. GABA-α-ketoglutarate transaminase has been purified to homogeneity from extracts of *Pseudomonas* sp. F.-126 and crystallized. See Yonaha and Toyama.[8]

[8] K. Yonaha and S. Toyama, *Arch. Biochem. Biophys.* **200,** 156 (1980).

[9] A. Schousboe, J.-Y. Wu, and E. Roberts, *Biochemistry* **12,** 2868 (1973).

[10] R. A. John and L. J. Fowler, *Biochem. J.* **155,** 645 (1976).

[11] M. Maitre, L. Ciesielski, C. Cash, and P. Mandel, *Eur. J. Biochem.* **52,** 157 (1975).

[12] M. Maitre, L. Ciesielski, C. Cash, and P. Mandel, *Biochim. Biophys. Acta* **522,** 385 (1978).

[13] I. A. Sytinsky and V. Y. Vasilijev, *Enzymologia* **39,** 1 (1970).

[14] R. A. Salvador and R. W. Albers, *J. Biol. Chem.* **234,** 922 (1959). The reaction of aldehydes possessing a methylene carbon with 3,5-diaminobenzaldehyde to form a fluorescent product was first described in 1948 [L. Velluz, G. Amiard, and M. Pesez, *Bull. Soc. Chim. Fr.* p. 678 (1948)].

[15] A. Waksman and M. Bloch, *J. Neurochem.* **15,** 99 (1968).

[16] P. Gonnard, A. Wicker, J.-C. Kouyoumdjian, and M. Bloch-Tardy, *Biochimie* **55,** 509 (1973).

[17] A unified single assay system has been developed to measure six enzymes of glutamate metabolism. Glutamate decarboxylase is measured directly. In the other five cases (glutamate dehydrogenase, glutaminase, glutamate-aspartate transaminase, glutamate-alanine

counter.[6,16,17] Schousboe *et al.* have measured glutamate formation [Eq. (2)] by a coupled enzymatic procedure.[9] The reaction is stopped with aminooxyacetate and glutamate is measured by oxidation with glutamate dehydrogenase–acetylpyridine–NAD^+. The increase in absorbance at 363 due to acytelpyridine–NADH formation is measured.[9] Yonaha and Toyama have measured glutamate formation [Eq. (2)] using an amino acid analyzer.[8]

^{14}C-Labeled GABA has also been used as a substrate of GABA transaminase. In the method of Hall and Kravitz, [1-^{14}C]GABA is quantitatively retained on a small Dowex 50 H^+ column, whereas the product [1-^{14}C]succinic semialdehyde (or [1-^{14}C]succinate if succinic semialdehyde dehydrogenase is present) is eluted in a water wash.[18] Other workers have used a similar assay except that [2,3-^{3}H]GABA is used in place of [1-^{14}C]GABA.[19] Sterri and Fonnum have used a liquid exchanger (tri-*n*-octylammonium phosphate) to extract ^{14}C-labeled succinate formed via transamination and oxidation of γ-amino [U-^{14}C]butyrate in mice cerebellar extracts.[20] The assay is very sensitive and is carried out in a total volume of 7 μl; succinate is extracted with 97% efficiency.

Finally, in tissue homogenates where succinic semialdehyde dehydrogenase is in excess, GABA transaminase may be estimated by using endogenous succinic semialdehyde dehydrogenase to convert formed succinic semialdehyde to succinate [Eq. (3)]. The increase in absorbance at 340 nm due to NADH production is used to estimate GABA transaminase activities. Thus, De Boer and Bruinvels showed that with rat brain homogenates the dehydrogenase activity is sufficient to prevent accumulation of succinic semialdehyde.[21] Jung *et al.* have described a similar assay for determining GABA transaminase in rat or mouse brain homogenates.[22]

transaminase, and GABA transaminase), the product formed is labeled glutamate which is then determined by decarboxylation with bacterial glutamate decarboxylase and estimation of labeled CO_2 [R. C. Dinwoodie and E. A. Boeker, *Anal. Biochem.* **96,** 24 (1978)].

[18] Z. W. Hall and E. A. Kravitz, *J. Neurochem.* **14,** 45 (1967); H. L. White and T. L. Sato, *ibid.* **31,** 41 (1978).

[19] L. Ossola, M. Maitre, J.-M. Blinderman, and P. Mandel, *J. Neurochem.* **34,** 293 (1980).

[20] S. H. Sterri and F. Fonnum, *Eur. J. Biochem.* **91,** 215 (1978).

[21] T. De Boer and J. Bruinvels, *J. Neurochem.* **28,** 471 (1977).

[22] M. J. Jung, B. Lippert, B. W. Metcalf, P. J. Schechter, P. Böhlen, and A. Sjoerdsma, *J. Neurochem.* **28,** 717 (1977).

[17] Glutamate-Aspartate Transaminase from Microorganisms

By TOSHIHARU YAGI, HIROYUKI KAGAMIYAMA, MITSUHIRO NOZAKI, and KENJI SODA

L-Aspartate + 2-oxoglutarate ⇌ oxalacetate + L-glutamate

Assay Method

Principle. Various methods have been developed for the assay of aspartate aminotransferase.[1] Most depend on the determination of oxalacetate or L-glutamate. Of these procedures the Karmen method[2] is used most widely, particularly in the field of clinical analysis, in which oxalacetate is determined with malate dehydrogenase (EC 1.1.1.37) and NADH by following a decrease in absorbance at 340 nm. Recently, a new method was reported; 2-oxoglutarate produced in the reverse reaction is determined with 2-hydroxyglutarate dehydrogenase (EC 1.1.99.2).[3] A method for the determination of aspartate aminotransferase activity after electrophoretic separation of the isoenzymes also has been developed.[4]

The procedure described here is essentially based on the Karmen method.[2]

Reagents

L-Aspartic acid (neutralized with NaOH), 0.5 *M*
2-Oxoglutaric acid (neutralized with NaOH before use), 0.05 *M*
NADH, 0.01 *M*
Malate dehydrogenase [2.5 mg/ml, dissolved in 50% glycerol, ≥1200 U/mg (25°), commercial preparation]
Tris–HCl buffer, 0.2 *M*, pH 8.0

Procedure. The enzyme reaction was carried out at 25°. To a microcuvette were added successively 100 μl of 2-oxoglutarate, 500 μl of Tris–

[1] H. U. Bergmeyer and E. Bernt, *in* "Methods of Enzymatic Analysis" (H. U. Bergmeyer, ed.), Vol. 2, p. 727. Academic Press, New York, 1962.

[2] A. Karmen, *J. Clin. Invest.* **34,** 131 (1955); see also this series, Vol. 17 [261C].

[3] T. Yagi, H. Kagamiyama, S. Ohtawara, K. Soda, and M. Nozaki, *Anal. Biochem.* **100,** 20 (1979).

[4] T. Yagi, H. Kagamiyama, and M. Nozaki, *Anal. Biochem.* **110,** 146 (1981).

Copyright © 1985 by Academic Press, Inc.
All rights of reproduction in any form reserved.

HCl buffer, 10 μl of NADH, 10 μl of malate dehydrogenase solution, 230 μl of water, 50 μl of enzyme solution, and 100 μl of L-aspartate. After mixing, a decrease in absorbance at 340 nm was determined. L-Aspartate was replaced by water in a blank.

Definition of Unit. One unit of enzyme was defined as the amount that catalyzes the formation of 1.0 μmol of NAD per minute. Protein concentration was determined by the method of Lowry *et al.*[5] with bovine serum albumin as a standard.

Culture Conditions. Escherichia coli B was grown in a medium composed of 0.5% succinate, 0.5% L-aspartate, 0.2% K_2HPO_4, 0.2% KH_2PO_4, 0.2% NaCl, and 0.01% $MgSO_4 \cdot 7H_2O$. The pH was adjusted to 7.5 with NaOH. After incubation at 28° for 18–20 hr under aeration, the cells were harvested, and washed with a 0.85% NaCl solution. Cell yield was about 4.5 g (wet weight) per liter of the medium. The cells could be stored at −20° for at least 2 years without loss of the enzyme activity.

Purification Procedures

Method A[6]

Step 1. Preparation of Cell-Free Extract. The cells (500 g) were suspended in 3 liters of 0.02 *M* potassium phosphate buffer (pH 7.0) containing 20 μ*M* pyridoxal 5′-phosphate and 0.01% 2-mercaptoethanol. The suspension (500 ml portion) was subjected to sonication at output control of 9 with Branson Sonifier W185 for 15 min, and centrifuged at 6000 *g* for 15 min. Temperature was kept below 15° during sonication. The supernatant solution was used as a cell-free extract for purification. The following operations were performed below 10° unless otherwise stated.

Step 2. Ammonium Sulfate Fractionation. The supernatant solution (2300 ml) was brought to 30% saturation with ammonium sulfate, and centrifuged at 10,000 *g* for 20 min. Ammonium sulfate was added to the supernatant solution to 75% saturation. The precipitate was collected by centrifugation, and was dissolved in 300 ml of 0.01 *M* potassium phosphate buffer (pH 7.0) containing 10 μ*M* pyridoxal 5′-phosphate (Buffer A), and dialyzed against the same buffer. The insoluble material formed during dialysis was removed by centrifugation.

[5] O. H. Lowry, N. J. Rosebrough, A. L. Farr, and R. J. Randall, *J. Biol. Chem.* **193,** 265 (1951).

[6] T. Yagi, H. Kagamiyama, K. Motosugi, M. Nozaki, and K. Soda, *FEBS Lett.* **100,** 146 (1979).

Step 3. First DEAE-Cellulose Column Chromatography. The supernatant solution was applied to a DEAE-cellulose column (5 × 40 cm) equilibrated with Buffer A. After the column was washed with 3 liters of the buffer, the enzyme was eluted with the buffer containing about 0.15 *M* NaCl by linear gradient elution (NaCl: 0–0.3 *M*, a total volume of 4 liters). The active fractions (350 ml) were pooled.

Step 4. Heat Treatment. After addition of 0.04 ml of 2-mercaptoethanol and 3.5 ml of 1 m*M* pyridoxal 5′-phosphate, the enzyme solution was heated in a water bath (80°) to bring its temperature to 58–63°. The enzyme solution was kept at the temperature for 5 min, and after cooling to about 4° was centrifuged at 10,000 *g* for 15 min to remove the precipitate. The supernatant solution was brought to 75% saturation with ammonium sulfate, and centrifuged at 10,000 *g* for 35 min. The precipitate was dissolved in a small volume of Buffer A, and dialyzed against 2 liters of Buffer A. The dialyzed enzyme was centrifuged to remove precipitate.

Step 5. Sephacryl S-200 Chromatography. The supernatant solution was divided into two portions, and applied on a Sephacryl S-200 column (2.0 × 100 cm) equilibrated with Buffer A. The active fractions eluted with Buffer A were collected, and concentrated by ammonium sulfate precipitation (75% saturation). The precipitate was dissolved in a small volume of 0.02 *M* sodium acetate buffer (pH 4.8) containing 10 μM pyridoxal 5′-phosphate, and dialyzed against the same buffer.

Step 6. Second DEAE-Cellulose Chromatography. After centrifugation, the enzyme solution was applied to a DEAE-cellulose (DE-52, Whatman) column (2.5 × 22 cm) equilibrated with the dialysis buffer. The elution was carried out with a linear gradient (0–0.3 *M* NaCl) in a total volume of 1.4 liters. The enzyme was eluted with the buffer containing about 0.15 *M* NaCl, and the solution was concentrated by ammonium sulfate precipitation (75% saturation). The precipitate was dissolved in a small volume of 1 m*M* potassium phosphate buffer (pH 6.0), and dialyzed against 100 volume of the buffer.

Step 7. Hydroxyapatite Chromatography. The dialyzed enzyme was applied to a hydroxyapatite column (2.5 × 23 cm) equilibrated with 1 m*M* potassium phosphate buffer (pH 6.0). The enzyme was eluted with about 0.01 *M* potassium phosphate buffer (pH 6.0), when the elution was performed with a linear gradient established with 0.001–0.1 *M* potassium phosphate buffer (pH 6.0) in a total volume of 1.4 liters. A part of the enzyme was sometimes eluted with the 0.005 *M* buffer, but identical with the enzyme eluted with the 0.01 *M* buffer.

A summary of the purification procedure is given in Table I.

TABLE I
PURIFICATION OF ASPARTATE AMINOTRANSFERASE FROM *Escherichia coli* B

	Total protein (mg)	Total units	Specific activity (units/mg protein)
1. Crude extract	26,000	23,000	0.9
2. Ammonium sulfate fractionation	15,900	21,000	1.3
3. First DEAE-cellulose chromatography	4,700	17,300	3.7
4. Heat treatment	1,500	17,000	11.3
5. Sephacryl S-200 chromatography	590	12,420	21.1
6. Second DEAE-cellulose chromatography	137	9,200	67.2
7. Hydroxyapatite chromatography	23	4,600	200.0

Method B[7]

Alternatively, the enzyme is purified in an apo form as follows, particularly when needs rapidly purifying, though on a small scale.

Step 1. Preparation of Cell-Free Extract. The cells (20 g) were suspended in 30 ml of 0.02 *M* potassium phosphate buffer (pH 6.0) (Buffer B). The suspension was subjected to sonication and centrifugation as described previously.

Step 2. DEAE-Cellulose Treatment. To the supernatant solution (32 ml) was added DEAE-cellulose (DE-52, Whatman) (15 g, wet weight) which was equilibrated with Buffer B. The suspension was centrifuged at 6000 *g* for 15 min. The precipitated DEAE-cellulose was washed twice with Buffer B, and loaded into a column (1.6 × 15 cm). The enzyme was eluted with 50 ml of the buffer containing 0.2 *M* NaCl.

Step 3. Ammonium Sulfate Precipitation. Ammonium sulfate (25 g) was added to the enzyme solution (50 ml). The precipitate collected by centrifugation (10,000 *g*, 20 min) was dissolved in a small volume of Buffer B, and dialyzed successively against two changes of Buffer B and two changes of 0.02 *M* Tris-acetate buffer (pH 7.0).

[7] T. Yagi, H. Sata, and S. Yamamoto, *Proc. Annu. Meet. Agric. Chem. Soc. Jpn.* p. 535 (1984).

TABLE II
PURIFICATION OF ASPARTATE APOAMINOTRANSFERASE FROM *Escherichia coli* B

	Total protein (mg)	Total units[a]	Specific activity (units/mg protein)
1. Crude extract	1300	1560	1.2
2. DEAE-Cellulose treatment	300	990	3.3
3. Ammonium sulfate precititation	290	840	2.9
4. Affinity chromatography			
Nonadsorbed	200	340	1.7
Adsorbed	1.9	400	210.0

[a] Samples were incubated with 10^{-4} *M* pyridoxal 5′-phosphate at 30° for 60 min prior to the determination of activity.

Step 4. Affinity Chromatography. The enzyme solution (15 ml) was applied to a column (1.2 × 10 cm) of pyridoxamine-Sepharose which was prepared by the method of Miller *et al.*,[8] and equilibrated with 0.02 *M* Tris-acetate buffer (pH 7.0). The column was washed successively with 60 ml of 0.02 *M* Tris-acetate buffer (pH 7.0) and 60 ml of the buffer containing 0.5 *M* NaCl, and 30 ml of 0.05 *M* potassium phosphate buffer (pH 7.0). The holo enzyme was not adsorbed on the column, and washed out with 0.02 *M* Tris-acetate buffer (pH 7.0). The apo enzyme was eluted with the phosphate buffer, and shown to be homogeneous.

A protocol of the purification is presented in Table II.

Properties

Stability. The enzyme purified by either method can be stored in Buffer A containing 75% ammonium sulfate at 4° for several months without loss of activity.

Physical and Chemical Properties. The purified enzyme was shown to be homogeneous by the criteria of ultracentrifugation and disc gel electrophoresis in the presence or absence of sodium dodecyl sulfate. The molecular weight was estimated to be 84,000 by the meniscus depletion sedimentation equilibrium method[9] assuming a partial specific volume of

[8] J. V. Miller, Jr., P. Cuatrecasas, and E. B. Thompson, *Biochim. Biophys. Acta* **276,** 407 (1972).

[9] D. A. Yphantis, *Biochemistry* **3,** 297 (1964).

TABLE III
AMINO ACID COMPOSITION OF ASPARTATE AMINOTRANSFERASE FROM *Escherichia coli* B

Amino acid	Residues[a] (M_r 42,000)	Amino acid	Residues[a] (M_r 42,000)
Asp	42.4	Met	7.9
Thr	22.3	Ile	17.3
Ser	18.1	Leu	35.9
Glu	47.6	Tyr	11.3
Pro	14.6	Phe	19.9
Gly	29.6	His	5.7
Ala	45.5	Lys	16.7
Cys[b]	6.3	Arg	21.8
Val	25.4	Trp[c]	4.4

[a] The average or extrapolated (threonine and serine) value from 24, 48, and 72 hr hydrolysis.
[b] Determined as cysteic acid after performic acid oxidation.
[c] Determined spectrophotometrically.

0.74. Polyacrylamide slab gel electrophoresis in the presence of sodium dodecyl sulfate gave a molecular weight of 43,000 ± 1,000, showing that the enzyme is composed of two subunits identical in molecular weight. The enzyme contains two moles of pyridoxal 5′-phosphate per mole, and shows absorption maxima at 280 nm (ε: 74,800) and 360 nm (ε: 13,500) at pH 8.5, and 280 nm and 440 nm (ε: 12,900) at pH 5.5. It shows an isoelectric point of pH 4.2.

pH Optimum and K_m. The enzyme shows the maximum activity in the pH range of 7.5–8.5. The K_m value determined by the method of Velick and Vavra[10] are 0.01 m*M* for oxalacetate, 0.24 m*M* for 2-oxoglutarate, 1.3 m*M* for L-aspartate, and 15 m*M* for L-glutamate.

Substrate Specificity. The amino donor specificity was determined by radioassay with 2-[^{14}C]oxoglutarate[6]: the relative activity is 100 for L-glutamate, 30 for L-aspartate, 144 for L-cysteine sulfinate, 11 for L-tryptophan, 6.2 for L-phenylalanine, 1.5 for L-tyrosine, and 1.0 for L-methionine. The amino acceptor specificity determined with malate dehydrogenase is as follows: 100 for 2-oxoglutarate, 20.2 for 4-hydroxyphenylpyruvate, 16.8 for indolpyruvate, 14.6 for phenylpyruvate, 1.7 for 2-oxomethionine, 0.17 for 2-oxobutyrate, 0.17 for 2-oxocaproate, and 0.05 for pyruvate. The reactivity of the following keto acids was determined with 2-hydroxyglutarate dehydrogenase[3]: 100 for oxalacetate, 16.5

[10] S. F. Velick and J. Vavra, *J. Biol. Chem.* **237,** 2109 (1962).

for 4-hydroxyphenylpyruvate, 16.1 for indolpyruvate, 1.12 for 2-oxomethionine.

Catalytic Activity of Apoenzyme. The apo form of enzyme catalyzes the pyridoxamine–oxalacetate transamination reaction with a specific activity of 1.87, but does not the pyridoxamine–2-oxoglutarate reaction.[11]

Amino Acid Composition[6] and Primary Structure.[12] Amino acid composition is shown in Table III. The primary structure of pyridoxal 5′-phosphate binding peptide is Ser-Lys (Pyx)-Asn-Phe. The complete amino acid sequence has been elucidated.[13]

Other Methods of Preparation

The similar *Escherichia* aspartate aminotransferases were purified from *coli* K12,[14–16] *coli* Crooks,[17] *coli* B,[6] and a strain of *coli.*[18] The enzyme has been purified also from brewer's yeast[19,20] and bakers' yeast.[18,21] Aspartate aminotransferases from *Rhizobium japonicum*[22] and *Pseudomonas striata*[23] also were purified. These microorganisms produce only one form of the enzyme, whereas plants and animals contain both cytosolic and mitochondrial isoenzymes.

[11] T. Yagi, H. Kagamiyama, and M. Nozaki, *Biochem. Biophys. Res. Commun.* **107,** 897 (1982).

[12] H. Kagamiyama and T. Yagi, *Biochem. Biophys. Res. Commun.* **89,** 1347 (1979).

[13] K. Kondo, S. Wakabayashi, T. Yagi, and H. Kagamiyama, *Biochem. Biophys. Res. Commun.* **122,** 62 (1984).

[14] S. Chesne and J. Pelmont, *Biochimie* **55,** 237 (1973).

[15] J. T. Powell and J. F. Morrison, *Eur. J. Biochem.* **87,** 391 (1978).

[16] E. Gout, S. Chesne, C. G. Beguin, and J. Pelmont, *Biochem. J.* **171,** 719 (1978).

[17] C. Mavrides and W. Orr, *J. Biol. Chem.* **250,** 4128 (1974).

[18] P. B. Porter, D. Barra, F. Bossa, G. Cantalupo, S. Doonan, F. Martini, D. Sheehan, and S. M. Wilkinson, *Comp. Biochem. Physiol.* **69B,** 737 (1981).

[19] G. Schreiber, M. Eckstein, A. Oeser, and H. Holzer, *Biochem. Z.* **340,** 13 (1964).

[20] I. Cozzani, M. Pellegrini, R. Barsacchi, G. Dibenedetto, and F. Sgarella, *Ital. J. Biochem.* **23,** 380 (1974).

[21] T. Yagi, H. Kagamiyama, and M. Nozaki, *J. Biochem.* (*Tokyo*) **92,** 35 (1982).

[22] E. Ryan, F. Bodley, and P. F. Fottrell, *Phytochemistry* **11,** 957 (1972).

[23] T. Yagi, M. Toyosato, and K. Soda, *FEBS Lett.* **61,** 34 (1976).

[18] L-Kynurenine Transaminase from *Hansenula schneggii*

By KATSUYUKI TANIZAWA, YASUHIKO ASADA, and KENJI SODA

L-Kynurenine + α-Ketoglutarate → [o-aminobenzoylpyruvic acid] + L-Glutamate; [o-aminobenzoylpyruvic acid] → H_2O + Kynurenic acid

L-Kynurenine-α-ketoglutarate aminotransferase [L-kynurenine transaminase, L-kynurenine : 2-oxoglutarate aminotransferase (cyclizing), EC 2.6.1.7] catalyzes the transfer of the α-amino group of L-kynurenine to α-ketoglutarate to yield L-glutamate and *o*-aminobenzoylpyruvic acid, which is immediately converted to the intramolecularly dehydrated and cyclized form, kynurenic acid. The enzyme has been demonstrated as the initial enzyme in the quinaldine pathway of kynurenine degradation in mammalian tissues,[1-4] *Pseudomonas fluorescens*,[5] and *Neurospora crassa*.[6] The present chapter describes the assay method, purification procedures, and general properties of L-kynurenine-α-ketoglutarate aminotransferase of a yeast, *Hansenula schneggii*.[7]

[1] M. Mason, *J. Biol. Chem.* **211,** 839 (1954).
[2] H. Okamoto and O. Hayaishi, *J. Biol. Chem.* **245,** 3603 (1970).
[3] J. Nakamura, T. Noguchi, and R. Kido, *Biochem. J.* **135,** 815 (1973).
[4] Y. Minatogawa, T. Noguchi, and R. Kido, *J. Neurochem.* **20,** 1479 (1973).
[5] I. L. Miller, M. Tsuchida, and E. A. Adelberg, *J. Biol. Chem.* **203,** 205 (1953).
[6] W. B. Jakoby and D. M. Bonner, *J. Biol. Chem.* **221,** 689 (1956).
[7] K. Soda, Y. Sawa, and K. Tanizawa, *in* "Current Developments in Yeast Research: Advances in Biotechnology" (G. G. Stewart and I. Russell, eds.), p. 441. Pergamon, Oxford, 1981.

METHODS IN ENZYMOLOGY, VOL. 113
Copyright © 1985 by Academic Press, Inc.
All rights of reproduction in any form reserved.

Assay Method

Principle. The assay method routinely employed is based on the spectrophotometric determination of kynurenic acid formed.

Reagents

L-Kynurenine sulfate, 0.05 *M*
Sodium α-ketoglutarate, 0.2 *M*
Pyridoxal 5′-phosphate, 0.5 m*M*
Potassium phosphate buffer, pH 8.0, 1.0 *M*
Hydrochloric acid, 2 *N*, 4 *N*

Procedure. The standard assay system consists of 0.2 ml of potassium phosphate buffer, 0.1 ml of L-kynurenine sulfate, 0.1 ml of sodium α-ketoglutarate, 0.2 ml of pyridoxal 5′-phosphate, and enzyme in a final volume of 1.0 ml. Sodium α-ketoglutarate or enzyme is replaced by water in a blank. After the mixture is incubated at 37° for 10 min, the reaction is terminated by addition of 1.0 ml of 4 *N* hydrochloric acid. The denatured protein is removed by brief centrifugation. To 0.2 ml of the supernatant solution is added 3.0 ml of 2 *N* hydrochloric acid, and the absorbance at 340 nm due to the formation of kynurenic acid is measured against the blank. The amount of the product formed is calculated from the difference between the absorbance coefficient of kynurenic acid and that of L-kynurenine ($\Delta\varepsilon = 7640$ in 2 *N* HCl).

Alternative Methods. (1) L-Kynurenine consumption is determined by measuring the rate of decrease in absorbance at 370 nm due to deamination of L-kynurenine ($\varepsilon = 4300$, at pH 8.0). (2) L-Glutamate formed from α-ketoglutarate is separated by paper chromatography with Toyo filter paper No. 51 (solvent system; *n*-butanol : acetic acid : water, 4 : 1 : 1 or ethanol : water : ammonium hydroxide, 18 : 1 : 1), and after the ninhydrin color of L-glutamate is developed and extracted with 95% ethanol containing 0.005% $CuSO_4$, the absorbance of the extract is determined at 500 nm. (3) Kynurenic acid formed is separated by paper chromatography as in (2), and the fluorescence of kynurenic acid in the water extract is measured by spectrofluorometer (excitation at 330 nm and emission at 378 nm).

Definition of Unit and Specific Activity. One unit of enzyme is defined as the amount that catalyzes the formation of 1 μmol of kynurenic acid per minute. Specific activity is expressed as units per mg of protein. The protein concentration is determined by the method of Lowry *et al.*,[8] with bovine serum albumin as a standard or from the absorbance coefficient of the purified enzyme at 280 nm ($E_{1\,\text{cm}}^{1\%} = 12.5$).

[8] O. H. Lowry, N. J. Rosebrough, A. L. Farr, and R. J. Randall, *J. Biol. Chem.* **193,** 265 (1951).

Purification Procedures[7]

Culture Conditions. *Hansenula schneggii* (IFO 0135) is grown in a medium containing 5% sucrose, 0.15% L-tryptophan, 0.2% K_2HPO_4, 0.1% $MgSO_4 \cdot 7H_2O$, 0.1% sodium L-glutamate, 0.1% urea, 0.1% $CaCO_3$, which is separately sterilized, and 0.01% yeast extract (pH 6.0–6.6) with a 200-liter fermentor at 28° for about 18 hr under aeration. The cells harvested by centrifugation are washed twice with 10 m*M* potassium phosphate buffer (pH 7.2). The washed cells are stored at −20° until use. All the subsequent operations are performed at 0–5°.

Step 1. Preparation of Crude Extract. The washed cells (1000 g, wet weight) are suspended in 2 liters of 10 m*M* potassium phosphate buffer (pH 7.2) containing 20 μ*M* pyridoxal 5′-phosphate and 0.01% 2-mercaptoethanol, and disrupted continuously by a DYNO-MILL apparatus (Willy A. Bachofen Maschinenfabrik) with glass beads (0.25–0.5 mm in diameter). The supernatant solution obtained by centrifugation is used as the cell-free extract.

Step 2. Polyethyleneimine Treatment. To the cell-free extract (2000 ml) is added 4 g of polyethyleneimine with stirring. The mixture is centrifuged and the bulky inactive precipitate is discarded. The enzyme solution (1700 ml) is concentrated by ammonium sulfate precipitation (70% saturation) and the precipitate collected by centrifugation is dissolved in 10 m*M* potassium phosphate buffer (pH 7.2) containing 20 μ*M* pyridoxal 5′-phosphate and 0.01% 2-mercaptoethanol. The enzyme solution (320 ml) is dialyzed overnight against 100 volumes of the same buffer.

Step 3. DEAE-Cellulose Column Chromatography. The dialyzed enzyme solution (610 ml) is placed on a DEAE-cellulose column (12 × 80 cm) equilibrated with the dialysis buffer. The column is washed thoroughly with the same buffer in order to remove the unadsorbed protein. After the column is washed with the buffer containing 0.08 *M* NaCl, the enzyme is eluted with the buffer supplemented with 0.16 *M* NaCl. The active fractions are pooled and concentrated by ammonium sulfate precipitation (70% saturation).

Step 4. BioGel P-300 Column Chromatography. The precipitated enzyme is dissolved in 10 m*M* potassium phosphate buffer (pH 7.2) containing 20 μ*M* pyridoxal 5′-phosphate and 0.01% 2-mercaptoethanol. A portion (7 ml) of the enzyme solution (35 ml) is applied to a BioGel P-300 column (2.5 × 120 cm), and eluted with the same buffer. The active fractions are pooled. The remainder of enzyme also is subjected to the gel filtration.

Step 5. Hydroxyapatite Column Chromatography. The enzyme solution (120 ml) is placed on a hydroxyapatite column (4 × 25 cm) equili-

PURIFICATION OF L-KYNURENINE-α-KETOGLUTARATE AMINOTRANSFERASE FROM *Hansenula schneggii*

Step	Total protein (mg)	Total activity (units)	Specific activity (units/mg)	Purification (fold)	Yield (%)
1. Cell-free extract	36,000	660	0.018	1	100
2. Polyethyleneimine treatment	14,000	480	0.033	1.8	73
3. DEAE-cellulose chromatography	682	472	0.700	39	71
4. BioGel P-300 chromatography	264	348	1.20	67	53
5. Hydroxyapatite chromatography	136	304	2.20	122	46
6. Crystallization	16	264	16.8	933	40

brated with the same buffer. After the column is washed with 50 m*M* potassium phosphate buffer (pH 7.2), the enzyme is eluted with 0.1 *M* potassium phosphate buffer (pH 7.2). The active fractions are pooled and concentrated by ammonium sulfate precipitation (70% saturation). The precipitate is collected by centrifugation and dissolved in a minimum volume of the above-mentioned buffer.

Step 6. Crystallization. Finely ground ammonium sulfate is added gradually to the enzyme solution until a faint turbidity is obtained. On standing at 4° overnight, crystal formation occurs. The crystals take the form of needles.

Approximately 930-fold purification is achieved with an overall yield of 40%. A summary of the purification procedures is presented in the table.

Properties[7]

Stability. The crystalline enzyme can be stored at 4° as a suspension in 70% saturated ammonium sulfate (pH 7.2) without loss of activity for periods of over 6 months. The purified enzyme can be stored also in a solution containing 20 μM pyridoxal 5′-phosphate and 0.01% 2-mercaptoethanol at −20° for several weeks without appreciable loss of activity when protein concentration is more than 0.6 mg/ml. The enzyme in a dilute solution (below 0.1 mg/ml), however, is irreversibly denatured by repeated freezings and thawings. The enzyme is stable up to 50°, when heated for 5 min in 0.1 *M* potassium phosphate buffer (pH 7.2). The enzyme is very stable at pH 6.0 to 10.0 when incubated at 50° for 5 min.

Physicochemical Properties. The crystalline enzyme is homogeneous by the criteria of ultracentrifugation and polyacrylamide disc gel electrophoresis. The sedimentation coefficient of the enzyme, calculated for water at 20° and zero protein concentration, is 5.95 S. The molecular weight of the enzyme is calculated as 106,000 ± 3000, 98,000 ± 4000, and 102,000 ± 2000, by the gel filtration method, sedimentation equilibrium method, and electrophoretic method with gradient polyacrylamide gels, respectively. The enzyme consists of two subunits identical in molecular weight (52,000). The enzyme exhibits absorption maxima at 280, 335, and 435 nm at pH 7.2 with molar absorption coefficients of 125,000, 18,200, and 4400, respectively. No appreciable spectrum shift occurs on varying the pH (4.5–9.0). An average pyridoxal 5′-phosphate content of 1 mol/47,000 g of protein shows that about 2 mol of pyridoxal 5′-phosphate are bound to 1 mol of enzyme protein. Circular dichroism of the enzyme shows the negative CD bands at 268, 332, and 408 nm, and the small positive CD bands at 363 and 450 nm. All the circular dichroism spectra of L-amino acid aminotransferases thus far studied are positive in visible wavelength region.[9] L-Kynurenine-α-ketoglutarate aminotransferase from *Hansenula schneggii* is unique in this respect.

Resolution and Reconstitution of the Enzyme. Incubation of the enzyme with 10 m*M* hydroxylamine (pH 7.2) at room temperature for 1.5 hr gives an inactive form of enzyme that can be reactivated with pyridoxal 5′-phosphate. The inactive form of the enzyme exhibits no absorption peak at 435 nm but has one at 335 nm. The active enzyme reconstituted with pyridoxal 5′-phosphate shows the same spectrum as the native holoenzyme. The Michaelis constants are 1.3 μM for pyridoxal 5′-phosphate and 3.2 μM for pyridoxamine 5′-phosphate.

Substrate Specificity and Michaelis Constants. The enzyme exhibits a very broad substrate specificity. In addition to L-kynurenine (relative activity, 100), which is the most preferred amino donor, various other L-amino acids, e.g., L-3-hydroxykynurenine (44), L-tryptophan (59), L-leucine (86), L-α-aminoadipic acid (59), and L-methionine (36), serve as good amino donors for α-ketoglutarate. D-Kynurenine, basic amino acids (L-lysine and L-arginine), ω-amino acids, and amines are inert. The enzyme also catalyzes transamination between L-kynurenine and a variety of α-keto acids such as α-ketoglutarate (relative activity, 100), α-ketoadipate (110), α-ketoisocaproate (57), β-phenylpyruvate (67), α-ketovalerate (55), β-indolepyruvate (37), and α-keto-γ-methylthiobutyrate (70). Appar-

[9] H. C. Dunathan, *Adv. Enzymol.* **35**, 79 (1971).

ent Michaelis constants for several amino donors and acceptors are 1.9 mM for L-kynurenine, 2.6 mM for L-tryptophan, 5.0 mM for L-leucine, 5.0 mM for L-methionine, 1.1 mM for α-ketoglutarate, 1.2 mM for β-phenylpyruvate, and 1.0 mM for α-ketoisocaproate.

Effect of pH. The enzyme shows the maximum reactivity at about pH 8.0 for the transamination of L-kynurenine, L-leucine, and L-cysteinesulfinate with α-ketoglutarate as an amino acceptor when examined in the presence of 0.1 M potassium phosphate, Tris–HCl, and glycine–KCl buffers.

Inhibitors. Various dicarboxylic acids inhibit the aminotransferase activity at the concentration of 1 mM. Adipate and pimelate inhibit the enzyme 39 and 53%, respectively. Glutarate (18%), dipicolinate (21%), phthalate (10%), and oxalacetate (19%) are also inhibitory. Mason reported that the mammalian L-kynurenine aminotransferase also is inhibited by a variety of dicarboxylic acids.[10] The enzyme is inhibited markedly by hydroxylamine (inhibition at 0.1 mM, 100%) and phenylhydrazine (81%), both of which are typical inhibitors for pyridoxal 5′-phosphate enzymes. The enzyme is also inhibited by $HgCl_2$ (at 0.1 mM, 89%), p-chloromercuribenzoate (27%), iodoacetate (23%), and N-ethylmaleimide (27%). Divalent metal ions and chelating reagents, e.g., ethylenediaminetetraacetic acid and α,α'-dipyridyl have no effect on the enzyme activity at the concentration of 1 mM.

Distribution in Yeasts. L-Kynurenine-α-ketoglutarate aminotransferase occurs in several strains of yeast: *Saccharomyces cerevisiae, S. marxianus, Pichia polymorpha, Debariomyces hansenii,* and *Hansenula schneggii*. Little activity is found in *S. fragilis, D. globosus, Cryptococcus albidus, Candida rugosa, Tricosporon cutaneum,* and *Schwanniomyces occidentalis*. The enzyme activity in *Hansenula schneggii* is induced severalfold by L-tryptophan.

[10] M. Mason, *J. Biol. Chem.* **234,** 2770 (1959).

[19] L-Lysine Transaminase from *Flavobacterium lutescens*

By KATSUYUKI TANIZAWA, TOHRU YOSHIMURA, and KENJI SODA

$$\underset{\text{L-Lysine}}{COOH-CHNH_2-(CH_2)_4-NH_2} + \underset{\alpha\text{-Ketoglutaric acid}}{COOH-CO-CH_2-CH_2-COOH} \longrightarrow \underset{\alpha\text{-Aminoadipate } \delta\text{-semialdehyde}}{COOH-CHNH_2-(CH_2)_3-CHO} + \underset{\text{L-Glutamic acid}}{COOH-CHNH_2-CH_2-COOH}$$

$$\alpha\text{-Aminoadipate } \delta\text{-semialdehyde} \xrightarrow{-H_2O} \Delta^1\text{-Piperideine-6-carboxylic acid}$$

Δ^1-Piperideine-6-carboxylic acid

L-Lysine-α-ketoglutarate 6-aminotransferase (transaminase) (EC 2.6.1.36) catalyzes the transamination of terminal amino group of L-lysine to α-ketoglutarate to produce Δ^1-piperideine-6-carboxylate (a cyclized form of α-aminoadipate δ-semialdehyde) and L-glutamate.[1] The enzyme has been purified in a crystalline form from *Flavobacterium lutescens* (formerly classified as *Achromobacter liquidum*).[2] The assay method, definition of unit and specific activity, purification procedures, and general properties of the enzyme have been already described in Vol. 17B.[3] We here describe an improved purification method of the enzyme by affinity chromatography, subunit structures, and the stereochemical analysis of the enzyme reaction.

Improved Purification by Affinity Chromatography[4]

Preparation of L-Lysylacetamidododecyl-Sepharose. The affinity adsorbent used for purification of the enzyme was L-lysylacetamidododecyl-

[1] K. Soda, H. Misono, and T. Yamamoto, *Biochemistry* **7,** 4102 (1968).
[2] K. Soda and H. Misono, *Biochemistry* **7,** 4110 (1968).
[3] K. Soda and H. Misono, this series, Vol. 17B, p. 222.
[4] T. Yagi, T. Yamamoto, and K. Soda, *Biochim. Biophys. Acta* **614,** 63 (1980).

Copyright © 1985 by Academic Press, Inc.
All rights of reproduction in any form reserved.

Sepharose 6B.[4] Sepharose 6B (Pharmacia, 200 ml) was activated in a 10% (w/v) solution of CNBr (pH 10.5–11.0) at 4° according to the method of Cuatrecasas and Anfinsen.[5] To the activated Sepharose 6B suspended in 250 ml of 0.1 *M* $Na_2B_4O_7$–Na_2CO_3 buffer (pH 10.5) was added 225 ml of 10% solution of dodecamethylenediamine dihydrochloride in 50% dioxane. The reaction was allowed to proceed with gentle stirring at 4° for 24 hr. After the ω-aminododecyl-Sepharose 6B was filtered off and washed thoroughly with 30% acetic acid and then with water, it was converted to the bromoacetyldodecyl-Sepharose 6B as described previously.[5] The bromoacetyldodecyl-Sepharose was suspended in 200 ml of 0.1 *M* potassium phosphate buffer (pH 7.5) and reacted with 1.0 *M* L-lysine (pH 7.3). The pH of the reaction mixture was kept at about neutral in order to allow the α-amino group of L-lysine to react with the matrix arm. The reaction was carried out with gentle stirring at room temperature for 24 hr. The adsorbent prepared was washed thoroughly with 1.0 *M* NaCl and then with water.

Purification of L-*Lysine 6-Aminotransferase. F. lutescens* (IFO 3084) was grown and disrupted as described previously,[2] and the resulting homogenate was brought to 30% saturation with $(NH_4)_2SO_4$ and the precipitate was removed by centrifugation. $(NH_4)_2SO_4$ was added to the supernatant solution to 70% saturation. The precipitate was dialyzed at 4° against 0.01 *M* potassium phosphate buffer (pH 7.2). The enzyme (500 mg protein) was applied to an L-lysylacetamidododecyl-Sepharose 6B column (1.5 × 20 cm) equilibrated with the phosphate buffer containing 0.1 *M* NaCl. The column was washed successively with the buffer supplemented with 0.3, 0.5, and 1.0 *M* NaCl. The enzyme activity was found in the fractions eluted with the buffer containing 1.0 *M* NaCl. The active fractions were combined and concentrated by addition of $(NH_4)_2SO_4$ (60% saturation). After the enzyme was dialyzed against 0.01 *M* potassium phosphate buffer (pH 7.2) containing 0.01 m*M* pyridoxal 5′-phosphate and 0.02% 2-mercaptoethanol, it was applied to a DEAE-cellulose column (1.2 × 20 cm) equilibrated with the same buffer. The column was washed thoroughly with the buffer containing 0.15 *M* NaCl, and the enzyme was eluted with the buffer containing 0.2 *M* NaCl. About 100-fold purification was achieved with an overall yield of 45%. A summary of the purification is presented in Table I. Although the specific activity of the purified enzyme was half that of the enzyme obtained by the previous procedure[2] which involved heat treatment, the enzyme was activated twice by the heat treatment at 65° for 5 min.[4]

[5] P. Cuatrecasas and C. B. Anfinsen, this series, Vol. 22, p. 345.

TABLE I
PURIFICATION OF L-LYSINE-α-KETOGLUTARATE 6-AMINOTRANSFERASE

Step	Total protein (mg)	Specific activity	Total units	Yield (%)
Cell-free extract	2292	4.8	11,000	100
$(NH_4)_2SO_4$ fractionation (30–70%)	1073	8.2	8,800	80
Affinity chromatography	26	279	7,250	66
DEAE-cellulose chromatography	10	500	5,000	45

Subunit Structure of L-Lysine 6-Aminotransferase[6]

L-Lysine-α-ketoglutarate 6-aminotransferase from *F. lutescens* has been shown to be composed of one each of four nonidentical subunits (A, B_1, B_2, and C)[6] as follows. Upon disc gel electrophoresis in the presence of 0.1% sodium lauryl sulfate of the enzyme carboxymethylated by the method of Crestfield *et al.*,[7] three bands of stained protein were observed, indicating that the enzyme consists of at least three nonidentical subunits (A, B, and C). Amino- and carboxy-terminal analyses of the enzyme also showed the heterogeneity of the subunit composition[6] (see Table II). To establish the subunit structure of L-lysine 6-aminotransferase and to characterize the subunits, isolation of each subunits was attempted with the enzyme oxidized with performic acid[8] and reduced with $NaBH_4$.[9] To avoid decomposition of the ε-*N*-phosphopyridoxyllysine residue, operations were carried out in the dark. The $NaBH_4$-reduced and performic acid-oxidized enzyme (30 mg) was dissolved in 1.5 ml of 0.05 *M* sodium acetate buffer (pH 4.8) containing 8 *M* urea and 1.0 m*M* EDTA, and chromatographed on a Sepharose 6B column (2.0 × 150 cm) equilibrated with the same buffer at a flow rate of 4.5 ml/hr. Three protein peaks (S-1, S-2, and S-3) were observed, which corresponded to subunits C, B, and A, respectively, upon sodium lauryl sulfate disc gel electrophoresis. Fluorescence, which is derived from the ε-*N*-phosphopyridoxyllysine residue, was found only in S-2 (subunit B). The peak area of S-2 measured by absorbance at 280 nm was approximately twice that of S-1 or S-3, suggesting that subunit B may account for two subunits in the enzyme. Hence, attempts were made to separate subunit B further. Subunit B (about 10

[6] T. Yagi, H. Misono, H. Kurihara, T. Yamamoto, and K. Soda, *J. Biochem. (Tokyo)* **87,** 1395 (1980).
[7] A. M. Crestfield, S. Moore, and W. H. Stein, *J. Biol. Chem.* **238,** 622 (1963).
[8] C. H. W. Hirs, this series, Vol. 11, p. 197.
[9] Y. Matsuo and D. M. Greenberg, *J. Biol. Chem.* **234,** 507 (1959).

TABLE II
MOLECULAR WEIGHTS AND TERMINAL AMINO ACIDS OF SUBUNITS

Subunits	Molecular weights determined by: Sedimentation equilibrium	Gel electrophoresis	Gel filtration	Amino terminals	Carboxyl terminals
Subunit A	24,000	24,000	24,000	Serine	Alanine
Subunit B_1	28,000	28,000	28,000	Arginine	Alanine
Subunit B_2	28,000	28,000	28,000	Glycine	Glycine
Subunit C	45,000	46,000	45,000	Arginine	Leucine

mg) was dissolved in 1.0 ml of 0.02 *M* Tris–HCl buffer (pH 8.0) containing 8 *M* urea and 0.1 m*M* EDTA, and applied to a DEAE-cellulose column (1.2 × 24 cm) equilibrated with the same buffer. The flow rate was 8 ml/hr, and 1.4-ml fractions were collected. After washing the column with the buffer, two protein peaks (subunits B_1 and B_2) were eluted successively with the buffer supplemented with 0.05 and 0.1 *M* NaCl. Subunits B_1 and B_2 showed similar mobility on sodium lauryl sulfate disc gel electrophoresis. Thus, they have the same molecular weight, but show different ionic behavior. Fluorescence determined at 390 nm with excitation at 330 nm was observed exclusively in subunit B_2: the pyridoxal 5′-phosphate absorbing at 415 nm binds to this subunit.

The molecular weights of the subunits were determined by three methods: sodium lauryl sulfate disc gel electrophoresis, Sephadex G-75 column chromatography in the presence of 6 *M* guanidine-HCl, and equilibrium centrifugation in the presence of 8 *M* urea. The results obtained by these three procedures are in good agreement (Table II). The amino-terminal amino acids of subunits A, B_1, B_2, and C determined by the dansylation method[10] and the carboxyl-terminal amino acids determined by the tritium labeling method[11] are also presented in Table II. The amino acid compositions of the enzyme and subunits are summarized in Table III. The sum of the numbers of each amino acid residue in the four subunits agrees well with the corresponding value for the whole enzyme. In conclusion, L-lysine-α-ketoglutarate 6-aminotransferase consists of one each of four nonidentical subunits, A, B_1, B_2, and C. Of the two molecules of bound pyridoxal 5′-phosphate, the one which absorbs at 415 nm is bound to subunit B_2 and participates in the catalytic action of the enzyme. Although the pyridoxal 5′-phosphate absorbing at 340 nm cannot

[10] A. M. Weiner, T. Platt, and K. Weber, *J. Biol. Chem.* **247,** 3242 (1972).

[11] H. Matsuo, Y. Fujimoto, and T. Tatsuno, *Biochem. Biophys. Res. Commun.* **22,** 69 (1966).

TABLE III
AMINO ACID COMPOSITIONS OF L-LYSINE-α-KETOGLUTARATE 6-AMINOTRANSFERASE AND ITS SUBUNITS[a]

Amino acids	Subunits A	Subunits B_1	Subunits B_2	Subunits C	Sum of four subunits	Native enzyme
Lysine	8.3	5.7	5.7	6.9	26.6	28.1
Histidine	9.1	8.2	8.2	10.6	36.1	35.6
Arginine	15.5	22.4	21.4	31.1	90.4	85.7
Aspartic acid	20.4	22.9	20.3	36.4	100.0	97.3
Threonine[b]	8.4	11.8	11.2	18.1	49.5	49.3
Serine[b]	6.8	10.4	8.2	14.6	40.0	38.0
Glutamic acid	17.2	29.7	31.6	43.5	122.0	127.6
Proline	15.3	12.9	15.7	29.1	73.0	74.9
Glycine	23.9	26.2	21.3	39.2	110.6	106.7
Alanine	40.9	33.0	29.4	59.7	163.0	159.0
Valine	9.7	16.3	19.1	26.5	71.6	68.5
Isoleucine	12.0	8.2	9.1	17.2	46.5	45.6
Leucine	22.8	27.1	28.6	45.2	123.7	122.5
Tyrosine	5.2	6.1	7.2	12.5	31.0	30.1
Phenylalanine	7.7	7.7	8.7	14.9	39.0	39.7
Half-cystine	2.9	3.6	3.0	5.6	15.1	17.4
Methionine	6.0	8.2	6.2	9.9	30.3	30.3

[a] Data are expressed in terms of the numbers of amino acid residues per mole of the enzyme and the subunits. The molecular weights used in the calculation were 24,000 for subunit A, 28,000 for subunits B_1 and B_2, 45,000 for subunit C, and 125,000 for the native enzyme. The values for the amino acids are averages of four sets of analyses.

[b] These values were extrapolated to zero time of hydrolysis.

be reduced with $NaBH_4$ and is easily released from the enzyme by denaturation in 8 *M* urea or by acid precipitation, a preliminary investigation suggested that it is probably bound to subunit C.[6]

Stereochemistry of Proton Abstraction during L-Lysine-α-Ketoglutarate 6-Aminotransferase Reaction[12]

Principle. Aminotransferase reactions proceed through the abstraction of a proton from the Schiff base complex formed between the enzyme-bound pyridoxal 5′-phosphate and a substrate to produce an anionic intermediate. Substrates of α-aminotransferases except glycine have only

[12] K. Tanizawa, T. Yoshimura, Y. Asada, S. Sawada, H. Misono, and K. Soda, *Biochemistry* **21,** 1104 (1982).

a single α proton to be removed during transamination, whereas those of ω-aminotransferases have two chemically identical hydrogen atoms on the prochiral distal carbon. Thus, in the ω-aminotransferase reactions, there are two stereochemical possibilities for the proton abstraction: the stereospecific abstraction of the *pro-R* or *pro-S* hydrogen and alternatively their indiscriminate removal. The stereochemistry of this proton abstraction in the L-lysine-α-ketoglutarate 6-aminotransferase reaction has been examined with two substrates tritiated stereospecifically on the C-6 carbon of L-lysine, (6*R*)-L-[6-^{3}H]lysine and (6*S*)-L-[6-^{3}H]lysine.

Procedure A. The reaction mixture contained 300 units of L-lysine 6-aminotransferase, 50 μmol of sodium α-ketoglutarate, 0.1 μmol of pyridoxal 5′-phosphate, 50 μmol of potassium phosphate buffer (pH 8.0), and 67.5 μmol (76,950 dpm) of (6*R*)-L-[6-^{3}H]lysine[12] or 5.88 μmol (33,600 dpm) of (6*S*)-L-[6-^{3}H]lysine[12] in a final volume of 1.0 ml. This was incubated at 37° for 1 hr, followed by addition of 0.1 ml of 50% trichloroacetic acid. After removal of trichloroacetic acid and the remaining α-ketoglutarate by extraction with ether, the reaction mixture was concentrated to dryness. The radioactivity of tritium released into the solvent was measured in an aliquot of the evaporated water. The residue was washed with water by repeated dissolutions and evaporations and finally dissolved into 50 μl of 1 *N* formate. The reaction products were separated by descending paper electrophoresis at 1500 V for 1 hr with 1 *N* formate as a solvent.[1] Δ^1-Piperideine-6-carboxylate and L-glutamate were determined with the *o*-aminobenzaldehyde reagent and the ninhydrin reagent, respectively,[1] and

TABLE IV
STEREOSPECIFICITY FOR PROTON ABSTRACTION BY L-LYSINE-α-KETOGLUTARATE 6-AMINOTRANSFERASE

Substrate	Specific activity (dpm/μmol)	Product	Total activity (dpm)	Specific activity (dpm/μmol)
(6*R*)-L-[6-^{3}H]Lysine	1140	Δ^1-Piperideine-6-carboxylate	43,600	1074
		L-Glutamate	520	13
		Water	204	
(6*S*)-L-[6-^{3}H]Lysine	5720	Δ^1-Piperideine-6-carboxylate	570	134
		L-Glutamate	510	117
		Water	31,480	
(6*R*)-L-[6-^{3}H]Lysine + L-[U-^{14}C]lysine	1083(^{3}H) 3738(^{14}C)[a]	Δ^1-Piperideine-6-carboxylate	41,470(^{3}H) 132,800(^{14}C)[b]	1168(^{3}H) 3741(^{14}C)[b]

[a] ^{14}C/^{3}H ratio = 3.45.
[b] ^{14}C/^{3}H ratio = 3.20.

their radioactivity was measured by counting a piece of the paper. As shown in Table IV, most of the radioactivity was found in an isolated product, Δ^1-piperideine-6-carboxylate, when (6*R*)-L-[6-^{3}H]lysine was used as a substrate for L-lysine 6-aminotransferase. In contrast, almost all of the radioactivity was released into the solvent, water, in the transamination between (6*S*)-L-[6-^{3}H]lysine and α-ketoglutarate. These results indicate that L-lysine-α-ketoglutarate 6-aminotransferase discriminates two hydrogens at the prochiral C-6 carbon of L-lysine, and abstracts the *pro-S* proton stereospecifically.

Procedure B. The above reaction system was employed except that both L-[U-^{14}C]lysine (3.2 μmol, 240,000 dpm) and (6*R*)-L-[6-^{3}H]lysine (61 μmol, 69,540 dpm) were incubated with the enzyme, and the radioactivity ratio (^{14}C/^{3}H) in Δ^1-piperideine-6-carboxylate was determined. Other procedures were as described for procedure A. Based on the result that no essential change in the radioactivity ratio (^{14}C/^{3}H) was observed between the substrate (L-[U-^{14}C]lysine plus (6*R*)-L-[6-^{3}H]lysine) and the product (Δ^1-piperideine-6-carboxylate) (see Table IV), it was again concluded that the *pro-S* hydrogen at the prochiral C-6 carbon of L-lysine is specifically abstracted by L-lysine-α-ketoglutarate 6-aminotransferase.

[20] Taurine-Glutamate Transaminase

By KAZUO YONAHA, SEIZEN TOYAMA, and KENJI SODA

$$\underset{\text{Taurine}}{NH_2CH_2SO_3H} + \underset{\alpha\text{-Ketoglutaric acid}}{HOOCCH_2CH_2COCOOH} \rightleftharpoons \underset{\text{Sulfoacetaldehyde}}{OHOCCH_2SO_3H} + \underset{\text{L-Glutamic acid}}{HOOCCH_2CH_2CH(NH_2)COOH}$$

Assay Method

Principles. Taurine : α-ketoglutrate aminotransferase (taurine-glutamate transaminase, taurine : 2-oxoglutarate aminotransferase, EC 2.6.1.55) catalyzes the transamination of taurine with α-ketoglutarate to yield sulfoacetaldehyde and L-glutamate, and the reverse reaction.[1] The assay method is based on the measurement of glutamate or sulfoacetaldehyde. Glutamate is determined with ninhydrin after the separation by

[1] S. Toyama and K. Soda, *J. Bacteriol.* **109,** 533 (1972).

Copyright © 1985 by Academic Press, Inc.
All rights of reproduction in any form reserved.

circular paper chromatography (procedure A). Sulfoacetaldehyde is determined by the reaction with 2,4-dinitrophenylhydrazine (procedure B), or *o*-aminobenzaldehyde in the presence of glycine (procedure C).

Reagents

Taurine 0.2 *M*
α-Ketoglutaric acid 0.2 *M*, neutralized with 0.4 *M* NaOH
Pyridoxal 5'-phosphate 1 m*M*
Potassium phosphate buffer 0.2 *M*, pH 8.0
Trichloroacetic acid 25% (w/v)
2,4-Dinitrophenylhydrazine 0.5% in 2 *N* HCl
o-Aminobenzaldehyde 0.025 *M* in 10% ethanol
Potassium hydroxide 0.8 *M*
Glycine–KCl–KOH buffer 1 *M*, pH 8.0
Ethyl acetate
Sodium carbonate 0.5 *M*

Procedure. The standard reaction system consists of 20 μmol of taurine, 20 μmol of α-ketoglutarate, 0.1 μmol of pyridoxal 5'-phosphate, 40 μmol of potassium phosphate buffer, pH 8.0, and enzyme, in a final volume of 1.0 ml. After the reaction mixture is incubated at 37° for 30 min, the reaction is terminated by addition of 0.1 ml of 25% trichloroacetic acid. The mixture is deproteinized by centrifugation at 17,000 *g* for 10 min. In a blank taurine is added to the mixture, after the reaction has been terminated.

A. Determination of glutamate. Glutamate formed from α-ketoglutarate is determined by the method of Katagiri *et al.*[2] The procedure is described in previous volume of this series in detail (Vol. 17B [169], p. 222).

B. 2,4-Dinitrophenylhydrazine method.[3] To the reaction mixture (0.5 ml) deproteinized as described above is added 0.5 ml of 0.5% 2,4-dinitrophenylhydrazine solution in 2 *N* HCl to produce the dydrazone of sulfoacetaldehyde. After incubation at 37° for 30 min, 2.0 ml of ethyl acetate is added to the solution, and the mixture is shaken for 2 min. To a 1.0-ml aliquot of the aqueous layer is added 2.0 ml of a mixture of 1 *N* NaOH and 0.5M Na_2CO_3 (1 : 1, v/v). After mixing the absorbance is measured against the blank at 435 nm within 3 min.

C. *o*-Aminobenzaldehyde method.[4] A 0.5-ml aliquot of the deproteinized supernatant solution obtained as described above is mixed with 1.0

[2] H. Katagiri, K. Soda, and T. Tochikura, *J. Agric. Chem. Soc. Jpn.* **34,** 814 (1960).
[3] S. Toyama, H. Misono, and K. Soda, *Biochim. Biophys. Acta* **523,** 75 (1978).
[4] S. Toyama, M. Yasuda, K. Miyasato, T. Hirasawa, and K. Soda, *Agric. Biol. Chem.* **38,** 2263 (1974).

ml of *o*-aminobenzaldehyde, 0.2 ml of 0.8 *M* KOH, 1.0 ml of 1 *M* glycine–KCL–KOH buffer (pH 8.0), and 0.3 ml of water. The mixture is incubated to develop a yellow color at 37° for 30 min. The absorbance at 440 nm is measured against the blank.

Definition of Unit and Specific Activity. One unit of enzyme is defined as the amount of enzyme which catalyzes the formation of 1.0 μmol of glutamate or sulfoacetaldehyde per min. The specific activity is expressed as units per mg of protein. The protein is determined by the method of Lowry *et al.*,[5] or estimated from the absorbance at 280 nm ($E_{1\,cm}^{1\%}$ = 7.20.)[3]

Purification Procedure[6]

Culture Conditions.[1] *Achromobacter superficialis* ICR B-89 (this organism can be obtained from Institutie for Chemical Research, Kyoto University, Uji, Kyoto 611, Japan) is grown in the medium composed of 0.2% β-alanine (as an inducer), 0.5% peptone, 0.2% glycerol, 0.1% KH_2PO_4, 0.1% yeast extract, and 0.01% $MgSO_4 \cdot 7H_2O$. The pH is adjusted to 7.2 with sodium hydroxide. The cultures are carried out in a 30-liter fermentor at 28° for 20–22 hr under aeration. The cells, harvested by centrifugation, are washed twice with 0.85% NaCl, and subsequently with 0.02 *M* potassium phosphate buffer, pH 8.0, containing 10^{-4}*M* pyridoxal 5′-phosphate and 0.01% 2-mercaptoethanol. All the subsequent operations are performed at 0–5°, and potassium phosphate buffer used contains 0.1% 2-mercaptoethanol.

Step 1. Preparation of Crude Extract. The washed cells (about 3.0 kg, wet weight) are mixed in 500-g portions with 500 g of levigated alumina and ground thoroughly in a mortar for 2 hr. To the mortar, 200 ml of 0.02 *M* potassium phosphate buffer, pH 8.0, containing 10^{-4} *M* pyridoxal 5′-phosphate is added and mixed well with the paste. The resultant slurry is centrifuged in order to separate the cell-free extract from the alumina and cellular debris. The cellular debris is again suspended in 1 liter of the same buffer, and subjected in 250-ml portions to sonication in a 19 kHz sonic oscillator for 30 min, followed by centrifugation. The supernatant solution is combined with the cell-free extract mentioned above, and then dialyzed against two changes of 100 volumes of 0.01 *M* potassium phosphate buffer, pH 7.0, containing 10^{-5} *M* pyridoxal 5′-phosphate. The inactive precipitate formed during dialysis is removed by centrifugation.

[5] O. H. Lowry, N. J. Rosebrough, A. L. Farr, and R. J. Randall, *J. Biol. Chem.* **193,** 265 (1951).

[6] S. Toyama, H. Misono, and K. Soda, *Biochem. Biophys. Res. Commun.* **46,** 1374 (1972).

Step 2. DEAE-Cellulose Column Chromatography. The enzyme solution is placed on a DEAE-cellulose column (10 × 90 cm) equilibrated with 0.01 *M* potassium phosphate buffer, pH 7.0, containing 10^{-5} *M* pyridoxal 5′-phosphate. The column is washed thoroughly with the same buffer and then with the buffer containing 0.27 *M* NaCl. The enzyme is eluted with the buffer supplemented with 0.35 *M* NaCl at the flow rate of 100 ml/hr, and 30-ml aliquots of the eluate are collected. The active fractions are pooled, and the enzyme is precipitated by addition of ammonium sulfate (80% saturation). The precipitate is dissolved in 0.05 *M* potassium phosphate buffer, pH 7.0, containing 10^{-4} *M* pyridoxal 5′-phosphate.

Step 3. First Sephadex G-150 Column Chromatography. The enzyme is applied to five Sephadex G-150 columns (3 × 150 cm) equilibrated with 0.01 *M* potassium phosphate buffer, pH 7.0, containing 10^{-5} *M* pyridoxal 5′-phosphate, and eluted with the same buffer. The active fractions are pooled and the enzyme is concentrated by addition of ammonium sulfate (80% saturation). The precipitate is dissolved in 10^{-3} *M* potassium phosphate buffer, pH 7.0, containing 10^{-4} *M* pyridoxal 5′-phosphate. The enzyme solution is dialyzed against three changes of 500 volumes of the same buffer.

Step 4. Hydroxylapatite Column Chromatography. The enzyme solution is applied to three hydroxylapatite columns (3 × 11 cm) equilibrated 10^{-3} *M* potassium phosphate buffer, pH 7.0, containing 10^{-5} *M* pyridoxal 5′-phosphate. The columns are washed with the same buffer, and then with 0.01 *M* potassium phosphate buffer, pH 7.0, containing 10^{-5} *M* pyridoxal 5′-phosphate. The enzyme is eluted with 0.04 *M* potassium phosphate buffer, pH 7.0, containing 10^{-5} *M* pyridoxal 5′-phosphate at the flow rate of 15 ml/hr and 5-ml fractions are collected.

Step 5. Ammonium Sulfate Fractionation. The enzyme solution is brought to 30% saturation with ammonium sulfate, and the resultant precipitate is centrifuged and discarded. Ammonium sulfate is added to the supernatant solution to 60% saturation, followed by centrifugation. The precipitate is dissolved in 0.01 *M* potassium phosphate buffer, pH 7.0, containing 10^{-4} *M* pyridoxal 5′-phosphate.

Step 6. Second Sephadex G-150 Column Chromatography. The enzyme solution is applied to three Sephadex G-150 columns (1.3 × 130 cm) equilibrated with 0.01 *M* potassium phosphate buffer, pH 7.0, containing 10^{-5} *M* pyridoxal 5′-phosphate, and eluted with the same buffer. The active fractions are combined and the enzyme is precipitated by addition of ammonium sulfate (60% saturation). The precipitate is dissolved in a small volume of 0.01 *M* potassium phosphate buffer, pH 7.0, containing 10^{-4} *M* pyridoxal 5′-phosphate.

PURIFICATION OF TAURINE:α-KETOGLUTARATE TRANSAMINASE FROM *Achromobacter superficialis*

Purification step	Total protein (mg)	Total activity[a] (units)	Specific activity[a] (units/mg)	Yield (%)
1. Crude extract	60,400	483	0.008	100
2. DEAE-cellulose	3,610	357	0.099	73.9
3. First Sephadex G-150	2,015	240	0.119	49.7
4. Hydroxylapatite	1,025	197	0.192	40.8
5. Ammonium sulfate fractionation (30–60%)	328	108	0.329	22.4
6. Second Sephadex G-150	174	74	0.427	15.3
7. Crystallization	112	58	0.513	12.0
8. Recrystallization	75	39	0.519	8.1

[a] The enzyme activity is determined by procedure A.

Step 7. Crystallization and Recrystallization (Step 8). Ammonium sulfate is added gradually to the enzyme solution until the solution becomes slightly turbid. The pH of solution is kept constant at 7.0 to 7.6 with 10% NH_4OH solution. On standing for 2 to 3 days, crystals with the form of hexagonal plates are obtained. The crystals are collected by centrifugation, and dissolved in a minimum volume of the above mentioned buffer. The enzyme is recrystallized in the same manner. The purification of the enzyme is summerized in the table. About 65-fold purification is achieved with an overall yield of 8 to 12%.

Properties[3,6]

Stability. The crystalline enzyme can be stored with little loss of activity at 0–5° in 0.01 *M* potassium phosphate buffer, pH 7.0, containing 60% saturation of ammonium sulfate, 10^{-4} *M* pyridoxal 5′-phosphate and 0.01% 2-mercaptoethanol, for periods of several months. Freezing and thawing of the enzyme cause its significant inactivation. Treatment of the enzyme with ammonium sulfate often results in loss of activity. The inactivated enzyme is fully activated by incubation with pyridoxal 5′-phosphate as described below. The enzyme is stable between pH 6.0 and 9.0 and at 60° for 60 min.

Physicochemical Properties. The crystalline enzyme is homogeneous by the criteria of ultracentrifugation and disc gel electrophoresis. The sedimentation coefficient ($s^0_{20,w}$) of the enzyme has been determined to be

9.1 S. Assuming a partial specific volume of 0.74, the molecular weight determined by the sedimentation equilibrium method[7] is approximately 156,000.

Absorption Spectrum and Coenzyme. When the enzyme is treated with ammonium sulfate (60% saturation) in 0.01 *M* potassium phosphate buffer (pH 7.0), the enzyme is converted to an inactive form. The inactivated enzyme is fully activated by incubation with pyridoxal 5′-phosphate at 60° for 10 min. The absorption spectrum of the inactivated enzyme shows maxima at 280 and 345 nm with molar absorption coefficients of 112,000 and 21,000, respectively. The activated enzyme exhibits a small new peak at 430 nm in addition to the peaks at 280 and 345 nm. The exhaustive dialysis of the activated enzyme against 0.01 *M* potassium phosphate buffer (pH 7.0) results in disappearance of the 430-nm peak with concomitant formation of a shoulder at 400 nm, though the enzyme activity is not affected. The pyridoxal 5′-phosphate contents are 2.8 and 3.9 mol per mole of the inactivated and activated enzymes, respectively. The absorption spectrum of the inactivated enzyme is not affected by varying the pH (6.0–9.0), and by addition of an amino donor or α-ketoglutarate. A decrease in absorbance at 400 nm and an increase in 345-nm absorption are observed, when an amino donor is added to the activated enzyme. One mole of pyridoxal 5′-phosphate can be resolved from the activated enzyme by incubation with taurine and 0.5 *M* potassium phosphate buffer (pH 5.0) at room temperature for 1 hr, followed by precipitation with ammonium sulfate (60% saturation). The enzyme has no activity in the absence or presence of pyridoxal 5′-phosphate, although the full activity is obtained after incubation with pyridoxal 5′-phosphate and pyridoxamine 5′-phosphate at 60° for 10 min. Pyridoxine, pyridoxal, pyridoxamine, pyridoxine 5′-phosphate, and some other vitamin B_6 compounds are inactive for the activation.

Substrate Specificity. Hypotaurine (relative activity, 601), DL-β-aminoisobutyrate (208), and β-alanine (184) show higher reactivity than taurine (100) in the transamination. In addition to them, γ-aminobutyrate (60), 3-aminopropanesulfonate (43), and DL-β-aminobutyrate (14) also serve as amino donors, whereas aminomethanesulfonate, glycine, δ-aminovalerate, L- and D-α-amino acids, and amines are not substrates. α-Ketoglutarate is the exclusive amino acceptor; pyruvate, phenylpyruvate, and oxaloacetate are inert.

Kinetics. The Michaelis constants are calculated; taurine, 12 m*M;* hypotaurine, 16 m*M;* DL-β-aminoisobutyrate, 11 m*M;* β-alanine, 17 m*M;* and α-ketoglutarate, 11 m*M*. The K_m value of 5 μ*M* is obtained for pyri-

[7] K. E. Van Holde and R. L. Baldwin, *J. Phys. Chem.* **62,** 734 (1958).

doxal 5′-phosphate. The pH optimum for the enzyme activity is 7.8–8.0. The maximum activity is obtained at 60°.

Inhibitors. Hydroxylamine, L-cycloserine, and $HgCl_2$ strongly inhibit the enzyme. Semicarbazide, D-cycloserine, phenylhydrazine, and 3-methyl-2-benzothiazolone hydrazone hydrochloride are weak inhibitors. *p*-Chloromercuribenzoate, monoiodoacetic acid, D- and L-penicillamine, and ethylenediaminetetraacetic acid have no effect on the enzyme activity.

Distribution.[1] The taurine transaminase activity has been found only in the extracts of *Achromobacter superficialis* ICR B-89 and *Achromobacter polymorph* ICR B-88. The enzyme activity is induced by β-alanine, but not by taurine.

[21] D-Glutamate-D-Amino Acid Transaminase from Bacteria

By WANDA M. JONES, THOMAS S. SOPER,
HIROSHI UENO, and JAMES M. MANNING

D-Amino acid transaminase catalyzes the transamination of a wide spectrum of α-amino acids and α-keto acids; however, it is absolutely specific for the D-isomer of amino acids. The enzyme was first described in *Bacillus subtilis* by Thorne *et al.*[1] and, subsequently, it has been found in many gram-positive bacteria. The probable physiological role of this enzyme in bacteria is the production of D-amino acids, especially D-glutamate, which is required for synthesis of the peptidoglycan layer of the bacterial cell wall.[2] The overall reaction is

$$\text{D-Amino acid} + \alpha\text{-ketoglutarate} \leftrightharpoons \alpha\text{-keto acid} + \text{D-glutamate}$$

The procedures described below are modifications[3] of those described by Martinez-Carrion and Jenkins[4] and by Yonaha *et al.*[5]

[1] C. B. Thorne, C. G. Gómez, and R. D. Housewright, *J. Bacteriol.* **69,** 357 (1954).
[2] J. Baddiley, *in* "β-Lactam Antibiotics: Mode of Action, New Developments, and Future Prospects" (M. R. J. Salton and G. D. Shockman, eds.), p. 13. Academic Press, New York, 1981.
[3] T. S. Soper, W. M. Jones, B. Lerner, M. Trop, and J. M. Manning, *J. Biol. Chem.* **252,** 3170 (1977).
[4] This article updates the earlier work of M. Martinez-Carrion and W. T. Jenkins, this series, Vol. 17A, p. 167.
[5] K. Yonaha, H. Misono, T. Yamamoto, and K. Soda, *J. Biol. Chem.* **250,** 6983 (1975).

Copyright © 1985 by Academic Press, Inc.
All rights of reproduction in any form reserved.

Methods of Assay

Principle. D-Amino acid transaminase can be assayed conveniently by measuring the production of pyruvate from D-alanine in the presence of an α-keto acid cosubstrate. Determination of the amount of pyruvate formed from D-alanine is made either with salicylaldehyde in a discontinuous assay[4] or with lactate dehydrogenase in a continuous assay.[3] The assay with salicylaldehyde is employed during the purification of the enzyme; the assay with lactate dehydrogenase cannot be used in crude extracts because of NADH oxidase activity, but it is very useful for investigating the kinetic properties of the purified enzyme. The former method is the less sensitive one and requires about 0.1–1.0 units of enzyme; the latter assay can measure as little as 0.005–0.05 units of D-amino acid transaminase. One unit of enzyme activity is defined as the amount of protein that produces 1 μmol of pyruvate per minute at 37°. The specific activity is defined as units of activity per mg of protein, where the latter is determined by the $E_{280}^{1\%}$ of the pure enzyme being equal to 12.5[5]

Salicylaldehyde Assay

Reagents

Potassium phosphate buffer, pH 8.5, 1 *M*
D-Alanine, 1 *M*
α-Ketoglutarate, 1 *M*, neutralized with NaOH
Pyridoxal phosphate, 10 m*M*
Dithiothreitol, 10 m*M*
Salicylaldehyde, 2% in absolute ethanol
Potassium hydroxide, 60% (w/v)

Procedure. The procedure of Martinez-Carrion and Jenkins[4] has been modified by increasing the concentration of substrates to ensure linear kinetics and saturating concentrations of substrates during the course of the assay over a 30-min period. The assay mixture contains 0.4 *M* potassium phosphate buffer, pH 8.5, 100 m*M* D-alanine, 100 m*M* α-ketoglutarate, 1 m*M* pyridoxal-P, and 1 m*M* dithiothreitol in a total volume of 1 ml. The reaction is initiated by addition of enzyme (0.1–1.0 unit) and the incubation is carried out at 37° for 10–30 min; the reaction is terminated by 1 ml of 60% KOH. The salicylaldehyde reagent (0.5 ml) is then added and this mixture is incubated at 37° for 10 min for development of the orange chromophore. The mixture is diluted with 2.5 ml of water and the absorbance is measured at 480 nm. The absorbance, which is proportional to the amount of pyruvate up to about 1.5 absorption units, should be measured shortly after the reaction is complete since the color slowly intensifies.

Lactate Dehydrogenase-Coupled Assay

Reagents

Potassium phosphate buffer, pH 8.5, 1 *M*
D-Alanine, 1 *M*
α-Ketoglutarate, 1 *M*, neutralized with NaOH
NADH, 10 m*M*
Lactate dehydrogenase, 700 μg/ml (from rabbit muscle, purchased from Sigma)

Procedure. The assay mixture contains 0.3 *M* potassium phosphate buffer, pH 8.5, 25 m*M* D-alanine, 25 m*M* α-ketoglutarate, 0.1m*M* NADH, and 70 μg/ml lactate dehydrogenase. One milliliter of this assay mixture is placed in a cuvette thermostated at 24° and the reaction is initiated by addition of 0.005 to 0.05 units of enzyme. The absorbance at 338 nm is followed for 3 to 5 min and the rate of production of pyruvate is calculated using a molar absorptivity of 6.2×10^4 M^{-1} cm^{-1} for NADH. The concentration of lactate dehydrogenase is significantly higher than usually employed with other coupled assays, since lower levels of lactate dehydrogenase result in a significant lag period which interferes with the determination of the rate of pyruvate formation under these conditions. The concentration of α-ketoglutarate is lower in this assay than that in the salicylaldehyde assay because this α-keto acid has a significant absorption at 330 nm. Since less than 0.4% of the total amount of α-ketoglutarate is consumed, any decrease in absorption due to keto acid utilized can be ignored.

This assay procedure can be easily modified to permit measurement of D-amino acid transaminase with other D-amino acids as substrates. However, since the K_m for D-aspartate is rather high, the concentration of D-aspartate should be above 100 m*M* to saturate the enzyme. The transamination of D-glutamate could be measured if glutamate dehydrogenase replaced lactate dehydrogenase in the above procedure. However, since D-glutamate is a potent inhibitor of glutamate dehydrogenase (K_i = 0.1 m*M*),[6] this assay is limited to subsaturating concentrations of this substrate; about 1 m*M* D-glutamate is the maximum concentration that can be used.

Purification Procedure

Step 1. Growth and Breakage of Cells. Bacillus sphaericus strain ATCC 14577 was grown in 550-liter batches at the New England Enzyme

[6] R. Sekura, P. Van Der Werf, and A. Meister, *Biochem. Biophys. Res. Commun.* **71,** 11 (1976).

Center, Boston, Mass. The specific activity of D-amino acid transaminase varies with the phase of growth of this organism with the maximum occurring as the cells enter stationary phase. If the cells are allowed to remain in stationary phase, enzyme activity is nearly abolished. The cells are harvested, washed, and suspended in 50 m*M* potassium phosphate buffer, pH 8.5, and 0.1 m*M* DTT. The cell suspension is disrupted using a French Pressure Cell. The crude extract retains activity for many months when stored frozen. Once thawed, the purification should be initiated since the enzyme activity is not very stable in the thawed extract.

Step 2. Removal of Nucleic Acids and Ammonium Sulfate Fractionation. The crude extract (approximately 5 kg) is thawed and centrifuged to remove any insoluble material. The extract is then treated with 20 mg of protamine sulfate per 100 mg of total protein, and the nucleic acid precipitate is removed by centrifugation at 16,300 *g*. Solid ammonium sulfate is added to 60% saturation and the insoluble material is collected by centrifugation at 16,300 *g*. The pellet is suspended in a minimal volume of 10 m*M* potassium phosphate buffer, pH 8.5, containing 0.1 m*M* DTT, and the suspension is dialyzed against several changes of the same buffer until most of the insoluble material is dissolved.

Step 3. Chromatography on DE-52. After centrifugation and concentration, this clear solution of protein is then applied to a 4 × 60 cm column of DE-52 (Whatman) equilibrated with the same buffer. When the protein has been absorbed by the cellulose, the eluent is changed to 50 m*M* potassium phosphate buffer, pH 8.5, with 0.1 m*M* DTT. The fractions containing activity are pooled and concentrated to as small a volume as possible using an Amicon concentration apparatus.

Step 4. Gel Filtration and Chromatography on Hydroxyapatite. The enzyme is next applied to a 4.5 × 75 cm column of Sephadex G-150 equilibrated with 1 m*M* potassium phosphate buffer, pH 7.5, and eluted with the same buffer. Fractions containing activity are pooled and concentrated by ultrafiltration. The enzyme is next applied to a 2.5 × 38 cm column of hydroxyapatite equilibrated with 1 m*M* potassium phosphate buffer, pH 7.5, and 0.1 m*M* DTT. After washing with 2-column volumes of this buffer, the enzyme is eluted with 5 m*M* potassium phosphate, pH 7.5, and 0.1 m*M* DTT. The fractions containing enzyme activity are pooled and concentrated by ultrafiltration to a minimal volume and applied to a 1.5 × 100 cm column of Sephadex G-100 equilibrated with 10 m*M* potassium phosphate buffer, pH 8.5, and 0.1 m*M* DTT. The enzyme is then concentrated, filtered, and stored in liquid nitrogen. The pure enzyme has a specific activity of 150 units per mg protein. If the specific activity is not near this value, the enzyme is rechromatographed either on hydroxyapatite or on DE-52 as described above, to remove any remaining impurities.

Properties of the Enzyme

D-Amino acid transaminase has the unique ability to catalyze the transamination of a wide variety of D-amino acids.[4,5] Apparently, the only requirement for a D-amino acid to be a substrate is that it possess at least two protons on the β-carbon of the molecule. Thus, D-norvaline and D-norleucine are substrates, but D-valine and D-leucine are not substrates. The necessity for the presence of at least two β-protons is not apparent, but it may be related to the enzymatic exchange of these protons with solvent water.[7]

The transaminase, purified from *B. subtilis*[4] or from *B. sphaericus*,[5] is a dimer of molecular weight of about 60,000. However, the enzymes do not behave similarly during purification. In addition, the enzyme from the latter source is inactivated by thiol-specific reagents whereas that from the former species is resistant to inactivation by such reagents. We have investigated the reactivity of the enzyme from *B. sphaericus;* there are 8 sulfhydryl groups per dimer and they exist in three distinct environments.[8] The first pair is fully exposed to solvent, and chemical modification of them does not result in inactivation. A second pair is partially exposed to solvent, and chemical modification results in partial inactivation of the enzyme; the remaining 4 sulfhydryl groups per dimer are inaccessible to DTNB and react only upon denaturation of the transaminase.

D-Amino acid transaminase contains two pyridoxal lysine linkages per dimer, but only one of the two is removed upon mild treatment with phenylhydrazine.[5] The remaining cofactor is involved in an, as yet, unidentified linkage on the enzyme.

β-Chloro-D-alanine, an inhibitor of bacterial growth,[9] is an efficient enzyme-activated inhibitor of D-amino acid transaminase.[3] This inhibitor undergoes a β-elimination reaction to form an α-aminoacrylate Schiff base with the cofactor. This intermediate partitions between solvolysis and irreversible alkylation of the enzyme. The completely inactivated enzyme contains 2 mol of inhibitor covalently attached to the dimeric enzyme.[10] Structural homologues of this inhibitor, including β-fluoro-D-alanine,[11] β-bromo-D-alanine,[12] *O*-carbamyl-D-serine, *O*-sulfo-D-serine,

[7] T. S. Soper and J. M. Manning, *J. Biol. Chem.* **257,** 13930 (1982).

[8] T. S. Soper, W. M. Jones, and J. M. Manning, *J. Biol. Chem.* **254,** 10901 (1979).

[9] J. M. Manning, N. E. Merrifield, W. M. Jones, and E. C. Gotschlich, *Proc. Natl. Acad. Sci. U.S.A.* **71,** 417 (1974).

[10] J. M. Manning and T. S. Soper, *in* "Enzyme-Activated Irreversible Inhibitors" (N. Seiler, M. J. Jung, and J. Koch-Weser, eds.), p. 163. Elsevier/North-Holland Biomedical Press, Amsterdam, 1978.

[11] T. S. Soper and J. M. Manning, *J. Biol. Chem.* **256,** 4263 (1981).

[12] T. S. Soper and J. M. Manning, *Biochemistry* **17,** 3377 (1978).

and β-cyano-D-alanine,[12a] inactivate this transaminase. Vinylglycine, on the other hand, undergoes transamination, and an α-keto acid cosubstrate is required for complete inactivation with this suicide substrate.[13] Other inhibitors such as gabaculine and D-cycloserine sequester the coenzyme without alklylating the enzyme.[11] D-Cycloserine displays half of the sites reactivity with the enzyme, suggesting cooperativity between the subunits of D-amino acid transaminase. Gabaculine acts via an aromatization mechanism.[14] A key step in this pathway involves removal of a proton from the β-carbon of the inhibitor; D-amino acid transaminase catalyzes such a reaction.[7] Since inhibitors of D-amino acid transaminase deprive bacteria of essential D-glutamate, they are potential antimicrobial agents.

Acknowledgments

This work was supported in part by NIH Grant AI-16190. The authors with to thank colleagues Nancy Merrifield, Emil Gotschlich, Brad Lerner, and Moshe Trop for their assistance during various aspects of this work.

[12a] H. Ueno, T. S. Soper, and J. M. Manning, *Biochem. Biophys. Res. Commun.* **122,** 485 (1984).

[13] T. S. Soper, J. M. Manning, P. A. Marcotte, and C. T. Walsh, *J. Biol. Chem.* **252,** 1571 (1977).

[14] R. R. Rando, *Biochemistry* **16,** 4604 (1977).

[22] Δ^1-Pyrroline-5-carboxylate Synthase from Rat Intestinal Mucosa

By JAMES J. KRAMER, JERRY G. HENSLEE, YASUO WAKABAYASHI, and MARY ELLEN JONES

Assay Method

Principle. The enzymatic conversion of L-glutamate to L-proline has been studied in microorganisms,[1–4] insects,[5,6] and mammalian tissue.[7–14]

[1] H. J. Vogel and B. D. Davis, *J. Am. Chem. Soc.* **74,** 109 (1952).

[2] H. J. Strecker and P. Mela, *Biochim. Biophys. Acta* **17,** 580 (1955).

[3] A. Baich, *Biochim. Biophys. Acta* **192,** 462 (1969).

[4] A. Baich, *Biochim. Biophys. Acta* **244,** 129 (1971).

[5] R. W. Brosemer and P. S. Veerabhadrappa, *Biochim. Biophys. Acta* **110,** 102 (1965).

[6] A. Wadano, *Experientia* **36,** 1029 (1980).

Copyright © 1985 by Academic Press, Inc.
All rights of reproduction in any form reserved.

This conversion involves the reduction of the γ-carboxyl group of glutamate to yield glutamate 5-semialdehyde, in which the latter spontaneously cyclizes to Δ^1-pyrroline-5-carboxylate (P5C).[15] P5C is then reduced in an irreversible reaction to L-proline by Δ^1-pyrroline-5-carboxylate reductase in the presence of either NADPH or NADH.[1,16]

The synthesis of P5C in bacteria requires two enzymes: γ-glutamyl kinase and γ-glutamate semialdehyde dehydrogenase.[3,4] In *Escherichia coli* the first enzyme is coded for by the *pro B* gene. This enzyme presumably catalyzes the formation of γ-glutamyl phosphate.[17] It has been suggested that this product exists as an enzyme-bound intermediate[3] not unlike that found in the synthesis of glutamine[18] and γ-glutamyl-cysteine[19] from glutamate, ATP and Mg^{2+}. However, γ-glutamyl phosphate is extremely labile to nucleophilic attack and must be measured indirectly. This is accomplished by the reaction of the acyl phosphate with hydroxylamine which results in the formation of γ-glutamyl hydroxamate.[18–21] Since hydroxamate formation is also observed when enzymes other than P5C synthase are present, the assay would measure these activities as well. However, in *E. coli* the γ-glutamyl kinase coded for by the *pro B* gene is inhibited by L-proline. Therefore duplicate assay vessels containing glutamate, ATP, and Mg^{2+}, either in the presence or absence of proline, are used as substrates. The difference between the amount of hydroxamate formed in these vessels is ascribed to the *pro B* gene product.

The second enzyme required for P5C synthesis in *E. coli* is the product of the *pro A* gene.[17] This enzyme is presumably γ-glutamate semialdehyde

[7] M. R. Stetten, *in* "Amino Acid Metabolism" (W. D. McElroy and B. Glass, eds.), p. 277. Johns Hopkins Press, Baltimore, Maryland, 1955.

[8] J. M. Phang, D. Valle, and E. M. Kowaloff, *Ann. Clin. Lab. Sci.* **5,** 298 (1975).

[9] A. Herzfeld and S. M. Raper, *Biochim. Biophys. Acta* **428,** 600 (1976).

[10] J. J. Wasmuth and C. T. Caskey, *Cell* **8,** 71 (1976).

[11] G. Ross, D. Dunn, and M. E. Jones, *Biochem. Biophys. Res. Commun.* **85,** 140 (1978).

[12] R. J. Smith, S. J. Downing, J. M. Phang, R. F. Lodata, and T. T. Aoki, *Proc. Natl. Acad. Sci. U.S.A.* **77,** 5221 (1980).

[13] Y. Wakabayashi and M. E. Jones, *J. Biol. Chem.* **258,** 3865 (1983).

[14] Y. Wakabayashi, J. G. Henslee, and M. E. Jones, *J. Biol. Chem.* **258,** 3873 (1983).

[15] C. Schöpf and H. Stener, *Justus Liebigs Ann. Chem.* **558,** 124 (1947).

[16] J. J. Rossi, J. Vender, C. M. Berg, and W. H. Coleman, *J. Bacteriol.* **129,** 108 (1977).

[17] D. J. Hayzer and T. Leisinger, *J. Gen. Microbiol.* **118,** 287 (1980).

[18] P. R. Krishnaswamy, W. Pamiljans, and A. Meister, *J. Biol. Chem.* **237,** 2932 (1962).

[19] A. Meister, *in* "The Enzymes" (P. D. Boyer, ed.), 3rd ed., Vol. 10, p. 671. Academic Press, New York, 1974.

[20] A. Katchalsky and M. Paecht, *J. Am. Chem. Soc.* **76,** 6042 (1954).

[21] F. Lipmann and L. C. Tuttle, *J. Biol. Chem.* **159,** 21 (1945).

dehydrogenase (EC 1.2.1.41) and is responsible for the reduction of γ-glutamyl phosphate to γ-L-glutamate 5-semialdehyde [Reaction (1)].

$$\gamma\text{-Glutamyl-P} + \text{NADPH} + \text{H}^+ \leftrightharpoons \gamma\text{-L-glutamate 5-semialdehyde} + \text{NADP}^+ + \text{P}_i \quad (1)$$

The product of this reaction is in chemical equilibrium with P5C [Reaction (2)].

$$\gamma\text{-L-Glutamate 5-semialdehyde} \leftrightharpoons \text{P5C} + \text{H}_2\text{O} \quad (2)$$

Since γ-glutamyl phosphate is unstable the reaction has been measured in the reverse direction.[22–24] The assay measures the stoichiometric conversion of P5C and $NADP^+$ to 5-oxopyrrolidine-2-carboxylic acid and NADPH, respectively.[22] This enzyme has been purified to homogeneity and requires inorganic phosphate for activation.[23]

These two enzymes have been separated from one another. However, there is some evidence[25,26] which indicates that aggregation of γ-glutamyl kinase and γ-glutamate semialdehyde dehydrogenase is necessary for P5C synthesis to occur. The concept of an enzyme complex was first suggested by Baich.[3]

The mammalian enzyme catalyzing the conversion of glutamate to P5C has recently been discovered in homogenates of rat intestinal mucosa.[13,14] Like the bacterial system, it requires ATP, Mg^{2+}, and NADPH for the synthesis of P5C. However, unlike the bacterial enzyme, the conversion of glutamate to P5C in mammalian tissue is not inhibited by proline. Although not purified to homogeneity, P5C synthase activity has been demonstrated in a mitochondrial membrane fraction. So far there are few or no data to establish whether the two enzymatic activities are separate or not.

The bacterial enzymes are very active in comparison to the mammalian system. Therefore the bacterial kinase can be followed by hydroxamate formation and dehydrogenase activity can be measured either by a colorimetric method measuring the loss of P5C[27] or spectrophotometrically by the formation of NADPH in the presence of P5C and phosphate.[23]

Due to the low activity in the mammalian P5C synthetic system, approximately 1 nmol product formed/min/mg protein, and the lack of sensitivity afforded by the spectrophotometric analysis used in the assay of

[22] R. V. Krishna, P. Beilstein, and T. Leisinger, *Biochem. J.* **181,** 223 (1979).
[23] D. J. Hayzer and T. Leisinger, *Eur. J. Biochem.* **121,** 561 (1982).
[24] D. J. Hayzer and T. Leisinger, *Biochim. Biophys. Acta* **742,** 391 (1983).
[25] H. Gamper and V. Moses, *Biochim. Biophys. Acta* **354,** 75 (1974).
[26] D. J. Hayzer and V. Moses, *Biochem. J.* **173,** 207, 219 (1978).
[27] H. J. Strecker, *J. Biol. Chem.* **235,** 2045 (1960).

the bacterial enzymes, a radioassay for P5C synthase has recently been developed in this laboratory.[28] It involves the use of [^{14}C]glutamate and its subsequent reduction to Δ^1-[^{14}C]pyrroline-5-carboxylate which can then be quantitated directly by use of liquid scintillation spectroscopy. Isolation of [^{14}C]P5C from [^{14}C]glutamate is accomplished by the use of ion-exchange chromatography. The efficiency of product isolation is followed by the addition of a known amount of [^{3}H]P5C after termination of enzymatic activity.

We will describe in detail the isolation of mammalian P5C synthase from rat intestinal mucosal cells and the radioactive assay procedure, since it should be applicable for extracts with low activity. The spectrophotometric assay for the phosphate dependent γ-glutamate semialdehyde dehydrogenase is also detailed.[27]

Δ^1-Pyrroline-5-carboxylate Synthase[28]

Reagents

(Hepes) *N*-2-Hydroxyethylpiperazine-*N'*-2-ethanesulfone: NaOH buffer, 0.5 *M*, pH 7.4
[^{14}C]L-Glutamic acid, 100 μCi/ml
Creatine phosphokinase, 500 units/ml
Nonidet P-40 (Polyethylene Glycol (9)*p*-Isooctylphenyl ether, 10% (v/v)
$MgCl_2$, 0.4 *M*
Perchloric acid, 70–72%
Δ^1-[^{3}H]Pyrroline-5-carboxylate[23]
Gabaculine (5-amino-1,3-cyclohexadienyl carboxylic acid), 10 m*M*

Substrate Mixture

Disodium ATP, 40 μmol; tetrasodium NADPH, 2.6 μmol; monosodium L-glutamic acid, 13.3 μmol; phosphocreatine, 200 μmol; per milliliter of Hepes-NaOH buffer, 0.1 *M*, pH 7.4.

Procedure

Solutions of Hepes buffer, L-[^{14}C]glutamic acid, creatine phosphokinase, the substrate mixture (ATP, NADPH, L-glutamic acid, and phosphocreatine), nonidet P-40, $MgCl_2$, and gabaculine, in aliquots of 185, 10, 20, 75, 25, 50, and 100 μl, respectively, are mixed together with an additional amount of distilled water such that the volume of the assay mixture after addition of enzyme solution is equal to 1 ml. This solution is then

[28] J. G. Henslee, Y. Wakabayashi, W. C. Small, and M. E. Jones, *Arch. Biochem. Biophys.* **226,** 693 (1983).

preincubated for 3 min in a room temperature water bath. The reaction is initiated by the addition of enzyme, equivalent to 1 mg of mitochondrial protein, and continues for 30 min at room temperature. The assay is terminated by the addition of 50 μl perchloric acid and the samples are placed in an ice bath.

Approximately 40,000 dpm of [^{3}H]P5C is added to the terminated assay mixture. The solution is then centrifuged at 21,000 g (4°) for 10 min after which the supernatant is transferred to a clean centrifuge tube and allowed to equilibrate to room temperature. The pH of the assay supernatant is adjusted to 5.7–6.3 with addition of approximately 70 μl of 10 M KOH and a dropwise addition of 0.1 N KOH. The solution is recentrifuged at 24,000 g (4°) for 10 min to pellet precipitable $KClO_4$. The protein-free perchlorate supernatant is allowed to come to room temperature and is then applied to AG1-X8 anion exchange resin (0.7 × 6.0 cm bed volume, acetate form).

Isolation of [^{14}C]P5C is accomplished by the use of ion-exchange chromatography. Five milliliters of H_2O is washed through the resin followed by 3 ml of 0.025 M acetic acid. Elution of Δ^1-pyrroline-5-carboxylate is achieved by subsequent additions of 1 ml aliquots of 0.025 M acetic acid. P5C elutes from the resin in a total of 4–7 ml of acetic acid. Each separate milliliter of this acid wash is mixed with 9 ml of liquid scintillation cocktail[29] and counted in a liquid scintillation counter.

Care should be taken in the isolation of Δ^1-pyrroline-5-carboxylate. P5C is most stable at 4°,[27] acidic pH and concentrations lower than 1 mM.[30]

Definition of Specific Activity. One unit of enzymatic activity is defined as that amount which produces 1 nmol of P5C per minute.

γ-Glutamate Semialdehyde Dehydrogenase[23]

Reagents

Potassium phosphate, 1 M
Δ^1-Pyrroline-5-carboxylate,[30] 20 mM
$NADP^+$, 10 mM
Imidazole buffer, 0.5 M, pH 7.0

Procedure

To each cuvette is added 100 μl each of potassium phosphate, P5C, $NADP^+$, and imidazole buffer followed by the addition of an appropriate

[29] S. V. Pande, *Anal. Biochem.* **74,** 25 (1976).

[30] J. Williams and L. Frank, *Anal. Biochem.* **64,** 85 (1975).

amount of water and enzyme to yield a final volume of 1 ml. Blank cuvettes included all reagents except phosphate. The reaction is started by the addition of enzyme and followed spectrophotometrically at A_{340} to determine the rate of NADPH formation. The molar extinction coefficient for NADPH is $6.22 \times 10^3\ M^{-1}\text{cm}^{-1}$ at A_{340}.

Definition of Specific Activity. A unit of enzymatic activity is that amount which results in the formation of 1 nmol NADPH/min/mg protein.

Preparation of Rat P5C Synthase[13,14]

Male, Sprague–Dawley CD rats, 250 g, are starved overnight and sacrificed by cervical dislocation. The small intestine is removed from the pyloric end to 5 cm proximal to the ileocecal valve. All operations are performed at 4°. The intestines are flushed with 0.9% (w/v) NaCl and cut into 11-cm segments on a glass plate placed on ice. The segments are inverted over a plastic Pasteur pipette, washed with saline, gently blotted with paper tissue, cut into 3-cm segments, and placed into a thick-walled shaker bottle into which has been added 50 ml of isolation buffer containing 0.14 *M* NaCl and 5 m*M* EDTA (pH 7.4 at 4°).

The shaker bottles, designed for the Braun mechanical shaker (B. Braun Melsungen Apparatebau), contain all the intestinal segments from one rat. The segments are then shaken for 1.5 min after which the segments are removed and placed in a 50 ml centrifuge tube. The mucosal epithelial cell suspension is poured into another centrifuge tube. The isolated intestinal segments are mixed with isolation buffer, parafilm is placed over the mouth, and the tube is vigorously shaken by hand. The segments are removed from the tube and the hand-shaken supernatant is added to the epithelial cell suspension. Typically, the mucosal suspension from one rat was diluted to 80 ml with isolation buffer to facilitate pelleting of epithelial cells.

The cell suspension is centrifuged at 600 *g* for 3 min, the supernatant aspirated, and the sides of the centrifuge tube are wiped with tissue to remove mucin. The cell pellet is resuspended in Buffer A (300 m*M* mannitol, 5 m*M* Hepes, 5 m*M* EDTA, pH 7.4 at 4°) by gentle hand shaking of the tube and is recentrifuged at 600 *g* for 3 min. The supernatant is again aspirated, the sides of the tube wiped with tissue and the cell pellet is resuspended in Buffer A. The cell suspension is centrifuged (600 *g* for 5 min at 4°), the supernatant is aspirated, and the cell pellet weighed.

The pellet is homogenized in approximately 30 ml of homogenization buffer (300 m*M* mannitol, 5 m*M* Hepes, 3 m*M* DTT, 0.2 m*M* EDTA, pH

7.4 at 4°) with two series of three strokes using a teflon pestle and a motor-driven Potter-Elvehjem tissue homogenizer set at about 600 rpm. The homogenate is then mixed with fine glass beads (0.2 mm in diameter) that have been acid cleaned and rinsed well with homogenization buffer. The pestle and homogenizer are rinsed in 20 to 30 ml of homogenization buffer which is then added to the cell homogenate. The homogenate and glass beads are stirred for 5 min, the supernatant is decanted into a large beaker, and an additional 40 ml of homogenization buffer is added to the glass beads and stirred for 5 min. The supernatant is again decanted and procedure is repeated once more. After decanting the second homogenate, the total volume was measured and enough 200 m*M* EDTA (pH 7.4 at 4°) is added to bring the final EDTA concentration to 5 m*M*. This is the modified homogenate.

The modified homogenate is centrifuged at 600 *g* for 10 min after which the supernatant is collected and the pellets are resuspended in fractionation buffer (300 m*M* mannitol, 5 m*M* Hepes, 5 m*M* EDTA, 3 m*M* DTT, pH 7.4 at 4°). The homogenate is centrifuged twice more, each time the supernatant is collected and the pellets resuspended in fractionation buffer. The combined supernatants are centrifuged at 12,000 *g* for 15 min. The resultant supernatant is aspirated, the pellet is resuspended in fractionation buffer and centrifuged at 600 *g* for 4 min. The supernatant is decanted into clean centrifuge tubes and centrifuged at 4000 *g* for 10 min. The supernatant is aspirated with the resultant pellet (the mitochondrial fraction) used as the enzyme source.

Properties

Stability. Rat P5C synthase is relatively stable for at least 2 weeks when kept at −70°. However, enzymatic activity is lost upon freeze/thawing of the preparation. *E. coli* γ-glutamate semialdehyde dehydrogenase is completely stable at −70° for at least 6 months.[23]

Specificity. P5C synthase catalyzes the reduction of L-glutamate to L-Δ^1-pyrroline-5-carboxylate.[14,28] Further metabolism of the γ-glutamate 5-semialdehyde by ornithine aminotransferase to ornithine is blocked by gabaculine.[31] The reduction of P5C to proline via P5C reductase is not possible since the enzyme is primarily located in the cytosol.[32,33] Oxidation of P5C to glutamate by P5C dehydrogenase does not occur under the

[31] M. J. Jung and N. Seiler, *J. Biol. Chem.* **253,** 7431 (1978).

[32] A. Meister, A. N. Radhakrishnan, and S. D. Buckley, *J. Biol. Chem.* **229,** 789 (1957).

[33] J. Peisach and H. J. Strecker, *J. Biol. Chem.* **237,** 2255 (1962).

assay conditions described here due to the dehydrogenase's greater affinity for NAD^+ over $NADP^+$.[34–36] γ-Glutamate semialdehyde dehydrogenase, like P5C dehydrogenase, exhibits greater activity in the presence of phosphate, but the former demonstrates a greater affinity for $NADP^+$.[23]

pH Optimum. Maximum rates of glutamate reduction to P5C are obtained at pH 7.0 in either bis(2-hydroxylethyl)imino-tris-(hydroxymethyl)methane–HCl or Hepes–KOH.[13] γ-Glutamate semialdehyde dehydrogenase exhibits a pH maximum at 7.0 in either phosphate or imidazole buffer.[23]

Inhibitors. Im mammalian tissue, 0.5 m*M* ornithine inhibits 50% of P5C synthase activity.[28,37] Inorganic phosphate, as low as 10 m*M*, is reported to inhibit as much as 50% of P5C synthase activity in rat intestinal mucosa.[28] Partially purified γ-glutamyl kinase from *Pseudomonas aeruginosa*[38] exhibits 50% inhibition with 5 m*M* proline,[3,38,39] 7–12% inhibition in the presence of 5 m*M* ADP and inorganic phosphate,[38] and 100% inhibition with 0.125 m*M* *p*-chloromercuribenzoate and *N*-ethylmaleimide.[22] Also, γ-glutamate semialdehyde dehydrogenase is inhibited by *p*-chloromercuribenzoate, copper, and nickel ions.[23]

Kinetic Properties. The apparent K_m values for P5C synthase activity are 2.5 m*M* for glutamate, 0.19 m*M* for ATP, and 6.5 μM for NADPH in the presence of 20 m*M* $MgCl_2$.[13] The K_m values for γ-glutamate semialdehyde dehydrogenase are 2.5 m*M* for DL-P5C, 0.05 m*M* for $NADP^+$, and 0.35 m*M* for phosphate.[24]

Distribution. As mentioned previously, P5C synthase activity has been demonstrated in many organisms. Recently,[40] a comprehensive review of P5C synthase activity in mammalian tissue has been reported.

[34] H. J. Strecker, *J. Biol. Chem.* **235,** 3218 (1960).

[35] G. Brunner and W. Neupert, *FEBS Lett.* **3,** 283 (1969).

[36] A. Herzfeld, V. A. Mezl, and W. E. Knox, *Biochem. J.* **166,** 95 (1977).

[37] R. F. Lodato, R. J. Smith, D. Valle, J. M. Phang, and T. T. Aoki, *Metab., Clin. Exp.* **30,** 908 (1981).

[38] R. V. Krishna and T. Leisinger, *Biochem. J.* **181,** 215 (1979).

[39] A. Baich and D. J. Pierson, *Biochim. Biophys. Acta* **104,** 379 (1965).

[40] M. E. Jones, *Trans. N.Y. Acad. Sci. Series II* **41,** 77 (1983).

[23] β-Methylaspartate-Glutamate Mutase from *Clostridium tetanomorphum*

By H. A. BARKER

$$\begin{array}{c} CO_2^- \\ | \\ CH_2 \\ | \\ CH_2 \\ | \\ H-C-NH_3^+ \\ | \\ CO_2^- \end{array} \quad \overset{\text{AdoCbl}}{\rightleftharpoons} \quad \begin{array}{c} CO_2^- \\ | \\ H_3C-C-H \\ | \\ H-C-NH_3^+ \\ | \\ CO_2^- \end{array}$$

L-Glutamate *threo*-3-Methyl-L-aspartate

The conversion of L-glutamate to threo-30-methyl-L-aspartate by glutamate mutase is the initial reaction in the anaerobic degradation of glutamate by *Clostridium tetanomorphum* and other Clostridium species.[1,2] Two readily separable proteins called Component E (MW 128,000) and Component S (MW 17,000), a cobamide coenzyme and a sulfhydryl compound, are required for activity. Component S is reversibly inactivated by oxygen and therefore must be reduced by a sulfhydryl compound for full activity. Stable complexes between the S and E components have not been detected and specific functions of the individual protein components are not known.

Assay Methods

Principle. Two assay methods are described: a spectrophotometric assay done in the presence of air and an anaerobic assay. The spectrophotometric assay,[2] which is generally used for estimating mutase activity, is based on the conversion of glutamate to mesaconate by the coupled action of mutase and β-methylaspartase (EC 4.3.1.2) and the measurement of the resulting increased absorbance at 250 nm.

$$\text{L-Glutamate} \xrightarrow{\text{mutase}} \text{threo-3-methyl-L-aspartate} \xrightarrow{\beta\text{-methylaspartase}} \text{mesaconate} + \text{ammonia}$$

This assay can be applied both to the estimation of mutase activity in relatively crude mutase preparations containing both E and S protein

[1] H. A. Barker, F. Suzuki, A. Iodice, and V. Rooze, *Ann. N.Y. Acad. Sci.* **112,** 644 (1964).
[2] H. A. Barker, V. Rooze, F. Suzuki, and A. Iodice, *J. Biol. Chem.* **239,** 3260 (1964).

Copyright © 1985 by Academic Press, Inc.
All rights of reproduction in any form reserved.

components (provided the background absorbance is not excessively high) and to the estimation of the individual components when the other is added in excess. However, since the calculated activities of the individual components vary somewhat with the quality and quantity of the second component, the assay results must be considered relative rather than absolute.

The anaerobic assay[2] is based upon the conversion of β-methylaspartate to glutamate and is applicable only to purified mutase preparations in which β-methylaspartase is either absent or fully inhibited by Ca^{2+}. The disappearance of β-methylaspartate is followed by stopping the reaction after a suitable period of incubation and estimating the remaining substrate by means of β-methylaspartase in a separate assay. This assay is useful for studying the equilibrium and certain other properties of the mutase reaction without complications resulting from the presence of β-methylaspartase.

Spectrophotometric Assay Reagents

Sodium L-glutamate 0.5 *M* pH 8.2
2-Mercaptoethanol, 0.1 *M*
$MgCl_2$, 0.2 *M*
KCl, 1.0 *M*
Tris–Cl, pH 8.2, 1.0 *M*
Potassium phosphate pH 8.0, 0.03 *M*
Benzimidazolyl-cobamide coenzyme[3] (preferably) 0.12 m*M* or adenosylcobalamin (AdoCbl) 3.0 m*M*
The coenzyme solutions must be stored in the dark and shielded from bright light during use
β-Methylaspartase,[4,5] 400 units/ml (100–200 units/mg) in 1% bovine serum albumin
Reduced Component S, 0.1 ml of S protein (8 mg/ml; 1 to 2 units/mg, purification step 4, below) is mixed with 0.1 ml of 0.1 *M* 2-mercaptoethanol and 0.3 ml 30 m*M* potassium phosphate buffer pH 8.2 in a 10 × 75 mm or smaller test tube and incubated for 45 at 0° or 15 min at 25°. Thereafter the activity remains almost constant for at least 2 hr at 0°C.
Glutamate mutase, Component E or Components S and E, 0.5 to 3 units/ml (see below)
Component E, 10 mg/ml, 1 to 2 units/mg (purification step 4, below).

[3] H. A. Barker, R. D. Smyth, H. Weissbach, J. I. Toohey, J. W. Ladd, and B. E. Volcani, *J. Biol. Chem.* **235,** 480 (1960).
[4] H. A. Barker, R. D. Smyth, R. M. Wilson, and H. Weissbach, *J. Biol. Chem.* **234,** 320 (1959).
[5] H. A. Barker, R. D. Smyth, H. J. Bright, and L. L. Ingraham, this series, Vol. 5, p. 827.

Procedure. For Component E assay[2] the solution in a 1.5-ml silica cuvette contains 50 μl Tris buffer, 20 μl sodium glutamate, 10 μl KCl, 5 μl $MgCl_2$, 5 μl benzimidazolyl cobamide coenzyme or 10 μl AdoCbl, 5 μl β-methylaspartase, 50 μl (0.1 unit) reduced Component S, 0.01 to 0.05 unit of Component E, and sufficient distilled water to give a final volume of 1.0 ml. The Component S is added next to last and the coenzyme is added last to start the reaction. The increase in $A_{240\ nm}$ is followed for 2–3 min at 25°. The maximal rate, usually reached within 45 sec after adding coenzyme, is used as a measure of Component E activity. The rate is generally proportional to the amount of enzyme in the range of 0 to 0.25 ΔA/min.

One unit of mutase activity causes a 3.85 change in absorbance units per minute at 240 nm, corresponding to 1 μmol of mesaconate per minute. Specific activity is defined as units per mg of protein determined by the method of Lowry *et al.* with bovine serum albumin as standard.

The same assay is used for Component S assay except that the added Component S is replaced by 10 μl (0.1 unit) of Component E.[6] The Component S is reduced by incubation with 2-mercaptoethanol as described above and is added last to the assay solution. The activity unit for Component S is the same as for Component E.

When neither protein component is added to the assay of crude mutase preparations, the observed activity is determined by the relative concentrations of the components. Extracts of *C. tetanomorphum* treated only with charcoal and protamine to remove cobamide coenzyme and nucleic acids appear to contain about equal levels of the two components.[7]

Anaerobic Assay[2]

Reagents. The following reagents are required in addition to those for the spectrophotometric assay.

Sodium *threo*-3-methyl-L-aspartate,[8,9] 0.2 *M*

$CaCl_2$, 0.05 *M*

Stock Tris, $CaCl_2$, 2-MET solution. Add 0.5 ml of 1.0 *M* Tris–Cl, pH 8.2, and 0.1 ml of 0.05 *M* $CaCl_2$ to 1.2 ml of water. Deaerate by bubbling argon or nitrogen through the solution for at least 15 min, then add 0.2 ml of 1 *M* 2-mercaptoethanol. Store in a stoppered tube under argon at 0°.

Oxygen-free water. Boil for 1 min, then cool under argon.

[6] R. L. Switzer and H. A. Barker, *J. Biol. Chem.* **242,** 2658 (1967).

[7] F. Suzuki and H. A. Barker, *J. Biol. Chem.* **241,** 878 (1966).

[8] H. A. Barker, R. D. Smyth, E. J. Wawszkiewicz, M. N. Lee, and R. M. Wilson, *Arch. Biochem. Biophys.* **78,** 468 (1958).

[9] H. A. Barker and R. D. Smyth, *Biochem. Prep.* **8,** 89 (1961).

Procedure. To a small test tube flushed with argon and kept at 0° add 25 μl of sodium *threo*-3-methyl-L-aspartate, 5 μl of benzimidazolyl cobamide coenzyme, 0.1 ml of the stock Tris–$CaCl_2$–MET solution, 2 to 20 μl (0.005 to 0.06 unit) of purified Component E, 50 μl (0.1 unit) of reduced Component S, and sufficient O_2-free water to give a final volume of 0.5 ml. The coenzyme is added last. The solution is mixed briefly and sucked into a cold 1-ml syringe (previously rinsed with argon), any gas bubbles are expelled and the tip of the syringe is sealed. The syringe is incubated in a water bath at 30°. After incubation for 10 and 20 min, 0.2-ml aliquots of the solution are removed from the syringe and mixed with 3.8 ml of cold 3 m*M* acetic acid to stop the reaction. After a few minutes the solutions are centrifuged, if necessary, and aliquots of the clear supernatant solution are used to estimate the remaining β-methylaspartate using β-methylaspartase.[8] The amount of β-methylaspartate decomposed is calculated by comparison with a control lacking enzyme.

The anaerobic assay gives a linear utilization of β-methylaspartate with both time and enzyme concentration until at least 80% of the substrate has been converted to glutamate.[2]

Purification Procedure for Component S[6]

Unless otherwise specified, all purification steps were performed at 0° to −4°.

Step 1. Sonic Extraction. Frozen *C. tetanomorphum* cell paste (197 g) is thawed in 800 ml of 20 m*M* potassium phosphate pH 7.4 and dispersed with a Teflon-glass homogenizer. The suspension is sonically treated in 75 ml batches with 2 g of grade FFF Corundum powder (Braun-Knecht-Heimann) per batch for 10 min in a Raytheon 10-KC sonic oscillator at full power (1.2 A). The resulting suspension is centrifuged for 30 min at 10,000 *g*. The precipitate is discarded and the supernatant liquid, including a slimy gray layer, is saved. The sonic extract can be stored overnight at 0° but better yields of Component S are obtained by continuing with the following steps through the beginning of Sephadex chromatography in rapid succession on the same day.

Step 2. Isoelectric Precipitation. The sonic extract (1080 ml) is diluted to 10 mg of proteins/ml with cold distilled water. The pH of the solution is then lowered from 7.2 to 4.60 by the addition of about 11 ml of 5 *N* acetic acid with continuous stirring during 5 min. A voluminous precipitate, removed by centrifugation at 10,000 *g* for 20 min, is discarded. The pH of the clear yellow supernatant solution is immediately raised to 7.6 by the addition of about 3 ml of conc. ammonia, with stirring. The total time of exposure of the extract to a pH below 5.0 should not exceed 60 min.

Step 3. DEAE-Cellulose Column Fractionation. To reduce the salt concentration, the *Step 2* solution (2000 ml; conductivity 1800 ppm as NaCl) is put in several 1-in.-diameter dialysis tubes and dialyzed against 30 liters of cold distilled water for 2 hr. The conductivity of the pooled dialysate should then correspond to 1000 ppm or less. The solution is made 1 m*M* in sodium EDTA pH 7.0 and 10 m*M* in 2-mercaptoethanol. The solution is then immediately allowed to flow into a DEAE-cellulose column, 4.8 cm (diam) × 12 cm, previously equilibrated with 10 m*M* potassium phosphate pH 7.5, containing 10 m*M* mercaptoethanol. The yellow pass-through solution is discarded. The column is then successively eluted with the following potassium phosphate buffers, pH 7.5, all containing 10 m*M* mercaptoethanol and 1 m*M* EDTA: 200 ml at 10 m*M*, 400 ml at 50 m*M*, and 300 ml at 0.1 *M*. Fractions containing 20 ml each are collected at 2 min intervals. Component S activity is eluted in the leading edge of a large trailing protein peak which comes off the column during elution with 50 m*M* buffer. Fractions 19 through 25 (126 ml), which are pink in color and often turbid, contain most of the activity and are pooled. The active fractions are largely resolved from a bright yellow zone (fractions 26–35) as well as from material eluting with 0.1 *M* buffer.

Step 4. Ammonium Sulfate Precipitation. The Step 3 preparation (126 ml) is immediately concentrated by addition of 70.6 g powdered ammonium sulfate with continuous stirring. After 20 min of stirring, the precipitate is collected by centrifugation at 12,000 *g* for 10 min, and is redissolved in 15 ml of 0.1 *M* potassium phosphate, pH 6.8, containing 10 m*M* mercaptoethanol and 1 m*M* EDTA, to give a final volume of 20 to 25 ml.

Step 5. Sephadex G-100. The Step 4 concentrate is immediately allowed to flow into a 2.4 cm (diam) × 98 cm column of Sephadex G-100 equilibrated with 0.1 *M* potassium phosphate pH 6.8 containing 10 m*M* mercaptoethanol and 1 m*M* EDTA. The protein is eluted with the same buffer and 5 ml fractions are collected at 18 min intervals. The major protein peaks are β-methylaspartase and Component S. Fractions 72 through 89, containing Component S, are pooled to give 72 ml of an almost colorless solution. Fractions 48 through 60 provide highly purified β-methylaspartase 208 units per mg, free of significant contamination by Component E or Component S.

Step 6. CM-Cellulose Column Fractionation. The Step 5 Component S pool is dialyzed against 2 liters of 5 m*M* sodium acetate pH 4.8, containing 10 m*M* mercaptoethanol and 1 m*M* EDTA, for at least 8 hr. After dialysis the solution is adjusted to pH 4.5 with 3 *N* acetic acid. The final conductivity must be less than that of 2000 ppm NaCl. The solution is then allowed to flow into a 2 cm (diam) × 12 cm column of CM-cellulose equilibrated with 5 m*M* sodium acetate pH 4.8 containing 10 m*M* mercaptoethanol and 1 m*M* EDTA. The column is eluted successively with the following buffers, all containing mercaptoethanol and EDTA: 100 ml of 5 m*M* sodium acetate, pH 4.8; 100 ml of 50 m*M* sodium acetate, pH 5.5; and 50 ml of 0.1 *M* potassium phosphate, pH 6.8. Fractions of 5 ml are collected at 2 to 3 min intervals. Component S activity elutes in a sharp peak with 50 m*M* buffer. The peak fractions are pooled (25 ml) and the pH adjusted to 6.8 with 1 *N* NaOH.

Step 7. Ammonium Sulfate Precipitation. The Step 6 pool is concentrated by adding 12.6 g powdered ammonium sulfate (to 90% saturation) with mechanical stirring (60 min). The colorless precipitate is collected by centrifugation at 12,000 *g* and redissolved in 3 ml of 0.1 *M* potassium phosphate pH 6.8 containing 1 m*M* EDTA but no mercaptoethanol. The solution is dialyzed against 1 liter of the same buffer for 18 hr and centrifuged to remove a small precipitate. The preparation is stored frozen at −10° or lower.

Typical results of Component S purification are shown in Table I. The enzyme is purified 350-fold to about 32 units per mg in 21% yield.

Properties of Component S[6]

Homogeneity and Molecular Weight. Polyacrylamide gel electrophoresis of Component S (36.7 units per mg) after exposure to 50 m*M* mercaptoethanol shows that it is at least 90% pure. Samples exposed to lower concentrations of thiol in the presence of air contain 20% or more of a partially oxidized dimer. The dimer is inactive in the absence of reducing

TABLE I
PURIFICATION OF COMPONENT S OF GLUTAMATE MUTASE[a]

Step	Volume (ml)	Protein (mg/ml)	Component S activity		
			Units/ml	Units/mg	Total units
1. Sonic extraction	1080	26	2.4	0.092	2590
2. Isoelectric precipitation	2000	0.61	0.57	0.94	1140
3. DEAE-cellulose column	126	3.5	5.84	1.67	736
4. Ammonium sulfate precipitation	24.5	19	30.4	1.60	745
5. Sephadex G-100 column	72	0.67	11.2	16.7	806
6. CM-cellulose column	19.5	1.1	30.4	28.6	593
7. Ammonium sulfate precipitation	3.8	4.4	142	32.2	540

[a] Switzer and Barker.[6]

agent, whereas the monomer is active in this condition. The monomer is reversibly oxidized to the dimer by oxygen; a slow irreversible oxidation also occurs at 0° or higher. The molecular weight of the monomer is 17,000 ± 1000 as determined by sedimentation equilibrium, sedimentation velocity, and gel filtration techniques. The monomer and dimer can be separated by gel filtration. The better preparations of Component S (28 to 31 units per mg) contain no detectable levels of β-methylaspartase, Component E, cobamide coenzyme, or vitamin B_6.

Stability. Component S is most stable at −10° or lower in presence of 0.1 *M* potassium phosphate pH 6.8 and in absence of thiols. Under these conditions, little or no activity is lost during several months. At 0°, 21 and 39% of the activity was lost during storage for 3 and 5.5 months, respectively.

Absorption Spectrum. The absorbance maximum is at 280 nm and the minimum at 254 nm. The A_{280}/A_{260} ratio is 1.44 and the absorbance of a 0.1% solution at 280 nm is 0.644. There is no absorbance in the visible region.

Amino Acid Composition and Sulfhydryl Groups. The amino acid composition has been determined.[6] The protein contains 1 mol of tryptophan and probably 5 mol of half-cysteine per mole. Fully reduced, native Component S appears to contain 5 sulfhydryl groups, one of which is deeply buried in the protein. Of the remaining 4 sulfhydryl groups, 2 are always available for reaction with sulfhydryl reagents but are not involved in disulfide bond formation. The remaining 2 sulfhydryl groups are reactive and readily form a disulfide bond with each other, with other thiols and probably with other protein molecules.

Activators and Inhibitors. Component S is activated by sulfhydryl compounds. 2-Mercaptoethanol, cysteine, and 1,2-dimercaptopropanol at 20 m*M* are about equally effective as activating agents, whereas reduced glutathione and sodium thioglycolate are less effective. Full activation by 20 m*M* mercaptoethanol requires about 15 min at 25° and 45 min at 0°; thereafter the activity remains almost constant for at least 2 hr. Lower concentrations of mercaptoethanol give only partial activation under these conditions. When fully activated Component S is assayed in a solution containing 1 m*M* mercaptoethanol and exposed to air, the activity remains constant for at least several minutes.

Component S is inhibited by reagents such as oxygen, mercuribenzoate, iodoacetate, and arsenite that oxidize or form stable compounds with sulfhydryl groups. Oxygen acts rather slowly, whereas iodoacetate and arsenite inhibit rapidly. Fully reduced Component S is completely inactivated by iodoacetate whereas partly oxidized protein retains some activity after exposure to iodoacetate and subsequent treatment with mercaptoethanol. Complete inactivation apparently results from carboxymethylation of a single sulfhydryl group.

Purification Procedure for Component E[7]

Step 1. Sonic Extraction. This step is the same as for Component S except that 130 g of frozen cells of *C. tetanomorphum* are used in the procedure described.

Step 2. Charcoal and Protamine Treatment. To 284 ml of extract from Step 1 (60 mg of protein per ml) is added 45 g of moist acid-washed charcoal (Nuchar C), equivalent to 2.7 g of charcoal per g of protein. The mixture is stirred gently for 10 min and then centrifuged for 30 min at 13,000 *g*. The precipitate is discarded. The supernatant solution (254 ml) is diluted with water to a protein concentration of 10 mg/ml. To the diluted solution is added 190 ml of 1% protamine sulfate (salmine, Lilly) dissolved in 40 m*M* potassium phosphate buffer, pH 7.0, equivalent to 0.15 g of protamine sulfate per g of protein. The protamine solution is added slowly while the enzyme solution is stirred mechanically. After stirring for an additional 10 min the solution is centrifuged for 20 min at 13,000 *g* and the precipitate discarded.

Step 3. Ammonium Sulfate Fractionation. To 1100 ml of protamine-treated extract (5 mg protein per ml) are added 55 ml of 1 *M* potassium phosphate buffer pH 7.0 and 1.1 ml of disodium EDTA. Then 320 g of powdered ammonium sulfate is added slowly with mechanical stirring to give a 0.45 saturated solution. After stirring for an additional 20 min, the precipitate is separated by centrifugation and discarded. Ammonium sulfate (223 g) is now added to the supernatant solution to bring the concentration to 0.70 saturation. The resulting precipitate containing the enzyme is separated by centrifugation and dissolved in 250 ml of 10 m*M* potassium phosphate pH 7.0 to give a protein concentration of 13 mg/ml. This solution is dialyzed in 1-in.-diameter Visking tubing for 6 hr against 6 liters of 2 m*M* potassium phosphate pH 6.5 with stirring. The dialyzed solution has a volume of 287 ml and a protein concentration of 9 mg/ml.

Step 4. Fractionation on DEAE-Cellulose. Enzyme solution from Step 3 (280 ml; 2.6 g protein) is dialyzed for 3 hr against 6 liters of 10 m*M* potassium phosphate pH 8.0 containing 10 m*M* mercaptoethanol. The dialyzed solution is then applied to column of DEAE-cellulose [4.6 cm (diam) × 26 cm] equilibrated with the same buffer. The column is eluted successively with the following potassium phosphate buffers, all containing 10 m*M* mercaptoethanol: 600 ml of 10 m*M* pH 8.0; 1 liter of 10 m*M* pH 6.65; 2 liters of 50 m*M*, pH 6.7; 1 liter of 0.1 *M*, pH 6.8; 1 liter of 0.2 *M*, pH 6.8; and 700 ml of 0.5 *M*, pH 6.8. Fractions of 80 ml are collected at a flow rate of 5 to 10 ml/min and stored at 4° before being assayed. Component E is eluted with 50 m*M* buffer and β-methylaspartase is eluted with 0.1 *M* buffer. Fractions containing Component E of higher specific activity are pooled and the enzyme is precipitated by 0.80 saturated ammonium sulfate in presence of 1 m*M* EDTA. The precipitate is recovered by centrifugation at 14,000 *g* for 20 min and dissolved in 10 ml of 10 m*M* phosphate buffer pH 6.8. This solution is desalted by passage through a column of Sephadex G-100 [2.4 cm (diam) × 20 cm]; both equilibration of the column and elution are done with the 10 m*M* phosphate buffer. The protein peak fractions ($A_{280\ nm}$) are pooled.

Step 5. Fractionation on Carboxymethyl Cellulose. Enzyme solution from Step 4 (20 ml; 162 mg of protein) is dialyzed for 3 hr against 6 liters of 5 m*M* sodium acetate, pH 5.2, and then immediately applied to a column of carboxymethyl cellulose [2.4 cm (diam) × 32.5 cm] equilibrated with 5 m*M* sodium acetate, pH 5.2 containing 5 m*M* mercaptoethanol. The column is eluted successively with the following phosphate buffers, pH 6.5, all containing 5 m*M* mercaptoethanol: 750 ml of 15 m*M* buffer; 150 ml of 50 m*M* buffer; and 200 ml of 0.2 *M* buffer. Fractions of 8 ml are collected at a flow rate of 1.0 ml/min. A little β-methylaspartase elutes with 5 m*M* buffer (fractions 16–19) and Component E elutes with 15 m*M* buffer (fractions 80–86). The peak Component E fractions are pooled.

Step 6. Fractionation on Brushite. The relatively dilute enzyme solution from Step 5 (60 ml; 0.35 mg of protein/ml) is applied directly to a column to brushite[10] [1.2 cm (diam) × 20 cm] equilibrated with 10 m*M* potassium phosphate, pH 6.1. After washing the column with 30 ml of the same buffer, it is eluted successively with the following phosphate buffers, pH 7.0: 40 ml of 70 m*M* and 40 ml of 0.2 *M* buffer. The first 100 ml of eluate is collected in four 25 ml fractions and then 1 ml fractions are collected at a flow rate of 1 ml/min. A small amount of β-methylaspartase elutes as soon as the 70 m*M* buffer is applied (fractions 13–28) and Component E elutes with the 0.2 *M* buffer (fractions 53–70). Component E is further concentrated by precipitating it with 0.80 saturated ammonium sulfate in the presence of 1 m*M* EDTA. After centrifugation, the precipitate is dissolved in 0.8 ml of 10 m*M* phosphate buffer. This solution is desalted by passage through a small column of Sephadex G-50 [1.0 cm (diam) × 10 cm] equilibrated with the same buffer. The protein peak fractions are combined to give 3 ml of a solution containing about 5 mg of protein per ml.

Table II summarizes the results of a typical purification and shows that Component E is purified 180-fold to a specific activity of 3.6 units/mg in

[10] W. T. Jenkins, *Biochem. Prep.* **9**, 83 (1962).

TABLE II
PURIFICATION OF COMPONENT E[a]

Step	Volume (ml)	Protein (mg)	Component E activity	
			Units	Units/mg
1. Sonic extraction	285	17,100	340	~0.02
2. Charcoal and protamine treatment	1,100	5,500	330	0.06
3. Ammonium sulfate precipitation	290	2,610	313	0.12
4. DEAE-cellulose column	22	170	187	1.1
5. Carboxylmethyl-cellulose column	60	21	67	3.2
6. Brushite column	15	16.5	60	3.6

[a] Suzuki and Barker.[7]

18% yield. Removal of most of the β-methylaspartase and Component S activities is achieved in the DEAE-cellulose and carboxymethyl-cellulose column steps. The final brushite column step is used mainly to concentrate Component E, although it also removes the last traces of β-methylaspartase activity. Attempts to crystallize purified Component E from ammonium sulfate solutions have been unsuccessful.

Properties of Component E[7]

Homogeneity and Molecular Weight. Polyacrylamide gel electrophoresis shows that the best Component E preparations contain several protein components. Sucrose density gradient centrifugation shows one major protein peak ($A_{280\ nm}$) and several minor peaks. Component E activity is associated only with the major peak. A sedimentation velocity experiment showed the presence of one major and two minor components. The major component gave a $s_{20,w}$ value of 6.90 S and an estimated molecular weight of 128,000. The best Component E preparations are free of Component S, β-methylaspartase, and glutamate : α-ketoglutarate transaminase activities.

Stability. Component E dissolved in 10 m*M* potassium phosphate buffer, pH 7.0 is stable for months when stored at −20° in absence of mercaptoethanol. In the presence of 10 m*M* mercaptoethanol 30–90% of the activity is lost during 18 hr at −20°. Repeated thawing and refreezing causes progressive small inactivation. At 0°, the enzyme lost 8% of its activity during storage for 21 days in either 10 m*M* phosphate, pH 6.7 or 0.3 *M* phosphate, pH 7.0. At 30° the enzyme retained full activity for at least 2 hr in 50 m*M* phosphate, pH 7.0. The enzyme is much less stable in presence of Tris–Cl and other cationic buffers; the rate of inactivation in Tris–Cl buffer increases markedly both with buffer concentration and pH. Inactivation by 50 m*M* Tris–Cl, pH 8.0 is largely prevented by 50 m*M* glutamate, β-methylaspartase, or aspartate.

Absorption Spectrum. Component E has a typical protein absorption spectrum with an absorbance maximum at 280 nm and a minimum at 251 nm. The $A_{280} : A_{260}$ ratio is 1.63. Most preparations have a faint pink color and show a small and progressively decreasing absorption between 300 and 550 nm. The color is probably caused by small amounts of corrinoid vitamins bound to the protein. Significant amounts of vitamin B_6 are not present.

Activators and Inhibitors. Unlike Component S, Component E does not require treatment with mercaptoethanol to show maximal activity when it is combined with reduced Component S in the assay. However, when combined with partially oxidized Component S, a higher, though submaximal mutase activity is obtained when Component E is previously exposed to mercaptoethanol. Unlike Component S, Component E is not inhibited by treatment with iodoacetate.

Interactions among Components of the Mutase System[6,7]

Effect of Relative Amounts of Components S and E on Reaction Rate. Neither protein component, when highly purified, shows activity in the absence of the other under conditions of the spectrophotometric assay. When Component E is kept constant, and Component S is progressively increased, a typical saturation curve is obtained; mutase activity rises rapidly at first and later approaches a maximum value which is proportional to the amount of Component E. However, with less pure Component S preparations, the addition of an excess may cause some inhibition. When Component S is kept constant and Component E is increased progressively, a saturation curve is also obtained. Half-maximal mutase activity is obtained under conditions of the spectrophotometric assay when the apparent molar ratio of Component E to Component S is 1.6 or lower.

Effect of Coenzyme Concentration on Mutase Activity. Benzimidazolyl cobamide coenzyme shows typical Michaelis equation kinetics in the mutase reaction. However, the apparent K_m values vary greatly with the Component S/E unit ratio. A value of 1.29 μM is observed at an S/E ratio of 0.35 units per unit. The apparent K_m decreases rapidly to 0.12 μM at a unit ratio of 3.6 and then declines more slowly to reach a value of 0.031 μM at a ratio of 17.8. This indicates that a complex of the two components binds coenzyme more strongly than does either one separately. Conversely at low S/E ratios, increasing the coenzyme concentration greatly increases mutase activity, presumably by increasing the association of the protein components.

Binding of Coenzyme to Components S and E. Determination of the binding of AdoCbl to the two protein components by a gel filtration technique has shown that Component S alone binds little coenzyme, Component E alone binds roughly 1 mol of coenzyme per mole of protein, and Component E in the presence of a 7-fold excess of Component S binds approximately 2 mol per mole of Component E.

Complex Formation between Components S and E. Attempts to demonstrate stable complex formation between the two protein components by a gel filtration technique either in the presence or absence of substrate and benzimideazolylcobamide coenzyme have been unsuccessful. Evidently the association and dissociation of the components is rapid, and their interaction is too weak to permit detection of a complex by this technique.

Coenzyme Specificity.[11] AdoCbl and at least seven analogs containing bases other than dimethylbenzimidazolyl can serve as coenzymes for glutamate mutase. The bases present in these compounds and the corresponding apparent K_m values ($M \times 10^7$) follow: benzimidazolyl, 2.3; 5(6)-trifluoromethylbenzimidazolyl, 4.0; 5(6)-methylbenzimidazolyl, 4.3; 5(6)-aminobenzimidazolyl, 4.6; 5(6)-nitrobenzimidazolyl, 5.3; adenine, 12.4; 2,6-diaminopurine, 16.4; and 5,6-dimethylbenzimidazolyl, 128. The V_{max} values for these coenzymes are similar, differing by 50% at most. The benzimidazolyl analog is the preferred coenzyme because of its low K_m value. A number of corrinoid compounds are inactive as coenzymes in the mutase reaction. These include the corrinoid vitamins (lacking a deoxadenosyl group), methyl, carboxymethyl and 5-deoxyuridyl cobalamins, and cobinamide coenzyme which contains the deoxyadenoxyl group but lacks the nucleotide moiety.

Substrate Specificity.[2] The only known substrates for the mutase are L-glutamate (K_m 1–2 mM) and *threo*-3-methyl-L-aspartate (K_m 0.5 mM). D-Glutamate, *threo*-3-ethyl-DL-aspartate, α-, β-, or γ-methyl-DL-glutamate, L-glutamine, the γ-methylester of L-glutamate, and DL-α-aminoadipate are not utilized at a significant rate.

Effect of pH and Buffers on Activity.[2] The pH optimum for activity is 8.5. pH 8.2 is usually used for the assay to reduce the background absorbance caused by mercaptoethanol. The enzyme is active between about pH 6.0 and 9.8. Although Component E is unstable during storage in Tris and other cationic buffers, no significant inactivation is caused by Tris under the assay conditions.

[11] J. I. Toohey, D. Perlman, and H. A. Barker, *J. Biol. Chem.* **236,** 2119 (1961).

Effect of Temperature on Activity.[2] The optimal temperature is at 38° and a half-maximal rate is obtained at about 27°.

Equilibrium of the Mutase Reaction.[2] The equilibrium favors the conversion of β-methylaspartase to glutamate. At 30° and pH 8.2, K_{eq} (L-glutamate)/(*threo*-3-methyl-L-aspartate) is about 10.7 corresponding to a $\Delta F°$ of −1.43 kcal.

[24] γ-Carboxyglutamic Acid

By JANE B. LIAN, CAREN M. GUNDBERG, PETER V. HAUSCHKA, and PAUL M. GALLOP

Introduction

In 1974, several laboratories[1–4] demonstrated the existence in prothrombin of a previously unknown amino acid, 4-carboxyglutamic acid (Gla). The biosynthesis of this compound is vitamin K dependent and involves the posttranslational carboxylation of specific glutamic acid residues in a precursor protein, such as acarboxyprothrombin.

To provide for the biosynthesis, microsomal enzymes and a vitamin K cycle exist which promote appropriate glutamate residue carboxylation with carbon dioxide (Fig. 1). Vitamin K hydroquinone is generated prior to the oxygen-dependent carboxylation reaction and undergoes two stages of oxidation, ultimately forming vitamin K epoxide. The hydroquinone is regenerated by a vitamin K epoxide reductase, and a vitamin K reductase. It has been documented that during the carboxylation, the vitamin is converted to the stable 2,3 epoxide and evidence supports the hypothesis that carboxylation and epoxidation are mediated at the same active site of the enzyme.[5] The carboxylase enzyme requires CO_2, O_2, and the hydroquinone form of vitamin K.[4–6] In this mixed function reac-

[1] J. Stenflo, P. Fernlund, W. Egan, and P. P. Roepstorff, *Proc. Natl. Acad. Sci. U.S.A.* **71,** 2730 (1974).

[2] G. L. Nelsestuen, T. H. Zytkovicz, and J. B. Howard, *J. Biol. Chem.* **249,** 6347 (1974).

[3] S. Magnusson, L. Sottrup-Jensen, T. E. Peterson, H. R. Morris, and A. Dell, *FEBS Lett.* **44,** 189 (1974).

[4] C. T. Esmon, J. A. Sadowski, and J. W. Suttie, *J. Biol. Chem.* **250,** 4744 (1975).

[5] J. W. Suttie, *CRC Crit. Rev. Biochem.* **8,** 191 (1980).

[6] J. P. Jones, E. J. Gardner, T. G. Cooper, and R. E. Olson, *J. Biol. Chem.* **252,** 7738 (1977).

Copyright © 1985 by Academic Press, Inc.
All rights of reproduction in any form reserved.

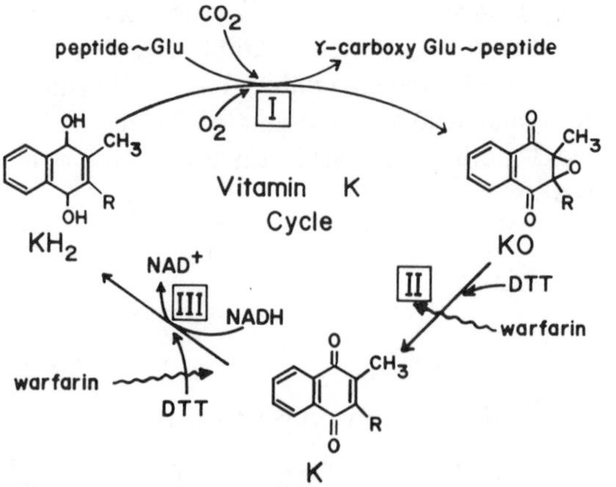

FIG. 1. Synthesis of γ-carboxyglutamic acid residues and the vitamin K cycles. A microsomal vitamin K-dependent carboxylase (or epoxidase, I), which requires the reduced vitamin K hydroquinone (KH_2), generates γ-carboxyglutamic acid (Gla) from peptide-bound glutamic acid, converting KH_2 to the 2,3 epoxide (KO). A vitamin K microsomal epoxide reductase, II, regenerates the quinone (K). Several quinone reductases, III, occur. These include a thiol requiring enzyme, as II, inhibited by warfarin and both microsomal and cytosolic NAD(P)H reductases. The reduction of K to the hydroquinone KH_2 is dependent in part on sulfhydryl groups that can be supplied *in vitro* by dithiothreitol (DTT) in a reaction inhibited by warfarin.

tion in which both a reducing compound and oxygen are required, an energetic intermediate, which might promote carbanion or radical carboxylation, could involve an oxygenated form of the vitamin, perhaps a 2- or 3-hydroperoxide, which is subsequently converted to the epoxide.[7] At physiologic levels of vitamin K, recycling of the epoxide to the hydroquinone is necessary for continued carboxylation. The enzymes, vitamin K epoxide reductase[8] and vitamin K quinone reductase, are involved in regeneration of the active hydroquinone. Multiple quinone reductases which utilize either reduced pyridine nucleotides[9–11] or thiols[12,13] have been described. Although not proven, there is evidence suggesting that the thiol-requiring enzyme may be of greater physiologic relevance with respect to carboxylation.[9,13]

[7] A. E. Larson and J. W. Suttie, *Proc. Natl. Acad. Sci. U.S.A.* **75,** 5413 (1978).

[8] A. K. Willingham and J. T. Matschiner, *Biochem. J.* **140,** 435 (1974).

[9] D. S. Whitlon, J. A. Sadowski, and J. W. Suttie, *Biochemistry* **17,** 1371 (1978).

[10] J. A. Sadowski, C. T. Esmon, and J. W. Suttie, *J. Biol. Chem.* **251,** 2770 (1976).

[11] R. Wallin, O. Gedhardt, and H. Prydz, *Biochem. J.* **169,** 95 (1978).

[12] P. A. Friedman and M. A. Shia, *Biochem. Biophys. Res. Commun.* **70,** 647 (1976).

[13] M. J. Fasco and L. M. Principe, *Biochem. Biophys. Res. Commun.* **97,** 1487 (1980).

Gla synthesis is inhibited directly by certain vitamin K antagonists, e.g., 2-chloro-3-phytyl-4-naphthoquinone,[14] which bind to the carboxylase enzyme. It is also inhibited indirectly by the 4-hydroxy-coumarin anticoagulants,[14] such as sodium warfarin, which appear to inhibit the epoxide reductase,[15] promoting accumulation of the inactive epoxide form of vitamin K (Fig. 1). Although the precise mechanism of inhibition is not known, recent experiments suggest that the vitamin K epoxide reductase is the physiologic important site of action[16] and that the 4-hydroxycoumarins can interact with the epoxide reductase only when the enzyme is in its inactive, oxidized, disulfide form.[17] These anticoagulants are also potent inhibitors of the thiol-requiring vitamin K quinone reductase, whereas the NADH-dependent quinone reductase is insensitive to warfarin (Fig. 1).

Formation of Gla in the vitamin K-dependent blood coagulation factors (prothrombin, factors VII, IX, and X) enables these proteins to specifically bond Ca^{2+} ions and phospholipids. This then allows conformational alteration leading to accelerated proteolytic activation and normal blood coagulation. The coumarin drugs promote synthesis of nonfunctioning Gla-deficient proteins; clinically, this results in an increased prothrombin time. Several recent reviews concerned with vitamin K are available.[5,18–21] Vitamin K-dependent carboxylase activity and Gla synthesis have been demonstrated to occur in microsomal cell fractions from a wide variety of tissues and isolated cells including those from bone,[22,23] kidney,[24] spleen,[25] lung,[26] placenta,[27] and certain tumors.[28] Gla-containing

[14] R. A. O'Reilly, *Annu. Rev. Med.* **27,** 245 (1976).

[15] J. A. Sadowski and J. W. Suttie, *Biochemistry* **13,** 3696 (1974).

[16] E. F. Hildebrandt and J. W. Suttie, *Biochemistry* **21,** 2406 (1982).

[17] M. J. Fasco, L. M. Principe, W. A. Walsh, and P. A. Friedman, *Biochemistry* **22,** 5655 (1983).

[18] J. W. Suttie, *in* "The Fat-Soluble Vitamins, Handbook of Lipid Research" (H. F. DeLuca, ed.). Vol. 2, Plenum, New York, 1978.

[19] J. Stenflo and J. W. Suttie, *Annu. Rev. Biochem.* **46,** 157 (1977).

[20] J. W. Suttie, *in* "Vitamin K Metabolism and Vitamin K-Dependent Proteins" (J. W. Suttie, ed.). University Park Press, Baltimore, Maryland, 1979.

[21] B. C. Johnson, *Mol. Cell. Biochem.* **38,** 77 (1981).

[22] P. V. Hauschka, J. B. Lian, and P. M. Gallop, *Proc. Natl. Acad. Sci. U.S.A.* **72,** 3925 (1975).

[23] P. A. Price, J. W. Poser, and N. Raman, *Proc. Natl. Acad. Sci. U.S.A.* **10,** 3374 (1976).

[24] P. A. Friedman, W. E. Mitch, and P. Silva, *J. Biol. Chem.* **257,** 1037 (1982).

[25] S. D. Buchthal and R. G. Bell, *Biochemistry* **22,** 1077 (1983).

[26] R. G. Bell, *Arch. Biochem. Biophys.* **203,** 58 (1980).

[27] P. A. Friedman, P. V. Hauschka, M. A. Shia, and J. K. Wallace, *Biochim. Biophys. Acta* **583,** 261 (1979).

[28] S. D. Buchthal, C. G. McAllister, D. C. Loux, and R. G. Bell, *Biochem. Biophys. Res. Commun.* **109,** 55 (1982).

proteins have been identified in mineralized tissues, bone, and tooth dentin,[29] and many types of ectopic calcifications.[30–32] Of these tissues, only the bone protein, osteocalcin, has been purified, sequenced, and structurally characterized.[23,33–36] It accounts for about 20% of noncollagenous bone protein and is, by far, the single largest reservoir of Gla in the body.

The demonstration that 4-carboxyglutamate occurs in proteins prompted development of several techniques for the direct identification of Gla and detection of the Gla-producing enzyme system. Current methods include (1) vitamin K-dependent CO_2 incorporation into endogenous precursor protein or added peptide substrate to form carboxyl-labeled and acid labile Gla-residues, (2) direct amino acid analysis of Gla in base-hydrolyzed protein samples, and (3) conversion of Gla in an intact protein to glutamate residues, which are tritiated at the γ-methylene position.

Biosynthesis of 4-Carboxyglutamic Acid Residues

In vitro synthesis of Gla residues can be carried out using microsomes which have accumulated endogenous precursor protein, as a result of treatment of the animal with either a vitamin K-deficient diet or warfarin. When detergent-solubilized microsomes are utilized, synthesis of Gla can also be demonstrated with synthetic peptide substrates, homologous to short sequences in prothrombin, one of which is Phe-Leu-Glu-Glu-Leu (FLEEL), the sequence of residues 5–9 in rat prothrombin.[37] These substrates can provide quantitation of vitamin K carboxylase activity in a tissue. The hepatic enzyme system has been studied most extensively for characterization of the microsomal vitamin K-dependent carboxylase enzyme complex and *in vitro* prothrombin formation. The general conditions described here for Gla synthesis in both endogenous and exogenous substrates derive largely from methodologies used in rat liver microsomal preparations,[5,21,38,39] but the same requirements have allowed for detection of Gla synthesis in other tissues.[24–28,40]

[29] A. Linde, M. Brown, W. C. Cothran, A. Hoglund, and W. T. Butler, *Biochim. Biophys. Acta* **704,** 235 (1980).

[30] J. B. Lian, M. Skinner, M. J. Glimcher, and P. M. Gallop, *Biochem. Biophys. Res. Commun.* **71,** 349 (1976).

[31] J. B. Lian, E. L. Prien, M. J. Glimcher, and P. M. Gallop, *J. Clin. Invest.* **59,** 1151 (1977).

[32] R. J. Levy, J. A. Zenker, and J. B. Lian, *J. Clin. Invest.* **65,** 563 (1982).

[33] A. Linde, M. Brown, and W. T. Butler, *J. Biol. Chem.* **255,** 5931 (1980).

[34] S. A. Carr, P. V. Hauschka, and K. Biemann, *J. Biol. Chem.* **255,** 5931 (1980).

[35] P. V. Hauschka, S. A. Carr, and K. Biemann, *Biochemistry* **21,** 63 (1982).

[36] J. W. Poser, F. S. Esch, N. C. Lenz, and P. A. Price, *J. Biol. Chem.* **255,** 8685 (1980).

[37] S. Magnussen, L. Sottrup-Jensen, T. E. Petersen, H. R. Morris, and A. Dell, *FEBS Lett.* **44,** 189 (1974).

Materials

Vitamin K_1 can be obtained as an oil preparation (Sigma Chemical Co.). The active vitamin K hydroquinone is freshly prepared from the K_1 oil by a modified dithionite reduction method.[10] To prepare the hydroquinone, 100 mg of oil is weighed in a disposable micropipette and then dissolved in 20 ml of ether. To this is added 60 ml of 8% sodium hydrosulfite in a separatory funnel which is then shaken and vented. The sodium hydrosulfite water layer (lower) is removed and the process repeated several times until the organic phase is clear. The ether layer is transferred to an Erlenmeyer flask containing $MgSO_4$ and flushed with nitrogen. The solution is removed to a preweighed glass culture tube and evaporate under nitrogen. The tube is reweighed and reconstituted to 20 mg/ml 100% ethanol, flushing the tube with nitrogen before capping. The reduced form of the vitamin will keep stored under nitrogen in the dark at −20°. Vitamin K concentration can be determined by UV absorbance at 248 nm in hexane with an absorptivity of 0.042 mg^{-1} ml^{-1}.[16]

Sodium [^{14}C]bicarbonate of a specific activity of 50–60 mCi/mmol (New England Nuclear or Amersham, Searle, Arlington Heights, IL) is generally employed for carboxylation assay from tissue microsomes. Recent studies in tumors and cell lines have employed sodium carbonate, Na_2CO_3, of specific activity 59 mCi/mmol.[28]

Many synthetic peptides have been studied[5] and those homologous to residues 5–9 of human and rat prothrombin, C-Phe-Leu-Glu-Glu-Val, or Phe-Leu-Glu-Glu-Leu, respectively, are available from Vega-Fox Chemical Co., Tucson, AZ. The synthetic peptides are checked for purity and may require P-2 gel filtration.

Preparation of Microsomes

Animals. To measure Gla synthesis in endogenous accumulated substrate *in vitro,* it is necessary to prepare microsomes from vitamin K-deficient or sodium warfarin-treated animals.[39] Using a commercial purified soy protein base-vitamin K-deficient diet (ICN Biochemicals, OHIO or Teklab Test Diet, Madison, WI), a 150–250 g rat can be made vitamin K deficient in about 7 days. Animals should be housed in coprophagy-preventing cages and fasted overnight before sacrifice. If warfarin is used instead of a vitamin K-deficient diet, the drug is administered at the level of 1–5 mg/kg body weight 18–24 hr before sacrifice. Although less carboxylation has been reported in microsomes prepared from warfarin-treated

[38] B. C. Johnson, this series, Vol. 67, p. 165.

[39] J. W. Suttie, L. M. Canfield, and D. V. Shah, this series, Vol. 67, p. 180.

[40] J. B. Lian and P. A. Friedman, *J. Biol. Chem.* **253,** 6623 (1978).

than from vitamin K-deficient animals, functional preparations are obtained.[39]

Tissue Homogenization. Microsomes are prepared from freshly excised tissues homogenized in either a Brinkmann Polytron or a Potter-Elvehjem tight-fitting Teflon homogenizer. One-half gram of wet weight tissue is homogenized per ml of Buffer A[39] which contains 0.25 *M* sucrose, 0.025 *M* imidazole hydrochloride, pH 7.2 at 0–4°. (Benzamidine hydrochloride at 1 m*M* has also been included in studies of purification of the carboxylase enzyme.[41]) The homogenate is centrifuged for 15 min at 10,000 *g*, 3–4°. The postmitochondrial supernatant is recentrifuged at 105,000 *g* for 1 hr, 3–4°. The fluffy top layer, if present, is removed from the pellet. The pellet is resuspended in a Dounce homogenizer with an A pestle, in one-half the volume of the original postmitochondrial supernatant. After recentrifugating at 105,000 *g* for 1 hr, 3–4°, the top layer is removed. At this point the crude microsomal pellet can be stored frozen at −80° or in liquid nitrogen without loss of activity.[42]

Microsomal Suspensions. For carboxylation of endogenous substrate, microsomes are prepared by resuspension of the crude pellet with a loose fitting Dounce homogenizer for 5–8 strokes in one-fifth the original volume in Buffer B[39] which contains 0.25 *M* sucrose, 0.025 *M* imidazole, 0.5 *M* KCl, and 2 m*M* dithiothreitol, pH 7.2. The protein concentration of the suspension is measured and should be in the range of 20 mg/ml. Soluble microsome preparations are required for use with synthesis in exogenous synthetic substrates, but will also retain the capacity to carboxylate the endogenous protein substrates. They are prepared by resuspension of the crude microsomal pellet in Buffer B containing 1.0% Triton X-100 using a Dounce homogenizer. The protein to detergent ratio can be critical and for suspensions of 20 mg/ml protein, a final concentration of 2% Triton X-100 is used. After centrifugation at 105,000 *g* for 1 hr, the supernatant contains the carboxylating system. An alternative procedure for disruption of microsomes for exogenous peptide carboxylation is the preparation of an acetone powder.[43] The intact microsomal pellet is homogenized in 30 ml of −20° acetone for 2–3 strokes, centrifuged for 15 min, and the procedure repeated after which the pellet is lyophilized and can be stored at 5° for several days, then resuspended in Buffer B for assay.

Incubation Conditions. The assays are carried out in triplicate in capped tubes using from 100–300 μl of microsomal suspension (10 mg/ml

[41] D. O. Mack, M. Wolfensberger, J. M. Girardot, J. Miller, and B. C. Johnson, *J. Biol. Chem.* **254,** 2656 (1979).

[42] L. Uotila and J. W. Suttie, *J. Infect. Dis.* **148,** 571 (1983).

[43] P. A. Friedman and M. A. Shia, *Biochim. Biophys. Acta* **616,** 362 (1980).

protein per tube). ^{14}C $NaHCO_3$ is present at 1–4 μC/ml, added in a 10-μl aliquot, and the reaction is initiated by the addition of 100 μg/ml of vitamin K hydroquinone added in 5–10 μl of 100% ethanol. A convenient reaction volume is 0.2 ml; however, the assay can be scaled up or down. Control tubes have no vitamin K but contain the vehicle. Maximum carboxylation of endogenous substrate is attained if incubations are carried out at 20–25° proceeding for 15–30 min with constant shaking.

For peptide carboxylation,[44] the solubilized microsomal suspension must be used in the presence of 1 to 4 m*M* peptide dissolved first in 100 μl of 1 m*M* NaOH or imidazole–HCl, pH 8.0. Pyridoxal-5-phosphate, a stimulator of exogenous peptide carboxylation, but not of endogenous substrate,[45] is added in a 10 μl volume to 1 m*M*; the reaction is initiated with [^{14}C]bicarbonate and vitamin K as above and run at 17 or 21°. Under these conditions linearity is maintained for 45 to 60 min. The soluble microsomal system is not inhibited by warfarin, although the direct vitamin K antagonists, e.g., 2-chloro-3-phtyl-1-4-naphthoquinone will inhibit Gla synthesis.

Assay of ^{14}C Incorporation. The reactions are terminated on ice and in a well-ventilated hood by the addition of acid, either 10% perchloric or 10% trichloroacetic acid (0.6 ml/0.2 ml reaction mix) to a final concentration of 7.5% for 15 min. Further removal of nonprotein bound [^{14}C]bicarbonate is accomplished by the addition of 0.4 ml of 1 *M* $NaHCO_3$ per 1 ml of reaction, placing the samples under reduced pressure with a KOH trap and shaking to remove liberated $^{14}CO_2$.

Endogenous microsomal protein is recovered from the reaction tube by the addition of 1 ml of 0.3 *M* Na_2CO_3 per 1 ml of reaction and transferring it to a scintillation vial. Then a second addition of 15 ml of cold 10% trichloracetic acid (TCA) is added to the scintillation vial. After 30 min at 4°, the precipitate is collected by low-speed centrifugation and the supernatant is discarded. Either Protosol (New England Nuclear, MA) or NCS (Nuclear Chicago, IL) solubilizer is added to the pellet in the vial. The degree of carboxylation is expressed as the ratio of counts in the sample of +K and −K incubation controls. If the −K control counts are greater than 50 cpm/tube in endogenous material, a repeated solubilization in 1 *M* sodium carbonate and reprecipitation in TCA will lower the background.

To measure carboxylated exogenous peptide, the reaction mixture is terminated as above. After the first addition of TCA or PCA, and standing in ice for 15 min, a major aliquot of supernatant is pipetted into a scintilla-

[44] J. W. Suttie, J. M. Hageman, S. R. Lehrman, and D. H. Rich, *J. Biol. Chem.* **251,** 5827 (1976).

[45] J. S. Suttie, S. R. Lehrman, L. O. Gewete, J. M. Hageman, and D. H. Rich, *Biochem. Biophys. Res. Commun.* **86,** 500 (1979).

tion vial and placed under vacuum for 15 to 20 min. Gassing the supernatant with CO_2 for 4 min has also proven useful in reducing the background in the supernatant.[39] Scintillation cocktail is then added for counting. In reactions where carboxylation is to be quantitated simultaneously in both endogenous and exogenous substrates, following the first precipitation with TCA or PCA and centrifugation, the samples are centrifuged and the supernatant (containing peptide) aliquoted into a scintillation vial. The pellet (endogenous protein) and supernatant are assayed as described above.

Confirmation of Incorporated $^{14}CO_2$. In examining other tissues for vitamin K-dependent carboxylase activity, it is necessary to confirm the formation of 4-carboxyglutamic acid. After the reaction is stopped with acetic acid, an aliquot is removed from the incubation and 1 ml of 1 *M* sodium bicarbonate is added. It is dialyzed against 0.05 *M* sodium bicarbonate in the cold for 24 hr, then against distilled water for the material is lyophilized and then hydrolyzed in KOH to release free 4-carboxyglutamic acid. At this point the material can be analyzed by amino acid analysis or Dowex-1 anion exchange for separation of Gla with collection for scintillation counting (see below). Alternatively, after precipitation of potassium with perchloric acid,[46] the hydrolysate can be divided equally with one aliquot being rehydrolyzed in 6 *N* HCl. This procedure converts 4-[^{14}C]carboxyglutamic acid to [^{14}C]glutamic acid, resulting in loss of 50% of the radioactivity as carbon dioxide during the decarboxylation procedure.[47]

Quantitation of 4-Carboxyglutamate

Several methods have been published for the direct analysis of Gla in protein hydrolysates. These methods employ either automated amino acid analysis,[46] high-pressure liquid chromatography,[48] or manual ion-exchange chromatography followed by fluorescent[49] or absorption spectroscopy.[50] The methods described here, which are the ones used in our laboratory, include automated cation exchange amino acid chromatography, and an isotope dilution method employing a single anion exchange column, and fluorescence detection.

[46] P. V. Hauschka, *Anal. Biochem.* **80,** 212 (1977).
[47] J. B. Lian, P. V. Hauschka, and P. M. Gallop, *Fed. Proc., Fed. Am. Soc. Exp. Biol.* **37,** 2615 (1978).
[48] M. Kuwada and K. Katayama, *Anal. Biochem.* **117,** 259 (1981).
[49] C. M. Gundberg, J. B. Lian, and P. M. Gallop, *Anal. Biochem.* **98,** 219 (1979).
[50] L. Pecci and P. Cavallini, *Anal. Biochem.* **118,** 70 (1981).

Materials. Alkai-resistant tubes (13 × 100 mm) are either glass (Corning No. 7280) or Teflon (Tefzel 3113-0065). Dowex-1 is obtained from Bio-Rad Laboratories as AG 1-XB, minus 400 mesh, chloride form. A number of methods for Gla synthesis have been developed.[51–57] 4-Carboxyglutamic acid is also commercially available from Sigma. 4-[^{14}C]Carboxyglutamic acid was prepared by New England Nuclear as a custom synthesis. *o*-Phthalaldehyde is purchased from Eastman Kodak Company. All other chemicals are reagent grade or better.

Base Hydrolysis. 4-Carboxyglutamate is a malonic acid derivative which readily decarboxylates to glutamic acid under the acidic conditions used for protein hydrolysis. Therefore, for quantitative recovery of Gla from proteins hydrolysates alkaline conditions are employed. Either pure protein, lyophilized tissues, or liquid samples are made 2.0 *N* in standardized KOH (150 μl/mg protein for solid samples) and placed in alkalai-resistant tubes. These in turn are placed inside of large Pyrex tubes containing sufficient 2 *N* KOH such that the level of liquid inside the outer tube exceeds that in the inner tube. After evacuation, the tubes are sealed with a torch under an atmosphere of nitrogen. Alternatively, a desiccator method which is more convenient for a large number of samples has been described.[58] The sealed samples or desiccator is heated for 22 hr at 100°. After hydrolysis, the sample tubes are removed from the desiccator or Pyrex tubes and placed on ice to cool. When they are cold, an equal volume of ice cold saturated $KHCO_3$ is added followed by an equal volume of 0.2 *M* sodium citrate, pH 2.2. Fifty-microliter aliquots of ice cold 3 *M* $HClO_4$ are added at several minute intervals until the solution has reached pH 7. The tubes are allowed to stand on ice for 30 min and then centrifuged for 15 min at 1000 *g* to remove the precipitate. At this point, the supernatant fluid can be stored at −20°.

Because the $KClO_4$ precipitate can trap some of the free Gla after hydrolysis, recovery can be monitored by adding an internal standard of

[51] N. T. Boggs, III, R. E. Gawley, K. A. Koehler, and R. Hiskey, *J. Org. Chem.* **40,** 2850 (1975).
[52] M. Opleger and R. Schwyzer, *Helv. Chim. Acta* **60,** 43 (1977).
[53] N. T. Boggs, III, B. Goldsmith, R. E. Gawley, K. A. Koehler, and R. G. Hiskey, *J. Org. Chem.* **44,** 2263 (1979).
[54] S. Danishefsky, E. Berman, L. A. Clizlue, and M. Hirama, *J. Am. Chem. Soc.* **101,** 4385 (1979).
[55] S. Bory, M. Gaudry, A. Marquet, and A. Azerod, *Biochem. Biophys. Res. Commun.* **87,** 85 (1979).
[56] A. Juhasz and S. Bajusz, *Int. J. Pept. Protein Res.* **15,** 154 (1980).
[57] K. Y. Zee Cheng and R. E. Olson, *Biochem. Biophys. Res. Commun.* **94,** 1128 (1980).
[58] H. T. Keutmann and J. T. Potts, *Anal. Biochem.* **29,** 175 (1969).

[^{14}C]Gla or [^{14}C]alanine, which are alkalai stable, either before or after hydrolysis.

Automated Amino Acid Analysis. Several procedures have been described to quantitate Gla in hydrolyzed samples. The method currently used in our laboratory and described here is an adaptation Hauschka's method[46]; this method is not suitable for total amino acid analysis because not all neutral and basic amino acids are well separated. However, it is the method of choice when analyzing for Gla in heterogeneous samples because it separates Gla from all known acidic, ninhydrin-positive interfering compounds. This procedure has been developed for a Beckman 121M amino acid analyzer, but is adaptable to other systems.

Hydrolyzed and neutralized samples for automated amino acid analysis are diluted in 0.2 *M* sodium citrate buffer, pH 2.2. The pH is checked on indicator paper (short range alkacid pH 0.0–3.0) and adjusted to pH 2.2 with 6 *M* HCl if necessary. Gla is stable for several days at 20° under these conditions.

A microbore column (2.8 × 330 mm) containing Beckman AA-20 resin is used in conjunction with nanobore tubing. The buffer flow rate is 8.0 ml/hr and the effluent stream is monitored by mixing with ninhydrin–DMSO (Beckman) reagent flowing at 4.0 ml/hr. All buffers contain 0.2 *M* citrate, 0.01% (v/v) octanoic acid, and 0.25% (v/v) thiodiglycol. The column is equilibrated with pH 2.58 citrate buffer. A 50 μl sample is injected and the following buffer program cycle is used: pH 2.58 equilibration (14.0 min), sample injection (0.1 min), pH 2.58 (8.0 min), pH 3.10 (25 min), 0.2 *M* NaOH, 0.1% EDTA column wash (5.0 min), reequilibration at pH 2.58 (13.1 min), and restart of the cycle. Elution time for Gla is 33.0 min after injection. Glutamic acid can also be determined by lengthening the pH 3.10 buffer cycle from 25.0 to 50.0 min and continuing as above. The glutamic acid elution time is 64.0 min after injection.

Several unidentified compounds found in alkaline hydrolysates of crude tissues elute in the Gla region. These may erroneously be assumed to be Gla. To prove that the putative peak is Gla, the peak must be present in alkaline hydrolysates, but not in acid hydrolysates; furthermore, the kinetics of decarboxylation from Gla to glutamic acid should match those with an authentic standard.[59] Acid conversion of 4-carboxyglutamic acid to glutamic acid results in a 2.38-fold increase in relative peak area based on the ninhydrin reaction.

Isotope Dilution and Anion Exchange. We have also developed a rapid, simple, and sensitive procedure for Gla analysis which employs isotope dilution coupled with a single anion exchange column and fluores-

[59] P. V. Hauschka, E. B. Henson, and P. M. Gallop, *Anal. Biochem.* **108,** 57 (1980).

cent detection. This method has been described in detail elsewhere but will be discussed briefly here.[49]

[^{14}C]Gla (2 nCi) (specific activity = 1.6 Ci/mmol) is added to 2.0 ml of neutralized base hydrolysate. The sample is adjusted to pH 11.5 with 6 *N* KOH and rapidly centrifuged to remove any precipitate. A small aliquot is removed and radioactivity determined. One milliliter of sample is applied to a 10 × 0.7 cm column filled with Dowex 1 × 8 (<400 mesh) previously equilibrated with 0.01 Hepes, pH 11.5. Fifty milliliters of 0.02 Hepes pH 5.0 is passed through the column and discarded. This eluate contains the basic and neutral amino acids. Fifty milliliters of 0.020 Hepes, 0.020 *M* $MgCl_2$, pH 4.5 is then passed through the column and collected in 5 ml fractions. These contain most of the acidic amino acids. Gla, free of any other primary amines, elutes at approximately 40 ml of this buffer. To determine precise location and recovery, a 0.5-ml aliquot of each fraction is removed for determination of radioactivity. Gla is quantitated by mixing a second 0.5-ml aliquot of each fraction with 0.5 ml 6.0 m*M* *o*-phthalaldehyde in 0.40 *M* boric acid, pH 9.7. Fluorescence is determined by use of a spectrophofluorometer and with excitation and emission wavelengths of 340 and 455 nm, respectively. Concentration is determined by comparison to a standard curve of Gla from 0.1 to 10 nmol/ml. Final Gla concentration is determined by correcting for recovery (generally >80%). Column fractions immediately adjacent to the Gla region are free of any fluorophore reactive components.

This method allows for the rapid determination of Gla in base hydrolysates. Comparison of results from this procedure and direct automated amino acid analysis yield values generally consistent to within 10%. This method can also be conveniently used for the isolation of radioactive Gla from the base hydrolysates of *in vitro* carboxylation experiments. The isolated Gla fraction can then be acid hydrolyzed and on rechromatography should yield a 50% recovery of the radioactivity as glutamic acid with a complete loss of Gla.

Decarboxylation and Tritium Exchange

Another useful method for the identification of 4-carboxyglutamate in proteins is by the specific incorporation of tritium into Gla residues in the protein. Thermal decarboxylation was first described in 1974 by Stenflo and in 1975 by Magnusson. By the restriction of water from the system, a decarboxylated protein was produced without introducing other changes.[60] The method has been further extended by Hauschka[61] to intro-

[60] J. W. Poser and P. A. Price, *J. Biol. Chem.* **254,** 431 (1979).

[61] P. V. Hauschka, *Biochemistry* **18,** 4992 (1979).

duce tritium into Gla proteins. Tritium-labeled proteins and peptides produced by this method are unmodified except for the substitution of 4-methylene tritiated glutamate residues for Gla residues. The method can also be useful in the identification and localization of Gla proteins in tissues without the requirement of extensive purification.

Isotopes. Tritiated water made up in 99.7% pure deuterated water (T_2O in D_2O) was obtained at a specific activity of 1.42 Ci/ml from New England Nuclear (Boston, MA). D_2O is obtained from the same source, and concentrated 38% DCl is purchased from ICN (Irvine, CA). Stock equilibration solutions of T_2O in 0.05 *M* DCl is prepared by volumetric dilution to final specific activities ranging from 85 to 426 mCi/ml or 1700 to 8520 dpm/nmol of deuterium. These solutions are referred to as "0.05 *M* TDCl," and at this specific activity the fraction of hydrogen atoms which are actually T ranges from 2.6×10^{-5} to 1.3×10^{-4}. The background of nonvolatile radioactivity in the 0.05 *M* TDCl is less than 10^{-10} of the total disintegrations per minute.

Liquid Scintillation Counting. Aqueous samples (0.02–1.0 ml) are dispersed in 10 ml of scintillation fluid (Formula 963; New England Nuclear) and counted at 0° in the tritium window of an Intertechnique liquid scintillation spectrometer. Efficiency is calibrated by internal standards of tritiated water (New England Nuclear). Stock solutions of 0.05 *M* TDCl are diluted 2×10^{-5} in 0.05 *M* HCl before counting for determination of specific activity.

Vacuum System. A vacuum pump operating at a pressure of 0.1 mm Hg in a well-vented hood is utilized for tritium exchange, and evacuation of samples prior to sealing for decarboxylation. Access of tritium to the pump is largely prevented by a glass U-tube trap in isopropanol-solid CO_2 followed by a small P_2O_5 trap. Because of the levels of radioactivity involved, all operations must be confined to the hood and extreme care and caution must be exercised.

Procedure for Tritium Incorporation at Gla Residues

Proteins are dissolved in cold 0.05 *M* HCl and dialyzed against this same solvent in Spectrapor 3 membrane tubing (Spectrum Industries, Los Angeles, CA) to remove metal ions. Aliquots of the clear solutions containing 0.05–1.6 mg of protein are dispensed into acid-washed breakseal vials (Virtis) with other samples being set aside for amino acid analysis and determination of protein concentration. The vials are lyophilized in a SpeedVac Concentrator (Savant). One milliliter of 0.05 *M* DCl in D_2O is added to each vial, and after 30 min at 20° the clear solution is again lyophilized, yielding the dry, largely deuterated protein. The rationale for this step is that tritium generally exchanges more effectively with deuterium than with hydrogen. Vials containing dry, deuterium-exchanged protein are then connected by 1 cm lengths of vacuum tubing to a small 8-port glass manifold on the vacuum system in the hood. Attached to one of the ports is a vial containing 1 ml of frozen 0.05 *M* TDCl. Evacuation of the manifold to below 0.1 mm Hg is achieved within 2 min, at which point the manifold is isolated from the vacuum pump by a valve, and tritium exchange is allowed to occur for 3 hr at 20°. Early studies by Linderstrom-Lang demonstrated the surprisingly rapid equilibration between exchangeable hydrogen atoms in a dry protein and deuterium from frozen D_2O in an adjacent connected vessel; a similar phenomenon occurs with tritium exchange.[61] During the exchange period, the pressure in the manifold rises gradually and the 0.05 *M* TDCl melts. The tritium solution is refrozen by contact with a small cup of isopropanol–solid CO_2 and then the manifold is reopened to the vacuum system. After about 15 min of evacuation the vials containing dry tritium-exchanged protein are sealed with a small torch, removed from the manifold, and placed in a 110° oven for 3–6 hr.[61] The vial containing 0.05 *M* TDCl is removed, sealed with a glass plug, and stored frozen for future use in a canning jar (Mason), tightly sealed with a metal lid. The cold trap of the vacuum system must be carefully opened and rinsed in the hood for recovery of the 5 to 50 mCi of tritium which has accumulated.

The vials containing thermally decarboxylated tritium-labeled protein which may contain 10^6 to 10^8 cpm at this point are opened carefully in the hood. Unincorporated and exchangeable tritium is removed by three or four successive lyophilizations after dissolution in 1 ml of 0.05 *M* HCl or 0.05 *M* NH_4CO_3. At each wash a small aliquot is counted to monitor the process. Occurrence of tritium-labeled glutamic acid at former Gla positions may be verified by acid hydrolysis and amino acid analysis.[61]

3-Carboxyaspartic Acid and Aminomalonic Acid

The presence of 4-carboxyglutamate in a variety of calcium binding proteins led to the search for other amino acids which contain a malonic acid component. Aminomalonic acid is easily prepared by the base hydrolysis of acetaminomalonic diethyl ester and related compounds. 3-Carboxyaspartic has been prepared by published procedures.[62–64] In the first of these methods, parabanic acid (oxalyl urea) is reduced with borohydride to 5-hydroxyhydantoin, which after treatment with thionyl chlo-

[62] E. B. Henson, P. M. Gallop, and P. V. Hauschka, *Tetrahedron* **37,** 2561 (1981).
[63] M. R. Christy and T. H. Koch, *J. Am. Chem. Soc.* **104,** 1771 (1982).
[64] N. E. Dixon and A. M. Sargeson, *J. Am. Chem. Soc.* **104,** 6716 (1982).

ride is converted to 5-chlorohydantoin. It is then reacted with sodium malonate ester to form the hydantoin 5-malonate diester which after base hydrolysis yields racemic 3-carboxyaspartic acid. In another published procedure, 1,1,2-tris (carbomethoxy) ethane is reflexed with benzyl alcohol to give tris (carbobenzyloxy) ethane. This is then brominated, reacted with triethylamine, then followed by treatment with hydrazoic acid to yield tribenzyl 1-azidoethane-1,2,2-tricarboxylate. Catalytic hydrogenation of this compound yields DL-3-carboxyaspartic acid.[63] Aminomalonic acid has been found in a small peptide called arcamine which is a fish attractant and has the structure, hypotaurylaminomalonate.[65] The antibiotic malonomycin also contains the aminomalonate moiety.[66]

Both aminomalonate and 3-carboxyaspartate can be separated from 4-carboxyglutamate by the amino acid analysis.[59] The kinetics of the decarboxylation of these three amino group-containing malonic acids has also been studied.[62] The occurrence of 3-carboxyaspartic has been reported in ribosomal proteins from *E. coli*,[67] but this remains to be confirmed.

[65] A. W. Sangster, S. E. Thomas, and N. L. Tingling, *Tetrahedron* **31,** 1135 (1975).
[66] D. Schipper, J. L. van der Baan, and F. Bickelhaupt, *J. Chem. Soc.* **1979,** 2017 (1979).
[67] M. R. Christy, R. M. Barkley, T. H. Koch, J. Van Buskirk, and W. M. Kirsh, *J. Am. Chem. Soc.* **103,** 3935 (1981).

[25] Capsular Poly-γ-D-glutamate Synthesis in *Bacillus licheniformis*

By FREDERIC A. TROY, II

Introduction

A membranous polyglutamyl synthetase complex (PGSC)[1-4] from *Bacillus licheniformis* 9945A catalyzes a sequence of membrane-associated enzymatic reactions in which L-glutamic acid is activated, racemized, and polymerized to form a poly(γ-D-glutamyl) capsule.[1] The molecular mass

[1] F. A. Troy, *J. Biol. Chem.* **248,** 305 (1973).
[2] F. A. Troy, *J. Biol. Chem.* **248,** 316 (1973).
[3] J. M. Gardner and F. A. Troy, *J. Biol. Chem.* **254,** 6262 (1979).
[4] Abbreviations used: PGSC, polyglutamyl synthetase complex; PGA, poly-γ-D-glutamic acid; NCA, N-carboxyanhydride; N-TFA, N-trifluoroacetyl.

Copyright © 1985 by Academic Press, Inc.
All rights of reproduction in any form reserved.

of these surface polymers exceeds 10^6 daltons.[2] The capsule is known to be important in the pathogenicity of *Bacillus anthracis* because of its ability to influence host-cell interactions.[5] Reactions (1)–(4) summarize the postulated mechanism of assembly:

$$\text{L-Glu} + \text{ATP} \rightleftharpoons [\gamma\text{-L-Glu-AMP}] + \text{PP} \quad (1)$$

$$[\gamma\text{-L-Glu-AMP}] \rightleftharpoons [\gamma\text{-x-Glu-S-protein}] + \text{AMP} \quad (2)$$

$$[\gamma\text{-x-Glu-S-protein}] \rightarrow (\gamma\text{-D-Glu})_n\text{-S-protein} \quad (3)$$

$$(\gamma\text{-D-Glu})_n\text{-S-protein} \xrightarrow[\text{Acceptor}]{\text{Poly}(\gamma\text{-D-glutamyl})} (\gamma\text{-D-Glu})_n\text{-acceptor} \quad (4)$$

The stereochemical configuration of the product of reaction (1), catalyzed by the membrane-associated L-glutamic acid activating enzyme, is L.[3] Thus, stereochemical inversion must occur after activation. Proposed reactions (2) and (3) are less well understood and only indirect evidence supports their existence. Since a step in polymerization subsequent to activation is inhibited by sulfhydryl group blocking agents, and since chemical treatments known to liberate thioesterified polypeptide chains release membrane-bound polyglutamic acid (reaction product 3), we conclude that a thioester is functionally involved.[3] What is less certain is the existence of a monoglutamyl thioesterified product shown in reaction (2). That AMP is inhibitory for polymer formation supports the existence of this intermediate and also the reversibility of the reaction. It is possible, however, that the activated glutamyl adenylate is transferred directly to a growing oligo- or polyglutamyl chain [reaction (3)] and that the stereochemical inversion occurs concomitantly with polymerization. The apparent high M_r of labeled γ-D-glutamyl polymers released by treatment of membranes with hydroxylamine or at pH 10 further supports the existence of reaction (3). That only partial release of radiolabeled glutamyl polymers from the membrane occurs with hydroxylamine is supportive of the product of reaction (4). Also compatible with reaction (4) is the observation that *in vitro*, polymer growth occurs by the addition of glutamyl residues to an endogenous acceptor that contains substantial amounts of D-glutamic acid.[3] Nothing is known concerning the nature of the thioprotein product of reactions (2) or (3) and there is no evidence that they may be two distinct proteins. A more definite answer to these questions awaits the isolation and characterization of the thioesterified protein(s).

The proposal that a thioester may serve as an acceptor of activated glutamyl moieties and thus function to anchor growing polyglutamyl chains in the multienzyme complex is analogous to the known role that enzyme-bound 4′-phosphopanthetheine plays in peptide antibiotic syn-

[5] F. A. Troy, *Annu. Rev. Microbiol.* **33,** 519 (1979).

thesis.[6,7] It is envisioned that such an intermediate may function, as seen in reaction (2) and (3), to facilitate the transfer of activated monomers to the growing polymeric chain. This would be in accord with the observation that these polymers are extended by the addition of new glutamyl residues to the amino terminus end of growing nascent chains.[3] This mechanism of elongation is in contrast to polypeptide antibiotic synthesis where the activated oligopeptide is transferred to an activated monomer. Another major difference between peptide antibiotic and γ-D-glutamyl capsular polymer synthesis is the latter is dependent upon the structural integrity of the cell membrane–cell wall complex.[3] Lysozyme and detergent sensitivity of the polyglutamyl polymerization reaction(s) is interpreted to mean that a particular conformation of the membrane-bound enzymes and endogenous acceptors is required. The appropriate domain has not yet been clarified. Thus, definitive characterization of the linkages involved and the interactions between the components of the membrane-bound polyglutamyl synthetase complex awaits further study. In this chapter, preparation and properties of the PGSC and the L-glutamic acid activating enzyme will be described. Methods for structural characterization of the products of these reactions will also be described.

Assay Methods

Principle. The *in vitro* determination of the overall reaction catalyzed by the membranous PGSC [reactions (1)–(4)] is based on measuring the rate of incorporation of L-[U-^{14}C]glutamic acid into high-molecular-weight polymeric products (poly-γ-D-glutamic acid, PGA). These polymers are chromatographically and electrophoretically immobile.[1] The membrane preparation is incubated with L-[U-^{14}C]glutamic acid, ATP and Mg^{2+}. Aliquots from the incubation mixture are applied to Whatman No. 3MM paper and either chromatographed or electrophoresed to separate glutamic acid from ^{14}C-labeled PGA that remains immobile. ^{14}C-labeled PGA which remains at the origin is determined by scintillation counting. It is important that subsequent chemical, physical, and enzymatic analyses be carried out to confirm that the enzymatically synthesized polymers are structurally identical with the capsular γ-D-glutamyl polymers synthesized *in vivo*.[2]

The activity of the L-glutamic acid activating enzyme [reaction (1)] is determined by measuring the rate of an ATP-dependent formation of

[6] F. Lipmann, W. Gevers, H. Kleinkauf, and R. Roskoski, Jr., *Adv. Enzymol.* **35,** 1–34 (1971).

[7] K. Kurahaski, *Annu. Rev. Biochem.* **43,** 445 (1974).

[^{14}C]glutamyl hydroxamate.[3] Glutamate activation is also studied by [^{32}P]pyrophosphate exchange with ATP.[3]

Procedure for Measuring Poly-γ-D-Glutamyl Synthesis [Reactions (1)–(4)]

Chromatographic Method. Incorporation of L-[U-^{14}C]glutamic acid into PGA is determined in standard reaction mixtures which contain the following components in a final volume of 0.25 ml: sodium phosphate buffer, pH 7.2, 24 μmol; $MgCl_2$, 10 μmol; ATP (pH 7.2), 1.0 μmol; L-[U-^{14}C]glutamic acid (pH 7.2), 1.0 μmol (1000 to 4000 dpm/nmol); dithiothreitol, 2.0 μmol; KCl, 50 μmol; glycerol, 25 μmol; dimethyl sulfoxide, 25 μmol; chloramphenicol, 20 μg; and membrane protein, 0.1 to 0.3 mg. After incubation at 37°, 20-μl aliquots are applied to Whatman No. 3MM paper and chromatographed in 1-butanol–acetic acid–water (40 : 8.8 : 20) for 16 to 18 hr. The chromatograms are dried and the ^{14}C-labeled PGA which remains at the origin is quantitatively determined by cutting out the origin segment and counting in a vial which contains scintillation fluid. For the purpose of isolating large quantities of the enzymatically synthesized product, the proportions of the standard incubation mixtures are increased by a factor of five and multiple incubation mixtures are prepared. This procedure resulted in greater yields of product than attempting to increase the volume by a factor greater than five.

Electrophoretic Method. An alternative assay for quantitation of the amount of PGA synthesized is to separate the labeled polymer from the L-[^{14}C]glutamic acid by high voltage electrophoresis. Twenty-microliter aliquots from the reaction mixture are applied to Whatman No. 3MM paper and subjected to electrophoresis at 60 V/cm in pyridinium acetate, pH 6.4. The electrophoretograms are dried and the amount of labeled polyglutamic acid at the origin is quantitated by scintillation counting as described above. Identical results are obtained with the chromatographic and electrophoretic assays.

Procedure for Assaying the L-Glutamic Acid Activating Enzyme (Reaction 1)

Hydroxamate Method. Glutamyl hydroxamate formation is measured essentially as described by Staudenbauer and Strominger[8] in a reaction mixture (200 μl) containing 100 m*M* Tris–HCl, pH 7.6; 50 m*M* $MgCl_2 \cdot 7H_2O$; 20 m*M* ATP · Na_2, pH 7.6; 800 m*M* salt-free hydroxylamine

[8] W. Staudenbauer and J. L. Strominger, *J. Biol. Chem.* **247,** 5095 (1972).

(NH_2OH), pH 7.6; 100 mM sodium L-[^{14}C]glutamate; 5 mM DL-methionine sulfoximine; 5 mM L-glutamine; and 50 to 250 μg of enzyme protein.[9] After incubation for 30 min at 37°, 0.3 ml of $FeCl_2$ reagent (10% $FeCl_2$, 5% trichloroacetic acid in 0.67 N HCl) is added. After centrifugation at 3000 g for 10 min, the absorbance of the supernatant is measured at 540 nm and that of the zero time control substracted. Authentic γ-L-glutamyl hydroxamate (Sigma, St. Louis, Mo.) is used as the standard reference compound.

Salt-free NH_2OH is prepared by adjusting hydroxylamine HCl to pH 7.6 with NaOH and passing the solution through a mixed bed ion exchange resin (Bio-Rad AG 501-X8, 20 to 50 mesh). This solution is stored at −20°.

Pyrophosphate Exchange Assay. Pyrophosphate exchange is measured by a procedure similar to that described by Calender and Berg.[10] The incubation mixture (0.5 ml) contains 0.1 M sodium cacodylate, pH 7.5, or 0.1 M glycylglycine, pH 8.6; 5 mM $MgCl_2 \cdot 7H_2O$; 2 mM ATP · Na_2, pH 7.5; 8 mM dithiothreitol; 10 mM sodium fluoride; 100 μg bovine serum albumin; 5 to 200 mM L-glutamic acid, pH 7.5 (or other amino acids, as indicated); 2 mM [^{32}P]pyrophosphate (500 to 800 cpm/nmol), and 50 to 250 μg of enzyme protein. Incubations are carried out at 37° for 30 min, and the reaction terminated by the addition of 1.0 ml of 15% perchloric acid containing 50 mM sodium pyrophosphate and 0.1 ml of 7.5% "acid washed" Norite A. The mixture is vortexed, filtered through glass fiber filters presoaked in saturated sodium pyrophosphate, washed with 30 ml of 5% trichloroacetic acid containing 20 mM sodium pyrophosphate and with 50 ml of water. Filters are dried and ^{32}P-labeled ATP determined directly by counting in a toluene-based scintillation fluid.

Enzyme Preparation

Growth of Bacterial Cells. A heavily encapsulated colony (smooth phenotype) of *B. licheniformis* ATCC 9945A (formerly designated *Bacil-*

[9] The membrane-bound L-glutamic acid activating enzyme is partially sensitive to the glutamine synthetase inhibitors glutamine and methionine sulfoximine. The product of the activating reaction, even in the presence of these inhibitors, was consistent with γ-L-glutamyl adenylate and not γ-L-glutamyl phosphate (see ref. 3). The nature of the methionine sulfoximine glutamine-sensitive activity, if distinct from the activation of L-glutamic acid related to the synthesis γ-D-glutamyl capsular polymers, is unknown. In order to minimize possible ambiguity due to the methionine sulfoximine glutamine-sensitive activity, 5 mM methionine sulfoximine and glutamine was added to relevant incubation mixtures.

[10] R. Calender and P. Berg, *Biochemistry* **5,** 1681 (1966).

lus subtilis 9945A[1]) is selected following growth on a chemically defined medium, Medium E,[11] supplemented with 1.5% agar. Stock cultures of the organism are maintained in Trypticase Soy Broth containing 15% glycerol as described previously.[12] The contents of one vial (1.5 ml) are thawed and used to inoculate 500 ml of Medium E contained in a 2-liter Erlenmeyer flask. Cultures are incubated on a rotary shaker (250 rpm) at 37° until growth reaches the late logarithmic or early stationary phase. The culture is cooled by the addition of an equal volume of ice and agitated for 2 min in a laboratory blender. This treatment shears capsular polymers from the surface of cells and facilitates harvesting of the cells by centrifugation.[12] Cells are recovered by centrifugation (16,000 *g* per 10 min) and washed by resuspension in 1 volume of cold 0.01 *M* Tris buffer, pH 7.1, containing 0.03 *M* NaCl. The cells at time of harvest must be actively engaged in the synthesis of PGA. This is readily assessed by the increase in the viscosity of the culture (Ref. 1, Fig. 1).

Preparation of Polyglutamyl Synthetase Complex. Membrane fractions capable of catalyzing the polymerization of L-glutamic acid are prepared by a modification of our original method.[1] Washed cell pellets (2 g, wet weight) are suspended in 10 ml cold 20 m*M* potassium phosphate buffer, pH 7.2 containing 0.1 *M* glycerol, 0.1 *M* dimethyl sulfoxide (KDG buffer), and sonicated 4 times in successive 5-sec pulses at 80 W with a 2-cm-diameter probe (Heat Systems-Ultrasonics, Inc.). Twenty milliliters of the same buffer is added and the homogenate centrifuged at 4000 *g* for 5 min to remove intact cells. The supernatant fraction is centrifuged at 48,000 *g* for 10 min and the middle layer (P-2) of the pellet, which contains all the polyglutamyl synthetase activity, is selectively removed as described previously[1] and washed 2 times in 30 ml KDG buffer by centrifugation at 48,000 *g* for 10 min. The washed cell envelope material is suspended in 5 ml of KDG buffer and used directly for polyglutamyl synthetase assays or for preparation of the lysozyme solubilized membrane fraction (see below). Approximately 80% of the polyglutamyl synthetase activity is lost upon storage for 24 hr at either −20 or 4°. For this reason, membranous enzyme complexes are prepared fresh daily.

Preparation of Membranous L-Glutamic Acid Activating Enzyme. The membrane fraction prepared above is treated with egg white lysozyme (500 μg/ml) and DNase (10 μg/ml) for 20 min at 37° and centrifuged at 48,000 *g* for 30 min. The supernatant fraction is then centrifuged in a 65 Ti rotor at 220,000 *g* for 4 hr. The pellet, which contains about 90% of the L-glutamic acid activating enzyme activity present in the original membrane

[11] C. G. Leonard, R. D. Housewright, and C. B. Thorne, *J. Bacteriol.* **76,** 499 (1958).

[12] F. A. Troy, F. E. Frerman, and E. C. Heath, *J. Biol. Chem.* **246,** 118 (1971).

preparation, is resuspended in 1 ml KDG buffer. An approximately 5-fold purification of L-glutamic acid activation activity is achieved by this procedure. No γ-D-polyglutamyl synthetase activity is present in either the pellet or the supernatant fraction following lysozyme treatment.

Isolation and Purification of γ-D-Glutamyl Polymers

γ-D-Glutamyl Polymers Synthesized in Vivo. A 2-liter Erlenmeyer flask, containing 500 ml of Medium E, is inoculated with 1.5 ml of a Trypticase soy-glycerol stock as described above. The flask is incubated at 37° on a rotary shaker (250 rpm) for approximately 96 hr after which time the viscosity reaches its maximum value (see Ref. 1, Fig. 1 for the relationship between growth and expression of the PGA capsule). Optical density measurements show there is no significant lysis of cells at this time although extensive sporulation has occurred. After removal of the cells by centrifugation, the supernatant fraction which contains the PGA is precipitated by the addition of 4 volumes of 95% ethanol previously cooled to −20°. After storage at −20° for 24 to 48 hr, the PGA precipitates as a gelatinous, viscid aggregate. This procedure yields nearly quantitative isolation of PGA. The ethanol is decanted, and the PGA dissolved in 250 ml of distilled water and dialyzed against distilled water at 4° for 48 hr. The dialyzed material is then centrifuged in the type 60 Ti rotor in the Spinco ultracentrifuge, model L2-65B, at 214,000 g_{av} for 6 hr at 4°. The clear, viscous supernatant fraction is then lyophilized. Amino acid analysis of the purified polymer is carried out following hydrolysis in 6 *N* HCl at 110° for 24 hr in deaerated, sealed tubes as described by Moore and Stein[13] on a Beckman automatic amino acid analyzer (model 121) using a column (0.9 × 55 cm) packed with Beckman PA-35 resin. The amount of material in the chromatographic peak is quantitatively determined by integration of the peak area with a Beckman integrator, model 125. PGA isolated by this procedure is characterized structurally as described below.

γ-D-Glutamyl Polymers Synthesized in Vitro. In order to characterize the product of the biosynthetic reaction, 20 1.25-ml incubation mixtures are prepared as described above and incubated at 37° for 2 to 4 hr. Aliquots are removed during the incubation period and assayed for incorporation of labeled glutamic acid into PGA by the chromatographic method. The enzymatically synthesized γ-D-glutamyl polymers are isolated free from essentially all of the protein components of the membrane because

[13] S. Moore and W. H. Stein, this series, Vol. 6, p. 819.

they are resistant to hydrolysis with Pronase.[1] Thus, the enzymatically synthesized polymers are "solubilized" by exhaustive digestion of the cell membrane proteins following treatment of the reaction mixture with Pronase by the following procedure. At the end of the incubation period, each incubation mixture receives 20 μl of a solution containing freshly prepared Pronase and sodium azide such that the final concentration in each 1.25-ml incubation mixture is 6.2 μg of Pronase and 0.02% sodium azide. Following incubation at 37° for 8 to 12 hr, 20 μg of RNase and 20 μg of DNase are added and incubation continued for an additional 2 hr. At the end of this period, the incubation mixtures are combined and centrifuged at 48,000 *g* for 20 min. The small pellet which remains does not contain any radioactivity. The supernatant fraction is dialyzed against distilled water at 4° for 16 to 18 hr and concentrated by lyophilization. The lyophilized product is resuspended in 3 ml of 1% NaCl, applied to a Sephadex G-200 column equilibrated with 1% NaCl, and eluted with 1% NaCl. Quantitative recovery (95 to 100%) of the labeled glutamic acid which is incorporated into PGA as measured by either the chromatographic or electrophoretic assays is obtained in a single peak which elutes from Sephadex G-100 in the column void volume.[1]

The enzymatically synthesized product can also be isolated from the incubation mixture by first removing the excess substrate by repeated washing of the 48,000 *g* pellet with 0.01 *M* Tris buffer, pH 7.2, containing 0.03 *M* NaCl or by exhaustive dialysis. The enzymatically synthesized polymers, which remain firmly attached to the particulate fraction, are solubilized by resuspending the washed particulate fraction in either 1 *M* KCl or 1 *M* NaCl and incubating at 37° for 60 min. The membranes are then removed by centrifugation at 48,000 *g* for 20 min and the solubilized PGA, which remains in the supernatant, is purified further by passage through Sephadex G-100 equilibrated with 1% NaCl as described above. The structure of these enzymatically synthesized polymers is determined as described below.

Structural Characterization of the γ-D-Glutamyl Polymers

Determination of Optical Configuration of Glutamic Acid: Synthesis and Separation of Diastereoisomeric Dipeptides. The optical configuration of glutamic acid isolated from the enzymatically synthesized polymers, the native polyglutamyl polymers synthesized *in vivo*, chemically synthesized α-D- and α-L-polyglutamic acid and a number of authentic glutamyl dipeptides is determined by chromatographic separation of diastereoisomeric dipeptides obtained by derivatization with L-leucine–NCA[4] using a procedure adapted from Manning and Moore.[14] Glutamic acid is isolated following hydrolysis of the polymers in 6 *N* HCl at 110° for 22 hr in evacuated, sealed tubes. The hydrolysates are evaporated to dryness on a rotary evaporator, redissolved in distilled water, and again evaporated to dryness three to four times. For the preparation of dipeptides, samples containing between 5 and 20 μmol of glutamic acid are dissolved in 2 ml of 0.45 *M* sodium borate, pH 10.2 at 0°. When necessary, the pH is adjusted to 10.2 with 1 *M* NaOH. A 20% molar excess of L-leucine–NCA is added and after 2 min of intermittent shaking on a Vortex mixer at 4°, the reaction is stopped by the addition of 0.85 ml of 1.0 *N* HCl. The samples are passed through a Millipore filter (type HA, 0.45-μm filter) and stored frozen. The yield of dipeptide is 90 to 96%. Dipeptide separations are carried out on a Beckman automatic amino acid analyzer (model 121) with a column, 0.9 × 55 cm, packed with Beckman PA-35 resin. Elution is with pH 3.10 sodium citrate buffer at a column temperature of 53° and a flow rate of 70 ml/hr. Peak areas are integrated with a Beckman model 125 Integrator. The operational ninhydrin color value for L-leucine-D-glutamic acid (0.59) and L-leucine-L-glutamic acid (0.75) relative to leucine (1.00) is used to correct for the true color value of each isomer in samples containing both antipodes, as described by Manning and Moore.[14]

Synthesis and Gas–Liquid Chromatographic Separation of N-Trifluoroacetyl-L-prolyl-D-glutamic Acid and N-Trifluoroacetyl-L-prolyl-L-glutamic Acid. Derivatization of glutamic acid isolated from the enzymatically synthesized polyglutamyl polymer and the native capsular polymers synthesized *in vivo* with N-TFA-L-prolyl chloride[4] and the subsequent separation of the resultant diastereoisomers by gas–liquid chromatography is carried out as follows. Glutamic acid is isolated from each polymer by acid hydrolysis, as described above, and the methyl ester prepared using ethereal diazomethane.[2] In this procedure, 1.02 to 5.5 μmol of dry glutamyl hydrochloride is methylated at 25° by the addition of an excess (3 ml) of freshly prepared ethereal diazomethane. After brief mixing and the disappearance of the yellow color, an additional 2.0 ml of diazomethane is added and the incubation continued at 25° for an additional 2 hr. The solvent is then evaporated with a flow of dry nitrogen gas. The extent of methylation is quantitatively determined by following the disappearance of labeled glutamic acid and the concomitant appearance of the labeled methyl ester by paper chromatography on Whatman No. 3MM in 1-butanol–acetic acid–water (40 : 8.8 : 20). The chromatograms are then scanned on a Packard model 7201 radiochromatographic scanner. Quantitative

[14] J. M. Manning and S. Moore, *J. Biol. Chem.* **243,** 5591 (1968).

conversion to the methyl ester is obtained when freshly prepared diazomethane is used.

For preparation of the *N*-TFA-L-prolyl derivatives, a 10-fold molar excess of *N*-TFA-L-prolyl chloride in 1.0 ml of chloroform is added to each sample of the glutamyl methyl ester. One milliliter of dichloromethane is then added and the pH adjusted to 9 by the addition of triethylamine. The reaction mixture is sealed and heated at 90° for 5 min. After cooling, 1.0 ml of 6 *N* HCl is added and mixed well on a Vortex shaker. The organic layer is removed, washed three times with distilled water, and dried by shaking with 0.15 g of anhydrous sodium sulfate and finally by passage through a column, 0.5 × 10 cm, of anhydrous sodium sulfate. The column is washed with 3 ml of dichloromethane and the solvent evaporated with a stream of dry nitrogen gas. *N*-TFA-L-prolyl-L-glutamic acid and *N*-TFA-L-prolyl-D-glutamic acid are resolved on a Varian model 2100 gas–liquid chromatograph (Varian Instruments) equipped with a 6-foot glass column containing 0.5% H1-EFF 2 AP on Gas-chrom Q (60 to 80 mesh) (Applied Science Laboratories). The oven temperature is programmed from 140 to 200°. The 13.6% of the D isomer present in the *N*-TFA-L-prolyl chloride reagent is corrected for precise quantitative work. Alternatively, labeled diastereoisomers are resolved by radio–gas–liquid chromatography on 1.5% OV-210 on Gaschrom Q at 200°. An Aerograph A-90P gas chromatograph (Wilkins Instrument and Research) is used which is coupled to a Nuclear Chicago Biospan proportional radiation detector (Nuclear Chicago Corp.). Identification of compounds is done by comparing the retention time and position of the radioactive peaks with standards.

Synthesis of γ-Amino-δ-hydroxyvaleric Acid. γ-Amino-δ-hydroxyvaleric acid is prepared by lithium borohydride reduction of the methyl ester of 5-oxo-2-pyrrolidine carboxylic acid by a procedure similar to that described by Chibnall *et al.*[15] Freshly prepared ethereal diazomethane, 2 ml, at −10° is added to 8 mg of the dry pyrrolidonecarboxylic acid. Following incubation at 28° for 2 min, in which there is considerable effervescence, the methyl ester of the pyrrolidone–carboxylic acid becomes soluble in the ethereal solution. The ethereal diazomethane is immediately evaporated to dryness with a gentle stream of dry nitrogen gas. The ester is dissolved in 1.0 ml of dry methanol and a 10-fold molar excess of lithium borohydride is added. This excess is necessary to ensure complete reduction since lithium borohydride does react with methanol. Other solvents without active hydrogen atoms including tetrahydrofuran and dichloro-

[15] A. C. Chibnall, C. Haselbach, J. L. Mangan, and M. W. Rees, *Biochem. J.* **68**, 122 (1958).

methane are less satisfactory due primarily to the decreased solubility of the ester. The reduced product is dissolved in 1.0 ml of distilled water and deionized by passage through a column, 0.5 × 10 cm, of mixed bed ion exchange resin (Dowex 50W-X8 and Dowex 1-X8). The column is washed with 1.0 ml of distilled water. The reduced ester, which does not adhere to the resin, is adjusted to 6 *N* HCl and hydrolyzed at 110° for 5 hr. The HCl is removed by repeated evaporation *in vacuo* and the resulting γ-amino-δ-hydroxyvaleric acid further purified by preparative thin-layer chromatography on Silica Gel F-254 (0.5 mm thickness) plates in phenol–water (100:20, w/v). It migrates as a single ninhydrin spot (R_f 0.28). The R_f for α-amino-δ-hydroxyvaleric acid is 0.32. The amino alcohol also gives a single ninhydrin spot ($R_f = 0.21$) on paper chromatography in 1-butanol–acetic acid–water (4:1:5). This mobility is identical with that reported by Chibnall *et al.*[15] for γ-amino-δ-hydroxyvaleric acid.

Characterization of the Product of the Membrane-Bound L-Glutamic Acid Activating Enzyme

AMP Exchange Assay. AMP exchange is assayed in 200 μl incubation mixtures containing 2 m*M* $MgCl_2 \cdot 7H_2O$; 100 m*M* sodium glutamate, pH 7.5; 100 m*M* glycylglycine, pH 8.6; 2.5 m*M* ATP · Na_2; 2.5 m*M* potassium fluoride; 0.5 m*M* dithiothreitol; 1 m*M* sodium pyrophosphate; 1 m*M* 5′-AMP and 0.25 μCi of [^{14}C]AMP (specific activity 61 mCi/mmol). After incubation for 30 min at 37°, the reaction mixture is spotted on Whatman No. 3 MM paper and electrophoresed for 2 hr at 2000 V in pyridine–acetic acid–water (6:60:1734) pH 3.6. ATP is localized by radiochromatographic scanning, cut out, and quantitated by scintillation counting. Alternatively, the reaction mixture can be chromatographed on PEI-cellulose thin layer plates as described below and radioactivity in ATP, ADP, or AMP quantitated.

Separation of ATP, ADP, and AMP. ATP, ADP, and AMP are separated on phosphoethyleneimine (PEI)-cellulose thin layer plates as described by Randerath and Randerath.[16] Separation is accomplished by a stepwise gradient elution procedure in which the plates are developed initially in 0.5 *M* sodium formate, pH 3.4 (30 sec) followed by a 2-min development in 2.0 *M* sodium formate, pH 3.4. Final development is carried out in 4 *M* sodium formate, pH 3.4 and terminated when the solvent front reaches the end of the plate.

Assay for Racemization of L-Glutamic Acid. Proof of the stereochemical configuration of glutamic acid in the glutamyl hydroxamate formed in

[16] K. Randerath and E. Randerath, this series, Vol. 12A, p. 323.

the activation assay is provided by chromatographic separation of diastereoisomeric dipeptides obtained by derivatization with L-leucine–NCA as described above. Glutamyl hydroxamate formed in the incubation mixture is first isolated by high-voltage electrophoresis on Whatman No. 3 MM paper in pyridine–acetic acid–water (3 : 3 : 194). The compound is eluted from the paper with 30% ethanol, Millipore filtered (HA, 0.45 μm), lyophilized, and hydrolyzed in 4 *N* HCl for 30 min at 110°. After evaporating to dryness, the free glutamic acid is isolated by preparative paper chromatography on Whatman No. 3MM paper in ethanol–1 *M* ammonium acetate, pH 7.5 (7 : 3). Glutamic acid, having a lower R_f than the hydroxamate (R_f, 0.45 for glutamic acid, 0.60 for glutamyl hydroxamate), is eluted with 30% ethanol, lyophilized, and suspended in 0.45 *N* $Na_2B_4O_7 \cdot 10H_2O$, pH 10.2. The diastereoisomeric dipeptides (L-Leu-D-Glu and L-Leu-L-Glu) are synthesized by reaction with L-leucine–NCA and resolved on a Beckman amino acid analyzer.[2] Alternatively, a simpler thin layer chromatographic assay for radiolabeled diastereoisomers of glutamic acid, as described by Barooshian *et al.*,[17] can be used. In this procedure, diastereoisomeric dipeptides are separated on Silica Gel HA plates (0.25 mm) in 1-butanol–acetic acid–water (4 : 1 : 1). L-Leu-L-Glu migrates ahead of L-Leu-D-Glu and is clearly resolved by autoradiography on Kodak medical X-ray film. The amount of radioactivity in the dipeptide is quantitated by scintillation counting after scraping 2-mm sections from the TLC plate.

Properties of the Membranous Polyglutamyl Synthetase

Kinetics, Effect of pH, Protein and Substrate Concentration. The rate of incorporation of L-glutamic acid into PGA is constant with time of incubation for ~2 hr and proportional to enzyme concentration. The optimum pH is 7.2 in both phosphate and Tris buffer. The apparent K_m for L-glutamic acid is 0.5 m*M*.[1]

Requirements for Incorporation of L-Glutamic Acid into γ-D-Glutamyl Polymers. Incorporation of L-glutamic acid into PGA requires ATP (4 m*M*), Mg^{2+} ion (10 m*M*) and to a lesser yet demonstrable extent, K^+ ion (0.2 *M* KCl) and dithiothreitol (8 m*M*).[1] The reaction is specific for ATP and the requirement for Mg^{2+} ion cannot be replaced by Mn^{2+} ion. Enzyme activity is stimulated 2-fold by 0.1 *M* glycerol and 0.1 *M* dimethyl sulfoxide. A 2- to 3-fold increase in activity is obtained when the mem-

[17] A. V. Barooshian, M. J. Lautenschlegger, J. M. Greenwood, and W. G. Harris, *Anal. Biochem.* **49,** 602 (1972).

brane fraction is prepared by either sonic oscillation or pressure disintegration in the presence of 0.1 *M* glycerol and 0.1 *M* dimethyl sulfoxide.

Substrate Specificity. The activity of the membranous polyglutamyl synthetase to incorporate radiolabeled amino acids into polymeric products is markedly specific for L-glutamic acid.[1] D-Glutamic acid is not incorporated into low- or high-molecular-weight products by the PGSC prepared by either alumina grinding, sonication, or pressure disintegration. The incorporation of L-glutamic acid is not related to peptidoglycan synthesis since no incorporation of L-[^{14}C]alanine is observed. A mechanism for the indirect conversion of L-glutamic acid to the D isomer in extracts from *B. licheniformis* was proposed earlier by Thorne *et al.*[18] to involve (1) a transamination between L-glutamic acid and pyruvate forming α-ketoglutarate and L-alanine, (2) a direct racemization of L-alanine to D-alanine, and (3) the formation of D-glutamic acid via a specific transaminase which catalyzes a transamination between α-ketoglutarate and D-alanine. This sequence of reactions is not involved in the membrane-mediated synthesis of γ-D-glutamyl polymers since α-ketoglutarate, a key intermediate in the proposed reaction sequence, has no effect on L-[^{14}C] glutamate incorporation, even when added to the incubation mixture in a 10-fold molar excess. Similarly, a 5-fold molar excess of D-glutamic acid also has no effect.[1]

Effects of Nucleotides and Thiol Reagents on Polyglutamyl Synthetase; Evidence for Thioester Linkage. The membrane-mediated synthesis of PGA is not influenced by the exogenous addition to the reaction mixture of DPN^+, TPN^+, TPNH (2 m*M*), or pyridoxyl phosphate (0.4 m*M*). No effect is observed with 3-phosphoglyceric acid, acetyl CoA, or CTP at 2 m*M*. *N*-Ethylmaleimide at 1.0 m*M* inhibits nearly completely PGA synthesis, suggesting that an essential thiol group is functionally involved in activation, racemization, or polymerization.[1] Subsequent studies have shown that L-glutamic acid activation is not inhibited by *N*-ethylmaleimide even at 10 m*M*. This places the active thiol group at a reaction step distal to activation. Evidence in support of the involvement of a high energy thioester is derived from the fact that labeled glutamic acid incorporated into the membrane fraction can be released by alkaline cleavage (pH 10, 90° for 20 min) and hydroxylamine (3 *M*, pH 6.1, 60° for 20 min). Both these treatments liberate the growing peptide chains in tyrocidine biosynthesis. It has been established[19,20] that these growing antibiotic peptide chains are covalently linked to the enzyme complex by thioester linkages. Thus, these results are consistent with the possible involvement of a high

[18] C. B. Thorne, C. G. Gomez, and R. D. Housewright, *J. Bacteriol.* **69,** 357 (1955).
[19] R. Roskoski, Jr., W. Gevers, H. Kleinkauf, and F. Lipmann, *Biochemistry* **9,** 4839 (1970).
[20] R. Roskoski, Jr., H. Kleinkauf, W. Gevers, and F. Lipmann, *Biochemistry* **9,** 4846 (1970).

energy thioester. In polyglutamyl synthesis, an enzyme-bound sulfhydryl may serve as the initial acceptor of the activated glutamyl moiety (possibly from adenylate) and thus function as the anchor or site within the multienzyme complex for subsequent polymerization of glutamyl residues.[3]

Effects of Glutamyl Dipeptides on Polyglutamyl Synthetase Activity. Incorporation of [^{14}C]glutamic acid into PGA is inhibited 85% by α-L-glutamic acid–L-glutamic acid at 8 m*M* and 70% γ-L-glutamic acid–L-glutamic acid (8 m*M*). Under these same conditions, 8 m*M* α-D-glutamic acid–D-glutamic acid inhibited incorporation 30%. The inhibition by α-L-glutamic acid–L-glutamic acid is not due to the dipeptide serving as a better substrate and, therefore, being preferentially polymerized. Rather it appears this dipeptide is a competitive inhibitor of the γ-D-glutamyl polymerase.[1]

Effect of Inhibitors of Protein Synthesis and Hydrolytic Enzymes on Membrane-Mediated Synthesis of Polyglutamic Acid. The membranous polyglutamyl synthetase is resistant to pretreatment with RNase, phospholipase D, and lipase while sensitive to pretreatment with lysozyme and Pronase. Slight (22%) loss in the biosynthetic capability is observed upon pretreatment of the membrane fraction with phospholipase C.[1] Synthesis is refractory to pretreatment of the membranes with RNase. This result is consistent with inhibitor studies which have shown that PGA is not synthesized by the regular mechanisms involved in protein synthesis. Thus, chloramphenicol (240 μg ml^{-1}), actinomycin D (100 μg ml^{-1}), puromycin (0.5 m*M*), or rifampicin (0.5 m*M*) have no effect on the enzymatic synthesis of PGA.[1]

Direction of Polyglutamyl Chain Elongation. Studies to determine if polymer extension occurs by addition of glutamyl moieties to the NH_2- or COOH-terminal end of the growing nascent chain are carried out by *in vitro* pulse-chase experiments. Radioactivity in the NH_2-terminal residues of PGA is determined after dansylation, acid hydrolysis, and isolation of DNS-Glu derivatives by chromatography on polyamide plates as described.[2] Results from these experiments indicate that only those residues of the polymeric chains that are added last contain the same amount of radioactivity in DNS-Glu as those that are synthesized entirely in the presence of [^{14}C]glutamic acid. The polyglutamyl chain is thus lengthened by sequential addition of new glutamyl residues to the NH_2 terminus of the growing nascent chain. *In vitro* construction of these polymers is initiated when the activated γ-carboxyl group is transferred to the NH_2 terminus of a glutamyl residue of an endogenous glutamyl acceptor. This mechanism of elongation is analogous to the synthesis of homopolysaccharides, mucopolysaccharides, and the inner core region of lipopolysaccharides and in contrast to the assembly of proteins, fatty acids, cyclic polypeptide antibiotics, and the O-antigen moiety of lipopolysaccharides.[5]

Endogenous Acceptor Involved in Synthesis of PGA. There is no evidence for the involvement of small-molecular-weight oligopeptides as intermediates in the membrane-mediated synthesis of polyglutamic acid.[1] Newly synthesized polymers remain associated with the membranous enzyme complex and can be released by alkaline cleavage or hydroxylamine. Although small amounts of glutamic acid or Glu-NHOH are observed, essentially all of the radioactivity is excluded from Sepharose 2B. These results are interpreted to imply the involvement of an endogenous acceptor of glutamyl residues in the biosynthesis of PGA. Two experimental observations support this idea. First, the apparent size of polyglutamyl polymers synthesized during a 45-sec incubation period appears identical in size to those synthesized for 30 min. Since the exclusion limits of Sepharose 2B is estimated at approximately 2×10^7 for polysaccharides and 4×10^7 for proteins, it is evident that polymers formed in 45 sec attain their apparent macromolecular size by transfer of activated glutamyl residues to a preformed, endogenous acceptor. *De novo* synthesis of a γ-D-glutamyl polymer of only 1×10^6 daltons would require polymerizing glutamic acid at a rate of ~10,000 residues/min. This rate is approximately an order of magnitude greater than the rate of which *E. coli*, growing under optimal conditions, can synthesize proteins.[2] Second, structural studies show that some γ-D-glutamyl polymers are present endogenously in the membranous PGSC.[2]

Properties of the Enzymatically Synthesized γ-D-Glutamyl Polymers

Rationale. The properties of the poly γ-D-synthetase complex described above support the contention that these capsular polymers are synthesized by a sequence of membrane-associated enzymatic reactions rather than a mechanism involving the concerted action of a several soluble transaminases, transamidases, or transpeptidases, as was originally postulated.[21] Of fundamental importance to the establishment of this concept is proof that the enzymatically synthesized polymers are structurally identical with the native poly(γ-D-glutamyl) capsule synthesized *in vivo*. The purified, *in vitro* synthesized, and native capsular polymers were characterized chemically, physically, and enzymatically, as described below.

[21] W. J. Williams and C. B. Thorne, *J. Biol. Chem.* **210**, 203 (1954).

Isolation and Compositional Analysis of Enzymatically Synthesized γ-D-Glutamyl Polymers. The product of the membranous polyglutamyl synthetase, after exhaustive digestion with Pronase, elutes as a large-molecular-weight component in the column void volume of Sephadex G-200 run in 1.0 *M* NaCl. Glutamic acid is the only labeled amino acid recovered after complete acid hydrolysis in 6 *N* HCl at 110° for 22 hr and chromatography on an amino acid analyzer.[1] The specific activity of glutamic acid in the enzymatically synthesized polymers is 90.2% that of the initial specific activity of glutamic acid used in the incubation mixture. These data indicate a considerable amount of *de novo* synthesis, in contrast to the attachment of a few glutamyl residues to preexisting PGA. They also indicate that a small amount of glutamic acid, also resistant to Pronase, is present endogenously in the membrane enzyme complex. This unidentified component may link PGA to the membrane.

Optical Configuration of Glutamyl Moieties. The optical configuration of glutamic acid recovered from the enzymatically synthesized polymers and the native capsular polymers synthesized *in vivo* is determined by chromatographic separation of diastereoisomeric dipeptides. At least 92.5% of the glutamyl residues isolated from the enzymatically synthesized polymers are of the D configuration. Similarly, the native capsular polymer synthesized *in vivo* contains at least 90% of the D isomer.[2] The extent to which the 7 to 10% L isomer contributes as an authentic component in these polymers is not known with certainty, although in the enzymatically synthesized polymers it may be derived in part from a small amount of endogenous glutamic acid. Alternatively, the L isomer is likely to result from racemization during hydrolysis since a chemically synthesized α-D-glutamyl polymer yields 7.2% of the L isomer.[2] Confirmation that at least 90% of the glutamyl residues in the enzymatically synthesized polymers are of the D configuration was obtained by gas liquid chromatographic separation of the *N*-TFA-L-proline–D-glutamic acid and *N*-TFA-L-proline–L-glutamic acid.[2]

Determination of the Amide Linkage. The amide linkage in both the enzymatically synthesized PGA and the native capsular polymers was determined to be γ by their susceptibility to acid-catalyzed hydrolysis.[2] As shown in Fig. 1, radiochromatographic and ninhydrin analysis of the hydrolysates following treatment with 2 *N* HCl at 110° for 30 min shows a loss of radioactivity in the polymer with a concomitant and stoichiometric increase of radioactivity in a series of glutamyl oligopeptides with varying degrees of polymerization. After 60 min of hydrolysis, essentially all of the radioactivity is present as glutamylglutamylglutamic, glutamylglutamic, and glutamic acid while after 120 min, the oligopeptides are com-

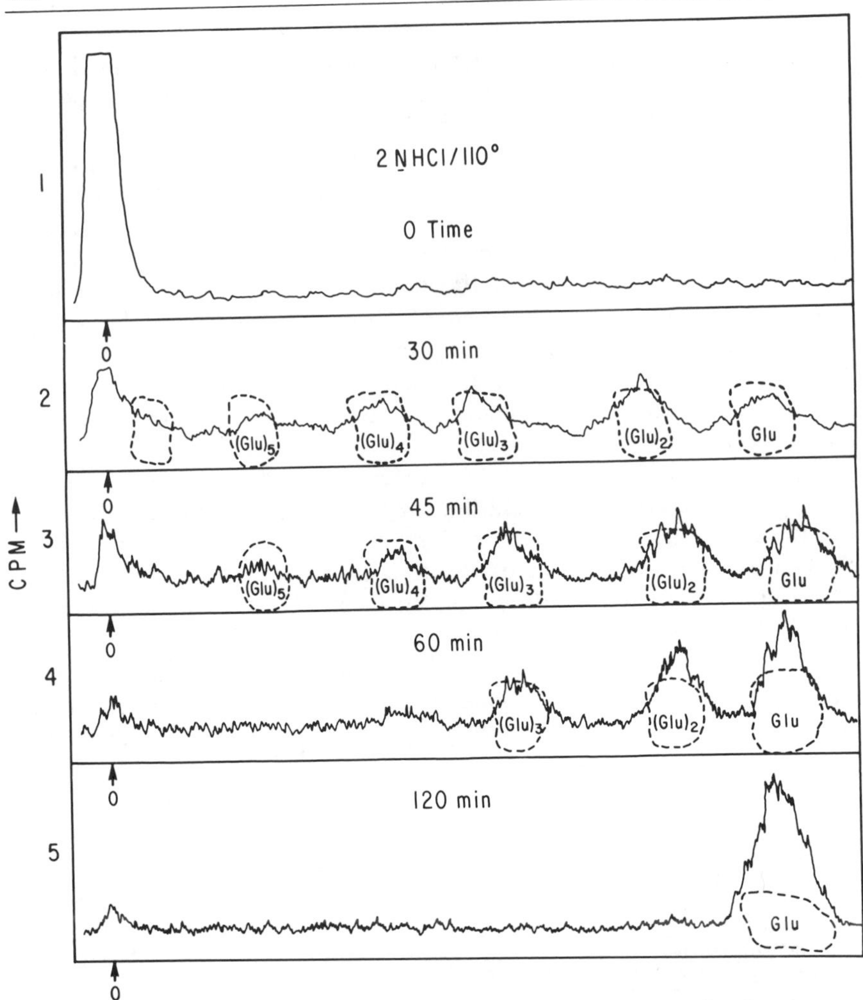

FIG. 1. Kinetics of acid hydrolysis of the enzymatically synthesized and native γ-D-glutamyl capsular polymers. Both polymers were subjected to acid-catalyzed hydrolysis in 2 *N* NCl at 110°. At the times indicated, aliquots were removed and chromatographed on Whatman 3MM in 1-butanol : acetic acid : water (40 : 8.8 : 20). The chromatograms were scanned for radioactivity (biosynthetic product, radioactivity profile) or stained with ninhydrin (native capsular polymer, dashed lines) as described in Ref. 2.

pletely hydrolyzed to glutamic acid. This lability of γ-amide glutamyl linkages to acid is in marked contrast to the stability shown by α-amide glutamyl linkages. Hydrolysis of authentic (chemically synthesized) poly(α-D-glutamic acid) (molecular weight 70,700) under identical conditions shows no ninhydrin-reactive products before 2 hr, a time after which only a small amount of glutamylglutamic is detected. After 4 hr of hydrolysis, free glutamic acid is detected while no glutamyl oligopeptides larger than glutamylglutamic are observed. These data establish that both the enzymatically synthesized polymers and the native capsular polymers produced *in vivo* have identical acid lability profiles consistent with the presence of γ-amide linkages.

Confirmation of the presence of only γ-amide linkages in both polymers is provided by chemical degradation and subsequent identification of the products. Lithium borohydride reduction of the methylated polymers followed by hydrolysis yields only γ-amino-δ-hydroxyvaleric acid. As illustrated by the sequence of reactions shown in Fig. 2, γ-amino-δ-hydroxyvaleric acid can only arise from an γ-amide-linked glutamyl polymer while an α-amide-linked glutamyl polymer would yield α-amino-δ-hydroxyvaleric acid. Indeed, identical treatment of the chemically synthesized α-D-glutamyl polymer (molecular weight 70,700) yields only α-amino-δ-hydroxyvaleric acid. While the two amino alcohols can be separated by thin layer chromatography on Silica Gel F-254 plates, greater resolution is obtained by electrophoretic separation as shown in Fig. 3. At pH 3.0 α-amino-δ-hydroxyvaleric acid is a dipolar zwitterion and has no electrophoretic mobility while γ-amino-δ-hydroxyvaleric acid

A: $\leftarrow NH-CH[(CH_2)_2-CO-O^-]-CO\rightarrow \xrightarrow{CH_2=N^+=N^-} \leftarrow NH-CH[(CH_2)_2-CO-O-CH_3]-CO\rightarrow \xrightarrow{LiBH_4} \leftarrow NH-CH[(CH_2)_2-CH_2OH]-CO\rightarrow \xrightarrow{H^+} H_3N^{\oplus}-CH[(CH_2)_2-CH_2OH]-CO-O^{\ominus}$

α-AMINO-δ-HYDROXY-VALERIC ACID

B: $\leftarrow NH-CH[(CH_2)_2-CO\rightarrow]-CO-O^- \xrightarrow{CH_2=N^+=N^-} \leftarrow NH-CH[(CH_2)_2-CO\rightarrow]-CO-O-CH_3 \xrightarrow{LiBH_4} \leftarrow NH-CH[(CH_2)_2-CO\rightarrow]-CH_2OH \xrightarrow{H^+} H_3N^{\oplus}-CH[(CH_2)_2-CO-O^{\ominus}]-CH_2OH$

γ-AMINO-δ-HYDROXY-VALERIC ACID

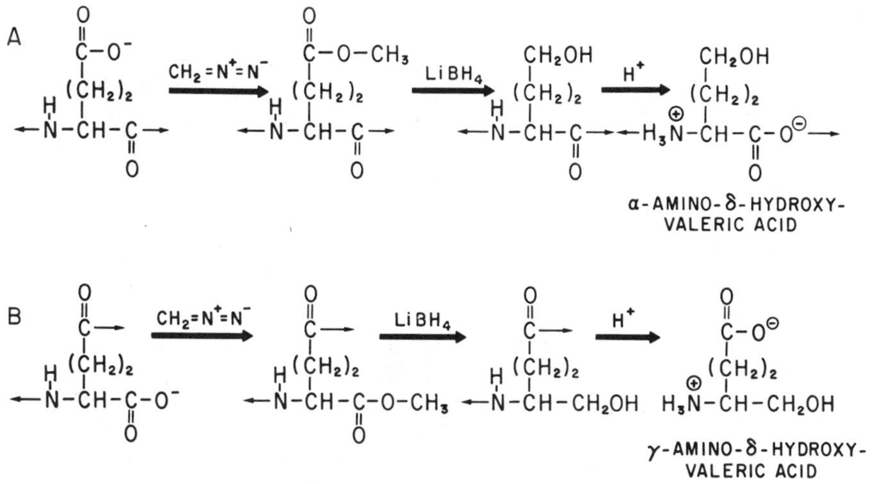

FIG. 2. Reaction scheme illustrating characterization of the amide linkages (A, α; B, γ) in polyglutamic acid. The enzymatically synthesized polymers and the native capsular polymers synthesized *in vivo* were methylated with diazomethane ($CH_2=N^+=N^-$) and reduced with lithium borohydride ($LiBH_4$). The reduced polymers were then hydrolyzed in 2 *N* HCl at 110° for 4 to 5 hr and the amino alcohol characterized by electrophoresis as shown in Fig. 3.

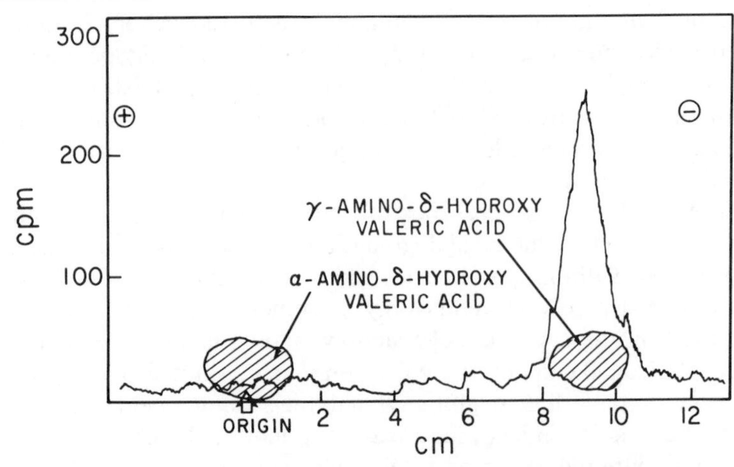

FIG. 3. Proof of γ-amide linkage in the enzymatically synthesized glutamyl polymers. The enzymatically synthesized capsular polymers which were methylated, reduced, and hydrolyzed as described in Fig. 2 were subjected to paper electrophoresis at pH 3.0 in pyridinium acetate at 60 V/cm. The distribution of radioactivity was then determined by radiochromatographic scanning and compared to known standards of the two amino alcohols which were subjected to coelectrophoresis and detected by ninhydrin staining. The unlabeled native capsular polymers synthesized *in vivo* (not shown) were treated in an identical manner and the γ-amino-δ-hydroxyvaleric acid was detected with ninhydrin.

possesses a net positive charge and, therefore, exhibits a cathodal mobility. That all of the radioactivity from the enzymatically synthesized polymers and the ninhydrin-reactive material from the native capsular polymers synthesized *in vivo* is associated with γ-amino-δ-hydroxyvaleric acid provides proof for the sole involvement of γ-amide glutamyl linkages in both polymers. A standard of γ-amino-δ-hydroxyvaleric acid is prepared by $LiBH_4$ reduction of the methyl ester of 5-oxo-2-pyrrolidine carboxylic acid followed by acid hydrolysis to open the ring as described in Ref. 2.

Sedimentation Equilibrium Analysis of Native and Biosynthetic γ-D-Glutamyl Polymers. Detailed ultracentrifugal studies have established that the biosynthetic polymers are polydisperse with weight-average molecular weight distribution values ranging from 1.72×10^5 to 3.63×10^5. Similar studies carried out on native γ-D-glutamyl capsular polymers synthesized *in vivo* show a much greater degree of polydispersity with apparent weight-average molecular weight distribution values ranging from 8.4×10^4 to 1.15×10^6.[2]

Digestion of Enzymatically Synthesized Polymers with Capsular Poly(γ-D-glutamyl) Depolymerase. The enzymatically synthesized polymers are susceptible to enzymatic degradation by a poly(γ-D-glutamyl) depolymerase isolated from the culture filtrate of *B. licheniformis* after 4 days of growth.[1] After this period of incubation, the viscosity of the culture filtrate drops to nearly that of the solvent indicating that the capsular material synthesized *in vivo* at earlier times has been nearly completely hydrolyzed by the extracellular enzyme. Although the depolymerase has not been extensively purified, it has no activity against authentic poly(α-D-glutamic acid). In addition, it is the only enzyme which we have found that catalyzes the hydrolysis of the poly(γ-D-glutamyl) capsular polymers. These results provide confirmatory evidence that the enzymatically synthesized polymers are structurally identical with the capsular poly(γ-D-glutamyl) polymer synthesized *in vivo*.

Properties of the Membranous L-Glutamic Acid Activating Enzyme

Introduction. The L-glutamic acid activating enzyme catalyzes Reaction (1). It is the only enzymatic activity within the PGSC that has been resolved. This has permitted us to determine that racemization of L-glutamic acid occurs after activation. It has not been possible, however, to reconstitute the activating enzyme back into a membrane complex that synthesizes γ-D-glutamic polymers.

Requirements for Activation of L-*Glutamic Acid*. The PGSC catalyzes the formation of glutamyl hydroxamate in the presence of ATP, Mg^{2+}, hydroxylamine, and L-glutamic acid.[3] The pH optimum for activation is 7.5. The K_m for L-glutamic acid is 11 mM and the V_{max}, 15 nmol/min/mg of protein.[22] Activation is dependent on Mg^{2+}, ATP, and glutamic acid. Only 5 to 10% of the ATP-dependent activation is observed with GTP or CTP. While the lysozyme solubilized glutamyl activating enzyme is inactive in synthesizing PGA, it is approximately 5 times more active in the ATP-dependent activation of glutamic acid than the membrane-associated enzyme. Glutamate activation can also be studied by [^{32}P]pyrophosphate exchange since both the membrane-bound and lysozyme solubilized activating enzyme catalyze [^{32}P]pyrophosphate exchange with ATP.[3] The apparent K_m for L-glutamic acid in the pyrophosphate exchange reaction is 5 mM.[22] Activation is partially sensitive to the glutamine and methionine sulfoximine. The product of the activation reaction, even in the presence of these two inhibitors of glutamine synthetase, is consistent with γ-L-glutamyl adenylate and not γ-L-glutamyl phosphate.[3]

[22] The K_m for L-glutamic acid in the poly(γ-D-glutamyl) synthetase assay is 0.5 mM. Thus, the reliability of the K_m values reported for the activation reaction by either the hydroxamate or [^{32}P]pyrophosphate exchange seems questionable. In the hydroxamate assay, the potentially inhibitory effect of 800 mM hydroxylamine cannot be properly assessed. Similarly, the 5 mM value obtained in the pyrophosphate exchange reaction may be related to the different experimental conditions under which this partial reaction is run.

Uncoupling of Poly(γ-D-glutamyl) Capsule Synthesis from Glutamic Acid Activation. N-Ethylmaleimide, a potent inhibitor of the membranous polyglutamyl synthetase,[1] can be used to distinguish features of the activation reaction from polymerization (Fig. 4). *N*-Ethylmaleimide at 1 m*M* does not inhibit L-glutamic acid activation yet inhibits by greater than 95% the incorporation of glutamate into PGA.[3] This result suggests than an active thiol group, possibly the thioester reported previously to be implicated in polymer synthesis,[1] is involved functionally in a reaction distal to activation. Uncoupling of activation and polymerization can also be accomplished by surfactants since Triton X-100 at 0.5% or sodium dodecyl sulfate at 0.2% have no effect on activation of L-glutamic acid but inhibit capsular polymer synthesis.

Identification of the Initial Activated Intermediate: Determination of the Stereochemical Configuration of γ-L-glutamyl Hydroxamate (*γ-L-Glu-NHOH*). A major question with respect to L-glutamic acid activation is whether racemization occurs simultaneously with or subsequent to activation. The PGSC cannot activate, racemize or polymerize D-glutamic acid.[1] In order to study this problem directly, the stereochemical configuration of the glutamyl residues isolated from the γ-glutamyl hydroxamate

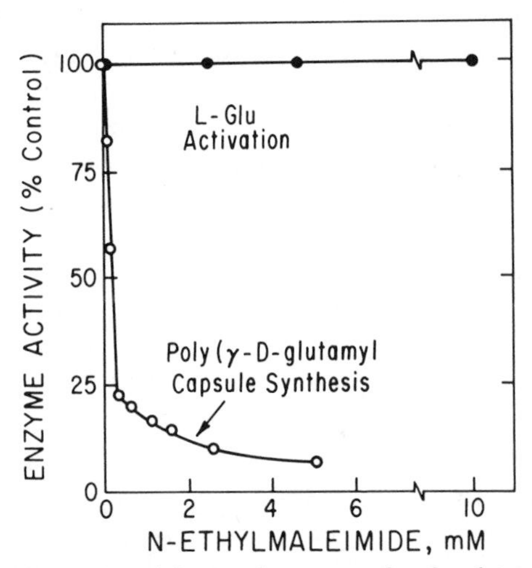

FIG. 4. Differential inhibition of the membrane-associated L-glutamyl activating enzyme and the polyglutamyl synthetase by *N*-ethylmaleimide. Standard incubation mixtures to measure poly(γ-D-glutamyl) capsule formation and L-glutamic acid activation (hydroxamate assay) were prepared and assayed at 37° for 60 min. Eash set of incubation mixtures contained *N*-ethylmaleimide at the concentration indicated. The 100% activity in the L-glutamic acid activation assay represented the formation of 475 nmol of γ-L-Glu · NHOH/mg of protein and, in the capsular polymer assay, the incorporation of 51 nmol of [U-^{14}C]glutamic acid/mg of protein.

was determined by dipeptide analysis and shown to be of the L-configuration. Thus, inversion of stereochemical configuration must occur after activation.[3]

Evidence that the γ-L-Glu-NHOH Is Derived from γ-L-Glu-AMP. As shown in Fig. 5, formation of glutamyl hydroxamate is inhibited by 5′-AMP. It is also inhibited by inorganic pyrophosphate. Since neither 5′-AMP nor pyrophosphate should inhibit the formation of glutamyl phosphate, these results supported the conclusion that the first activated intermediate catalyzed by the PGSC is glutamyl adenylate and not glutamyl phosphate. The fact that PGA formation is inhibited by 2 m*M* *N*-ethylmaleimide (Fig. 4) and that this sulfhydryl group blocking agent does not inhibit activation of L-glutamic acid suggests that a glutamyl thioester may occur as an intermediate subsequent to the activation reaction. Thioesters are also sensitive to hydroxylamine but since Glu-NHOH formation also occurs in the presence of 2 m*M* *N*-ethylmaleimide, this establishes that it is the glutamyl adenylate and not the putative glutamyl thioester that is the activated intermediate from which the glutamyl hydroxamate is derived.

Characterization of the Amide Linkage in L-Glu-NHOH. The amide linkage in the enzymatically synthesized polymer was shown to be γ, as described above. If the Glu-NHOH formed above is a relevant intermedi-

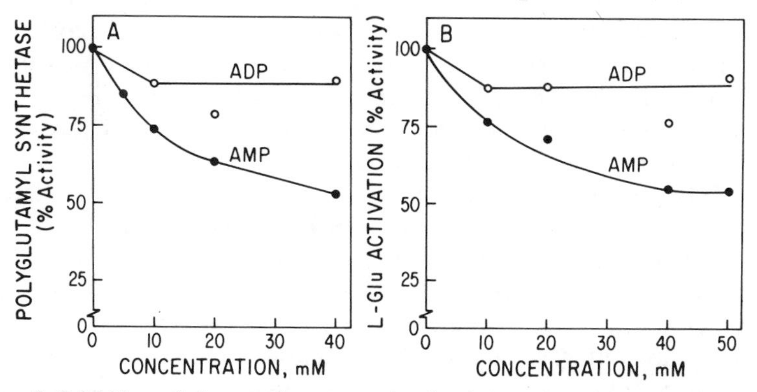

FIG. 5. Inhibition of the membrane-associated poly(γ-D-glutamyl) synthetase and L-glutamic acid activating activity by 5′-AMP. (A) Polyglutamic acid synthetase activity was assayed at 37° for 30 min in standard incubation mixtures. The 100% activity represented the incorporation of 46 nmol of [U-^{14}C]glutamic acid. This corresponded to a specific activity of 161 nmol incorporated/mg of membrane protein. (B) L-Glutamic acid activation was measured in the same membranous enzyme preparation after incubation at 37° for 30 min. The 100% activity corresponded to the synthesis of 225 nmol of γ-L-glutamyl hydroxamate representing a specific activity of 832 nmol of γ-L-Glu · NHOH/mg of membrane protein. The effect of ADP may be due to hydrolysis of ADP to AMP.

ate in PGA synthesis, it would also be expected to be of the γ configuration. To examine this possibility directly, enzymatically synthesized Glu-NHOH was isolated and the amide linkage characterized by its electrophoretic properties at pH 3.6. At this pH, γ-Glu-NHOH is a dipolar zwitterion and has no electrophoretic mobility while the α-amide derivative possesses a net positive charge and exhibits a cathodal mobility. Eighty-nine percent of the labeled glutamyl hydroxamate isolated exhibits a mobility identical to that of authentic γ-glutamyl hydroxamate, providing proof that it is the γ-carboxyl of L-glutamic acid that is activated.[3] This is consistent with the formation of subsequent γ-linked residues in polymer synthesis[2] and is also in accord with the conclusion that the first activated intermediate in polymer synthesis is γ-glutamyl adenylate.

Purification and Characterization. The L-glutamic acid activating enzyme has been purified 40-fold from membranes of *B. licheniformis*.[23] The native enzyme has an estimated molecular weight of 200,000 and SDS–PAGE yields subunits of 51,000.

The enzyme is released from the cell wall–membrane complex with lysozyme or sonication. Purification has involved molecular sieve, hydroxylapatite, cellulose, hydrophobic (butylamine Sepharose), and affinity chromatography on a glutamyl-alkyl-Sepharose. The enzyme is a relatively hydrophobic protein and is quite stable to both ionic and nonionic detergents. It yields one polypeptide band on SDS gels. The partially purified enzyme is heat stable, losing only 30% activity after 3 hr at 60° in buffer. During further purification, however, a protective factor is apparently removed and the enzyme becomes very labile under a variety of conditions. The exogenous addition of membrane phospholipids from *B. licheniformis* appears to stabilize the solubilized enzyme. Unexpectedly, the partially purified enzyme is fully active in 10% Triton X-100 and 0.5% SDS. This stability is critically dependent on the weight ratio of detergent to protein. At a weight ratio of 1.0 g SDS to 1.0 g protein, the enzyme loses only 10% of its activity but at a ratio of 1.5 g detergent to 1.0 g protein, 80% of the activity is lost.

Enzyme activity is extremely sensitive to mercurials and this inhibition is completely reversed by dithiothreitol. The enzyme is inactivated by high concentrations (~100 m*M*) of *N*-ethylmaleimide. Inactivation by *N*-ethylmaleimide, but not by mercury, is prevented by prior formation of glutamyl adenylate. Presumably glutamyl adenylate protects the active site against *N*-ethylmaleimide but not mercury. The purified enzyme does not catalyze racemization of L-Glu.

[23] J. M. Gardner and F. A. Troy, unpublished results.

[26] Synthesis of Folylpolyglutamates

By LINDA D'ARI and JESSE C. RABINOWITZ

Although folate functions in cellular processes as a derivative of tetrahydrofolyl*poly*glutamate, the monoglutamate derivatives are the compounds most commonly used in experimental studies. The monoglutamate form of the folate coenzyme will generally function in place of the polyglutamyl forms. However, the affinity constants of the polyglutamyl forms of the coenzyme are from 10 to 1000 times less than that of the monoglutamyl form for some enzymes. The polyglutamyl derivatives are not available from commercial sources. The procedures that have been described for the synthesis of the folylpolyglutamates include solid phase synthesis[1,2] or procedures employing anhydrous solution chemistry for synthesis of the protected peptides of oligo-γ-glutamic acid.[3–6] It is our impression that extensive purification of products formed by solid phase synthesis is required, and that highly specialized skills in synthetic organic chemistry are required in the alternate methods. We believe that the procedure described here offers several advantages over those previously published. These include chromatographic methods for the purification of intermediates, and the use of water-soluble carbodiimide synthesis that makes the procedures potentially accessible to investigators with ordinary skills in synthetic organic chemistry. Folylpolyglutamates are obtained in yields comparable to the previous methods described and are of at least comparable purity.

Principle

The overall approach to the synthesis of the pteroylpolyglutamates is shown in Figs. 1 and 2:

Figure 1. The N^{10}-trifluoroacetyl derivative of pteroic acid (**I**) is coupled with the γ-glutamyl-polypeptide-*tert*-butyl ester containing from two

[1] C. L. Krumdieck and C. M. Baugh, *Biochemistry* **8,** 1568 (1969).

[2] C. L. Krumdieck and C. M. Baugh, this series, Vol. 66, p. 523.

[3] J. Meienhofer, M. J. Paula, H. A. Godwin, and I. H. Rosenberg, *J. Org. Chem.* **35,** 4137 (1970).

[4] H. A. Godwin, I. H. Rosenberg, C. R. Ferenz, P. M. Jacobs, and J. Meienhofer, *J. Biol. Chem.* **247,** 2266 (1972).

[5] J. K. Coward, K. N. Parameswaran, A. R. Cashmore, and J. R. Bertino, *Biochemistry* **13,** 3899 (1974).

[6] C. N. C. Drey and G. P. Priestley, *J. Chem. Soc., Chem. Commun.* p. 144 (1977).

Copyright © 1985 by Academic Press, Inc.
All rights of reproduction in any form reserved.

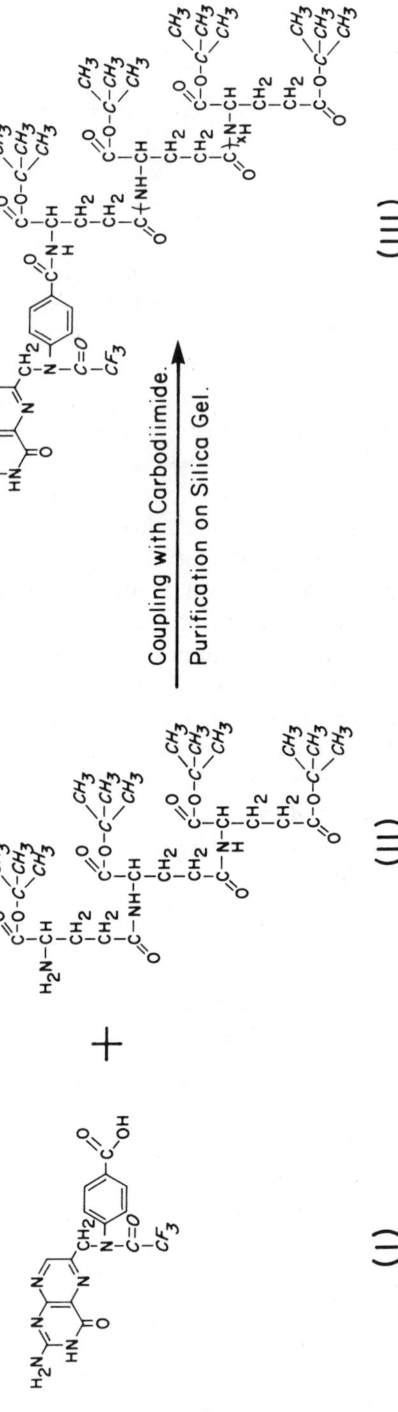

FIG. 1. Condensation to form folate derivatives.

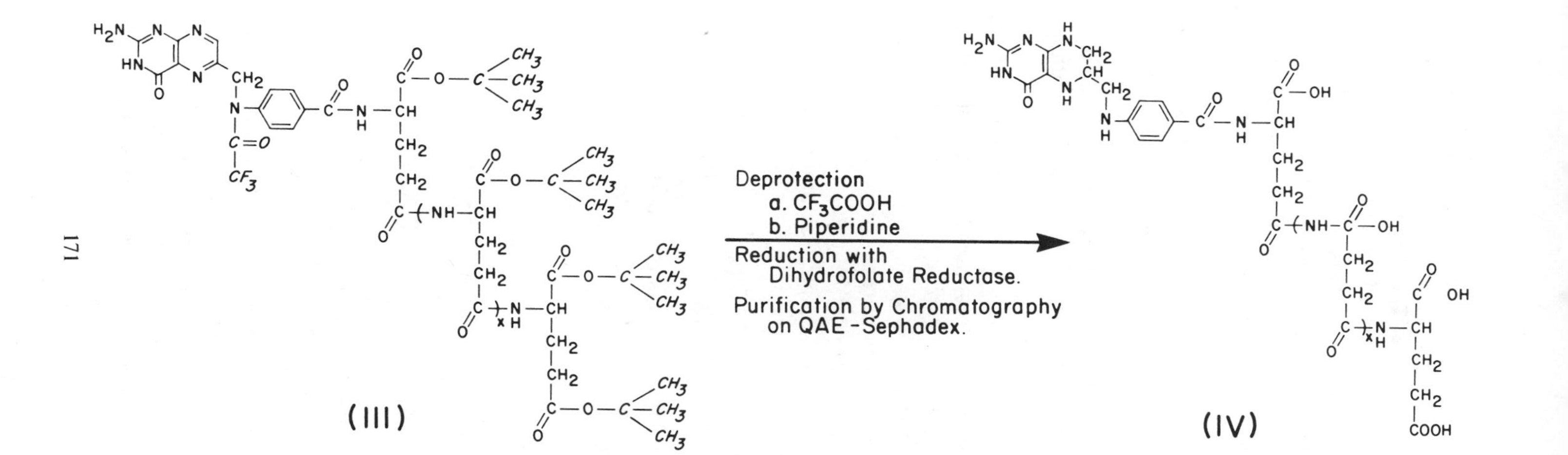

FIG. 2. Formation of tetrahydrofolates.

to five glutamic acid residues (**II**) to form the pteroylpolyglutamate derivative (**III**). A water-soluble carbodiimide is used for this process.

Figure 2. Removal of the trifluoroacetyl and *tert*-butyl protecting groups to yield the pteroylpolyglutamates and their reduction to the 5,6,7,8-tetrahydro form. The trifluoroacetate group, protecting the N-10 position, is removed by treatment with piperidine, and the *tert*-butyl groups, protecting the carboxyl groups, are removed by treatment with trifluoroacetic acid. Reduction is carried out enzymatically at pH 5.5 with dihydrofolate reductase, and the reduced product is purified by column chromatography.

The starting materials for the reactions illustrated in Fig. 1 are not available from commercial sources. They can be synthesized by the reactions shown in Figs. 3 and 4:

Figure 3. Derivatives of the γ-glutamylpeptides of varying length are synthesized according to the reactions shown here. Benzyloxycarbonyl-L-glutamic acid di-*tert*-butyl ester (**V**) and L-glutamic acid di-*tert*-butyl ester (**VI**) are coupled with a water soluble carbodiimide to yield compound (**VII**). This product is purified on a silica gel column. The carbobenzoxy group is then removed by catalytic hydrogenation to yield the dipeptide (**VIII**) containing a single free amino group. This compound is condensed with benzyloxycarbonyl-L-glutamic acid-α-*tert*-butyl ester (**V**) in another synthetic round to yield the tripeptide. This procedure is repeated two more times to obtain the tetrapeptide and pentapeptide derivatives.

Figure 4. Pteroic acid (**X**) was obtained by fermenting folic acid (**IX**) with a glutamic acid-utilizing *Pseudomonas* species. The pteroic acid (**X**) was converted to the 10-trifluoroacetyl derivative (**I**).

Materials

Since all of the solvents were fresh and of a high grade, they were not distilled prior to use.

The following materials were obtained from the commercial sources indicated:

Benzyloxycarbonyl-L-glutamic acid α-*tert*-butyl ester dicyclohexylamine salt was purchased from Bachem Fine Chemicals (P.O. Box 3426, Torrance, CA. 90510).

1-Ethyl-3(3-dimethylaminopropyl)carbodiimide-HCl was from Sigma Chemical Co., St. Louis, Mo.

Glass supported thin layer plates are preferable. Those used were made by EM Manufacturing Laboratories, and were supplied by MC/B Manufacturing Chemists, Inc., Cincinnati, Ohio.

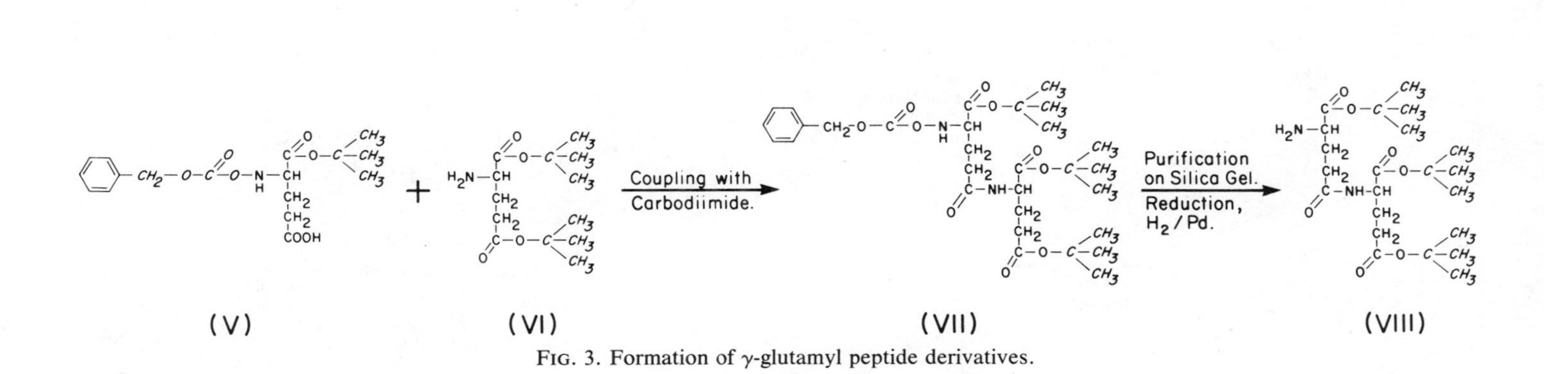

FIG. 3. Formation of γ-glutamyl peptide derivatives.

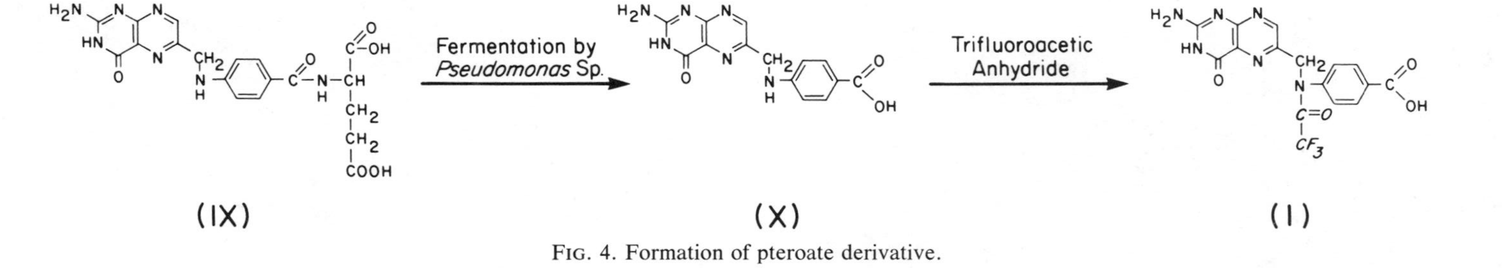

FIG. 4. Formation of pteroate derivative.

L-Glutamic acid di-*tert*-butyl ester (**VI**) was purchased from Bachem Fine Chemicals.

N-Methyl-morpholine, trifluoroacetic acid, and trifluoroacetic anhydride were purchased from Pierce Chemical Co., Rockford, Ill.

Methylene chloride and ethyl acetate were from Matheson, Coleman and Bell Manufacturing Chemists, Norwood, Ohio.

Palladium (5%) on barium sulfate was from Matheson, Coleman and Bell.

Piperidine and *N*,*N*-dimethylformamide were from Eastman Organic Chemicals, Rochester, N.Y.

QAE-Sephadex (A-25) was from Pharmacia Fine Chemical, Piscataway, N.J.

Silica Gel 60, particle size 0.040–0.063 (230–400 ASTM) was from EM.

4,4′-Methylenebis-*N*,*N*(dimethylaniline) was from Fluka Chemical Corp., Hauppauge, N.Y. This was used as a substitute for toluidine, which is a carcinogen.

The following materials were prepared by published procedures:

Dihydrofolate reductase prepared from *Lactobacillus casei*[7] was a generous gift from Dr. R. L. Kisliuk. The preparation contained 33 μU/mg.

Pteroic acid (**X**) was produced from a commercial grade of folic acid (**IX**) according to the fermentative method described by Goldman and Levy[8] and Scott[9] involving the use of a glutamate utilizing *Pseudomonas* species (ATCC #25301). The product was obtained in roughly 50% yield starting with 11 g of folic acid. It was stored in a desiccator in the dark.

N^{10}-Trifluoroacetylpteroic acid (**I**) was prepared by the procedure described by Godwin *et al.*[4] in order to convert the pteroic acid to a form soluble in the condensation reactions and to minimize side reactions. The procedure was carried out on a 3 g scale, and the product was stored in a desiccator in the dark.

Methods

Purity Control

The purity of intermediates and final products of the chemical syntheses was determined by thin layer chromatography. The most satisfactory

[7] E. J. Pastore, L. T. Plante, and R. L. Kisliuk, this series, Vol. 34B, p. 281.

[8] P. Goldman and C. C. Levy, *Biochem. Prep.* **13,** 79 (1971).

[9] J. M. Scott, this series, Vol. 66, p. 657.

TABLE I
R_f OF INTERMEDIATES ON SILICA GEL[a]

Compound	$CHCl_3$/EtOAc = 1/2	$CHCl_3$/MeOH = 5/1
Z-Glu_2-$(OtBu)_3$	0.60	0.75
Z-Glu_3-$(OtBu)_4$	0.53	0.77
Z-Glu_4-$(OtBu)_5$	0.44	0.80
Z-Glu_5-$(OtBu)_6$	0.37	0.79
$Glu(OtBu)_2$	0.2	
$Glu_2(OtBu)_3$	0.15	0.54
$Glu_3(OtBu)_4$	<0.1	0.57
$Glu_4(OtBu)_5$	<0.1	0.58
$Glu_5(OtBu)_6$	<0.1	0.60
TFA-pteroic acid	0	0
TFA-pteroyl-γ-$glu_2(OtBu)_3$	0	0
TFA-pteroyl-γ-$glu_3(OtBu)_4$	0	0.45
TFA-pteroyl-γ-$glu_4(OtBu)_5$	0	0.47
TFA-pteroyl-γ-$glu_5(OtBu)_6$	0	0.50
Z-glu-α-OtBu	0.23	
DCHA	0	
N-Methyllmorpholine	<0.1	
carbodiimide	0	

[a] Z, benzyloxycarbonyl; (OtBu), *tert*-butyl ester; TFA, N^{10}-trifluoroacetyl.

solvent systems found among the many suggested in the literature are listed in Table I. Three methods of visualization were used. Examination under ultraviolet light reveals the aromatic compounds. The carbobenzoxy derivatives and deprotected pteroyl compounds show only faint absorption, but the trifluoroacetylpteroyl derivatives appear as brightly fluorescent spots. Ninhydrin spray detects compounds with free amino groups. Toluidine spray[10] detects both protected and unprotected amino groups. Toluidine is now known to be a carcinogen and should therefore be replaced by 4,4′-methylenebis-*N*,*N*(dimethylaniline). These three methods can be used sequentially on the same plate. In general, glass plates give more sharply defined spots than plastic plates. The only exception in Table I is the chromatography of TFA-pteroic acid on plastic supported cellulose plates.

γ-Glutamyl Peptide Synthesis (*Fig. 3*)

Benzyloxycarbonyl-L-glutamic Acid α-tert-Butyl Ester. The dicyclohexylamine was removed as follows from the salt form of benzyloxycar-

[10] D. E. Nitecki and J. W. Goodman, *Biochemistry* **5**, 665 (1966).

bonyl-L-glutamic acid-α-*tert*-butyl ester that was obtained commercially: a mixture of the salt, 20–50% excess of $KHSO_4$ relative to the dicyclohexylamine, and ethyl acetate/water = 2/1 was stirred vigorously at room temperature for 1 hr. The phases were then separated. The organic phase was extracted three times with water and then the combined aqueous phases were back extracted with ethyl acetate. The combined organic phases were dried over $MgSO_4$, filtered, and concentrated on a rotoevaporator. The oil obtained was dried overnight in an evacuated desiccator containing P_2O_5.

Coupling with Carbodiimide. Synthesis of the dipeptide was carried out in a ground glass stoppered vessel containing 1 equivalent each of benzyloxycarbonyl-L-glutamic acid α-*tert*-butyl ester (**V,** Fig. 3), L-glutamic acid di-*tert*-butyl ester (**VI,** Fig. 3), 1-ethyl-3(3-dimethylaminopropyl)carbodiimide hydrochloride, and methylene chloride as the solvent. When done on a small scale (<1 mmol), volumes were kept as small as feasible; when carried out with 1 mmol or more of substrate, the volume was adjusted so that the concentration of each reactant was 1 mmol/ml. Two equivalents of *N*-methyl-morpholine were added to neutralize the hydrochlorides used. The carbodiimide is not soluble in methylene chloride. It was added as a solid to the reaction mixture as the last component, and dissolved as the reaction proceeded at 4° with stirring. The course of the reaction was determined by examining the reaction mixture by thin-layer chromatography. It was complete after several hours, and the mixture could then be stored overnight in the refrigerator, if necessary.

The reaction mixture was diluted to 50 ml with CH_2Cl_2 (20 ml for a small scale reaction) and washed with equivalent volumes of 1 *M* citric acid, saturated NaCl, water saturated $NaHCO_3$, and saturated NaCl. Any emulsions that form can readily be cleared by centrifugation. The washes were backextracted with CH_2Cl_2. The combined organic layers were then dried over $MgSO_4$, filtered, and concentrated to an oil on a rotary evaporator. The oil was dried overnight over P_2O_5 in an evacuated desiccator. Yields of the benzyloxycarbonyl-*tert*-butyl ester of the dipeptide are about 75% and this value can be taken as representative of the yield obtained for the addition of a single glutamic acid residue in γ-peptide linkage in the coupling reaction.

Purification on Silica Gel. The protected peptide was purified on a silica gel column. A 150 g column was poured in the following solvent mixture: $CHCl_3$/EtOAc = 1/2. The sample was put on the column in a minimal amount of solvent. Fractions (5 ml) were collected and analyzed by TLC. The peak fractions were pooled and concentrated to an oil on a rotary evaporator. The protected peptide seems fairly stable at this point and may be stored for several days without any degradation.

Carbobenzoxy Group Removal. The carbobenzoxy group was removed through the use of catalytic hydrogenation with 5% palladium on $BaSO_4$ in methanol.[3] The amount of catalyst used was 30% by weight of the protected peptide. The catalyst was prewet with ethyl acetate since methanol vapors are potentially explosive in the presence of palladium dust. The protected peptide dissolved in methanol (7–15 ml) was added to the prewet catalyst. Nitrogen was bubbled through the solution for 15 min to flush out oxygen. Then hydrogen was bubbled through the solution. The palladium was kept suspended by continuous stirring with a magnetic stir bar. The CO_2 produced by the decomposition of the carbobenzyoxy groups was monitored by bubbling the outlet tube through a solution of 0.1 *M* $Ba(OH)_2$. The reaction was continued for 15 min after the cessation of $BaCO_3$ formation. The time required for the reaction depends on the amount of Z-peptide. For a preparation of less than 1 mmol, it takes about 0.5 hr; for 5 mmol, it takes 1 to 2 hr. To minimize methanol evaporation during the reaction, the gas is first bubbled through water. At the end of the reaction, nitrogen is bubbled through the mixture for 15 min. The catalyst was removed by suction through GF/B filter paper. The filtrate was quickly concentrated to an oil on a rotoevaporator; dissolved in the appropriate solvent, and used immediately in the next coupling step. Occasionally, the Pd forms a colloid that is impossible to remove by filtration. In this event, a pinch of Norit added to the suspension will trap the Pd. The benzyloxycarbonyl-L-glutamic acid-α-*tert*-butyl ester does not appear to stick to the charcoal in appreciable amounts, and near quantitative yields of the product of this reaction (**VIII**) or the larger peptides are obtained.

Coupling the Peptide to Trifluoroacetyl Pteroic Acid (Fig. 1)

Coupling with Carbodiimide. The protected pteroic acid was coupled to the desired peptide derivative in essentially the same manner used for the synthesis of the peptides. The reaction involved 1 equivalent of each of the four components of the reaction: the pteroic acid derivative (**I**), the *tert*-butyl ester derivative of the γ-glutamyl-polypeptide-*tert*-butyl ester (**II**), carbodiimide hydrochloride, and *N*-methylmorpholine. Because of the insolubility of the trifluoroacetyl pteroic acid, dimethylformamide was used as a solvent. The trifluoroacetyl pteroic acid was suspended in the dimethylformamide, and was gradually solubilized as the reaction proceeded. The reaction was carried out in an ice bath, with stirring, away from direct light.

The procedure for isolating the product was modified from that used in isolating the peptides formed in the carbodiimide reaction in order to remove the dimethylformamide. The reaction mixture was added to 50 ml of ethyl acetate with stirring. The unreacted trifluoroacetyl pteroic acid is insoluble and is removed by centrifugation. The precipitate was washed once with 50 ml of ethylacetate. This wash was saved for the back extraction. The ethylacetate supernatant solution was extracted with 25 ml each of saturated sodium bicarbonate, water, 1 *M* citric acid, water, and saturated NaCl. After back extraction, the organic phases were combined, dried over $MgSO_4$, filtered, and concentrated to an oil on a rotary evaporator. If the product precipitates during concentration, it should be dissolved in methanol, precipitated by the slow addition of water, and collected by centrifugation. The oil is dried overnight over P_2O_5 in an evacuated desiccator.

Purification on Silica Gel. The product (**III**) obtained in the previous reaction was dissolved in methanol and purified on a silica gel column developed with chloroform/methanol = 5/1. The fractions containing the product were pooled and concentrated on a rotary evaporator. The concentrate was then dissolved in methanol and precipitated by the slow addition of water. The product was collected by centrifugation and dried in an evacuated desiccator with P_2O_5.

The small amount of the *tert*-butyl ester of the amino-peptide (**VIII**) that does not react in the coupling comes off this column just before the product (**III**) and complete separation of the two compounds is difficult. The separation is sometimes more successful after a second column. We were unable to find a better solvent system. Another opportunity for removing the peptide derivative occurs during chromatography on QAE-Sephadex, a step included in the preparation of the reduced folylpolyglutamates. The fractions containing the product were pooled and concentrated on a rotary evaporator. The yield of product (**III**) in this reaction is approximately 50–60%. A similar yield is obtained in the comparable reaction carried out with the longer peptides.

Removal of Protecting Groups. The deprotection procedures described by Meienhofer and associates[4] were followed, with a few modifications, for removal of the trifluoroacetyl and *tert*-butyl groups from the product isolated in the previous step.

Removal of *tert*-butyl esters was accomplished with trifluoroacetic acid according to the following procedure and gave near quantitative yields. Approximately 1 ml of trifluoroacetic acid was added per 0.1 mmol of the dried product (**III**). The reaction was carried out at room temperature with occasional swirling for 15 min. At the end of the reaction, the mixture was poured into 10 ml of ice cold ether. The precipitate formed was centrifuged for 10 min at 8000 rpm and washed twice with 10 ml of ice cold ether. The precipitate was then dried for 30–60 min in an evacuated desiccator over P_2O_5.

Removal of trifluoroacetyl groups was accomplished with piperidine

TABLE II
CALCULATED OVERALL YIELDS OF PTEROYL-GLU$_x$ AND TETRAHYDRO-PTEROYL-GLU$_x$ IN CARBODIIMIDE SYNTHESIS AND ENZYMATIC REDUCTION[a]

	Yield (%)	
x	Oxidized	Reduced
2	30	12
3	23	9
4	17	7
5	13	5
6	9	4
7	7	3

[a] Based on yields for individual steps described in the text.

according to the following procedure and gave near quantitative yields. Approximately 10 ml of 0.1 *M* aqueous piperidine were used per 0.1 mmol of product. The reaction was carried out at 15°, away from direct light, with continuous stirring and nitrogen blown gently over the surface, for 1 hr. At the end of the reaction, 2 *N* HCl was carefully added to the solution until the mixture was between pH 2 and 3. The deprotected product precipitated at low pH and was removed by centrifugation. It was washed once with a small amount of ice cold 0.02 *N* HCl. The final product was then dried by lyophilization. During the acidification, the solution often becomes quite gelatinous. If this occurs, remove the precipitate that has formed by centrifugation, and then continue to acidify the supernatant solution.

The final product is stored in a desiccator in the dark. The deprotection steps go to completion. There is no appreciable pteroic acid formed during these reactions. The overall yields of the pteroyl-diglutamic acid to pteroyl-heptaglutamic acid calculated from the yields reported here for the individual steps involved in the syntheses are shown in Table II.

Reduction of the Folylpolyglutamates (Fig. 2)

The folylpolyglutamates were reduced to the 5,6,7,8-tetrahydro derivatives through the use of *Lactobacillus casei* dihydrofolate reductase[7] at pH 5.5. This procedure gave about a 55% yield of (*S*)-isomer as determined by enzymatic analysis. The procedure used is described below in detail. It is a modification of a procedure described in a personal commun-

TABLE III
R_f OF INTERMEDIATES ON CELLULOSE DEVELOPED WITH 5% Na_2HPO_4

Compound	R_f	Micromoles/spot
Pte	<0.1	5
Pte-glu_1	0.21	5
Pte-glu_2	0.37	5
Pte-glu_3	0.52	5
Pte-glu_4	0.61	5
Pte-glu_5	0.70	5
H_2-pte-glu[a]	0.17	10–20
H_4-pte-glu	0.54	10–20
Glu	0.93	50
p-Aminobenzoyl-glu	0.86	10
TFA-pte[b]	0.77	

[a] Reduced compounds do not give very discreet spots.
[b] Must use plastic plate.

ication from R. L. Kisliuk. The reduction is carried out in relatively dilute concentrations of the folate derivative at pH 5.5. To reduce 20 μmol of pteroyl-x-glutamate, 200 ml of deaerated reaction components containing 20 mg dihydrofolate reductase (650 units), 0.1 m*M* pteroyl-x-glutamate, 0.13 m*M* TPNH, 0.05 *M* sodium acetate buffer, pH 5.5, 0.2 *M* mercaptoethanol, in an evacuated filter flask was incubated at room temperature in the dark. The reaction progress is followed by determination of the decrease in A_{340} of an aliquot of the reaction mixture. The reaction is considered complete when the A_{340} is constant. Accurate determinations of the amount of reduced folylpolyglutamate formed is carried out by enzymatic determination with 10-formyltetrahydrofolate synthetase.[11] The reaction is finished in about 40 min. Mercaptoethanol is added to a concentration of 0.5 *M* and the medium is adjusted to pH 7.5 with TRIZMA base.

Purification of the reduced product is carried out by chromatography on a QAE-Sephadex column. Elution is carried out through the use of a linear gradient of Tris-buffer, pH 7.5. For the chromatography of products containing from one to 3 glutamate residues, the gradient used was 200 ml of 0.05 to 1.5 *M* Tris-chloride buffer, pH 7.5 of 0.05 *M* Tris-chloride buffer, pH 7.5. For the products containing 4 and 5 glutamate residues, chromatography was carried out with 1.0 *M* buffer in place of the 1.5 *M* buffer. All buffers contained 0.2 *M* mercaptoethanol. The

[11] J. C. Rabinowitz and W. E. Pricer, Jr., *J. Biol. Chem.* **229**, 321 (1957).

TPN^+, TPNH, and *p*-aminobenzoyl-x-glutamate elute before the tetrahydropteroyl-x-glutamates; the unreduced pteroyl-x-glutamates elute afterward. The tetrahydropteroyl-x-glutamates were all eluted with approximately 0.6–0.7 *M* Tris chloride buffer, pH 7.5. The concentration of the tetrahydropteroyl-diglutamate recovered was approximately 1.0 m*M* while that of the tetrahydropteroyl-pentaglutamate was about 0.4 m*M*.

About 70% of the tetrahydropteroyl-x-glutamates applied to the column were recovered in the peak. The spectrum was used as a criterion of purity and quantitation was based on the enzymatic assay noted above. Another useful criterion for characterization of the folylpolyglutamates is their mobility on cellulose TLC. The mobilities of compounds of interest are shown in Table III. The final product was stored frozen in a test tube covered with a serum cap and flushed with argon. The overall yield of the reduced (*S*)-isomer calculated from the yields reported here for the individual steps including the enzymatic reduction are shown in Table II. These low values emphasize the cumulative effect of the successive application of individual synthetic steps that are not quantitative.

Acknowledgment

The authors are indebted to Dr. Danute E. Nitecki of the Department of Microbiology of the University of California School of Medicine, San Francisco, CA, who originally suggested the use of a water-soluble carbodiimide coupling procedure and was a continuous source of help and advice as we worked out the details of this synthesis.

Section II

Glutamine

[27] Glutamine Synthetase from Mammalian Tissues

By ALTON MEISTER

$$\text{Glutamate} + NH_3 + \text{ATP} \xrightleftharpoons{Mg^{2+}(Mn^{2+})} \text{glutamine} + \text{ADP} + P_i \quad (1)$$

The ω-amidation of glutamate to form glutamine was first observed in slices of animal tissues and it was proposed that this reaction requires energy.[1] Later work showed that the energy requirement could be fulfilled by ATP, and studies with cell-free preparations of glutamine synthetase established the stoichiometry of the reaction.[2–4] The glutamine synthesis reaction was shown to be freely reversible,[5] and from the experimentally determined equilibrium constant, a good estimate of the standard free energy of hydrolysis of ATP was obtained.[5,6] When the enzyme is incubated with 10 m*M* concentrations each of L-glutamate, ammonium ions, and ATP in the presence of Mg^{2+} at pH 7.0 and 37°, equilibrium is attained when about 90% of the L-glutamate is converted to L-glutamine.[5] Substitution of hydroxylamine for ammonia in this system leads to a reaction that goes to greater than 99% of completion. These results are in accord with the fact that the free energy change associated with the hydrolysis of glutamine is about 3500 cal/mol whereas that of γ-glutamyl hydroxamate is about 700 cal/mol.[7]

Glutamine synthetase was first isolated in essentially homogeneous form from sheep brain[8]; a procedure for the isolation of this enzyme has been published in a previous volume of this series[9] (see also ref. 10). The enzyme has also been isolated from the brains of other species including man,[11,12] rat,[12] and pig.[13–15] Its isolation has also been described from

[1] H. A. Krebs, *Biochem. J.* **29,** 1951 (1935).
[2] W. H. Elliot, *Nature (London)* **161,** 128 (1948).
[3] J. F. Speck, *J. Biol. Chem.* **168,** 403 (1947).
[4] F. Leuthardt and E. Bujard, *Helv. Med. Acta* **14,** 274 (1947).
[5] L. Levintow and A. Meister, *J. Biol. Chem.* **209,** 265 (1954).
[6] J. T. Edsall and J. Wyman, *Biophys. Chem.* **1,** 210 (1958).
[7] E. Ehrenfeld, S. J. Marble, and A. Meister, *J. Biol. Chem.* **238,** 3711 (1963).
[8] V. Pamiljans, P. R. Krishnaswamy, G. Dumville, and A. Meister, *Biochemistry* **1,** 153 (1962).
[9] W. B. Rowe, R. A. Ronzio, V. P. Wellner, and A. Meister, this series, Vol. 17A, p. 900.
[10] R. A. Ronzio, W. B. Rowe, S. Wilk, and A. Meister, *Biochemistry* **8,** 2670 (1969).
[11] A. Meister, *in* "The Enzymes" (P. D. Boyer, ed.), 3rd ed., Vol. 10, p. 699. Academic Press, New York, 1974.
[12] S. S. Tate and A. Meister, *in* "The Enzymes of Glutamine Metabolism" (S. Prustner and E. R. Stadtman, eds.), p. 77. Academic Press, New York, 1973.

Copyright © 1985 by Academic Press, Inc.
All rights of reproduction in any form reserved.

chicken liver,[16,17] chinese hamster liver,[18] chick neural retina,[19] rat spleen,[20] and rat liver.[21–23] A method for isolation of the enzyme from rat liver is given below; this procedure is based on the work of Tate *et al.*[21] and that of Ronzio *et al.*[10]

The glutamine synthetases of mammalian origin closely resemble each other with the respect to amino acid composition, subunit structure, and molecular weight. They differ substantially in these respects from bacterial glutamine synthetases, some of which exist in adenylylated forms. The glutamine synthetase of *E. coli* is considered in another chapter of this volume.[24]

Determination of Enzymes Activity

Principles

Glutamine synthetase activity may be followed by measuring the rate of formation of inorganic phosphate, ADP, or glutamine. In some instances, it may be convenient to determine the rate of disappearance of glutamate. Inorganic phosphate may be determined by the method of Fiske and SubbaRow[25] or by a modification of this procedure. ADP may be determined by coupling the glutamine synthetase reaction with those catalyzed by pyruvate kinase and lactate dehydrogenase. By use of reaction mixtures containing labeled glutamate, the disappearance of glutamate and the formation of glutamine may be determined (after separation of these compounds). A commonly used procedure for determining glutamine synthetase activity involves replacement of ammonia by hydroxylamine. In this reaction γ-glutamylhydroxamate is formed, which gives a characteristic color reaction on addition of ferric chloride.[26]

---

[13] L. Jaenicke and J.-C. Jesior, *FEBS Lett.* **90,** 115 (1978).
[14] J. Stahl and L. Jaenicke, *Eur. J. Biochem.* **29,** 401 (1972).
[15] D. R. Rao, K. Beyreuther, and L. Jaenicke, *Eur. J. Biochem.* **35,** 582 (1973).
[16] S. Seyama, Y. Kuroda, and N. Katunuma, *J. Biochem. (Tokyo)* **72,** 1017 (1972).
[17] J. E. Vorhaben and J. W. Campbell, *J. Biol. Chem.* **247,** 2763 (1972).
[18] D. C. Tiemeier and G. Milman, *J. Biol. Chem.* **247,** 2272 (1972).
[19] P. K. Sarkar, D. A. Fischman, E. Goldwasser, and A. A. Moscona, *J. Biol. Chem.* **247,** 7743 (1972).
[20] C. Wu, *Can. J. Biochem.* **55,** 332 (1977).
[21] S. S. Tate, F.-Y. Leu, and A. Meister, *J. Biol. Chem.* **247,** 5312 (1972).
[22] T. F. Deuel, M. Louis, and A. Lerner, *J. Biol. Chem.* **253,** 6111 (1978).
[23] K. Iqbal and C. Wu, *Enzyme* **12,** 553 (1971).
[24] S. G. Rhee, P. B. Chock, and E. R. Stadtman, this volume [29].
[25] C. H. Fiske and Y. SubbaRow, *J. Biol. Chem.* **66,** 375 (1925).
[26] F. Lipmann and L. C. Tuttle, *J. Biol. Chem.* **159,** 21 (1945).

Enzyme activity may also be followed by use of the γ-glutamyl transfer reaction in which glutamine synthetase catalyzes conversion of L-glutamine to the corresponding hydroxamate in the presence of hydroxylamine, catalytic quantities of nucleotide (ADP or ATP), and inorganic phosphate or arsenate [reaction (2)]. This reaction also requires divalent cations.

$$\text{L-Glutamine} + NH_2OH \xrightarrow[As_i(P_i)]{ATP(ADP);\ Mn^{2+}(Mg^{2+}, Co^{2+})} \text{L-}\gamma\text{-glutamyl-hydroxamate} + NH_3 \quad (2)$$

Reagents

Ferric chloride solution (0.37 *M* $FeCl_3$, 0.67 *M* HCl, and 0.20 *M* trichloroacetic acid)
Sodium L-glutamate
ATP
ADP
$MgCl_2$
$MnCl_2$
2-Mercaptoethanol
Imidazole–HCl buffer, pH 7.2
L-Glutamine
Potassium phosphate buffer, pH 7.2
Hydroxylamine hydrochloride (adjusted to pH 7.2 by addition of sodium hydroxide or salt-free hydroxylamine)
Ammonium chloride

Assay Procedures

The assay mixture (final volume, 0.5 ml) contains 0.1 *M* imidazole–HCl buffer (pH 7.2), 25 m*M* 2-mercaptoethanol, 50 m*M* sodium L-glutamate (pH 7.2), 10 m*M* NaATP, 20 m*M* $MgCl_2$, and 125 m*M* hydroxylamine or 40 m*M* ammonium chloride. During purification of the enzyme, the concentration of ATP should be 20 m*M*. For the determination of Mn^{2+}-dependent synthetase activity, 2 m*M* $MnCl_2$ may be used in place of $MgCl_2$. After incubation at 37°, the formation of inorganic phosphate may be determined by the method of Fiske and SubbaRow.[25] It is often more convenient to use the γ-glutamylhydroxamate procedure. In this method 0.75 ml of ferric chloride reagent is added after incubation, and the precipitated protein is removed by centrifugation. The absorbance of the solution at 535 nm is read against a reagent blank. Blanks should also be run in which enzyme, ATP, and glutamate are separately omitted. Such blanks are essential, especially when crude tissue preparations are examined, to ensure that the observed hydroxamate formation is a true function of glutamine synthetase. Several other enzyme activities are known that catalyze γ-glutamylhydroxamate formation; these include certain glutaminases and glutamine amidotransferases. It should also be noted that hydroxamates other than γ-glutamylhydroxamate may be formed in crude tissue preparations after incubation with hydroxylamine. Artifactual hydroxamate formation may thus occur.[27] Protein is determined as described.[28] A unit of glutamine synthetase activity is defined here as the amount of enzyme that catalyzes the synthesis of 1 μmol of product in 15 min at 37° under the conditions given above. Specific activity is expressed as units per mg of protein.

Glutamine synthetase also catalyzes the formation of L-γ-glutamylhydroxamate and ammonia from L-glutamine and hydroxylamine according to the γ-glutamyl transfer reaction [reaction (2)]. This reaction may be used for the determination of glutamine synthetase activity (but see above for caution regarding determination of hydroxamates). The assays are carried in mixtures (final volume, 0.5 ml) containing 0.1 *M* imidazole–HCl buffer (pH 7.2), 50 m*M* L-glutamine, 0.2 m*M* $MnCl_2$, 62.5 m*M* hydroxylamine, 10 m*M* sodium arsenate, and 0.4 m*M* ADP. After incubation at 37° for 15 min 0.5 ml of ferric chloride reagent is added and the formation of hydroxamate is carried out as described above.

Purification of Glutamine Synthetase from Rat Liver

The isolation procedure[21] is similar to that described for the isolation of ovine brain glutamine synthetase.[9,10] Sprague–Dawley rats (250 to 300 g) are used. Unless otherwise stated, all procedures are carried out at 4°.

Step 1. Liver Extract. The rats are killed by decapitation and exsanguinated. The livers are excised and homogenized in a Waring blendor with 4 volumes of 0.15 *M* potassium chloride containing 5 m*M* 2-mercaptoethanol and 1 m*M* EDTA (pH 7.2). The homogenate is centrifuged at 0° for 30 min at 16,000 *g*.

Step 2. Acid Precipitation. The extract obtained in Step 1 is placed in an ice bath and sufficient 1 *M* acetic acid is added with mechanical stirring over a period of 1 hr to bring the pH to 4.3. The precipitate is collected by centrifugation at 27,000 *g* for 20 min. It is then suspended in 100 ml of Solution A (0.15 *M* potassium chloride, 5 m*M* 2-mercaptoethanol, and 1 m*M* EDTA, adjusted to pH 7.2 by adding sodium hydroxide). The pH of the enzyme-containing solution is rapidly adjusted to 7.2 by adding 1 *M*

[27] J. E. Vorhaben, L. Wong, and J. W. Campbell, *Biochem. J.* **135,** 893 (1973).

[28] O. H. Lowry, N. J. Rosebrough, A. L. Farr, and R. H. Randall, *J. Biol. Chem.* **193,** 265 (1951).

ammonium hydroxide and the resulting suspension is centrifuged at 43,000 *g* for 20 min. Both the pellet and the supernatant solution are saved. The pellet is suspended in 125 ml of Solution A and then centrifuged at 43,000 *g* for 20 min. The two supernatant solutions are combined and centrifuged again at 55,000 *g* to remove turbidity. The clarified supernatant solution is then dialyzed for 18 hr against 2 liters of 10 m*M* potassium phosphate buffer (pH 7.2) containing 10 m*M* 2-mercaptoethanol. The next step is done immediately.

Step 3. Chromatography on Hydroxylapatite. Dry hydroxylapatite (Bio-Rad, 100 g) is suspended in 400 ml of 10 m*M* potassium phosphate buffer (pH 7.2) and allowed to stand for at least 4 hr at 4°. The sintered glass plate at the bottom of a column (4.5 × 30 cm) is covered with a plastic filter cloth or a piece of filter paper and the column is filled to about two-thirds of its volume by adding buffer. The suspension of hydroxylapatite is poured into the column and allowed to settle to achieve a bed depth of 10 cm. The column is equilibrated with 200 ml of 10 m*M* potassium phosphate buffer (pH 7.2) containing 10 m*M* 2-mercaptoethanol. A flow rate no greater than 100 ml/hr is used.

The dialyzed solution obtained in Step 2 is carefully applied to the column, the top of which should be protected by a plastic filter cloth or a piece of filter paper. Elution is then carried out at a rate no greater than 100 ml/hr with 0.1 *M* potassium phosphate buffer (pH 7.2) containing 2 m*M* 2-mercaptoethanol. When the absorbance of the effluent solution at 280 nm decreases to about 0.05, elution is begun with a linear gradient established between 300 ml of 0.15 *M* potassium phosphate buffer (pH 7.2) and 300 ml of 0.3 *M* potassium phosphate buffer (pH 7.2); both buffers containing 10 m*M* 2-mercaptoethanol. The enzyme emerges from the column when the phosphate concentration is about 0.22 *M*. The fractions containing the enzyme activity are pooled.

Step 4. DEAE-Cellulose Chromatography. The pooled fractions obtained in Step 3 are treated with solid ammonium sulfate to achieve 50% of saturation and stirring is continued for 30 min. The solution is then centrifuged at 16,000 *g* for 40 min. The pelleted protein is dissolved in 15 ml of a buffer containing 10 m*M* imidazole–HCl, 2 m*M* EDTA, and 5 m*M* 2-mercaptoethanol (pH 7.9). This solution is dialyzed for 18 hr against 2 liters of this buffer and then run into a DEAE-cellulose column (Whatman-52, microgranular, previously swollen) (2.5 × 8 cm), which is equilibrated with the same buffer. The column is washed with 100 ml of buffer followed by elution with a solution containing 20 m*M* imidazole–HCl buffer containing 2 m*M* EDTA and 5 m*M* 2-mercaptoethanol (adjusted to pH 6.2 at 4° by addition of citric acid). Fractions of 4 to 5 ml are collected and the enzymatically active fractions are pooled. Glycerol is added to achieve a final concentration of 10% (v/v). The solution is concentrated by ultrafiltration through dialysis tubing *in vacuo* to about 6 or 7 ml.

Step 5. Chromatography on Sephadex G-100. The concentrated solution obtained in Step 4 is applied to the top of a column of Sephadex G-100 (2.5 × 90 cm) previously equilibrated with 0.15 M potassium chloride buffer containing 1 mM EDTA and 5 mM 2-mercaptoethanol (pH 7.2). The column is eluted with the same buffer at a flow rate of about 20 ml/hr and fractions of 5 ml are collected. Fractions exhibiting specific activities greater than 120 are pooled and concentrated by ultrafiltration through dialysis tubing *in vacuo*.

A summary of the purification is given in the table.

Properties of the Isolated Enzyme

The isolated enzyme is homogeneous on polyacrylamide gel electrophoresis at pH values of 7.2, 8.0, and 8.5. It is also homogeneous in the ultracentrifuge ($s_{20,w}$ 15.0 S). The enzyme has a molecular weight of about 360,000 and dissociates in the presence of sodium dodecyl sulfate to yield monomeric units of molecular weight about 44,000.

The apparent K_m value for L-glutamate is about 5 mM, and that for ATP in the presence of Mg^{2+} is about 2.3 mM. The enzyme is inhibited by L-methionine-S-sulfoximine. However, complete inhibition of the enzyme was found to be associated with the binding of about 4 mol of methionine sulfoximine.[21] Studies on the binding of ATP indicated that a maximum of

PURIFICATION OF GLUTAMINE SYNTHETASE FROM RAT LIVER[a]

Step	Volume (ml)	Protein: Concentration (mg/ml)	Protein: Total (mg)	Total units	Activity: Specific activity (units/mg)	Activity: Yield (%)
1. Liver extract	960	25.1	24,100	11,400	0.47	[100]
2. Acid precipitation	277	4.91	1,368	8,100	5.96	71
3. Hydroxylapatite chromatography	370	0.46	170	5,900	34.7	52
4. DEAE-cellulose chromatography	55	0.51	28.0	2,720	97	24
5. Chromatography on Sephadex G-100	9.5	1.77	16.8	2,260	134	20

[a] From 220 g of rat liver.

5 mol of ATP bind to the enzyme. In this respect the isolated rat liver enzyme differs from the purified sheep brain glutamine synthetase, which can bind tightly about 8 mol of ATP per mole of enzyme, and whose inhibition is associated with binding of substantially more than 4 mol of methionine sulfoximine. Several possible explanations for these apparent differences between the rat liver and sheep brain enzymes have been considered.[21]

The rat liver enzyme is activated by α-ketoglutarate and is inhibited by glycine, L-alanine, L-serine, and carbamyl phosphate.[21] Rat liver glutamine synthetase resembles the glutamine synthetases of mammalian brain and other tissues in terms of its physical and structural characteristics, and catalytic behavior; these properties are briefly reviewed below.

Physical and Chemical Properties

The glutamine synthetases of animal origin are composed of eight apparently identical subunits of molecular weight 44,000–49,000. The brain and liver enzymes were found to exhibit cube-like appearance in the electron microscope.[29] The enzyme is considered to possess D_4 symmetry. The octameric brain enzyme is formed by isologous association of 2 heterologously bonded tetramers. In the presence of 2 *M* urea at 25° (and under certain other conditions), the enzyme dissociates reversibly to yield a tetramer which exhibits a sedimentation coefficient of 8.6 S. Such dissociation is inhibited by MgATP, and reassociation of the tetramer to the octamer is favored by MgATP.[30,31] Thus far, studies on the further dissociation of the tetramer to the monomer have shown that this process is irreversible.

The amino acid compositions of the several glutamine synthetases of mammalian origin are similar.[11,12] It is of interest that the enzymes from pea and *E. coli* contain, respectively, 2 and 4 half-cystine residues per subunit as compared to 12 for the mammalian enzymes. All of the half-cystine residues of the brain enzyme react with 5,5′-dithiobis(2-nitrobenzoate) in the presence of 5 *M* guanidine hydrochloride indicating the absence of disulfide linkages.[10]

Catalytic Properties

Under the usual conditions of assay, maximal activity is observed in the pH range 7.0–7.4. It is interesting to note, however, that the pH

[29] R. H. Haschemeyer, *Adv. Enzymol.* **33,** 71 (1970).
[30] S. Wilk, A. Meister, and R. H. Haschemeyer, *Biochemistry* **8,** 3168 (1969).
[31] S. Wilk, A. Meister, and R. H. Haschemeyer, *Biochemistry* **9,** 2039 (1970).

optimum varies from about pH 4.8 to 8.5 depending upon the nature and concentration of the divalent cations present. At pH 7.2, Mg^{2+} are more effective than Mn^{2+} or Co^{2+}, but at their respective pH optima equivalent activity has been found with these divalent metal ions.[32,33] The apparent K_m values determined under usual assay conditions in the presence of Mg^{2+} for the sheep brain enzyme are 2.3, 2.5, and 0.18 m*M*, respectively for ATP, L-glutamate, and ammonium ion.

Many studies have been carried out on the specificity of glutamine synthetase. The enzyme interacts with a wide variety of analogs of L-glutamate (see below). Of the various nucleoside triphosphates tested, only ATP and deoxy-ATP are appreciably active. Ammonia and hydroxylamine are about equally active in reaction mixtures containing L-glutamate. Hydrazine, methylhydrazine, methylamine, ethylamine, and glycine ethyl ester are active in place of ammonia; in their presence the corresponding γ-glutamyl derivatives are formed. When the enzyme is incubated in reaction mixtures containing methylhydrazine in place of ammonia, 1-(5-glutamyl)-2-methylhydrazine is formed.[34] This is converted nonenzymatically to 1-(5-glutamyl)-1-methylhydrazine. This indicates the methylhydrazine attaches to the ammonia-binding site of the enzyme in such a manner that only the nonalkylated nitrogen atom of methylhydrazine is available for reaction with enzyme-bound γ-glutamyl phosphate.

Mechanism

Much effort has been devoted to the understanding of the mechanism by which this enzyme catalyzes the coupling of the endergonic synthesis of glutamine from glutamate and ammonia to the exergonic cleavage of ATP to ADP and inorganic phosphate; several reviews and recent publications are available.[11,35–37] The finding that the enzyme does not catalyze exchange reactions between (1) ATP and inorganic phosphate, (2) ATP and ADP, (3) glutamine and glutamate, or (4) ammonia and glutamine in the absence of the other components of the synthetase system was initially considered as strong evidence that the reaction operates by means

[32] J. Greenberg and N. Lichtenstein, *J. Biol. Chem.* **234,** 2337 (1959).

[33] C. Monder, *Biochemistry* **4,** 2677 (1965).

[34] M. L. Rueppel, S. L. Lundt, J. D. Gass, and A. Meister, *Biochemistry* **11,** 2839 (1972).

[35] A. Meister, *in* "Glutamine: Metabolism, Enzymology, and Regulation" (J. Mora and R. Palacios, eds.), p. 1. Academic Press, New York, 1980.

[36] R. D. Allison and D. L. Purich, *Biochem. Biophys. Res. Commun.* **115,** 220 (1983).

[37] R. D. Allison, J. A. Todhunter, and D. L. Purich, *J. Biol. Chem.* **252,** 6046 (1977).

of a concerted mechanism.[38–40] However, there is now compelling evidence that the reaction takes place through formation of enzyme-bound intermediates including a complex consisting of enzyme, γ-glutamyl phosphate, and ADP, from which ADP does not dissociate readily.[41] Important information about the mechanism of the reaction has come from studies of the partial reactions catalyzed by the enzyme, investigations on the mapping of the active site with a number of substrate analogs, and a variety of studies which show that enzyme-bound γ-glutamyl phosphate is an intermediate.

The partial reactions catalyzed by glutamine synthetase include the γ-glutamyl transfer reaction [reaction (2)]. It is notable that only catalytic amounts of nucleotides are required for this reaction. This is also true of the arsenolysis reaction catalyzed by the enzyme [reaction (3)]. The finding that both L-glutamate and D-glutamate are substrates of the enzyme, and that synthesis of the corresponding γ-glutamyl hydroxamates occurs at a similar rate for both isomers of glutamate suggested that the enzyme catalyzes an initial activation of glutamate in a reaction of low optical specificity.[42] This was thought to be followed by a more specific reaction of the activated intermediate with ammonia, which becomes rate limiting in the case of D-glutamate. An interesting partial reaction was discovered after it was postulated that an enzyme-bound γ-carboxyl-activated glutamate derivative would exhibit a marked tendency to cyclize in the absence of ammonia. Thus, the enzyme catalyzes conversion of glutamate to 5-oxoproline in the absence of ammonia [reaction (4)]:

$$\text{L-Glutamine} + H_2O \xrightarrow{\text{ADP, } M^{2+}\text{, As}_i} \text{L-glutamate} + NH_3 \quad (3)$$

$$\text{L- (or D-)Glutamate} + \text{ADP} \overset{M^{2+}}{\rightleftharpoons} \text{5-oxo-L- (or D-)proline} + \text{ADP} + P_i \quad (4)$$

The finding that this reaction proceeds at about equal rates with L-glutamate and D-glutamate provided evidence that the initial step in the reaction is formation of an activated glutamate derivative.

A "pulse-chase" method was developed to explore the nature of the activated glutamate intermediate.[41,43] In this approach, the enzyme was briefly incubated with ATP, Mg^{2+}, and [^{14}C]glutamate; a mixture containing hydroxylamine and a large excess of unlabeled glutamate was then

[38] J. M. Buchanan and S. C. Hartman, *Adv. Enzymol.* **21,** 199 (1959).
[39] J. M. Buchanan, *Adv. Enzymol.* **39,** 91 (1973).
[40] F. C. Wedler and P. D. Boyer, *J. Biol. Chem.* **247,** 984 (1972).
[41] P. R. Krishnaswamy, V. Pamiljans, and A. Meister, *J. Biol. Chem.* **237,** 2932 (1962).
[42] L. Levintow and A. Meister, *J. Am. Chem. Soc.* **75,** 3039 (1953).
[43] P. R. Krishnaswamy, V. Pamiljans, and A. Meister, *J. Biol. Chem.* **235,** 39 (1960).

added, and, after deproteinization, the formation of labeled γ-glutamylhydroxamate was determined. Since this experiment showed preferential conversion of labeled glutamate to the hydroxamate, it was clear that equilibration between the labeled and unlabeled glutamates did not occur. A similar study was carried out with labeled ATP, and experiments involving the use of ultracentrifugation and ultrafiltration showed that (1) glutamate attaches to the enzyme only when both ATP and Mg^{2+} are present, and (2) binding of glutamate to the enzyme is associated with cleavage of ATP to ADP.[41] Of much interest, the phosphate formed in this cleavage remained attached to the enzyme. This experiment provided evidence for the intermediate formation of an enzyme complex containing an activated form of glutamate, ADP, and phosphate.

The finding that both D-glutamate and L-glutamate are substrates of the enzyme[42] led to experiments which showed when the amino group is attached to the third possible position of the glutarate carbon chain, a molecule with substrate activity is obtained.[44,45] Thus, the enzyme amidates β-aminoglutarate to form β-glutamine, and it was also found that the enzyme uses β-glutamyl phosphate for synthesis of ATP from ADP [reaction (5)].[45] In addition to these partial reactions, the enzyme can also catalyze the phosphorylation of the cyclic analog of glutamate, *cis*-1-amino-1,3-dicarboxycyclohexane (cycloglutamate) [reaction (6)].[46]

$$\beta\text{-Glutamyl-P} + \text{ADP} \xrightarrow{M^{2+}} \text{ATP} + \beta\text{-glutamate} \quad (5)$$

$$\text{Enzyme} + \text{cycloglutamate} + \text{ATP} \xrightarrow{M^{2+}} \text{enzyme [cycloglutamyl-P] [ADP]} \quad (6)$$

The γ-glutamyl phosphate hypothesis is strongly supported by the finding that methionine sulfoximine, an irreversible inhibitor of the enzyme, is phosphorylated by the enzyme in the presence of ATP [reaction (6)].[47–51] Chemically prepared methionine sulfoximine phosphate also inhibits. The intermediate formation of an acyl phosphate is also in accord with studies in which γ-glutamyl phosphate was trapped by borohydride treatment of the enzyme complex in the absence of ammonia to yield α-amino-δ-hydroxyvalerate.[52] An elegant experiment which shows that γ-glutamyl phosphate is a kinetically competent intermediate was carried

[44] E. Khedouri and A. Meister, *J. Biol. Chem.* **240,** 3357 (1965).
[45] E. Khedouri, V. P. Wellner, and A. Meister, *J. Biol. Chem.* **239,** 2553 (1964).
[46] Y. Tsuda, R. A. Stephani, and A. Meister, *Biochemistry* **10,** 3186 (1971).
[47] R. Ronzio and A. Meister, *Proc. Natl. Acad. Sci. U.S.A.* **59,** 164 (1968).
[48] R. A. Ronzio, W. B. Rowe, and A. Meister, *Biochemistry* **8,** 1066 (1969).
[49] W. B. Rowe, R. A. Ronzio, and A. Meister, *Biochemistry* **8,** 2674 (1969).
[50] J. M. Manning, S. Moore, W. B. Rowe, and A. Meister, *Biochemistry* **8,** 2681 (1969).
[51] W. B. Rowe and A. Meister, *Proc. Natl. Acad. Sci. U.S.A.* **66,** 500 (1970).
[52] J. A. Todhunter and D. L. Purich, *J. Biol. Chem.* **250,** 3505 (1975).

out in which the enzyme was incubated with glutamate and ATP labeled with ^{18}O in the $\beta\gamma$-bridge oxygen atom. Reversible cleavage of the P_β–O–P_γ-bond was shown by appearance of ^{18}O in the β-nonbridge oxygen atom of ATP. The rate of such ^{18}O scrambling is about the same as the maximum rate of the enzyme-catalyzed reaction.[53]

The active site of glutamine synthetase has been mapped by studies in which many glutamate analogs were used.[11,54,55] Although the enzyme acts on both L- and D-glutamate, it accepts only the L-isomer of α-methyl-DL-glutamate.[56] Curiously, the product formed from β-glutamate (β-aminoglutarate) has the D-configuration.[44] These findings led to further work on the interaction of the enzyme with glutamate analogs.

Since the enzyme does not interact significantly with monocarboxylic amino acids or with glutarate derivatives that lack an amino group, it appears that the enzyme has binding sites for both carboxyl groups of glutamate as well as for its amino group. Since aspartate interacts poorly with the enzyme, it appears that the distance between the carboxyl binding sites on the enzyme must be greater than the maximal possible intercarboxyl distance for aspartate. These considerations suggest that L-glutamate, when attached to the enzyme, must be in the fully extended, or almost fully extended form. If the enzyme sites that interact with the carboxyl and amino groups of L-glutamate also interact with the corresponding groups of D-glutamate, it is necessary to postulate that D-glutamate can bind only in a rotated manner. In other words, the α-hydrogen atom of D-glutamate must be directed in an essentially opposite manner to that of the α-hydrogen atom of L-glutamate. Thus, if L-glutamate attaches to the enzyme so that its α-hydrogen atom is directed away from the enzyme, the α-hydrogen atom of D-glutamate must be directed toward the enzyme. If the α-hydrogen atom of L-glutamate is directed away from the enzyme, it would be expected that substitution of this hydrogen atom by a methyl group would not interfere with its binding to the active site. However, such a substitution on D-glutamate would be expected to be accompanied by substantial interference to attachment of this substrate to the active site. This reasoning seems to explain why the enzyme can use L- and D-glutamates, but only the L-isomer of α-methylglutamate. Since both of the β-hydrogen atoms of L-glutamate and the erythro β-hydrogen atom of D-glutamate are on the same side of the glutamate molecule as the α-hydrogen atom of D-glutamate, whereas the threo-β-hydrogen of D-glutamate is in a position close to that of the α-hydrogen of L-glutamate, it can

[53] C. F. Midelfort and A. Rose, *J. Biol. Chem.* **254,** 5881 (1976).

[54] A. Meister, *Harvey Lect.* **63,** 139 (1969).

[55] A. Meister, *Adv. Enzymol.* **31,** 183 (1968).

[56] H. M. Kagan, L. R. Manning, and A. Meister, *Biochemistry* **4,** 1063 (1965).

be predicted that substitution by a methyl group of either of the β-hydrogen atoms of L-glutamate or of the erythro β-hydrogen atom of D-glutamate would interfere with binding to the enzyme. These considerations suggest that substitution of the threo-β-hydrogen atom of D-glutamate by a methyl group would not lead to loss of substrate activity. These predictions proved to be correct[57]; thus, of the four stereoisomers of β-methylglutamate, only threo-β-methyl D-glutamate is a substrate. Analogous reasoning regarding the 4 isomers of γ-methyl glutamate led to the prediction that only threo-γ-methyl-L-glutamate would be a substrate; this was demonstrated experimentally.[58] Thus, of the 10 possible monomethyl substitutions of the glutarate carbon chain, only 3 are substrates of glutamine synthetase: α-methyl-L-glutamate, threo-β-methyl-D-glutamate, and threo-γ-methyl-L-glutamate. The methyl group of each of these substrates is directed away from the enzyme. *cis*-1-Amino-1,3-dicarboxycyclohexane, a molecule that can be thought of as containing carbon atoms 2, 3, and 4 of L-glutamate and a chain of 3 carbon atoms attached to carbon atoms 2 and 4 of glutamate, was synthesized and found to be an excellent substrate.[59] This is in accord with the considerations discussed above.

Thus, the data show that L-glutamate attaches to the active site in an extended conformation in which the α-hydrogen atom of L-glutamate is directed away from the enzyme. Application of a computer approach to this problem, in which the three-dimensional coordinates of the substrates and of inhibitors of the enzyme were calculated, led to additional conclusions.[60] For example, it is possible to calculate the structure of the tetrahedral addition compound formed in the reaction of γ-glutamyl phosphate with ammonia. The computer study allowed identification of five points on the enzyme derived from the positions of particular atoms of L-glutamate, i.e., the α-carboxyl carbon atom, the α-amino nitrogen atom, the γ-carboxyl oxygen atom which is phosphorylated, the γ-carboxyl oxygen atom which is not phosphorylated, and the nitrogen atom of the tetrahedral intermediate formed in reaction of γ-glutamyl phosphate with ammonia.

A similar approach elucidates interaction of the irreversible inhibitor L-methionine-*S*-sulfoximine with the enzyme. This isomer (but not the other three isomers of methionine sulfoximine) is phosphorylated by the enzyme and binds tightly to the active site thus inhibiting it irreversibly.[50,51] Computer studies show that the sulfoximine oxygen atom of L-methionine-*S*-sulfoximine binds to the enzyme site that binds the gluta-

[57] H. M. Kagan and A. Meister, *Biochemistry* **5**, 725 (1966).
[58] H. M. Kagan and A. Meister, *Biochemistry* **5**, 2423 (1966).
[59] J. D. Gass and A. Meister, *Biochemistry* **9**, 842 (1970).
[60] J. D. Gass and A. Meister, *Biochemistry* **9**, 1380 (1970).

mate γ-carboxyl oxygen atom that is not phosphorylated, that the sulfoximine nitrogen atom binds to the binding site of the enzyme that binds the glutamate γ-carboxyl oxygen atom that is phosphorylated, and that the methyl group of L-methionine-*S*-sulfoximine binds to the ammonia binding site of the enzyme.[60] Thus, L-methionine-*S*-sulfoximine is bound to the enzyme in such a manner that its sulfoximine nitrogen atom is in a favorable position to be phosphorylated. These studies indicate that L-methionine-*S*-sulfoximine is an inhibitory analog of the enzyme-bound intermediate or transition state formed in the normal catalytic reaction.

The conclusion that the methyl group of methionine sulfoximine binds to the ammonia binding site of the enzyme proved to be important because it suggested a way in which the methionine sulfoximine molecule might be modified so as to achieve a selective inhibitor of γ-glutamylcysteine synthetase. The latter enzyme also functions by a mechanism involving enzyme-bound γ-glutamyl phosphate and it is also inhibited by L-methionine-*S*-sulfoximine, which is also phosphorylated on the enzyme by ATP.[61] Although the ammonia binding site of glutamine synthetase can accommodate the *S*-methyl group of L-methionine-*S*-sulfoximine, it cannot accommodate a molecule in which the methyl group is replaced by a bulkier moiety such as propyl or butyl. Thus, glutamine synthetase is not affected by prothionine sulfoximine or buthionine sulfoximine (which effectively inhibit γ-glutamylcysteine synthetase.[62,63]

The observation that α-methylglutamate is a much better substrate of glutamine synthetase than of γ-glutamylcysteine synthetase suggested that there is more space in the region of glutamine synthetase that is occupied by the α-hydrogen atom of glutamate than there is in the comparable region of γ-glutamylcysteine synthetase. This finding was important because it suggested that introduction of an α-ethyl moiety into a substrate or inhibitor that binds to the glutamate site of the enzyme might lead to a molecule that would be too large to fit into the glutamate site of γ-glutamylcysteine synthetase, but which might be accepted by the glutamate site of glutamine synthetase. This line of reasoning proved to be useful.[64] Thus, α-ethylmethionine sulfoximine was found not to inhibit γ-glutamylcysteine synthetase, but it did inhibit glutamine synthetase.

Since L-methionine-*S*-sulfoximine (but not the other 3 isomers of this compound) inhibits glutamine synthetase and γ-glutamylcysteine synthetase, the convulsant activity of L-methionine-*S*-sulfoximine (which is not shared by the other 3 isomers[51]) might be due to an effect either on

[61] P. G. Richman, M. Orlowski, and A. Meister, *J. Biol. Chem.* **248,** 6684 (1973).

[62] O. W. Griffith, M. E. Anderson, and A. Meister, *J. Biol. Chem.* **254,** 1205 (1979).

[63] O. W. Griffith and A. Meister, *J. Biol. Chem.* **254,** 7558 (1979).

[64] O. W. Griffith and A. Meister, *J. Biol. Chem.* **253,** 1799 (1978).

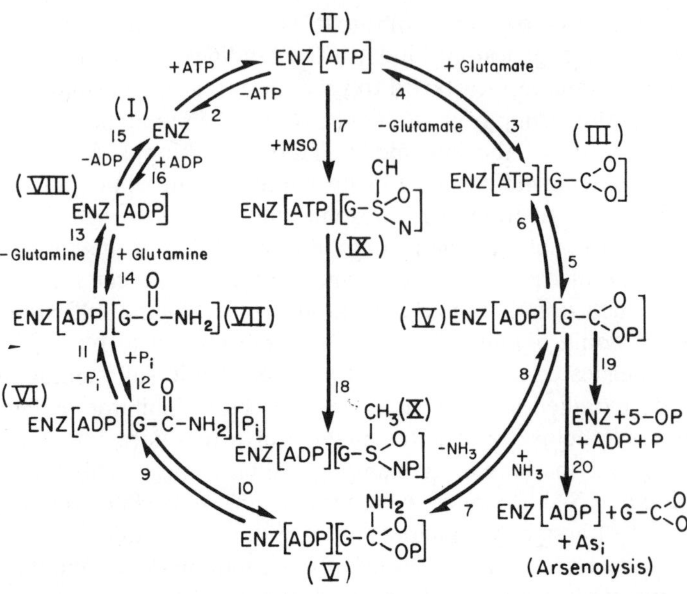

FIG. 1. Scheme of the mechanism of action of glutamine synthetase. G, $^{-}OOCCH(NH_3^+)(CH_2)_2^-$. Glutamine synthesis: 1, 3, 5, 7, 9, 11, 13, 15. Reversal of synthesis: 16, 14, 12, 10, 8, 6, 4, 2. Partial reactions: γ-Glutamyl transfer: 16, 14, 12, 10, 8, 7 (+NH_2OH), 9, 11, 13. Arsenolysis of glutamine: 16, 14, 12 (+As_i), 10, 8, 20. Phosphorylation of L-methionine-*S*-sulfoximine (MSO): 1, 17, 18. Cyclization; formation of 5-oxoproline (5-OP): 1, 3, 5, 19. Phosphorylation of glutamate analog (cycloglutamate): 1, 3, 5. Acyl phosphate (β-glutamyl ~ P, carbamyl ~ P, acetyl ~ P) + ADP → ATP: 6, 4, 2.

glutathione synthesis or on glutamine synthesis. That α-ethyl methionine sulfoximine, but not buthionine sulfoximine, produces convulsions indicates that the convulsant effect of methionine sulfoximine is most likely associated with its effect on glutamine synthetase. The possibility that convulsions are due to metabolic products of L-methionine-*S*-sulfoximine seems unlikely because α-alkyl derivatives of methionine sulfoximine, like other α-alkyl amino acids, are not metabolized to a significant extent.

A scheme for the mechanism of action of glutamine synthetase that accounts for the experimental observations cited above and also explains the partial reactions catalyzed by the enzyme has been proposed[11] (Fig. 1). According to this scheme, the reaction involves several enzyme-bound intermediates. A full discussion is given elsewhere.[11] Although there is no direct chemical evidence for the binding of glutamate or of glutamine to the enzyme in the absence of divalent metal ions and nucleotides, kinetic studies are consistent with the possibility that this may occur. However, the predominant mechanism for glutamate binding seems to be nucleotide dependent.[37] Studies purporting to show a difference between the kinetic behavior of the pea and *E. coli* enzymes as compared to brain glutamine synthetase, in which equilibrium exchange measurements were used,[65] have not been confirmed.[36] The reported studies[65] on the brain enzyme were apparently not carried out under equilibrium conditions.[36] The available findings on glutamine synthetases from various sources are consistent with the scheme given in Fig. 1.

[65] F. C. Wedler, *J. Biol. Chem.* **249,** 5080 (1974).

[28] Glutamine Synthetase from Muscle

By W. BRUCE ROWE

$$\text{Glutamate} + \text{ammonia} + \text{ATP} \xrightleftharpoons{Mg^{2+}} \text{glutamine} + \text{ADP} + P_i$$

Assay Procedures

The assay for glutamine synthetase activity in muscle measures the rate of synthesis of radioactive glutamine from radioactive glutamate.[1] In this assay, samples were incubated in a mixture (0.1 ml) containing imidazole–HCl (0.05 *M*, pH 7.2), mercaptoethanol (10 m*M*), EDTA (1 m*M*), L-[U-^{14}C]glutamate (10 m*M*, 14,000 cpm/μmol), NH_4Cl (20 m*M*), NaATP (10 m*M*), $MgCl_2$ (20 m*M*), phosphoenolpyruvate (2 m*M*), and pyruvate kinase (1 IU) and enzyme solution. The samples were incubated at 37° for 30 min. The reaction was stopped by placing the samples on ice and adding 0.4 ml of a cold solution containing 4 μmol of imidazole–HCl (pH 7.2) and 10 μmol of glutamine (quenching solution). The treated samples were placed on 0.5 × 7 cm columns of Dowex-1(A$\bar{c}$), the incubation tubes were washed with 0.5 ml of quenching solution which was placed on the columns, and the columns eluted with 1.0 ml of 0.05 *M* acetic acid. The combined sample, washing, and eluate from the column was collected in a scintillation counting vial and assayed for radioactivity in 10 ml of a scintillation fluid prepared by dissolving 8 g of Omnifluor in a liter of dioxane. A unit of enzyme activity is defined as the amount of enzyme required to catalyze the synthesis of 1 μmol of labeled glutamine per hour at 37°.

Tissues used in the survey of glutamine synthetase activity reported in Table I were taken from fasted 230–260 g male Sprague–Dawley rats.

[1] S. Prusiner and L. Milner, *Anal. Biochem.* **37,** 429 (1970).

Copyright © 1985 by Academic Press, Inc.
All rights of reproduction in any form reserved.

TABLE I
GLUTAMINE SYNTHETASE ACTIVITY IN RAT TISSUES[a]

Tissue	Enzymatic activity: Units/mg of soluble protein, +ATP regenerating system	Enzymatic activity: Units/mg of soluble protein, −ATP regenerating system	Enzymatic activity: Units/g of tissue
Brain	0.81 ± 0.07	0.76	89
Liver	0.86 ± 0.08	0.80	106
Kidney	0.28 ± 0.05	0.12	56
Heart	0.13 ± 0.5	<0.01	6.4
Muscle	0.07 ± 0.03	<0.01	3.8

[a] Activity was determined by the radioactive assay described in the methods section. In order to determine the activity in the absence of the ATP regenerating system, phosphoenolpyruvate and pyruvate kinase were omitted. The values for the complete system represent the mean of five determinations each representing the pooled tissues of two animals ± standard deviation.

Tissues were removed and placed on ice, and then homogenized within 15 min in 5 volumes of 0.15 *M* KCl, 2 m*M* in EDTA, and 10 m*M* mercaptoethanol using a Teflon pestle homogenizer.

The activities in brain and liver as determined using the radioactive assay described above are, within experimental error, identical to activities determined using the colorimetric hydroxamate assay system.[2] In the absence of a phosphoenol pyruvate and pyruvate kinase ATP regenerating system, the activity in heart and skeletal muscle extracts was less than 0.01 units/mg of soluble protein. In the absence of the ATP regenerating system, activity in the kidney extracts was less than 50% of the activity measured in the presence of the ATP regenerating system. Glutamine synthetase activity was also determined in rabbit heart and skeletal muscle, mouse heart and skeletal muscle, and in beef, pig, and sheep heart. The specific enzymatic activities determined in these tissues were comparable to those reported in Table I for rat tissues.

Purification of Pig Heart Glutamine Synthetase

Pig hearts were obtained from a local slaughterhouse and were kept on ice until used (usually within 3 hr). Pig heart was chosen as the source of heart glutamine synthetase solely on the basis of the availability of fresh tissue. Hearts were trimmed to remove excess connective tissue, ground in a meat grinder, then homogenized in a blender in five volumes

[2] W. B. Rowe, R. A. Ronzio, V. P. Wellner, and A. Meister, this series, Vol. 17A, p. 900.

of cold 0.15 *M* KCl, 2 m*M* in EDTA, and 10 m*M* in mercaptoethanol. The extract was centrifuged for 20 min at 15,000 *g* and the precipitate discarded. Glutamine synthetase activity precipitated from the supernatant between 30 and 45%, 45 and 60%, and 60 and 75% saturation with ammonium sulfate. The fraction precipitated between 45 and 60% saturation was chosen for further purification as it had a higher specific activity than the other enzymatically active fractions. This precipitate was dissolved in 0.005 *M* potassium phosphate buffer, pH 7.2, 10 m*M* mercaptoethanol, and dialyzed overnight against 20 volumes of the same buffer. Chromatography of this extract on hydroxylapatite was performed as for the purification of glutamine synthetase from sheep brain.[2] The activity from pig heart eluted from the hydroxylapatite at a phosphate concentration of about 0.35 *M*, significantly higher than the concentration required to elute the brain enzyme (0.225 *M*). The enzymatically active fractions from hydroxylapatite chromatography were combined and the protein precipitated at 90% saturated ammonium sulfate. The precipitate was redissolved in 10 m*M* potassium phosphate buffer, pH 7.2, 2 m*M* in EDTA, and 10 m*M* in mercaptoethanol, and dialyzed for a minimum of 5 hr against 5 liters of 5 m*M* potassium phosphate buffer, pH 7.2, 2 m*M* in EDTA, and 10 m*M* in mercaptoethanol. The dialyzed extract was adjusted to pH 7.9 using dilute NH_4OH and applied to a 1 × 15 cm column of DEAE-cellulose (DE-52) equilibrated with 5 m*M* potassium phosphate buffer, pH 8.0, 2 m*M* in EDTA, and 10 m*M* in dithiothreitol. Glutamine synthetase activity eluted as a symmetrical peak near the change in pH of the eluant from pH 8.0 to 7.3. The fractions containing enzyme activity were combined and concentrated to about 0.4 mg protein per ml by ultrafiltration. This enzyme preparation had a specific enzymatic activity of about 20 units/mg of protein and contained three protein components when visualized after acrylamide slab gel electrophoresis. This preparation contained an active non-glutamate-dependent ATPase activity. Further purification on a 2.4 × 45 cm Sephadex G-200 column (equilibrated with and eluted with 10 m*M* potassium phosphate, pH 7.2, 2 m*M* in EDTA, and 5 m*M* in dithiothreitol) yielded an enzyme preparation with a specific enzymatic activity of about 60 units/mg of protein. This preparation had no glutamate-independent ATPase activity. On gel electrophoresis, a single major band and two minor low-molecular-weight components were present. A summary of the results of this purification is given in Table II.

Purification of Rat Skeletal Muscle Glutamine Synthetase

Rat skeletal muscle was obtained from fasted 220–350 g breeder reject rats. The larger muscles of rat hind limbs were grossly dissected, freed of

TABLE II

PURIFICATION OF GLUTAMINE SYNTHETASE FROM PIG HEART AND RAT SKELETAL MUSCLE

Step	Volume (ml)	Protein concentration (mg/ml)	Total (mg)	Activity concentration (units/ml)	Total (units)	Specific (units/mg)	Yield (%)
Pig heart: 401 g of tissue							
1. 0.15 *M* KCl extract	1915	19	36,400	2.3	4370	0.12	(100)
2. 45–60% $(NH_4)_2SO_4$	185	12	2,200	6.5	1200	0.54	28
3. Hydroxylapatite chromatography	212	1.2	250	3.1	650	2.6	15
4. DEAE Cellulose chromatography	65	0.38	25	7.7	500	20.1	11
5. Sephadex G-200 chromatography	25	0.20	5	12.2	300	61.0	7
Rat skeletal muscle: 314 g of tissue							
1. 0.15 *M* KCl extract	1638	23.4	38,300	1.4	2300	0.06	(100)
2. 45–60% $(NH_4)_2SO_4$	136	28.4	3,900	7.85	1070	0.28	46
3. Hydroxylapatite chromatography	165	1.2	200	3.5	580	2.9	25
4. Sephadex G-200 chromatography	65	0.38	25	4.9	500	20.2	22
5. DEAE Cellulose chromatography	25	0.2	5	13.1	3	65.6	14

excess connective tissue, and ground in a meat grinder. The ground muscle was homogenized in a blender in five volumes of cold 0.15 *M* KCl, 2 m*M* in EDTA, and 10 m*M* in mercaptoethanol, centrifuged at 19,000 *g* for 20 min, and the precipitate discarded. Glutamine synthetase activity precipitated from the supernatant at 30 to 45%, and 45 to 60% ammonium sulfate saturation. The 45 to 60% precipitate which had a higher specific activity was retained for further purification. The precipitate was dissolved and dialyzed as described above for the pig heart preparation. Hydroxylapatite chromatography of this extract performed as described above resulted in elution of glutamine synthetase activity at about 0.35 *M* potassium phosphate. The combined concentrated, and dialyzed enzyme from hydroxylapatite did not bind to DEAE-cellulose (DE-52) under a variety of conditions. The extract was applied to a 2.4 × 45 cm column of Sephadex G-200 equilibrated with 10 m*M* potassium phosphate buffer, pH 7.2, 2 m*M* in EDTA and 5 m*M* in dithiothreitol. The column was then eluted with the equilibration buffer. Glutamine synthetase activity was eluted with a slightly asymmetric protein peak near the excluded volume of the column. Only a 2-fold purification was attained in this procedure and chromatography on DEAE-cellulose was then performed on the enzyme obtained from gel filtration. DEAE-cellulose chromatography on the concentrated, dialyzed enzyme fraction from gel filtration was performed successfully as described above for the heart enzyme to yield an enzyme with a specific activity of about 60 units/mg of protein. This preparation had no glutamate-independent ATPase activity. A single major protein component and a minor higher molecular weight component were seen on slab gel acrylamide electrophoresis. A summary of the results of this purification is given in Table VII.

Kinetic Properties of the Enzyme

Data on the reaction requirements for purified rat muscle and pig heart glutamine synthetase are presented in Table III. The enzymatic activity in the absence of the phosphoenolpyruvate and pyruvate kinase ATP regenerating system was less than one-half of the activity in the presence of the ATP regenerating system. Creatine phosphate and creatine kinase were also effective as an ATP regenerating system. EDTA (1 m*M*) and mercaptoethanol (10 m*M*) both stimulated the reactions. The rate of reaction catalyzed by these enzymes is essentially identical with hydroxylamine and NH_4Cl as substrates. The rate of synthesis of glutamine in the presence of optimum concentrations of $MnCl_2$ and ATP (2 and 10 m*M*, respectively) was about 25% of the rate in the presence of $MgCl_2$ and ATP (20 and 10 m*M*, respectively). Both enzymes are strongly inhibited by

TABLE III
REACTION REQUIREMENTS: MUSCLE GLUTAMINE SYNTHETASE

Reaction condition	Relative activity[a]	
	Rat muscle	Pig heart
Complete assay system	100	100
-ATP regenerating system	44	47
-EDTA	88	90
-Mercaptoethanol	69	72
$-NH_4Cl + NH_2OH$	103	105
-MgATP + MnATP	24	22
+Methionine sulfoximine (5 m*M*)	8	7

[a] Activity was determined on the purified enzymes with a specific activity of about 60 units/mg.

L-methionine sulfoximine. The pH optima of the muscle enzymes was approximately 7.2 in imidazole–HCl, Tris–HCl, and Hepes buffers. At a buffer concentration of 0.05 *M* the activities of the enzymes were essentially identical in these three buffers. The enzymes are inhibited by phosphate.

The apparent K_m values for ATP, L-glutamate, and NH_4 were determined. The results of these experiments are summarized in Table IV. The substrate concentration versus reaction velocity curves for all three substrates were hyperbolic for both the rat muscle and pig heart enzymes. The apparent K_m values for the brain enzyme determined under the conditions of these experiments do not differ significantly from those previously reported.[3] The muscle enzymes appear to have a significantly higher K_m for L-glutamate than do the enzymes from brain and spleen.[4] The apparent K_m for ATP for the muscle enzymes is lower than the values for brain, liver, and spleen.

The optimum ratio of $MgCl_2$ to ATP was approximately 2 : 1 at 1 m*M* ATP for the muscle enzymes and the enzymes were strongly inhibited by $MgCl_2$ at greater than a 20 m*M* excess of $MgCl_2$ over ATP. As noted in Table III the enzyme activity at the optimum ratio of $MnCl_2$ to ATP is about 25% of the activity with $MgCl_2$ and ATP. CaATP is not a substrate for muscle glutamine synthetase when tested with $CaCl_2$ to ATP ratios

[3] V. Pamiljans, P. R. Krishnaswamy, G. Dumville, and A. Meister, *Biochemistry* **1,** 153 (1962).

[4] C. Wu, *Can. J. Biochem.* **55,** 332 (1976).

TABLE IV
APPARENT K_m VALUES FOR PIG HEART AND RAT MUSCLE TO GLUTAMINE SYNTHETASES[a]

Substrate	Enzyme	K_m (m*M*)
ATP	Rat muscle	1.1
	Pig heart	1.6
	Sheep brain	2.6
L-Glutamate	Rat muscle	5.8
	Pig heart	5.3
	Sheep brain	2.5
NH_4^+	Rat muscle	0.19
	Pig heart	0.14
	Sheep brain	0.17
NH_2OH	Rat muscle	0.17
	Pig heart	0.15

[a] The reaction mixtures contained the non-varied substrates at the concentration used in the standard radioactive assay. The K_m values for L-glutamate and ATP were determined in the presence of NH_4Cl. When L-[U-^{14}C]glutamate was the variable substrate a constant specific radioactivity of 14,000 cpm/μmol was maintained.

ranging from 0.05 : 1 to 2 : 1. Both $CaCl_2$ and $MnCl_2$ are inhibitory at concentrations greater than 2 m*M* when tested in the presence of 20 m*M* $MgCl_2$ and 10 m*M* ATP. MgGTP was not a substrate for muscle glutamine synthetase.

Inhibitors of Muscle Glutamine Synthetases

A number of compounds previously reported as inhibitors of glutamine synthetase activity were evaluated as inhibitors of muscle glutamine synthetase (Table V). The enzymes were inhibited by phosphate and pyrophosphate. This inhibition was not competitive with either ATP or glutamate. Carbamyl phosphate, but not creatine phosphate or acetyl phosphate, also inhibited the enzyme. A number of amino acids inhibited the muscle enzymes less than 10% at 20 to 25 m*M*; however, glycine, L-alanine, D-alanine, and β-alanine were all effective inhibitors (>20% inhibition at 20 m*M*). These four amino acids have been reported to inhibit the activity of rat liver glutamine synthetase but only when assayed with

TABLE V
EFFECT OF VARIOUS COMPOUNDS ON MUSCLE GLUTAMINE SYNTHETASE ACTIVITY

Compound added	Concentration (m*M*)	Inhibition (%)	
		Rat muscle	Pig heart
Potassium phosphate	10	23	27
	35	48	49
Sodium pyrophosphate	10	50	52
Carbamyl phosphate	10	23	29
Creatine phosphate	10	11	7
Acetyl phosphate	10	8	4
Glycine	20	21	21
L-Alanine	20	27	24
D-Alanine	20	48	53
β-Alanine	20	36	41
L-Methionine	25	6	4
L-Histidine	20	4	3
L-Serine	20	7	4
L-Leucine	10	2	4
L-Tryptophan	10	3	6

MnATP as a substrate.[5] Only D-alanine was found to inhibit the sheep brain enzyme.

Methionine sulfone and methionine sulfoximine are effective inhibitors of muscle glutamine synthetase (Table VI), while L-methionine-*S*,*R*-sulfoxide (25 m*M*) is no more effective as an inhibitor than is L-methionine and several other α-amino acids. L-Methionine sulfone and L-methionine sulfoximine (Fig. 1) and D-methionine sulfone are competitive with L-glutamate.

At low concentrations, L-methionine-*S*,*R*-sulfoximine appears to be an inhibitor competitive with L-glutamate (Fig. 1). At a higher concentration (5 m*M*), a pronounced curvature in the double reciprocal plot is observed that is consistent with the irreversible inhibition previously observed with this inhibitor on the brain enzyme.[6] Preincubation of the rat muscle enzyme in the presence of MgATP and methionine sulfoximine (2 m*M*) resulted in nearly complete inactivation of the enzyme when small aliquots of the preincubation mixture were assayed by the standard radioactive assay. Preincubation of the enzyme with methionine sulfoximine in

[5] S. Tate and A. Meister, *Proc. Natl. Acad. Sci. U.S.A.* **68,** 781 (1975).
[6] R. A. Ronzio, W. B. Rowe, and A. Meister, *Biochemistry* **8,** 1066 (1969).

TABLE VI
EFFECT OF METHIONINE DERIVATIVES ON MUSCLE GLUTAMINE SYNTHETASE ACTIVITY

Amino acid added	Concentration (m*M*)	Inhibition (%) Rat muscle	Inhibition (%) Pig heart
L-Methionine	25	6	4
L-Methionine-*S*,*R*-sulfoxide	25	12	10
L-Methionine sulfone	10	80	76
D-Methionine sulfone	10	82	74
L-Methionine-*S*,*R*-sulfoximine	2	75	81
L-Methionine-*S*-sulfoximine	1	71	72
L-Methionine-*R*-sulfoximine	2	5	ND[a]

[a] Not determined.

the absence of MgATP or in the presence of MgATP, L-glutamate, and NH_4Cl resulted in no appreciable inhibition of enzyme activity. The L-methionine-*S*-sulfoximine diastereoisomer was inhibitory, while L-methionine-*R*-sulfoximine was ineffective as an inhibitor at 2 m*M*.

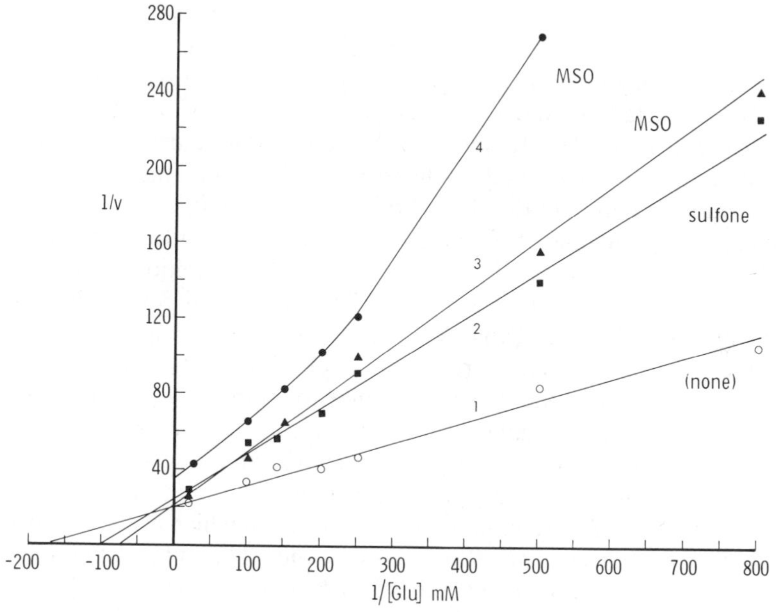

FIG. 1. Inhibition of rat muscle glutamine synthetase by L-methionine sulfoximine (MSO) and L-methionine sulfone. Activity was determined by the radioactive assay. Curve 1, no inhibition (control); curve 2, L-methionine sulfone (1 m*M*); curve 3, L-methionine sulfoximine (1 m*M*); and curve 4, L-methionine sulfoximine (5 m*M*).

TABLE VII
INHIBITION OF GLUTAMINE SYNTHETASE BY NUCLEOSIDE DIPHOSPHATE[a]

Enzyme source	Nucleotide	Concentration (mM)	Inhibition (%)	K_i (mM)
Pig heart	ADP	1	40	1.1
		5	63	
	GDP	1	32	1.35
		5	69	
Rat muscle	ADP	1	32	0.8
		5	67	
	GDP	1	41	0.4
		5	70	
Sheep brain	ADP	1	0	<5
		5	23	
	GDP	1	4	<5
		5	12	

[a] Activities were determined using the radioactive assay in the absence of the ATP regenerating system. K_i values were determined graphically by plotting l/v vs [I] at two concentrations of ATP (5 and 10 mM). All three enzymes were inhibited less than 10% at 5 mM of UDP, CDP, TDP, AMP, GMP, cAMP, guanine, and adenine.

The dependence of enzyme activity in both crude muscle extracts and the purified enzyme preparations on the presence of an ATP regenerating system suggested that the muscle enzymes are inhibited by ADP. Studies, summarized in Table VII, indicate that both ADP and GDP are strong inhibitors of rat muscle and pig heart glutamine synthetase but are relatively poor inhibitors of the sheep brain enzyme. UDP, CDP, TDP, AMP, GMP, cAMP, guanine, and adenine are not significantly inhibitory at 5 mM. Data presented in Fig. 2 indicate that the inhibition by ADP and GDP is competitive with respect to ATP. The values for the inhibition constant (K_i) given in Table VII were determined graphically by the method of Dixon[7] at concentrations of ATP of 5 and 10 mM.

The sensitivity of the muscle enzymes to inhibition by ADP is probably of physiological significance in the *in vivo* regulation of glutamine metabolism. In addition to being subject to inhibition by ADP, these enzymes appear to have a significantly lower apparent K_m for MgATP than do the enzymes from brain, liver, and spleen which are less sensitive to inhibition by ADP. This is taken as evidence that an important mechanism of control of glutamine synthesis in muscle may be by the adenylate energy charge[8] or by the ATP/ADP ratio in the cell. At prevailing con-

[7] M. Dixon, *Biochem. J.* **55,** 170 (1953).

[8] D. E. Atkinson, "Cellular Energy Metabolism and its Regulation," p. 88. Academic Press, New York, 1977.

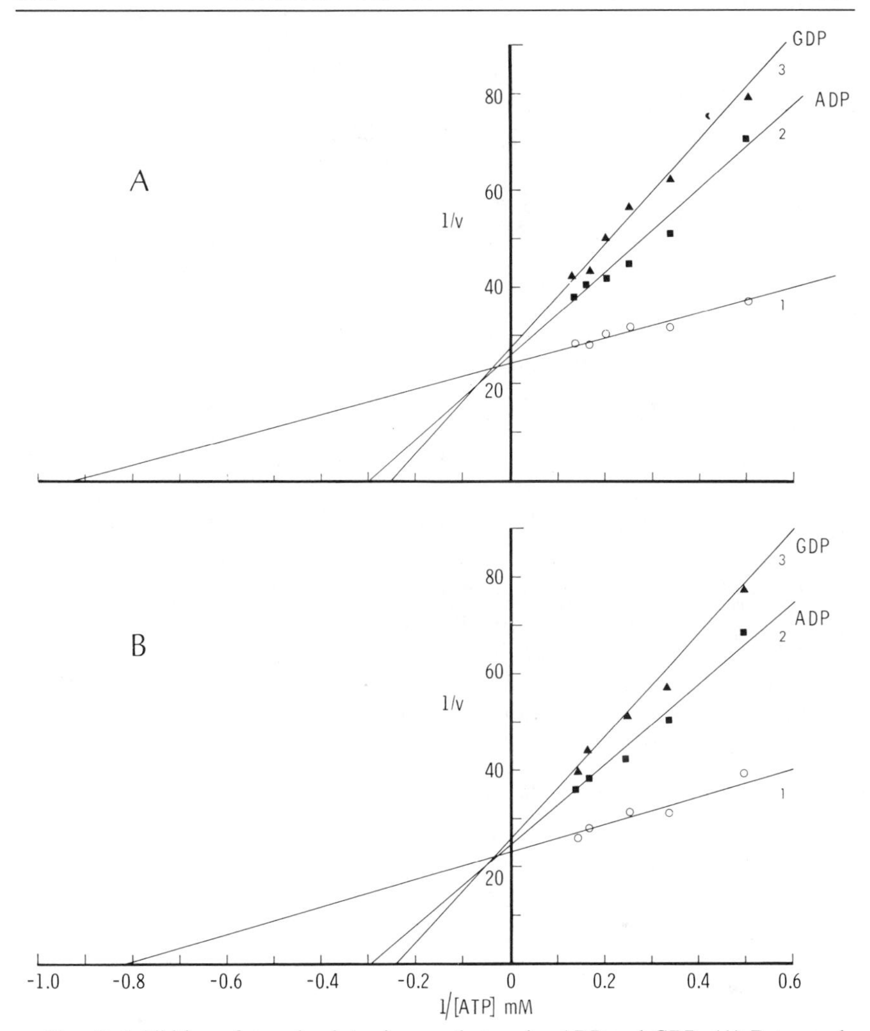

FIG. 2. Inhibition of muscle glutamine synthetase by ADP and GDP. (A) Rat muscle glutamine synthetase; (B) pig heart glutamine synthetase. Curve 1, no added nucleoside diphosphate; Curve 2, ADP (1 mM); Curve 3, GDP (1 mM). Activity was determined by the radioactive assay with no ATP regenerating system present.

centrations of adenine nucleotides in muscle [perfused rat heart; 7.5 mM ATP, 1.6 mM ADP, and 0.19 mM AMP, adenylate energy charge = 0.89, ATP/ADP = 4.7[9]; skeletal muscle; 8 mM ATP, 0.95 mM ADP, and 0.09 mM AMP, adenylate energy charge = 0.94, ATP/ADP = 8.4[10]] glutamine

[9] A. Parmeggiani and H. E. Morgan, *Biochem. Biophys. Res. Commun.* **9,** 252 (1962).

[10] R. Piras and R. Staneloni, *Biochemistry* **8,** 2153 (1969).

synthetase activity could be significantly altered by relatively small and physiologically reasonable changes in adenylate energy charge or ATP/ADP ratio. Data on the effect of ADP concentration on *in vitro* glutamine synthetase activity of pig heart and sheep brain at 10 mM ATP are plotted in Fig. 3 to show the effect of adenylate energy charge and ATP/ADP ratio on enzyme activity. A change in ATP/ADP ratio within a reasonable physiological range from 5.0 to 2.0[9] would result in a 30% decrease in enzymatic activity of the muscle enzyme. By contrast, a similar change in ATP/ADP ratio would result in only a 10% decrease in activity of the brain enzyme. In terms of *in vivo* regulation by adenine nucleotides as represented by a parameter such as the adenylate energy charge, a change in adenylate energy charge from 0.90 to 0.80 would decrease the rate of reaction catalyzed by muscle glutamine synthetase 35% and the rate catalyzed by the brain enzyme 14%. At the point of maximum difference between the two enzymes (high adenylate energy charge) the muscle enzyme is 4 to 5 times more sensitive to changes in adenylate energy charge than is the brain enzyme. This difference in sensitivity to regulation by adenine nucleotides is probably a reflection of the different role this enzyme plays in the metabolism of brain and muscle. If one considers the primary roles of glutamine synthetase in brain to be ammonia detoxification and conversion of glutamate as a neurotransmitter to an inactive product, one could reason that glutamine synthetase activity should be

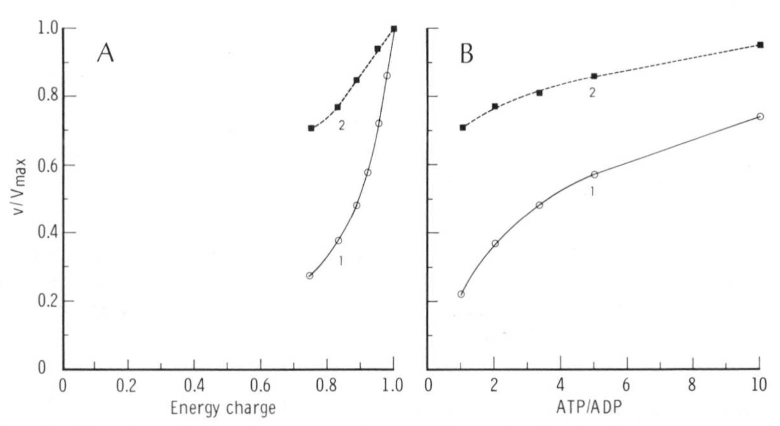

FIG. 3. The effect of adenylate energy charge and ATP/ADP ratio on pig heart and sheep brain glutamine synthetase. (A) Fractional activity plotted against adenylate energy charge = ([ATP] + 0.5 [ADP])/([ATP] + [ADP] + [AMP]); (B) fractional activity plotted against ATP/ADP at 10 mM ATP. Curve 1, pig heart glutamine synthetase; curve 2, sheep brain glutamine synthetase. These data are replotted from the data at 10 mM ATP used in determining the K_i of ADP.

maintained at a high level, which is relatively insensitive to changes in the levels of adenine nucleotides. In muscle, glutamine synthetase can be considered to play a biosynthetic role in maintaining the availability of glutamine as a source of nitrogen and as a respiratory or gluconeogenic substrate when dietary sources of nitrogen or carbon are limited. In addition, one could consider that regulation of glutamine synthetase by the level of adenine nucleotides to be a mechanism to minimize utilization of high energy phosphate bond energy for glutamine synthetase when energy demands are high for muscle contraction. It thus appears reasonable that glutamine synthetase of muscle should be under tighter metabolic control by ATP and ADP than is the brain enzyme.

Physiologic Significance of Muscle Glutamine Synthetase

Skeletal muscle is capable of releasing both glutamine and alanine at a rate in excess of that expected from release of these amino acids as a consequence of muscle protein breakdown.[11–15] The magnitude of this amino acid release from muscle suggests that glutamine and alanine represent an important source of carbon for gluconeogenesis and energy metabolism as well as a source of nitrogen in nonmuscle tissue.[13,16–18] The enzymatic capacity of muscle to form alanine by transamination is well established. However, the enzymatic capacity for glutamine synthesis in muscle has been the subject of some disagreement in the literature. Wu[19] and Trush[20] reported that avian heart had a relatively high level of glutamine synthetase activity. Trush presented evidence that glutamine synthetase activity occurred in both heart and skeletal muscle of several mammalian species, but others have reported no activity or very low levels of activity in skeletal muscle.[19,21] Iqbal and Ottaway[22] found no

[11] D. R. London, T. H. Foley, and C. B. Webb, *Nature (London)* **208,** 588 (1965).
[12] T. Pozefsky, P. Felig, J. D. Tobin, J. S. Soeldner, and G. F. Cahill, Jr., *J. Clin. Invest.* **48,** 2273 (1969).
[13] E. B. Marliss, T. T. Aoki, T. Pozefsky, A. S. Most, and G. F. Cahill, Jr., *J. Clin. Invest.* **50,** 814 (1971).
[14] N. B. Ruderman and P. Lund, *Isr. J. Med.* **8,** 295 (1972).
[15] A. J. Garber, I. E. Karl, and D. M. Kipnis, *J. Biol. Chem.* **251,** 8236 (1976).
[16] D. S. Parsons, *Top. Gastroenterol.* **7,** 253 (1979).
[17] H. Schrock and L. Goldstein, *Am. J. Physiol.* **240,** E519 (1981).
[18] H. G. Windmueller and A. E. Spaeth, *J. Biol. Chem.* **253,** 69 (1978).
[19] C. Wu, *Comp. Biochem. Physiol.* **8,** 335 (1963).
[20] G. P. Trush, *Ukr. Biokhim. Zh.* **35,** 713 (1963), translated in *Fed. Proc., Fed. Am. Soc. Exp. Biol. (Trans. Suppl.)* **23,** 1305.
[21] A. Herzfeld, *Biochem J.* **133,** 49 (1973).
[22] K. Iqbal and J. H. Ottaway, *Biochem. J.* **119,** 145 (1970).

detectable activity in rat heart but demonstrated activity in skeletal muscle and partially purified the enzyme from muscle. Using an assay similar to the method described above, King *et al.* have clearly demonstrated glutamine synthetase activity in rat skeletal muscle.[23]

The data presented here show that both skeletal muscle and heart tissue from small laboratory mammals and heart tissue from larger animals contain glutamine synthetase activity. The specific enzymatic activity of muscle is low compared to tissues such as brain and liver even on the basis of activity per mg of soluble protein. However, it is evident that the large mass of skeletal muscle in the body represents a significant source of glutamine formation. In a 250 g rat in which skeletal muscle constitutes approximately 45% of body weight, the maximum rate of synthesis of glutamine by glutamine synthetase is estimated to be 7 μmol of glutamine synthesis in liver (20 μmol/min) and kidney (4 μmol/min).

A comparison of the maximum rate of muscle glutamine synthesis by glutamine synthetase with reported rate of glutamine release by isolated muscle preparations may be made. In the absence of added amino acid substrates, glutamine release by three muscle preparations from the rat ranged from 25 to 34 nmol/min/g of tissue.[15] The rate of release of glutamine by the epitrochlaris muscle in the presence of the best amino acid precursors, valine and tyrosine in the presence of NH_4Cl, was about 50 nmol/min/g.[24] Using the results from *in vitro* assays of crude muscle extracts under optimal assay conditions, the maximum rate of glutamine synthesis catalyzed by glutamine synthetase is about 60 nmol/min/g. If some substantial fraction of glutamine released (one-quarter to one-third) is considered to be derived from muscle proteolysis, it appears reasonable that there is in muscle sufficient glutamine synthetase activity to account for the rate of glutamine release by muscle. Muscle glutamine synthetase probably does not function at optimal rates under most physiological conditions due in substantial part to the sensitivity of this activity to inhibition by ADP.

Acknowledgment

This work was supported by a grant from the American Heart Association, Grant 74-743.

[23] P. A. King, L. Goldstein, and E. A. Newsholme, *Biochem. J.* **216,** 523 (1983).
[24] A. J. Garber, I. E. Karl, and D. M. Kipnis, *J. Biol. Chem.* **251,** 836 (1976).

[29] Glutamine Synthetase from *Escherichia coli*

By SUE GOO RHEE, P. BOON CHOCK, and EARL R. STADTMAN

Glutamine synthetase in *Escherichia coli* and other gram-negative bacteria is subject to rigorous cellular control. This is because glutamine produced by the enzyme serves as a source of nitrogen atoms in the biosynthesis of all amino acids, purines and pyrimidine nucleotides, of glucosamine-6-phosphate, *p*-aminobenzoic acid, and of nicotinamide derivatives. In essence, glutamine synthetase (GS) links the assimilation of NH_3 with biosynthetic pathways leading to the formation of proteins, nucleic acids, complex polysaccharides, and different coenzymes.[1] To accommodate such diverse functions, GS is regulated by varying the enzyme level through protein synthesis,[2] irreversible inactivation[3] and degradation, by cummulative feedback inhibition on the enzymic activity,[4] and by a closed bicyclic cascade[5] which is capable of sensing simultaneously changes in the concentration of many metabolites and integrating their effects such that the catalytic activity of the enzyme can be adjusted to meet the everchanging demand for glutamine.

The Closed Bicyclic Cascade

Glutamine synthetase activity is modulated by a closed bicyclic covalent interconvertible cascade. It consists of two protein nucleotidylation cycles. One involves the cyclic adenylylation and deadenylylation of glutamine synthetase; the other involves the uridylylation and deuridylylation of Shapiro's regulatory protein, P_{II}.

The adenylylation and deadenylylation of GS are catalyzed at separate, and essentially noninteractive sites (designated AT_a and AT_d, respectively) on a single adenylyltransferase.[6–8] The adenylylation reaction

[1] E. R. Stadtman, *in* "The Enzyme of Glutamine Metabolism" (S. Prusiner and E. R. Stadtman, eds.), p. 1. Academic Press, New York, 1973.

[2] B. Magasanik, *Annu. Rev. Genet.* **16,** 135 (1982).

[3] C. N. Oliver, R. L. Levine, and E. R. Stadtman, *in* "Experiences in Biochemical Perception" (L. N. Ornston and S. Sligar, eds.), p. 233. Academic Press, New York, 1982.

[4] E. R. Stadtman and A. Ginsburg, *in* "The Enzymes" (P. D. Boyer, ed.), 3rd ed., Vol. 10, p. 755. Academic Press, New York, 1974.

[5] E. R. Stadtman and P. B. Chock, *Curr. Top. Cell. Regul.* **13,** 53 (1978).

[6] W. B. Anderson, S. B. Hennig, A. Ginsburg, and E. R. Stadtman, *Proc. Natl. Acad. Sci. U.S.A.* **61,** 1417 (1970).

[7] S. B. Hennig and A. Ginsburg, *Arch. Biochem. Biophys.* **144,** 611 (1971).

[8] S. G. Rhee, R. Park, P. B. Chock, and E. R. Stadtman, *Proc. Natl. Acad. Sci. U.S.A.* **75,** 3138 (1978).

Copyright © 1985 by Academic Press, Inc.
All rights of reproduction in any form reserved.

involves the attachment of an adenylyl group from ATP, through phosphodiester linkage, to the hydroxyl group of a specific tyrosyl residue in each subunit.[9-11] The amino acid sequence of the adenylylation site has been identified to be Ile-His-Pro-Gly-Glu-Ala-Met-Asp-Lys-Asn-Leu-Tyr-Asp-Leu-Pro-Pro-Glx-Gln-Ala-Lys.[12,13]
|
AMP

Because GS is composed of 12 identical subunits[14,15] (M_r = 50,000) up to 12 adenylyl groups can be attached to each enzyme molecule. Since adenylylated subunits are catalytically inactive under most physiological conditions, the specific activity of the enzyme is inversely proportional to the average number, $\bar{n}$, of adenylylated subunits per dodecamer. Deadenylylation is achieved by phosphorylysis of the adenylyl-*O*-tyrosyl bond to yield ADP and the active unmodified GS.[16] Because the catalytic sites for both adenylylation and deadenylylation reaction are present on the same polypeptide chain, without proper regulation, the two reactions will be coupled and result in senseless phosphorylysis of ATP to form ADP and PP_i. Such futile cycle is prevented by the coupling of the adenylylation cycle with another nucleotidylation cycle in which P_{II} undergoes interconversion between uridylylated and unmodified forms.[17,18]

Uridylylation of P_{II} involves the attachment of a uridylyl group from UTP, through phosphodiester linkage to the hydroxyl group of a specific tyrosyl residue in each subunit (M_r = 11,000) of the tetramer.[18-20] The reaction is catalyzed by uridylyltransferase (UT) and the amino acid sequence at the uridylylation site has been shown to be Gly-Ala-Glu-Tyr-Met-Val-Asp-Phe-Leu-Pro-Lys.[21] A separate uridylyl removing enzyme (UR) activity catalyzes the hydrolytic cleavage of the uridylyl-*O*-tyrosyl
UMP
|
(the UMP is attached to the Tyr)

[9] H. S. Kingdon, B. M. Shapiro, and E. R. Stadtman, *Proc. Natl. Acad. Sci. U.S.A.* **58,** 1703 (1967).

[10] K. Wulff, D. Mecke, and H. Holzer, *Biochem. Biophys. Res. Commun.* **28,** 740 (1967).

[11] B. M. Shapiro and E. R. Stadtman, *J. Biol. Chem.* **243,** 3769 (1968).

[12] R. L. Heinrickson and H. S. Kingdon, *J. Biol. Chem.* **246,** 2246 (1971).

[13] L. M. Keefer, P. S. Keim, H. S. Kingdon, C. Noyes, and R. L. Heinrikson, *Fed. Proc., Fed. Am. Soc. Exp. Biol.* **38,** 325 (1979).

[14] R. C. Valentine, B. M. Shapiro, and E. R. Stadtman, *Biochemistry* **7,** 2143 (1968).

[15] B. M. Shapiro and A. Ginsburg, *Biochemistry* **7,** 2153 (1968).

[16] W. B. Anderson and E. R. Stadtman, *Biochem. Biophys. Res. Commun.* **41,** 704 (1970).

[17] B. M. Shapiro, *Biochemistry* **8,** 659 (1969).

[18] M. S. Brown, A. Segal, and E. R. Stadtman, *Proc. Natl. Acad. Sci. U.S.A.* **68,** 2949 (1971).

[19] J. H. Mangum, G. Magni, and E. R. Stadtman, *Arch. Biochem. Biophys.* **158,** 514 (1973).

[20] S. P. Adler, D. Purich, and E. R. Stadtman, *J. Biol. Chem.* **250,** 6264 (1975).

[21] S. G. Rhee and P. B. Chock, *Isozymes: Curr. Top. Biol. Med. Res.* **8,** 141 (1983).

bond to form UMP and unmodified P_{II}.[20,22] Like the adenylylation/deadenylylation reaction, both UT and UR activities are located at a separate site on a single polypeptide chain of $M_r = 95,000$.[23] Because the same polypeptide catalyzes both the uridylylation and deuridylylation reactions, the two reactions are coupled to yield cyclic interconversion of P_{II} between its uridylylated and unmodified forms with concomitant hydrolysis of UTP to UMP and PP_i. This interconversion is regulated by metabolites.

Linkage of the GS adenylylation cycle with the P_{II} uridylylation cycle is derived from the fact that the unmodified form of P_{II} (also referred to as P_{IIA}) stimulates the capacity of AT_a to catalyze the adenylylation of GS, whereas the uridylylated form of P_{II} (also referred to as P_{IID}) is required to activate the deadenylylation activity of AT_d.[5,8,18,21] The coupling of the adenylylation cycle with the uridylylation cycle produces a closed bicyclic cascade for regulating the GS activity. The cyclic interconversion of P_{II} and GS is controlled by various metabolites which affect the activities of the converter enzymes; i.e., AT_a, AT_d, UR, and UT. In fact, almost 40 different metabolites have been shown to affect one or more of these enzymes.[5] Among them, α-ketoglutarate (α-Kg) and L-glutamine play a dominant role in the regulation of GS. L-Glutamine stimulates the adenylylation of GS and the deuridylylation of $P_{II}(UMP)_n$, whereas it inhibits the deadenylylation of $GS(AMP)_n$ and the uridylylation of P_{II}. Conversely, α-ketoglutarate inhibits the adenylylation of GS but it stimulates the deadenylylation of $GS(AMP)_n$ and the uridylylation of P_{II}. In other words, those reactions which lead to the inactivation of GS are stimulated by L-glutamine and inhibited by α-ketoglutarate, whereas those leading to the reactivation of GS are inhibited by L-glutamine and stimulated by α-ketoglutarate, a precursor of L-glutamine. Consequently, the ratio of L-Gln to α-Kg, which varies in response to the availability of ammonia, determines the state of adenylylation of GS, and hence the GS activity.

Theoretical Analysis of Glutamine Synthetase Cascade

In order to illustrate the superiority of cyclic cascade as a metabolite regulator for glutamine synthetase a theoretical analysis of this system has been carried out.[5,24] The schematic representation of the closed bicyclic cascade of GS is given in Fig. 1. In the forward cascade, UR is

[22] S. P. Adler, J. H. Mangum, G. Magni, and E. R. Stadtman, *in* "Metabolic Interconversion of Enzymes" (E. H. Fischer, E. G. Krebs, and E. R. Stadtman, eds.), p. 221. Springer, New York, 1974.

[23] E. Garcia and S. G. Rhee, *J. Biol. Chem.* **258,** 2246 (1983).

[24] E. R. Stadtman, P. B. Chock, and S. G. Rhee, *in* "From Gene to Protein: Informative Transfer in Normal and Abnormal Cells" (T. R. Russell, K. Brew, J. Schultz, and H. Faber, eds.), p. 521. Academic Press, New York, 1979.

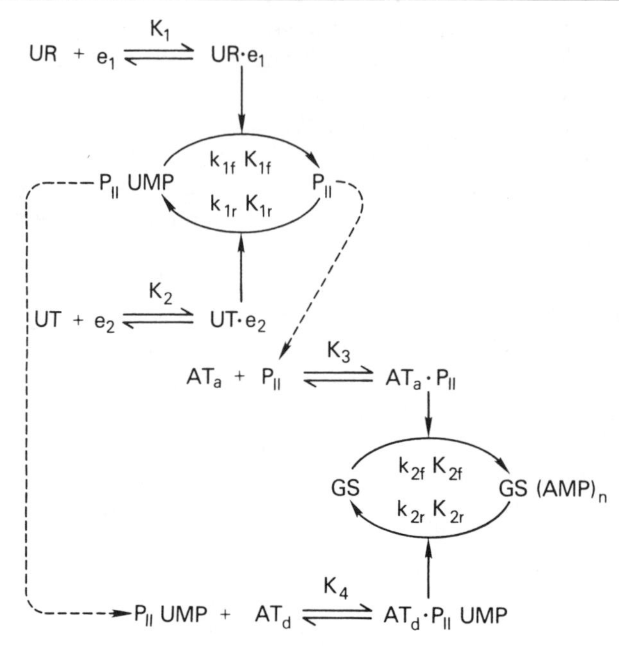

FIG. 1. The glutamine synthetase bicyclic cascade.

activated by an allosteric effector e_1. The active $UR \cdot e_1$ catalyzes the hydrolytic cleavage of UMP from $P_{II}(UMP)_n$ to form P_{II}, which in turn activates AT_a. The activated AT_a P_{II} then catalyzes the adenylylation of GS. The regeneration cascade is triggered by the interaction of an allosteric effector e_2 with UT to form an active $UT \cdot e_2$, which in turn catalyzes the uridylylation of P_{II}. The uridylylated P_{II} activates AT_d to catalyze the deadenylylation of $GS(AMP)_n$. Dynamic coupling of these forward and regeneration cascades results in the cyclic interconversion between GS and $GS(AMP)_n$, P_{II}–(P_{IIA}) and $P_{II}(UMP)_n$ (P_{IID}) and the concomitant consumption of UTP and ATP. When the ATP and UTP concentrations are maintained in excess relative to the enzymes involved and at reasonably constant levels, a steady state will be established in which the rate of $GS(AMP)_n$ formation will be equal to the rate GS regeneration. Under these steady state conditions, it is possible to derive an equation to show how the state of adenylylation, $\bar{n}$, varies as a function of other cascade parameters. For simplicity, both ATP and UTP dependency are ignored in deriving the steady-state expression because they can be treated as constants due to the conditions described above. Further simplification was made by assuming that[5,25,26] (1) enzyme–enzyme and enzyme–effec-

[25] E. R. Stadtman and P. B. Chock, *Proc. Natl. Acad. Sci. U.S.A.* **74,** 2761 (1977).
[26] P. B. Chock and E. R. Stadtman, *Proc. Natl. Acad. Sci. U.S.A.* **74,** 2766 (1977).

tor complexes are formed via a rapid equilibrium mechanism; (2) the concentrations of the enzyme–enzyme complexes are negligible compared to the concentrations of the modified and unmodified enzymes; and (3) the concentrations of the allosteric effectors e_1 and e_2 are maintained at constant levels for any metabolic state. With these assumptions, the simplified steady-state expression for the state of adenylylation, $\bar{n}$, is shown in Eq. (1).

$$\bar{n} = 12 \left\{ \left[\left(\frac{k_{1r}}{k_{1f}}\right)\left(\frac{k_{2r}}{k_{2f}}\right)\left(\frac{K_{1f}}{K_{1r}}\right)\left(\frac{K_{2f}}{K_{2r}}\right)\left(\frac{K_3}{K_4}\right)\left(\frac{[\mathrm{UT}]}{[\mathrm{UR}]}\right) \left(\frac{[\mathrm{AT_d}]}{[\mathrm{AT_a}]}\right)\left(\frac{[e_2]}{[e_1]}\right) \frac{(K_1 + [e_1])}{(K_2 + [e_2])} \right] + 1 \right\}^{-1} \qquad (1)$$

In this equation, K_1, K_2, K_3, and K_4 are the dissociation constants for activated complexes, $\mathrm{UR} \cdot e_1$, $\mathrm{UT} \cdot e_2$, $\mathrm{AT_aP_{IIA}}$, and $\mathrm{AT_d} \cdot \mathrm{P_{IID}}$, respectively; k_{1f}, k_{1r}, k_{2f}, and k_{2r} are catalytic rate constants for the interconversion reactions indicated in Fig. 1; K_{1f}, K_{1r}, K_{2f}, and K_{2r} are dissociation constants for the activated converter enzyme-interconvertible protein substrate complexes formed by the reactions shown in Fig. 1.

The simplified Eq. (1) shows that in addition to the two allosteric effectors, e_1 and e_2, the state of adenylylation of GS can be regulated by altering one or more of the 16 parameters through interactions between metabolite(s) and cascade enzyme(s). Furthermore, because $\bar{n}$ is a multiplicative function of multiple variables, GS cascade is endowed with unique regulatory properties. Theoretical analysis of this cascade reveals that (1) it possesses an enormous capacity for signal amplification, i.e., a relatively small fractional activation of UR enzyme can promote an extensive adenylylation of GS.[5,24] The concentration of the primary allosteric stimuli, e_1, required to attain a given value of $\bar{n}$ varies with the value of the overall parameter ratios shown in Eq. (1), and it can be much lower than the concentration of e_1 needed to obtain a comparable activation of the UR enzyme with which it reacts directly. For comparative purposes, this signal amplification is defined[5] as the ratio of the concentration of effector required to produce 50% activation of the converter enzyme, UR, to that required to obtain 50% activation of the interconvertible enzyme, GS. Stadtman *et al.*[24] have shown that a 2-fold and a 4-fold change in each of the 16 parameters in favor of the adenylylation reaction would yield a 400-fold and a 10,000-fold increase in signal amplification. It should be pointed out that the signal amplification described here is a time-independent parameter. It is distinctly different from the catalytic amplification which is time dependent and is a function of the relative concentrations and the maximal catalytic efficiencies of the converter enzyme and interconvertible enzyme. (2) It can modulate the maximum value of $\bar{n}$ that can be accomplished with saturating levels of primary effector, e_1.[5,25] By changing the magnitude of the parameters in Eq. (1), the values of $\bar{n}$ can be varied smoothly from 0 to 12. Therefore, even at saturating levels of primary effector, interconvertible enzymes do not function as an On-Off switch. Instead, they allow the cells to maintain the GS activity at various levels of activation in response to their physiological needs. (3) It can modulate the sensitivity of GS adenylylation in response to changes in the concentration of primary allosteric effectors; i.e., it is capable of eliciting a positive cooperative type of response to increasing concentrations of an allosteric effector.[5] This positive cooperativity can be obtained when the same effector serves as a positive modifier in two forward steps in the bicyclic cascade, or when the same effector is a positive modifier of a forward step and also a negative modifier of a regeneration step. It should be pointed out that positive cooperativity can also be attained for a monocyclic cascade when the converter enzyme forms a tight complex with its interconvertible substrate.[27,28] (4) Cyclic cascades are extremely flexible systems for allosteric regulation. They are capable of exhibiting both quantitatively and qualitatively, various responses to allosteric stimuli.[5,24] In addition, because each of the 16 parameters in Eq. (1), except e_1 and e_2, can be altered by allosteric interaction of one or more of the cascade proteins with a single or multiple allosteric effectors, the GS bicyclic cascade can function as a biological integration device which can sense simultaneous fluctuations in the intracellular concentrations of numerous metabolites and vary the specific activity of GS accordingly. (5) Like other cyclic cascades, the GS cascade can serve as rate amplifier.[29] Therefore, GS cascade is capable of responding very rapidly to changes in metabolite levels.

In the theoretical analysis described above the roles of ATP and UTP as nucleotidyl group donors in the two interconversion cycles have been ignored. This does not invalidate the results, because for a given physiological state the concentrations of ATP and UTP are maintained at nearly constant levels which are several orders of magnitude higher than the concentrations of the proteins involved. Nevertheless, for each complete interconversion cycle one equivalent of nucleoside triphosphate is consumed. This decomposition of nucleoside triphosphate is an essential

[27] E. Shacter-Noiman, P. B. Chock, and E. R. Stadtman, *Philos. Trans. R. Soc. London Ser. B* **302,** 157 (1983).

[28] A. Goldbetter and D. E. Koshland, Jr., *Proc. Natl. Acad. Sci. U.S.A.* **76,** 6840 (1981).

[29] P. B. Chock and E. R. Stadtman, *in* "Modulation in Protein Function" (D. E. Atkinson and C. F. Fox, eds.), p. 185. Academic Press, New York, 1979.

feature of the cyclic cascade regulatory mechanism because it provides the free energy required to maintain the steady-state distribution between the nucleotidylated and unmodified forms of the interconvertible proteins at various metabolite-specified levels which are different from that specified by thermodynamic consideration. The nucleoside triphosphate consumption is therefore the price the cell must pay to maintain such an effective control mechanism. It should be pointed out that the rate of nucleoside triphosphate consumption is also regulated by the parameters that regulate the fractional modification of the interconvertible proteins.[5]

Validity of the Cascade Model

The cyclic cascade model described above is based on the assumption that for any metabolic condition a steady-state is achieved in which the rates of GS adenylylation and deadenylylation are equal. This assumption is supported by the studies of Segal *et al.*[30] showing that a steady state was established when purified unadenylylated GS was incubated in a mixture containing all the components needed to sustain the cascade system. The steady-state level of $\bar{n}$ varied as a function of effector concentrations. These results confirm that GS cascade is not merely a metabolic switch for turning On or Off the GS activity, instead, it functions as a dynamic processing device which can integrate large amounts of metabolic signals and accordingly adjust the specific activity of GS in a smooth and continuous manner.

Attempt to reconstitute the GS cascade with purified proteins has not been accomplished because of the instability of the UR/UT enzymes and the inability to obtain them in a reasonable quantity. However, because the sole function of the uridylylation/deuridylylation cycle is to modulate the steady-state ratios of P_{IIA} to P_{IID}, one can substitute this cycle by varying the mole fraction of P_{IIA}. As a consequence, the bicyclic cascade of GS is converted to a monocyclic cascade system. Study of this monocyclic cascade system using the purified proteins has been reported by Rhee *et al.*[8,31] Their results demonstrate that the GS cascade is much more complex than that predicted by the simplified monocyclic cascade model because the effector, Gln, activates the adenylylation reaction and inactivates the deadenylylation reaction, while another effector, α-ketoglutarate, activates the deadenylylation reaction and inactivates the

[30] A. Segal, M. S. Brown, and E. R. Stadtman, *Arch. Biochem. Biophys.* **161,** 319 (1974).

[31] S. G. Rhee, P. B. Chock, and E. R. Stadtman, *in* "The Enzymology of Post-Translational Modification of Proteins" (R. Freedman, ed.). Vol. II, p. 273. Academic Press, Orlando, 1985.

adenylylation reaction; furthermore, the kinetic analysis shows that four AT_a–effector complexes are active in catalyzing the adenylylation reaction and four AT_d–effector complexes are capable of catalyzing the deadenylylation reaction. Therefore, in order to quantitate this cascade it is necessary to evaluate 10 different binding constants, 14 synergistic or antagonistic coefficients and 8 rate constants. These constants were determined by detailed kinetic analysis. Despite the complexity observed for the adenylylation/deadenylylation cycle, the data show that the GS cascade is remarkably flexible to allosteric regulation. Further, signal amplification and the increase in sensitivity with respect to the variation of effector concentrations were observed as predicted by the cyclic cascade model.

The predicted properties of bicyclic cascade have also been verified by studying the GS cascade in the permeabilized *E. coli* cells.[32] Following the freeze-thaw cycle, treatment of *E. coli* with Lubrol WX causes the cells to be permeable to small molecules while retaining the cytosolic proteins.[33] When cells were cultured with 10 m*M* Gln as the sole nitrogen source, all components of the bicyclic cascade are functional after Lubrol treatment. However, with 5 m*M* Gln-grown cells, Lubrol treatment seems to selectively inactivate the UR/UT bifunctional enzyme thus reducing the bicyclic cascade into a monocyclic cascade. This is revealed by the observation that the state of adenylylation in the 5 m*M* Gln-grown cells does not respond to UTP, a substrate for UT, or to CMP and Mn^{2+}, the allosteric effectors of the UR enzyme, while the GS cascade in the 10 m*M* Gln-grown cells does response to these metabolites. The data show that in both 5 and 10 m*M* Gln-grown cells, the state of adenylylation is dependent upon the concentration of allosteric effectors of AT_a/AT_d. As predicted by the cyclic cascade model, the 5 and 10 m*M* Gln-grown permeabilized cells exhibit characteristics of mono- and bicyclic cascades, respectively. As expected, the cells with both cycles functioning possess a higher signal amplification than the cells with only 1 cycle functioning. In addition, the data reveal that the bicyclic cascade system possesses higher potential for increasing sensitivity in response to changes in the concentration of effector than that of the monocyclic cascade system. Results of these studies show that GS adenylylation in the permeabilized cells is regulated by the mechanism established in *in vitro* studies using purified cascade proteins and that the principles disclosed by theoretical analysis of the cyclic cascade model operate *in situ* in the permeabilized cells.

[32] U. Mura, P. B. Chock, and E. R. Stadtman, *J. Biol. Chem.* **256,** 13022 (1981).
[33] U. Mura and E. R. Stadtman, *J. Biol. Chem.* **256,** 13014 (1981).

Purification and Characterization of Protein Components in Glutamine Synthetase Cascade

Glutamine Synthetase Purification Procedure

Since the purification method of Woolfolk and Stadtman[34] as modified by Shapiro and Stadtman[35] had been described, several new procedures were developed based on the unique properties of the enzyme. These new procedures are described here. Through all purification steps, enzyme activity was measured by γ-glutamyltransferase assay.

Zinc Precipitation Method

This procedure depends on the fact that GS from *E. coli* forms highly insoluble paracrystalline aggregates in the presence of Zn^{2+} and Mg^{2+}. Under optimal conditions (pH 5.85, 25°, 1 m*M* $ZnSO_4$ and 50 m*M* $MgCl_2$), over 95% of the enzyme is precipitated from crude extracts.[36] In the original procedure by Miller *et al.*[36] GS was extracted from the precipitate with Zn^{2+}-free buffer. Later, the extraction procedure was improved by using a filter and celite, and two more steps (acetone fractionation step and acid ammonium sulfate precipitation step) from the procedure of Woolfolk and Stadtman[34] were added to ensure homogeneity. With these procedures, overall yields were high (>80%), even when large quantities of cells (>2 kg) were used. Unless otherwise stated, the purification was performed at 4°. pH adjustments were made with 1 *N* acetic acid and 1 *N* NH_4OH.

Step 1. Crude Extract. Frozen *E. coli* cells were suspended in 2 ml of buffer (10 m*M* imidazole–HCl, pH 7.4, 1 m*M* $MgCl_2$) per g of cell. The suspension was sonicated or passed through a French pressure cell and then cellular debris was removed by centrifugation.

Step 2. Streptomycin Precipitation. Streptomycin sulfate, 1 g/100 ml, was added to the supernatant solution and the pH was adjusted to 5.85. After stirring at room temperature for 15 min, the mixture was centrifuged at 30,000 *g* for 30 min.

Step 3. Zinc Precipitation. To the supernatant solution, $MgCl_2$ and $ZnSO_4$ were added to final concentrations of 50 and 1.5 m*M*, respectively.

[34] C. A. Woolfolk, B. M. Shapiro, and E. R. Stadtman, *Arch. Biochem. Biophys.* **116,** 177 (1966).

[35] B. M. Shapiro and E. R. Stadtman, this series, Vol. 17A, p. 910.

[36] R. E. Miller, E. Shleton, and E. R. Stadtman, *Arch. Biochem. Biophys.* **163,** 155 (1974).

The pH was adjusted to 5.85 and the mixture was stirred at room temperature for 30 min. For large scale preparations, the mixture was allowed to stand overnight at 4°. Dry, unwashed celite (Hyflo, Fischer Scientific Company), 1 g/10 ml of supernatant, was stirred into the suspension at room temperature. This was filtered on a sintered glass funnel (medium porosity) to which a base of celite (0.7 g/10 ml supernatant solution) was previously added. The celite was washed with buffer (10 m*M* imidazole–HCl, pH 7.0, 50 m*M* $MgCl_2$, and 1 m*M* $ZnSO_4$) in a quantity 2 ~ 3 times that of the supernatant solution after streptomycin precipitation. GS was eluted batchwise with 10 m*M* imidazole buffer not containing $MgCl_2$ and $ZnSO_4$. Each batch was assayed for GS activity and the most active batches were pooled.

Step 4. Second Zinc Precipitation. $MgCl_2$ was added to the pooled eluate to a final concentration of 50 m*M* and the pH was adjusted to 7.0. The concentration of GS in the pooled eluate was estimated based upon γ-glutamyltransferase activity and then $ZnSO_4$ was added to a final stoichiometry of 1.5 equivalents per subunit. The suspension was allowed to stand overnight at 4°. The supernatant solution was decanted and discarded. The remaining suspension was centrifuged and the precipitate was resuspended to a protein concentration of 2–10 mg/ml in buffer containing 10 m*M* imidazole–HCl, pH 7.0, 100 m*M* KCl, 2 m*M* Mg-Titriplex (E. Merck, Germany) and dialyzed against the resuspending buffer and finally against 10 m*M* imidazole–HCl, pH 7.0, 10 m*M* $MnCl_2$. The dialyzed protein solution was cleared by centrifugation.

Step 5. Acetone Fractionation. The pH of the supernatant from step 4 was lowered to 5.7 and acetone at room temperature was added to a final concentration of 45% by volume. After 5 min at room temperature, the mixture was centrifuged and the precipitate was resuspended in the same volume of 10 m*M* imidazole–HCl, pH 7.0, 10 m*M* $MnCl_2$ buffer. The carefully suspended precipitate was then centrifuged.

Step 6. Acid Ammonium Sulfate Precipitation. To the supernatant solution thus obtained, 10% (v/v) of saturated $(NH_4)_2SO_4$ (4°) was added; the pH was adjusted to 5.15 and the mixture was then centrifuged. The supernatant was then adjusted to pH 4.4, equilibrated at that pH for 15 min, then centrifuged. The precipitate was resuspended in buffer containing 10 m*M* imidazole–HCl, pH 7.0, 100 m*M* KCl, 1 m*M* $MnCl_2$, and stored at 4°.

Affinity Chromatography on Affi-Gel Blue

This new approach was suggested by the demonstration that Cibacron Blue binds to the nucleotide substrate site on the enzyme and can be displaced by ADP.[37,38] Affi-gel Blue (Bio-Rad Laboratories) is a cross-linked agarose with covalently attached Cibacron Blue. Combining this affinity chromatography with several steps from the procedure of Woolfolk and Stadtman, GS from *A. vinelandii*,[39] *P. putida*,[40] *P. aeruginosa*,[40] Cyanobacteria,[41] *E. coli*,[42,43] and *S. cerevisiae*[44] were purified to homogeneity. The following are procedures used by Burton *et al.*[42,43] for the purification of GS from *E. coli* to give a yield of 80%.

Step 1. Cell Lysis. Frozen cells were resuspended in preparation buffer (20 m*M* imidazole, pH 7.15, 10 m*M* $MnCl_2$, 0.1 m*M* EDTA, 20 μl/100 ml 2-mercaptoethanol, 0.03% NaN_3) at a concentration of 1 g cell paste/3 ml preparation buffer. Cells were lysed by sonication, and cell debris pelleted 30 min at 15,000 rpm in a Sorvall SS-34 rotor or 45 min at 9000 rpm in a GS-3 rotor.

Step 2. Streptomycin Sulfate Precipitation. A 20% (v/v) streptomycin sulfate solution, freshly made, was added to the supernatant from step 1 to a final 1.5%. The solution was stirred 10 min and centrifuged. The pellet, which contains nucleic acids, was discarded. If the supernatant is not completely clear, a second streptomycin sulfate precipitation is recommended, because nucleic acids interfere with the precipitation of proteins by polyethyleneimine.

Step 3. Polyethyleneimine (PEI) Precipitation. A PEI stock solution was made up as follows: 4.3 g 50% PEI solution (ICN Biochemicals) was diluted in H_2O, the pH adjusted to 7.0 with concentrated HCl, and the volume brought to 100 ml (referred to as a "2.15%" PEI stock solution). The 2.15% PEI stock solution was added to the supernatant from step 2 to a final "0.12%." After it was stirred for 10 min, the solution was centrifuged. Burton and Eisenberg[42] suggest that the concentration of PEI necessary for complete precipitation of GS from the streptomycin sulfate supernatant be determined for each new PEI solution purchased. The PEI pellet was resuspended in high salt preparation buffer (preparation buffer with 1.2 *M* NaCl) using a tissue grinder with a Teflon pestle. The volume of resuspension was one-fifth to about one-tenth the volume of the whole cell homogenate. The solution was cleared by centrifugation.

[37] E. R. Stadtman, R. J. Hohman, J. N. Davis, M. Wittenberger, P. B. Chock, and S. G. Rhee, *Mol. Biol. Biochem. Biophys.* **32,** 144 (1980).

[38] E. R. Stadtman and M. Federici, unpublished results.

[39] J. Siedel and E. Shelton, *Arch. Biochem. Biophys.* **192,** 214 (1979).

[40] J. M. Meyer and E. R. Stadtman, *J. Bacteriol.* **146,** 705 (1981).

[41] G. Stacey, C. V. Baalen, and R. Tabita, *Arch. Biochem. Biophys.* **194,** 457 (1979).

[42] Z. F. Burton and D. Eisenberg, *Arch. Biochem. Biophys.* **205,** 478 (1980).

[43] Z. F. Burton, T. Sutherland, and D. Eisenberg, *Arch. Biochem. Biophys.* **211,** 509 (1981).

[44] A. P. Mitchell and B. Magasanik, *J. Biol. Chem.* **258,** 119 (1983).

Step 4. Affi-Gel Blue (Blue Dextran) Column Chromatography. The resuspended PEI pellet was loaded on a Affi-gel Blue column (1 × 4 cm) preequilibrated with 1.2 *M* NaCl preparation buffer. This solution was passed over the column twice to allow GS to interact fully with the affinity resin. The column was washed with 3–5 column volume of 1.2 *M* NaCl preparation buffer, and the GS was eluted with 1.2 *M* NaCl preparation buffer containing 5 m*M* ADP.

Step 5. Acid Ammonium Sulfate Precipitation. The GS peak eluted from the column was concentrated by acid ammonium sulfate precipitation. A concentrated ammonium sulfate stock solution was added with stirring to a final 0.3 *M*, and 1.0 *M* acetic acid was added to adjust the pH to 4.4. The solution was stirred for 15 min and centrifuged. The pellet was resuspended in preparation buffer, dialyzed against preparation buffer, and stored under refrigeration.

Burton *et al.*[43] also described a modified procedure in which instead of treating the whole cell homogenate with streptomycin sulfate, PEI was added directly to the whole cell homogenate to a final concentration of 0.3%.

Affinity Chromatography on Anthranilic Acid Bound to Sepharose

Anthranilic acid is an effective feedback inhibitor of *Neurospora crassa* GS.[45] Thus, when Palacios[46] prepared an affinity gel by cross-linking anthranilic acid to CH-Sepharose 4B (Pharmacia Fine Chemicals) by a carbodiimide coupling procedure, the anthranilate bound Sepharose was able to bind GS. The following are procedures[46] used by Palacios for the purification of GS from *N. crassa*. This method resulted in a homogeneous preparation of GS with a recovery of about 40% of the original enzyme activity.

Step 1. Crude Extract. Acetone powder of *N. crassa* (12 g) was ground with dry ice in a mortar until a very find powder was obtained. To each gram of powder, 10 ml of extraction buffer (5 m*M* phosphate, 0.5 m*M* EDTA, 50 m*M* K_2SO_4, pH 7.2) was added, the suspension was stirred for 15 min and then homogenized with two strokes in a motor-driven Teflon pestle homogenizer at 5000 rpm. The preparation was centrifuged for 15 min at 15,000 *g* and the supernatant was filtrated through a Sephadex G-25 column equilibrated with extraction buffer.

Step 2. DEAE-Cellulose Column Chromatography. The filtrate was applied to a DEAE-cellulose column (2.5 × 20 cm) equilibrated in extrac-

[45] M. Kapoor and D. Bray, *Biochemistry* **7,** 3583 (1968).

[46] R. Palacios, *J. Biol. Chem.* **251,** 4787 (1976).

tion buffer. The column was washed with 500 ml of extraction buffer and enzyme activity was then eluted with 400 ml linear gradient of 0 to 0.4 M NaCl in extraction buffer. The peak of activity is found at about 150 mM NaCl. Fractions with enzyme activity were pooled, solid ammonium sulfate was added to 55% saturation, stirred for 15 min, and centrifuged for 15 min at 15,000 g. The resulting pellet was resuspended in 10 ml of extraction buffer and dialyzed overnight against the same buffer.

Step 3. Chromatography on Anthranilic Acid Bound to Sepharose. The affinity gel was packed in a column (1.5 × 20 cm) and equilibrated with extraction buffer containing 2.25 mM $MnCl_2$. The dialyzed sample from step 2 was made 2.25 mM $MnCl_2$ and passed through the column at a flow rate of about 0.5 ml/min. The affinity gel binds about 0.5 to 1 mg of GS per ml of bed. The column was washed with extraction buffer containing 2.25 mM $MnCl_2$ until A_{280} was below 0.05. GS activity was then eluted by a linear gradient of 0 to 40 mM AMP in the same buffer. Fractions containing activity were pooled and the enzyme was precipitated with ammonium sulfate at 55% saturation. After centrifugation, the pellet was resuspended in 1 to 2 ml of extraction buffer, dialyzed against the same buffer, and stored at −70°.

Polyethylene Glycol (PEG) Precipitation

This method developed by Streicher and Tyler[47] takes advantage of the fact that GS precipitates from crude extracts in the presence of 10% PEG and 10 mM $MgCl_2$. Under this condition, DNA and some other protein also precipitate. However, GS is solubilized by adding a high concentration of NaCl (final concentration of 3 to 4 M) to the 10% PEG solution which still results in the precipitation of DNA and some protein contaminants. The PEG procedure was used to purify GS[47] from *K. aerogenes, K. pneumoniae, E. coli, S. typhimurium, Rhizobium* sp. strain 32HI, *C. crescentus, R. meliloti, R. capsulata, A. vinelandii,* and *P. putida.* Although the yield of GS (20 to 90%) and the degree of purity varied depending on bacteria, nearly homogeneous GS was obtained with several strains.

Step 1. Crude Extract. Frozen cells were resuspended in cold standard buffer (20 mM imidazole, pH 7.5, 10 mM $MgCl_2$, 2 mM 2-mercaptoethanol) at a concentration ranging from 0.05 to 0.5 g of cells per ml of buffer. Cells were ruptured either by two passages through a French pressure cell or by sonic oscillation, and cell debris was removed by centrifugation at 35,000 g for 45 min.

[47] S. L. Streicher and B. Tyler, *J. Bacteriol.* **142,** 69 (1980).

Step 2. Polyethylene Glycol (PEG) Precipitation. To one volume of crude extract was added, dropwise with stirring, an equal volume of 20% PEG prepared in standard buffer. Ten minutes after the final addition of the PEG solution, the mixture was centrifuged for 15 to 20 min at 27,000 *g*. This usually resulted in a clear supernatant fluid and a large pellet. In several experiments, especially when concentrated extracts were used, the supernatant fractions remained slightly turbid. The residual precipitate was pelleted by a second centrifugation, and the additional pellet was combined with the initial one.

Step 3. Solubilization. The pellet was suspended in 4 *M* NaCl prepared in standard buffer at a volume equal to one-half the original crude extract. To obtain maximal solubilization of the GS from the pellet, the suspension was homogenized in a glass-and-Teflon tissue grinder. This usually resulted in the formation of a clear solution that contained 80 to 90% of the GS activity. To the solubilized fraction was then added, dropwise with stirring, an equal volume of 20% PEG-2 *M* NaCl prepared in standard buffer. Ten minutes after the final addition, the mixture was centrifuged for 15 min at 27,000 *g*.

Step 4. Acetone Fractionation. The resulting clear supernatant fluid from step 3 was placed in an ice bath, and one-half volume of very cold (−20°) acetone was added rapidly with stirring. A heavy precipitate formed and was immediately removed by centrifugation for 15 min at 27,000 *g*. The supernatant and fluid were carefully decanted. The small amount of residual supernatant fluid and acetone was removed by gently blowing a stream of sterile air into the tube until the pellet surface appeared to be dry.

Step 5. Sepharose 6B Column Chromatography. The pellet was resuspended in a small volume of column buffer (20 m*M* imidazole, pH 7.5, 1 m*M* $MgCl_2$, 10% glycerol). Very little material was solubilized at this stage and it was always necessary to thoroughly homogenize the suspension to ensure maximal recovery of GS. The homogenate was centrifuged for 15 min at 27,000 *g*, and the supernatant was further purified by gel filtration chromatography, using a Sepharose 6B column and an elution buffer containing 20 m*M* imidazole, pH 7.5, 1 m*M* $MgCl_2$. Enzyme preparations were stored at 4° in column buffer or −20° after the glycerol concentration was brought to 50%. Other purification procedures reported in the literature include a differential sedimentation method[47] which depends upon the association of GS with DNA in cell extracts and an affinity chromatography method using ADP–hexane–agarose matrix.[48]

[48] S. Soliman, S. Nordlund, B. C. Johansson, and H. Baltscheffsky, *Acta Chem. Scand., Ser. B* **B35**, 63 (1981).

Regulatory Protein P_{II}

The regulatory protein P_{II} in *E. coli* exists in two interconvertible forms; an unmodified form (P_{IIA}) which activates the adenylylation of glutamine synthetase (GS) and a uridylylated form (P_{IID}) which is required for the deadenylylation of GS.[19]

Assay Method

The degree of activation of adenylylation by P_{IIA} is dependent on the presence of glutamine and divalent cation.[49] Under the condition in which the stimulation by P_{IIA} is maximal, the basal activity due to AT_a-catalyzed reaction is not negligible.[50] However, the requirement of P_{IID} for the deadenylylation reaction catalyzed by AT_d is absolute. Therefore, the P_{II} assay following the P_{IID}-dependent deadenylylation reaction is simpler than following the adenylylation reaction activated by P_{IIA}.

If necessary, P_{IID} can be quantitated before and after incubation of P_{II} samples under an uridylylating condition, and from their differences P_{IIA} concentration can be determined. P_{IID} is quantitated by either the colorimetric[20,51] or the radioassay.[19] In the colorimetric assay, unadenylylated GS subunits formed from adenylylated GS is measured by γ-glutamyltransferase assay; the measurement of [^{14}C]ADP released from [^{14}C]adenylyl-GS was the basis of the radioassay. Recently we developed a fluorometric assay in which an extrinsic fluor was introduced into the adenylylated GS by adenylylating the enzyme with 2-aza-1,N^6-etheno-ATP (aza-ε-ATP), a fluorescent analog of ATP.[52] Upon deadenylylating the aza-ε-ATP adenylylated GS (aza-ε-GS), the fluorescence intensity of aza-ε-GS at the excitation wavelength of 300 nm and the emission wavelength of 470 nm decreased by 39%. This fluorescence intensity was used to monitor the deadenylylation of aza-ε-GS. Unlike the other two methods, with this fluorometric method a continuous recording of catalytic activity is possible. Under the assay condition described below, the fluorescence decrease due to the deadenylylation is linear with respect to time up to 70% completion of the reaction.

Fluorometric Assay Procedure

Two stock solutions were prepared; *the aza-ε-GS solution* contained 15 μM aza-ε-ATP adenylylated subunit, 20 mM Hepes, pH 7.5 100 mM

[49] C. E. Caban and A. Ginsburg, *Biochemistry* **15,** 1569 (1976).
[50] S. G. Rhee, R. Park, and M. Wittenberger, *Anal. Biochem.* **88,** 174 (1978).
[51] S. G. Rhee, C. Y. Huang, P. B. Chock, and E. R. Stadtman, *Anal. Biochem.* **90,** 752 (1978).
[52] S. G. Rhee, G. A. Ubom, J. B. Hunt, and P. B. Chock, *J. Biol. Chem.* **256,** 6010 (1981).

KCl, and 20 m*M* $MgCl_2$, 50 μM EDTA, and *the effector mixture* contained 20 m*M* ATP, 120 m*M* α-ketoglutarate, 200 m*M* KPO_4, pH 7.2. For each assay, 400 μl of the aza-ε-GS solution and 50 μl of the effector mixture were placed in a 1.2 ml quartz cuvette and equilibrated at 37°. Immediately after the addition of 20 μl of P_{IID}, 30 μl of adenylyltransferase (5 ~ 10 μg) was added to initiate the reaction and the fluorescence decrease was recorded at the excitation wavelength of 300 nm with the emission wavelength set at 470 nm. The rate of decrease was linearly dependent on the concentration of either adenylyltransferase or P_{IID} as long as the other protein was present in excess. Therefore, this fluorometric assay can be used to quantitate both P_{IID} and adenylyltransferase.

Purification Procedure

Because the concentration of P_{II} in *E. coli* is very low it was difficult to prepare sufficient quantities of P_{II} for structural studies. The problem was overcome by the use of molecular cloning technique to increase P_{II} production. A strain [YMC 26 (pglnB5)] bearing the multicopy plasmid vector (pglnB5) carrying the *glnB* gene (structural gene for P_{II}) was constructed by cloning a PVUI generated fragment of the recombinant plasmid PGSl in pBR322. This new strain overproduced P_{II} by 70-fold.[53] It was known that the plasmid PGSl contained *glyA* gene[54] and the *glnB* gene is located closely to *glyA* gene. The use of this P_{II} overproducing strain allowed us to prepare larger quantities of P_{II}. Furthermore, finding that in the presence of 26% 2-mercaptoethanol most contaminating proteins precipitate while P_{II} remains active in the supernatant made it possible to simplify the purification procedure of Adler *et al.*[20] All steps in the following purification procedure were performed at 4° unless otherwise specified.

Step 1. Streptomycin Precipitation. Strain YMC26 (pglnB5) was grown on medium E of Vogel and Bonner supplemented with 0.4% glucose, 0.2% glutamine, 0.0003% thiamine, and 30 μg/ml ampiciline. The cells were harvested in stationary phase and stored frozen at −15°. Typically, 1.5 kg of frozen cells was suspended in a Waring blender in 3 liters of standard buffer (20 m*M* 2-methylimidazole, pH 7.5, 0.1 m*M* Mg EDTA, 1 m*M* 2-mercaptoethanol). The suspension was passed through a French pressure cell (9000 psi) twice and then centrifuged. The supernatant was brought to 1% streptomycin sulfate by slow addition of 10% streptomycin sulfate. After 15 min of slow stirring, the suspension was

[53] S. G. Rhee, T. Hunt, G. Pahel, and B. Magasenik, unpublished results.

[54] G. V. Stauffer, M. D. Plamann, and L. T. Stauffer, *Gene* **14,** 63 (1981).

centrifuged at 15,000 *g* for 30 min and the precipitate was discarded. If necessary, the supernatant was clarified by a second centrifugation.

Step 2. DEAE-Cellulose Ion Exchange Chromatography. The clear supernatant was loaded onto a DEAE column (10 × 80 cm) equilibrated previously with standard buffer containing 70 m*M* KCl. The column was washed with 8 liters of standard buffer containing 200 m*M* KCl and eluted with a linear KCl gradient from 200 to 400 m*M* in a total volume of 40 liters of standard buffer at a flow rate of 1 liter/hr. Fractions of 700 ml were collected and assayed for P_{IID} activity by using the colorimetric assay. The fractions containing high specific activity (at least one-half of the specific activity of the peak fraction) were combined and the remaining side fractions were saved to be combined with future preparations. The pooled protein solution (3.7 liters) was brought to 70% by adding solid $(NH_4)_2SO_4$ and stirred for 30 min. The precipitate was dissolved and dialyzed against standard buffer.

Step 3. 2-Mercaptoethanol Treatment. To the dialyzed protein solution (170 ml), 60 ml of 2-mercaptoethanol was added at room temperature. After stirring for 20 min at room temperature, the precipitate was centrifuged off. Then the supernatant was dialyzed against 2 × 6 liters of standard buffer. The dialyzed solution was centrifuged again to remove precipitate formed during dialysis. At this stage, the P_{II} preparation was more than 80% pure.

Step 4. Second DEAE-Cellulose Ion Exchange Chromatography. The clear supernatant from step 3 (260 ml) was loaded on a DEAE column (2.5 × 18 cm) equilibrated with standard buffer containing 70 m*M* KCl. Then the column was eluted with a linear gradient from 150 to 400 m*M* KCl in a total volume of 1 liter of standard buffer. The fractions containing most of the enzyme activity were combined, and the pooled protein solution was brought to 70% by adding solid $(NH_4)_2SO_4$. The precipitate was dissolved, dialyzed against standard buffer.

Step 5. BioGel Filtration. (This step may be omitted unless a highly pure enzyme preparation is desired.) The dialyzed P_{II} solution (20 ml) from step 4 was divided into two halves which were run consecutively over a 2.6 × 90 cm column of BioGel A 0.5M by elution with standard buffer at a flow rate of 0.6 ml/min. Fractions of 3.7 ml were collected. The fraction containing most of the enzyme activity was pooled, and to the pooled fraction (40 ml), 90 ml of saturated $(NH_4)_2OS_4$ was added. The precipitate was dissolved in 15 ml of standard buffer and stored frozen at −80°. There was no loss of P_{IID} activity during several months storage at −80°.

The results of this purification are summarized in the table.

P_{II} Purification Summary[a]

Step	Total (g)	Yield (%)	Relative activity
1. Streptomycin	77	100	1
2. DEAE cellulose chromatography	2.8	47	13
3. Supernatant from 26% 2-mercaptoethanol ppt	0.754	29	30
4. DEAE cellulose chromatography	0.547	26	36
5. Agarose (0.5 *M*) chromatography	0.459	21	36

[a] Purified from 1.5 kg of YMC 26 (pglnB5).

Preparation of P_{IIA} and P_{IID}

P_{II} obtained in a particular preparation is a mixture of P_{IIA} and P_{IID}. To prepare P_{IIA} free from P_{IID}, the mixture of P_{IIA} and P_{IID} was treated with snake venom phosphodiesterase as follows. One hundred units of snake venom phosphodiesterase (Worthington Diagnostic Systems, Inc.) was purified on a concanavalin A-Sepharose (Pharmacia) column (0.9 × 5 cm) by the procedure of Sulkowski and Laskowski.[55] The P_{II} mixture (20 mg) was incubated at 37° with purified snake venom phosphodiesterase (50 units) in a 5 ml of buffer containing 20 m*M* Tris–HCl, pH 8.0, 10 m*M* $MgCl_2$. The reaction was monitored by assaying P_{IID}. When P_{IID} activity became negligible, the reaction mixture was stopped by adding 5 ml of 0.4 *M* sodium acetate and adjusting the pH to 6.0. Then the reaction mixture was loaded very slowly (0.5 ml/min) onto the concanavalin A-Sepharose column preequilibrated with 0.2 *M* sodium acetate at room temperature. The flow was stopped for ~20 min to ensure a complete retention of the diesterase by the affinity column. P_{IIA} was eluted with 0.2 *M* sodium acetate buffer, the fractions containing P_{IIA} were concentrated with an Amicon concentrator, and dialyzed against the standard buffer used in the P_{II} purification. To prepare a fully uridylylated P_{II}, P_{II} (mixture of P_{IIA} and P_{IID}) was incubated with a partially purified UT-UR preparation, usually from the post-DEAE step (see UT-UR purification procedure). The uridylylation mixture containing 20 mg of P_{II}, UT-UR (enzyme quantity depended on its activity), 200 m*M* KCl, 10 m*M* $MgCl_2$, 2 m*M* UTP, 10 m*M* α-ketoglutarate, and 0.5 m*M* ATP in a 10 ml volume of 50 m*M* 2-methylimidazole buffer, pH 7.6, was incubated at 26°. When no further increase in the P_{IID} activity was observed, the uridylylation was stopped by adding 3.5 ml of 2-mercaptoethanol. The impure proteins

[55] E. Sulkowski and M. Laskowski, Jr., *Biochem. Biophys. Res. Commun.* **57,** 463 (1974).

introduced together with UT-UR were eliminated by repeating purification steps 3 and 4 described in the P_{II} purification procedure.

Properties

The regulatory P_{II} protein is composed of four identical subunits, each with a molecular weight of 11,000.[20] Each subunit contains two tyrosines, five phenylalanines, but no tryptophan.[20] Only one of the two tyrosines is uridylylated by uridylyltransferase activity.[20] Since all four of the subunits can be uridylylated, five different hybrid molecules exist which differ from each other with respect to the number ($0 \sim 4$) of the UMP group attached. An empirical formula for calculating the average number ($\bar{n}$) of uridylylated subunits per tetramer and protein concentration ($[P_{II}]$) was derived from UV spectra of P_{IIA}, P_{IID}, and their difference spectrum. The difference spectrum was essentially the same as that of UMP. Since the UMP moiety absorbed through the entire region of unmodified P_{II} spectrum, from 250 to 300 nm, absorption at any given wavelength is a function of both protein concentration and the state of uridylylation ($\bar{n}$). Therefore, calculation of n and $[P_{II}]$ requires solving two simultaneous equations:

$$A_{260}(\text{OD/cm}) = 1.995([\text{A}] + [\text{D}]) + 9.90[\text{D}] + 2.92\,A_{340} \quad (2)$$

$$A_{290}(\text{OD/cm}) = 0.7011([\text{A}] + [\text{D}]) + 0.277[\text{D}] + 2.17\,A_{340} \quad (3)$$

$$[P_{II}] = [\text{A}] + [\text{D}] \quad (4)$$

$$n = \frac{4[\text{D}]}{[\text{A}] + [\text{D}]} \quad (5)$$

where [A] and [D] represent the subunit concentration (m*M*) of P_{IIA} and P_{IID}, respectively. The first, second, and third terms in Eqs. (2) and (3) represent the absorption due to the protein concentration, the uridylylated tyrosine, and light scattering correction, respectively. The absorbancy of P_{II} at 290 nm is low compared to other proteins containing tryptophan. Therefore, practically, it is difficult to measure n and $[P_{II}]$ accurately from the above equations unless the P_{II} preparation is very pure.

UTP : P_{II} Uridylyltransferase and Uridylyl-Removing Activities: Assay Method

Uridylyltransferase (UT) and uridylyl-removing (UR) activities are measured by either of the colorimetric assays,[51] or the radioassay.[20] The colorimetric assay procedures were described in a previous volume of this series.[56]

[56] S. G. Rhee, this series, Vol. 107, p. 183.

Purification Procedure

A bifunctional enzyme which carries both UT and UR activities was purified to near homogeneity.[23] The purification was facilitated by the use of an *E. coli* strain JA200/pLC 39-39 which carries multiple copies of *ColE1 glnD*$^+$ hybrid plasmid. This strain produces UT and UR at levels 25-fold higher than the wild type *E. coli*. All steps were performed at 4°. Unless indicated otherwise, standard buffer (20 m*M* 2-methylimidazole buffer, pH 7.6, 0.1 m*M* K_2Mg EDTA, 1 m*M* DTT) containing various concentrations of KCl was used.

Step 1. Streptomycin Precipitation. Typically, 2 kg of the frozen cells was resuspended in 4 liters of standard buffer containing 200 m*M* KCl and homogenized in a Waring blender for 2 min. Crude extracts of cells were made by passing the cell suspension twice through a French press at 9000 psi and removing the cell debris by centrifugation for 45 min at 15,000 g. All subsequent steps were carried out at 0–4°. The supernatant was brought to 1% streptomycin sulfate by slow addition of a 10% solution of streptomycin sulfate. After 15 min of slow stirring, the suspension was centrifuged at 15,000 *g* for 30 min and the precipitate was discarded.

Step 2. Polyethylene Glycol Precipitation. The pH of the supernatant was adjusted to 6.2 with 1 *N* acetic acid and made to 4.5% polyethylene glycol (PEG M_r = 6000) by slow addition of 40% w/v PEG with stirring. Stirring was continued for 30 min and the suspension was centrifuged at 8000 *g* for 10 min and the supernatant discarded. The pellet was resuspended in one-third the volume of the streptomycin step in standard buffer containing 30 m*M* KCl. Insoluble material remaining after resuspension was removed by centrifugation at 29,000 *g* for 1 hr. This step enriched the UT activity approximately 2-fold, gave a recovery of greater than 80%, and constitutes a clear advantage over ammonium sulfate precipitation because it permits recovery of the UT activity under conditions which allow immediate loading onto the subsequent DEAE cellulose column without an overnight dialysis which can lead to considerable loss in activity.

Step 3. DEAE-Cellulose Ion Exchange Chromatography. The supernatant from the previous step (1.5 liters) was immediately applied to a DEAE-cellulose column (DE52, Whatman, 10 × 60 cm) previously equilibrated with standard buffer containing 30 m*M* KCl. UT was eluted at a flow rate of 800 ml/hr with a linear gradient generated from 15 liters each of buffer containing 50 and 300 m*M* KCl. After several 2-liter fractions were collected, the fraction size was reduced to 450 ml for the remainder of the chromatography. A broad peak of UT activity elutes at a conductivity of around 8–10 mmho. The peak fractions were pooled, precipitated with 65% ammonium sulfate. After resuspending in one-tenth the original volume, dialyzed overnight against 100 volumes of standard buffer containing 200 m*M* KCl and divided into six equal portions. (This post-DEAE fraction is stable for several months when stored at −70°.) This step gives approximately a 10-fold enrichment of UT activity.

Step 4. Matrex Gel Blue A Affinity Chromatography. One-sixth of the supernatant from the DEAE-cellulose step was loaded onto a column (30 ml bed volume) of Matrex Gel Blue A (Amicon Corp.) equilibrated previously with standard buffer containing 200 m*M* KCl. Fractions of 15 ml were collected at a rate of 90 ml/hr. The column was washed with 200 ml of UT buffer containing 500 m*M* KCl and eluted with a linear salt gradient of 150 ml each of UT buffer containing 50 m*M* $MgCl_2$, 5 m*M* ATP, 0.75 *M* KCl, and standard buffer containing 50 m*M* $MgCl_2$, 5 m*M* ATP, 2.5 *M* KCl, respectively.

UT and UR activities coeluted as two broad peaks of activity. The first peak, which was associated with the bulk of the protein, was not retained on the column and was eluted during the column wash. The second peak of UT-UR activity eluted at a conductivity of ~40 mmho. Although UT and UR clearly coeluted, the UR activity was sometimes found associated more with the front portion of the peak. This is probably due to the fact that ATP is a potent inhibitor of UR. Efforts to eliminate the leaking by decreasing the size of the sample applied to the column or by reusing the column or recycling the activity of this peak were unsuccessful.

Although Matrex Gel Blue A produced broad UT-UR peaks which had a very dilute protein concentration and a low recovery yield, it was still a very powerful step. This affinity step resulted routinely in a 3- to 6-fold increase in UT specific activity after correcting for inhibition by KCl and ATP. We believe that this value is a gross underestimate caused by the inherent lability of the enzyme; a 10- to 15-fold enrichment of UT-UR protein is a better estimate for purification by Matrex Gel Blue A chromatography.

Step 5. Phenyl-Sepharose CL-B4 Chromatography. Since UT and UR activities are highly unstable, the pooled fraction (~200 ml) from the previous step was immediately applied to a phenyl-sepharose column (1.0 × 10 cm) equilibrated with buffer containing 2.5 *M* KCl. Fractions of 2 ml were collected at a rate of 0.6 ml/min. After washing with 10 ml of buffer lacking KCl, the column was eluted with a linear gradient of 0–2% Lubrol WX in the same buffer. The UT and UR activities coeluted as abroad peak. As in the Blue Gel A step, a small amount of protein activity did not bind to the column. Although only a modest increase in the UT-UR specific activity was obtained with this step, it was valuable since it provided a rapid and efficient way to concentrate fractions containing UT-

UR with recovery greater than 80%. Other concentration procedures such as ultrafiltration and negative pressure dialysis led invariably to large losses in activity and the formation of insoluble aggregates of UT-UR. The fractions containing UT-UR activity were pooled and further concentrated by negative dialysis against standard buffer containing 200 m*M* KCl. At this stage, the UT-UR preparation was 80 ~ 90% pure based on gel electrophoresis and high-pressure liquid chromatography results.

Step 6. Gel Chromatography. Pooled, concentrated fractions from the phenyl-sepharose column were applied to a Sephacryl-300 column (1.6 × 80 cm) or a BioGel A 1.5M column (1.6 × 80 cm), treated with 2 m*M* phenylmethylsulfonyl fluoride and equilibrated with standard buffer containing 200 m*M* KCl. Elution was achieved with the same buffer at a flow rate of 10 ~ 20 ml/hr. The elution profiles were not very reproducible, even with the same gel medium. This irreproducibility is caused mainly by a process of oligomerization of UT-UR which occurs to different extents through a mechanism not well understood at the present time.

Properties and Molecular Weight

Despite the irregularities in the gel filtration step, when proper fractions (fractions preceding the peak of highest activity) were pooled, a nearly homogeneous band (>90% of the protein), whose electrophoretic mobility was similar to that of phosphorylase *b* (M_r = 94,000), was obtained under denaturing conditions. However, when the same samples were subjected to native gel (4–20% acrylamide gradient) electrophoresis, the results were variable. Frequently, three or four broad bands with molecular weights corresponding to different oligomeric states of the protein were observed. Occasionally, a single band could be seen in a native gel when fractions with the highest UT specific activity were pooled. The molecular weight of this band, determined from a plot of the logarithm of molecular weight versus mobility, is approximately 100,000. Estimation of the molecular weight of UT was also made by HPLC analysis. Gel filtration was carried out on a Hewlett-Packard HPLC Model 1084B equipped with a TSK-gel 3000 SW column (600 × 7.5 mm, Toyo Soda, Tokyo, Japan). The eluants used were 10 m*M* potassium phosphate buffer, pH 6.5, containing 200 m*M* potassium sulfate for nondenaturing conditions, and 100 m*M* sodium phosphate buffer, pH 7.0, containing 0.1% SDS for denaturing conditions, which were pumped at a flow rate of 0.5 ml/min. Absorbance of the effluent was monitored at 280 nm. To prevent precipitate formation with SDS, samples containing KCl were dialyzed rapidly (<2 hr) by placing a small volume of sample (50 ~ 100 μl) on a pellicon membrane filter (Millipore, type PSED) floating on a KCl-free buffer. On a TSK-Gel 3000 SW column, a linear relationship between the logarithm of molecular weight of standard proteins and retention volume was obtained under native and denaturing conditions. Under denaturing conditions, the pooled phenyl-Sepharose CL-B4 fractions eluted in a peak with minor side peaks. The molecular weight estimated for the major peak is 94,000. However, under native conditions, the pooled phenyl-Sepharose CL-B4 fractions eluted into three major peaks. Molecular weights estimated from the calibration curve for the protein peaks are 550,000, 185,000, and 94,000, respectively. They correspond to the molecular weights of hexamer, dimer, and monomer.

Stability

The stability of UT-UR at the post DEAE stage of purification was exceptionally good in the presence of 200 m*M* KCl at −70° as long as the protein concentration was >5 mg/ml; there was virtually no loss of activity after 2 ~ 3 months of storage. However, freezing and thawing in the absence of KCl resulted in complete loss of activity.

The stability of the enzyme was drastically reduced at later stages of purification; the post-Blue Gel A fractions containing 70 μg/ml of protein in the presence of 200 m*M* KCl and 5 m*M* ATP lost 35 ~ 50% of UT activity upon freezing and thawing, and lost 15 ~ 20% of activity upon incubating at 4° for 10 hr, when compared to the UT fractions kept at −5° (200 m*M* KCl and 5 m*M* ATP prevented freezing). It is unlikely that the instability of the post-Blue Gel A fractions is caused by proteolysis because incubation with various protease inhibitors did not improve the stability. We have been unable to solve the instability problem satisfactorily. As mentioned above, the purified UT-UR undergoes oligomerization to dimer and hexamer forms and this oligomerization might be responsible for the instability.

Feedback Inhibition

Woolfolk and Stadtman[4,57] reported that *E. coli* is inhibited by nine end products of glutamine metabolism. They are CTP, AMP, glucosamine-6-P, histidine, tryptophan, carbamyl-P alanine, glycine, and serine. Inhibition of GS by these metabolites constitutes a feedback regulation system in which each end product of a highly branched metabolic pathway inhibits the first common enzyme in the pathway.[58] Many of these nine

[57] C. A. Woolfolk and E. R. Stadtman, *Arch. Biochem. Biophys.* **118,** 736 (1967).

[58] E. R. Stadtman, *in* "The Enzymes" (P. D. Boyer, ed.), 3rd ed., Vol. 1, p. 397. Academic Press, New York, 1970.

metabolites caused only partial inhibition when tested individually, but in combination, their effects were cumulative.[57] Therefore, it was proposed that different feedback inhibitors serve to separate allosteric sites. However, an alternate mechanism has been proposed in which all feedback inhibitors of GS bind in a competitive fashion with substrates at substrate sites.[59] In an effort to distinguish these two mechanisms binding of some feedback inhibitors to the unadenylylated GS, particularly L-Ala, has been studied extensively by means of fast reaction kinetic,[60] fluorometric,[61] calorimetric,[62] and NMR[61] techniques. The results from these studies support the existence of an allosteric site for L-Ala which is distinct from the L-Glu substrate site. Similar studies[60,61] also suggest the presence of an allosteric site for Gly. In addition, direct binding measurement[63] and calorimetric[64] study show that AMP and tryptophan binds at separate sites on GS and binding of AMP is not affected by Ala, Gly, His, and CMP.

The feedback inhibition mechanism was further investigated using kinetic methods[60] where the effect of L-Ala, Gly, and D-Val on the catalytic cycle of unadenylylated GS catalyzed biosynthetic reaction was studied. All the data obtained either from the initial rate measurements or from the transient kinetic studies suggest that GS contains a minimum of three binding sites for nonaromatic amino acids. In addition to the substrate L-Glu site, there exist two allosteric sites for the binding of L-amino acid and D-amino acid. When L-Ala binds to GS, it does not alter significantly the L-Glu binding affinity; instead, it inhibits the products formation from a reaction intermediate derived from MgGs · ATP · Glu · NH_3 complex. However, binding of D-Val on the D-amino acid site exerts a strong antagonistic effect on Glu binding but does not inhibit products formation from the reaction intermediate. Gly, which does not have D,L-enantiomer, exhibits an inhibition pattern of both L-Ala and D-Val. In addition, the inhibition pattern observed with D-Ala is very similar to that shown by D-Val, and the inhibition pattern exhibited by L-Val appeared to be similar to that of L-Ala.

[59] F. W. Dahlquist and D. L. Purich, *Biochemistry* **14,** 1980 (1975).

[60] S. G. Rhee, P. B. Chock, and E. R. Stadtman, *in* "Frontiers of Biological Energetics" (P. L. Dutton, J. S. Leigh, and A. Scarpa, eds.), Vol. 1, p. 725. Academic Press, New York, 1978.

[61] S. G. Rhee, J. J. Villafranca, P. B. Chock, and E. R. Stadtman, *Biochem. Biophys. Res. Commun.* **78,** 244 (1977).

[62] A. Shrake, R. Park, and A. Ginsburg, *Biochemistry* **17,** 658 (1978).

[63] A. Ginsburg, *Biochemistry* **8,** 1726 (1969).

[64] P. D. Ross and A. Ginsburg, *Biochemistry* **8,** 4690 (1969).

Oxidative Inactivation

It has been shown that GS is susceptible to inactivation by a number of mixed-function oxidation systems.[3,65–70] Among them are (1) dialyzed cell-free extract from *K. aerogenes* or *E. coli* (after removal of catalase) and NADPH; (2) purified components of a rabbit microsomal cytochrome *P*-450 reductase and cytochrome *P*-450, and NADPH; (3) purified components of a putidaredoxin reductase and redoxin from *P. putida* and NADH; (4) NAD(P)H oxidase and NAD(P)H; (5) xanthine oxidase, hypoxanthine, and ferredoxin or redoxin. The GS inactivation reaction catalyzed by these mixed-function oxidation systems required the presence of O_2, and in general, is stimulated by Fe^{3+} and is inhibited by catalase, by metal ion chelating agents, and by Mn^{2+}. Activated oxygen scavengers such as superoxide dismutase, dimethyl sulfoxide, and histidine exert no effect on the inactivation systems except with the putidaredoxin reductase–redoxin system where the inactivation reaction is inhibited by these agents. However, when this reductase–redoxin system is supplemented with $P\text{-}450_{cam}$ (a cytochrome *P*-450 from *P. putida* when grown with camphor as the carbon and energy source), the inactivation reaction is no longer sensitive to activated oxygen scavengers. In addition, several nonenzymatic systems, namely (1) ascorbate, Fe^{3+} and O_2, (2) Fe^{2+} and H_2O_2 or O_2, and (3) dihydroxyfumarate, Fe^{3+} and O_2, were also found to cause GS inactivation.[65,69] Based on the fact that Fe^{2+} and H_2O_2 alone can inactivate GS and because catalase and Mn^{2+} each inhibit the inactivation reaction, it is suggested[62,67,69] that the inactivation mechanism involves the oxidation of GS bound Fe^{2+} by H_2O_2 to form Fe^{3+} and an activated oxygen species, likely a OH radical, which in turn oxidizes a single histidine residue near or at the metal ion binding site. The proposed inactivation mechanism is supported by the fact that all the mixed-function oxidation systems studies are known to catalyze the production of H_2O_2 and can reduce Fe^{3+} to Fe^{2+}. In the case of ascorbate, a strong reducing agent,

[65] R. L. Levine, C. N. Oliver, R. M. Fulks, and E. R. Stadtman, *Proc. Natl. Acad. Sci. U.S.A.* **78,** 2120 (1981).

[66] C. N. Oliver, Ph.D. Dissertation, Johns Hopkins University, Baltimore, Maryland (1981).

[67] L. Fucci, C. N. Oliver, M. J. Coon, and E. R. Stadtman, *Proc. Natl. Acad. Sci. U.S.A.* **80,** 1521 (1983).

[68] C. N. Oliver, L. Fucci, R. Levine, M. Wittenberger, and E. R. Stadtman, *in* "Cytochrome P-450 Biochemistry, Biophysics and Environmental Implication" (E. Hietanen, M. Laitinen, and O. Hänninen, eds.), p. 531. Elsevier/North-Holland Biochemical Press, Amsterdam, 1982.

[69] R. Levine, *J. Biol. Chem.* **258,** 11823 (1983).

[70] R. Levine, *J. Biol. Chem.* **258,** 11828 (1983).

the enzyme bound Fe^{3+} is reduced to form GS · Fe^{2+} complex and oxidation of ascorbate by O_2 in the presence of transition metal ion forms H_2O_2.[71] The reduced Fe^{2+}–enzyme complex in turn reacts with H_2O_2 to form OH radical which oxidizes a specific histidine residue. Evidence for the site-specific oxidation of a unique histidine residue in each GS subunit is derived from the observations[65,69] that only 1 out of 16 histidine residues was oxidized; in addition an as yet unidentified carbonyl derivative is formed which can be quantitated by reacting the oxidized GS with 2,4-dinitrophenylhydrazine to form a 2,4-dinitrophenylhydrazone derivative. The production of 2,4-dinitrophenylhydrazine was monitored spectrophotometrically at 387 and 400 nm.[69]

It is believed that the oxidative inactivation of GS may constitute the "marking" step for intracellular GS turnover.[65–67,69] When oxidatively inactivated GS and native GS were each incubated with catalase-free partially purified *E. coli* extract, the inactivated GS was rapidly degraded while the native GS remained intact. However, in the presence of NADPH or ATP, proteolysis of native GS occurs. Addition of catalase, an inhibitor for oxidative inactivation of GS, protects the native enzyme from degradation, but exerts no effect on the proteolysis of inactivated GS. These observations indicate that both GS inactivation and degradation are catalyzed by the catalase-free *E. coli* extract and suggest that the oxidative inactivation converts the native GS to a form which is susceptible to proteolytic attack. In other words intracellular degradation of GS occurs in two steps. In the first step GS is "marked" by oxidative inactivation, followed by proteolytic digestion of the "marked" enzyme by a specific protease(s).

The rate of oxidative inactivation of GS is both a function of the state of adenylylation and the degree of substrate saturation.[65,66] In the presence of ATP and Glu, the physiologically active unadenylylated GS is protected from inactivation, whereas the physiologically inactive adenylylated GS becomes more susceptible to oxidative inactivation. Therefore, when glutamine synthesis is required in the cell, during which the concentrations of ATP and Glu are high, the physiologically active unadenylylated GS is not converted to the "marked" form. However, the inactive adenylylated enzyme is readily "marked" for proteolysis. It should be pointed out that using the 2,4-dinitrophenylhydrazone formation to monitor oxidative inactivation of GS, Levine showed[69] that the variability in specific activity for various batches of highly purified GS can be correlated to the amount of copurified oxidative inactivation enzyme. Because the buffer used in the first step of purification contained 10 m*M*

[71] A. R. Morgan, R. L. Cone, and T. M. Elgert, *Nucleic Acids Res.* **3,** 1139 (1976).

Mn^{2+}, an inhibitor for the oxidative inactivation, it is believed that the oxidative inactivation probably occurred during the growth of the *E. coli* or during storage of the cells.

Catalytic Cycle and Topographic Studies

The catalytic cycle for the biosynthesis of Gln catalyzed by both the unadenylylated[72] and the adenylylated[73] GS has been established through transient kinetic studies. Protein fluorescence intensity changes were used to monitor the unadenylylated GS during the catalysis.[72,74,75] The results show that either ATP or L-Glu can bind independently to the unadenylylated GS in the presence of Mg^{2+}, and in the catalytic cycle there exist three fluorometrically distinct intermediates. The first fluorescence intermediate is attributed to the formation of enzyme–substrate complex and the second intermediate in the enzyme-bound γ-glutamylphosphate, a reaction intermediate first proposed by Meister and coworkers for sheep brain GS.[76,77] The enzyme-bound NH_3 will then react with γ-glutamylphosphate to form enzyme-bound products. Based on the rate constant for the dissociation of ADP from MgGS · ADP · P_i complex which was estimated to be ~100 sec^{-1}, the rate-limiting step in the catalytic cycle is therefore the formation of intermediate and not the off-rate of ADP.[78] However, a value of 14 sec^{-1} has been assigned for the ADP off-rate based on rapid-quench and isotope partitioning experiments.[79] The results of this study confirm the existence of reaction intermediates which include enzyme-bound γ-glutamylphosphate, and suggest the release of ADP is rate limiting.

Unlike the unadenylylated GS, the adenylylated enzyme exhibits only negligible fluorescence changes upon substrates binding. Therefore an extrinsic fluor was introduced by adenylylating the unadenylylated GS with a fluorescence analog of ATP, 1,N^6-etheno-2-aza-ATP. The modified enzyme (aza-ε-GS) exhibits catalytic and kinetic properties similar to those of the naturally adenylylated enzyme.[52] However, the extrinsic fluor provides large fluorescence changes due to binding of various li-

[72] S. G. Rhee and P. B. Chock, *Proc. Natl. Acad. Sci. U.S.A.* **73,** 476 (1976).

[73] S. G. Rhee, G. A. Ubom, J. B. Hunt, and P. B. Chock, *J. Biol. Chem.* **257,** 289 (1982).

[74] R. B. Timmons, S. G. Rhee, D. L. Luterman, and P. B. Chock, *Biochemistry* **13,** 4479 (1974).

[75] S. G. Rhee, P. B. Chock, and E. R. Stadtman, *Biochimie* **58,** 35 (1976).

[76] A. Meister, *in* "The Enzymes" (P. D. Boyer, ed.), 3rd ed., Vol. 10, p. 699. Academic Press, New York, 1974.

[77] P. R. Krishnaswamy, V. Pamiljans, and A. Meister, *J. Biol. Chem.* **237,** 2932 (1962).

[78] S. G. Rhee and P. B. Chock, *Biochemistry* **15,** 1755 (1976).

[79] T. D. Meek, K. A. Johnson, and J. J. Villafranca, *Biochemistry* **21,** 2158 (1982).

gands. The results of fluorometric studies on this aza-ε-GS indicate that L-Glu and ATP bind to both Mn^{2+} and Mg^{2+} forms of the enzyme in a random order, but only the Mn^{2+} form is capable of catalyzing the biosynthesis of Gln. Stopped flow kinetic studies on the biosynthesis of Gln catalyzed by Mn^{2+} · aza-ε-GS reveal that five fluorometrically distinct intermediates are formed in the catalytic cycle.[80] The mechanism is very similar to that established for the unadenylylated enzyme. Substrates bind to the enzyme via a rapid equilibrium random mechanism, and the formation of an enzyme-bound γ-glutamylphosphate from ATP and L-Glu is the rate-limiting step. Comparison of the catalytic cycles of both unadenylylated GS and adenylylated GS suggests that these two enzyme forms catalyze the biosynthesis of Gln by a very similar pathway in spite of their different divalent cation requirements and specific activities.

Because GS can be adenylylated with various ATP analogs to yield an adenylylated enzyme which exhibits divalent cation requirement, pH profile, catalytic and inhibitor properties similar to those of naturally adenylylated enzyme,[52,81–83] ATP analogs have greatly facilitated topographical studies on the regulatory, catalytic, and the divalent metal ion activating sites. By monitoring the paramagnetic effect of Mn^{2+} on ^{13}C and ^{32}P nuclear magnetic resonance (NMR) signals from the [2-^{13}C]ATP adenylylated GS, Villafranca *et al.*[82] have determined the distances from the 2-C position and the phosphorus of covalently bound AMP to the two Mn^{2+} binding sites, n_1 and n_2. Binding of Mn^{2+} to the n_1 site converts an inactive apo-enzyme to its active form, while the metal ion bound at n_2 occupies the metal–nucleotide substrate site.[4] The distances from Mn^{2+} at n_1, and n_2 sites to phosphorus are ~10 and ~7 Å and to the 2-C position of the adenine ring are 12 and 11 Å, respectively. The fluorescence energy transfer method was also used to determine distances between Co^{2+} at n_1, and n_2 and the adenylyl site.[82] For this experiment, GS was adenylylated with a fluorescence analog of ATP, 1-N^6-etheno-ATP (εATP). The distances between ε-adenine and Co^{2+} at n_1 and n_2 are 13 and 11 Å, respectively. In addition, the paramagnetic effect of Mn^{2+} on the electron spin resonance (ESR) signal of the spin-labeled Tempo-ATP (2,2,6,6-tetramethylpiperidine-1-oxyl adenosine triphosphate) adenylylated GS was used to determine the distances from the nitroxyl moiety of the covalently bound

[80] S. G. Rhee, G. A. Ubom, J. B. Hunt, and P. B. Chock, *J. Biol. Chem.* **257,** 289 (1982).

[81] P. B. Chock, C. Y. Huang, R. B. Timmons, and E. R. Stadtman, *Proc. Natl. Acad. Sci. U.S.A.* **70,** 3134 (1973).

[82] J. J. Villafranca, S. G. Rhee, and P. B. Chock, *Proc. Natl. Acad. Sci. U.S.A.* **75,** 1255 (1978).

[83] P. B. Chock, J. J. Villafranca, S. G. Rhee, G. A. Ubom, and E. R. Stadtman, *in* "NMR and Biochemistry" (S. J. Opella and P. Lu, eds.), p. 405. Dekker, New York, 1979.

Tempo-AMP to n_1 and n_2 sites. The distances from the Mn^{2+} at n_1 and n_2 sites to the nitroxyl moiety are 19 and 17 Å, respectively.[83,84] Considering the fact that the nitroxide group of the enzyme-bound Tempo-AMP is ~5 Å away from the 2-C position of adenine, the distances obtained by the ESR method are in good agreement with those attained from the ^{13}C NMR and fluorescence energy transfer studies. All these data show that the adenylyl regulating site is located closely to the catalytic site (12–20 Å). In addition, the rotational correlation time calculated from both NMR and ESR data indicates that the adenylylation site is located on the surface of GS. The ESR signal was also used to monitor protein conformational change due to substrate binding.[84] The results show that binding of L-Glu induces a protein conformational change such that the distances between the nitroxyl radical and n_1 site is shortened by ~2 Å while the distance between the nitroxyl radical and n_2 is lengthened by ~2 Å. However, in the presence of both ATP and L-Glu, these 2 distances, nitroxide–n_1 site and nitroxide–n_2 site are lengthened by ~4 and ~3 Å respectively, from those observed with L-Glu bound enzyme. These stepwise protein conformational changes are consistent with the fluorescence changes induced by substrate binding to the adenylylated enzyme,[52] and in accord with the reports that L-Glu enhances Mn^{2+} binding affinity at n_1 site,[85] and binding of L-methionine-*S*-sulfoximine phosphate and ADP strengthens intra- and intersubunit bonding domains in the dodecameric GS.[86]

In essence, detailed studies of *E. coli* GS reveal some basic principles in metabolic regulation of enzymes, namely, the cyclic cascade regulation, control in protein turnover, and regulation by allosteric effectors. In addition, topographic and mechanistic studies provide better understanding on how enzyme functions as a catalyst.

[84] G. A. Ubom, S. G. Rhee, J. B. Hunt, and P. B. Chock, unpublished results.
[85] A. Shrake, D. M. Powers, and A. Ginsburg, *Biochemistry* **16,** 4372 (1977).
[86] M. R. Maurizi and A. Ginsburg, *J. Biol. Chem.* **257,** 7246 (1982).

[30] Glutaminase from Mammalian Tissues

By ELLING KVAMME, INGEBORG AA. TORGNER, and GERD SVENNEBY

Glutaminase (EC 3.5.1.2, L-glutamine amidohydrolase) catalyzes the reaction: L-glutamine + $H_2O \rightarrow$ L-glutamate + ammonia.

In the following the mitochondrial phosphate activated glutaminase (PAG), which is the most important glutaminase in mammalian tissues, will be described.

Copyright © 1985 by Academic Press, Inc.
All rights of reproduction in any form reserved.

Assay Methods

Measurement of both the reaction products ammonia or glutamate may be used for assay of PAG. Measurement of ammonia suffers from two disadvantages: The great water solubility of this compound makes it difficult to keep the background values sufficiently low, and some undissociated ammonia may escape, particularly at prolonged incubations at 37° and when the pH exceeds 8.0. Ammonia can be monitored, e.g., with Nessler's reagent after microdistillation,[1] by the ammonia electrode,[2] or by coupling to the glutamate dehydrogenase (GDH) reaction,[3] whereby NADH oxidation is measured. The two latter methods permit measurement of enzyme rates. The GDH-coupled method, which is the more sensitive, will be described below in detail.

When assaying crude enzyme preparations, determination of glutamate formed in a fixed time is the method of choice. Glutamate can be monitored by chromatography (on paper,[3,4] column,[5,6] or high-performance liquid chromatography[7]), or by using enzymatic methods. Here glutamate is measured either by coupling to GDH (and monitoring the NADH formed by fluorophotometric[8] or colorimetric[9,10] methods) or by measuring [^{14}C]CO_2 released from glutamate in the glutamate decarboxylase reaction.[11,12]

Using L-[^{14}C]glutamine as the substrate, the paper chromatographic method is reliable, sensitive, and simple, but suffers from the disadvantage that it is time consuming. The colorimetric GDH coupled technique according to Davis and Prusiner[9] is the method of choice in detecting PAG, following separation by gel electrophoresis. The colorimetric method described by Gella and Pascual[10] is simple to use when assaying a great number of fractions (e.g., column chromatography, sucrose gradients). The fluorophotometric GDH coupled assay[8] is versatile, easy to perform, and more sensitive than these methods. Therefore this method also will be described below in detail.

[1] R. Ballentine, this series, Vol. 3, p. 984. See also this volume, p. 350.
[2] Y.-Z. Huang, *Anal. Biochem.* **61,** 464 (1974).
[3] E. Kvamme and G. Svenneby, *Res. Methods Neurochem.* **3,** 277 (1975).
[4] E. Kvamme and B. E. Olsen, *J. Neurochem.* **36,** 1916 (1981).
[5] S. Prusiner and L. Milner, *Anal. Biochem.* **37,** 429 (1970).
[6] R. A. Shapiro, R. F. Morehouse, and N. P. Curthoys, *Biochem. J.* **207,** 561 (1982).
[7] F. Martin, A. Auzuki, and B. Hirel, *Anal. Biochem.* **125,** 24 (1982).
[8] N. P. Curthoys and O. H. Lowry, *J. Biol. Chem.* **248,** 162 (1973).
[9] J. N. Davis and S. Prusiner, *Anal. Biochem.* **54,** 272 (1973).
[10] J. Gella and M. A. Pascual, *Anal. Biochem.* **127,** 322 (1982).
[11] E. G. McGeer and P. L. McGeer, *J. Neurochem.* **32,** 1071 (1979).
[12] R. C. Dinwoodie and E. A. Boeker, *Anal. Biochem.* **96,** 24 (1979).

Assay of Purified PAG Preparations

Measurement of Rate of Ammonia Formation[3]

Principle. Ammonia, which is formed in the PAG reaction, is utilized in the GDH reaction and assayed by monitoring the rate of NADH oxidation at 340 nm. PAG is incubated with excess of glutamine, phosphate, 2-oxoglutarate, NADH, and GDH. This method is particularly useful if the initial rate of PAG is wanted.

Reagents

100 m*M* L-glutamine, pH 8.0 (freshly prepared)
50 m*M* 2-oxoglutarate Na, pH 8.0
1.0 *M* potassium phosphate buffer, pH 8.0
2 m*M* EDTA, pH 8.0
GDH (EC 1.4.1.3) ammonia free[13]
NADH

Procedure. For 10 determinations 1 ml each of L-glutamine, 2-oxoglutarate, phosphate buffer, and EDTA is mixed with 2.2 mg solid NADH and 30 μl GDH. Add in the cuvette (1 ml, 10 mm, light path) 403 μl of this mixture, PAG, and H_2O to give a total volume of 1 ml. In order to remove contaminating ammonia in the reagents, the sample is preincubated for 5 min at 25° before the addition of PAG. The decrease in absorbance at 340 nm is registered by a recorder connected to the spectrophotometer.

The rate is usually linear with time for several minutes and proportional to the amount of PAG in the assay, but under certain conditions a lag period of 1–3 min duration is observed before the maximal rate is obtained.

Endpoint Determination of Glutamate

Principle. PAG is incubated for a prefixed time in the presence of high concentrations of glutamine and phosphate, and the amount of glutamate formed is measured.

Reagents

0.1 *M* L-glutamine, pH 8.0 (freshly prepared)
0.5 *M* potassium phosphate buffer, pH 8.0
96% v/v ethanol

[13] We have found L-glutamic dehydrogenase (EC 1.4.1.3) sold by the Sigma Chemical Co. (No. G-2626) to be satisfactory. The enzyme (~ 40 units/mg protein and 140 mg protein/ml) is suspended in 50% glycerol and essentially free of ammonium ions.

Procedure. In an Eppendorf tube (1 ml) add 20 μl of potassium phosphate buffer, 10 μl of L-glutamine, enzyme, and H_2O to give a total volume of 50 μl. The enzyme is the last addition and added to reagents which have been preincubated for 5 min at 25°. The tube is capped and the incubation performed for 2 min at 25°. The reaction is stopped by adding 100 μl of cold ethanol and cooling in ice, followed by centrifugation (18,000 *g*, 2 min) (Ole Dich microcentrifuge, Ole Dich Instrumentmakers, Denmark). Blanks are incubated with no PAG, but the enzyme is added after the ethanol.

Assay of PAG in Isolated Mitochondria, Synaptosomes, Cultured Neurons, Granulocytes, and Astrocytes

Principle. Endpoint determination of glutamate. Glutamate formed in a prefixed time is assayed following incubation of the cellular material with L-glutamine (preferably in the physiological concentration range), two concentrations of phosphate (to ensure that PAG is assayed), and inhibitors to prevent the metabolism of glutamate.

Reagents

Substrate solution: Krebs–Ringer buffer[14] in double concentration, but with no addition of $NaHCO_3$. The solution is added in final concentrations either 2, 10, or 40 m*M* sodium phosphate buffer pH 7.4, 1 or 4 m*M* L-glutamine, pH 7.4, antimycin A (Sigma) 1.2 mg/liter, and oligomycin (Sigma) 20 mg/liter. The final solution is made up just before use. The antimycin and oligomycin must be dissolved in a small volume of 96% v/v ethanol (1 mg/5 ml and 1 mg/ml ethanol, respectively), before the addition to the substrate solution.

Solution for suspension of synaptosomes and mitochondria:
- 0.25 *M* D-mannitol
- 0.07 *M* sucrose
- 0.01 *M* Hepes (*N*-2 hydroxyethylpiperazine-*N*′-2-ethane sulfonic acid), pH 7.4

Solution for suspension of cultured cells:
- 0.1 *M* Tris–HCl, pH 7.4

96% v/v ethanol

Procedure. In Eppendorf tubes (1 ml) add 25 μl of the substrate solution, cellular material (0.02–1.25 mg of protein), and H_2O to give a total

[14] P. P. Cohen, *in* "Manometric Techniques" (W. W. Umbreit, R. H. Burris, and J. F. Stauffer, eds.), p. 147. Burgess, Minneapolis, Minnesota, 1959.

volume of 50 μl. (If desired, all the volumes and the amount of protein added may be scaled up, e.g., 5-fold.) Each sample is incubated with two phosphate concentrations (either 1 and 5 m*M* or 5 and 20 m*M* phosphate, final concentration) to ensure the phosphate activation of PAG. If no phosphate activation is observed, the enzymatic activity is not due to PAG. The cellular preparation is the last addition, and is added to reagents which have been temperature equilibrated by preincubating at the temperature chosen for 5 min. The tubes are capped and the incubation is performed for 2–10 min at 25 or 37°. The reaction is stopped by adding 100 μl of cold ethanol and cooling in ice. The cooled reaction mixtures are centrifuged at 18,000 *g* for 2 min. The supernatants are generally well suited for determination of glutamate by enzymatic or chromatographic methods. Blanks are incubated with cellular material added after the ethanol. Due to the variation in enzyme activity it is necessary to perform preliminary experiments to ensure that the rate of hydrolysis is linear for the experimental conditions chosen. There are two main reasons for nonlinear rates, inhibition of the reaction by accumulated products and depletion of the substrate glutamine.

Measurement of Glutamate

Principle. Glutamate produced is assayed fluorophotometrically by monitoring NADH formed in the GDH reaction, following removal of the reaction product 2-oxoglutarate with H_2O_2[8] to ensure that the reaction goes to completion.

Reagents

0.1 *M* Tris–HCl pH 9.4
24 m*M* NAD
3 m*M* ADP
0.4% H_2O_2
GDH[13] (EC 1.4.1.3)

Procedure. Add 100 μl of NAD, 100 μl of ADP, 100 μl of Tris–HCl, 100 μl of H_2O_2, 5 μl of GDH, ethanolic extract, and H_2O to 1.2 ml. Incubate for 30 min at room temperature. The fluorescence is read in a fluorophotometer (photometer "Eppendorf" Netheler + Hinz, Hamburg) using a primary filter of 265–366 nm and secondary filter of 430–470 nm. Blank values are subtracted from the measured values, and the glutamate concentrations calculated from a calibration curve of glutamate (in the range 0–4 μ*M* in the cuvette).

Purification Procedure

Principle. The purification is based on the property of PAG to solubilize and polymerize in a reversible manner, when exposed to Tris–HCl and phosphate–borate buffer, respectively.[15]

Reagents

Homogenate buffer: 0.1 m*M* sodium phosphate, 3 m*M* EDTA, pH 8.0
2 *M* Na_2SO_4 (stored at 37° to keep the salt in solution)
Borate buffer: 10 m*M* sodium borate, 3 m*M* EDTA, pH 8.0
Dialyzing buffer: 50 m*M* Tris–HCl, 3 m*M* EDTA, 10 m*M* mercaptoethanol, pH 8.0
Polymerizing buffer: 1 *M* potassium phosphate, 0.3 *M* sodium borate, pH 8.0
Diluting buffer: 0.1 *M* sodium phosphate, 3 m*M* sodium borate, 3 m*M* EDTA, 10 m*M* mercaptoethanol, pH 8.0

Pig Kidney PAG[15] (Table I)

The kidneys are collected at the slaughter house and stored in the freezer prior to use.

Homogenization. The kidneys are cut in two halves, the membranes are removed, and the pieces are frozen and can be stored at −20° for several weeks. The cortices are cut into small pieces and homogenized for 1 min at 2000 rpm (ATO-MIX) followed by 2 min at 6000 rpm in 3 volumes of homogenate buffer. The homogenate is frozen and can be stored at −20° for 2 to 3 weeks. For further purification homogenate is thawed for 24 hr in the cold and centrifuged at 17,000 *g* for 30 min at 5° (Sorvall centrifuge type RCB-2, rotor GSA). It is important that the supernatant (S1) is well separated without intermediate fluffy layer. S1 cannot be stored without inactivation of PAG. The purification procedure must be continued on the same day.

Na_2SO_4 Fractionation. All Na_2SO_4 precipitations are performed at room temperature, by addition of 2 *M* Na_2SO_4 slowly under constant stirring. The precipitated proteins are collected by centrifugation at 17,000 *g* for 30 min at 17°. The precipitates are then homogenized in cold borate buffer using a Potter Elvehjem homogenizer.

The supernatant (S1) after centrifugation of the homogenate is precipitated with Na_2SO_4 in a final concentration of 1.0 *M*. Two liters of the precipitate suspension in borate buffer (B2) is made up from 10 liters of

[15] E. Kvamme, B. Tveit, and G. Svenneby, *J. Biol. Chem.* **245,** 1871 (1970).

TABLE I
SUMMARY OF THE PURIFICATION PROCEDURE OF PAG FROM PIG KIDNEY

Fraction	Total protein (mg)	Total activity[a]	Specific activity[b]	Percentage recovery	Purification (fold)
Homogenate	1,670,000	50,100	0.030	100.0	1
S1	688,666	20,660	0.030	41.2	1
Na_2SO_4 precipitation 0–1 *M* (B2)	351,633	21,098	0.060	42	2
Na_2SO_4 precipitation 0–0.95 *M* (B4)	25,461	20,369	0.80	41	27
Na_2SO_4 precipitation 0.7–0.9 *M* (B6)	6,680	12,760	1.91	26	64
First Tris–HCl solubilization S10	3,784	1,665	0.44	3.3	15
First P-B[c] precipitation B11	193	3,337	17.3	6.7	577
Reprecipitation B12	92	2,828	30.7	5.6	1,023
Second Tris–HCl solubilization S13	10	614	63.7	1.2	2,123
Second P-B precipitation B14	3	1,116	417.5	2.2	13,917
Third Tris–HCl solubilization S15	1.2	142	121.1	0.3	4,037

[a] Activity (units): μmol NH_3 formed/min.

[b] Specific activity: units/mg protein. PAG activity in all fractions is determined by the GDH coupled measurement of ammonia as described in the text.

[c] P-B, phosphate–borate.

homogenate. The B2 preparations are stored at $-20°$. For refractioning, B2 is thawed in the cold room overnight and centrifuged at 16,000 *g* for 20 min at 5°. The supernatant (S3) is collected. The freezing and recentrifugation of the homogenized B2 preparation is important for successful further purification.

The amount of Na_2SO_4 added to this supernatant (S3) and in later fractionation steps is based on measurements by flame photometry of the sodium concentration in the protein suspension before the addition. To the supernatant S3 which contains about 0.3 *M* Na_2SO_4, more salt is added to a final concentration of 0.95 *M*.

After centrifugation, the precipitate (B4) from 6 liters B2 suspension is homogenized in 1 liter borate buffer. This preparation may be stored for 3 to 4 weeks at $-20°$. The protein concentration of the B4 suspension is adjusted to 20 mg/ml and the suspension is precipitated (B6) between 0.7 and 0.9 *M* Na_2SO_4. If necessary, the precipitation between 0.7 and 0.9 *M* Na_2SO_4 is repeated until a specific activity of at least 1.6 units/mg protein is obtained in the precipitate (B8). The final preparations may be stored for at least 1 year at −20°.

Repeated Solubilizations and Precipitations in Tris–HCl Buffer and Phosphate–Borate Buffer, Respectively

This part of the purification procedure is performed at 0–2° using B6 and/or B8 preparations containing at least 10,000 units (μmol NH_3 formed/min) of the enzyme with a specific activity of approximately 1.6 units/mg protein. The combined fractions are by convention called B9.

To solubilize the enzyme the Na_2SO_4 fractionated PAG is first diluted with dialyzing buffer, to give a protein concentration of about 10–15 mg protein/ml, and thereafter dialyzed overnight against the same buffer. Usually, 4 portions of 55 ml have to be each dialyzed against 8000 ml of buffer. After centrifugation for 6 hr in a Spinco Beckman ultracentrifuge (model L, rotor 30) at 65,000 *g* (27,000 rpm), the supernatant (S10) is collected. The solubilized enzyme is precipitated by adding polymerizing buffer to a final concentration of 50 m*M* phosphate and 15 m*M* borate. The enzyme suspension is left for at least 30 min or might be stored overnight in ice, and is then centrifuged at 65,000 *g* for 3.5 hr. The supernatants have to be carefully removed with a fine-tipped pipette before the precipitates (B11) can be resuspended in a total of 12 ml diluting buffer. The suspension is left in ice for at least 10 hr and is then homogenized by freezing and thawing (in the cold) once. Alternatively, 3 ml portions are homogenized with a Branson sonifier, equipped with a microtip. The power switch is set at position 3, and the meter detection should be about 2 A dc. The sonication is carried out for periods of 10 sec until the preparation appears homogeneous. It is important that foaming be avoided, and the enzyme should be cooled in ice during the procedure. B11 should contain about 10 mg protein/ml and have a specific activity of about 15 units/mg protein. It may be stored in ice overnight or at −75° for 2 weeks.

About 25 ml of B11 suspension is thawed in an icebath, added diluting buffer to obtain a protein concentration of 6 mg/ml, and centrifuged for 3 hr at 65,000 *g*. The precipitate (B12) is suspended in 10–12 ml of diluting buffer, kept on ice for at least 10 hr, and homogenized as described above (for B11). At this stage, a specific activity of about 25 units/mg protein of B12 should have been achieved. The enzyme suspension is diluted with the dialyzing buffer to a protein concentration of 6 mg/ml and dialyzed for 20 hr against 500 vol of the same buffer. The dialyzing buffer has to be changed after the first hour of dialyzation. The enzyme is centrifuged for 6 hr at 65,000 *g*. Only the clear supernatant fluid (S13) should be used. To S13 is added polymerizing buffer to a final concentration of 50 m*M* phosphate and 15 m*M* borate. The preparation is left for 30 min and centrifuged for 1.5 hr at 65,000 *g*. The supernatant fluid is carefully removed as described above, and the precipitate (B14) is suspended in 0.5 ml diluting buffer. Following storage of the B14 suspension overnight in ice (important), the preparation is homogenized by freezing and thawing, or by sonication as described above.

The final yield is 3–6 mg of about 10,000-fold purified PAG protein in 0.5 ml (the phosphate–borate form). No further purification could be achieved by repeating the dialysis of this preparation against Tris–HCl, followed by precipitation with phosphate–borate, although a considerable loss of activity invariably occurred.

The solubilized form (the Tris–HCl form) of highly purified PAG is obtained by dialyzing the B14 suspension overnight against 2 changes of 2000 volumes of dialyzing buffer, followed by centrifugation at 65,000 *g* for 1 hr.

The PAG in supernatant S13 is almost homogeneous as judged by polyacrylamide gel electrophoresis and ultracentrifugation methods, and may be used for kinetic studies.

Pig Brain PAG[3] (Table II)

Acetone Powder Preparation. The brains are taken out immediately after sacrifice of the pigs, cooled, and membranes and cerebellum are removed. The brains may be kept frozen at −20° for some weeks. Portions (200 g wet wt) of frozen brains are cut into slices of less than 1 mm thickness using a meat slicer, transferred to a precooled glass container (2 liters), and homogenized (starting carefully) in 1000 ml cold acetone (−40°) with a Sorvall Omnimixer at half maximal speed until the slices are cut into pieces of less than peanut size (usually after 2–5 min). Two-thirds of the acetone is then decanted off and saved. The residual one-third, with the precipitate, is transferred to a 600-ml steel container, homogenized at maximal speed for 1 min, and recombined with the main portions of acetone. Following stirring for about 15 min, the homogenate is filtered using a Büchner funnel. The process is repeated once by rehomogenizing the filter-cake in another 1000 ml portion cold acetone. After being dried for some minutes by suction, the acetone powder is allowed to dry com-

TABLE II
SUMMARY OF THE PURIFICATION PROCEDURE OF PAG FROM PIG BRAIN

Fraction	Total protein (mg)	Total activity[a]	Specific activity[b]	Percentage recovery	Purification (fold)
Homogenate	1,427,000	21,812	0.03	100	1
Acetone-powder in borate buffer	1,500,000	22,500	0.015	53	—
S1	127,000	4,700	0.037	21	1.2
Na_2SO_4 precipitation 0–1 *M* (B2)	79,486	16,692	0.21	39.0	7
Na_2SO_4 precipitation 0.6–0.8 *M* (B5)	6,451	17,740	2.75	41.4	92
First Tris–HCl solubilization S10	2,600	1,846	0.71	4.3	—
First P-B[c] precipitation B11	236	5,427	23.0	12.7	767
Reprecipitation B12	84	4,092	48.6	9.6	1,620
Second Tris–HCl solubilization S13	20.4	853	40.9	2.0	1,363
Second P-B precipitation B14	2.8	1,296	471	3.0	15,700
Third Tris–HCl solubilization S15	1.8	239	130	0.6	4,333

[a] Activity (units): μmol NH_3 formed/min.
[b] Specific activity: units/mg protein. PAG activity in all fractions is determined as for the kidney enzyme (see Table I).
[c] P-B, phosphate–borate.

pletely at room temperature, and finally stored at −20°. The yield is about 40 g acetone powder from 200 g (wet wt) of brain tissue, and it is necessary to work up acetone powder from about 9000 g (wet wt) brain tissue, representing some 150 pig brains, before purification.

Portions of 40 g of acetone powder are homogenized in a total volume of 800 ml cold borate buffer. The suspensions are centrifuged at 16,000 *g* for 1 hr at 5°, and the supernatant (S1) carefully removed by aspiration as the precipitate is somewhat fluffy.

Na_2SO_4 Fractionation. Precipitation with Na_2SO_4, centrifugation, and homogenization of precipitated proteins and preparation of B2 and S3 are performed as described above for pig kidney PAG. The total volume of B2 suspension is 60–70 ml per 40 g portion of acetone powder. The specific activity of this preparation is usually 0.1–0.2 units/mg protein with a protein concentration of 15–25 mg/ml.

From S3 the protein fraction precipitated between 0.6 and 0.8 *M* Na_2SO_4 is collected (B5). The specific activity of this preparation is usually 1.6–4.0 units/mg protein, with a protein concentration of 20–30 mg/ml. Occasionally, when this desired purification is not obtained, the procedure from B2 to B5 has to be repeated, and the last precipitate is called B7. If the protein precipitated at 0.6 *M* Na_2SO_4 (B4), which ordinarily is discarded, contains a high enzyme activity, this precipitate is resolubilized and worked up similarly. The final preparation may be stored for at least 1 year at −20°.

Repeated Solubilizations and Precipitations in Tris–HCl Buffer and Phosphate–Borate Buffer, Respectively. This part of the purification procedure is slightly modified from that of pig kidney PAG, as follows: Supernatant S10 is centrifuged for 6 hr following addition of polymerization buffer. The homogenized B11 suspension is centrifuged for 5 hr.

Similarly to pig kidney PAG, pig brain PAG can be kept for months at −75° without appreciable loss of activity. However, PAG stored in dilution buffer (called phosphate–borate form) may be difficult to solubilize.

Purified pig brain PAG has about the same specific activity as that of pig kidney, and both glutaminases have to be at least 10,000-fold purified as assayed in the phosphate–borate form, to yield an apparently homogeneous preparation.

A modification of this method for purifying pig brain PAG has been published[16] (Table III).

Rat Kidney PAG

Principle. This method by Curthoys *et al.*[17] is based on the same principle as for pig kidney and brain PAG, alternate solubilization of PAG in Tris–HCl buffer and precipitation in phosphate–borate buffer, but isolated mitochondria are used as a starting material.

Reagents

Homogenate buffer: 25 m*M* Tris–HCl, 0.2 m*M* EDTA, 0.33 *M* sucrose, pH 7.5
Digitonin
BPP buffer: 10 m*M* sodium tetraborate, 100 m*M* potassium phosphate, 100 m*M* potassium pyrophosphate, pH 8.9
$(NH_4)_2SO_4$

[16] G. A. Nimmo and K. F. Tipton [*Biochem. Pharmacol.* **29,** 359 (1980)] have modified the purification method of soluble pig brain PAG.

[17] N. P. Curthoys, T. Kuhlenschmidt, and S. S. Godfrey, *Arch. Biochem. Biophys.* **174,** 82 (1976).

TABLE III
MODIFICATION OF THE PURIFICATION PROCEDURE OF PAG FROM PIG BRAIN[a]

Reagents

Buffer T1: 50 m*M* Tris–HCl, 1 m*M* EDTA, 0.05% v/v mercaptoethanol, pH 8.0

Buffer T2: 100 m*M* Tris–HCl, pH 7.4, 0.1 m*M* EDTA

2 *M* Na_2SO_4

Buffer P-B: 0.5 *M* phosphate, 15 m*M* borate–NaOH, 1 m*M* EDTA, 0.05% v/v mercaptoethanol, pH 8.0

0.2 *M* Na_2HPO_4

4 *M* $CaCl_2$

$(NH_4)_2SO_4$

Frozen pig brain mitochondria (also containing synaptosomes) from 1 kg brain, suspended in 1 liter buffer T1 are thawed, sonicated, and centrifuged (20,000 *g*, 30 min). The pellet is resuspended in 500 ml buffer T1, sonicated, and centrifuged as before. To the combined supernatants is added 2 *M* Na_2SO_4 at room temperature to a final concentration of 1 *M* and the floating cake obtained following centrifugation (20,000 *g*, 40 min) is suspended in 500 ml buffer P-B. The suspension is frozen overnight, thawed, and centrifuged (20,000 *g*, 30 min). Thereafter the lipids are removed from the supernatant using calcium phosphate. The supernatant is mixed with an equal volume of 0.2 *M* Na_2HPO_4. pH is adjusted to 8.5–8.1 and kept in this range while 4 *M* $CaCl_2$ gradually is added to a final concentration of 40 m*M*. The mixture is stirred for 1 hr, followed by centrifugation at 20,000 *g* for 5 min. The enzyme protein is then precipitated from the supernatant with 20% $(NH_4)_2SO_4$, and thereafter centrifuged (20,000 *g*, 30 min) and resuspended in 15 ml buffer T1. Following dialysis against 2 liter buffer T1 (3–4 hr), and centrifugation (35,000 *g*, 20 min), the supernatant is applied to a Sepharose 4B column (3 × 85 cm), which is equilibrated with buffer T1. The active fractions are pooled and concentrated solutions of sodium phosphate and sodium borate pH 8.0 added to a final concentration of 50 and 15 m*M*, respectively. The protein is precipitated with 20% $(NH_4)_2SO_4$, resuspended in 3 ml of buffer P-B, then dialyzed against 1 liter P-B buffer for 3–4 hr, and centrifuged. The supernatant is filtered through a Sepharose 4B column (2 × 50 cm), which is equilibrated with P-B buffer. The active fractions from the column are pooled and centrifuged (160,000 *g*, 4 hr). The pellet is resuspended in 1 ml of buffer T2 and dialyzed overnight against 2 × 1 liter of the same buffer. To the enzyme which is obtained in the supernatant after centrifugation (195,000 *g*, 20 min), 1 m*M* dithiothreitol is added and stored at 4°; 0.22 mg of enzyme protein is obtained. According to the authors, sodium dodecyl sulfate electrophoresis of the purified enzyme suggests that the preparation is essentially homogeneous.

[a] From Nimmo and Tipton.[16]

TPD buffer: 10 m*M* Tris–HCl, 10 m*M* potassium pyrophosphate, 1 m*M* dithiothreitol, pH 8.6

Purification Procedure. Mitochondria from 40–50 acidotic rats are prepared and resuspended in a volume of homogenate buffer which is equivalent to the initial wet weight of the kidneys. This preparation is slowly stirred (30 min, 3°) in an equal volume of homogenate buffer containing a final concentration of 0.15 mg digitonin/mg protein. The pellet obtained by centrifugation (10,000 *g*, 15 min) is homogenized in 2 vol of homogenate buffer and recentrifuged. The final pellet is homogenized in 1 vol homogenate buffer and then is added 1/10 vol of a 10-fold concentrated BPP buffer. This sample is allowed to stand at room temperature for 30 min and then lyophilized overnight. After resuspension in 300 ml water, standing at room temperature for 30 min, and centrifugation (25,000 *g*, 30 min) the supernatant is collected. The pellet is resuspended in a total volume of 200 ml in BPP buffer, homogenized, and centrifuged. The combined supernatants are made 40% saturated with $(NH_4)_2SO_4$, allowed to stand for 30 min, at 3°, centrifuged (10,000 *g*, 15 min), suspended in up to 40 ml BPP buffer, dialyzed against 2 liter of the same buffer (4 hr, two changes of buffer), and centrifuged (20,000 *g*, 10 min). The supernatant is applied to a Sepharose 4B column (4 × 100 cm), equilibrated with BPP buffer. The active fractions are pooled and precipitated in 40% $(NH_4)_2SO_4$. The precipitate is suspended in up to 10 ml of TPD buffer and dialyzed against 2 × 2 liter of the same buffer. Following centrifugation, the supernatant is filtered through a Sepharose 4B column (2.5 × 50 cm), which is equilibrated with TPD buffer. Thereafter, fractions with PAG activity are precipitated with 60% $(NH_4)_2SO_4$. The precipitate is suspended in 1–2 ml BPP buffer and dialyzed overnight against 1 liter of the same buffer. According to the authors 2.2 mg of the purified protein is obtained, and it yields a single protein band after polyacrylamide gel electrophoresis in the presence of 0.1% Triton.

Properties

Molecular Properties. Purified pig kidney PAG is composed of two subunits with MW of 64,000 and 57,000, respectively, whereas pig brain PAG has one subunit with MW of 64,000.[3,18] However, if PAG is prepared from acetone powder of kidney cortices, the enzyme obtained is composed of apparently identical subunits with MW of 57,000. Recent findings indicate that the 57,000 subunit is a degradation product of that of 64,000. Others[16] report that the subunit of pig brain PAG has a MW of 73,000. Rat kidney PAG contains subunits of MW of 65,000, but subunits with lower molecular weights are formed, probably due to proteolysis.[19] Recently, the 64,000 PAG subunit has been detected by immunological

[18] B. R. Olsen, I. Aa. Torgner, T. B. Christensen, and E. Kvamme, *J. Mol. Biol.* **74,** 239 (1973).

[19] W. G. Haser, R. Shapiro, and N. P. Curthoys, *Hoppe-Seylers Z. Physiol. Chem.* **364,** 1238 (1983).

technique in brain and kidney of a variety of mammalian species, including man, monkey, cow, pig, rabbit, rat, mouse, and hen.[20,21]

Pig kidney and pig brain PAG solubilized in Tris–HCl buffer at a protein concentration below 0.1 mg/ml have molecular weights of 130,000–150,000 and 120,000–135,000, respectively. If the protein concentration is higher than about 0.1 mg/ml, PAG will dimerize in the presence of phosphate, but polymers are also present.[3,16–18] Insoluble polymers with MWs of 1.5–2.0 mill are formed upon addition of borate to the phosphate buffer. The monomeric form of rat kidney PAG has a MW of 160,000,[17] it dimerizes in the presence of phosphate, and polymer forms are formed in the presence of phosphate and borate.[22]

The Tris–HCl enzyme (monomeric form), the phosphate enzyme (dimeric form), and the phosphate–borate enzyme (polymeric form) have been distinguished. In addition, a membrane-bound, insoluble form of pig brain PAG has been suggested.[23]

By electron microscopy, pig kidney and brain PAG show similar pictures.[18,24] Electron microscopy of the Tris–HCl enzyme after negative staining demonstrates the presence of small rounded, square, or rectangular particles. In high magnification micrographs a number of molecules show a stain-filled cleft which appeared to divide each molecule into two parts. After addition of phosphate to the Tris–HCl enzyme, an elongated dimer composed of two Tris–HCl enzyme molecules becomes a major component of this preparation. On addition of phosphate–borate buffer, large rope-like helical polymers are formed.

Activators and Inhibitors. The PAG catalyzed hydrolytic reaction is for practical purposes irreversible. It is fast, energy independent, and independent of any known cofactor. Nevertheless, PAG is subject to a very complicated regulation. Phosphate is a potent activator, and the enzyme is also stimulated by a variety of different compounds and particularly by anions. PAG prepared as described above from pig kidney, and which consists of two different subunits, as well as PAG from pig brain exhibit pH-dependent nonlinear kinetics against both the substrate (glutamine), and the activator phosphate.[3] Activators are members of the

[20] G. Svenneby and J. Storm-Mathisen, *in* "Glutamine, Glutamate, and GABA in the Central Nervous System" (L. Hertz, E. Kvamme, E. G. McGeer, and A. Schousboe, eds.), p. 69. Alan Liss, Inc., New York, 1983.

[21] G. Svenneby, I. Aa. Torgner, and E. Kvamme, *Acta Neurol. Scand.* **69,** 356 (1984).

[22] S. Godfrey, T. Kuhlenschmidt, and N. P. Curthoys, *J. Biol. Chem.* **252,** 1927 (1977).

[23] G. A. Nimmo and K. F. Tipton, *Eur. J. Biochem.* **117,** 57 (1981).

[24] B. R. Olsen, G. Svenneby, E. Kvamme, B. Tveit, and T. Eskeland, *J. Mol. Biol.* **52,** 239 (1970).

citric acid cycle,[15,25–27] thyroxine,[28] various phosphorylated compounds,[29] acetyl-CoA,[30] and some acyl-CoA derivatives.[31,32] The enzyme is therefore not phosphate dependent so that the old term phosphate-activated glutaminase is preferred. Maleate also activates the purified PAG to the same extent as do succinate and citrate. This activation disappears when PAG is inactivated by heat (50°, 10 min) and is for that reason not due to contamination with "maleate activated glutaminase" [γ-glutamyl transpeptidase (EC 2.3.2.2)] which is resistant to such heat treatment.[4]

The citric acid cycle intermediates and acetyl-CoA produce no additional activation to that of phosphate at high concentrations, hence they appear to compete for the same site.[30,33,34] On the other hand, some acyl-CoA derivatives increase the activation by phosphate.[31]

The activation by anions differs for the Tris–HCl enzyme and the polymeric enzyme, as the polymeric form of pig kidney PAG is more susceptible than the Tris–HCl enzyme to activation by anions, such as succinate and citrate.[15,25]

The specific activity of the dimeric and polymeric forms of PAG is three times that of the monomeric Tris–HCl form.[3,25] Therefore, in addition to the described activation of the various enzyme forms, the Tris–HCl form is characterized by time-dependent activation. This occurs when the protein concentration is sufficiently high to permit dimerization or polymerization on incubation with anionic activators, such as phosphate, phosphate–borate, or acyl-CoA derivatives.[23,25,30,31,35] Half time ($t_{1/2}$) of the time-dependent activation at room temperature with phosphate or phosphate plus borate has been found to be about 53 sec for both the pig kidney and pig brain enzyme.[25,35] Time-dependent activation may initially be accompanied by a conformational change of PAG, followed by polymerization. This increase in specific activity seems to be a hysteretic effect.[36]

[25] G. Svenneby, B. Tveit, and E. Kvamme, *J. Biol. Chem.* **245,** 1878 (1970).
[26] J. P. Greenstein and F. M. Leuthardt, *Arch. Biochem. Biophys.* **17,** 105 (1948).
[27] J. O'Donovan and W. D. Lotspeich, *Nature (London)* **212,** 930 (1966).
[28] V. S. Hovhannessian, H. C. Buniatian, G. S. Mkrdumova, and L. L. Badalian, *Vopr. Biokhim. Mozga. Akad. Nauk. Armjan. SSR* **6,** 5 (1970).
[29] H. Weil-Malherbe and G. D. Beall, *J. Neurochem.* **17,** 1101 (1970).
[30] E. Kvamme and I. Aa. Torgner, *Biochem. J.* **137,** 525 (1974).
[31] E. Kvamme and I. Aa. Torgner, *Biochem. J.* **149,** 83 (1975).
[32] E. Kvamme, *Contr. Nephrol.* **31,** 60 (1982).
[33] H. Weil-Malherbe, *J. Neurochem.* **16,** 855 (1969).
[34] E. Kvamme and K. Lenda, *Neurochem. Res.* **7,** 667 (1982).
[35] G. Svenneby, *J. Neurochem.* **19,** 165 (1972).
[36] C. Frieden, *J. Biol. Chem.* **245,** 5788 (1970).

The reaction product, glutamate, is a potent inhibitor of PAG, but the enzyme is also inhibited by cAMP, cGMP,[37] protons,[15] palmityl-CoA, and stearyl-CoA (0.02 m*M*).[31] Recently, the kinetic properties of human brain[38] and kidney[39] PAG have been investigated and found to be very similar to the kinetics of these enzymes in the pig.

PAG is predominantly localized in the inner mitochondrial membrane.[32,40,41] Membrane bound PAG, which appears to be the dominating PAG in intact mitochondria, has different properties than the soluble PAG that may be purified. Thus membrane bound PAG in intact mitochondria is more susceptible to activation by phosphate and inhibition by glutamate than the purified forms.[23,34] Moreover, calcium enhances the activation by phosphate of pig kidney mitochondrial PAG and brain synaptosomal PAG,[42–44] but has no effect on purified PAG. Also the other reaction product, ammonia, exerts similar differential effects. Ammonia in concentrations 0.5–1.0 m*M* inhibits PAG in intact mitochondria and synaptosomes,[34,45,46] but has no effect on purified PAG, whereas 10–20 m*M* ammonia activates the purified enzyme.[15] *N*-Ethylmaleimide, which inhibits mitochondrial-bound pig kidney and brain PAG strongly, also has no effect on the purified enzyme.[4]

Using metabolic inhibitors, which are impermeable to the inner mitochondrial membrane, it has been shown that the enzyme is localized to the outer face of the membrane,[47] and is thus accessible to regulation by external compounds with restricted permeability of the inner membrane.[4,34]

[37] H. Weil-Malherbe, *J. Neurochem.* **19,** 2257 (1972).

[38] E. Kvamme, I. Aa. Torgner, and G. Svenneby, *Acta Neurol. Scand.* **69,** 328 (1984).

[39] E. Kvamma, G. Svenneby, K. J. Tveter, and I. Aa. Torgner, *Contr. Nephrol.* **47** (1985), in press.

[40] N. P. Curthoys and R. F. Weiss, *J. Biol. Chem.* **249,** 3261 (1974).

[41] Z. Kovaćević, *Biochim. Biophys. Acta* **430,** 399 (1976).

[42] E. Kvamme and B. E. Olsen, *FEBS Lett.* **107,** 33 (1979).

[43] A. M. Benjamin, *Brain Res.* **208,** 363 (1981).

[44] E. Kvamme, G. Svenneby, and I. Aa. Torgner, *Neurochem. Res.* **8,** 23 (1983).

[45] D. P. Simpson and W. Adams, *J. Biol. Chem.* **250,** 8148 (1975).

[46] H. F. Bradford and H. K. Ward, *Brain Res.* **110,** 115 (1976).

[47] E. Kvamme, *in* "Glutamine Metabolism in Mammalian Tissues" (D. Häussinger and H. Sies, eds.), p. 32. Springer-Verlag, Berlin and New York, 1984.

[31] Glutaminase from *Acinetobacter glutaminsificans*

By JOHN S. HOLCENBERG

Heightened interest has developed in certain bacterial glutaminase enzymes with the demonstration that they have antitumor activity against some animal and human cancers.[1] In contrast to the *E. coli* glutaminase previously described,[2] these enzymes have optimal activity at physiologic pH and most catalyze the hydrolysis of both glutamine and asparagine. This chapter will present general assay methods and the purification and properties of a representative, well-characterized enzyme from an *Acinetobacter* soil organism.[3]

Assay Methods

Four assay methods have been commonly used to measure the hydrolysis of glutamine or asparagine: (1) release of ammonia and direct nesslerization; (2) release of ammonia, distillation, and assay with a phenol-hypochlorate reagent; (3) isolation of the radiolabeled dicarboxylic acid on ion-exchange columns; (4) coupling the formation of dicarboxylic acid to disappearance of NADH.[4,5] Asparagine is usually used as a substrate because of its greater stability. We have used the first 3 assays; in general the first one for samples with enzyme activity greater than 1 IU/ml, the second for activity of 0.05 to 1 IU/ml, and the third for enzyme activity as low as 0.0005 IU/ml.

Reagents. All materials are stored at 5° unless indicated.

Assay 1

L-Asparagine monohydrate 0.025 *M* in Tris–HCl buffer 0.05 *M*, pH 7.5

Tris–HCl 0.05 *M*, ph 7.5, containing 5 mg/ml bovine serum albumin

H_2SO_4, 0.5 *M*; store at room temperature

Nessler's reagent for ammonia; dissolve in 1 liter distilled water and filter. Store at room temperature

[1] J. S. Holcenberg, *in* "Enzymes as Drugs" (J. S. Holcenberg and J. Roberts, eds.), p. 25. Wiley (Interscience), New York, 1981.

[2] See also S. Hartman, this series, Vol. 17A, p. 941.

[3] J. Roberts, J. S. Holcenberg, and W. C. Dolowy, *J. Biol. Chem.* **247,** 84 (1972).

[4] See also J. C. Wriston, Jr., this series, Vol. 17A, p. 732.

[5] H. N. Jayaram, D. A. Cooney, S. Jayaram, and L. Rosenblum, *Anal. Biochem.* **50,** 327 (1974).

Copyright © 1985 by Academic Press, Inc.
All rights of reproduction in any form reserved.

Assay 2

Phenol, 12.4 g and sodium nitroprusside, 0.05 g, dissolved to a total volume of 200 ml distilled water

NaOH, 20 g, and 43 ml household bleach (5.25% hypochlorite) dissolved and diluted to final volume of 100 ml with distilled water

K_2CO_3, saturated solution at room temperature

Assay 3

Tris–HCl, 0.1 *M*, pH 7.5

L-Asparagine monohydrate, 1 m*M*

[^{14}C]- or [^{3}H]Asparagine, uniformly labeled, purified by passage through a AG-1 × 8 column equilibrated and washed with water

Bovine serum albumin, 5 mg/ml

Buffer 1, imidazole, 20 m*M* and asparagine, 75 m*M*, pH 7.0

Buffer 2, imidazole, 10 m*M* and asparagine, 30 m*M*, pH 7.0

Procedure Assay 1. Enzyme solution is diluted to approximately 1 IU/ml with Tris-albumin. The reaction is started by pipetting 0.1 ml of the diluted enzyme solution into 0.4 ml of the asparagine–Tris solution that had been previously warmed to 37°. The reaction is stopped after 15 min by addition of 0.5 ml of 0.5 *M* H_2SO_4. The acid is added at zero time for blanks. The tubes are centrifuged if necessary. Two-tenths milliliter of supernatant is transferred to a 7-ml glass tube containing 2.8 ml distilled water and mixed. Nessler's solution (0.5 ml) is added and mixed. The absorbance at 420 nm is read 2 to 20 min after addition of the Nessler's solution. This is compared with standard curves of ammonium chloride, run with each batch of Nessler's solution. Results are expressed as IU, micromoles ammonia released per minute under these conditions.

Procedure Assay 2.[6] The enzyme reaction mixture, conditions, and H_2SO_4 acidification are the same as in assay 1 except that the reaction time can be extended to 4 hr at 37° for assay of very low enzyme activity. Four-tenths milliliter of reaction mixture is transferred to a 30-ml serum bottle with 1 ml of distilled water. One milliliter of saturated K_2CO_3 is added and the bottle is immediately closed with a rubber stopper containing a glass rod with a 1-cm abraded tip that has been dipped in 0.5 *M* H_2SO_4. Care must be taken to avoid excess H_2SO_4 on the glass rod and to avoid touching the rod to the solution or sides of the bottle. The bottles are rotated on a wheel for 20 min at room temperature. The rod is carefully removed and rinsed with 1 ml of water into a 7-ml glass tube. Ammonia is completely distilled in this period. One milliliter phenol–nitroprusside and 0.2 ml NaOH–bleach is added, with mixing after each addition.

[6] A. L. Chaney and E. P. Marbach, *Clin. Chem.* (*Winston-Salem, N.C.*) **8,** 130 (1962).

The tube is warmed 15 min at 37° and then cooled for 15 min to room temperature. The absorbance is measured at 640 nm and compared with distilled ammonium chloride standards.

Procedure Assay 3.[7] Reaction mixtures of 240 μl containing 125 μl of 0.1 *M* Tris–HCl, 25 μl of 1 m*M* L-asparagine, 5 μl of radiolabeled asparagine, 25 μl of 5 mg/ml bovine serum albumin, and 60 μl distilled water are pipetted into small glass tubes and warmed to 37°. Ten microliters of enzyme, standard, or water is added and the mixtures incubated for 20 min. The reactions are stopped by adding 1.5 ml of ice cold Buffer 1. The mixtures are applied to 0.5 × 3.5 cm columns of AG 1 X8 (200–400 mesh) in Pasteur pipettes stoppered with glass wool. The columns are washed with 6 ml of Buffer 2. This procedure removes all the unreacted asparagine. The aspartic acid is eluted with 4 ml of 30 m*M* HCl. A 1-ml aliquot of this eluate is mixed with 10 ml of Aquasol (New England Nuclear) scintillation cocktail and radioactivity determined. The amount of enzyme was adjusted so that less than 25% of the substrate was converted to product. Enzyme activity is calculated by the percentage conversion of radioactive × total moles asparagine/time.

Purification Procedure

All steps were carried out at 5°.

Step 1. Preparation of the Crude Extracts

Cells (ATCC 27197) were grown by Grain Processing Company, Museative, Iowa, in the media described by Roberts *et al.*[3] and supplied as frozen paste weighing approximately 1 pound per plastic bag. In a typical purification procedure two 1-lb bags containing 60,000–80,000 IU were thawed overnight and suspended in 700 ml 0.05 *M* sodium phosphate buffer, pH 7.2 containing 0.02% sodium azide. The pH was adjusted to 7.2 with dilute HCl and the suspension sonicated in 300- to 400-ml aliquots for 5 min with a 2-cm-diameter probe. The sonicate was centrifuged 90 min at 18,000 rpm. This supernatant (SN1) was assayed, used in the next step or lyophilized. Each batch contained a mean of 31,000 IU with a specific activity of about 2 IU/mg protein. The pellet was suspended in 400 ml 30% saturated ammonium sulfate. The pH was adjusted to 7.0 and the mixture sonicated twice for 5-min periods. After centrifugation for 90 min at 18,000 rpm, the pellet was discarded. This supernatant (SN2) was dialyzed in 4 liters 0.1 *M* sodium phosphate buffer, pH 7.2 and then twice in 4 liters each of 0.02 *M* sodium phosphate buffer, pH 7.2. The dialyzed

[7] S. Prusiner and L. Milner, *Anal. Biochem.* **37,** 429 (1970).

material was assayed, used in the next step or lyophilized. It contained a mean of 26,000 IU.

Step 2. CM-Sephadex Chromatography

SN1, SN2, and/or lyophilized powder (20,000–40,000 IU) was dissolved in distilled water and adjusted to a pH of 6.5 with dilute HCl and conductivity of <3 μmho. This material was applied by gravity to a 5×80 cm column of C-50 Sephadex equilibrated with 0.04 *M* sodium phosphate buffer, pH 6.5 containing 0.02% sodium azide (conductivity of <2 μmho). The column was washed with 400 ml of this buffer and then eluted with a linear gradient of 1 liter of this buffer and 1 liter buffer containing 0.5 *M* NaCl, pH 6.5. Twelve milliliter fractions were collected and assayed. The enzyme activity appeared in the last third of the gradient. The active fractions were pooled and concentrated with an Amicon P10 filter to approximately 200 ml. Concentration is needed for optimal ammonium sulfate fractionation. This concentrate contained a mean of 18,500 IU with a mean specific activity of 28 IU/mg protein.

Step 3. Ammonium Sulfate Fractionation

The pH of the concentrate was adjusted to 7 and crystalline, enzyme grade ammonium sulfate added to 53% saturation (31.5 g/100 ml initial volume). After 30 min incubation this solution was centrifuged at 10,000 rpm. The pellet was dissolved in approximately 70 ml of 0.1 *M* potassium phosphate buffer, pH 7.2. Solid ammonium sulfate (31.4 g/100 ml) was added to the supernatant to make a 98% saturated solution. After 30 min incubation, this solution was centrifuged at 10,000 rpm (16,000 *g*). The precipitate was dissolved in about 70 ml 0.1 *M* potassium phosphate buffer, pH 7.2. The two dissolved pellets and 10 ml of the supernatant were dialyzed in 2 liters of this buffer followed by 2 changes of 0.005 *M* potassium phosphate buffer, pH 8.0. The 53–98% saturated ammonium sulfate fraction contained a mean of 14,000 IU with a mean specific activity of 81 IU/mg protein.

Step 4. DEAE-Sephadex Chromatography

The 53–98% saturated ammonium sulfate fraction was adjusted to pH 8.0 and a conductivity <1.0 mho and applied to a 2.5×35 cm column of DEAE Sephadex A-50 equilibrated with 0.005 *M* potassium phosphate buffer, pH 8.0 and a conductivity <1.0 μmho. The enzyme eluted near the solvent front when washed with the same buffer; yellow impurities were retained on the column. The fractions with enzyme activity were combined, adjusted to pH 7.5, and lyophilized. It is important to use potassium phosphate rather than sodium phosphate as the pH remains above 7 with the former buffer during freezing. This fraction contained a mean of 10,000 IU with a mean specific activity of 130 IU/mg protein. The overall yield averaged 34%.

Properties

Purity. The enzyme preparations after the DEAE-Sephadex step have been shown to be homogeneous by disc gel electrophoresis in SDS gels; isoelectric focusing in acrylamide gels; moving boundary, zone, and equilibrium centrifugation; and analysis of N-terminal amino acid sequence.[3,8–10]

Stability. The enzyme is stable indefinitely stored at 5° as a lyophilized powder. It is also stable in sterile solutions at 1 mg/ml protein or greater for at least 10 days. Freezing and thawing produces a substantial loss of activity.

Physical Properties. Acinetobacter glutaminase-asparaginase consists of 4 identical subunits with molecular weight of 33,000. Analytic sedimentation equilibrium showed that the enzyme reversibly dissociates to a dimer of 69,000 ± 3000 MW. Moving boundary velocity sedimentation showed that most of the enzyme exists as a tetramer ($s_{20,w}$ = 7.42 ± 0.03 S). Equivalent boundary calculations showed a smaller sedimentation coefficient. Hydrolysis of 5-diazo-4-oxo-L-norvaline by the enzyme is associated with the loss of absorbance at 274 nm. Zone sedimentation through a solution of this substrate showed that the active enzyme species has a $s_{20,w}$ of 7.6 ± 0.2 S. By matching zone sedimentation experiments with and without this substrate, enzyme species less than tetramers were shown to have little, if any, activity.[8]

Specificity. The enzyme catalyzes the hydrolysis of L- and D-glutamine and -asparagine. At 20 m*M* the relative activities are L-glutamine, 1:00; L-asparagine, 0.77; D-glutamine, 0.35; D-asparagine 0.26; and β-cyanoalanine 0.002. Glutamic acid γ-monohydroxamate is hydrolyzed and synthesized by the enzyme. Dipeptide amides and various aliphatic and aromatic amides are not hydrolyzed. 5-Diazo-4-oxo-L-norvaline is degraded at a low relative rate (0.0008 at 0.1 m*M*).[3] The next larger homolog, 6-diazo-5-oxo-L-norleucine, is hydrolyzed only when incubated with stoichiometric amounts of enzyme.[10]

[8] J. S. Holcenberg, D. C. Teller, J. Roberts, and W. C. Dolowy, *J. Biol. Chem.* **247,** 7750 (1972).

[9] J. S. Holcenberg, G. Schmer, D. C. Teller, and J. Roberts, *J. Biol. Chem.* **250,** 4165 (1975).

[10] J. S. Holcenberg, E. Ericsson, and J. Roberts, *Biochemistry* **17,** 411 (1978).

PROPERTIES OF THERAPEUTIC GLUTAMINASE–ASPARAGINASE ENZYMES

Enzyme	Ratio maximal activity L-Gln/L-Asn	K_m (μM) L-Gln	K_m (μM) L-Asn	p*I*
Acinetobacter[a]	1.2	5.8	4.8	8.2
Pseudomonas 7A[b]	2.0	4.6	4.4	5.8
E. coli-2[c]	0.03	3000	12	5.0
Erwinia carotovora[d]	0.05	1000	10	8.7

[a] J. Roberts, J. S. Holcenberg, and W. C. Dolowy, *J. Biol. Chem.* **247,** 84 (1972).
[b] J. Roberts, *J. Biol. Chem.* **251,** 2119 (1976).
[c] J. C. Wriston and T. O. Yellin, *Adv. Enzymol.* **39,** 185 (1973).
[d] J. B. Howard and F. H. Carpenter, *J. Biol. Chem.* **247,** 1020 (1972).

Kinetic Properties. The K_m for L-glutamine and L-asparagine are 5.8 ± 1.5 and 4.8 ± 1.4 μM, respectively. The isoelectric point with both substrates is 8.4 to 8.5. Optimal activity is seen at pH 7 for L-asparagine and L-glutamine.[3]

Inhibitors. The products of the reaction, ammonia, glutamate, or aspartate are not inhibitors. EDTA, *p*-mercuribenzoate, 1,10-phenanthroline, diisopropylphosphorofluoridate, and phenylmethanesulfonylfluoride do not effect the enzyme activity. Methylene blue (0.005%) or rose bengal (0.01%) dyes irreversibly inactivated the enzyme in the presence of light. 6-Diazo-5-oxo-L-norleucine rapidly and irreversibly inactivated the enzyme; nearly complete inhibition occurred when 3.2 mol of inhibitor was bound per mole of enzyme subunit. This inhibition was prevented by coincubation with asparagine or glutamine.[3,10]

Sequence Analysis and X-Ray Diffraction Analyses. The amino acid sequence has been determined for a 60-residue N-terminal segment and a 16-residue C-terminal segment of the 96-residue N-terminal peptide cleaved by cyanogen bromide. When the enzyme was treated with [6-C^{14}]diazo-5-oxonorleucine, reduced with sodium borohydride and cleaved with cyanogen bromide, all the radioactivity was present in this N-terminal peptide. During automatic sequencing radioactivity was released with a threonine in residue 12. Crystals have been obtained from *Acinetobacter* and *Pseudomonas 7A* glutaminase-asparaginase. Both contain 1 subunit per asymmetric unit with 222 point group symmetry.[11]

Chemical Modification. The *Acinetobacter* enzyme has been altered by succinylation and glycosylation. These modifications decrease the p*I*

[11] H. L. Ammon, K. C. Murphy, L. Sjolin, A. Wlodawer, J. S. Holcenberg, and J. Roberts, *Acta Crystallogr. Sect. B39,* 250 (1983).

to as low as 5.0 and prolong the plasma half-life in animals and humans. Succinylation does not alter the kinetic properties, physical properties, or antigenicity.[1,9]

Comparison with Other Enzymes. Several enzymes with glutaminase activity, optimal activity at physiologic pH, and antitumor activity have been isolated.[1] The properties of four of these enzymes are listed in the table. All these enzymes have molecular weights between 138,000 and 141,000. The first three have similar crystal symmetry and considerable homology in the N-terminal amino acid sequence. They also have an identical 8 amino acid segment that contains the threonine residue binding site for 6-diazo-5-oxonorleucine. The *Pseudomonas 7A* enzyme has an extra 8 amino acid at the N-terminus not present in the other two enzymes. The *Acinetobacter* enzyme was the only one that dissociates to inactive dimers. In contrast, the *Pseudomonas 7A* enzyme appears to polymerize during zone sedimentation when the initial protein concentration is greater than 1 mg/ml and the buffer contained its substrates.[12,13]

[12] J. S. Holcenberg, D. C. Teller, and J. Roberts, *Arch. Biochem. Biophys.* **161,** 306 (1974).
[13] J. S. Holcenberg, D. C. Teller, and J. Roberts, *J. Biol. Chem.* **251,** 5375 (1976).

[32] Glutamine Amidotransferases

By HOWARD ZALKIN

Glutamine amidotransferases utilize the amide of glutamine for the biosynthesis of several amino acids, purine and pyrimidine nucleotides, two coenzymes, glucosamine and antibiotics. A comprehensive review article[1] and the proceedings of a symposium[2] have summarized the properties of many of these enzymes. Most glutamine amidotransferases exhibit several common properties: (1) capacity to utilize NH_3 in place of glutamine, (2) selective inactivation of glutamine-dependent activity by alkylation of an active site cysteine residue with the glutamine affinity analogs, 6-diazo-5-oxonorleucine,[3] or L-2-amino-4-oxo-5-chloropentanate,[4] (3) glutaminase activity. These properties imply a common func-

[1] J. M. Buchanan, *Adv. Enzymol. Relat. Areas Mol. Biol.* **39,** 91 (1973).
[2] S. Prusiner and E. R. Stadtman, eds., "The Enzymes of Glutamine Metabolism." Academic Press, New York, 1973.
[3] B. Levenberg, I. Melnick, and J. M. Buchanan, *J. Biol. Chem.* **225,** 163 (1957).
[4] E. Khedouri, P. M. Anderson, and A. Meister, *Biochemistry* **5,** 3552 (1966).

Copyright © 1985 by Academic Press, Inc.
All rights of reproduction in any form reserved.

tional domain for glutamine amide transfer integrated with one or more domains for the covalent attachment of N and the specialized catalytic features of each enzyme. Where information on functional domains is available, it is included under the properties of individual enzymes. In the case of anthranilate synthase and *p*-aminobenzoate synthase, amino acid sequence homology has been detected in glutamine amide transfer domains.[5] As a consequence of an active site cysteine required for glutamine-dependent activity, precautions are usually required to prevent oxidative inactivation. Thiol reagents are often employed but in the case of amidophosphoribosyltransferase deoxygenation of buffers is more effective.

Glutamine amidotransferases involved in the synthesis of purine[6,7] and pyrimidine[8] nucleotides, asparagine[9] and glucosamine[10] have been purified from avian and mammalian tissues but most studies with homogeneous enzymes have used bacteria and yeast. In chapters [33]–[40] procedures are described for the assay and purification of glutamine amidotransferases from the best studied or most advantageous sources and references to alternative sources or procedures are included. The glutamine amidotransferases involved in the biosynthesis of histidine,[11] chloramphenicol,[12] and candicidin[13] are not included since these enzymes have not yet been purified.

[5] J. B. Kaplan and B. P. Nichols, *J. Mol. Biol.* **168,** 451 (1983).
[6] E. W. Holmes, *Adv. Enzyme Regul.* **19,** 215 (1981).
[7] S. Ohnoki, B.-S. Hong, and J. M. Buchanan, *Biochemistry* **16,** 1065 (1977).
[8] R. P. McPartland and H. Weinfeld, *J. Biol. Chem.* **251,** 4372 (1976).
[9] B. Horowitz and A. Meister, *J. Biol. Chem.* **247,** 6708 (1972).
[10] P. J. Winterburn and C. F. Phelps, *Biochem. J.* **121,** 701 (1971).
[11] D. W. E. Smith and B. N. Ames, *J. Biol. Chem.* **239,** 1848 (1964).
[12] A. Jones and D. W. S. Westlake, *Can. J. Microbiol.* **20,** 1599 (1974).
[13] J. A. Gil, P. Liras, G. Naharro, J. R. Villanueva, and J. F. Martin, *J. Gen. Microbiol.* **118,** 189 (1980).

[33] Amidophosphoribosyltransferase

By HOWARD ZALKIN

Amidophosphoribosyltransferase catalyzes the initial reaction in the *de novo* pathway for purine nucleotide synthesis, the reaction of PP-ribose-P and glutamine to yield phosphoribosylamine, glutamate, and

Copyright © 1985 by Academic Press, Inc.
All rights of reproduction in any form reserved.

PP_i.[1] The enzyme has been studied from a variety of microbial, avian, and mammalian sources.[2,3] The enzyme from all species is subject to end product inhibition by purine nucleotides. Homogeneous preparations which are stable and retain feedback inhibition have been obtained from *E. coli*[4,5] and *B. subtilis*.[6] Reports of homogeneous enzymes from chicken[7] and pigeon liver[8] have appeared but rigorous criteria for purity were lacking and feedback inhibition was labile. Careful chemical and physical studies have identified a diamagnetic [4Fe-4S] cluster in *B. subtilis* amidophosphoribosyltransferase.[9] The avian and mammalian enzymes appear to contain Fe-S[2] but *E. coli* amidophosphoribosyltransferase clearly lacks Fe-S. Procedures are described here for the purification of amidophosphoribosyltransferase from *E. coli*.

Assay Methods for Amidophosphoribosyltransferase

The glutamine- and NH_3-dependent reactions are shown.

$$\text{PP-Ribose-P} + \text{glutamine} \xrightarrow{Mg^{2+}} \text{phosphoribosylamine} + \text{glutamate} + PP_i$$

$$\text{PP-Rib-P} + NH_3 \xrightarrow{Mg^{2+}} \text{phosphoribosylamine} + PP_i$$

The glutamine-dependent activity is assayed by the rate of glutamate formation. Glutamate can be determined using glutamate dehydrogenase or by a DEAE filter-binding method. These assays are taken from Messenger and Zalkin.[4] For all procedures an enzyme unit is defined as the amount of enzyme catalyzing the formation of 1 μmol of product per min. Specific activity is expressed as units per mg of protein. Protein is determined by either the methods of Lowry *et al.*[10] or Bradford[11] using bovine

[1] The abbreviations used are PP-Ribose-P, 5-phosphoribosyl-1-pyrophosphate; PMSF, phenylmethylsulfonylfluoride; chloroketone, L-2-amino-4-oxo-5-chloropentanoate; DON, 6-diazo-5-oxonorleucine; DTT, dithiothreitol, GSH, glutathione.

[2] E. W. Holmes, *Adv. Enzyme Regul.* **19,** 215 (1981).

[3] H. Zalkin, *Adv. Enzyme Regul.* **21,** 225 (1983).

[4] L. J. Messenger and H. Zalkin, *J. Biol. Chem.* **254,** 3382 (1979).

[5] J. Y. Tso, M. A. Hermodson, and H. Zalkin, *J. Biol. Chem.* **257,** 3532 (1982).

[6] J. Y. Wong, D. A. Bernlohr, C. L. Turnbough, and R. L. Switzer, *Biochemistry* **20,** 5669 (1981).

[7] S. C. Hartman, *J. Biol. Chem.* **238,** 3024 (1963).

[8] P. B. Rowe and J. B. Wyngaarden, *J. Biol. Chem.* **243,** 6373 (1968).

[9] B. A. Averill, A. Dwivedi, P. Debrunner, S. J. Vollmer, J. Y. Wong, and R. L. Switzer, *J. Biol. Chem.* **255,** 6007 (1980).

[10] E. Layne, this series, Vol. 3, p. 447.

[11] M. M. Bradford, *Anal. Biochem.* **72,** 248 (1976).

serum albumin as a standard. The relationship between methods is 1.28 mg of protein, Lowry method, equals 1.15 mg protein, Bradford method, equals 1.0 mg protein according to amino acid analysis.

Procedure 1. Measurement of Glutamate Using Glutamate Dehydrogenase

This is a two step end point assay in which the glutamate formed by the amidophosphoribosyltransferase reaction in Step 1 is determined in Step 2 using glutamate dehydrogenase.

Reagents

PP-Ribose-P, 0.3 *M*
Glutamine, 0.13 *M*
Tris–HCl buffer pH 8.0, 0.5 *M*
MgCl, 0.5 *M*
Bovine serum albumin, 10 mg/ml
KCl, 1.0 *M*
Glutamate dehydrogenase, bovine liver, 3 times crystallized. A 10 mg/ml solution in 50% glycerol containing sodium phosphate is diluted 1 : 1 with 0.25 *M* potassium phosphate pH 7.5
Acetyl pyridine NAD^+, 0.06 *M*

Step 1. Pipet into a 1.5-ml polypropylene self-capped tube in ice, 25 μl Tris–HCl buffer pH 8.0, 10 μl $MgCl_2$, 5 μl PP-Ribose-P, 50 μl 0.13 *M* glutamine, 50 μl bovine serum albumin, and water to 0.49 ml. A control reaction is prepared which lacks PP-Ribose-P. Tubes are placed at 37° and the reactions are initiated by addition of 10 μl of enzyme solution. Incubation is at 37° for 10 to 30 min. Reactions are stopped by transferring to a boiling water bath for 2 min. If denatured protein is visible, shake at high speed using a mechanical mixer and then centrifuge at room temperature in an Eppendorf microfuge to sediment the protein precipitate.

Step 2. Glutamate is determined in a 1.0 ml reaction in a 12 × 75-mm test tube containing 50 μl KCl, 100 μl Tris–HCl buffer, 50 μl of glutamate dehydrogenase, 5 μl acetyl pyridine NAD^+, 0.3 ml of supernatant from Step 1, and 0.5 ml H_2O. A reagent blank is included which lacks the 0.3 ml sample of Step 1 supernatant. Incubation is for 30 min at 37°. The absorbance of acetylpyridine NADH is determined at 363 nm. A standard curve should be prepared and used to calculate the glutamate content. Linearity is obtained from 2.5–50 nmol of glutamate per 1.0 ml. An A_{363} reading of 0.460 was obtained for 50 nmol glutamate per ml.

Procedure 2. Measurement of [^{14}C]Glutamate by Filter Binding

Reagents

PP-Ribose-P, 25 m*M*
Glutamine, 50 m*M* (sp. act. 500 cpm/nmol)
Tris–HCl buffer pH 8.5, 0.5 *M*
Bovine serum albumin, 10 mg/ml
$MgCl_2$, 0.05 *M*
Ethanol, 95%

Purification of [^{14}C]Glutamine. It is essential to purify the [^{14}C]glutamine used in the filter binding assay of amidophosphoribosyltransferase since [^{14}C]glutamate and [^{14}C]pyrrolidone carboxylic acid interfere. The [^{14}C]glutamine is purified prior to mixing with unlabeled carrier glutamine. A 0.5 × 5.0 cm column of Bio-Rad AG1X2 (100–200 mesh) resin is washed with H_2O. Up to 1.0 ml of [^{14}C]glutamine is added to the column which is then eluted with H_2O. Glutamine does not bind and is collected in the first 2–3 ml. The [^{14}C]glutamine is freeze dried and then made up to its original volume with H_2O. Recovery should be greater than 90%. When the background levels of binding to DE-81 squares become unacceptable, purification of [^{14}C]glutamine must be repeated.

Washing of DE-81 Paper. Sheets of Whatman DE-81 paper are washed in pyridine acetate (200 ml pyridine, 8 ml glacial acetic acid) by descending chromatography. Wash at least 9 hr or until the pyridine acetate drips off the bottom. The pH (using pH indicator paper) at the bottom of the paper should be 6.5–7. Dry paper 1 day in a hood. Cut into 2-cm squares and dry in a vacuum oven for 1–2 hr at 50–60°. The pH of wet paper when pressed against pH indicator paper should be 7–8.

Step 1. Conversion of [^{14}C]Glutamine to [^{14}C]Glutamate. The following reagents are pipeted into a 1.5-ml self-capped polypropylene tube kept in ice: 20 μl Tris–HCl, 10 μl each of PP-Ribose-P, $MgCl_2$, and [^{14}C]glutamine. The volume is made to 90 μl with H_2O and 10 μl of enzyme solution is added to initiate the reaction. A control reaction is included which lacks PP-Ribose-P. Incubation is for 5–15 min at 37°. Reactions are terminated with 0.2 ml cold 95% ethanol. Precipitated protein is sedimented by centrifugation in an Eppendorf microfuge.

Step 2. Determination of [^{14}C]Glutamate by Filter Binding.[12] Filter squares are numbered with a soft pencil. Duplicate samples of 30 μl are applied to filters. Filters are washed in 1 liter of H_2O in a larger beaker. Up to 100 filters can be processed in 1 liter. Washing is continued for 5

[12] D. W. Martin, Jr., *Anal. Biochem.* **46,** 239 (1972).

min with gentle stirring. The water is carefully decanted, 1 liter of fresh water added, and the filters rinsed a second time for 2 min. The rinsing procedure is repeated 2 additional times. Filters are initially dried on top of filter paper on a large Buchner funnel with gentle suction followed by drying under a heat lamp. In addition to the washed samples, several controls are necessary. Samples of the reaction mixture should be applied to filters which are not washed. These will yield the input [^{14}C]glutamine. In addition, samples of [^{14}C]glutamate should be included to determine the binding efficiency. Thus, duplicate samples of [^{14}C]glutamate are prepared, one is washed, one not washed. After drying, the filters are placed in 3–4 ml scintillation fluid and counted for radioactivity. The binding efficiency for [^{14}C]glutamate should be greater than 85%.

Procedure 3. Measurement of Phosphoribosylamine[13]

This is a two-step assay in which phosphoribosylamine formed in Step 1 is determined electrophoretically as a complex with [^{35}S]cysteine in the second step. Hydrolysis of phosphoribosylamine to ribose-5-P is inconsequential as these two products react equally well. This assay can be used for NH_3-dependent, as well as glutamine-dependent amidophosphoribosyltransferase.

Reagents

PP-Ribose-P, 0.1 *M*
Glutamine, 0.05 *M*
Tris–HCl buffer pH 8.5, 0.5 *M*
$MgCl_2$, 0.1 *M*
NH_4Cl pH 8.5, 0.75 *M*
Carrier ribose 5-P/cysteine, 0.5 *M* Tris–HCl pH 8.5/0.25 *M* cysteine/0.1 *M* dithiothreitol/12.5 m*M* ribose 5-P. Store frozen and incubate 30 min 37° before use to reduce the cysteine
[^{35}S]Cysteine, specific activity > 300 Ci/mmol, stored frozen in 0.1 *N* HCl
[^{35}S]Cysteine stop solution, 10 μl [^{35}S]cysteine is mixed with 240 μl 0.01 *N* HCl and to this is added 750 μl of 0.5 *M* Tris–HCl, pH 9.0/500 m*M* cysteine/40 m*M* DTT/500 m*M* EDTA. [^{35}S]Cysteine is adjusted to yield 500 cpm/nmol
Pyridine acetate pH 3.5, 10 ml pyridine, 100 ml glacial acetic acid, 1890 ml of H_2O

Step 1. Formation of Phosphoribosylamine. The following are pipeted into a 1.5 ml polypropylene tube kept in ice: 14 μl Tris–HCl buffer, 7 μl

[13] G. L. King and E. W. Holmes, *Anal. Biochem.* **75**, 30 (1976).

PP-Ribose-P, 7 μl $MgCl_2$, 14 μl of either glutamine or NH_4Cl, 18 μl H_2O. The tubes are placed at 37° and the reactions are initiated with addition of 10 μl of enzyme. Control reactions are included in which glutamine or NH_4Cl is omitted. Incubation is for 5–15 min. Reactions are quenched by addition of 10 μl of stop solution containing [^{35}S]cysteine.

Step 2. Electrophoretic Separation of [^{35}S]Cysteine Adducts. Carrier ribose 5-P/cysteine solution (15 μl) is spotted on a 46 × 57-cm sheet of Whatman 3 MM paper at 2.7 cm intervals along a line 12.5 cm from one end. This arrangement allows 14 aliquots to be placed along the narrow dimension of the chromatography sheet. The spots of carrier ribose 5-P/cysteine are dried by a stream of cool air. Samples of each reaction mixture from Step 1 are then applied to spots containing carrier, using 2 applications of 10 μl. Drying is with a steam of cool air. The sheet of chromatography paper is wet with pyridine acetate pH 3.5 and placed in an electrophoresis apparatus. Electrophoresis is for 5 min at 300 V then 30 min at 3500 V. After thorough drying in a hood the paper is sprayed with ninhydrin and again dried at room temperature. The ribose 5-P/[^{35}S]cysteine adduct (relative mobility 1) is readily separated from [^{35}S]cysteine (relative mobility 0.14). The spots containing the ribose 5-P/cysteine adduct are cut out and counted for radioactivity in 3–4 ml of scintillation solution. A sample of the [^{35}S]cysteine stop solution or the reaction mixture on a filter paper strip is counted for radioactivity to calculate the [^{35}S]cysteine specific activity.

Procedure for Assay of PP-Ribose-P[14]

Due to chemical instability and variable quality of some commercial preparations it is frequently necessary to assay solutions of PP-Ribose-P.

Reagents

- Tris–HCl pH 8.5, 1.0 *M*
- $MgCl_2$, 0.2 *M*
- 5-Fluororotate, 0.6 m*M* (dissolved in 0.03 *M* Tris-HCl pH 8.5)
- Orotidine 5′-P pyrophosphorylase, 30 U/ml

The conversion of 5-fluororotate to 5 F-UMP at pH 8.5 results in a molar absorbance decrease of 3.82×10^3. Into a 1.2-ml cuvette is pipeted 20 μl Tris–HCl pH 8.5, 10 μl $MgCl_2$, 0.3 ml fluororotate, 0.1 ml enzyme, and H_2O to 0.99 ml. The initial absorbance at 295 nm is recorded. The sample of PP-Ribose-P (10 μl containing less than 0.12 μmol PP-Ribose-P) to be assayed is added and the decrease in absorbance is recorded. The assay is repeated using one-half the amount of PP-Ribose-P to verify

[14] J. L. Dahl, J. L. Way, and R. E. Parks, Jr., *J. Biol. Chem.* **234,** 2998 (1959).

linearity. An absorbance decrease of 0.382 is equivalent to 0.1 μmol PP-Ribose-P.

Purification Procedure for *E. coli* Amidophosphoribosyltransferase

E. coli amidophosphoribosyltransferase is purified from a strain bearing the cloned *E. coli* gene *purF*.[5] *E. coli* TX158/pSB2 is grown in a fermentor in media containing salts,[15] trace minerals,[16] 1 g/liter ammonium sulfate, 3.32 g/liter K_2HPO_4, 1 g/liter glucose, 1 g/liter acid casein hydrolysate, 1 mg/liter thiamin, and 4 mg/liter ampicillin. Cells are grown for about 12 hr at 37° with vigorous aeration to a cell density giving a turbidity reading of approximately 380 with a Klett colorimeter using a 66 red filter. Cells are harvested by centrifugation in a Sharples centrifuge. The cell yield is approximately 1 kg/100 liters. Cell paste is stored at −20°.

Step 1. Preparation of Crude Extract. All steps are conducted in a cold room at 6° or at ice bath temperature. Where noted buffer solutions are flushed with nitrogen to remove dissolved oxygen. Mercaptoethanol and DTT are less effective for enzyme stabilization. Approximately 200 g of cells is suspended in 480 ml of buffer A (0.05 *M* Tris–HCl pH 7.5/5 m*M* $MgCl_2$/1 m*M* EDTA/1 m*M* glutamine/0.5 m*M* PMSF) and the cells are disrupted by a French press at 12,000 psi. A few flakes of DNase are added and DNA digestion is allowed to proceed for 15 min. To the disrupted cell suspension 0.1 vol of 10% streptomycin–SO_4 in buffer A is slowly added with gentle stirring. Stirring is continued for 1 hr after the last addition. The suspension is centrifuged at 19,000 *g* for 1 hr to obtain the crude extract fraction.

Step 2. Ammonium Sulfate Fractionation. A solution of 3.8 *M* ammonium sulfate (pH 7.5, 4°) is added with stirring to give a final concentration of 1.267 *M*. The solution is centrifuged for 1 hr at 19,000 *g* and the precipitate is discarded. Ammonium sulfate solution is added to the supernatant to a final concentration of 2.067 *M*. The solution is centrifuged at 19,000 *g* for 1 hr and the precipitates containing amidophosphoribosyltransferase are dissolved in approximately 120 ml of buffer solution A containing 7.5 mg catalase. Dialysis, as described above, is repeated with fresh buffer for 5 hr.

Step 3. Heat Treatment. GMP and AMP are each added to the dialyzed ammonium sulfate fraction to a concentration of 2.5 m*M*. Approximately 100 ml of enzyme solution in a 250 ml stainless-steel container is placed in a 69° water bath. The solution is stirred with a thermometer and

[15] H. J. Vogel and D. M. Bonner, *J. Biol. Chem.* **218,** 97 (1956).

[16] R. E. Miller and E. R. Stadtman, *J. Biol. Chem.* **247,** 7407 (1972).

within 1.5 min when the temperature reaches 60° the container is placed in a 60° water bath for 2.5 min with continued stirring. The container is then transferred to an ice bath and stirring continued until the temperature is less than 10°. Denatured protein is removed by centrifugation at 19,000 *g* for 1 hr.

Step 4. DEAE Sepharose Chromatography. The enzyme solution from Step 3 is diluted with buffer A to approximately 18 mg protein/ml. The enzyme solution is applied to a 2.7 × 45 cm column of DEAE Sepharose equilibrated with buffer A. The column is washed successively with 50 ml of buffer A, 1.2 liters of 0.1 *M* potassium phosphate pH 7.0/1 m*M* EDTA/5 m*M* $MgCl_2$/1 m*M* glutamine and 20 ml of 0.1 *M* potassium phosphate pH 7.0/1 m*M* EDTA/1 m*M* glutamine. Amidophosphoribosyltransferase is eluted with a 3 liter linear gradient of 0–0.3 *M* $(NH_4)_2SO_4$ in 0.1 *M* potassium phosphate pH 7.0/1 m*M* EDTA/1 m*M* glutamine. Amidophosphoribosyltransferase elutes at approximately 0.16 *M* $(NH_4)_2SO_4$ immediately after glutamate dehydrogenase. Fractions with maximal activity (approximately 200 ml) are pooled and dialyzed against 6 liters of deoxygenated 0.01 *M* Tris–HCl pH 7.4. The buffer is changed 3 times over a 24-hr period. The enzyme is concentrated to 30 ml by pressure filtration using an Amicon PM-10 membrane. Insoluble protein is removed by centrifugation and the enzyme is stored in liquid nitrogen. A sample purification is summarized in the table.

Comments on Purification Procedure

The specific activity in *purF* plasmid-bearing strain TX158/pSB2[5] was approximately 90-fold higher than in the wild-type strain lacking the plasmid. As a result of the higher initial enzyme content in plasmid-bearing cells the final 3 steps of the older procedure[4] were unnecessary. Incorporation of the deleted steps may be required in specific cases. Chromatography on Blue dextran-Sepharose is effective in removing glutamate dehydrogenase. Chromatography on hydroxylapatite is effective in

SUMMARY OF AMIDOPHOSPHORIBOSYLTRANSFERASE PURIFICATION

Fraction	Volume (ml)	Activity (units)	Protein (mg)	Specific activity (units/mg)	Recovery (%)
Extract	858	16,310	22,650	0.72	100
Ammonium sulfate	185	17,290	8,730	1.98	100
Heat treatment	165	12,580	3,700	3.40	76
DEAE Sepharose	32	9,010	530	17.0	55

removing other contaminating proteins. The enzyme as purified in the table is homogeneous by the criteria of sodium dodecyl sulfate–polyacrylamide gel electrophoresis and automated Edman degradation.

Properties

Fe-S Centers. Fe was not detected by either atomic absorption[4] or X-ray fluorescence spectrometry.[5] The following additional elements were also undetected (<1 ppm): K, Ca, Sc, Ti, V, Cr, Mn, Fe, Co, Ni, Cu, Zn, Ga, Ge, As, Se, Rb, Sr, Mo. Although sulfide analyses were not done, the metal analyses clearly indicate that Fe-S centers are not a constituent of *E. coli* amidophosphoribosyltransferase. The lack of Fe-S distinguishes the *E. coli* enzyme from that in *B. subtilis*, avian liver and mammalian tissues. Interestingly, Fe-S centers were correctly assembled into fully active *B. subtilis* amidophosphoribosyltransferase made from the cloned *B. subtilis* gene in *E. coli*.[17]

Quaternary Structure. *E. coli* amidophosphoribosyltransferase is an oligomer of 3 or 4 subunits of M_r 56,395. The subunit M_r was estimated by sodium dodecyl sulfate–polyacrylamide gel electrophoresis and also calculated from the amino acid sequence. Molecular weights of 194,000 and 224,000 for the native enzyme were determined by sedimentation equilibrium centrifugation and gel filtration.[4]

Amino Acid Sequence. The amino acid sequence of amidophosphoribosyltransferase was deduced from the nucleotide sequence of cloned *purF*.[18] The protein subunit is a chain of 504 amino acids in which Met-1 is processed off and the NH_2-terminal residue of the isolated enzyme is Cys.

Glutamine Site. Glutamine-dependent activity is specifically inactivated by the glutamine affinity analogs, chloroketone or DON. Inactivation results from incorporation of 1 eq of DON per enzyme subunit. Cys-1 was identified as the residue alkylated by DON.[5]

Relationship between Glutamine and Ammonia Sites. Glutamine-dependent activity exhibits a broad pH optimum between 7.8 and 8.6 in Tris–HCl buffer whereas a pH optimum of 8.5 was determined for NH_3-dependent activity in Tris–HCl.[4] The rate of the NH_3-dependent reaction was approximately 2.8-fold greater than that of the glutamine-dependent reaction. The glutamine and NH_3 activities are not additive. K_m values of 1.7 mM, 8.8 mM, and 67 μM were calculated for glutamine, NH_3, and PP-Ribose-P, respectively.

[17] C. A. Makaroff, H. Zalkin, R. L. Switzer, and S. J. Vollmer, *J. Biol. Chem.* **258,** 10,586 (1983).

[18] J. Y. Tso, H. Zalkin, M. van Cleemput, C. Yanofsky, and J. M. Smith, *J. Biol. Chem.* **257,** 3525 (1982).

Glutaminase. The rate of glutamine hydrolysis in the absence of other substrates was 4 to 5% that of the overall reaction. This rate was stimulated 4-fold by ribose 5-P, Pi and Mg^{2+}.

Endproduct Inhibition. Purine nucleotides inhibited enzyme activity. The nucleoside 5′-monophosphates were most inhibitory. AMP + GMP gave synergistic inhibition. Inhibition exhibited positive cooperativity, n' 2.0 for AMP and 4.6 for GMP. GMP promotes cooperativity for PP-Ribose-P ($n = 2.5$) but AMP does not. Endproduct inhibition is stable and attempts to desensitize the enzyme to inhibition were unsuccessful.

[34] GMP Synthetase

By HOWARD ZALKIN

GMP synthetase catalyzes the final reaction in the *de novo* pathway for purine nucleotide synthesis, the conversion of XMP to GMP. The enzyme has been partially purified from pigeon liver,[1] calf thymus,[2] and *Klebsiella aerogenes*,[3] and purified to homogeneity from *E. coli*.[4] The procedure described here is a modified version[5] of that reported by Sakamoto *et al.*[4] for GMP synthetase from *E. coli*.

Assay Method for GMP Synthetase

The glutamine- and NH_3-dependent syntheses of GMP are shown below.

$$\text{XMP} + \text{glutamine} + \text{ATP} \xrightarrow{Mg^{2+}} \text{GMP} + \text{glutamate} + \text{AMP} + \text{PP}_i$$

$$\text{XMP} + \text{NH}_3 + \text{ATP} \xrightarrow{Mg^{2+}} \text{GMP} + \text{ATP} + \text{PP}_i$$

The rate of formation of GMP is determined by a continuous spectrophotometric assay at 290 nm. Molar extinction coefficients at 290 nm of 3.3×10^3 for GMP and 4.8×10^3 for XMP at pH 8.5 were reported by Moyed and Magasanik.[6] The difference in extinction coefficients of 1500 at 290 nm is the basis for a convenient assay. Assays are conducted at room temperature (22°).

[1] U. Lagerkvist, *J. Biol. Chem.* **233,** 143 (1958).
[2] R. Abrams and M. Bentley, *Arch. Biochem. Biophys.* **79,** 91 (1959).
[3] B. Magasanik, this series, Vol. 6, p. 106.
[4] N. Sakamoto, G. W. Hatfield, and H. S. Moyed, *J. Biol. Chem.* **247,** 5880 (1972).
[5] H. Zalkin and C. D. Truitt, unpublished.
[6] H. S. Moyed and B. Magasanik, *J. Biol. Chem.* **226,** 351 (1957).

Copyright © 1985 by Academic Press, Inc.
All rights of reproduction in any form reserved.

Reagents

Tris–HCl pH 8.3[7], 0.9 *M*
ATP, 0.2 *M*
XMP, 0.1 *M*
$MgCl_2$, 0.4 *M*
Glutamine, 0.2 *M*
Ammonium sulfate, 1.0 *M*
DTT, 10 m*M*
EDTA, 10 m*M*

Procedure. Stock solutions of nucleotides are stored frozen at pH 6.8–7.0. To assay the glutamine-dependent activity pipet into a 1.2-ml cuvette, 0.1 ml of Tris–HCl pH 8.3, 0.1 ml glutamine, 10 μl DTT, 10 μl EDTA, 3 μl XMP, 5 μl ATP, and 10 μl $MgCl_2$. Water is added to bring the volume to 0.99 ml and the reaction is initiated by addition of 10 μl of enzyme solution. The linear decrease in absorption at 290 nm is recorded at room temperature. A decrease of 0.1 at 290 nm corresponds to a conversion of 0.067 μmol of XMP to GMP. NH_3-dependent activity is assayed by replacing glutamine with 40 μl of 1 *M* ammonium sulfate, adjusted to pH 7.5. One unit of activity is defined as the amount of enzyme giving 1 μmol of GMP per min at 22°. Specific activity is defined as units per mg of protein. Protein is determined by the biuret procedure[9] through the DEAE cellulose purification step and by absorbance at 280 nm for the latter stages of purification. An absorbance of 1.00 at 280 nm is equivalent to 0.893 mg/ml of *E. coli* GMP synthetase.[4]

Purification Procedure

E. coli strain B96 (ATCC No. 13473), a purine auxotroph lacking inosinicase and transformylase (encoded by *purJ* and *purH*, respectively) is grown in a fermentor under conditions allowing derepression of GMP synthetase. The growth media contains 0.4% glucose, salts,[10] 0.2% acid casein hydrolysate, and 40 mg AMP per liter (filter sterilized). For a typical preparation of cells, 200 liters of media is inoculated with 8 liters of cells grown overnight in the same media in flasks. The fermentor is operated at 37° with vigorous aeration. Growth is for approximately 16 hr. Growth ceases upon exhaustion of AMP yielding a turbidity of approxi-

[7] In view of the reported inhibition of glutamine-dependent GMP synthetase by Tris[8] substitution with HEPES should be considered.

[8] N. Patel, H. S. Moyed, and J. F. Kane, *Arch. Biochem. Biophys.* **178,** 652 (1977).

[9] E. Layne, this series, Vol. 3, p. 447.

[10] H. J. Vogel and D. M. Bonner, *J. Biol. Chem.* **218,** 97 (1956).

mately 140–150 units using a Klett colorimeter (66 filter). This turbidity corresponds to 1.8–2.0 g (wet wt) of cells per liter. Cells are harvested by centrifugation and stored at −20°.

Step 1. Preparation of Crude Extract. All steps are conducted at ice bath temperature or in a cold room at 6°. To 200 g of thawed cells are added 400 ml of buffer A containing 7.5 m*M* potassium phosphate pH 7.4/0.1 m*M* DTT/0.1 m*M* EDTA. A few flakes of DNase are added. The cell suspension is passed two times through a French pressure cell at 10,000 psi. Buffer A is added to bring the volume up to 640 ml and the broken cell suspension is centrifuged at 27,000 *g* for 60 min.

Step 2. Streptomycin Sulfate. To the crude extract from Step 1, 0.1 volume of a 10% streptomycin sulfate solution adjusted to pH 7–7.5 is slowly added with stirring. Stirring is continued for 30 min and the suspension is then centrifuged at 23,000 *g* for 50 min. The supernatant solution of approximately 600 ml is dialyzed overnight against 8 liters of buffer A. The precipitate that forms is removed by centrifugation.

Step 3. Ammonium Sulfate Precipitation. The enzyme solution is slowly brought to 43% saturated ammonium sulfate by addition of 263 mg ammonium sulfate per ml. NH_4OH (1.0 *M*) is added to maintain the pH at 7.0. Stirring is continued for 30 min after the last addition of ammonium sulfate and the solution is then centrifuged at 23,000 *g* for 50 min. The pellet is dissolved in buffer B (20 m*M* potassium phosphate pH 7.4/1.0 m*M* dithiothreitol/1.0 m*M* EDTA/0.1 *M* KCl) and saved for assay to ensure that less than 8% of the activity is in this fraction. The enzyme solution at 43% saturated ammonium sulfate is then brought to 55% saturation by the further slow addition of 78 mg ammonium sulfate per ml. As before the pH is monitored and kept at neutrality by addition of 1 *M* NH_4OH. Gentle stirring is continued for 30 min after the last addition of ammonium sulfate and the solution is then centrifuged at 23,000 *g* for 50 min. The pellet is dissolved in 60 ml of buffer B and dialyzed overnight against 2 liters of the same buffer.

Step 4. DEAE Cellulose Chromatography. The dialyzed enzyme from Step 3 is adjusted to a protein concentration of 20 mg/ml with buffer B. The diluted enzyme solution (about 330 ml) is applied to a 5 × 80 cm column of DEAE cellulose (DE-52) that is equilibrated with buffer B. Elution is with a 4 liter gradient of 0.1–0.45 *M* KCl in buffer B. Fractions containing the enzyme, usually eluting between 2500 and 2900 ml, are combined and the solution is concentrated to approximately 90 ml by pressure filtration using an Amicon XM-50 membrane at 60 psi. The enzyme solution is dialyzed overnight against 2 liters of buffer C (0.02 *M* Tris–HCl pH 7.8/1 m*M* DTT/1 m*M* EDTA/0.2 *M* KCl). In this and subsequent column chromatographies the side fractions containing enzyme activity are saved, stored at −20°, and added to the next enzyme preparation at the step preceding the one where the side fraction was collected.

Step 5. DEAE Sepharose Chromatography. The dialyzed enzyme solution from Step 4 is applied to a 2.5 × 90 cm column of DEAE Sepharose CL6B equilibrated with buffer C. Elution is with a 2 liter gradient of 0.2–0.45 *M* KCl in buffer C. Fractions with greater than 20% of the peak activity are pooled and the solution is concentrated to 10–15 ml by pressure filtration with an Amicon XM-50 membrane. The concentrated enzyme solution is dialyzed overnight against 2 liters of buffer solution B. Side fractions are saved.

Step 6. Gel Filtration. Gel filtration is conducted with tandem 2.5 × 90 cm columns. The enzyme solution is applied to the first column of Sephacryl S-200. The eluate from the Sephacryl column is pumped into a column of BioGel A 1.5-m (200–400 mesh). Fractions of 3 ml are collected. Enzyme activity is associated with the major peak of protein. Fractions containing greater than 25% of the peak fraction are pooled. Side fractions are saved. The main pooled enzyme fraction of approximately 50 ml is concentrated to approximately 8 ml using pressure filtration. The enzyme is stored at −20°. A sample purification is summarized in the table.

Comments on Purification Procedure

The procedure described here is a modification[5] of that reported by Sakamoto *et al.*[4] We have changed the concentration of ammonium sulfate used in Step 3, substituted DEAE cellulose and DEAE Sepharose for two DEAE Sephadex steps and added a final gel filtration step. The procedure as described is highly reproducible. It yields somewhat less enzyme than reported by Sakamoto *et al.* Sakamoto *et al.* reported a specific

SUMMARY OF GMP SYNTHETASE PURIFICATION

Fraction	Volume (ml)	Activity (units)	Protein (mg)	Specific activity (units/mg)	Recovery (%)
1. Crude extract	520	3188	27,560	0.12	100
2. Streptomycin sulfate[a]	640	4608	23,040	0.20	—
3. Ammonium sulfate[a]	98	2626	6,566	0.40	82
4. DEAE-cellulose[b]	84	1777	1,932	0.92	56
5. DEAE-Sepharose[b]	12	1908	636	3.0	60
6. Gel filtration	8.4	1828	259	7.1	57

[a] After dialysis.

[b] After concentration and dialysis.

activity of 9.1 for homogeneous GMP synthetase. It is unclear whether the activity was determined at 38 or 22°. Enzyme of specific activity 7.0 is greater than 90% homogeneous by the criterion of sodium dodecyl sulfate–polyacrylamide gel electrophoresis. Cloning of *guaBA* on a multicopy plasmid should increase the GMP synthetase level in cells and facilitate purification of the enzyme.

Properties

Quaternary Structure. Molecular weights of 63,000 and 126,000 for the subunit and native enzyme, respectively, were determined by sedimentation equilibrium measurements.[4] The native enzyme is thus a dimer of subunits encoded by *guaA*.

pH Optima. pH optima of 8.3–8.5 were obtained for the glutamine- and NH_3-dependent activities.[11] The maximal rate of the glutamine-dependent activity is about 33% greater than the NH_3-dependent activity.

Glutamine Site. Affinity labeling with DON or chloroketone inactivated the glutamine-dependent activity.[8] Concomitant but lesser inactivation of NH_3-dependent GMP synthetase occurred. The maximal rate of affinity labeling by DON or chloroketone required ATP, XMP and Mg^{2+}. The stoichiometry for affinity labeling with chloroketone was 0.44 eg per subunit. On the other hand, selective inactivation of glutamine-dependent GMP synthetase was obtained by alkylation with iodoacetamide. The NH_3-dependent activity was unaffected. Approximately 1.1 eq carboxamidomethyl groups were incorporated into completely inactivated enzyme. Iodoacetamide and chloroketone appeared to react with the same residue which was identified as a cysteine.[12] *E. coli gua A* encoding GMP synthetase was recently cloned and sequenced. The derived protein chain of 525 amino acids has a calculated M_r of 58,604. Residues 1–198 are homologous with As II and thus constitute the glutamine amide transfer domain.

Relationship between Glutamine and Ammonia Sites. Calculations indicated that upon release, the amide of glutamine must remain enzyme bound prior to incorporation into XMP. The glutamine and NH_3 activities were not additive suggesting that a common site for NH_3 may participate in both reactions.

Glutaminase. GMP synthetase exhibits glutaminase activity similar to other glutamine amidotransferases. Hydrolysis of glutamine requires

[11] H. Zalkin and C. D. Truitt, *J. Biol. Chem.* **252,** 5431 (1977).

[12] C. D. Truitt, M. A. Henderson, and H. Zalkin, *J. Biol. Chem.* **253,** 8470 (1978).

[12a] A. A. Tiedeman, J. M. Smith, and H. Zalkin, *J. Biol. Chem.* **260,** in press.

[12b] H. Zalkin, P. Argos, S. V. L. Narayana, A. A. Tiedeman, and J. M. Smith, *J. Biol. Chem.* **260,** 3350 (1985).

XMP, Mg^{2+}, and either adenosine + PP_i or the adenosine analog psicofuranine. The maximal rate of glutaminase was found to be 1.8-fold greater than GMP synthetase.[11]

XMP Aminase. It was previously reported that the activity responsible for GMP synthesis in *E. coli*[13] and *Klebsiella aerogenes*[6] was a strictly NH_3-dependent activity designated XMP aminase. It is now known that the glutamine-dependent activity of these enzymes was in some way inactivated. Patel *et al.*[8] reported that the glutamine- but not the NH_3-dependent activity of GMP synthetase was inactivated by incubation in Tris buffer. It is possible that glutamine-dependent GMP synthetase was inactivated by Tris in the previously used 30-min end point assay.

[13] S. Udaka and H. S. Moyed, *J. Biol. Chem.* **238,** 2797 (1963).

[35] Glucosamine-6-Phosphate Synthase

By HOWARD ZALKIN

Glucosamine-6-phosphate synthase has been partially purified from *E. coli,*[1] *B. subtilis,*[1] *N. crassa,*[2] and animal tissues.[2,3] The lability of the enzyme, from all sources examined, is a serious problem which has precluded purification to homogeneity. No satisfactory purification procedure has been reported. The procedure described here for rat liver glucosamine synthase[3] is representative and might be a good starting point for future attempts to improve the enzyme purification.

Assay Method

Glucosamine synthase catalyzes the following reaction.

D-Fructose 6-P + glutamine → D-glucosamine 6-P + glutamate

Reagents

Fructose 6-P, 0.06 *M*
Glutamine, 0.12 *M*
Sodium phosphate, pH 7.5, 0.4 *M*

[1] R. Kornfeld, *J. Biol. Chem.* **242,** 3135 (1967).
[2] S. Ghosh, H. J. Blumenthal, H. J. Davidson, and S. Roseman, *J. Biol. Chem.* **235,** 1265 (1960).
[3] P. J. Winterburn and C. F. Phelps, *Biochem. J.* **121,** 701 (1971).

Copyright © 1985 by Academic Press, Inc.
All rights of reproduction in any form reserved.

EDTA, 0.05 *M*
Glucosamine 6-P, 20 m*M*
$NaHCO_3$, saturated solution
Acetic anhydride, 5% aqueous solution prepared fresh and kept cold
Sodium borate (pH 9), 0.8 *M*
Ehrlich's reagent, 1.0 g *p*-dimethylaminobenzaldehyde is added to 1.25 ml 10 *N* HCl and then diluted to 100 ml with gl acetic acid.

Procedure.[1] A 0.5 ml reaction mixture contains 0.05 ml fructose 6-P, 0.05 ml glutamine, 0.05 ml sodium phosphate, 12.5 μl EDTA, 0.29 ml H_2O, and 0.05 ml enzyme solution. Incubation is for 30–60 min at 37°. Reactions are terminated in a boiling water bath. The protein precipitate is removed by centrifugation and a 0.4 ml sample of the supernatant is used for assay of glucosamine 6-P by a modification of the Morgan–Elson method described by Ghosh *et al.*[2] To the supernatant is added 0.5 ml of $NaHCO_3$ followed by 0.05 ml of cold 5% acetic anhydride. The contents are shaken vigorously and then incubated at room temperature for 3 min. Tubes are transferred to a boiling water bath for 3 min to destroy excess acetic anhydride. After cooling to room temperature, 0.1 ml of sodium borate is added, the contents mixed, and heated in a boiling water bath for 3 min. Tubes are cooled at room temperature and 3.0 ml of Ehrlich's reagent added. Tubes are incubated for 20 min at 37° and color is measured at 585 nm. Glucosamine 6-P standards are included along with complete reaction mixtures that are boiled at zero time. A unit of activity is defined as the amount of enzyme catalyzing the synthesis of 1.0 μmol of glucosamine 6-P per hr. Specific activity is units per mg of protein. Protein is determined by the biuret procedure after precipitation with an equal volume of 12% trichloroacetic acetic acid to avoid interference by Tris and GSH. Trichloroacetic acid-precipitated protein is dissolved in a minimal volume of 0.7 *M* NaOH and brought to a convenient volume with water.

Purification Procedure for Rat Liver Glucosamine Synthase[3]

This preparation employs the stabilizing properties of fructose 6-P but is unsuitable for some studies since no adequate method for substrate removal was found that did not result in substantial loss of enzyme activity. An alternative procedure for purification in the absence of substrate is available and yields enzyme of somewhat lower specific activity.[3] In Steps 1–3, glucose 6-P added to buffer solutions is converted to fructose 6-P by phosphoglucose isomerase present in these fractions.

Step 1. Preparation of Homogenate. All operations are conducted at 4° unless noted otherwise. Care should be taken to avoid freezing the less pure fractions, since this results in complete inactivation. Rat livers are rapidly removed from sacrificed animals and chilled to 2° in buffer A (50 mM Tris/5 mM EDTA/5 mM GSH/5 mM glucose 6-P adjusted to pH 7.8 with 2 M HCl) plus 100 mM KCl, pH 7.8. The livers are homogenized in 3 vol of the above buffer in a Potter-Elvehjem homogenizer.

Step 2. High-Speed Supernatant. The homogenate is centrifuged at 18,000 g for 30 min. The supernatant solution is centrifuged at 105,000 g for 60 min and the supernatant retained.

Step 3. First DEAE-Cellulose Chromatography. Because of the need for a rapid separation of this unstable enzyme, a 5.5 × 5 cm column of DEAE-cellulose equilibrated in buffer A plus 100 mM KCl pH 7.8 is used. A flow rate of 600–800 ml/hr is maintained using gentle suction. A column of these dimensions is used to fractionate up to 1 g of high-speed supernatant. After application of the supernatant, the column is washed with 3–4 bed volumes of the equilibrating buffer and the enzyme is eluted with 150 ml of buffer A containing 150 mM KCl pH 7.8.

Step 4. Second DEAE-Cellulose Chromatography. This step is intended to concentrate as well as to further purify the enzyme. The eluate from Step 3 is diluted with 3 vol of 5 mM EDTA/5 mM GSH/5 mM glucose 6-P, pH 7.0 and adjusted to pH 6.7 with 0.1 M cacodylic acid. This solution is applied to a column (2 × 7 cm) of DEAE-cellulose equilibrated in buffer B/10 mM cacodylic acid, pH 6.7. Buffer B is identical to buffer A except that 5 mM glucose 6-P is replaced by 1 mM fructose 6-P. The enzyme is applied to the column at a flow rate of 300 ml/hr and is then eluted with 2 bed vol of buffer B/10 mM cacodylic acid/100 mM KCl, pH 6.7. The enzyme solution (about 15 ml) is adjusted to pH 7.5 by the careful addition of 0.1 M KOH.

Step 5. Hydroxyapatite Fractionation. To the solution from Step 4 is added 10 mg of hydroxyapatite per mg of protein. The solution is gently agitated for 10 min prior to collecting the hydroxyapatite by brief low speed centrifugation. The hydroxyapatite is washed for 10 min with 10 ml of 150 mM KH_2PO_4/5 mM EDTA/5 mM GSH/1 mM fructose 6-P adjusted

SUMMARY OF PURIFICATION OF GLUCOSAMINE SYNTHASE

Fraction	Specific activity (μmol/hr/mg)	Purification	Yield (%)
Homogenate	0.026	1	100
High-speed supernatant	0.058	2.2	91
First DEAE-cellulose	0.57	22	67
Second DEAE-cellulose	1.78	69	36
Hydroxyapatite	9.7	374	27

to pH 7.5 with 2 *M* KOH. After low-speed centrifugation the supernatant is discarded and the enzyme desorbed by gentle stirring for 10 min with 7 ml of 220 m*M* KH_2PO_4/5 m*M* EDTA/5 m*M* GSH/1 m*M* fructose 6-P, adjusted to pH 7.5 with 2 *M* KOH. The hydroxyapatite is removed by low speed centrifugation. The purification procedure is summarized in the table.

Comments on Purification Procedure

The entire procedure can be completed in 6 hr. If the 105,000 *g* supernatant is prepared the previous day and stored at 4°, the purification time is reduced to 3 hr. The enzyme can be concentrated by precipitation with 2.3 *M* ammonium sulfate with some attendant loss in activity. The enzyme can be stored at 4 or −15° with 50% activity loss in 8 days.

After the second DEAE-cellulose column, the following enzymes were not detected: phosphomannose isomerase, phosphoglucose isomerase, phosphofructokinase, glucosamine 6-P *N*-acetyltransferase, glutaminase, and glucosamine 6-P isomerase. Estimates of purity were not reported but the enzyme should require extensive further purification before reaching homogeneity. Further purification is dependent on improved techniques for stabilization.

Properties

Glutamine and Ammonia Sites. Glucosamine synthase was reported[2] to be inhibited (or inactivated) by DON and in this respect is similar to other glutamine amidotransferases. The enzyme apparently cannot utilize ammonia as a substrate.[2,3]

Catalytic Properties. K_m values of approximately 0.24 m*M* for fructose 6-P and 0.69 m*M* for glutamine were reported for the rat liver enzyme.[3] The enzyme from rat liver but not *E. coli* is subject to allosteric inhibition by UDP-*N*-acetylglucosamine, the presumed end product of the pathway.[1,3] The pH optimum is between pH 7 and 8.[3]

Cloning and Sequencing. *E coli glmS* was recently localized downstream of the *unc* operon in closed DNA.[4] The derived sequence of glucosamine synthase contains 609 amino acids. Residues 1–186 are homologous to a corresponding segment of *E. coli* amidophosphoribosyltransferase and constitute a glutamine amide transfer domain. The active site cysteine, identified by homology, is residue 2 in the derived sequence. Availability of the cloned gene on a multicopy plasmid should facilitate enzyme purification.

[4] J. E. Walker, N. J. Gay, M. Saraste, and A. N. Eberle, *Biochem. J.* **224,** 799 (1984).

[36] CTP Synthetase

By HOWARD ZALKIN

CTP synthetase catalyzes the terminal reaction in the pathway for *de novo* synthesis of pyrimidine nucleotides. The enzyme from *E. coli* B has been purified to homogeneity.[1] From eukaryotic organisms the most highly purified CTP synthetase, approximately 50% pure, was obtained from calf liver.[2] The purification and properties of CTP synthetase from *E. coli* B are described in this section. Methods for assay and purification are taken from publications by Long and Pardee[3] and by Levitzki and Koshland.[1]

Assay Method for CTP Synthetase

The glutamine- and NH_3-dependent syntheses of CTP are shown below.

$$\text{UTP} + \text{glutamine} + \text{ATP} \xrightarrow[\text{GTP}]{\text{Mg}^{2+}} \text{CTP} + \text{glutamate} + \text{ADP} + \text{P}_i$$

$$\text{UTP} + \text{NH}_3 + \text{ATP} \xrightarrow{\text{Mg}^{2+}} \text{CTP} + \text{ADP} + \text{P}_i$$

GTP is a positive allosteric effector for the glutamine-dependent reaction but is not required for NH_3-dependent synthesis of CTP. The simplest assay measures the amination of UTP by the increase in absorbance at 291 nm. The extinction coefficient at 291 nm of UTP at pH 7.0 is 182, while that of CTP is 1520. This difference of 1338 in extinction coefficient is the basis for a suitable assay. More sensitive assays have been described and are useful when the activity in crude microbial cell or tissue preparations is low.[4,5] In such cases the separation of [^{14}C]CTP from unreacted [^{14}C]UTP by thin-layer chromatography appears to be the method of choice.[5] The spectrophotometric assay is used for purification of CTP synthetase from *E. coli*.

Reagents

UTP, 0.1 *M*
ATP, 0.1 *M*

[1] A. Levitzki and D. E. Koshland, Jr., *Biochim. Biophys. Acta* **206**, 473 (1970).
[2] R. P. McPartland and H. Weinfeld, *J. Biol. Chem.* **251**, 4372 (1976).
[3] C. W. Long and A. B. Pardee, *J. Biol. Chem.* **242**, 4715 (1967).
[4] C. R. Savage and H. Weinfeld, *J. Biol. Chem.* **245**, 2429 (1970).
[5] J. C. Williams, H. Kizaki, E. Weiss, and G. Weber, *Anal. Biochem.* **91**, 46 (1978).

Copyright © 1985 by Academic Press, Inc.
All rights of reproduction in any form reserved.

GTP, 0.02 *M*
Glutamine, 0.2 *M*
Ammonium sulfate, 1.0 *M*
$MgCl_2$, 1.0 *M*
Tris–HCl pH 7.1, 1.0 *M*
Tris–HCl pH 8.2, 1.0 *M*

Procedure. Stock solutions of nucleotides are stored frozen at pH 6.8–7.0. To assay the glutamine-dependent activity, pipet into a 1.2 ml cuvette, 20 μl of 1.0 *M* Tris–HCl pH 7.1, 5 μl of 0.1 *M* UTP, 5 μl of 0.1 *M* ATP, 5 μl of 0.02 *M* GTP, 10 μl of 1.0 *M* $MgCl_2$, and H_2O to a final volume of 0.95 to 0.99 ml. The contents of the cuvette are equilibrated for 3–5 min to a temperature of 38° or room temperature. The reaction is initiated by addition of 10–50 μl of enzyme solution. The linear increase in absorbance at 291 nm is measured as a function of time at room temperature or 38° in a recording spectrophotometer. A control reaction is conducted by omitting UTP from the incubation mixture and any rate of absorbance increase in subtracted from that obtained with the complete mixture. The NH_3-dependent activity is determined by a similar procedure except that the buffer solution is Tris–HCl pH 8.2 and glutamine is replaced by 10 μl of 1.0 *M* $(NH_4)_2SO_4$. A small portion of the $(NH_4)_2SO_4$ stock solution is adjusted to pH 7.5 and any remaining solution is discarded at the end of the day. One unit of activity is defined as the amount of enzyme giving 1 μmol of CTP per min. Specific activity is defined as units per mg of protein. Protein is determined according to the method of Lowry *et al.*[6] or, for highly purified preparations, using an $A^{280}_{0.1\%}$ value of 0.89.[7]

Purification Procedure

E. coli B grown in minimal media and harvested in middle log phase was obtained from a commercial supplier. *E. coli* K12 strains grown in the laboratory in minimal media[8] appear to be equally suitable. The cell paste can be stored frozen at −20°.

Step 1. Preparation of Extract. To 850 g of thawed cells is added 1000 ml of buffer solution A (20 m*M* sodium phosphate, pH 7.0/1 m*M* EDTA/2 m*M* glutamine). The cells are disrupted by sonic oscillation. Alternatively a cell suspension containing 3 ml buffer A per g cell paste can be broken by passage through a French Press at 10,000 psi. The temperature is maintained below 6° throughout subsequent steps. Cell debris is removed by centrifugation at 13,000 *g* for 45 min yielding the crude extract.

[6] E. Layne, this series, Vol. 3, p. 447.
[7] A. Levitzki and D. E. Koshland, Jr., *Biochemistry* **11**, 247 (1972).
[8] H. J. Vogel and D. M. Bonner, *J. Biol. Chem.* **218**, 97 (1956).

Step 2. Streptomycin Sulfate Treatment. To 1650 ml of crude extract, 800 ml of 10% streptomycin sulfate is added, with stirring over a period of 2 hr. The suspension is centrifuged at 13,000 *g* for 30 min and the supernatant fraction is saved.

Step 3. First Ammonium Sulfate Treatment. To 2300 ml of Step 3 supernatant, 575 g of ammonium sulfate is added with stirring, over a 60-min period. The precipitate obtained by centrifugation at 13,000 *g* for 30 min is dissolved in 1000 ml of buffer solution A (Fraction III).

Step 4. Second Ammonium Sulfate Treatment. To 1050 ml of Fraction III, 200 g of ammonium sulfate is added with stirring over a period of 45 min. The precipitate obtained after centrifugation at 13,000 *g* for 30 min is discarded. An additional 60 g of ammonium sulfate is added to the supernatant solution, with stirring, over a period of 20 min. The precipitate obtained following centrifugation is dissolved in 150 ml of buffer solution A. The enzyme solution is desalted in 5 batches using a column of Sephadex G-50 (6 × 40 cm). Buffer solution B (20 m*M* sodium phosphate, pH 7.4/1 m*M* EDTA/40 m*M* glutamine/70 m*M* 2-mercaptoethanol) is used in the desalting step (Fraction IV) and all subsequent steps of the purification.

Step 5. DEAE-Sephadex Chromatography. The desalted enzyme is applied to a DEAE-Sephadex A-50 column (6 × 30 cm) and eluted with a gradient of 0–0.19 *M* $(NH_4)_2SO_4$ in 2.4 liters of buffer solution B. The enzyme elutes at approximately 0.08–0.09 *M* $(NH_4)_2SO_4$. It is likely that substitution of DEAE-Sepharose for DEAE-Sephadex would facilitate chromatography because of superior mechanical properties. Pooled fractions are treated with ammonium sulfate (351 mg/ml) to concentrate the protein. After centrifugation the precipitated protein is dissolved in buffer B to yield a solution of less than 20 mg/ml. Alternatively, the protein solution may be concentrated by pressure filtration using an Amicon PM-30 or XM-50 membrane.

Step 6. First Gel Filtration. This is the first of a series of gel filtration steps which are conducted in the absence or presence of nucleotide substrates. ATP and UTP synergistically promote a dimer to tetramer conversion that is utilized to purify the enzyme from contaminating proteins. Enzyme from Step 5 is applied to a column of Sephadex G-100 (4.2 × 75 cm). Elution in the absence of nucleotides is at a flow rate of 60 ml/hr and yields enzyme of M_r approximately 105,000. It is likely that a 2.5 × 95 cm column of Sephacryl S-200 would facilitate this gel filtration because of superior rigidity and flow rate compared to Sephadex G-100.

Step 7. Second Gel Filtration. This gel filtration is conducted in the presence of ATP plus UTP which induce a dimer to tetramer conversion. A 2.5 × 95 cm column of BioGel A-0.5m is equilibrated with buffer B containing 0.75 mM ATP/0.75 mM UTP/10 mM $MgCl_2$. Enzyme from the preceding step is concentrated to approximately 15 ml by pressure filtration using an Amicon PM-30 or XM-50 membrane. ATP, UTP, and $MgCl_2$ are added to give the concentrations noted above and the enzyme is applied to the column and eluted with buffer B containing nucleotides and $MgCl_2$. In the presence of nucleotides the enzyme elutes at a position corresponding to M_r approximately 210,000. Protein in pooled fractions is precipitated with ammonium sulfate (351 mg/ml) and following centrifugation the enzyme is dissolved in approximately 15 ml of buffer B.

Step 8. Third Gel Filtration. This gel filtration is conducted on a BioGel A-0.5m column (2.5 × 95 cm) equilibrated and eluted with buffer B. In the absence of nucleotides the enzyme elutes at a position corresponding to M_r approximately 105,000. The pooled fractions are concentrated to approximately 10 ml by precipitation with $(NH_4)_2SO_4$ (390 mg/ml) or pressure filtration.

Step 9. Fourth Gel Filtration. Although the original procedure calls for a final gel filtration on a 2.5 × 90 cm column of Sephadex G-200 equilibrated with buffer B, it is likely that a similar column of Sephacryl S-200 or S-300 would yield a comparable result and is preferred because of increased rigidity. The enzyme is concentrated to approximately 10 mg/ml by pressure filtration and stored at −20° in buffer solution containing 20 mM sodium phosphate pH 7.2/2 mM EDTA/70 mM 2-mercaptoethanol/4 mM glutamine/20% glycerol. A sample purification is summarized in the table.

SUMMARY OF CTP SYNTHETASE PURIFICATION[a]

Fraction	Activity (units)	Protein (mg)	Specific activity (units/mg)	Recovery (%)
1. Crude extract	185[a]	73,000[a]	<0.01	
2. Streptomycin sulfate	390	38,000[a]	0.01	100
3. First ammonium sulfate	339[a]	14,000[a]	0.02	87
4. Second ammonium sulfate	260	5,200	0.05	67
5. DEAE-Sephadex	160	533	0.30	41
6. First gel filtration	150	115	1.30	38
7. Second gel filtration	150	75	2.00	38
8. Third gel filtration	112	24	4.60	29
9. Fourth gel filtration	100	17	5.8–6.1	26

[a] Data except as noted are taken from ref. 1. Data for Fractions 1 and 3 are extrapolated from ref. 3. Assay temperature was 38°.

Comments on Purification Procedure

A single protein band was obtained upon polyacrylamide gel electrophoresis at pH 9.5 and pH 8.0 in 8 *M* urea.[9] Gel electrofocusing in 8 *M* urea also gave a single protein band.

With crude extract, enzyme activity is sometimes not detected using the A_{291} assay. In these cases it is advisable to carry out Steps 2 and 3. After Step 3, the first ammonium sulfate precipitation, the activity should be detected. Alternatively, a more sensitive assay may be used.

This purification procedure was worked out prior to the availability of materials such as DEAE-Sepharose, Blue Sepharose, nucleotide affinity resins, and gel filtration media exhibiting high porosity and rigidity (such as Sephacryl S-200 and S-300). These materials should be tested as replacements for those used originally. Furthermore it is probable that cloning of *E. coli pyrG* on a multicopy plasmid would result in elevated levels of CTP synthetase in crude extracts thus allowing several of the purification steps to be omitted. A simplified purification procedure employing hydrophobic column chromatography was recently reported.[9a]

Properties

pH Optima. The glutamine-dependent activity exhibits a broad optimum between pH 7.5 and 9.3, whereas the pH dependence of the ammonia activity rises linearly in the pH region 7–10.[10] A maximum is observed at pH 10.3–10.4 and is coincidental with rapid denaturation. The maximal activity with ammonia is at least 20% greater than with glutamine.

Glutamine Site. Incubation of [^{14}C]glutamine with glutamine-free CTP synthetase in 0.02 *M* sodium phosphate pH 7.4/1 m*M* EDTA led to formation of a radioactive enzyme–ligand intermediate which was isolated by gel filtration.[10] The radioactivity was concluded to represent a covalent glutamyl–enzyme intermediate. The stoichiometry was 0.9 eq of intermediate per subunit of 52,000 M_r. The intermediate was discharged when incubated with ATP, UTP, and GTP. There was, however, no GTP requirement for formation of the intermediate. Sulfhydryl titrations indicated involvement of a cysteinyl residue suggesting that the glutamyl–enzyme is a thioester.

The glutamine affinity analog DON inactivates glutamine-dependent CTP synthetase by specific alkylation of a cysteinyl residue. Ammonia-dependent activity is completely retained in the DON-treated enzyme. Affinity labeling by DON is complex and differs significantly from gluta-

[9] A. Levitzki and D. E. Koshland, Jr., *Biochemistry* **10,** 3371 (1971).

[9a] P. M. Anderson, *Biochemistry* **22,** 3285 (1983).

[10] A. Levitzki and D. E. Koshland, Jr., *Biochemistry* **10,** 3365 (1971).

myl–enzyme intermediate formation. (1) The rate of affinity labeling is increased 8-fold by GTP + Mg^{2+}. (2) The stoichiometry of DON incorporation is 0.5 eq per subunit (half of the sites reactivity). Levitzki and Koshland[9] postulated that DON induces a conformation that is different from that induced by glutamine.

Relationship between Glutamine and Ammonia Sites. Investigation of the effect of pH on the glutamine- and ammonia-dependent activities indicated that glutamylation of the enzyme liberates NH_3 which reacts with UTP as nascent ammonia without leaving the enzyme. Since only a single ammonia site could be detected it was suggested that ammonia released from glutamine may bind to the same site as exogenous ammonia from solution. Levitzki and Koshland suggest that the ammonia-dependent reaction has been preserved only because exogenous ammonia is able to utilize the same site as ammonia generated from glutamine hydrolysis.[10]

Dimer–Tetramer Interconversion. E. coli CTP synthetase is a dimer of M_r 108,000. The dimer associates to form a tetramer in the presence of either ATP or UTP.[7] ATP and UTP are synergistic. The dimer to tetramer conversion is ligand induced. Calf liver CTP synthetase also exhibits an ATP- and UTP-dependent dimer ($M_r \sim$ 133,000) to tetramer ($M_r \sim$ 263,000) interconversion.[2]

[37] Anthranilate Synthase

By HOWARD ZALKIN

Anthranilate synthase catalyzes the initial reaction in the tryptophan biosynthetic pathway. The enzyme from a number of bacteria,[1] *Neurospora crassa*,[2] and *Euglena*[3] has been purified to homogeneity. Anthranilate synthase contains two functional domains designated anthranilate synthase Components I (AS I) and II (AS II). AS I catalyzes the NH_3-dependent reaction and AS II confers glutamine amide transfer capacity. AS II exhibits the interesting property of existing in different fused arrangements with other enzymes of the tryptophan pathway (multifunctional proteins).[1] It is the fusion pattern that largely distinguishes anthranilate synthase from different organisms. Briefly, the following AS II

[1] H. Zalkin, *in* "Multifunctional Proteins" (H. Biswanger and E. Schmincke-Ott, eds.), p. 123. Wiley, New York, 1980.

[2] J. Keesey, Jr., J. Paukert, and J. A. DeMoss, *Arch. Biochem. Biophys.* **207,** 103 (1981).

[3] C. N. Hankins and S. E. Mills, *J. Biol. Chem.* **251,** 7774 (1976).

Copyright © 1985 by Academic Press, Inc.
All rights of reproduction in any form reserved.

arrangements have been detected: unfused in *Serratia marcescens*, fused to anthranilate-5′-phosphoribosyl-1-pyrophosphate phosphoribosyltransferase in *E. coli, Salmonella typhimurium*, and *Klebsiella aerogenes*, fused to indole-3-glycerol-P synthase and phosphoribosylanthranilate isomerase in *N. crassa*, fused to indole-3-glycerol-P synthase in *Saccharomyces cerevisiae*, fused to AS I in *Euglena*. Anthranilate synthase from *S. marcescens* is the easiest of these enzymes to purify and is described here along with a procedure for the AS I subunit from the same organism. Methods for purification of bifunctional anthranilate synthase-anthranilate-5′-phosphoribosyl-1-pyrophosphate phosphoribosyltransferase from *K. aerogenes*, partially degraded anthranilate synthase from *S. typhimurium* in which anthranilate-5′-phosphoribosyl-1-pyrophosphate phosphoribosyltransferase is removed by proteolysis and the trifunctional enzyme from *Neurospora* have been described in an earlier volume of this series.[4]

Assay Method

The glutamine- and NH_3-dependent reactions are shown below.

$$\text{Chorismate} + \text{glutamine} \xrightarrow{Mg^{2+}} \text{anthranilate} + \text{pyruvate} + \text{glutamate}$$

$$\text{Chorismate} + NH_3 \xrightarrow{Mg^{2+}} \text{anthranilate} + \text{pyruvate}$$

The formation of anthranilate is measured fluorimetrically with excitation at 325 nm and emission at 400 nm. The assay procedure for the glutamine-dependent activity is described in a previous volume.[4] The procedure for assay of NH_3-dependent anthranilate synthase is given here.

Reagents

Triethanolamine–HCl, pH 8.9, 0.5 *M*
$MgCl_2$, 0.1 *M*
$(NH_4)_2SO_4$ pH 8.0, 0.5 *M*
Chorismic acid, 10 m*M* (solution in H_2O is stored at −20°)
Anthranilic acid, 1 m*M* (in 95% ethanol, stored at −20° in a brown bottle)

Procedure. The glutamine-dependent activity is routinely used to assay the native enzyme. Assay of NH_3-dependent activity is used with AS I, but can also be used with the native enzyme. For NH_3-dependent assay, pipet into a fluorimeter cuvet 0.68 ml H_2O, 0.1 ml of triethanolamine buffer, $MgCl_2$, and $(NH_4)_2SO_4$ and 10 μl of chorismic acid. The reaction is

[4] A. F. Egan and F. Gibson, this series, Vol. 17 [47]; F. H. Gaertner and J. A. DeMoss, *ibid*. [48]; H. Tamir and P. R. Srinivasan, *ibid*. [48A].

initiated with 10 μl of enzyme solution and the initial linear increase of anthranilate formation at 22° is recorded. The fluorescence of 10 μl of anthranilic acid (10 nmol) in the 1.0 ml reaction mixture is used to determine the relationship between fluorescence and concentration. For assay of the glutamine-dependent activity, the buffer is 50 m*M* potassium phosphate pH 7.5 and the $(NH_4)_2SO_4$ is replaced by 5 m*M* glutamine. A unit of activity is the amount of enzyme giving 1 nmol anthranilate per min. Specific activity is units per mg of protein. Protein is determined by the method of Lowry *et al.*[5] For the purified enzyme 1.0 mg protein, Lowry determination, equals 0.87 mg dry wt.

Purification Procedure for Anthranilate Synthase

The procedure of Zalkin and Hwang[6] was modified by J. Yun Tso. *S. marcescens* strain HY150 (ATCC 27143), a tryptophan auxotroph lacking indole-3-glycerol-P synthase, is grown in a fermentor in media containing salts,[7] 4 g/liter glucose, 1 g/liter acid casein hydrolysate, 4.5 mg/liter tryptophan. Growth at 34° with vigorous aeration is for 20–22 hr. Cells are harvested in a Sharples centrifuge and stored at −20° until used. Yield is approximately 3–3.5 g cell paste per liter.

Step 1. Preparation of Extract. All steps are performed in a cold room at 6° or at ice bath temperature. Suspend approximately 700 g cells in 2 liters of buffer A (0.1 *M* potassium phosphate pH 7.4/0.1 m*M* EDTA/0.2 m*M* dithiothreitol), add a few flakes of DNase, and pass two times through a French press at 10,000 psi. The broken cell suspension is centrifuged at 23,000 *g* for 1 hr to obtain the crude extract.

Step 2. Protamine Sulfate. Protamine sulfate (20 mg/ml in buffer A) is added dropwise to a final concentration of 0.16 mg protamine sulfate per mg of protein in the crude extract. Stir for 15 min after the last addition and then centrifuge at 23,000 *g* for 30 m in.

Step 3. Ammonium Sulfate. To the supernatant from Step 2, 230 mg/ml $(NH_4)_2SO_4$ is added. Stirring is continued for 20 min after the last addition and the suspensions are then centrifuged at 23,000 *g* for 45 min. The pellets are dissolved in approximately 300 ml of buffer B (0.05 *M* potassium phosphate/0.1 m*M* EDTA/0.2 m*M* dithiothreitol) and dialyzed overnight against 4 liters of buffer B. The dialyzed enzyme is centrifuged at 23,000 *g* to remove insoluble protein and diluted to 10 mg/ml with buffer B.

[5] E. Layne, this series, Vol. 3, p. 447.

[6] H. Zalkin and L-H. Hwang, *J. Biol. Chem.* **246,** 6899 (1971).

[7] H. J. Vogel and D. M. Bonner, *J. Biol. Chem.* **218,** 97 (1956).

TABLE I
SUMMARY OF ANTHRANILATE SYNTHASE PURIFICATION

Fraction	Volume (ml)	Activity[a] (units) ($\times 10^6$)	Protein (mg)	Specific activity (units/mg)	Recovery (%)
Extract	1840	3.5	36,800	95	100
Protamine sulfate	2050	3.8	26,240	145	100
Ammonium sulfate	330	3.6	8,410	428	100
DE-52	23	2.4	1,317	1820	69
Gel filtration	12.8	1.9	853	2230	54

[a] Glutamine-dependent activity was assayed.

Step 4. DEAE-Cellulose Chromatography. The dialyzed enzyme is applied to a 5.5 × 85 cm column of DE-52 cellulose equilibrated with buffer B. The column is washed with 2 liters of buffer B prior to elution with a 4 liter 0–0.5 *M* KCl gradient in buffer B. Fractions containing greater than 20% of the activity in the peak tube are pooled and the enzyme is precipitated with $(NH_4)_2SO_4$ (277 mg/ml). The precipitate is dissolved in less than 20 ml of buffer B and then dialyzed overnight against this buffer. Denatured proteins are removed by centrifugation.

Step 5. Gel Filtration. Enzyme from Step 4 is applied to the first of 3 gel filtration columns (2.5 × 90 cm) in series. The first column is packed with BioGel A-1.5 m (200–400 mesh) and the other two with Sephacryl S-200. Fractions containing the major peak of protein are pooled and the enzyme is precipitated and dialyzed as in Step 4. The enzyme is exceptionally stable when stored at −20° in buffer B. The purification procedure is summarized in Table I.

Purification Procedure for AS I[8]

Serratia marcescens AS I is purified from *E. coli* strain JA221 (Δ*trpE5 thr leu thi recA*) carrying plasmid pGM14.[9] This plasmid contains *S. marcescens trpE* on a *Sal*I fragment inserted into the *Sal*I site of pBR322. Bacteria are grown in a fermentor at 30° in media containing salts,[7] 4 g/liter glucose, 1 g/liter acid casein hydrolysate, 20 mg/liter threonine, 20 mg/ml leucine, 4 mg/liter thiamin, and 20 mg/liter 3-indoleacrylic acid. Growth is for approximately 16 hr to give 3.5–4.5 g cell paste per liter corresponding to turbidity readings of 180–225 with a Klett colorimeter (66 filter). Cells are harvested in a Sharples centrifuge and stored at −20°.

[8] H. Zalkin, unpublished.

[9] G. F. Miozzari and C. Yanofsky, *Nature (London)* **277**, 486 (1979).

Step 1. Preparation of Extract. Approximately 350 g of cells are suspended in 1.3 liter of buffer B (0.05 *M* potassium phosphate pH 7.5/0.1 m*M* EDTA/0.2 m*M* dithiothreitol) and are disrupted in a French press at 10,000 psi. The suspension is centrifuged at 23,000 *g* for 1 hr and the supernatant is diluted to 15 mg protein per ml.

Step 2. Streptomycin Sulfate. To the extract is added 0.15 vol of 20% streptomycin sulfate in buffer B. If necessary pH 7.4 is maintained by addition of NH_4OH. The solution is stirred for 15 min after the last addition of streptomycin sulfate and is then centrifuged at 23,000 *g* for 30 min.

Step 3. Ammonium Sulfate. AS I is precipitated by addition of 250 g/liter ammonium sulfate. Ammonium hydroxide (5 *N*) is added as needed to maintain pH 7.5. The suspension is gently stirred 1 hr after the last addition of ammonium sulfate and is then centrifuged 23,000 *g* for 1 hr. The pellet is dissolved in 100 ml of buffer B and is dialyzed overnight against 5 liters of buffer C (buffer B containing 5% glycerol).

Step 4. DEAE-Sepharose Chromatography. The dialyzed fraction from Step 3 is diluted to 10 mg protein per ml with buffer C and is applied to a 2.5 × 95 cm column of DEAE-Sepharose CL6B equilibrated with the same buffer. The column is washed until the A_{280} of the effluent is less than 0.200. Enzyme is eluted using a 2 liter linear gradient of 0–0.5 *M* KCl in buffer C. Fractions containing the bulk of activity are pooled and the enzyme is concentrated to less than 20 ml by precipitation with 313 mg/ml ammonium sulfate. The concentrated fraction is dialyzed overnight against 500 ml of buffer B and any insoluble material is removed by centrifugation.

Step 5. Gel Filtration. Gel filtration is on 3 tandem columns of Sephacryl S-200 (2.5 × 95 cm) equilibrated and eluted with buffer B. Fractions with maximal activity are pooled and enzyme is concentrated by precipitation with ammonium sulfate as in Step 4. The enzyme is stable at −20°. The purification is summarized in Table II.

TABLE II
SUMMARY OF AS I PURIFICATION

Fraction	Volume (ml)	Activity (units)	Protein (mg)	Specific activity (units/mg)	Recovery (%)
Extract	1470	864,000	63,000	13.7	100
Streptomycin sulfate	3780	754,000	60,000	12.6	87
Ammonium sulfate	620	559,000	8,870	63.0	65
DEAE-Sepharose	32	452,000	2,240	202	52
Gel filtration	12.1	167,000	294	568	19

Comments on Purification Procedures for Anthranilate Synthase and AS I

Native anthranilate synthase and AS I, purified by these procedures, are homogeneous by the criterion of sodium dodecyl sulfate–polyacrylamide gel electrophoresis. The purification of anthranilate synthase could likely be improved by using a cloned *Serratia trpEG* fragment to increase the enzyme content of cells. It is also possible that higher initial AS I levels could be obtained from pGM14 in another *E. coli* host that could be more efficiently starved for tryptophan. It should be possible to increase the AS I yield by improving the gel filtration step. The present procedures yield 120 mg anthranilate synthase and 84 mg AS I per 100 g of cell paste.

Properties

Quaternary Structure. Serratia anthranilate synthase is a tetramer containing two ASI subunits (M_r 60,000)[6] and two AS II subunits (M_r 20,956).[10]

Structure–Function Relationship. AS I catalyzes NH_3-dependent synthesis of anthranilate. Glutamine is not a substrate for AS I. AS II provides glutamine amide transfer function. Affinity labeling with DON selectively inactivates glutamine-dependent activity and alkylates an AS II cysteine residue.[6,10] AS II is a protein chain of 192 residues and Cys-83 is the active site residue.[10] It is important to emphasize that *E. coli* AS I which is similar to the *Serratia* AS I subunit functions *in vivo* for the NH_3-dependent synthesis of tryptophan.[11] Formation of the covalent glutaminyl intermediate with AS II Cys-83 requires prior binding of chorismate to AS I.[6] It is interesting to note that the ratio of the glutamine- and NH_3-dependent activities is unity in the native enzyme,[6] yet the specific activity of the tetramer is 4-fold higher than NH_3-dependent AS I (Tables I and II). Thus, AS II activates the AS I NH_3-dependent reaction.

Catalytic Properties.[6] The pH optima for glutamine- and NH_3-dependent reactions are 7.6 and 8.5, respectively. K_m values of 0.5 mM (glutamine), 5.9 mM ($(NH_4)_2SO_4$), and 4 μM (chorismate) were reported. The enzyme is subject to allosteric inhibition by tryptophan.[12] Glutaminase activity is stimulated 7-fold by chorismate and is inhibited by tryptophan.

[10] J. Y. Tso, M. A. Hermodson, and H. Zalkin, *J. Biol. Chem.* **255,** 1451 (1980).
[11] H. Zalkin and T. Murphy, *Biochem. Biophys. Res. Commun.* **67,** 1370 (1975).
[12] H. Zalkin and S. H. Chen, *J. Biol. Chem.* **24,** 5996 (1972).

[38] *p*-Aminobenzoate Synthase

By HOWARD ZALKIN

p-Aminobenzoate synthase catalyzes the first specific reaction in the folic acid biosynthetic pathway, the formation of *p*-aminobenzoate (PABA) from chorismate and glutamine. PABA is also a likely intermediate in the biosynthesis of candicidin by *Streptomyces griseus*.[1] Crude enzyme preparations have been obtained from *E. coli*,[2] *B. subtilis*,[3] and *N. crassa*.[4] The enzyme is an oligomer containing dissimilar subunits. The oligomeric enzyme has not been purified due to the very low levels in cell extracts and problems with stability. The small subunit, designated G, has been purified from *B. subtilis*.[5] The two genes for *E. coli* PABA synthase *pabA* and *pabB* have recently been cloned and overproducing *E. coli* strains are anticipated. In this chapter, the assay and some properties of PABA synthase from *E. coli* and *B. subtilis* are described along with purification of *B. subtilis* subunit G. *B. subtilis* PABA synthase subunit G also functions with AS I in the synthesis of anthranilate.

Assay Method

PABA synthase catalyzes the following reactions.

$$\text{Chorismate} + \text{glutamine} \xrightarrow[\text{guanosine}]{} \text{PABA} + \text{pyruvate} + \text{glutamate}$$

$$\text{Chorismate} + NH_3 \xrightarrow[\text{guanosine}]{} \text{PABA} + \text{pyruvate}$$

Guanosine was identified[3] as an activator of the *B. subtilis* enzyme and may be required for PABA synthase from other organisms as well. Formation of PABA is determined by an end point fluorimetric assay.[3]

Reagents

Chorismate, 0.01 *M*
Glutamine, 0.2 *M*
$MgCl_2$, 0.2 *M*

[1] J. A. Gil, P. Liras, G. Naharro, J. R. Villanueva, and J. F. Martin, *J. Gen. Microbiol.* **118,** 189 (1980).
[2] M. Huang and F. Gibson, *J. Bacteriol.* **102,** 767 (1970).
[3] J. F. Kane and H. D. O'Brien, *J. Bacteriol.* **123,** 1131 (1975).
[4] S. Hendler and P. R. Srinivasan, *Biochim. Biophys. Acta* **141,** 656 (1967).
[5] W. M. Holmes and J. F. Kane, *J. Biol. Chem.* **250,** 4462 (1975). In this report AS I and subunit G are designated subunits E and X, respectively.

Copyright © 1985 by Academic Press, Inc.
All rights of reproduction in any form reserved.

Ammonium chloride, 1.0 *M*
Guanosine, 0.2 *M*
N-2-Hydroxyethylpiperazine-*N'*-2'-ethanesulfonic acid (Hepes) pH 7.8, and pH 8.6, 0.6 *M*
EDTA, 1.0 *M*/2-mercaptoethanol, 60 m*M*/glycerol, 30%
HCl, 1 *N*
Ethyl acetate

Procedure. For glutamine-dependent activity, 0.5 ml reaction mixtures contain 50 μl Hepes pH 7.8, 100 μl glutamine, 20 μl chorismate, 50 μl $MgCl_2$, 50 μl guanosine, 50 μl EDTA/2-mercaptoethanol/glycerol, 30 μl H_2O, and 150 μl enzyme (4–6 mg extract protein). The NH_3-dependent assay is modified by replacing the pH 7.8 buffer with pH 8.6 Hepes buffer and replacing glutamine with 100 μl of NH_4Cl. Incubation is at 37° for 30 min. Reactions are terminated by the addition of 0.1 ml of HCl. Ethyl acetate, 1.5 ml, is added and the samples are mixed vigorously for 20 sec. The layers are separated by low-speed centrifugation and the fluorescence of the ethyl acetate layer is determined at an excitation wavelength of 290 nm and an emission wavelength of 340 nm (both uncorrected). The fluorescence is compared to a standard curve prepared by extraction of authentic PABA from an acid-precipitated reaction mixture. Reactions terminated at zero time serve as controls.

Purification Procedure for B. subtilis Subunit G.[5] In this procedure PABA synthase subunit G is isolated in association with anthranilate synthase Component I (AS I) and is then further purified as isolated subunit G. For this purification, glutamine-dependent anthranilate synthase is used to assay subunit G in the presence of excess AS I. A unit of activity is the amount of enzyme which catalyzes the formation of 1 μmol of anthranilate per min at 37°. *B. subtilis* mutant NP100, genetically derepressed for the tryptophan biosynthetic enzymes, is grown in a fermentor at 37° in Spizizen minimal media supplemented with 50 mg/liter of phenylalanine. The minimal media contains, in percent: $(NH_4)_2SO_4$ 0.2, K_2HPO_4 1.4, KH_2PO_4 0.6, $MgSO_4 \cdot 7H_2O$ 0.02, sodium citrate $\cdot$ $2H_2O$ 0.1, glucose 0.5. Cells are grown to late log phase and harvested in a Sharples centrifuge. Cells are frozen at −20°.

Step 1. Preparation of Extract. Cell paste (100 g) is suspended in 500 ml of buffer A (0.04 *M* potassium phosphate pH 7.75/20 m*M* glutamine/1 m*M* histidine/6 m*M* 2-mercaptoethanol/0.1 m*M* EDTA), 50 mg lysozyme added, and the mixture incubated 1 hr at 37°. Subsequently, 50 ml of a solution of 1.0 *M* $MgCl_2$ containing 700 μg of DNase (Sigma, type II) is added and incubation continued for 15 min at 37° to reduce the viscosity. All subsequent steps are conducted at 4° unless indicated otherwise. The extract is clarified by centrifugation for 1 hr at 19,000 *g*. The supernatant is decanted and recentrifuged as before. The supernatants are saved and the two pellets are resuspended in 300 ml of buffer C (0.04 *M* potassium phosphate pH 7.75/0.1 m*M* EDTA/6 m*M* 2-mercaptoethanol/30% glycerol). The resuspended material is centrifuged for 2 hr at 19,000 *g*. The supernatants are pooled and the volume adjusted to 800 ml with buffer A.

Step 2. Ammonium Sulfate Precipitation. Solid ammonium sulfate is added to 0.4 saturation (24.3 g/100 ml). The suspension is stirred for 1 hr and then centrifuged at 13,000 *g* for 30 min. The supernatant is decanted and brought to 0.8 saturation by the further addition of 28.5 g ammonium sulfate per 100 ml. After stirring for 1 hr the suspension is centrifuged as before and the precipitate is dissolved in buffer B containing 0.04 *M* potassium phosphate pH 7.75/20 m*M* glutamine/1 m*M* histidine/6 m*M* 2-mercaptoethanol/0.1 m*M* EDTA, 10 m*M* $MgCl_2$/30% glycerol. The volume is reduced by ultrafiltration with an Amicon XM-50 membrane to approximately 100 ml.

Step 3. Gel Filtration. The enzyme solution from Step 2 is applied to a 5 × 130 cm column of Sephadex G-100 equilibrated with buffer B. Elution with buffer B is at a flow rate of 22 ml/hr. Fractions containing glutamine-dependent anthranilate synthase (ASI · subunit G complex) are pooled and concentrated to 50 ml as described above. The sample is applied to a 5 × 130 cm column of Sephadex G-100 equilibrated with buffer C. Under these conditions the ASI · subunit G complex dissociates into free subunits. AS I subunit is detected by assaying for NH_3-dependent anthranilate synthase (chapter [37]) and free subunit G is detected by assaying for glutamine-dependent anthranilate synthase in the presence of excess AS I subunit. The pooled AS I fraction is concentrated to approximately 50 ml by ultrafiltration and stored at −20° for use in assays of subunit G.

Step 4. DEAE-Sephadex Chromatography. The pooled fraction of subunit G obtained by gel filtration in buffer C is chromatographed on a 2.5 × 30 cm column of DEAE-Sephadex equilibrated with buffer C. Subunit G is eluted with a linear 140 ml gradient of 0.1 *M* to 0.4 *M* KCl in buffer C.

Step 5. Preparative Polyacrylamide Gel Electrophoresis. The pooled fraction from Step 4 is concentrated to approximately 10 ml by ultrafiltration and then is equilibrated with Tris buffer by gel filtration through a 2 × 25 cm column of Sephadex G-25 equilibrated with 0.2 *M* Tris-acetate pH 8.6/5 m*M* 2-mercaptoethanol/30% glycerol. A solution containing 7.5% acrylamide (w/v), 0.25% bisacrylamide, and 30% glycerol is prepared in 0.1 *M* Tris–acetate pH 8.6. To 70 ml of this mixture are added 25 mg of ammonium persulfate and 0.15 ml of *N,N,N′,N′*-tetramethylenediamine. The solution is poured into the upper barrel assembly of a Buchler model 100 preparative electrophoresis apparatus and allowed to polymerize. The

SUMMARY OF PABA SYNTHASE SUBUNIT G PURIFICATION

Fraction	Volume (ml)	Protein (mg)	Activity units	Specific activity (units/mg)	Recovery (%)
Extract	800	13,900	66	0.005	100
Ammonium sulfate	100	7,600	110	0.015	166
Gel filtration (ASI/subunit G complex)	370	2,800	92	0.033	139
Gel filtration (subunit G)	392	81	60	0.740	89
DEAE-Sephadex	28	26	54	2.0	80
Electrophoresis	5	5	15	3.0	21

lower electrode chamber is filled with 0.25 *M* Tris–acetate pH 8.6/5 m*M* 2-mercaptoethanol/30% glycerol. The elution chamber is filled with 0.07 *M* Tris–acetate pH 8.6/5 m*M* 1-mercaptoethanol. The temperature of the water jacket is maintained at 4°. Buffers are pumped through the chambers at a rate of 1.0 ml/min. A regulated potential of 250 V is applied to the gel and maintained for 4 hr. The upper electrode reservoir is filled with fresh buffer and the sample containing subunit G and bromophenol is layered over the gel and a potential of 250 V applied. Fractions of 7.5 ml are collected and assayed for subunit G. Subunit G is stored in buffer C at −20°. A summary of the purification procedure is in the table.

Properties of *B. subtilis* Subunit G

PABA synthase subunit G from *B. subtilis* is a protein chain of M_r approximately 19,000. Subunit G interacts with PABA synthase subunit A (M_r approximately 31,000) to form glutamine-dependent PABA synthase or with AS I (M_r approximately 67,000) to form glutamine-dependent anthranilate synthase. AS I and subunit A each exhibit an NH_3-dependent activity. Thus subunit G provides glutamine amide transfer function to the subunits catalyzing NH_3-dependent synthesis of anthranilate and PABA. Subunit G contains an essential cysteine residue which is alkylated by DON. By analogy with anthranilate synthase, glutamine presumably forms a covalent intermediate with the subunit G active site cysteine. Glutamine also promotes association of subunit G with AS I[5] and presumably also with subunit A.[3]

Guanosine is essential for PABA synthase activity following partial purification.[3] GMP, GDP, and GTP can replace guanosine but are somewhat less effective. The subunit A subunit G complex is most stable at pH 6.8 and exhibits maximal activity in the presence of 30% glycerol/6 m*M* 2-mercaptoethanol/0.1 m*M* EDTA.[3]

E. coli PABA Synthase

E. coli PABA synthase contains dissimilar subunits which are the product of genes *pabA* and *pabB*.[2] Subunits A and B, isolated from crude extracts, were reported to have molecular weights of 9,000 and 48,000, respectively. The calculated M_r of cloned *pabA* subunit having 187 amino acids is however 20,752.[6] The deduced *pabA* amino acid sequence exhibits extensive homology to AS II from enteric bacteria.[6] Structural similarity between *E. coli pabA* subunit and *B. subtilis* subunit G is likely. *pabB* was recently cloned and sequenced.[7] The derived amino acid sequence of 453 residues (M_r 50,958) is homologous to As I.

[6] J. B. Kaplan and B. P. Nichols, *J. Mol. Biol.* **168,** 451 (1983).
[7] P. Goncharoff and P. B. Nichols, *J. Bacteriol.* **159,** 57 (1984).

[39] NAD Synthetase

By HOWARD ZALKIN

In 1958, Preiss and Handler described the biosynthesis of NAD from nicotinic acid in extracts of red blood cells, yeast, and liver.[1] The final reaction in the pathway is the conversion of nicotinate adenine dinucleotide (N^aAD) to NAD. NAD synthetase has been highly purified from *E. coli*[2] and *S. cerevisiae*.[3] Ammonia was reported to be far superior to glutamine as nitrogen donor for *E. coli* NAD synthetase in contrast to nearly equal reactivity of these substrates with the yeast enzyme. Procedures for the purification of NAD synthetase from *E. coli* and yeast are herein described.

Assay Method for NAD Synthetase

The glutamine- and NH_3-dependent reactions for synthesis of NAD are shown.

$$N^aAD + \text{glutamine} + ATP \xrightarrow{Mg^{2+}} NAD + \text{glutamate} + AMP + PP_i$$

$$N^aAD + NH_3 + ATP \xrightarrow{Mg^{2+}} NAD + AMP + PP_i$$

[1] J. Preiss and P. Handler, *J. Biol. Chem.* **233,** 493 (1958).
[2] R. L. Spencer and J. Preiss, *J. Biol. Chem.* **242,** 385 (1967).
[3] C. K. Yu and L. S. Dietrich, *J. Biol. Chem.* **247,** 4794 (1972).

Copyright © 1985 by Academic Press, Inc.
All rights of reproduction in any form reserved.

The NAD synthesized is measured enzymatically with alcohol dehydrogenase or fluorimetrically. The more sensitive fluorimetric method is described here.

Reagents

Tris–HCl pH 8.5, 1.0 *M*
ATP, 0.1 *M*
KCl, 1.0 *M*
$MgCl_2$, 0.5 *M*
Glutamine, 0.2 *M*
NH_4Cl_2, 0.1 *M*
N^aAD,[4] 0.05 *M*
Bovine serum albumin, 10 mg/ml
NaOH, 7 *N*

Procedure. The NH_3-dependent reaction is used to assay *E. coli* NAD synthetase.[2] The 0.5 ml reaction mixture contains 10 μl ATP, 10 μl KCl, 20 μl $MgCl_2$, 30 μl Tris–HCl buffer, 10 μl NH_4Cl, 10 μl N^aAD, 10 μl bovine serum albumin, enzyme solution, and H_2O to 0.5 ml. Yeast NAD synthetase is most conveniently assayed using the glutamine-dependent activity.[3] The 0.5 ml reaction mixture contains 10 μl ATP, 28 μl KCl, 5 μl $MgCl_2$, 25 μl Tris–HCl, 50 μl glutamine, 10 μl N^aAD, 10 μl bovine serum albumin, enzyme solution, and H_2O to 0.5 ml. In each case incubation is for 30 min at 37°. Reactions are stopped by heating in a boiling water bath for 1 min.

To determine the NAD synthesized, a 20 to 50 μl aliquot is withdrawn from the reaction mixture and added to 0.2 ml of 7 *N* NaOH and immediately mixed. The fluorescent product is developed by heating for 10 min at 100° or by incubating at 37° for 30 min. After incubation, 1.8 ml of H_2O is added and the fluorescence measured using 398 nm for activation and 460 nm for emission. The fluorescence is compared to that obtained using standard NAD solutions to calculate the amount of NAD. A unit of enzyme activity is defined as that amount synthesizing 1 μmol of NAD per min at 37°.

Purification Procedure for *E. coli* NAD Synthetase[2]

E. coli B cells are grown in a fermentor in media containing 1.1% K_2HPO_4, 0.85% KH_2PO_4, 0.6% yeast extract, and 1% glucose. In early stationary phase the cells are harvested in a Sharples centrifuge and

[4] N^aAD is synthesized enzymatically by the exchange of nicotinate with NAD catalyzed by beef spleen microsomal NADase (EC 3.2, 2.5) [T. Honjo, M. Ikeda, A. J. Andreoli, Y. Nishizuka, and O. Hayaishi, *Biochim. Biophys. Acta* **89,** 549 (1964)]. Modifications are described in refs. 2 and 3.

stored at $-20°$ prior to use. All subsequent procedures are carried out at 0–4°.

Step 1. Preparation of Extract. Frozen cells, 200 g, are thawed and suspended in 800 ml of 0.05 *M* Tris–HCl pH 7.5/0.01 *M* GSH/0.01 *M* EDTA. The cells are broken by sonic oscillation in a Brownwell Biosonik at 20 kc for 8 min in 50 ml batches. The resulting suspension is centrifuged at 32,000 *g* for 15 min. NAD synthetase activity is in the supernatant fluid.

Step 2. Protamine Sulfate Fractionation. Protamine sulfate solution, 1%, is added slowly with stirring in the ratio 50 ml of protamine sulfate solution per 100 ml of supernatant fluid. After 10 min the suspension is centrifuged at 12,000 *g* for 15 min and the precipitate is discarded.

Step 3. Ammonium Sulfate Fractionation 1. To 1.2 liter of supernatant solution from Step 2, 290 g of $(NH_4)_2SO_4$ is added slowly with stirring. Stirring is continued for 10 min and the suspension is then centrifuged at 12,000 *g* for 10 min. The precipitate is discarded and 178 g of $(NH_4)_2SO_4$ is slowly added with stirring. After stirring for an additional 10 min the suspension is centrifuged for 15 min at 12,000 *g*. The precipitate is dissolved in 0.05 *M* potassium phosphate pH 7.2/0.001 *M* GSH to yield a solution of 62 ml.

Step 4. Ammonium Sulfate Fractionation 2. The pH of the supernatant from Step 3 is carefully adjusted to 5.0 with 2 *N* HCl followed by the slow addition of 16 ml of saturated $(NH_4)_2SO_4$ solution. After stirring for 10 min the precipitate is collected by centrifugation, as above, and dissolved in 18 ml of buffer solution A containing 0.01 *M* potassium phosphate pH 7.2/1 m*M* EDTA. The solution is dialyzed for 4 hr against 2 liters of the same buffer.

Step 5. DEAE-Cellulose Fractionation 1. The enzyme from Step 4 is applied to a 2 × 15 cm column of DEAE-cellulose equilibrated with buffer solution B containing 0.01 *M* potassium phosphate pH 7.6/1 m*M* GSH/1 m*M* EDTA. The column is washed with 90 ml of the same buffer and the enzyme is then eluted with a salt and pH gradient established with 450 ml of buffer B in the mixing flask and buffer containing 0.1 *M* potassium phosphate pH 6.8/1 m*M* GSH/1 m*M* EDTA in the reservoir. After 320 ml of eluting buffer is collected the enzyme is obtained in the next 175 ml. The enzyme is precipitated with 98 g of $(NH_4)_2SO_4$, dissolved in 3 ml of 0.05 *M* potassium phosphate pH 7.2/1 m*M* GSH/1 m*M* EDTA and dialyzed for 4 hr against 1 liter of buffer A.

Step 6. DEAE Cellulose Fractionation 2. The enzyme solution from Step 5 is applied to a 1 × 5 cm column of DEAE cellulose equilibrated with buffer B. The column is washed with 20 ml of buffer B and the enzyme then eluted as in Step 5 using 50 ml each in the mixing chamber and reservoir. The enzyme is collected in approximately 50 ml and is precipitated, dissolved, and dialyzed as in Step 5. Protein is determined

TABLE I
PURIFICATION OF *E. coli* NAD SYNTHETASE

Fraction	Activity (units)	Specific activity (units/mg)	Recovery (%)
Extract	280	0.015	100
Protamine sulfate	261	0.037	93
Ammonium sulfate I	281	0.21	100
Ammonium sulfate II	164	0.56	58
DEAE-cellulose I	74	5.9	26
DEAE-cellulose II	29	6.9	10

by absorption at 215 and 225 nm. The purification is summarized in Table I.

Purification Procedure for Yeast NAD Synthetase[3]

All purification steps are carried out at 4°. All of the buffers contain 1 m*M* DTT and 1 m*M* EDTA.

Step 1. Preparation of Extract. Fleischmann bakers' yeast cake (2.7 kg) purchased locally is washed twice with 0.9% NaCl solution. The yeast cells are suspended in 50 m*M* Tris–HCl pH 7.4 (1 g wet wt/ml of buffer) and are disrupted in a French press at 16,000 psi yielding 4960 ml of disrupted cell suspension. The suspension is centrifuged at 14,000 *g* for 40 min and the pellet washed once with the same buffer. The resulting turbid supernatant material is centrifuged at 40,000 *g* for 1 hr in a Spinco centrifuge. NAD synthetase is recovered in the slightly turbid supernatant solution.

Step 2. Ammonium Sulfate Fractionation. Ammonium sulfate (28.2 g/100 ml) is added slowly with stirring to the extract. The suspension is stirred for 30 min after the last addition and then kept at 4° overnight. The precipitate is collected by centrifugation at 27,000 *g* for 30 min and dissolved in 0.05 *M* Tris–HCl pH 7.4. Any insoluble material is removed by centrifugation. Most of the NAD synthetase is recovered in this 0–40% ammonium sulfate fraction. The preparation is dialyzed overnight against 40 vol of 0.04 *M* Tris–HCl pH 7.2/0.04 *M* KCl with one buffer change.

Step 3. DEAE-Cellulose Chromatography. The fraction from Step 2 is applied to a 2.5 × 56 cm column of DEAE cellulose equilibrated with 40 m*M* Tris–HCl pH 7.2/40 m*M* KCl. Elution is with a linear gradient of KCl (0.04–0.26 *M*) in the same buffer. Fractions containing maximal activity are pooled and concentrated to 20 ml by pressure filtration using an Amicon PM-10 or PM-30 membrane. The concentrated enzyme solution is dialyzed against 100 vol of 25 m*M* potassium phosphate pH 7.5 for 3–4 hr and insoluble material removed by centrifugation.

Step 4. Hydroxylapatite Chromatography. Enzyme solution from Step 3 is applied to a 1.6 × 15 cm column of hydroxylapatite equilibrated with 0.1 *M* potassium phosphate pH 7.4. The column is washed with 200 ml of the same buffer and the enzyme then eluted with a linear gradient of potassium phosphate pH 7.4 (0.025–0.2 *M*). The active fractions are pooled and concentrated to 4 ml by pressure filtration using an Amicon PM-10 membrane. The concentrated material is dialyzed 4 hr against 250 vol of 50 m*M* Tris–HCl pH 7.6 containing 100 or 350 m*M* KCl. Insoluble material is removed by centrifugation.

Step 5. Gel Filtration. The enzyme solution is applied to a Sepharose 4B column equilibrated with 50 m*M* Tris–HCl pH 7.6 and 100 or 350 m*M* KCl. Elution is with the same buffer. Fractions containing a single protein band coincident with NAD synthetase activity are pooled and concentrated to 1–2 ml by pressure filtration using an Amicon PM-10 membrane. The purification procedure is summarized in Table II.

Comments on Purification Procedures

The homogeneous yeast enzyme is stored at 4° and loses approximately 60% activity in 10 days. Addition of bovine serum albumin at a concentration of greater than 10 mg/ml greatly decreases inactivation. Inactivation is not prevented by storage at −70°, −20°, lyophilization, changes in ionic strength, pH, DTT, 30% glycerol, 70% ammonium sulfate. Partially inactivated enzyme can be fully reactivated by the addition of 1–2% mercaptoethanol. Once activity is totally lost mercaptoethanol has no effect.

TABLE II
PURIFICATION OF YEAST NAD SYNTHETASE

Fraction	Volume (ml)	Protein (mg)	Specific activity (units/mg)	Recovery (%)
Extract	1236	7220	0.001	100
Ammonium sulfate	100	4080	0.012	63
DEAE-cellulose	20	294	0.111	42
Hydroxylapatite	4	26.8	0.43	15
Gel filtration	1.2	1.14	2.10	3.4

The stability and conditions for storage of *E. coli* NAD synthetase purified as described here were not reported. The purity of *E. coli* NAD synthetase was not specified.[2] Yeast NAD synthetase was reportedly homogeneous by the criteria of polyacrylamide gel electrophoresis and high-speed equilibrium centrifugation.[3] Subunits of M_r approximately 80,000 and 65,000 were obtained by sodium dodecyl sulfate–polyacrylamide gel electrophoresis indicating either that the homogeneous enzyme contains dissimilar subunits or that the enzyme was partially nicked by proteolysis.

Properties

Glutamine and Ammonia Sites. The major differences between the bacterial and yeast enzymes are in the amide specificity donor and the effect of inhibitors. Glutamine- and NH_3-dependent activities are nearly equal for the yeast enzyme at optimal pH values of approximately 6.2 to 8.4 and 8.4 to 8.8, respectively. For *E. coli* NAD synthetase the K_m for NH_3 is reportedly 250-fold lower than that for glutamine. Apparent K_m values for the *E. coli* enzyme are 2.0×10^{-4} *M* (ATP), 1.1×10^{-4} *M* (N^aAD), 6.5×10^{-5} *M* (NH_3), 1.6×10^{-2} *M* (glutamine). For yeast NAD synthetase apparent K_m values are 1.7×10^{-4} *M* (ATP), 1.9×10^{-4} *M* (N^aAD), 6.4×10^{-3} *M* (NH_3), 5×10^{-3} *M* (glutamine).

Azaserine and DON inactivated glutamine- but not NH_3-dependent yeast NAD synthetase. The *E. coli* enzyme was not inactivated under similar conditions. Whether these properties indicate a fundamental difference between the two enzymes or whether the glutamine site of *E. coli* NAD synthetase was inactivated is not known. Possibly the glutamine-dependent activity of *E. coli* NAD synthetase was inactivated by Tris buffer as previously noted for *E. coli* GMP synthetase.[6] A glutaminase activity has not been described.

The adenosine analogs psicofuranine and decoyinine are potent inhibitors of *E. coli* NAD synthetase but have little effect on the yeast enzyme. these compounds are also strong inhibitors of *E. coli* GMP synthetase.

Oligomeric Structure. Yeast NAD synthetase has a M_r of approximately 630,000. Based on subunit M_r of 80,000 and 65,000 the enzyme may be an $\alpha_2\beta_2$ tetramer. If the 65,000 subunit arose from proteolytic nicking an α_4 tetramer is suggested. Information for the *E. coli* enzyme is not available.

[6] N. Patel, H. S. Moyed, and J. F. Kane, *Arch. Biochem. Biophys.* **178,** 652 (1977).

[40] Glu-tRNA^{Gln} Amidotransferase

By HOWARD ZALKIN

A survey of a number of gram-positive bacteria including *Bacillus subtilis, Bacillus megaterium, Lactobacillus acidophilus, Streptococcus faecalis,* and *Streptomyces antibioticus* did not detect Gln-tRNA synthetase activity.[1] Instead these organisms charge glutamic acid to tRNA^{Gln} to form missense Glu-tRNA^{Gln} which is subsequently amidated to Gln-tRNA^{Gln} by a specific glutamine amidotransferase. The limited procedures previously described[2,3] for the assay and partial purification of Glu-tRNA^{Gln} amidotransferase are summarized in this section.

Assay Method for Glu-tRNA^{Gln} Amidotransferase

Glu-tRNA^{Gln} amidotransferase catalyzes the reaction shown.

$$[^{14}C]glu\text{-}tRNA^{Gln} + ATP + glutamine \xrightarrow{Mg^{2+} \text{ or } Mn^{2+}} [^{14}C]Gln\text{-}tRNA^{Gln} + ADP + P_i + glutamate$$

[^{14}C]Glutamine is isolated by ion exchange chromatography after deacylation.

Reagents

Potassium cacodylate pH 6.5, 0.5 *M*
2-Mercaptoethanol, 0.06 *M*
$MnCl_2$, 3.5 m*M*
Glutamine, 2 m*M*
[^{14}C]Glu-tRNA^{Gln}, 60 A_{260} units/ml[4]
KOH, 0.02 *N*
HCl, 0.1 *N*
Dowex-1 (Cl-form)

Procedure. Complete reactions contain in a final volume of 50 μl: 5.0 μl each of potassium cacodylate, 2-mercaptoethanol, $MnCl_2$, ATP, glutamine, and [^{14}C]Glu-tRNA^{Gln}. Water and enzyme are added to bring the

[1] M. Wilcox, *Cold Spring Harbor Symp. Quant. Biol.* **34,** 521 (1969).
[2] M. Wilcox and M. Nirenberg, *Proc. Natl. Acad. Sci. U.S.A.* **61,** 229 (1968).
[3] M. Wilcox, *Eur. J. Biochem.* **11,** 405 (1969).
[4] tRNA can be isolated, acylated with [^{14}C]glutamate, and fractionated by benzoylated DEAE-cellulose chromatography as described in ref. 2.

Copyright © 1985 by Academic Press, Inc.
All rights of reproduction in any form reserved.

volume to 50 μl. Incubation is at 24° for 5 min. Reactions are stopped by addition of 0.5 ml of KOH and are incubated for 2 min at 24° to deacylate tRNA. Reactions are neutralized with 0.1 ml of HCl and are applied to 0.5 × 2 cm of columns of Dowex 1 (Cl^- form) previously equilibrated with H_2O. Each column is washed twice with 0.5 ml of H_2O. The effluent containing [^{14}C]glutamine is collected and counted for radioactivity.

Purification Procedure for *B. subtilis* Glu-tRNA^{Gln} Amidotransferase[3]

B. subtilis Marburg strain is grown in a fermentor at 37° with vigorous aeration to mid log phase in Spizizin minimal media and is harvested with a Sharples centrifuge. Cells are frozen until used. All steps in the purification are carried out at approximately 3°.

Step 1. Preparation of Extract. Cells are suspended in buffer A (0.04 *M* Tris–HCl pH 7.2/0.01 *M* mercaptoethanol/1 m*M* $MgCl_2$/0.5 m*M* $MnCl_2$) in a ratio of approximately 3 ml of buffer per g of cells. Cell disruption is in a French press at 18,000 psi. The lysate is centrifuged for 2 hr at 150,000 *g* and the pellet is discarded.

Step 2. Ammonium Sulfate Fractionation. Solid ammonium sulfate is added slowly with stirring to 60% of saturation (39 g/100 ml). The suspension is centrifuged and the pellet is dissolved in 1/5 vol of buffer A and dialyzed against the same buffer solution. A second ammonium sulfate precipitation is carried out to obtain protein precipitating between 46 and 60% of saturation. This fraction is dissolved in 1/5 vol of buffer B (0.02 *M* imidazole–HCl 6.5/6 m*M* mercaptoethanol/0.5 m*M* $MnCl_2$/10% glycerol).

Step 3. Gel Filtration. The material from Step 2 is applied to a 3 × 149 cm column of BioGel A-0.5m equilibrated with buffer B. Tubes containing activity are pooled and concentrated by dialysis against buffer B containing 15% polyethylene glycol-6000.

Step 4. DEAE-Cellulose Chromatography. KCl is added to the fraction from Step 3 to a final concentration and 0.01M and the solution is applied to a 0.9 × 7 cm column of DE-52 cellulose equilibrated with buffer B containing 0.01 *M* KCl. Protein is eluted with a linear 0.01–0.4 *M* gradient of KCl in buffer B. The amidotransferase activity elutes at approximately 0.21 *M* KCl. Tubes containing the bulk of the activity are pooled, dialyzed, and concentrated as described above and finally stored at −70°.

Comments on Purification Procedure

A purification summary was not reported. The enzyme was purified approximately 25-fold with respect to the activity in the 150,000 *g* supernatant fraction. In contrast to the procedure described above, ion-exchange chromatography of an extract from *B. megaterium* yielded two fractions of amidotransferase.

Properties

Glu-tRNA^{Gln} amidotransferase appears to replace Gln-tRNA^{Gln} synthetase in gram-positive organisms. The amidation of Glu-tRNA^{Gln} proceeds via a γ-phosphorylated intermediate:

$$\text{Glu-tRNA}^{Gln} + \text{ATP} \xrightarrow{Mn^{2+}} \text{P-}\gamma\text{-Glu-tRNA}^{Gln} + \text{ADP}$$
$$\text{P-}\gamma\text{-Glu-tRNA}^{Gln} + \text{glutamine} \rightarrow \text{Gln-tRNA}^{Gln} + \text{glutamate} + P_i$$

Conflicting statements have appeared regarding the capacity of the amidotransferase to utilize NH_3.[2,3] Of interest is the report that asparagine can replace glutamine as amide donor.[2]

[41] Carbamyl Phosphate Synthetase (Glutamine-Utilizing) from *Escherichia coli*

By DEBORAH S. KASEMAN and ALTON MEISTER

$$\text{L-Glutamine} + 2\ \text{ATP} + HCO_3^- + H_2O \xrightarrow{Mg^{2+},\ K^+} \text{carbamyl phosphate} + 2\ \text{ADP} + P_i + \text{L-glutamate}$$

This activity was discovered in *Agaricus bisporus*, the common edible mushroom.[1] Later it was found in *E. coli*[2] and the stoichiometry of the reaction was established.[3] In *E. coli*, this enzyme (carbamyl phosphate synthetase II) catalyzes the synthesis of carbamyl phosphate that is used for the formation of both arginine and of the pyrimidines. Carbamyl phosphate synthetase II, which can utilize either glutamine or ammonia as the nitrogen-donating substrate, is different from carbamyl phosphate synthetase I, which is found in the liver mitochondrial matrix of eukaryotes and functions in the urea cycle.[4] Carbamyl phosphate synthetase I is active only with ammonia and requires *N*-acetyl-L-glutamate as a cofactor.

[1] B. Levenberg, *J. Biol. Chem.* **237,** 2590 (1962).
[2] A. Pierard and J. M. Wiame, *Biochem. Biophys. Res. Commun.* **15,** 76 (1964).
[3] P. M. Anderson and A. Meister, *Biochemistry* **4,** 2803 (1965).
[4] S. Ratner, *Adv. Enzymol.* **39,** 1 (1973).

Copyright © 1985 by Academic Press, Inc.
All rights of reproduction in any form reserved.

Eukaryotic cells also contain a cytosolic, glutamine-utilizing carbamyl phosphate synthetase II that supplies carbamyl phosphate for pyrimidine biosynthesis.[5-8] This enzyme, which exists as a multifunctional protein together with aspartate transcarbamylase and dihydroorotase, is similar to the *E. coli* enzyme with regard to stoichiometry, substrate specificity, and feedback inhibition. Carbamyl phosphate synthetase III, which also uses glutamine as a nitrogen donor and which requires *N*-acetyl-L-glutamate as a cofactor, has been purified from *Squalus acanthias* (spiny dogfish) and is also present in other elasmobranchs, where it provides carbamyl phosphate for the synthesis of urea used in osmoregulation, and in freshwater teleosts.[9-11] In contrast to carbamyl phosphate synthetase II, carbamyl phosphate synthetase III is apparently not subject to allosteric regulation. Carbamyl phosphate synthetases have also been found in other prokaryotes, fungi, protozoa, angiosperms, *Drosophila*, and amphibians.[12] Human carbamyl phosphate synthetase I has been purified and characterized.[13]

The *E. coli* enzyme was first purified to apparent homogeneity nearly 20 years ago[3]; a detailed description of its isolation has been published in an earlier volume of this series.[14] The purification procedure given here is a modified and improved method.

Determination of Carbamyl Phosphate Synthetase Activity

Activity may be determined by measuring the rate of formation of ADP, glutamate, inorganic phosphate, or carbamyl phosphate.[3,14] The rate of carbamyl phosphate formation may also be determined by coupling the synthetase reaction to those catalyzed by either ornithine transcarbamylase[6] or aspartate transcarbamylase,[15] and measuring the rate of citrulline or carbamyl aspartate formation, respectively. Carbamyl aspartate and citrulline may be measured spectrophotometrically.[15-17] Alterna-

[5] M. E. Jones, *Annu. Rev. Biochem.* **49,** 253 (1980).
[6] S. E. Hager and M. E. Jones, *J. Biol. Chem.* **242,** 5667 (1967).
[7] P. F. Coleman, D. P. Suttle, and G. R. Stark, *J. Biol. Chem.* **252,** 6379 (1977).
[8] M. Mori and M. Tatibana, *J. Biochem.* (*Tokyo*) **78,** 239 (1975).
[9] C. A. Casey and P. M. Anderson, *J. Biol. Chem.* **258,** 8723 (1983).
[10] P. M. Anderson, *J. Biol. Chem.* **256,** 12228 (1981).
[11] P. M. Anderson, *Science* **208,** 291 (1980).
[12] A. J. Makoff and A. Radford, *Microbiol. Rev.* **42,** 307 (1978).
[13] V. Rubio and S. Grisolia, *Enzyme* **26,** 233 (1981).
[14] P. M. Anderson, V. P. Wellner, G. A. Rosenthal, and A. Meister, this series, Vol. 17A, p. 235.
[15] J. C. Gerhart and A. B. Pardee, *J. Biol. Chem.* **237,** 891 (1962).
[16] D. Hunninghake and S. Grisolia, *Anal. Biochem.* **16,** 200 (1966).
[17] L. M. Prescott and M. E. Jones, *Anal. Biochem.* **32,** 408 (1969).

tively, by adding [^{14}C]bicarbonate to the reaction mixture, [^{14}C]citrulline, or [^{14}C]carbamyl aspartate may be quantitated by measurement of radioactivity following ion-exchange chromatography. Carbamyl phosphate and cyanate (a decomposition product of carbamyl phosphate) can be trapped as hydroxyurea after reaction with hydroxylamine, and the hydroxyurea may be quantitated using an assay for hydroxamic acids,[18] or by a modified spectrophotometric assay specific for hydroxyurea.[19]

In the first assay system described below, the reaction mixture contains $H^{14}CO_3^-$; after incubation, labeled carbamyl phosphate is determined by converting it to labeled urea.[3] This conversion is accomplished by treating the reaction mixture with an alkaline ammonia solution to convert labeled carbamyl phosphate to labeled cyanate. Cyanate is then converted to urea by heating to 100° in the presence of ammonium chloride. Unreacted $H^{14}CO_3^-$ is removed from the mixture by ion-exchange chromatography, and [^{14}C]urea is quantitated by measurement of radioactivity.

In the second assay, the formation of ADP is measured with a coupled assay system containing pyruvate kinase and lactate dehydrogenase.

Assay by Conversion of [^{14}C]Carbamyl Phosphate to [^{14}C]Urea

Reagents

ATP, 0.2 *M*, pH 8
$MgCl_2$, 0.2 *M*
KCl, 1 *M*
Tris–HCl buffer, 0.5 *M*, pH 7.9
L-Glutamine, 0.1 *M*
Sodium [^{14}C]bicarbonate, 0.1 *M* (specific activity = 20,000 cpm/μmol)
KOH, 4 *M*
NH_4OH, 2 *M*
NH_4Cl, 4 *M*, pH 8.5

Procedure. An amount of enzyme (0.075 ml), sufficient to catalyze the synthesis of 0.08–0.4 μmol of carbamyl phosphate in 10 min] is added to 0.175 ml of an assay mixture (at 37°) containing 5 μmol of ATP, 5 μmol of $MgCl_2$, 25 μmol of KCl, 2.5 μmol of L-glutamine, 5 μmol of $Na[^{14}C]HCO_3$, and 12.5 μmol of Tris–HCl buffer (pH 7.9). After incubation at 37° for 10 min, 0.025 ml of a solution containing 2.7 *M* KOH and 0.7 *M* NH_4OH (prepared immediately before use by mixing 2 parts of 4 *M*

[18] F. Lipmann and L. C. Tuttle, *J. Biol. Chem.* **159,** 21 (1945).
[19] R. L. Levine and N. Kretchmer, *Anal. Biochem.* **42,** 324 (1971).

KOH with 1 part 2 *M* NH_4OH) is added to stop the reaction. After thorough mixing with a Vortex mixer, the solution is incubated for an additional 10 min at 37° to convert the [^{14}C]carbamyl phosphate to [^{14}C]cyanate. The [^{14}C]cyanate is then converted to [^{14}C]urea by adding 0.1 ml 4 *M* NH_4Cl (pH 8.5), and heating at 100° for 12 min. The mixture is cooled in ice and then applied to a 0.5 × 6 cm column of Dowex 1 X 8 (OH^- form, 200–400 mesh). The [^{14}C]urea is eluted by adding 4 ml of distilled water, 1.0 ml of the eluate is transferred to a liquid scintillation vial, and 10 ml of scintillation solution is added. Radioactivity is determined by liquid scintillation counting. This assay may also be used to measure ammonia-dependent carbamyl phosphate synthetase activity by substituting 50 μmol of NH_4Cl for L-glutamine in the assay mixture.

Carbamyl Phosphate Synthetase Assay: ADP Formation

This assay is based on the measurement of ADP by a coupled enzyme assay using pyruvate kinase and lactate dehydrogenase.[3]

Reagents

ATP
$MgCl_2$
$NaHCO_3$
L-Glutamine
$MgSO_4 \cdot 7H_2O$
Tris base
KCl
Phosphoenolpyruvate, monocyclohexylamine or sodium salt
β-NADH, disodium salt
Pyruvate kinase, $(NH_4)_2SO_4$ suspension
Lactate dehydrogenase, $(NH_4)_2SO_4$ suspension
HCl, 4 *M*

Procedure. The enzyme (0.075 ml; sufficient to catalyze the synthesis of 0.1–1 μmol ADP per 10 min) is added to 0.075 ml of an assay mixture solution of the following composition: 40 m*M* ATP, 20 m*M* L-glutamine, 200 m*M* Tris base, 200 m*M* KCl, 40 m*M* $MgSO_4 \cdot 7H_2O$, and 40 m*M* $NaHCO_3$. The pH of the assay mixture solution is adjusted to pH 8.2 by adding 4 *M* HCl before the addition of $MgSO_4 \cdot 7H_2O$ and $NaHCO_3$. Ammonium chloride (400 m*M*) may be substituted for L-glutamine to assay for ammonia-dependent synthetase activity and the bicarbonate-dependent ATPase activity may be assayed by omitting L-glutamine from the assay mixture. HEPES buffer may be substituted for Tris, if desired; the enzyme is not stable in Tris–HCl buffer for extended periods (i.e., longer than a few hours). The assay solution is stored frozen, and is stable for several months if not frozen and thawed repeatedly.

After incubation at 37° for 10 min, 0.05 ml of 1 *M* HCl is added to stop the reaction, and the mixture is placed in ice for 5 min. The mixture is neutralized by adding 0.05 ml of 1 *M* Tris base, and 1.0 ml of the coupling assay solution is added. The coupling solution contains 150 m*M* Tris base, 50 m*M* KCl, 20 m*M* $MgCl_2$, 1 m*M* phosphoenolpyruvate, 0.2 m*M* NADH, 1–3 U/ml pyruvate kinase, and 1–2 U/ml of lactate dehydrogenase. The pH of the coupling solution is adjusted to pH 7.6 by adding 4 *M* HCl before addition of phosphoenolpyruvate, NADH, and the enzymes. (This solution may be stored at 4° for about 1 week.) After addition of the coupling solution, the mixture is incubated for 5 min at 37°, and the absorbance at 340 nm is measured. The amount of ADP produced is calculated (using $\varepsilon = 6.22 \times 10^{-3}$ for NADH). A unit of activity is defined as one-half of the micromoles of ADP formed per hour.

Protein Determinations

Enzyme concentration may be determined by the method of Lowry *et al.*,[20] using crystalline bovine serum albumin as a standard, or by direct measurement of absorbance at 280 nm for the purified enzyme ($E_{280}^{1\%}$ = 7.0.[21] Protein assays based on the binding of Coomassie blue[22,23] are unsuitable because, for unknown reasons, carbamyl phosphate synthetase gives erroneously high values (using crystalline bovine serum albumin as a standard) compared to the Lowry method, which agrees closely with protein quantitation based on measurement of absorbance at 280 nm.

Purification of Carbamyl Phosphate Synthetase

Frozen *E. coli* B (ATCC 11303, 3/4 log phase, minimal medium) may be obtained from Grain Processing Corp, Muscatine, Iowa. Except where indicated, all steps in the purification are carried out at 4°. A summary of the purification is given in the table.

Step 1. Cell Lysis. The frozen cell paste (1 kg) is broken into small pieces, suspended in 2.5 l of 0.2 *M* potassium phosphate buffer (pH 7.7) containing 0.5 m*M* EDTA, and stirred overnight. The cell suspension is cooled to 2–3° in an ice-salt bath and the cells are lysed by mechanical disruption in a Gaulin mill, Model 15 M (Gaulin Corp., Everett, Mass.) at

[20] O. H. Lowry, N. J. Rosebough, A. L. Farr, and R. J. Randall, *J. Biol. Chem.* **193,** 265 (1951).

[21] V. P. Wellner, P. M. Anderson, and A. Meister, *Biochemistry* **12,** 2061 (1973).

[22] M. M. Bradford, *Anal. Biochem.* **72,** 248 (1976).

[23] S. Spector, *Anal. Biochem.* **86,** 140 (1978).

SUMMARY OF PURIFICATION

No.	Step	Volume (ml)	Total protein[a] (mg)	Activity[b] Total units	Specific activity (U/mg)	Recovery (%)
1.	Crude lysate supernatant[c]	3730	94,000	84,600	0.9	(100)
2.	Heat step supernatant	3460	66,600	79,900	1.2	94
3.	Protamine sulfate supernatant	4655	60,000	54,000	0.9	64
4.	Ammonium sulfate precipitation	290	17,000	74,800	4.4	88
5.	G-25 Sephadex[d]	840	15,000	ND[d]	—	—
6.	DEAE-Sephadex					
	Pool I	120	324	23,700	73	28
	Pool II	242	605	31,500	52	37
7.	Agarose A-0.5m					
	Pool IA	14.2	54.0	11,400	211	13
	Pool IB	25.6	48.6	5,700	117	6.7
	Pool II	19.5	109	17,500	160	21
8.	Ammonium sulfate fractionation[e]					
	Pool IA					
	0–50%	4.6	44.2	5,260	119	6.2
	50–65%	1.5	4.4	930	212	1.1
	Pool IB					
	0–65%	3.6	36.0	3,300	92	3.9
	Pool II					
	0–50%	3.8	21.3	1,360	64	1.6
	50–65%	2.8	73.6	13,800	188	16.3

[a] In steps 1–4, protein was determined by the Lowry method.[20] In steps 6–8, protein was determined by measurement of absorbance at 280 nm.[21]

[b] In steps 1–4, activity was determined by measurement of glutamine-dependent [^{14}C]carbamyl phosphate formation. In steps 6–8, activity was determined by measurement of glutamine-dependent ADP formation.

[c] From 1 kg frozen *E. coli* B.

[d] Desalted protein solution from the G-25 Sephadex column was not assayed for activity because the enzyme is unstable in 0.14 *M* potassium phosphate buffer at pH 6.8.

[e] Total yield of carbamyl phosphate synthetase from all ammonium sulfate fractions = 180 mg (29% of initial activity in crude lysate supernatant).

8000 psi. The cell suspension is passed through the mill twice; the mixture is cooled in an ice-salt bath to <10° between passes. The temperature of the lysate should not exceed 25° after the second pass through the mill. Potassium phosphate buffer (0.2 *M*; pH 7.7, containing 0.5 m*M* EDTA) is used to wash the cell suspension through the mill. After lysis, the mixture is centrifuged at 16,000 *g* for 30 min and the pellet is discarded.

Step 2. Heat Denaturation of Impurities. The supernatant solution obtained in Step 1 is divided into 800–900 ml portions in 1 liter Erlenmeyer flasks, 5 g of L-glutamine is added to each flask and allowed to dissolve. The flask is placed in a water bath at 100° and the solution is heated to 53° with constant mechanical stirring; this takes about 4 min. The temperature is maintained at 53° for an additional 6 min. (The flask is removed from the water bath when the temperature reaches 53° and allowed to stand at room temperature without stirring.) If the temperature drops below 52°, the flask is briefly immersed (with stirring) to reheat to 53°. After a total heating time of 10 min (timed from initial immersion of the flask), the solution is cooled rapidly in an ice-salt bath to $<10°$ with stirring, then centrifuged at 16,000 *g* for 20 min and the pellet is discarded.

Step 3. Protamine Sulfate Treatment. A 2% solution of protamine sulfate (Sigma Grade X, or equivalent) is adjusted to pH 5.5 by adding 10 *M* KOH. This solution must be clear and colorless, and should be filtered, if necessary, before use. Each liter of the supernatant solution from the heat step is treated with 375 ml of 2% protamine sulfate. The protamine sulfate is added dropwise from a separatory funnel over a period of 1–2 hr, and the solution is stirred for an additional 15–20 min after all the protamine sulfate is added. The mixture is centrifuged at 16,000 *g* for 20 min, and the pellet discarded.

Step 4. Ammonium Sulfate Precipitation. Each liter of the protamine sulfate supernatant solution is treated with 350 g of ammonium sulfate (Schwarz-Mann, Enzyme Grade) and 200 mg of EDTA. Ammonium sulfate is added with stirring over a period of 1 hr, and the solution is stirred for 30 min following $(NH_4)_2SO_4$ addition. The suspension is centrifuged at 16,000 *g* for 30 min, and the supernatant is carefully decanted and discarded. The $(NH_4)_2SO_4$ precipitate is dissolved in 200–250 ml of 0.2 *M* potassium phosphate buffer (pH 7.65) containing 0.5 m*M* EDTA. Any undissolved material is removed by centrifugation at 16,000 *g* for 20 min and discarded.

Step 5. Gel Filtration (Desalting) on Sephadex G-25. A 5 × 87 cm column of Sephadex G-25 (coarse) is equilibrated with 0.14 *M* potassium phosphate buffer (pH 6.85) containing 0.5 m*M* EDTA. The flow rate is adjusted to about 25 ml/min, and one-half (100–125 ml) of the resuspended ammonium sulfate precipitate is applied to the column. A volume (about 700 ml) equivalent to the void volume of the column is collected and discarded; then 400–450 ml of desalted protein solution is collected. The appearance of the salt peak [containing $(NH_4)_2SO_4$] is monitored by testing the column effluent for ammonia with Nessler's reagent on a spot plate. The column is washed with 2–3 liters of the elution buffer after the salt peak has eluted. The procedure given above is repeated with the other half of the resuspended $(NH_4)_2SO_4$ precipitate.

It should be noted that the enzyme is *not* stable in 0.14 *M* potassium phosphate at this stage of purification, and, therefore, it is advantageous to begin loading the desalted protein solution on the DEAE-Sephadex column (Step 6) as soon as possible, i.e., the first half may be loaded while the second half is being desalted.

Step 6. DEAE-Sephadex Chromatography. The DEAE-Sephadex column should be prepared in advance. DEAE-Sephadex A-50-120 (35 g) is swelled in 3 liters of 0.7 *M* potassium phosphate buffer (pH 6.8) for 48 hr at 25°. The buffer is decanted and the resin is resuspended in 1.5 liters of 0.14 *M* potassium phosphate buffer (pH 6.8). The mixture is allowed to stand at least 30 min at 25°. The buffer is decanted, and the 0.14 *M* potassium phosphate buffer wash is repeated 5 times. The resin is deaerated under vacuum, and a 5 × 25 cm column is poured and equilibrated with at least 2 liters of 0.14 *M* potassium phosphate (pH 6.8) containing 0.5 m*M* EDTA.

The desalted protein solution is loaded onto the column, which is then washed with 1–2 liters of 0.14 *M* potassium phosphate buffer (pH 6.8) at a flow rate of about 100 ml/hr. A linear gradient is established between 2 liters of 0.3 *M* potassium phosphate buffer (pH 7.65) and 2 liters of 0.14 *M* potassium phosphate buffer (pH 6.85); both buffers contain 0.5 m*M* EDTA. The flow rate is adjusted to 60–80 ml/hr and 15 ml fractions are collected. The enzyme is found 2300–2500 ml from the start of the 4 liter gradient (at ~0.24 *M* potassium phosphate). The carbamyl phosphate synthetase-containing fractions (which may now be assayed by measuring ADP formation using the coupled enzyme assay, since contaminating ATPases have been removed at this stage of the purification) are combined into 2 pools, and the protein present in each pool is precipitated by adding 370 g of $(NH_4)_2SO_4$ per liter of enzyme solution. After centrifugation at 16,000 *g* for 30 min, the $(NH_4)_2SO_4$ precipitate is resuspended in about 25 ml of 0.2 *M* potassium phosphate buffer (pH 7.65) containing 0.5 m*M* EDTA. The resuspended precipitate is concentrated to about 5 ml in a collodion bag apparatus (Schleicher and Schuell, Inc., Keene, New Hampshire) against 0.2 *M* potassium phosphate buffer (pH 7.65) containing 0.5 m*M* EDTA and 40 m*M* NH_4Cl. The presence of NH_4Cl causes the enzyme to oligomerize to a high molecular weight species (M_r ~400,000) and thus elute in a tight peak near the void volume of the Agarose A-0.5m column (Step 7).

Step 7. Gel Filtration on Agarose A-0.5m (200–400 Mesh). The concentrated solution obtained in the preceding step is loaded on a 2.5 × 94 cm column of BioGel A-0.5m (Bio-Rad, Inc., Rockville Centre, N.Y.) that has been previously equilibrated with 0.2 *M* potassium phosphate buffer (pH 7.65) containing 40 m*M* NH_4Cl and 0.5 m*M* EDTA. This buffer is also used for elution. The flow rate is adjusted to about 15 ml/hr, and 2–3 ml fractions are collected, assayed for activity, and combined into 2 pools. A shoulder on the leading edge of the carbamyl phosphate synthetase peak (which is centered at an elution volume of about 200 ml) contains the yellow flavoenzyme glutamate synthase[24] which copurifies with CPS up to this step. Similar results may be obtained by gel filtration on Ultrogel AcA 34 (LKB, Inc., Gaithersburg, MD). This step is repeated with the second pool (of lower specific activity) from the DEAE-Sephadex column.

Step 8. Ammonium Sulfate Fractionation. The two enzyme pools obtained from the Agarose A-0.5m column (Pool I, specific activity ~200 U/mg; Pool II, specific activity ~160 U/mg) are further purified by $(NH_4)_2SO_4$ fractionation. A 0–50% $(NH_4)_2SO_4$ cut is performed on each pool, followed by a 50–65% cut. The $(NH_4)_2SO_4$ precipitate from each fractionation is collected by centrifugation at 25,000 *g* for 30 min and redissolved in 3–5 ml of 0.2 *M* potassium phosphate buffer (pH 7.65) containing 0.5 m*M* EDTA. The enzyme is stored at 4° in this buffer [containing $(NH_4)_2SO_4$] at a protein concentration of 10–20 mg/ml. More dilute solutions of the enzyme [<1 mg/ml, without $(NH_4)_2SO_4$] are much less stable and lose activity in 24–48 hr. Freezing the enzyme solution also leads to partial (10–20%) loss of activity. Long-term storage stability may also be improved by adding 2 m*M* IMP and/or by removal of bacteria from the enzyme solution by filtration through a 0.22-μm filter.

Enzyme stored under these conditions exhibits a rapid, irreversible decrease in the glutamine-dependent synthetase activity to about 50% of the initial activity within 1–2 weeks, followed by a slow decline in the remaining activity over 1–2 months. About 25% of the initial glutamine-dependent activity remains after 2 months at 4°. The ammonia-dependent synthetase activity does not show a rapid initial decrease in activity, but decreases slowly over several months at a slower rate than the glutamine-dependent synthetase activity. The bicarbonate-dependent ATPase activity remains essentially constant for as long as 6 months, and some preparations up to 2 years old still retain about 50% of their original ATPase activity, but do not catalyze carbamyl phosphate synthetase from either L-glutamine or ammonia. Concentrated solutions (notably those stored at >20 mg/ml) exhibit increased bicarbonate-dependent ATPase activity

[24] P. P. Trotta, K. E. B. Platzer, R. H. Haschemeyer, and A. Meister, *Proc. Natl. Acad. Sci. U.S.A.* **71,** 4607 (1974).

upon extended (>6 months) storage, as well as a tendency to precipitate. Discontinuous SDS gel electrophoresis[25] of this precipitate shows an increased proportion of light subunit to heavy subunit as compared to the unprecipitated enzyme solution. These observations suggest a possible selective denaturation and precipitation of the light subunit during long-term storage, which causes disruption of the native quaternary structure of the enzyme, and results in uncoupling of the ATPase activity (which is catalyzed by the heavy subunit alone) from the glutamine-dependent synthetase activity (which requires both subunits). This hypothesis is supported by previous studies[26-28] on the reconstitution of native enzyme from catalytically active subunits separated by KSCN solvent perturbation. These studies showed that a 4-fold molar excess of isolated light subunit to heavy subunit was required to restore maximal glutamine-dependent synthetase activity, which suggests the presence of a population of damaged light subunits.

Subunit Separation of Catalytically Active Subunits of Carbamyl Phosphate Synthetase by Solvent Perturbation with Potassium Thiocyanate

This procedure is modified from that described in a previous volume of this series.[28] A column (1.6 × 80 cm) of Ultrogel AcA 34 (LKB, Inc) is equilibrated with 0.1 *M* potassium phosphate buffer (pH 7.7) containing 1 *M* KSCN and 5 m*M* EDTA. The buffer is deoxygenated during column equilibration and gel filtration by flushing the reservoir with N_2 through a sintered glass filter. Buffer deoxygenation is essential to reduce aggregation of the heavy subunit during chromatography. Carbamyl phosphate synthetase (5–7 mg) is applied to the column, the flow rate is adjusted to about 5 ml/hr and 1–2 ml fractions are collected. Fractions containing the heavy subunit are pooled, dialyzed to remove KSCN against 0.2 *M* potassium phosphate (pH 7.7) containing 5 m*M* 2-mercaptoethanol, and concentrated in a collodion bag apparatus to a concentration of 0.5–1.0 mg/ml. The addition of 2-mercaptoethanol or 5 m*M* dithiothreitol reduces aggregation and precipitation of the heavy subunit. The fractions containing the light subunit are pooled, dialyzed and concentrated similarly, but

[25] U. K. Laemmli, *Nature (London)* **227,** 680 (1970).

[26] P. P. Trotta, L. M. Pinkus, R. H. Haschemeyer, and A. Meister, *J. Biol. Chem.* **249,** 492 (1974).

[27] P. P. Trotta, M. E. Burt, R. H. Haschemeyer, and A. Meister, *Proc. Natl. Acad. Sci. U.S.A.* **68,** 2599 (1971).

[28] P. P. Trotta, M. E. Burt, L. M. Pinkus, L. F. Estis, R. H. Haschemeyer, and A. Meister, this series, Vol. 51, p. 21.

sulfhydryl compounds are not added, because the H_2O_2 formed by oxidation of thiols causes irreversible inactivation of the glutaminase activity of the light subunit.[29] The separated subunits are stored at 4°, and remain catalytically active for several days.

The use of Ultrogel AcA 34 results in better resolution of the two subunits than Sephadex G-200, which was used in previously published procedures.

The enzyme may be reconstituted by mixing solutions of the heavy and light subunits which have been dialyzed to remove KSCN. Maximal glutamine-dependent carbamyl phosphate synthetase activity of the renatured enzyme is achieved when a 4-fold molar excess of light subunit to heavy subunit is used.[26–28] Hybrid carbamyl phosphate synthetase containing subunits from both *E. coli* and *Enterobacter aerogenes* has been prepared by this method.[24]

Separation of the Subunits of Denatured Carbamyl Phosphate Synthetase

The two subunits of carbamyl phosphate synthetase, after guanidine denaturation, may be separated by gel filtration using a modified procedure originally described by Matthews and Anderson.[30] A denaturing solution containing 6 *M* guanidine hydrochloride, 0.05 *M* KCl, 0.01 *M* EDTA, and 0.05 *M* 2-mercaptoethanol, pH 6.5[31] is used to equilibrate and elute a 1 × 95 cm column of BioGel A-5m (Bio-Rad, Inc.). Carbamyl phosphate synthetase (4–6 mg) is dialyzed against the denaturing solution for 24–48 hr at 25°, and is then applied to the BioGel A-5m column. The flow rate is adjusted to 2–3 ml/hr and 1 ml fractions are collected.

The protein-containing fractions are assayed by a turbidimetric procedure[32]: 0.25 ml of the protein solution in denaturing solution is added to 1.0 ml of 4% trichloroacetic acid, and the solution is mixed thoroughly on a Vortex mixer and allowed to stand at 25° for 30 min. The solution is then mixed thoroughly and the absorbance at 450 nm is measured.

This procedure is preferable to the KSCN subunit separation protocol (if catalytically active subunits are not required), because the heavy subunit does not aggregate in the denaturing solution as it does during the KSCN procedure. Omission of 2-mercaptoethanol from the denaturing solution does not affect the subunit separation on BioGel A-5m, and permits protein determination by direct measurement of absorbance at

[29] P. P. Trotta, L. M. Pinkus, and A. Meister, *J. Biol. Chem.* **249,** 1915 (1974).
[30] S. M. Matthews and P. M. Anderson, *Biochemistry* **11,** 1176 (1972).
[31] P. F. Davison, *Science* **161,** 906 (1968).
[32] W. W. Fish, K. G. Mann, and C. Tanford, *J. Biol. Chem.* **244,** 4989 (1969).

280 nm, which is not possible in solutions containing 2-mercaptoethanol due to the high ultraviolet absorbance of 2-mercaptoethanol disulfide.

Physical and Chemical Properties

The enzyme is homogeneous by discontinuous SDS–polyacrylamide gel electrophoresis[25] and in the analytical ultracentrifuge. The weight average molecular weight of the enzyme monomer, as determined by sedimentation equilibrium ultracentrifugation, is 163,000 $\pm$ 4,000[33]; discontinuous SDS–gel electrophoresis indicates a monomer molecular weight of ~172,000. The enzyme monomer is composed of two nonidentical subunits, a heavy subunit (M_r ~130,000) which has the binding sites for ammonia, bicarbonate, ATP and the allosteric effectors, and a light subunit (M_r ~40,000).[26,27] The light subunit has the binding site for the γ-glutamyl moiety of glutamine; this site has a sulfhydryl group that interacts with glutamine to form a γ-glutamyl enzyme. Both subunits contribute to the binding of glutamine. The amino acid compositions of both subunits have been determined.[26,30] The two subunits of *E. coli* carbamyl phosphate synthetase are encoded by the genetically linked *Car A* and *Car B* genes, respectively.[34,35] These two genes have recently been cloned,[36,37] and the amino acid sequence of the heavy subunit has been deduced from the nucleotide sequence of a cloned fragment of *E. coli* with genetic determinants for *Car B*.[38] The amino acid sequence derived from *Car B* shows a highly significant homology between the amino- and carboxyl-terminal halves of the heavy subunit. This suggests that the *Car B* gene may have been formed by an internal duplication of a smaller ancestral gene; the identity of this ancestral gene is unknown at present, but *Car B* may be related to the gene for carbamate kinase, an enzyme that catalyzes ATP synthesis from ADP and carbamyl phosphate formed in the arginine dihydrolase pathway. There is substantial evidence for two separate but nonidentical ATP binding sites on the heavy subunit (see section below), and the duplicated amino acid sequence could reflect the existence of these two sites.

[33] P. P. Trotta, L. F. Estis, A. Meister, and R. H. Haschemeyer, *J. Biol. Chem.* **249,** 482 (1974).

[34] A. Pierard, N. Glansdorff, M. Mergeay, and J. M. Wiame, *J. Mol. Biol.* **14,** 23 (1965).

[35] M. Mergeay, D. Gigot, J. Beckmann, N. Glansdorff, and A. Pierard, *Mol. Gen. Genet.* **133,** 299 (1974).

[36] N. Glansdorff, C. Dambly, S. Palchaudhuri, M. Crabeel, A. Pierard, and P. Halleux, *J. Bacteriol.* **127,** 302 (1976).

[37] M. Crabeel, D. Charlier, G. Weyens, A. Feller, A. Pierard, and N. Glansdorff, *J. Bacteriol.* **143,** 921 (1980).

[38] H. Nyunoya and C. J. Lusty, *Proc. Natl. Acad. Sci. U.S.A.* **80,** 4629 (1983).

Carbamyl phosphate synthetase can exist in monomer, dimer, and higher oligomeric forms, which are associated with its regulation by allosteric effectors. At least two different monomer conformations exist, each of which can undergo self-association to yield dimer and higher oligomers.[39] Reversible self-association is promoted by allosteric effectors, inorganic phosphate, and ATP. The specific activity of the enzyme increases with increasing enzyme concentration; thus, there is a direct relation between allosteric regulation and self-association. Sedimentation velocity ultracentrifugation and covalent cross-linking studies show that the enzyme exists in monomeric, dimeric, and tetrameric forms. In barbital buffer, the enzyme is present as a monomer. In sodium phosphate buffer, the enzyme exists as a partially dissociating dimer. Another type of association, distinct from the phosphate-induced dimerization, occurs in the presence of positive allosteric effectors. In the presence of phosphate and a positive allosteric effector, the maximum association state of the enzyme is a tetramer with two types of intersubunit contacts and probable D_2 symmetry. UMP promotes conversion to a dimer which closely resembles the phosphate-induced dimer, but UMP prevents further self-association to the tetramer. A detailed model for the self-association of the enzyme has been proposed.[39]

Catalytic Properties

The enzyme exhibits optimal activity in the pH range 7.8–8.2. The apparent K_m values for bicarbonate and L-glutamine are, respectively, 1.2 and 0.38 mM. The K_m value for NH_4^+ is 93 mM. Half-maximal activity is observed with an ATP concentration of about 7 mM. The relationship between the concentration of ATP and the initial rate of the reaction is described by a sigmoidal curve.[40] The enzyme is inhibited by UMP and to a lesser extent by UDP and UTP. Its activity is enhanced by IMP, XMP, GMP, GDP, GTP, AMP, ITP, and IDP. Ammonia and ornithine are also activators.[41] Potassium ions are required for maximal enzyme activity; the optimal concentration of K^+ is about 0.1 M. Ammonium ions at the same concentration can replace potassium ions.

The enzyme can be reversibly dissociated into catalytically active heavy and light subunits (see above).[26,27] Studies on the separated subunits indicate that the heavy subunit can catalyze ammonia-dependent carbamyl phosphate synthetase and two partial reactions, i.e., bicarbonate-

[39] S. G. Powers, A. Meister, and R. H. Haschemeyer, *J. Biol. Chem.* **255,** 1554 (1980).
[40] P. M. Anderson and A. Meister, *Biochemistry* **5,** 3157 (1966).
[41] P. M. Anderson and A. Meister, *Biochemistry* **5,** 3164 (1966).

dependent ATPase and the synthesis of ATP from ADP and carbamyl phosphate. The only reaction catalyzed by the separated light subunit is the hydrolysis of glutamine. The affinity of the separated light subunit for glutamine is substantially lower than that of the native enzyme.

Carbamyl phosphate synthetase was the first glutamine amidotransferase found to be inactivated by treatment with L-2-amino-4-oxo-5-chloropentanoate.[42] This reagent binds stoichiometrically to a cysteine residue on the light subunit.[43] The enzyme is also inactivated in an apparently similar manner by treatment with azaserine and 6-diazo-5-oxonorleucine. Covalent modification by these glutamine analogs destroys the glutamine-dependent carbamyl phosphate synthetase activity; however, the treated enzyme preparations are active in catalyzing carbamyl phosphate synthesis with ammonia as the nitrogen donor. Treatment of the enzyme with cyanate also reversibly inactivates the glutamine-dependent activity.[44,45] It is of interest that after treating the enzyme with L-2-amino-4-oxo-5-chloropentanoate, the bicarbonate-dependent ATPase activity of the enzyme is stimulated by about 2-fold and that the apparent K_m value for ammonia in the ammonia-dependent carbamyl phosphate synthesis reaction is decreased.

Interactions between the Subunits

The findings cited above on the selective inactivation of the glutamine binding site and the experiments on the dissociation of the enzyme into enzymatically active subunits indicate that there are significant interactions between the subunits. These interactions appear to stabilize the enzyme and also to facilitate catalysis. It is notable that the activities of the separated heavy subunit in catalyzing ammonia-dependent carbamyl phosphate synthesis, bicarbonate-dependent ATPase and ATP synthesis from carbamyl phosphate and ADP are somewhat lower than the corresponding activities exhibited by the isolated enzyme and the reconstituted enzyme (obtained by mixing the heavy and light subunits). These activities increase appreciably when the subunits are mixed together; this suggests that the light subunit contributes to the catalytic activity of the heavy subunit by either stabilizing it or by facilitating catalysis, or both. The relatively low affinity of the isolated light subunit for glutamine (apparent K_m = 130–180 mM)[26] suggests that the heavy subunit may contrib-

[42] E. Khedouri, P. M. Anderson, and A. Meister, *Biochemistry* **5,** 3552 (1966).

[43] L. M. Pinkus and A. Meister, *J. Biol. Chem.* **247,** 6119 (1972).

[44] P. M. Anderson, J. D. Carlson, G. A. Rosenthal, and A. Meister, *Biochem. Biophys. Res. Commun.* **55,** 246 (1973).

[45] P. M. Anderson and J. D. Carlson, *Biochemistry* **14,** 3688 (1975).

ute to the binding of the amide moiety of glutamine; alternatively, intersubunit interactions may alter the conformation of the light subunit so as to enhance its ability to bind glutamine. The binding of the chloroketone or of cyanate to the light subunit leads to a significant decrease in the apparent K_m value for ammonia, (from 90–150 to 20–60 m*M*)[43,44] suggesting that the binding of these compounds to the glutamine active site on the light subunit may be associated with conformational changes that promote the binding of ammonia to the heavy subunit. The substantial increase in bicarbonate-dependent ATPase activity (a reaction catalyzed by the heavy subunit) produced by the binding of substrate analogs to the glutamine binding site of the light subunit also provides evidence for significant interactions between the subunits. Conversely, the binding of substrates to the heavy subunit greatly enhances the rate of glutamine cleavage by the light subunit; thus, hydrolysis of glutamine in the absence of the other substrates occurs at a rate that is only about 1% of that observed in the presence of all of the substrates. If these effects occur during the normal catalytic reaction, it would appear that the conformational changes induced by the binding of glutamine facilitate glutamine cleavage by promoting the formation or utilization of activated carbon dioxide by the heavy subunit.

The interaction between heavy and light subunits is also evident in the pH dependence of the glutaminase activity catalyzed by the native enzyme as compared to the separated light subunit. The glutaminase activity of the native enzyme exhibits two pH optima (at 4.2 and 9.5)[46]; in contrast, the separated light subunit exhibits a single optimum at pH 6.7. Reconstitution of the enzyme by mixing of the light and heavy subunits leads to restoration of the two pH optima at 4.2 and pH 9.5. Thus, it appears that one active site (on the light subunit) is involved in glutamine hydrolysis at both acid and alkaline values of pH and the glutaminase activity of the enzyme is affected by intersubunit interactions. The dual pH optima of the native enzyme may reflect the participation of different catalytic groups of the enzyme which affect maximal velocity rather than binding of substrate.

Sulfhydryl group modification with *N*-ethylmaleimide increases the glutaminase activity of the enzyme about 250-fold and converts the pH profile to one that is similar to that exhibited by the separated light subunit.[47] Under these conditions, the activities exhibited by the heavy subunit are irreversibly destroyed. The effects produced by *N*-ethylmaleimide

[46] V. P. Wellner and A. Meister, *J. Biol. Chem.* **250,** 3261 (1975).

[47] P. P. Trotta, V. P. Wellner, L. M. Pinkus, and A. Meister, *Proc. Natl. Acad. Sci. U.S.A.* **70,** 2717 (1973).

modification seem to be associated with structural changes that alter the intersubunit relationships so as to uncouple glutamine hydrolysis from carbamyl phosphate synthesis.

Inhibition by Oxidation

The glutamine-dependent functions of carbamyl phosphate synthetase are inactivated by incubating the enzyme in air with low concentrations of dithiothreitol; other mercaptans such as glutathione, cysteine, homocysteine, or 2-mercaptoethanol produce similar effects.[48,49] Exclusion of oxygen reduces inactivation by thiols. Enzyme preparations inactivated in this way cannot be reactivated by treatment with dithiothreitol or other mercaptans, suggesting that disulfide bond formation is not the cause of the inactivation. When the enzyme is incubated with 0.2 m*M* hydrogen peroxide, the glutamine-dependent activities of the enzyme are irreversibly destroyed without affecting ammonia-dependent carbamyl phosphate synthesis activity. It is notable that [^{14}C]chloroketone binds to the dithiothreitol-inhibited enzyme to the same extent that it does to the native enzyme and that such binding is reduced in the presence of glutamine. The inactivation produced by mercaptans may be prevented by the addition of catalase. Thus, it is likely that the inhibition observed when the enzyme is incubated in air in the presence of free thiol compounds is due to the metal ion-catalyzed sulfhydryl oxidation accompanied by the formation of hydrogen peroxide, which oxidizes one or more critical residues in the glutamine binding site of the light subunit. However, this oxidation does not appear to involve the sulfhydryl group that reacts with the chloroketone, nor does it involve a sulfhydryl group on the light subunit whose reaction with *N*-ethylmaleimide in the presence of ATP, magnesium ions, and bicarbonate leads to loss of glutamine-dependent synthetase activity and to enhanced glutaminase activity.[47]

Hydroxylamine and certain alkyl hydrazines also inactivate the glutamine-dependent activity of carbamyl phosphate without affecting the synthesis of carbamyl phosphate from ammonia.[50] The relative order of effectiveness of several compounds in inactivating the glutamine-dependent synthetase activity was found to be[51]

$$H_2O_2 \gg NH_2OH, CH_3NHNH_2 > CH_3NHNHCH_3 > (CH_3)_2NNH_2 > CH_3ONH_2, NH_2OCH_3$$

[48] P. P. Trotta, L. M. Pinkus, and A. Meister, *J. Biol. Chem.* **249,** 1915 (1974).

[49] D. S. Kaseman, A. J. L. Cooper, and A. Meister, unpublished observations.

[50] D. S. Kaseman, *Fed. Proc., Fed. Am. Soc. Exp. Biol.* **39,** 2032, Abstr. 2252 (1980).

[51] D. S. Kaseman and A. Meister, *Abstr. Pap., 181st Meet., Am. Chem. Soc.* Abstr. 101 (1981).

Treatment of the enzyme with 25 m*M* hydroxylamine leads to a 2-fold increase in ATPase activity but has no effect on the synthesis of ATP from ADP and carbamyl phosphate catalyzed by the enzyme. Amino acid hydroxamates do not inactivate, with the notable exception of γ-glutamylhydroxamate, which is effectively cleaved by the enzyme to yield glutamate and hydroxylamine.

It may be postulated that treatment of the enzyme with low concentrations (10–25 m*M*) of hydroxylamine could lead to reaction of hydroxylamine with an activated carboxyl group of the enzyme to form a hydroxamate. Recent studies on the mechanism of this phenomenon are not consistent with this hypothesis, since it has been shown that hydroxylamine does not form a stable covalent bond with the enzyme. This was demonstrated conclusively by direct binding studies in which the enzyme was incubated with ^{13}N-labeled hydroxylamine.[52,53] Hydroxylamine is known to oxidize spontaneously in aqueous alkaline solution by a mechanism that involves attack of oxygen on the deprotonated species NH_2O^- to yield nitroxyl ion (NO^-), which is further oxidized to peroxonitrite.[54,55] The addition of glutathione peroxidase plus glutathione prevents NH_2OH inactivation of the glutamine-dependent activities of the enzyme.[56] This observation suggests that the inactivation by hydroxylamine is due to oxidative damage of the light subunit by hydrogen peroxide or peroxonitrite, or both. The inhibition of the glutamine-dependent activities of the enzyme by treatment with alkyl hydrazines also seems to involve the generation of peroxides. This inactivation may be prevented by the addition of EDTA or catalase. Studies with [^{14}C]methylhydrazine showed that inactivation of the glutamine-dependent synthetase activity is accompanied by the binding of about 0.7 mol of labeled compound per mole of monomeric enzyme. Subsequent dissociation of the enzyme into its subunits by treatment with guanidine hydrochloride or KSCN led to nearly complete loss of the bound radioactivity. However, the binding of methyl hydrazine was found to occur even when inactivation was prevented by carrying out the reaction in the presence of catalase; thus, methylhydrazine binding is evidently not directly related to the inactivation, which appears to be due to oxidative modification of the enzyme.

[52] D. S. Kaseman, A. J. L. Cooper, A. Meister, A. S. Gelbard, and R. E. Reiman, *J. Labelled Compd. Radiopharm.* **21,** 803 (1984).
[53] D. S. Kaseman, Doctoral dissertation, Cornell University (1985).
[54] M. N. Hughes and H. G. Nicklin, *J. Chem. Soc. A* p. 925 (1970).
[55] M. N. Hughes and H. G. Nicklin, *J. Chem. Soc. A* p. 164 (1971).
[56] D. S. Kaseman and A. Meister, *Biochemistry* **22,** 7A (1983).

Mechanism

The carbamyl phosphate synthetases are unusual in that they catalyze a reaction in which the formation of 1 mol of product requires the cleavage of 2 mol of ATP to 2 mol of ADP. Early consideration of the mechanism of this reaction led to the suggestion that carboxyphosphate is an enzyme-bound intermediate.[57–59] This hypothesis was in accord with observations[60] on the ammonia-utilizing carbamyl phosphate synthetase I of frog liver in which it was shown that ^{18}O is transferred from $HC^{18}O_3^-$ to inorganic phosphate in the course of the reaction. The idea that carboxy phosphate is an intermediate had also been considered for reactions catalyzed by certain biotin-containing enzymes.[61] However, other mechanisms also seemed possible, since the ^{18}O data could be explained by pathways not involving carboxyphosphate, including mechanisms involving concerted reactions.

Evidence that the reaction catalyzed by carbamyl phosphate synthetase involves intermediate formation of an activated form of bicarbonate was obtained in studies on the *E. coli* enzyme in which the "pulse-chase" technique[62] was used.[3] When a relatively large amount of enzyme was incubated with $H^{14}CO_3^-$ and ATP, and this reaction mixture was then treated with a solution containing glutamine and a large excess of unlabeled HCO_3^-, it was found that the [^{14}C]carbamyl phosphate formed contained much more radioactivity than could be accounted for if the labeled and unlabeled bicarbonates had equilibrated. These studies showed that bicarbonate binds to the enzyme in a reaction requiring ATP and that binding facilitates subsequent reaction of labeled bicarbonate with glutamine. Analogous "pulse-chase" studies with [^{32}P]ATP showed that the ATP-dependent binding of bicarbonate to the enzyme is associated with cleavage of ATP to ADP. These studies supported the view that the first step involves activation of bicarbonate on the enzyme, and that the activated bicarbonate reacts with glutamine to form enzyme-bound carbonate. In this scheme, a second molecule of ATP is used to phosphorylate enzyme-bound carbamate:

(a) Enzyme + ATP + HCO_3^- $\rightleftharpoons$ enzyme–($HCO_3PO_3^{2-}$) + ADP

(b) Enzyme–($HCO_3PO_3^{2-}$) + L-glutamine $\rightleftharpoons$ enzyme–($HCO_3PO_3^{2-}$)(L-glutamine)

[57] M. E. Jones, *Annu. Rev. Biochem.* **34,** 381 (1965).
[58] A. Meister, "Biochemistry of the Amino Acids," p. 692. Academic Press, New York, 1965.
[59] M. E. Jones, *Science* **140,** 1373 (1963).
[60] M. E. Jones and L. Spector, *J. Biol. Chem.* **235,** 2897 (1960).
[61] Y. Kaziro, L. F. Hass, P. D. Boyer, and S. Ochoa, *J. Biol. Chem.* **237,** 1460 (1962).
[62] P. R. Krishnaswamy, V. Pamiljans, and A. Meister, *J. Biol. Chem.* **237,** 2932 (1962).

(c) Enzyme–($HCO_3PO_3^{2-}$) (L-glutamine) → enzyme–($NH_2CO_2^-$) + L-glutamate + P_i

(d) Enzyme–($NH_2CO_2^-$) + ATP ⇌ enzyme + $NH_2CO_2PO_3^{2-}$ + ADP

The first step (a) seems to explain the bicarbonate-dependent ATPase activity exhibited by the enzyme, since enzyme-bound carboxyphosphate would be expected to break down rapidly to yield bicarbonate, inorganic phosphate, and ADP.

Evidence that enzyme-bound carboxy phosphate is an intermediate is supported by several experiments. In one approach, the enzyme-bound intermediate was converted to formate by reduction with KBH_4.[63,64] In these studies, the enzyme was incubated briefly with $H^{14}CO_3^-$ and ATP. Two volumes of dimethyl sulfoxide were then added, after which unlabeled carbon dioxide was bubbled through the solution to remove unreacted [^{14}C]bicarbonate. In the presence of dimethyl sulfoxide the enzyme, although catalytically inactive, retains the labeled carboxy phosphate, which may then be reduced to formate by the addition of potassium borohydride. Using this protocol, ^{14}C label may be selectively introduced into the enzyme-bound activated intermediate and thus also into the formate derived from it by borohydride reduction. The key to success in this experiment lies in the stabilization of the enzyme-bound carboxy phosphate afforded by dimethyl sulfoxide. This is important because free carbon dioxide is also reduced to formate on treatment with KBH_4. In another approach, the trimethyl ester of carboxy phosphate was isolated from the enzyme by treating a reaction mixture containing the enzyme, $H^{14}CO_3^-$, and ATP with diazomethane. Studies with labeled bicarbonate and ^{32}P-labeled ATP showed that the molar ratio of incorporation of ^{14}C and ^{32}P into product was close to unity. The amount of carboxy phosphate trapped as the trimethyl ester approached values that were stoichiometric with the enzyme. In the course of this work, it was found that a small amount of trimethyl carboxy phosphate is formed in the absence of enzyme; this probably results from metal ion catalyzed transphosphorylation of ATP to bicarbonate. The enzyme does not catalyze ATP–ADP or ATP–P_i exchanges; this suggests that the ADP formed in the interaction of ATP with bicarbonate remains bound to the enzyme and is part of a stabilized enzyme–carboxy phosphate–ADP complex. Analogous complexes are formed in the reactions catalyzed by glutamine synthetase,[65] γ-glutamylcysteine synthetase,[66] and glutathione synthetase.[67]

[63] S. G. Powers and A. Meister, *Proc. Natl. Acad. Sci. U.S.A.* **73,** 3020 (1976).

[64] S. G. Powers and A. Meister, *J. Biol. Chem.* **253,** 1258 (1978).

[65] See this volume [27].

[66] See this volume [47, 48].

[67] See this volume [49].

Evidence that carboxy phosphate is a kinetically competent intermediate in the reaction was obtained by application of the ATP $\beta\gamma$-bridge: β-nonbridge ^{18}O scrambling method.[68] The enzyme was found to catalyze reversible cleavage of ATP to bound ADP in the presence of bicarbonate and in the absence of ammonia or glutamine at a rate that is 1.4 to 1.7 times the rate of net ATP cleavage. In these studies it was also found that bicarbonate oxygen is not incorporated into the γ-phosphoryl moiety of ATP.[69]

Since the carbamyl phosphate reaction requires the cleavage of two molecules of ATP, it was important to ascertain whether the enzyme has a single binding site for ATP that is successively used for carboxy phosphate formation and for phosphorylation of carbamate, or whether the enzyme has two separate binding sites for the two molecules of ATP that are utilized in the reaction. Several indirect approaches suggested that there are two different binding sites for ATP.[70] Direct evidence in support of this hypothesis was obtained by studying the binding of a series of α,ω-adenosine-5′-polyphosphates (Ap_nA; $n = 2–6$)[71] to the enzyme. With the exception of adenylate kinase, which has two nucleotide binding sites, compounds of the type Ap_nA do not inhibit a variety of ATP-utilizing kinases, hydrolases, and synthetases which catalyze reactions that require only a single molecule of ATP.[72] Of several compounds of this type, only Ap_5A is an effective inhibitor of carbamyl phosphate synthetase, indicating that Ap_5A binds to the sites on the enzyme that normally interact with the two molecules of ATP. Since other Ap_nA compounds do not inhibit, it appears that the length of the pentaphosphate chain is a measure of the distance between the binding sites on the enzyme for the two adenosine moieties.

Additional evidence that the enzyme can bind two molecules of ATP simultaneously was obtained by an approach in which the "pulse-chase" method was used.[73] In these studies, the enzyme was incubated briefly with [γ-^{32}]ATP and bicarbonate and then a solution containing both glutamine and an excess of unlabeled ATP was added. The results expected if the [^{32}P]ATP binds initially only to the first ATP binding site would be formation, from the mixture of labeled and unlabeled ATP, of carbamyl phosphate containing relatively little label. On the other hand, if both

[68] C. F. Midelfort and I. A. Rose, *J. Biol. Chem.* **251,** 5881 (1976).

[69] M. J. Wimmer, I. A. Rose, S. G. Powers, and A. Meister, *J. Biol. Chem.* **254,** 1854 (1979).

[70] A. Meister, *Trans. N.Y. Acad. Sci.* [2] **41,** 117 (1983) (A Festschrift for Sarah Ratner, edited by M. E. Pullman).

[71] S. G. Powers, O. W. Griffith, and A. Meister, *J. Biol. Chem.* **252,** 3558 (1977).

[72] G. E. Lienhard and I. I. Secemski, *J. Biol. Chem.* **248,** 1121 (1973).

[73] S. G. Powers and A. Meister, *J. Biol. Chem.* **253,** 800 (1978).

ATP sites bind labeled ATP initially, an amount of carbamyl [^{32}P]phosphate equivalent maximally to the [^{32}P]ATP bound at the second site would be expected to be formed immediately when glutamine and excess unlabeled ATP are added, followed by formation of carbamyl phosphate containing much less label. It was found that there is an initial burst of labeled carbamyl phosphate formation, followed by a linear rate of formation of carbamyl phosphate containing much less label. It was also found that the enzyme catalyzes stoichiometric formation of unlabeled plus labeled P_i and carbamyl phosphate. From this evidence, it may be concluded that the enzyme initially binds two molecules of ATP, one of which is used for carboxy phosphate formation. The second molecule of bound ATP is not directly involved in the activation of bicarbonate, but is used for phosphorylation of enzyme-bound carbamate. This is consistent with the finding that the ATP bound to the second ATP site does not equilibrate with subsequently added unlabeled ATP and is preferentially used for the phosphorylation reaction.

An overall mechanism based on the intermediate formation of enzyme-bound carboxy phosphate and enzyme-bound carbamate has been proposed (see ref. 74 for details). According to this scheme, two molecules of ATP, bicarbonate, and glutamine are bound to the enzyme at the start of the catalytic cycle. Carboxy phosphate is formed by interaction of one specific bound ATP molecule with bicarbonate. (This reaction is reversible so that in the absence of glutamine and ammonia, carboxy phosphate breaks down to ADP and inorganic phosphate, thus accounting for the bicarbonate-dependent ATPase activity exhibited by the enzyme.) The enzyme binds glutamine in the absence of the other substrates; glutamine hydrolysis is facilitated by conformational changes associated with carboxy phosphate formation. Cleavage of glutamine is concomitant with formation of a γ-glutamyl-S-enzyme and subsequent release of ammonia. Ammonia reacts with carboxy phosphate on the enzyme to form enzyme-bound carbamate, with release of glutamate and inorganic phosphate. This step of the reaction is not reversible and therefore accounts for the irreversibility of the overall synthesis reaction. On the other hand, phosphorylation of enzyme-bound carbamate is reversible. (This accounts for the ability of the enzyme to catalyze ATP synthesis from carbamyl phosphate and ADP.) In this reaction, carbamate decomposes to ammonia and bicarbonate. It is important to note that ADP does not readily dissociate from the carboxy-phosphate-containing enzyme complex; this explains the inability of the enzyme to catalyze ADP–ATP exchange. The reaction mechanism involves two highly unstable interme-

[74] A. Meister and S. G. Powers, *Adv. Enzyme Regul.* **16,** 289 (1978).

diates, carboxy phosphate and carbamate. Both of these compounds are stabilized by their binding to the enzyme, but such stabilization is limited. Thus, formation of carbamate from carbamyl phosphate and ADP leads to formation of ammonia and bicarbonate, and enzyme-bound carboxy phosphate decomposes rapidly in the absence of glutamine and ammonia.

Further direct evidence for the presence of two functionally different ATP-binding sites was subsequently obtained in studies on the covalent modification of the active site of the enzyme by reaction with 5′-*p*-fluorosulfonyl-benzoyladenosine.[75] These studies also indicate that one site alone does not form carbamyl phosphate or does so inefficiently.

Electron paramagnetic resonance,[76] NMR,[77] and fluorescence energy transfer[78] measurements have been used to map distances between and to define the topographical relationships among substrate, metal ion, and allosteric sites on the enzyme.

Recent studies have begun to elucidate the nature of the binding sites for the allosteric effectors.[79,80] It was found that although UMP is an allosteric inhibitor, the dialdehyde derivative of UMP (prepared by periodate oxidation of UMP) is a potent activator. Dialdehyde UMP, like IMP, activates the enzyme. The corresponding dialcohol derivatives of UMP and IMP do not affect activity. The arabinose analog of UMP, in contrast to UMP, is an activator. dUMP neither activates nor inhibits, but binds to the enzyme in a manner similar to UMP. dUMP decreases inhibition by UMP and activation by IMP, but has no effect on the activation of the enzyme by L-ornithine. These results suggest that IMP and UMP bind to the same region of the enzyme, and a model has been proposed for such overlapping binding sites.

Other studies indicate that inorganic phosphate may function in the modulation of the allosteric regulation of the enzyme by nucleotides. Phosphate markedly increases inhibition by UMP, decreases activation by IMP, but does not affect activation by L-ornithine. Since the extent of activation by IMP and by L-ornithine, and that of inhibition by UMP are decreased when Mg^{2+} concentrations are increased relative to fixed concentration of ATP, it appears that the allosteric effectors may alter the affinity of the enzyme for divalent metal ions in addition to affecting the affinity of the enzyme for MgATP.[80]

[75] B. R. Boettcher and A. Meister, *J. Biol. Chem.* **255,** 7129 (1980).

[76] F. M. Raushel, C. J. Rawding, P. M. Anderson, and J. J. Villafranca, *Biochemistry* **18,** 5562 (1979).

[77] F. M. Raushel, P. M. Anderson, and J. J. Villafranca, *Biochemistry* **22,** 1872 (1983).

[78] P. G. Kasprzyk, P. M. Anderson, and J. J. Villafranca, *Biochemistry* **22,** 1877 (1983).

[79] B. R. Boettcher and A. Meister, *J. Biol. Chem.* **256,** 5977 (1981).

[80] B. R. Boettcher and A. Meister, *J. Biol. Chem.* **257,** 13971 (1982).

[42] Glutamate Synthase from *Escherichia coli, Klebsiella aerogenes,* and *Saccharomyces cerevisiae*

By ALTON MEISTER

Glutamate synthase catalyzes the following reaction:

$$\text{L-Glutamate} + \alpha\text{-ketoglutarate} + \text{NAD[P]H} + \text{H}^+ \rightleftharpoons 2\ \text{L-glutamate} + \text{ADP} + \text{P}_i + \text{NAD[P]}^+$$

This enzyme activity was discovered in *Klebsiella aerogenes* by Tempest *et al.,*[1-3] who found that this organism can grow when intracellular ammonia levels were less than 0.5 m*M*, a value below that for the K_m value for ammonia for glutamate dehydrogenate. When the organisms were grown on limiting levels of ammonia they exhibited very low levels of glutamate dehydrogenase and increased levels of glutamine synthetase. It was shown that ammonia assimilation involves glutamine synthetase and the glutamine amidotransferase glutamate synthase. Glutamate synthase has been found in many bacteria that can grow on media containing simple salts including nitrates,[3-10] nitrogen fixing bacteria,[11] the root nodules of legumes,[12] certain higher plants,[13-16] and some yeasts.[17-22] Thus far, it has

[1] D. W. Tempest, J. L. Meers, and C. M. Brown, *J. Gen. Microbiol.* **64,** 171 (1970).
[2] D. W. Tempest, J. L. Meers, and C. M. Brown, *Biochem. J.* **117,** 405 (1970).
[3] K. Adachi and I. Suzuki, *J. Bacteriol.* **129,** 1173 (1977).
[4] R. H. Dainty, *Biochem. J.* **126,** 1055 (1972).
[5] C. Elmerich and J. P. Aubert, *Biochem. Biophys. Res. Commun.* **42,** 371 (1971).
[6] R. E. Miller and E. R. Stadtman, *J. Biol. Chem.* **247,** 7407 (1972).
[7] D. W. Tempest, J. L. Meers, and C. M. Brown, *in* "The Enzymes of Glutamine Metabolism" (S. Prusiner and E. R. Stadtman, eds.), p. 167. Academic Press, New York, 1973.
[8] C. M. Brown, D. S. MacDonald-Brown, and S. O. Stanley, *J. Mar. Biol.* **52,** 793 (1972).
[9] P. J. Lea and B. J. Miflin, *Biochem. Biophys. Res. Commun.* **64,** 856 (1975).
[10] P. J. Lea and B. J. Miflin, *Biochem. Soc. Trans.* **3,** 381 (1975).
[11] H. Nagatani, M. Shimizu, and R. C. Valentine, *Arch. Mikrobiol.* **79,** 164 (1971).
[12] J. G. Robertson, M. P. Warburton, and K. J. F. Farnden, *FEBS Lett.* **55,** 33 (1975).
[13] D. K. Dougall, *Biochem. Biophys. Res. Commun.* **58,** 639 (1974).
[14] M. W. Fowler, W. Jessup, and G. S. Sarkissian, *FEBS Lett.* **46,** 340 (1974).
[15] P. J. Lea and B. J. Miflin, *Nature (London)* **251,** 614 (1974).
[16] B. J. Miflin and P. J. Lea, *Biochem. J.* **149,** 403 (1975).
[17] C. M. Brown, V. J. Burn, and B. Johnson, *Nature (London)* **246,** 115 (1973).
[18] V. J. Burn, P. R. Turner, and C. M. Brown, *Antonie van Leeuwenhoek* **40,** 93 (1974).
[19] B. Johnston and C. M. Brown, *J. Gen. Microbiol.* **85,** 169 (1974).
[20] B. Johnson, *J. Gen. Microbiol.* **89,** 295 (1975).
[21] R. J. Roon, H. L. Even, and F. Larrimore, *J. Bacteriol.* **118,** 89 (1974).
[22] D. S. Masters, Jr., Doctoral Dissertation, Graduate School of Medical Sciences, Cornell University (1979).

Copyright © 1985 by Academic Press, Inc.
All rights of reproduction in any form reserved.

not been found in animal tissues (which typically have very low levels of free ammonia and relatively high levels of glutamine synthetase). In this chapter, the purification of glutamate synthase from *E. coli* and *K. aerogenes* is described; these glutamate synthases are active with NADPH. The isolation of glutamate synthase from the eucaryote *Saccharomyces cerevisiae* is also described; this glutamate synthase uses only NADH.

Determination of NADPH-Dependent Glutamate Synthase Activity

Reagents

Tris–HCl buffer (pH 7.8 at 25°)
L-Glutamine
NADPH
Na α-ketoglutarate
Na α-keto[^{14}C]glutarate (80,000–100,000 cpm/μmol)
1 *M* HCl
1 *M* Tris
Dowex-1 acetate
0.3 *M* acetic acid

Glutamate synthase activity may be assayed in a reaction mixture (final volume, 0.5 ml) containing 0.35 m*M* NADPH, 5 m*M* sodium α-ketoglutarate, 100 m*M* Tris–HCl (pH 7.8), and 10 m*M* L-glutamine at 25°. Initial velocities are obtained from the decrease in absorbance at 340 nm and are given as units (micromoles/hour). Assays may also be performed in which α-keto[^{14}C]glutarate (80,000 to 100,000 cpm/μmol) is used in the assay mixture and in which the formation of labeled glutamate is determined. After incubation for 1.5–2 min, the reaction is stopped by adding 0.1 ml of 1 *M* HCl and the solution is placed at 0° for 10 min. After neutralization with 0.1 ml of 1 *M* Tris, the mixture is placed on a Dowex-1 acetate column; labeled glutamate is eluted with 0.3 *M* acetic acid, which does not elute α-ketoglutarate. Units of enzyme activity are expressed as μmoles of ^{14}C product formed per hour; specific activity is expressed as units per mg of protein.

Purification of Glutamate Synthase from K. aerogenes (or E. coli)

The procedure given below is based on that of Geary and Meister[23]; see also ref. 24. All steps are carried out at 4° and the centrifugations are done at 16,000 *g* for 30 min, unless otherwise stated.

[23] L. E. Geary and A. Meister, *J. Biol. Chem.* **252,** 3501 (1977).

[24] P. P. Trotta, K. E. B. Platzer, R. H. Haschemeyer, and A. Meister, *Proc. Natl. Acad. Sci. U.S.A.* **71,** 4607 (1974).

Step 1. Sonication. The frozen cell paste (400 g) is suspended in 1500 ml of 0.2 *M* potassium phosphate buffer (pH 7.6) containing 0.5 m*M* Na_2EDTA for 18 hours. The cells are ruptured in batches of 350 ml by sonication with a Branson sonifier for 15 min; the mixture is then centrifuged.

Step 2. Heat Denaturation of Impurities. Solid L-glutamine is added to the supernatant solution obtained in Step 1 to yield a final concentration of 50 m*M*. The solution is then heated with continuous stirring in a water bath at 90°. The temperature of the solution was increased to 53° within 5 min and this temperature is maintained for an additional 5 min. The solution is then cooled and the precipitated protein is removed by centrifugation.

Step 3. Protamine Sulfate Precipitation. The supernatant obtained in Step 2 is treated with 0.25 volume of protamine sulfate solution (2%; pH 5.5). After stirring for 10 min, it is centrifuged.

Step 4. Ammonium Sulfate Precipitation. The supernatant obtained in Step 3 is treated with 350 g of solid ammonium sulfate and 0.2 g Na_2EDTA per liter. The suspension is stirred for 10 min and then centrifuged. The pellet is dissolved in 280 ml of 0.2 *M* potassium phosphate (pH 7.6) containing 0.5 m*M* Na_2EDTA.

Step 5. Gel Filtration on Sephadex G-50. The solution from Step 4 is applied to a column (5 × 75 cm) of Sephadex G-50 equilibrated with 0.14 *M* potassium phosphate (pH 6.8) containing 0.5 m*M* Na_2EDTA. Effluent corresponding to the void volume of the column (about 500 ml) is collected. The protein appears in the next 500 ml, and the solution is immediately processed as described in Step 6.

Step 6. Chromatography on DEAE-Sephadex. The solution (420 ml) is applied to a column (5 × 25 cm) of DEAE-Sephadex (A-50) equilibrated with 0.14 *M* potassium phosphate buffer (pH 6.8) containing 0.5 m*M* Na_2EDTA, and the column is eluted with a linear gradient established between 3 liters of 0.14 *M* potassium phosphate (pH 7.6) containing 0.5 m*M* Na_2EDTA and 3 liters of 0.3 *M* potassium phosphate (pH 7.6) containing 0.5 m*M* Na_2EDTA. The fractions containing activity are pooled and solid ammonium sulfate (35 g/100 ml) is added. The precipitate is separated by centrifugation and dissolved in 0.2 *M* potassium phosphate buffer (pH 7.6) containing 0.5 m*M* Na_2EDTA to yield a final volume of 15 ml.

Step 7. Gel Filtration on Sephadex G-200. The solution obtained in Step 6 is applied to a column (3.3 × 100 cm) of Sephadex G-200 equilibrated with 0.2 *M* potassium buffer (pH 7.6) containing 0.5 m*M* Na_2EDTA. Elution is carried out with this buffer at a flow rate of 15 ml/hr; fractions of 5 ml are collected. Glutamate synthase elutes just after the void volume of the column (250 ml) is collected. The elution profile of ammonia-dependent activity exhibits a shoulder under the peak of glutamine-dependent activity, and a maximum appears several fractions later. The fractions containing glutamine-dependent glutamate synthase of the highest specific activity are pooled and concentrated in an Amicon cell fitted with an XM-50 membrane.

Step 8. Gel Filtration on Sepharose 6B. The concentrate obtained in Step 7 is applied (in two batches) to a column (2 × 65 cm) of Sepharose 6B equilibrated with 0.2 *M* potassium phosphate buffer (pH 7.6) containing 0.5 m*M* Na_2EDTA. Elution is carried out with this buffer at a flow rate of 10 ml/hr; fractions of 1.2 ml are collected. The fractions containing enzyme with the highest glutamine-dependent activity are pooled (fractions 54–65) and concentrated in an Amicon cell. In this chromatography, the elution of glutamine-dependent activity closely corresponds to that of the protein, but the elution profile for ammonia-dependent activity shows two peaks, the first of which corresponds to the major glutamate synthase peak. The second peak appears to be glutamate dehydrogenase. After the preparation is rechromatographed on Sephadex G-200 (i.e., Step 7 is repeated), the glutamine- and ammonia-dependent activities coincide closely. Table I gives a summary of the purification procedure.

The procedure described above is also suitable for purification of glutamate synthase from *E. coli B*. The enzyme obtained after Step 8 may be stored at 4° in 0.2 *M* potassium phosphate buffer (pH 7.6) containing 0.5

TABLE I
PURIFICATION OF GLUTAMATE SYNTHASE FROM *K. aerogenes*

Step	Volume (ml)	Protein (g)	Activity[a] Total (units)	Specific (units/mg)
1. Sonicate[b]	1,340	33.5	203,000	6.05
2. After heat treatment	1,040	18.1	187,000	10.3
3. Protamine sulfate precipitate	1,380	15.1	224,000	14.8
4. Ammonium sulfate precipitate	310	6.21	255,000	41.1
5. Gel filtration on G-50	420	5.54	208,000	37.5
6. Chromatography on DEAE-Sephadex	15	0.259	59,000	228
7. Gel filtration Sephadex G-200	3.0	0.024	21,000	875
8. Gel filtration Sepharose 6B	2.0	0.012	15,000	1250

[a] Glutamine-dependent glutamate synthase.
[b] From 400 g of cell paste.

mM Na_2EDTA. The enzyme isolated by this procedure from *E. coli* exhibits glutamine-dependent activity of about 800 μmol/mg/hr at 25°.[23] The enzyme, isolated from *E. coli B* by a different procedure, exhibited an activity of 1570 μmol/mg/hr at 37°.[25]

The purified enzyme gives a single band on polyacrylamide gel electrophoresis in 0.1 M Tris-acetate (pH 8.0). It exhibits a symmetrical boundary in sedimentation velocity runs in the analytical ultracentrifuge.[24] The sedimentation coefficient ($s_{20,w}$), determined at relatively low protein concentration, is 17.6 S. In the presence of sodium dodecyl sulfate, 2 bands are seen on polyacrylamide gel electrophoresis that corresponds in molecular weight to about 175,000 and 51,500. Dissociation of the enzyme has been achieved by treating the enzyme with 1.2 M KSCN for 1 hr at 4°. Such dissociation is accompanied by loss of catalytic activity, which is not restored by removal of thiocyanate.[24]

The absorption spectrum of the enzyme, which is to similar to that reported for *E. coli* glutamate synthase,[6] exhibits maxima at 278, 375, and 447 nm, as well as a broad shoulder of low absorbance extending above 500 nm. Glutamate synthase contains iron (mainly ferrous), sulfur, and both FMN and FAD. The available data indicate that most of the iron-sulfide and flavin binding sites are located on the heavy subunit.[24]

Catalytic Behavior

The enzyme catalyzes the formation of 2 mol of glutamate for each mole of NADPH and α-ketoglutarate utilized; studies with labeled α-ketoglutarate show that 1 mol of glutamate is formed from this substrate. The apparent K_m values for NADPH, L-glutamine, and α-ketoglutarate are 0.012, 0.3, and 0.3 mM, respectively. Neither NADH, nor deamido-NADH substitute for NADPH. Ammonium chloride *appears* to substitute for glutamine as a nitrogen donor. The pH optimum for the ammonia-dependent reaction is about 9, whereas that for the glutamine-dependent reaction is about 7.8. The apparent K_m value for ammonium chloride varied from 0.5 to 2 mM with different preparations of the enzyme which were considered to be free of glutamate dehydrogenase.

Inhibition by L-2-Amino-4-oxo-5-chloropentanoate

When the isolated enzyme is preincubated with 55 μM L-2-amino-4-oxo-5-chloropentanoate[26] there is a rapid loss of glutamine-dependent ac-

[25] R. E. Miller, *in* "Microbial Iron Metabolism: A Comprehensive Treatise" (J. B. Neilands, ed.), p. 283. Academic Press, New York, 1974.

[26] E. Khedouri, P. M. Anderson, and A. Meister, *Biochemistry* **5**, 3552 (1966).

tivity.[24] L-Glutamine (10 mM) provides substantial protection against inactivation. The glutaminase activity of the enzyme is similarly inhibited by the chloroketone and protected by glutamine. It is notable that the chloroketone has no effect on the ammonia-dependent activity. When the enzyme was incubated with ^{14}C-labeled chloroketone, and the enzyme then subjected to polyacrylamide gel electrophoresis in the presence of sodium dodecyl sulfate, 90% of the radioactivity was found in the heavy subunit. The amount of radioactivity incorporated corresponds to the binding of about 1.1 mol of analog per 227,000 g of enzyme. These studies indicate that the glutamine binding site of the enzyme is located on the heavy subunit of the enzyme.

Although highly purified preparations of glutamate synthase catalyze reductive amination in the presence of either L-glutamine or ammonia, preparations of the enzyme that lack flavins (FAD and FMN) or the flavins and iron sulfide, catalyze ammonia-mediated glutamate synthesis, but not glutamine-mediate glutamate synthesis.[23] The participation of the enzyme-flavin in the reductive amination reaction with glutamine, but not in that with ammonia is also indicated by studies in which the enzyme is reduced by treatment with dithionite. Thus, when the enzyme was incubated with $Na_2S_2O_4$, α-keto[^{14}C]glutarate, and glutamine, substantial formation of labeled glutamate (which was shown to be of the L configuration) was found.[23] However, only trace amounts of labeled glutamate were formed when glutamine was replaced by ammonia. A study with stereospecifically labeled [^3H]NADPH shows that glutamate synthase, like glutamate dehydrogenase, uses the hydrogen atom on the B side of C_4 of the nicotinamide ring of NADPH. In studies with [^3H]NADPH(B), it was found that the products formed in the reductive amidation of α-ketoglutarate with glutamine and ammonia are different. In the reaction with glutamine, ^3H is transferred from [^3H]NADPH(B) to water; ^3H_2O and glutamate are formed stoichiometrically. On the other hand, in the reaction with NH_3, [^3H]glutamate is formed. The enzyme catalyzes rapid exchange of tritium from [^3H]NADPH(B) with H_2O, and a much slower exchange of tritium of [α-^3H]glutamate with water.[23]

These observations indicate that reductive amidation with glutamine is flavin-mediated whereas reductive amidation with ammonia is not and therefore closely resembles the reaction catalyzed by glutamate dehydrogenase in which ^3H is transferred from [^3H]NADPH(B) to [^3H]glutamate. These observations suggest that highly purified glutamate synthase preparations from *E. coli* and *K. aerogenes* may contain two different catalytic entities, i.e., glutamate synthase and glutamate dehydrogenase. Although preparations of glutamate synthase can evidently use either glutamine or ammonia for the reductive amidation of α-ketoglutarate, there is good evidence that the mechanism of glutamine-dependent glutamate formation is fundamentally different from that involved in the ammonia-dependent formation of glutamate. It is notable, for example, that apoglutamate synthase (preparations that lack flavins and iron sulfide) as well as the deflavoenzyme can form glutamate from ammonia and α-ketoglutarate, but do not form glutamate from glutamine and α-ketoglutarate. These and other[23] considerations suggest that the reductive amidation reaction with ammonia is not a partial reaction of glutamate synthase. In should be mentioned, however, that other investigators[27-29] concluded that the ammonia-dependent activity is not due to the presence of glutamate dehydrogenase; further work in this area would be desirable.

Although a number of glutamine amidotransferases have been found to have separate subunits that serve the function of binding glutamine and facilitating transfer of its amide nitrogen, the subunits of *E. coli* and *K. aerogenes* glutamate synthase do not seem to function in this way. It thus appears that the heavy subunit contains the flavins, iron sulfide, and the glutamine binding site. The function of the light subunits of *K. aerogenes* and *E. coli* glutamate synthases is not yet known.

Determination of NADH-Dependent Glutamate Synthase Activity (S. cerevisiae)

Reagents

Tris–HCl buffer (pH 7.6 at 37°)
Na α-keto[^{14}C]glutarate
L-Glutamine
NADH
1 *M* Trichloroacetic acid
Dowex 50 (H^+)
3 *M* ammonium hydroxide

The composition of the reaction mixture used for assay is as follows: 10 m*M* Tris–HCl (7.6), 0.5 m*M* sodium α-keto[^{14}C]glutarate (70,000 cpm), 5 m*M* glutamine, 0.1 m*M* NADH, and enzyme in a final volume of 1 ml. The reactions are initiated by adding enzyme and are incubated at 37° for 15 min. The reactions are stopped by adding 0.2 ml of 1 *M* trichloroacetic acid. After centrifugation, portions (0.5 ml) of the deproteinized supernatant solutions are added to columns (1 × 3 cm) of Dowex-50 (H^+). The columns are washed with 6 ml of water; [^{14}C]glutamate is then eluted with

[27] P. Mantsala and H. Zalkin, *J. Biol. Chem.* **251,** 3294 (1976).
[28] P. Mantsala and H. Zalkin, *J. Biol. Chem.* **251,** 3300 (1976).
[29] P. Mantsala and H. Zalkin, *J. Bacteriol.* **126,** 539 (1976).

TABLE II
SUMMARY OF PURIFICATION OF GLUTAMATE SYNTHASE FROM *Saccharomyces cerevisiae*

Step	Volume (ml)	Protein (g)	Activity[a] Total (units)	Specific (units/mg)
1. Extract, supernatant[b]	2000	32.5	175,000	5.38
2. Ammonium sulfate precipitate	200	3.30	61,000	18.5
3. First DEAE-Sepharose chromatography	140	0.560	96,300	172
4. Phenyl-Sepharose chromatography	100	0.260	55,900	215
5. Second DEAE-Sepharose chromatography	18	0.133	50,500	380
6. Blue Sepharose chromatography	10	0.007	25,000	3,570
7. AcA 22 chromatography	5	0.0003	12,500	41,700

[a] Glutamine-dependent glutamate synthase activity.
[b] From 2280 g of yeast cakes.

3 ml of 3 *M* ammonium hydroxide and the radioactivity present is determined by scintillation counting. Controls in which NADH, glutamine, and enzyme are separately omitted are carried out. A unit of enzyme activity is defined as the amount of enzyme that catalyzes the formation of 1 μmol of glutamate per minute under these conditions. Protein is determined by the method of Lowry *et al.*[30] using crystalline bovine serum albumin as standard.

Purification of Glutamate Synthase from Saccharomyces cerevisiae

This procedure is that of Masters and Meister.[31] Unless otherwise stated, all procedures are carried out at 4° and centrifugations are performed at 16,000 *g* for 30 min. A summary of the purification is given in Table II.

Step 1. Preparation of Extracts. Commercial yeast cakes (2280 g) are suspended in ice-cold 100 m*M* potassium phosphate buffer (pH 7.6) in a ratio of 1 g of yeast per ml of buffer. The cells are disrupted by 2 passages through a Manton-Gaulin model 15 M laboratory homogenizer fitted with a cell disruption valve at 8000 pounds pressure. Cell debris is removed by

[30] O. H. Lowry, N. J. Rosebrough, A. L. Farr, and R. J. Randall, *J. Biol. Chem.* **193,** 265 (1951).
[31] D. S. Masters, Jr. and A. Meister, *J. Biol. Chem.* **257,** 8711 (1982).

centrifugation and the pH of the supernatant solution is adjusted to pH 7.6 by adding 1 *M* KOH. Protamine sulfate (1 g/100 g of yeast) is dissolved in the minimal volume of 100 m*M* potassium phosphate buffer (pH 7.6). The protamine solution is then added with stirring to the supernatant solution obtained above. After standing for 25 min the mixture is centrifuged, and the pellet is discarded.

Step 2. Precipitation with Ammonium Sulfate. The supernatant solution obtained in Step 2 is brought to 40% of ammonium sulfate saturation by adding 243 g of solid ammonium sulfate per liter, and the pH is adjusted to 7.6 by adding concentrated NH_4OH. After stirring for 30 min, the mixture is centrifuged, and the pellet obtained is resuspended in 200 ml of 20 m*M* Tris–HCl (pH 7.6) buffer containing 20 m*M* KCl, 1 m*M* EDTA, and 20% glycerol (Buffer A). This solution is dialyzed against 40 volumes of Buffer A for 18 hr, after which the insoluble material is removed by centrifugation.

Step 3. First Chromatography on DEAE-Sepharose. The solution obtained in Step 2 is added to the top of a column (2.5 × 30 cm) of DEAE-Sepharose previously equilibrated with Buffer A. The column is washed with Buffer A until no material absorbing at 280 nm appears in the effluent. The enzyme is then eluted with a linear gradient established between 50 ml of Buffer A and 500 ml of Buffer A containing 500 m*M* KCl. Enzyme activity elutes in a peak centered at a conductivity of 8.2 mS. The fractions containing activity are pooled and the protein present is precipitated by adding 32 g of solid ammonium sulfate per 100 ml. The precipitate is collected by centrifugation and dissolved in Buffer A containing 88 g/liter of ammonium sulfate.

Step 4. Chromatography of Phenyl-Sepharose. The solution obtained in Step 3 is added to the top of a column (2.5 × 30 cm) of Phenyl-Sepharose previously equilibrated with Buffer A containing 88 g/liter of ammonium sulfate. The column is washed with this buffer until no material absorbing at 280 nm appears in the effluent. The enzyme was then eluted from the column using Buffer A; the fractions containing the enzyme are combined and the solution is treated with 32 g of solid ammonium sulfate per 100 ml. The precipitated protein is collected by centrifugation and dissolved in Buffer A and then dialyzed against Buffer A for 18 hr.

Step 5. Second Chromatography on DEAE-Sepharose. The solution obtained in Step 4 is added to the top of a column (1 × 30 cm) of DEAE-Sepharose previously equilibrated with Buffer A. The enzyme is eluted using a linear gradient established between 500 ml of Buffer A containing 20 m*M* KCl and 500 ml of Buffer A containing 50 m*M* KCl. The fractions containing the enzyme are combined and the protein is precipitated by adding 32 g of solid ammonium sulfate per 100 ml. The precipitate is collected by centrifugation and dissolved in the minimal volume of Buffer A containing 50 m*M* 2-mercaptoethanol and then dialyzed against this buffer for 18 hr.

Step 6. Chromatography on Blue-Sepharose. The enzyme solution obtained in Step 5 is added to the top of a column (1 × 20 cm) of Blue-Sepharose equilibrated with Buffer A containing 50 m*M* 2-mercaptoethanol. The column is washed with this buffer until no further material absorbing at 280 nm appears in the effluent. The enzyme is eluted with Buffer A containing 50 m*M* 2-mercaptoethanol and 100 μM NADH. The fractions containing activity are combined.

Step 7. Chromatography on Ultrogel AcA 22. The solution obtained in Step 6 is added to the top of a column (2 × 100 cm) of Ultrogel AcA 22 equilibrated with Buffer A containing 50 m*M* 2-mercaptoethanol. The enzyme activity eluted as a single peak; the fractions containing activity are combined.

Comments on the Purification Procedure

It should be noted that the enzyme is relatively unstable at the end of Step 5. Thus, losses of about two-thirds of the activity are observed when the enzyme is stored for 2 hr at 4° in Tris-buffer after the second chromatography on DEAE. Addition of 2-mercaptoethanol leads to substantial but incomplete stabilization of the enzyme; the enzyme is dissolved in buffer containing 50 m*M* 2-mercaptoethanol, and the loss of enzyme activity is about 8% after storage at 4° for 18 hr. The glutamate dehydrogenase activity of *S. cerevisiae* is about 10 times greater than that of glutamine synthase activity present. More than 95% of the glutamate dehydrogenase activity is removed by the first chromatography on DEAE-Sepharose (Step 3), in which glutamate dehydrogenase elutes after the bulk of the glutamate synthase. In the second DEAE-Sepharose chromatography (Step 5), glutamate synthase is completely separated from the small amount of remaining glutamate dehydrogenase activity.

Molecular Weight and Subunit Composition

The purified enzyme inhibits a single band when subjected to polyacrylamide gel electrophoresis at pH 8.9. The molecular weight of the enzyme as determined by gel filtration is 265,000 ± 26,500. When the enzyme is treated with 1% sodium dodecyl sulfate and 1% 2-mercaptoethanol, it migrates as two components on polyacrylamide gel electrophoresis carried out in the presence of 0.1% sodium dodecyl sulfate. The apparent molecular weight of the large subunit is 169,000 ± 16,900 and that of the small subunit is 61,000 ± 6,100. When the isolated enzyme was treated with dimethylsuberimidate and then subjected to polyacrylamide gel electrophoresis in the presence of sodium dodecyl sulfate, three protein bands are found whose positions are consistent with components equivalent to the large subunit, small subunit, and the holoenzyme.[31]

The spectrum of the purified enzyme exhibits a maximum at 278 nm and shoulders at 380, 430, and 470 nm.

Catalytic Properties

In contrast to the bacterial enzymes, this enzyme uses NADH exclusively, and exhibits low but significant activity when ammonia is substituted for glutamine. NADH is not oxidized in the absence of α-ketoglutarate. The apparent K_m values for L-glutamine, α-ketoglutarate, and NADH are 280, 40, and 7 μM, respectively. The enzyme is not active when α-ketoglutarate is replaced by oxaloacetate, α-keto-*n*-valerate, or α-ketoadipate. No activity is found when L-glutamine is replaced by D-glutamine, the optical isomers of asparagine and homoglutamine, α-methyl-L-glutamine, DL-β-glutamine, L-isoglutamine, glutaramide, or glutaramate. The activity with ammonia is about 2.7% of that found with L-glutamine. The glutamine-dependent activity of the enzyme exhibits a broad optimal range between pH 7 and 7.5, whereas the rate of the ammonia-dependent reaction increases steadily as the pH is increased from 6 to 9.

In contrast to glutamate synthases from *K. aerogenes* and *E. coli,* the ammonia-dependent activity appears to be a property of the synthase itself rather than to glutamate dehydrogenase.

The glutamate synthase activity of *S. cerevisiae* is markedly inhibited by DL-homocysteine sulfonamide and L-methionine sulfone. In contrast, D-methionine sulfone is a poor inhibitor; some inhibition was observed with DL-ethionine sulfone, L-methionine-*SR*-sulfoximine, L-albizziin, DL-homocysteic acid, L-methionine-*SR*-sulfoxide, and D-glutamine. The findings suggest that homocysteine sulfonamide is a transition state inhibitor.

[43] Glutamine Transaminase L from Rat Liver

By ARTHUR J. L. COOPER and ALTON MEISTER

Assay Method

Principles. The original assay procedure required coupling of the transaminase and ω-amidase activities.[1,2] Ammonia liberated from amide cleavage of α-ketoglutaramate was measured by diffusion into acid followed by Nesslerization. Later, a radiochemical assay was devised which did not require the simultaneous presence of ω-amidase.[3] Thus, labeled glutamine is converted to labeled α-ketoglutaramate (or to labeled α-ketoglutarate if ω-amidase is present). The labeled α-keto acid is separated from labeled glutamine on small Dowex 50 (H^+) columns. However, the procedure is tedius and lacks sensitivity due to high blanks. Labeled glutamine is slowly converted to labeled 5-oxoproline in the assay mixture; labeled 5-oxoproline elutes with α-ketoglutaramic acid. Glycine formed by glyoxylate can be determined colorimetrically, but the reaction is nonlinear and only of limited applicability.[3] Glycine produced in the L-glutamine-glyoxylate transaminase assay can be rapidly (~15 min) quantitated by means of precolumn derivatization with *o*-phthalaldehyde, followed by HPLC analysis on a C_{18} column (method 1). An alternative and sensitive assay can be used to screen a large number of assay samples. This method (assay 2) depends on the fact that L-albizziin (a structural analog of glutamine) is a good substrate of the L-type transaminase but not of the K type.[4-8] The α-keto acid analog of L-albiziin cyclizes to 4-

[1] A. Meister, P. E. Fraser, and S. V. Tice, *J. Biol. Chem.* **206,** 561 (1954).
[2] A. Meister and P. E. Fraser, *J. Biol. Chem.* **210,** 37 (1954).
[3] A. J. L. Cooper and A. Meister, *Biochemistry* **11,** 661 (1972).
[4] A. J. L. Cooper and A. Meister, *J. Biol. Chem.* **248,** 8499 (1973).
[5] A. J. L. Cooper and A. Meister, *J. Biol. Chem.* **249,** 2554 (1974).
[6] A. J. L. Cooper and A. Meister, *Comp. Biochem. Physiol. B* **69B,** 137 (1981).
[7] A. J. L. Cooper and A. Meister, "Glutamine Transaminases. International Union of Biochemistry, Symposium 121 on Chemical and Biological Aspects of Vitamin B_6 Catalysis, May 1983, Athens, Greece." p. 3. Alan R. Liss, Inc., New York, 1984.
[8] The V_{max} with L-albizziin is slightly less than with L-glutamine and slightly greater than with methionine (20 m*M* glyoxylate as substrate).[2,4] However, the affinity toward L-albizziin is ~10 times lower than toward L-glutamine or L-methionine. For example, addition of 20 m*M* L-glutamine or of 40 m*M* L-methionine to the standard 40 m*M* L-albizziin–5 m*M* α-keto-γ-methiolbutyrate transaminase assay mix results in ~75 and 90% inhibition, respectively, of 2-imidazolinone-4-carboxylate formation catalyzed by both soluble and mitochondrial glutamine transaminase L.

Copyright © 1985 by Academic Press, Inc.
All rights of reproduction in any form reserved.

hydroxy-2-imidazolidone-4-carboxylic acid which in turn is readily dehydrated in acid or base to 2-imidazolinone-4-carboxylic acid. 2-Imidazolinone-4-carboxylic acid absorbs strongly in the UV whereas most α-keto acids do not. The L-albizziin-glyoxylate or L-albizziin-α-keto-γ-methiolbutyrate transamination reactions can be used to monitor glutamine transaminase activity in rat organ preparations.[6,7,9] The ratio of the glutamine-glyoxylate/L-albizziin-α-keto-γ-methiolbutyrate transaminase activities is 1.2 and 1.6, respectively, for the soluble and mitochondrial glutamine transaminase L.

Glutamine Transaminase L Assay 1

Reagents

500 m*M* sodium borate buffer, pH 8.5
100 m*M* L-glutamine
200 m*M* sodium glyoxylate
400 m*M* sodium acetate buffer, pH 5.0

OPT Reagent. *o*-Phalaldehyde (270 mg), 5 ml methanol, 0.2 ml 2-mercaptoethanol, made up to 50 ml with 0.4 *M* potassium borate, pH 9.6. Liquid phases for HPLC: (A) 50 m*M* sodium acetate, pH 6.2; (B) 100% methanol.

Procedure. The reaction mixture contains 100 m*M* sodium borate buffer, 20 m*M* L-glutamine, 20 m*M* sodium glyoxylate, and glutamine transaminase L. After incubation at 37° for 30–60 min the reaction is stopped by addition of 50 μl of 3 *M* perchloric acid. The blank contains assay mixture to which enzyme is added after addition of 3 *M* perchloric acid. Precipitated protein is removed by centrifugation. The supernatant is neutralized with 2 *M* potassium bicarbonate and the precipitate is removed by centrifugation. The supernatant is mixed with OPT reagent (20 μl) for 30 sec followed by addition of 10 μl of sodium acetate buffer. After 45 sec, 10 μl is injected into the HPLC system. The C_{18} column is eluted for 2 min with 25% B : 75% A. Between 2 and 10 min a gradient is established between 25% B : 75% A and 35% B : 65% A. Glutamine is eluted at 7 min and glycine is eluted at 9 min. The flow rate is 0.85 ml/min and detection is by fluorescence (excitation, 330 nm; emission, 418 nm). The

[9] A small amount of ammonia production was noted in reaction mixtures containing L-albizziin, glyoxylate, purified glutamine transaminase, and ω-amidase. However, no ammonia production was noted on incubating homogenates in assay system 2, suggesting that transamination of L-albizziin results in stoichiometric formation of 4-hydroxy-2-imidozolidone-4-carboxylate.

glycine peak is compared to 1 mM standard taken through the extraction procedure.[10]

Glutamine Transaminase L Assay 2

Reagents

1 M ammediol–HCl, pH 8.5
100 mM L-albizziin
100 mM sodium α-keto-γ-methiolbutyrate
100 mM 2-mercaptoethanol

Procedure. The reaction mixture contains 50 mM ammediol, 40 mM L-albizziin, 5 mM α-keto-γ-methiolbutyrate, 10 mM 2-mercaptoethanol, and glutamine transaminase L, pH 8.5, 37°. After incubation for 5–60 min, the reaction is terminated by addition of 0.9 ml of 1 M KOH. After a further 20 min at 37° the absorbance at 280 nm due to 2-imidazolinone-4-carboxylate is determined ($\varepsilon = 10{,}000$), The blank contains assay mixture to which enzyme is added after addition of base.[11]

Purification of Soluble Glutamine Transaminase L[12]

The previously published method based on the glutamine-glyoxylate transamination reaction is reproduced here. Note that this purification will be easier to follow using the L-albizziin-α-keto-γ-methiolbutyrate transaminase assay in place of the L-glutamine-glyoxylate assay.

In the following procedure all steps are carried out at 0–4° unless otherwise stated and all centrifugations are carried out at 45,000 g.

Step 1. Male Sprague–Dawley rats are decapitated and exsanguinated. The livers (200 g) are removed and homogenized in a Waring Blender in 3 volumes of 0.05 M Tris–HCl, pH 8.4. The homogenate is centrifuged and the precipitate is discarded.

[10] The OPT analysis method is based on that of B. N. Jones, S. Pääbo, and S. Stein [*J. Liq. Chromatogr.* **4,** 565 (1981)]. We thank Dr. Sol N. Mora for providing us with the routine for separating the OPT derivatives of glutamine and glycine.

[11] It was previously reported that high levels of α-keto-γ-methiolbutyrate (>5 mM) inhibited the enzyme.[3] Part of this inhibition was due to substrate inhibition but some activity was lost even after dialysis. Some preparations of α-keto-γ-methiolbutyrate apparently contain a contaminant that oxidizes the enzyme; this loss of activity due to oxidation cannot be reversed by subsequent addition of 2-mercaptoethanol but can be completely prevented by inclusion of 2-mercaptoethanol in the assay mix.

[12] A unit of enzyme activity is defined as the amount that catalyzes the formation of 1 μmol of α-ketoglutaramate (or glycine) per hour at 37° in the standard glutamine-glyoxylate reaction mixture. Specific activity is expressed as units/mg protein. Protein was determined by the method of Lowry *et al.* [O. H. Lowry, N. J. Rosebrough, A. L. Farr and N. J. Randall, *J. Biol. Chem.* **193,** 265 (1951)].

Step 2. The supernate solution from step one is treated with solid sodium pyruvate to yield a final concentration of 10 m*M*. This solution is then heated with constant stirring to 63–65° at which temperature it is kept for 20 min. The solution is cooled in ice and centrifuged. The pellet is discarded. This step effectively destroys the ω-amidase activity.

Step 3. The supernatant solution from step 2 is adjusted to pH 5.0 by addition of 2 *M* acetic acid. Solid ammonium sulfate (25 g/100 ml) is slowly added with constant stirring and after standing for 20 min the precipitate which formed is removed by centrifugation and discarded.

Step 4. The active fraction from step 3 (fraction 1) is treated with additional solid ammonium sulfate (18 g/100 ml) and after standing for 20 min the precipitate which forms is collected by centrifugation and dissolved in approximately 20 ml of 5 m*M* potassium phosphate buffer, pH 7.2, and dialyzed against 2 changes of 5 liters each of the same buffer. An inactive precipitate which forms is removed by centrifugation.

Step 5. The dialyzed solution from step 4 (fraction 2) is applied to the top of a column (3.5 × 20 cm) of DE-52 previously equilibrated with 5 m*M* potassium phosphate buffer, pH 7.2. The column is eluted with a linear gradient established between 1 liter of 5 m*M* and 1 liter of 70 m*M* potassium phosphate, pH 7.2, at a flow rate of 20 ml/hr. Both buffers contain 1 m*M* 2-mercaptoethanol.

Step 6. The active fractions are pooled, concentrated by ultrafiltration to 10 ml with a Diaflo XM 50 membrane, and dialyzed against 15 m*M* potassium phosphate buffer, pH 7.2. The dialyzed solution is then applied to the top of a hydroxylapatite column previously equilibrated with the same buffer used for dialysis. The column is eluted with this buffer until very little protein appears ($A_{280\ nm} < 0.01$) in the effluent. The column is then further eluted with a linear gradient established between 250 ml of 0.15 m*M* and 250 ml of 80 m*M* potassium phosphate buffer, pH 7.2, containing 1 m*M* 2-mercaptoethanol. The flow rate is 20 ml/hr. A single protein peak containing the enzyme is eluted at 30 m*M* phosphate. The enzyme solution is concentrated with a Diaflo XM 50 membrane to 10 ml.

The purification procedure is summarized in Table I.

Comments on the Purification Procedure

The isolation procedure yields a preparation of glutamine transaminase that is devoid of glutamate-aspartate transaminase,glutamate-alanine transaminase, and ω-amidase. The enzyme is moderately stable and may be stored for at least 3 months at 4° in 10 m*M* potassium phosphate buffer, pH 7.2, containing 10% glycerol, 1 m*M* 2-mercaptoethanol, and 10 μ*M* pyridoxal 5′-phosphate.

TABLE I
PURIFICATION OF SOLUBLE GLUTAMINE TRANSAMINASE L FROM RAT LIVER

Step	Volume (ml)	Protein (mg)	Units	Specific activity	Recovery (%)	Purification (fold)
1. Crude extract from 200 g of liver	600	47,800	16,500	0.345	100	1
2. After heat treatment	580	19,800	15,200	0.768	92	2.2
3. Ammonium sulfate fraction 1	190	540	10,800	2.00	65	5.8
4. Ammonium sulfate fraction 2	25	906	8,250	9.10	50	26.4
5. DE-52 column chromatography	500	47.3	7,250	153	44	443
6. Hydroxylapatite column chromatography	120	22.0	6,600	300	40	870

Properties of Soluble Glutamine Transaminase L. The enzyme has an estimated molecular weight of 110,000 and is composed of two identical subunits. It contains tightly bound pyridoxal 5′-phosphate which is difficult to remove even under denaturing conditions.[3] The enzyme is inhibited by hydroxylamine, cycloserine, and aminooxyacetate. Soluble glutamine transaminase L is active with a large number of α-keto acids[1-3] and amino acids.[3] The best amino acid substrates are glutamine (and various glutamine analogs) and methionine (and various methionine analogs). The most active α-keto acid substrates are α-keto-γ-methiolbutyrate, glyoxylate, 3-mercaptopyruvate, and α-ketoglutaramate (open-chain from). The enzyme may be regarded as a fully reversible glutamine-methionine transaminase.

Purification of Mitochondrial Glutamine Transaminase L

The starting material is the mitochondrial pellet obtained as described in the section on the purification of mitochondrial ω-amidase. Mitochondrial glutamine transaminase L copurifies with ω-amidase during ammonium sulfate fractionation but is much more strongly retained on DE-52. After eluting the ω-amidase with 5 mM potassium phosphate buffer, pH 7.2, the transaminase is eluted with a gradient established between 1 liter of this buffer and 1 liter of 120 mM potassium phosphate buffer, pH 7.2. The activity is eluted in a broad peak centered at about 60 mM phosphate.

TABLE II
PURIFICATION OF MITOCHONDRIAL GLUTAMINE TRANSAMINASE L FROM RAT LIVER

Step	Volume (ml)	Protein (mg)	Units	Specific activity	Recovery (%)	Purification (fold)
1. Crude extract from disrupted mitochondria	550	13,500	15,350	1.14	100	1
2. After ammonium sulfate fractionation	360	7,610	15,800	2.07	103	1.8
3. DE-52 chromatography	880	716	4,860	6.79	32	6.8

(For comparison in a separate experiment the soluble enzyme was found to be eluted with 30 m*M* phosphate buffer.)

A summary of the purification is given in Table II.

Comments on the Occurrence of Isozymic Forms of Rat Liver Glutamine Transaminase L

Yoshida[13] noted that glutamine-glyoxylate transaminase activity was present in the soluble and mitochondrial fractions of rat liver and presented evidence for two isozymic forms of the enzyme. Thus the mitochondrial enzyme is more strongly retained on DEA-cellulose than the cytosolic form. The two activities migrate with different mobilities during zone electrophoresis.[13] We later confirmed that rat liver mitochondria contain glutamine transaminase L activity[3] and now show that the mitochondrial glutamine transaminase L activity (L-albizziin-α-keto-γ-methiolbutyrate transaminase assay) is more strongly retained on DE-52 than is the soluble form in agreement with the previous observation of Yoshida.

Although the present procedure provides only a 7-fold purification it is devoid of ω-amidase activity and of glutamine transaminase K activity. The available evidence suggests distinct cytoplasmic and mitochondrial glutamine transaminase L-ω-amidase pathway in the rat, but more work is needed to characterize the mitochondrial pathway.

[13] T. Yoshida, *Bitamin* **35,** 227 (1967).

[44] Glutamine Transaminase K from Rat Kidney

By ARTHUR J. L. COOPER and ALTON MEISTER

Introduction

Early studies suggested that glutamine transaminase activity is higher in rat liver than in rat kidney and that, of the tissues investigated, activity is greatest in the liver.[1] Subsequently, Kupchik and Knox concluded that, in contrast to the earlier reports, rat kidney has a much higher level of glutamine transaminase than does the rat liver.[2,3] However, Cooper and Meister noted that most of the early reports described transamination between glutamine and pyruvate, whereas Kupchik and Knox had used the glutamine-phenylpyruvate reaction.[4] It was later shown that rat liver homogenates are indeed more active than kidney homogenates in catalyzing transamination between glutamine and pyruvate (or glyoxylate), but that kidney homogenates are about 5 times more active than liver homogenates in catalyzing transamination between glutamine and phenylpyruvate.[4] It was found that kidney possesses a glutamine transaminase activity distinct from that previously purified from the liver.[4] The two activities were named glutamine transaminase K and glutamine transaminase L, respectively.[4] Later studies showed that the L-type activity is present in the soluble and mitochondrial fractions of rat liver and that L-type activity is low in rat kidney and brain.[5] Conversely, the K-type activity appears to be more widespread; it is highest in the soluble fraction of kidney with smaller amounts in the soluble liver fractions, the kidney mitochondria, liver mitochondria, and brain mitochondria.[5] The L-type activity was purified about 155-fold from rat kidney homogenates and shown to be about 80% pure as judged by the criterion of polyacrylamide gel electrophoresis; however, no attempt was made to separate the mitochondrial and soluble form of the enzyme.[4] Later, the enzyme was purified 200-fold from the soluble fraction of rat liver homogenates and about 400-fold from the mitochondrial fraction.[6] By the criterion of polyacrylamide gel electrophoresis in three separate systems the soluble enzyme was 90% pure; the mitochondrial preparation contains two proteins of

[1] A. Meister, H. A. Sober, S. V. Tice, and P. E. Fraser, *J. Biol. Chem.* **197,** 319 (1952).
[2] H. Z. Kupchik and W. E. Knox, *Arch. Biochem. Biophys.* **136,** 178 (1970).
[3] H. Z. Kupchik and W. E. Knox, this series, Vol. 17A, p. 951.
[4] A. J. L. Cooper and A. Meister, *J. Biol. Chem.* **249,** 2554 (1974).
[5] A. J. L. Cooper and A. Meister, *Comp. Biochem. Physiol. B* **69B,** 137 (1981).
[6] A. J. L. Cooper, *Anal. Biochem.* **89,** 451 (1978).

Copyright © 1985 by Academic Press, Inc.
All rights of reproduction in any form reserved.

which the glutamine transaminase protein was the smaller component. Both the soluble and mitochondrial glutamine transaminase K are of similar molecular weight (100,000; 2 identical subunits) but the mitochondrial enzyme binds more tightly to DE-52 and is relatively more active with short-chain α-keto acids.[5]

Assay Methods

Principles. The original assay required coupling of the transaminase to ω-amidase; the liberated ammonia from amide cleavage of α-ketoglutaramate was measured by diffusion into acid followed by Nesslerization.[1] Later, Kupchik and Knox described an assay that did not require the presence of ω-amidase.[2,3] The procedure measures the rate of disappearance of phenylpyruvate-enol in the presence of 300 mM borate buffer, pH 8.5 (ε = 3570 at 300 nm) and 60 mM L-glutamine. The method is somewhat insensitive and has the drawback that disappearance of substrate, rather than appearance of product is measured. An 8-fold more sensitive assay that relies on measurement of product appearance has been described.[6] It has been shown that, in ammediol buffer, pH 9.0, L-phenylalanine-α-keto-γ-methiolbutyrate transaminase activity [Eq. (1)] in rat kidney is solely due to glutamine transaminase K.[6] The product of the reaction, phenylpyruvate, absorbs strongly in 3 M NaOH (ε = 24,000 at 322 nm) whereas the absorbance due to α-keto-γ-methiolbutyrate is negligible.[6]

$$\text{L-Phenylalanine} + \alpha\text{-keto-}\gamma\text{-methiolbutyrate} \rightleftharpoons \text{phenylpyruvate} + \text{L-methionine} \quad (1)$$

Standardization of Phenylpyruvate and α-Keto-γ-methiolbutyrate Solutions. Phenylpyruvate and α-keto-γ-methiolbutyrate are substrates of lactate dehydrogenase although the activity is lower than that with pyruvate.[7] Standardization is achieved by noting the disappearance of NADH at 340 nm (ε = 6230) due to complete reduction of the α-keto acid [Eq. (2)].

$$\alpha\text{-Keto acid} + \text{NADH} + \text{H}^+ \xrightarrow[\text{dehydrogenase}]{\text{lactic}} \alpha\text{-hydroxy acid} + \text{NAD}^+ \quad (2)$$

The reaction mixture contains 50 mM potassium phosphate buffer, pH 7.0, 100 μM NADH (at least 20% excess), 5 μg of lactic dehydrogenase (beef heart), and α-keto acid ($\leq$80 nmol of α-keto acid to be standardized) in a final volume of 1.0 ml. The reaction is over in 1 hr at 25°. The α-keto acid concentration is calculated from the decrease in absorbance at 322 nm relative to a blank lacking α-keto acid.

[7] A. Meister, *J. Biol. Chem.* **197,** 309 (1952).

L-*Phenylalanine–α-Keto-γ-methiolbutyrate Transaminase Assay*

Reagents

1 *M* ammediol–HCl, pH 9.0 (filtered through activated charcoal)
100 m*M* L-phenylalanine
100 m*M* sodium α-keto-γ-methiolbutyrate
3.3 *M* NaOH

Procedure. The reaction mixture contains 200 m*M* ammediol–HCl buffer, pH 9.0, 10 m*M* L-phenylalanine, 5 m*M* α-keto-γ-methiolbutyrate, and enzyme in a final volume of 0.1 ml. After incubating at 37° for 10 min, 0.9 ml of 3.33 *M* NaOH is added and the absorbance of 322 nm is determined. The blank contains the assay mixture in which the enzyme is omitted (or added just prior to addition of base).[8,9] The absorbance due to phenylpyruvate is stable for at least 15 min.[10]

Purification of Soluble Mitochondrial Glutamine Transaminase K[6]

All steps are carried out at 0–4°.

Step 1. Seventy adult Wistar rats are killed by decapitation. The kidneys are removed, cleaned, weighed, and homogenized in 200 ml of 0.25 *M* sucrose in a Waring Blender for 20 sec. The homogenate is centrifuged at 16,000 *g* for 30 min.

Step 2. The supernate from step 1 is applied directly to a DE-52 column (16 × 2.5 cm) equilibrated with 2 m*M* potassium phosphate buffer, pH 7.2, and eluted with 1 liter of 7 m*M* potassium phosphate buffer, pH 7.2. The enzyme is eluted with 25 m*M* potassium phosphate buffer, pH 7.2.

Step 3. The active fractions from step 3 are combined, dialyzed against

[8] The reaction is only linear to about 10% conversion of phenylalanine to phenylpyruvate after which, presumably, the reverse reaction becomes significant.

[9] A unit of enzyme activity is defined as the amount that catalyzes the formation of 1 μmol of phenylpyruvate/hr at 37°. Specific activity is expressed as units/mg of protein. Protein was measured by the method of Lowry *et al.* [O. H. Lowry, N. J. Rosebrough, A. L. Farr, and R. J. Randall, *J. Biol. Chem.* **193,** 265 (1951)].

[10] The reported extinction coefficients of phenylpyruvate in base have been somewhat variable. Scandurra and Cannella reported ε = 15,700 (318 nm) in 1 *M* NaOH [R. Scandurra and C. Cannella, *Anal. Biochem.* **27,** 253 (1969)]. Goldstein reported an ε value of 10,400 (322 nm) in 1.1 *M* NaOH [F. B. Goldstein, *J. Biol. Chem.* **236,** 2656 (1961)]. Cooper reported an ε value of 24,000 (322 nm) in 3 *M* NaOH.[6] Part of the explanation for the differences may be due to the fact that absorbance at 322 nm increases with increasing NaOH, reaching a maximum value in 3 *M* NaOH. Also, we have noted that concentrations of enzymatically standardized solutions of phenylpyruvate are slightly lower than expected from weighed samples, suggesting that commercial preparations contain an impurity or are somewhat hydrated.

distilled water so that the final concentration of buffer is 5 mM, and applied to a hydroxylapatite column (8 × 2.5 cm) equilibrated with the same buffer. The column is eluted with 500 ml of 5 mM potassium phosphate buffer, pH 7.2, followed by a linear gradient established between 250 ml of 5 mM and 250 ml of 150 mM potassium phosphate buffer, pH 7.2. The enzyme is eluted in the first 160 ml of the gradient. The active fractions are combined, diluted 9-fold with deionized water, and added to a small DE-52 column (1 × 1.5 cm). The enzyme is eluted with 1 ml of 1 M potassium phosphate buffer, pH 7.2.

Step 4. The concentrated enzyme from step 3 is applied directly to the top of a Sephadex G-150 column (108 × 2.5 cm) equilibrated with 5 mM potassium phosphate buffer, pH 7.2, and eluted with the same buffer. The active fractions are combined with concentrated as directed in step 3.

The purification procedure is summarized in Table I.

Purification of Mitochondrial Glutamine Transaminase K

All steps are carried out at 0–4°, and all centrifugations are carried out at 40,000 g for 30 min unless otherwise stated.

Step 1. Seventy male Wistar rats, weighing approximately 300 g each, are decapitated and their kidneys are removed. The kidneys are homogenized in a Teflon-glass homogenizer in 7 vol of 50 mM Tris–HCl buffer, pH 7.4 containing 150 mM KCl, 50 mM $MgCl_2$, and 1 mM EDTA. After centrifugation at 100,000 g for 60 min, the supernate is discarded. The pellet is resuspended in 200 ml of water, freeze-thawed twice, and then sonicated twice, each time for a period of 1 min. After centrifugation, the inactive pellet is discarded.

Step 2. Solid sodium pyruvate is added to bring the final concentration to 20 mM. The solution is heated at 60° for 20 min, cooled, and centrifuged.

TABLE I
PURIFICATION OF SOLUBLE GLUTAMINE TRANSAMINASE K FROM RAT KIDNEY[a]

Step	Volume (ml)	Protein (mg)	Units	Specific activity	Yield (%)	Purification (fold)
1. Homogenate	700	14,000	27,000	1.93	100	1
2. DE-52 chromatography	500	758	21,600	28.5	80	15
3. Hydroxylapatite chromatography	160	142	10,800	76.2	40	40
4. Sephadex G-150 chromatography	30	10.8	4,200	389	16	202

[a] From 180 g of rat kidney.

Step 3. The supernate from step 2 is applied to the top of a DE-52 column (36 × 2.5 cm) previously equilibrated with 5 mM potassium phosphate buffer, pH 7.2. The column is eluted with 400 ml of the same buffer and then with a linear gradient established between 1 liter of 10 mM and 1 liter of 250 mM potassium phosphate buffer, pH 7.2. The active fractions are pooled and dialyzed against several volumes of distilled water in order to bring the concentration of potassium phosphate to approximately 5 mM.

Step 4. The dialyzed sample is added to a column of hydroxylapatite (11.2 × 2.5 cm) previously equilibrated with 5 mM potassium phosphate buffer, pH 7.2. The column is eluted with 100 ml of the same buffer and then with a linear gradient established between 1 liter of 5 mM and 1 liter of 80 mM potassium phosphate buffer, pH 7.2. The active fractions are pooled and concentrated to 4 ml by ultrafiltration through a Diaflo XM-50 membrane.

Step 5. The concentrated solution is then added to a Sephadex G-150 column (119 × 2.5) previously equilibrated with 5 mM potassium phosphate buffer; the enzyme is eluted with this buffer.

The purification is summarized in Table II.

Comments on the Purification Schemes

At no time during the purification does addition of pyridoxal 5′-phosphate result in an increase in transaminase activity. The mitochondrial enzyme has been stored for over a year in 200 mM potassium phosphate buffer, pH 7.2, containing 20% glycerol and 5 mM 2-mercaptoethanol. The soluble enzyme has been stored for up to 7 years under similar conditions.

TABLE II

PURIFICATION OF MITOCHONDRIAL GLUTAMINE TRANSAMINASE K FROM RAT KIDNEY[a]

Step	Volume (ml)	Protein (mg)	Units	Specific activity	Yield (%)	Purification (fold)
1. Sonicated mitochondria	240	9070	3440	0.38	100	1
2. After heat treatment	190	2120	2430	1.15	71	3
3. DE-52 chromatography	1100	176	1630	9.26	47	24
4. Hydroxylapatite chromatography	820	26.7	1390	52.0	40	137
5. Sephadex G-150 chromatography	80	6.6	1050	159	31	418

[a] From 180 g of rat kidney.

The microsomal fraction of rat kidneys appears to contain very little glutamine-phenylpyruvate transaminase activity. Thus, in one experiment the mitochondrial, microsomal and high-speed supernates contained 12, 1.8, and 86%, respectively, of the total activity.[6]

Comments on the Specificity

At least 12 amino acids and sixteen α-keto acids are known to be substrates of the soluble enzyme.[4,6] At least 8 amino acids and 11 α-keto acids are known to be substrates of the mitochondrial form.[4-6] Noguchi *et al.* showed that rat tissues possess two enzymes (I and II) capable of catalyzing the histidine-pyruvate transaminase reaction.[11] Enzyme I was shown to be present only in liver and to be induced by glucagon.[11-13] Enzyme II (noninducible) was shown to be present in liver, kidney, heart, and brain. Enzyme I is probably asparagine transaminase.[14] Enzyme II was later shown by Noguchi *et al.* to copurify with glutamine transaminase in rat kidney preparations[15] and to be present in both soluble and mitochondrial fractions.[11-13] It was later shown that the glutamine-phenylpyruvate/phenylalanine-α-keto-γ-methiolbutyrate/histidine-pyruvate transaminase activities copurified in a constant ratio for both soluble and mitochondrial forms of the enzyme.[6] The ratio of activities is approximately 0.9 : 1.0 : 0.09, respectively, for both enzymes. The highest activity measured, thus far, is with the glutamine (20 mM)–α-keto-γ-methiolbutyrate (10 mM) pair ($\sim$2.5 times the phenylalanine-α-keto-γ-methiolbutyrate transaminase activity).[4,5,16] Both enzymes should be regarded as fully reversible glutamine (methionine)-phenylalanine transaminases.[4,5]

[11] T. Noguchi, Y. Minatogawa, E. Okuno, and R. Kido, *Biochem. J.* **157,** 635 (1976).
[12] T. Noguchi, E. Okuno, Y. Minatogawa, and R. Kido, *Biochem. J.* **159,** 607 (1976).
[13] T. Noguchi, E. Okuno, and R. Kido, *Biochem. J.* **159,** 607 (1976).
[14] T. Noguchi, Y. Takada, and Y. Oota, *Hoppe-Seyler's Z. Physiol. Chem.* **360,** 919 (1979).
[15] T. Noguchi, Y. Minatogawa, Y. Takada, E. Okuno, and R. Kido, *Biochem. J.* **170,** 173 (1978).
[16] Unfortunately, the glutamine-α-keto-γ-methiolbutyrate transaminase activity is more difficult to assay and is less sensitive than the phenylalanine-α-keto-γ-methiolbutyrate transaminase activity.[4,5]

[45] α-Keto Acid ω-Amidase from Rat Liver

By ARTHUR J. L. COOPER, THOMAS E. DUFFY, and ALTON MEISTER

Introduction

In the 1940s Greenstein and co-workers discovered two glutaminases in rat liver. Glutaminase I was activated by phosphate and to a lesser extent by arsenate; glutaminase II was enhanced by pyruvate (see Greenstein[1] for an early review). Later work established that the glutaminase II reaction was due to the combined action of two enzymes, a glutamine transaminase (EC 2.6.1.15) which catalyzes the transfer of the amino group of glutamine to a suitable α-keto acid acceptor [Eq. (1)] and an amidohydrolase, designated α-keto acid ω-amidase (EC 3.5.1.3), which catalyzes the deamidation of α-ketoglutaramate [Eq. (2)].[2–11]

$$\text{L-Glutamine} + \alpha\text{-keto acid} \rightleftharpoons \alpha\text{-ketoglutaramate} + \text{L-amino acid} \quad (1)$$
$$\alpha\text{-Ketoglutaramate} + H_2O \rightarrow \alpha\text{-ketoglutarate} + \text{ammonia} \quad (2)$$

Glutamine transaminase activity is widespread.[12] There are now known to be two major glutamine transaminases in the rat. A liver type (designated L) and a kidney type (designated K).[13–18] In addition, each

[1] J. P. Greenstein, *Adv. Enzymol.* **8,** 117 (1949).
[2] A. Meister and S. V. Tice, *J. Biol. Chem.* **187,** 173 (1950).
[3] A. Meister, H. A. Sober, S. V. Tice, and P. E. Fraser, *J. Biol. Chem.* **197,** 319 (1952).
[4] A. Meister, *J. Biol. Chem.* **200,** 571 (1953).
[5] A. Meister, *J. Biol. Chem.* **206,** 587 (1954).
[6] A. Meister, *J. Biol. Chem.* **210,** 17 (1954).
[7] A. Meister, *Science* **120,** 43 (1954).
[8] A. Meister, P. E. Fraser, and S. V. Tice, *J. Biol. Chem.* **206,** 561 (1954).
[9] A. Meister and P. E. Fraser, *J. Biol. Chem.* **210,** 37 (1954).
[10] A. Meister, L. Levintow, R. E. Greenfield, and P. A. Abendschein, *J. Biol. Chem.* **215,** 441 (1955).
[11] A. Meister, *Adv. Enzymol.* **16,** 185 (1955).
[12] For a review, see A. J. L. Cooper and A. Meister, *in* "The Transaminases" (D. E. Metzler and P. Christen, eds.), p. 401. Wiley, New York, 1985.
[13] A. J. L. Cooper and A. Meister, *Biochemistry* **11,** 661 (1972).
[14] A. J. L. Cooper and A. Meister, *J. Biol. Chem.* **248,** 8489 (1973).
[15] A. J. L. Cooper and A. Meister, *J. Biol. Chem.* **249,** 2554 (1974).
[16] A. J. L. Cooper and A. Meister, *CRC Crit. Rev. Biochem.* **4,** 281 (1977).
[17] A. J. L. Cooper and A. Meister, *Comp. Biochem. Physiol. B* **69B,** 137 (1981).
[18] A. J. L. Cooper and A. Meister, "Glutamine Transaminases. International Union of Biochemistry Symposium 121 on Chemical and Biological Aspects of Vitamin B_6 Catalysis, May 1983, Athens, Greece." p. 3. Alan R. Liss, Inc., New York, 1984.

Copyright © 1985 by Academic Press, Inc.
All rights of reproduction in any form reserved.

type exists in soluble and mitochondrial forms. (The purification and characteristics of these enzymes are described in the preceding chapters.)

ω-Amidase activity is also widespread. For example, the activity has been detected in various rat organs, Novikoff tumors, mouse liver, yeast, bacteria, and plants.[4] Meister obtained a 40-fold purification of a rat liver enzyme and later Hersh obtained a 146-fold purification from the soluble fraction of rat liver homogenates.[19] Hersh assumed that the ω-amidase is a cytosolic enzyme, but later work with freeze-thawed rat liver mitochondria established the presence of ω-amidase in rat liver mitochondria.[15] Cooper and Meister[15] showed that the activity in a disrupted mitochondrial preparation migrates on discontinuous gel electrophoresis with a different R_f value from that of the activity in the soluble fraction and suggested that rat liver contains isozymic forms of ω-amidase. This suggestion has been confirmed.[20] The metabolic importance of the glutamine transaminase–ω-amidase pathway has been reviewed.[12,18]

Principles of the Assay Methods. The assay method first used by Meister and co-workers involved measurement of ammonia released from α-ketoglutaramate by diffusion into acid followed by Nesslerization.[4] Later Hersh measured α-ketoglutarate formation [Eq. (2)] in a continuous assay by noting the rate of disappearance of absorbance at 340 nm due to oxidation of NADH in a reaction mixture containing ammonium chloride, excess glutamate dehydrogenase, and NADH.[19] Vergara *et al.*[21] used the same coupled assay but included 0.1 m*M* ADP to stabilize glutamate dehydrogenase. An alternative assay of ω-amidase makes use of the fact that α-ketoglutarate forms a 2,4-dinitrophenylhydrazone which absorbs strongly at 430 nm in base, whereas α-ketoglutaramate does not readily form a hydrazone. In neutral or acid solution, α-ketoglutaramate exists >99% in a cyclic (lactam) configuration (5-hydroxy-2-pyrrolidone-5-carboxylate)[4,19,22,23] which is unreactive toward carbonyl reagents. The hydrazone method is more convenient than the coupled assay when analyzing large numbers of samples, such as during purification of ω-amidase.

Preparation of Barium α-Ketoglutaramate. L-Glutamine is oxidized with L-amino acid oxidase in the presence of catalase.[4,24] Following removal of the enzyme by dialysis or by ultrafiltration, ammonia and unreacted amino acid are removed on a Dowex 50 (H^+) column. The effluent

[19] L. B. Hersh, *Biochemistry* **10,** 2884 (1971).

[20] K. G. MacPhee and S. M. Schuster, *Fed. Proc., Fed. Am. Soc. Exp. Biol.* **42,** 1967 (1983) (abstr.).

[21] F. Vergara, F. Plum, and T. E. Duffy, *Science* **183,** 81 (1974).

[22] A. J. L. Cooper and M. Gross, *J. Neurochem.* **28,** 771 (1977).

[23] T. T. Otani and A. Meister, *J. Biol. Chem.* **224,** 137 (1957).

[24] A. Meister, this series, Vol. 3, p. 404.

is decolorized with charcoal and concentrated by flash evaporation or by lyophilization to a volume one-tenth that of the original volume. The pH is raised to 5.0 by addition of saturated barium hydroxide (previously filtered to remove barium carbonate), and the barium salt of α-ketoglutaramic acid is precipitated by addition of five volumes of ice-cold ethanol. Nonenzymatic cyclization of glutamine to 2-pyrrolidone-5-carboxylate (5-oxoproline) and ammonia is minimized by carrying out the incubation with L-amino acid oxidase/catalase at pH 6.8 and 25°.[25]

Standardization of Solutions of α-Ketoglutaramate. There are two available methods. Ammonia released on hydrolysis of α-ketoglutaramic acid with 1 *N* HCl (1 hr, 100°) or with ω-amidase at pH 8.0 can be standardized, spectrophotometrically.[26,27] α-Ketoglutarate liberated from the ω-amidase reaction can be standardized spectrophotometrically.[27,28]

ω-Amidase Assay Method 1[21]

Reagents

1 *M* Tris–HCl, pH 8.5
10 m*M* adenosine disphosphate, sodium salt
100 m*M* α-ketoglutaramate, barium salt
1 *M* ammonium acetate
5 m*M* NADH
100 m*M* dithiothreitol
Glutamate dehydrogenase (10 mg/ml in 50% glycerol; specific activity of 120 U/mg at 25°)

Procedure. The reaction mixture (1.0 ml) contains 100 m*M* Tris–HCl buffer, 20 m*M* α-ketoglutaramate, 40 m*M* ammonium acetate, 0.1 m*M* adenosine disphosphate, 1 m*M* dithiothreitol, and 0.1 mg of glutamate dehydrogenase. The reaction is initiated by addition of ω-amidase (2–5 μl) and the rate of loss of absorbance at 340 nm due to oxidation of NADH is continuously recorded; $\varepsilon = 6.23 \times 10^3$. This assay method yields activity values 30–40% lower than values obtained by assay method 2 due to inhibition of ω-amidase by ammonium ions (see below).

[25] A sample of the barium salt of α-ketoglutaramate prepared in this manner was heated at 100° in 1 *N* HCl to convert 2-pyrrolidone-5-carboxylate to glutamate. Amino acid analysis revealed <0.1% 2-pyrrolidone-5-carboxylate in the preparation.

[26] T. E. Duffy, A. J. L. Cooper, and A. Meister, *J. Biol. Chem.* **249,** 7603 (1974).

[27] T. E. Duffy, A. J. L. Cooper, and F. Vergara, *Bioorg. Chem.* **5,** 351 (1976).

[28] A. J. L. Cooper, A. K. Dahr, H. Kutt, and T. E. Duffy, *Anal. Biochem.* **103,** 118 (1980).

ω-Amidase Assay Method 2[22,28]

Reagents

1 *M* Tris–HCl, pH 8.5
100 m*M* α-ketoglutaramate, barium salt
100 m*M* 2-mercaptoethanol
0.1% 2,4-dinitrophenylhydrazine in 2 *M* HCl
1 *M* KOH

Procedure. The reaction mixture (20 μl in a small stoppered tube) contains 100 m*M* Tris–HCl buffer, 20 m*M* α-ketoglutaramate, 5 m*M* 2-mercaptoethanol, and 1–5 μl of ω-amidase. After incubation of the mixture for 5–30 min, the reaction is terminated by addition of 20 μl of 2,4-dinitrophenylhydrazine reagent. After a further incubation for 10 min, 0.96 ml of 1 *M* KOH is added and the absorbance at 430 nm is determined within 5 min after addition of base. The blank contains assay mixture which is incubated without enzyme but to which enzyme is added after addition of 2,4-dinitrophenylhydrazine reagent. ω-Amidase activity is directly proportional to the increase in absorbance up to 0.8 optical density units. If the enzyme preparation contains considerable inorganic orthophosphate (such that the concentration of phosphate in the assay mixture is ≥5 m*M*), sodium α-ketoglutaramate should be used in place of barium α-ketoglutaramate to prevent precipitation of barium phosphate. Sodium α-ketoglutaramate is prepared by treating a solution of barium α-ketoglutaramate with an equivalent amount of sodium sulfate; the precipitated barium sulfate is removed by filtration or centrifugation.

Purification of Soluble ω-Amidase

The purification procedure is essentially that of Hersh[19] but with several modifications which result in a higher yield and a preparation with a higher specific activity. The most important modifications are the use of hydroxylapatite packed into a column, rather than added directly to the enzyme solution, and the elimination of the Sephadex G-75 chromatography step.

All steps are carried out at 0–4°.[29]

[29] A unit of enzyme activity is defined as the amount that catalyzes the formation of 1 μmol of α-ketoglutarate per hour at 37°. Specific activity is expressed as units/mg protein. Protein was determined using the Bio-Rad Protein Assay and bovine serum albumin as a standard.

Step 1. Twenty male Wistar rats weighing 350–500 g are decapitated and exsanguinated. Livers are homogenized in 2.4 volumes of a solution containing 70 m*M* sucrose, 220 m*M* mannitol, and 2 m*M* potassium HEPES (pH 7.4), using a Potter-Elvehjem homogenizer. The liver homogenate is centrifuged at 14,000 *g* for 1 hr.

Step 2. To the supernatant solution from step 1 is added solid ammonium sulfate (200 g/liter). After gentle stirring for 12 hr the precipitate is removed by centrifugation at 14,000 *g* for 30 min and discarded. A further 250 g/liter of ammonium sulfate is added and after 5 hr the precipitate is removed by centrifugation and resuspended in the minimum volume (~200 ml) of 5 m*M* potassium phosphate buffer (pH 7.2). After extensive dialysis against 3 changes of 4 liters of 5 m*M* phosphate buffer, the solution is centrifuged to remove an inactive precipitate.

Step 3. The supernatant solution from step 2 is applied to a DE-52 column (28 × 2.5 cm) equilibrated with 5 m*M* potassium phosphate buffer (pH 7.2). The cloudy pass-through is discarded and the enzyme is eluted with 2 liters of the same buffer. Soluble glutamine transaminase L is retained on the column.[30]

Step 4. The active fractions from step 3 are combined and concentrated by ultrafiltration to ~5 ml using a Millipore Pellicon membrane (cut off 10,000 molecular weight). The concentrate is added to a Sephadex CM-50 column (105 × 2.5 cm), previously equilibrated with 5 m*M* potassium phosphate buffer, pH 7.2, and eluted with the same buffer.

Step 5. The active fractions from step 4 are combined and added directly to a hydroxylapatite column (9 × 2.5 cm) equilibrated with 5 m*M* potassium phosphate, pH 7.2. The column is eluted with 650 ml of 10 m*M* potassium phosphate, pH 7.2. The enzyme is then eluted with 1.5 l of 50 m*M* potassium phosphate, pH 7.2, concentrated to ~250 ml, and stored in 20% glycerol containing 5 m*M* 2-mercaptoethanol. (Sephadex G-150 column chromatography of the concentrated enzyme from this step did not increase the specific activity of the preparation.)

The purification scheme is summarized in Table I.

Comments on the Purified Enzyme

The enzyme activity is sensitive to oxidation but full activity is easily regenerated in a few minutes by addition of 5 m*M* 2-mercaptoethanol. The enzyme is quite stable. A preparation with a high specific activity and stored in 30% glycerol (0.8 mg protein/ml)[21] is still active after 11 years.

[30] The stronger binding of glutamine transaminase compared to ω-amidase on DEA-cellulose was first noted by T. Yoshida [*Bitamin* **35,** 227 (1967)].

TABLE I
PURIFICATION OF SOLUBLE ω-AMIDASE FROM RAT LIVER

Step	Volume (ml)	Protein (mg)	Units	Specific activity	Recovery (%)	Purification (fold)
1. Crude extract from 347 g of liver	740	20,800	373,000	17.9	100	1
2. Ammonium sulfate fractionation and dialysis	340	11,400	318,000	27.9	85	1.6
3. DE-52 chromatography	1,150	2,020	260,000	129	70	7.2
4. Sephadex CM-50 chromatography	200	964	207,000	215	55	12.0
5. Hydroxylapatite chromatography	1,500	77	196,000	2,545	53	142

The present procedure gives a 5-fold higher recovery of enzyme than that originally described by Hersh.[19] The specific activity of 42.4 μmol/min/mg at 37° (Table I) is also higher than that reported previously (11.7 μmol/min/mg at 30°).[19] Vergara *et al.*[21] have obtained a specific activity of 15.8 μmol/min/mg at 25° in which the exact procedure of Hersh was followed except that the last step (hydroxylapatite adsorption) was omitted (the ratio of activities at 25, 30, and 37° are 1.0 : 1.3 : 2.0, respectively). Hersh stated that his preparation contained a single protein band on discontinuous gel electrophoresis. Analysis of the ω-amidase obtained from the present procedure by polyacrylamide slab gel electrophoresis[31] under nondenaturing conditions also revealed a major protein band (R_f, 0.41 relative to bromophenol blue) which coincided with enzyme activity. However, electrophoresis of the enzyme preparation under the same conditions, except that 0.1% SDS was included in the preparation and added to the gel, revealed 4 major bands and many (>10) minor bands in the molecular weight region of 30,000–60,000.

Properties of Cytoplasmic ω-Amidase. The enzyme has an estimated molecular weight of 58,000 and is composed of 2 subunits of molecular weight ~28,000.[19] It hydrolyzes α-ketoglutaramate, α-ketosuccinamate, glutaramate, and succinamate to the corresponding dicarboxylate and ammonia.[4] ω-Amidase hydrolyzes the monomethyl and monoethyl esters of α-ketoglutarate (γ-ester), glutarate, and succinate,[19] and the *p*-chloro, *p*-methyl, and unsubstituted phenyl esters of glutarate.[32] ω-Amidase also

[31] U. K. Laemmli, *Nature (London)* **277,** 680 (1970).
[32] L. B. Hersh, *Biochemistry* **11,** 2251 (1972).

catalyzes hydroxaminolysis and transamidation reactions.[4,32] Apparently, the catalytic sequence involves two steps and an acyl enzyme intermediate.[19,32]

The enzyme is inhibited by sulfhydryl reagents, such as *p*-mercuribenzoate, 5,5′-dithiobis-(2-nitrobenzoic acid) (Ellman's reagent), *N*-ethylmaleimide, iodoacetate, and iodoacetamide.[19] Ammonium ions have been reported to inhibit ω-amidase.[19] We have confirmed this finding with the present preparation and estimate that the K_i for NH_4^+ is ~10 m*M*. MacPhee and Schuster[20] reported that Ca^{2+} activates soluble ω-amidase-catalyzed hydrolysis of α-ketosuccinamate. Hersh noted no effect of calcium on ω-amidase activity toward α-ketoglutaramate and we were unable to detect activation of ω-amidase, with 50 m*M* $CaCl_2$, toward either α-ketoglutaramate or α-ketosuccinamate. Glycylglycine is a competitive inhibitor of ω-amidase (K_i = 5 m*M*).[33] The apparent K_m values of α-ketoglutaramate and α-ketosuccinamate have been reported as 3 m*M*[19] and 3.8 m*M*,[16] respectively. However, only 0.3% of α-ketoglutaramate exists in the open-chain form at pH 8.5; the true K_m of ω-amidase for α-ketoglutaramate open-chain structure may be of the order of 10 μ*M*.[19]

Use of Soluble ω-Amidase-Glutamate Dehydrogenase to Determine α-Ketoglutaramate in Biological Samples

α-Ketoglutaramate has been shown to occur in rat liver, kidney, and brain and in human cerebrospinal fluid.[26] In patients with liver disease and encephalopathy there is a good correlation between the degree of neurological dysfunction and increase in α-ketoglutaramate in cerebrospinal fluid.[34] The ω-amidase prepared by the above procedure or that previously[21] is suitable for the determination of α-ketoglutaramate.

The samples to be analyzed (15 to 50 μl) are added to 100 μl of a solution containing 50 m*M* Tris–HCl buffer (pH 8.5), 10 m*M* 2-mercaptoethanol, and 5 μg of ω-amidase. As controls, samples are mixed with 100 μl of a similar solution lacking ω-amidase. After incubation at 37° for 3 hr, the mixture is treated with 1 ml of a solution containing 50 m*M* Tris–HCl, buffer (pH 7.5), and 0.1 m*M* sodium ADP, 10 m*M* ammonium acetate, and 2 μ*M* NADH. After determination of the initial fluorescence (excitation, 360 nm; fluorescence emission, 460 nm), 0.1 mg of glutamate dehydrogenase is added, and the decrease in fluorescence is followed until the reaction is complete as judged by a constant level of fluorescence. The concentration of α-ketoglutaramate in the sample is calculated

[33] A. J. L. Cooper, *J. Biol. Chem.* **252,** 2032 (1977).

[34] T. E. Duffy, F. Vergara, and F. Plum, *Res. Publ.—Assoc. Res. Nerv. Ment. Dis.* **53,** 39 (1974).

TABLE II
PURIFICATION OF MITOCHONDRIAL ω-AMIDASE FROM RAT LIVER

Step	Volume (ml)	Protein (mg)	Units	Specific activity	Recovery (%)	Purification (fold)
1. Crude extract from disrupted mitochondria	550	13,500	114,000	8.4	100	1
2. Ammonium sulfate fractionation	360	7,610	110,000	14.5	96	1.7
3. DE-52 chromatography	440	—	82,500	—	75	—
4. Hydroxylapatite chromatography	78	50	37,100	742	33	88
5. Sephadex CM-50 chromatography	160	12	12,700	1060	11	126

from the net change in NADH fluorescence, i.e., after subtraction of the value obtained on the control sample that is not treated with ω-amidase.[35]

Purification of Mitochondrial ω-Amidase

All steps are carried out at 0–4°.

Step 1. The pellet from the 14,000 *g* centrifugation of the rat liver homogenate obtained in step 1 of the purification procedure for soluble ω-amidase is frozen at −20°. The frozen pellet is slowly thawed, suspended in 3 volumes of ice-cold 10 m*M* potassium phosphate buffer, pH 7.2, and sonicated for 60 sec. The homogenate is centrifuged at 14,000 *g* for 1 hr and the pellet is discarded.

Steps 2 and 3. These steps are identical to those described for the soluble enzyme.

Step 4. The active fractions from step 3 are added directly to a hydroxylapatite column (2.5 × 8 cm) equilibrated with 5 m*M* potassium phosphate buffer, pH 7.2. The column is eluted with 1 liter of this buffer, followed by 1 liter of 50 m*M* potassium phosphate buffer, pH 7.2.

Step 5. The active fractions from step 4 are combined and concentrated to ~3 ml by ultrafiltration using a Diaflo XM 50 membrane. The concentrate is added directly to a column of Sephadex CM-50 (2.5 × 110

[35] α-Ketoglutaramate has also been assayed in spinal fluid by gas chromatography of the tris trimethylsilyl derivative of the lactam.[28] Recent studies suggest that α-ketoglutaramate can be estimated in deproteinized human cerebrospinal fluid by HPLC analysis using a Whatman SAX column, 10 m*M* phosphate–HCl (pH 4.5) as eluting buffer, and UV detection at 205 nm (S. M. Fitzpatrick, A. J. L. Cooper, and T. E. Duffy, unpublished results).

cm) equilibrated with 5 mM potassium phosphate buffer, pH 7.2. The enzyme is eluted with this same buffer, and stored in the presence of 20% glycerol containing 5 mM 2-mercaptoethanol.

The purification scheme is summarized in Table II.

Comments on the Purified Enzyme

The preparation of mitochondrial ω-amidase exhibits a major band of activity on discontinuous polyacrylamide slab gel electrophoresis (R_f relative to bromophenol blue, 0.38) corresponding to enzyme activity. The mitochondrial enzyme is more heat labile than the soluble enzyme.[20] An investigation of the heat lability of the soluble ω-amidase and mitochondrial ω-amidase prepared by the above procedures shows that at 60° both enzymes lose all activity within 10 min. However, at 50° the purified mitochondrial enzyme (10 μg in 10 mM potassium phosphate buffer, pH 7.2) loses activity with a $t_{1/2}$ of ~10 min; under the same conditions the purified soluble enzyme loses activity with a $t_{1/2}$ of ~3 hr.

[46] Transglutaminases[1]

By J. E. Folk and Soo Il Chung

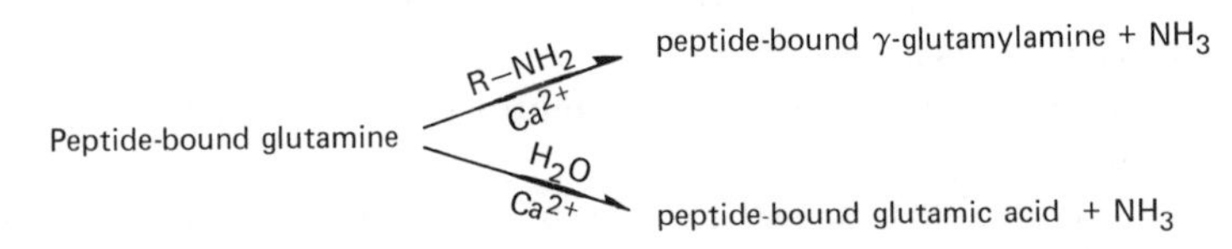

There are several enzymes that catalyze the calcium-dependent acyl transfer reaction in which γ-carboxamide groups of peptide-bound glutamine residues serve as acyl donors. These enzymes, the transglutaminases, display specificity differences that are expressed as variations in reactivity toward glutamine residues[2] and that are predicated by the sequences of amino acid residues surrounding the glutamines.[3] The specificity of each of the transglutaminases toward acceptor substrates is

[1] EC 2.3.2.13; Protein-glutamine: amine γ-glutamyltransferase (hydrolyzing).

[2] J. J. Gorman and J. E. Folk, *J. Biol. Chem.* **255,** 419 (1980).

[3] J. J. Gorman and J. E. Folk, *J. Biol. Chem.* **256,** 2712 (1981).

Copyright © 1985 by Academic Press, Inc.
All rights of reproduction in any form reserved.

broad. Indeed, many primary amines function *in vitro* as acceptors[4,5] with the subsequent formation of peptide-bound γ-glutamylamines.[6] At lower than adequate levels of a primary amine, water acts as an acceptor and peptide-bound glutamic acid is a product.[6,7] The acceptor role of ε-amino groups of peptide-bound lysine residues is of prime biological importance. Intermolecular ε-(γ-glutamyl)lysine crosslinking, catalyzed through exchange of the ε-amino groups of a lysine residue in one molecule of protein for ammonia at the carboxamide groups of a glutamine residue of another protein molecule, is known to be the basic reaction underlying a number of physiological processes.[8] These include the covalent polymerization of fibrin during hemostasis,[9,10] the production of the vaginal plug by postejaculatory clotting of rodent seminal plasma,[11] and the formation of the chemically resistant envelope of the stratum corneum during terminal differentiation of keratinocytes.[12] Each of these events is catalyzed by a different transglutaminase and in each the proper and efficient disposition of ε-(γ-glutamyl)lysine bonds between protein molecules reflects the specificity of the enzyme involved.

Guinea Pig Liver Transglutaminase[13]

Assay

The activity of this enzyme has been determined by measurement of the incorporation of any of several radiolabeled primary amines (e.g., putrescine, methylamine, or glycine ethyl ester) into mixed[14] or pure[2] caseins, the acetylated B chain of oxidized insulin[15] or the simple glutaminyl peptide derivative, benzyloxycarbonyl(*Z*)-glutaminylglycine.[16,17]

[4] D. D. Clarke, M. J. Mycek, A. Neidle, and H. Waelsch, *Arch. Biochem. Biophys.* **79,** 338 (1959).

[5] J. H. Pincus and H. Waelsch, *Arch. Biochem. Biophys.* **126,** 44 (1968).

[6] J. E. Folk and P. W. Cole, *J. Biol. Chem.* **240,** 2951 (1965).

[7] M. J. Mycek and H. Waelsch, *J. Biol. Chem.* **235,** 3513 (1960).

[8] J. E. Folk and J. S. Finlayson, *Adv. Protein Chem.* **31,** 1 (1977).

[9] J. J. Pisano, J. S. Finlayson, and M. P. Peyton, *Science* **160,** 892 (1968).

[10] S. Matacic and A. G. Loewy, *Biochem. Biophys. Res. Commun.* **30,** 356 (1968).

[11] H. G. Williams-Ashman, A. C. Notides, S. S. Pabalan, and L. Lorand, *Proc. Natl. Acad. Sci. U.S.A.* **69,** 2322 (1972).

[12] R. H. Rice and H. Green, *J. Cell Biol.* **76,** 705 (1978).

[13] Less purified preparations of this enzyme are described in this series, Vol. 5 [112] and Vol. 17 [127].

[14] H. Walsch and M. J. Mycek, this series, Vol. 5 [112].

[15] S. I. Chung and J. E. Folk, *J. Biol. Chem.* **247,** 2798 (1972).

[16] J. E. Folk, *J. Biol. Chem.* **244,** 3707 (1969).

[17] T. Abe, S. I. Chung, R. P. DiAugustine, and J. E. Folk, *Biochemistry* **16,** 5495 (1977).

When hydroxylamine is employed as the acceptor substrate with Z-glutaminylglycine, the product is a peptide derivative of γ-glutamylhydroxylamine:

$$C_6H_5{-}CH_2{-}O{-}\overset{O}{\overset{\|}{C}}{-}\overset{H}{\overset{|}{N}}{-}\underset{C(=O){-}NH{-}CH_2{-}COOH}{\overset{CH_2CH_2C(=O){-}NH_2}{C{-}H}} \longrightarrow C_6H_5{-}CH_2{-}O{-}\overset{O}{\overset{\|}{C}}{-}\overset{H}{\overset{|}{N}}{-}\underset{C(=O){-}NH{-}CH_2{-}COOH}{\overset{CH_2CH_2C(=O){-}NHOH}{C{-}H}}$$

The colored complex of this derviative which is formed upon addition of acid ferric chloride reagent is sufficiently soluble in acid to allow its direct measurement. This is the principle of the assay method described here for guinea pig liver transglutaminase and routinely used in its purification. Although this method is limited in sensitivity, it provides the specificity needed for purification in a simple and straightforward colorimetric assay.

The purified enzyme may be assayed by continuous recording of the *p*-nitrophenol release during the calcium-dependent hydrolysis of *p*-nitrophenyl acetate.[18] Procedures are available for active site titration of purified guinea pig liver transglutaminase with the use of iodoacetamide,[19] α-bromo-4-hydroxy-3-nitroacetophenone,[20] and methyl *N*-(2 hydroxy-5-nitrophenylacetyl)-L-2-amino-4-oxo-5-chloropentanoate.[20]

Reagents

Tris-acetic acid buffer, pH 6.0, 1 *M* in Tris

Z-L-glutaminylglycine solution, 0.2 *M*, 67.5 mg of the peptide derivative is dissolved in 0.5 ml of 0.4 *M* NaOH and the solution is adjusted to pH 6.0 and to a final volume of 1 ml

$CaCl_2$, 0.1 *M*

Hydroxylamine hydrochloride, 2 *M*

Ethylenediaminetetracetic acid (EDTA) tetra sodium salt, 0.02 *M*

Ferric chloride reagent. Prepared by mixing equal parts of 15% trichloroacetic acid, 5% $FeCl_3$ in 0.1 *N* HCl, and 2.5 *N* HCl.[21] This mixture may be stored indefinitely

Synthesis of Z-L-Glutaminylglycine.[22] A solution of 28 g of Z-L-glu-

[18] J. E. Folk, P. W. Cole, and J. P. Mullooly, *J. Biol. Chem.* **242,** 2615 (1967).

[19] J. E. Folk and P. W. Cole, *J. Biol. Chem.* **241,** 5518 (1966).

[20] J. E. Folk and M. Gross, *J. Biol. Chem.* **246,** 6683 (1971).

[21] F. Lipmann and L. C. Tuttle, *J. Biol. Chem.* **159,** 21 (1945); see also this series, Vol. 3 [39].

[22] J. E. Folk, unpublished; see J. E. Folk and P. W. Cole [*Biochim. Biophys. Acta* **122,** 244 (1966)] for an alternative procedure used for the preparation of Z-D-glutaminylglycine.

tamine and 14 ml of triethylamine in 140 ml of dimethylformamide is cooled to −7°. To this solution is added slowly with stirring 13 ml of isobutyl chloroformate. Stirring is continued at −7° for 30 min at which time is added a mixture of 14 g of finely powdered glycine ethyl ester HCl and 14 ml of triethylamine in 70 ml of dimethylformamide. The mixture is allow to warm to room temperature and is stirred overnight at this temperature. A portion of 2.1 liters of 0.05 *N* $NaHCO_2$ solution is added. Upon cooling to 5° a heavy crystalline precipitate forms. This precipitate is collected after overnight cooling, is washed well with cold water, and is dried *in vacuo* with gentle heat. Yield, 22 g; mp 160–161°. A solution of 22 g of this *Z*-dipeptide ester in 750 ml of ethanol is made by warming. The solution is cooled quickly to room temperature and 65 ml of 1 *N* NaOH is added. After 90 min at room temperature the mixture is treated with 66 ml of 1 *N* HCl. The solution is filtered and taken to about one-tenth volume by rotary evaporation. The crystalline product that forms upon cooling is recrystallized from boiling water. Yield, 13 g; mp 181°.

Procedure. A mixture is prepared by adding together 1 ml of buffer, 0.75 ml of *Z*-L-glutaminylglycine, and 0.25 ml each of $CaCl_2$, hydroxylamine, and EDTA. The pH of the mixture is brought to 6.0 with 5 *N* NaOH and its volume is adjusted to 4.0 ml. For assay a 0.4 ml portion of this mixture is warmed to 37° and enzyme solution and water are added to give a final volume of 0.5 ml. After incubation for 10 min at 37° the reaction is stopped and the color is developed by addition of 0.5 ml of $FeCl_3$ reagent. The solution is centrifuged at low speed in order to remove any precipitate and the absorbance of the clear supernatant is measured at 525 nm. An extinction coefficient of 340 M^{-1} cm^{-1} at 525 nm for the product, *Z*-L-glutamyl(γ-hydroxylamine)-glycine, has been determined with the use of synthetic material.[23]

Definition of Unit and Specific Activity. One unit of enzyme is defined as the amount that catalyzes formation of 1 μmol of the peptide derivative of γ-glutamylhydroxylamine per minute. Specific activity is given as the units per mg of protein.

Isolation of Enzyme[24]

Unfrozen guinea pig livers are used as the starting material for purification of enzyme. Only unfrozen livers yield satisfactorily pure enzyme and livers kept in crushed ice and used as long as 5 days after removal from the animals have given good yields of pure enzyme. All operations are carried out at 4°.

[23] M. Gross and J. E. Folk, *J. Biol. Chem.* **249,** 3021 (1974).

[24] J. M. Connellan, S. I. Chung, N. K. Whetzel, L. M. Bradley, and J. E. Folk, *J. Biol. Chem.* **246,** 1093 (1971).

Step 1. Preparation of Crude Extract. A 200 g portion of liver in enough 0.25 *M* sucrose to give a final volume of 450 to 500 ml is homogenized for about 2 min with a Polytron PT 20 homogenizer (Brinkmann Instruments) at intermediate speed. The homogenate is centrifuged at 105,000 *g* for 1 hr.

Step 2. Chromatography on DEAE-Cellulose. After filtration of the supernatant fluid through four layers of cheesecloth, it is pumped rapidly into a column, 3.5 × 10 cm, of DEAE-cellulose equilibrated with 5 m*M* Tris–chloride, pH 7.5 containing 2 m*M* EDTA. A 200 ml wash of equilibrating buffer is passed through the column and the protein is then eluted at a flow rate of 5 to 10 ml/min by the use of a 1.5-liter linear gradient of 0 to 1.0 *M* NaCl in the same buffer. The enzyme emerges from the column between 0.25 and 0.40 *M* NaCl. Those fractions rich in transglutaminase activity are combined.

Step 3. Protamine Precipitation and Extraction. To these pooled fractions is added gradually with stirring a 20 ml portion of a freshly prepared 1% (w/v) solution of protamine sulfate (Eli Lilly and Company). The precipitate which contains all of the enzyme is collected by centrifugation for 15 min at 14,600 *g* and is washed by suspending it in 10 ml of 0.2 *M* Tris-acetate buffer, pH 6.0, and homogenizing for 1 min at low speed with the Polytron homogenizer. The precipitate is collected by centrifugation and is extracted 3 times with 20 ml portions of 0.05 *M* ammonium sulfate in 5 m*M* Tris–chloride buffer, pH 7.5, containing 2 m*M* EDTA. Each extraction is carried out using three homogenizations of 0.5 min duration at low speed in the Polytron homgenizer. After each extraction the precipitate is removed by centrifugation. To the combined extracts are added a 1.2 ml portion of 1 *M* EDTA and 23.7 g of solid ammonium sulfate.

Step 4. Exclusion Chromatography. The precipitate that is formed upon stirring is collected after 30 min by centrifugation and is dissolved in 1 to 2 ml of 10 m*M* Tris–acetate buffer, pH 6.0 containing 1 m*M* EDTA and 0.16 *M* KCl. Any insoluble material is removed by centrifugation for 30 min at 27,000 *g* and the clear supernatant is placed on a column, 2.5 × 100 cm, of 10% agarose (BioGel A-0.5m, fine, Bio-Rad) equilibrated with the Tris-EDTA-KCl buffer. The chromatogram is developed using the same buffer and 4 ml fractions are collected at a flow rate of 30–40 ml/hr. The enzyme emerges from the column after about one-half of the column volume is collected. The enzyme activity in and the absorbance of each fraction are determined. The first fractions of enzyme activity to elute exhibit lower specific activity than the major portion of the active peak, indicating the presence of some inactive material that elutes slightly earlier than the enzyme. The fractions that have relatively constant ratios of activity to absorbance at 280 nm and the enzyme specific activities of

TABLE I
PURIFICATION OF TRANSGLUTAMINASE FROM GUINEA PIG LIVER

Step	Protein (mg)	Specific activity (units/mg)	Yield (%)
1. Homogenate	25,000[a]	0.05	
Supernatant fluid	10,000[a]	0.11	88
2. DEAE-cellulose chromatography	250[a]	2.5	50
3. Protamine extracts	56[a]	8.0	36
4. Agarose chromatography	18[b]	14.0	20

[a] Protein estimated by the method of O. H. Lowry, N. J. Rosebrough, A. L. Farr, and R. J. Randall [*J. Biol. Chem.* **193,** 265 (1951)] with bovine serum albumin as standard; see also this series, Vol. 3 [73].

[b] Protein estimated by the use of $E_{280}^{1\%}$, 15.8 [J. E. Folk and P. W. Cole, *J. Biol. Chem.* **241,** 5518 (1966)].

which are calculated to be above 12 are combined. The volume of the pooled fractions is reduced by ultrafiltration with the use of a Diaflo UM-10 membrane to give a final protein concentration of between 10 and 20 mg per ml.

Table I summarizes the purification procedure.

Properties

Stability. Enzyme preparations have been stored frozen at $-20°$ in concentrated solutions for several months without detectable loss in activity. Repeated freezing and thawing results in some losses in activity. Hence, the enzyme solutions are generally stored in small aliquots. Much of the lost activity can be restored upon incubation at room temperature after addition of 1 to 5 m*M* dithiothreitol.

Physical Properties. The purified enzyme appears homogeneous by several criteria.[19,24] It exists in a single-chain monomeric form of $M_r = \sim 75,000$, contains no disulfide bonds and no carbohydrate, and, thus, seems to be unbridged and unbranched. Only one of its 16 to 18 –SH groups is essential for catalytic activity. The sequence around this cysteine residue is Tyr-Gly-Gln-Cys-Trp.[25] Because Ca^{2+} which is essential for enzymatic activity induces conformational alterations in the enzyme protein, but does not effect gross changes in the size or shape of the molecule, there is reason to believe that the catalytically active form of

[25] J. E. Folk and P. W. Cole, *J. Biol. Chem.* **241,** 3238 (1966).

the liver enzyme is the monomeric metal–enzyme complex.[24] The purified enzyme exhibits an $E_{280}^{1\%}$ of 15.8.[19]

Specificity and Mechanism. In addition to its activity toward peptide-bound glutamine residues, the enzyme catalyzes the hydrolysis and aminolysis of certain aliphatic amides,[26] of active esters (e.g., *p*-nitrophenyl esters),[18,26,27] and certain esters that are not active.[23] The kinetics of each of these reactions has been found to follow a modified double displacement mechanism (modified ping pong reaction).[15,16,23,26,27] Evidence for a stable enzyme–substrate intermediate and isolation of a covalent trimethylacetyl-enzyme are in accordance with the postulated reaction mechanism.[28]

Human Coagulation Factor XIIIa[29]

This transglutaminase exists in inactive zymogen (Factor XIII) form. The zymogen of blood platelets differs in subunit composition from that found in blood plasma[30,31] (see *Properties* section below). For this reason and because the starting materials vary widely in composition, there are necessary differences in the procedures for isolation of the two zymogens. Both procedures are described here. Each of the zymogens is activated proteolytically by thrombin and the enzymes formed, following addition of Ca^{2+} to the level needed for full catalytic activity, are indistinguishable by several criteria.[32]

Assay

The assay used for purification of the zymogens is based upon measurement of incorporation of radiolabeled putrescine into a casein mixture as catalyzed by factor XIIIa. Activation to factor XIIIa by thrombin is carried out in a separate incubation prior to assay. Many modifications of

[26] M. Gross and J. E. Folk, *J. Biol. Chem.* **248,** 1301 (1973).

[27] S. I. Chung, R. I. Shrager, and J. E. Folk, *J. Biol. Chem.* **245,** 6424 (1970).

[28] J. E. Folk, P. W. Cole, and J. P. Mullooly, *J. Biol. Chem.* **242,** 4329 (1967); see also this series, Vol. 87 [3].

[29] An alternative method for purification of the zymogen of this enzyme is described in this series, Vol. 19 [59A], and in Vol. 45 [15] and in Vol. 80 [27] are given modifications of this method. No data are given on enrichment or yield at various purification steps. It is stated (this series, Vol. 45 [15]), however, that enzyme from different preparations of zymogen isolated by this procedure varied in purity over a wide range (42 to 85% based on active site titration).

[30] M. L. Schwartz, S. V. Pizzo, R. L. Hill, and P. A. McKee, *J. Biol. Chem.* **248,** 1395 (1973).

[31] H. Bohn and H. G. Schwick, *Arzneim. Forsch.* **21,** 1432 (1971).

[32] S. I. Chung, M. S. Lewis, and J. E. Folk, *J. Biol. Chem.* **249,** 940 (1974).

this assay have been given. These include the use of purified caseins, derivatized caseins, and amines other than putrescine. Glutamine 167 has been identified as the site of the most rapid incorporation of amine into β-casein.[2] Peptides modeled on the sequence surrounding this glutamine are excellent substrates for factor XIIIa[3] and offer a basis for design of specific low-molecular-weight substrates for the enzyme.

The purified enzyme may be assayed by measurement of free thiol release during the hydrolysis[33] and aminolysis[34] of thiol esters. Titrations of the active site–SH group of factor XIIIa with iodoacetamide[32] and with α-bromo-4-hydroxy-3-nitroacetophenone are described.[32]

Reagents

Thrombin, human or bovine, 250 NIH units/ml in 0.05 *M* sodium citrate, pH 6.5, containing 0.15 *M* NaCl
Tris-acetate buffer, pH 7.5, 1 *M*
Hammersten casein, 2% in 0.1 *M* Tris-acetate buffer, pH 7.5, containing 0.1 m*M* EDTA
$CaCl_2$, 1 *M*
EDTA, 0.1 *M*
Lubrol, 10%
Dithiothreitol, 0.01 *M*
[^{14}C]Putrescine · 2HCl, 0.2 m*M* (~20 Ci/mol)

Procedure. To the zymogen in a volume of 0.04 ml is added 0.01 ml of Tris buffer and 0.005 ml of thrombin. After 20 min at 25° to this zymogen activation mixture is added a solution (0.445 ml) composed of 0.25 ml of casein, 0.01 ml each of $CaCl_2$ and EDTA, 0.05 ml each of Tris buffer, dithiothreitol, and [^{14}C]putrescine, and 0.025 ml of lubrol. Incubation is carried out for 15 to 60 min at 25° and the reaction is stopped by addition of 0.5 ml of 10% trichloracetic acid. The precipitate is collected on a Whatman fiber-glass paper (GF/A) and washed free of unincorporated radioactive amine with 5% trichloroacetic acid. Radioactivity on the paper is measured in a scintillation fluid by the use of a liquid scintillation spectrometer.

Definition of Unit and Specific Activity. One unit of zymogen is defined as the amount that, after activation with thrombin, catalyzes the incorporation of 1 nmol of putrescine into casein per minute. Specific activity is given as the units per mg of protein.

[33] C. G. Curtis, P. Stenberg, K. L. Brown, A. Baron, K. Chen, A. Gray, I. Simpson, and L. Lorand, *Biochemistry* **13,** 3257 (1974); see also this series, Vol. 45 [15].

[34] P. Stenberg, C. G. Curtis, D. Wing, Y. S. Tong, R. B. Gredo, A. Gray, and L. Lorand, *Biochem. J.* **147,** 155 (1975); see also this series, Vol. 45 [15].

Isolation of the Zymogens[15,32]

Several essential steps in the procedures outlined here are performed as described in earlier purifications.[35–37] All manipulations are carried out at 4° unless otherwise stated.

Plasma Factor XIII

Step 1. $BaCl_2$ Treatment and Glycine Precipitation. To 4 liters of fresh acid citrate dextrose (USP Formula A) human plasma[38] is added portionwise with stirring 1.25 g of benzamidine hydrochloride and 200 ml of 1 *M* $BaCl_2$. Stirring is continued for 30 min and the precipitate is removed by centrifugation. The clear supernatant is made 0.1 *M* in ε-aminocaproic acid and 2.1 *M* in glycine by slow addition of these materials over a period of 30 min with stirring. The precipitated fibrinogen fraction containing most of the factor XIII is collected by centrifugation for 20 min at 23,000 *g*. The precipitate is dissolved in 1 liter of 0.05 *M* sodium citrate buffer pH 6.8, containing 2 m*M* benzamidine, 0.02% sodium azide, and 20,000 units/liter of aprotinin (protease inhibitor). The solution is clarified by centrifugation at 23,000 *g* for 20 min, is warmed to room temperature, and is made 2.1 *M* in glycine. The fibrinogen fraction, essentially free of cold-insoluble globulin, is obtained by centrifugation for 30 min at 23,000 *g* and 20°. The precipitate dissolved in 500 to 700 ml of the above buffer is dialyzed at 4° against 10 volumes of 0.05 *M* sodium citrate buffer, pH 6.8, containing 0.15 *M* NaCl and 0.02% sodium azide. The dialyzed solution, clarified by centrifugation at 23,000 *g* for 30 min at 2°, should contain protein that is greater than 92% clotable.[39] If the clotability is lower than this, glycine precipitation at room temperature and dialysis must be repeated.

Step 2. 56° Heat Treatment. Portions (~200 ml) of the protein solution from step 1 in 1-liter Erlenmeyer flasks are rapidly heated to 56° with

[35] L. A. Kazal, S. Amsel, O. P. Miller, and L. M. Tocantins, *Proc. Soc. Exp. Biol. Med.* **113,** 989 (1963).

[36] A. G. Loewy, C. Veneziale, and M. Forman, *Biochim. Biophys. Acta* **26,** 670 (1970).

[37] M. L. Schwartz, S. V. Pizzo, R. L. Hill, and P. A. McKee, *J. Biol. Chem.* **246,** 5851 (1971).

[38] When outdated human plasma is used as a source of zymogen it is necessary to employ an ammonium sulfate precipitation step between the $BaCl_2$ treatment and the glycine precipitation. In this case solid ammonium sulfate (176 g/liter) is added slowly with stirring. The resulting precipitate is collected by centrifugation and is dissolved in 1 liter of 0.05 *M* sodium citrate buffer, pH 6.8, containing 2 mmol of benzamidine and 20,000 units of aprotinin. Significant losses in zymogen occur at this stage of purification.

[39] The clotability may be estimated by comparing the absorbance at 280 nm of a 1 to 10 dilution of the fibrinogen solution in 0.05 *M* sodium phosphate buffer, pH 7.4, and the supernatant portion of the same solution obtained 30 min after addition of 2 NIH units/ml of human thrombin.

vigorous shaking in an 80° waterbath. After exactly 3 min at 56° the solutions are cooled rapidly in ice and filtered through cheesecloth. The combined filtrates are dialyzed against three changes of 20 volumes each of distilled water. After clarification by centrifugation, the dialyzed solution is taken to dryness by lyophilization.

Step 3. Exclusion Chromatography. Chromatography is carried out at room temperature on a 2.5 × 100 cm column of 6% agarose (BioGel A-5m, fine, Bio-Rad) that was equilibrated with the sodium citrate–NaCl buffer. The sample is applied in a total volume of ~ 6 ml of the same buffer and is clarified by centrifugation before application if needed. When 4 ml fractions are collected the zymogen emerges from the column between fractions 68 and 72. Fractions that display the highest specific activity upon thrombin activation are combined and stored at 5° or are dialyzed against distilled water and dried by lyophilization.[40]

Table IIA summarizes the purification.

Platelet Factor XIII[41]

Step 1. Preparation of Crude Extract. Platelet-concentrated plasma, obtained within 6 days after removal of blood from donors, is centrifuged for 10 min at 450 *g* to remove residual erythrocytes. The platelet-rich supernatant is made 1 m*M* in EDTA and the platelets are collected by centrifugation for 10 min at 14,000 *g*. The platelets obtained from 25 to 30 units of blood are washed 3 times by suspending them in 100 ml of 0.01 *M* Tris-acetate buffer, pH 7.4, containing 0.15 *M* NaCl and 1 m*M* EDTA and centrifuging. The cell homogenate is prepared in 100 ml of 0.01 *M* Tris acetate buffer, pH 7.4, containing 1 m*M* EDTA with the use of the Polytron PT 20 homogenizer at low speed for 3 min. The homogenate is made

[40] Before combining, the individual fractions are examined by polyacrylamide gel electrophoresis in sodium dodecyl sulfate on gels polymerized in 6 *M* urea [M.L. Schwartz, S. V. Pizzo, R. L. Hill, and P. A. McKee, *J. Clin. Invest.* **50,** 1506 (1971)]. Those fractions in which impurities are present may be combined and subjected to an additional chromatography step on DEAE-cellulose in 0.05 *M* Tris acetate, pH 7.5, containing 1 m*M* EDTA. Elution is with a linear gradient of NaCl in this buffer. The zymogen elutes at ~0.1 *M* NaCl as a single protein. The losses of zymogen in this step are always large, however.

[41] A zymogen indistinguishable from platelet factor XIII in physical, chemical, and enzymatic properties may be prepared from human placental tissue [H. Bohn, *Ann. N.Y. Acad. Sci.* **202,** 256 (1972)]. This tissue is a particularly rich source of the zymogen which is precipitated with 40% ammonium sulfate from the 105,000 *g* supernatant. The crude zymogen is dissolved in the Tris-EDTA buffer of Step 1 of the platelet factor XIII preparation and is precipitated with 6,9-diamino-2-ethoxyacridine lactate, H_2O. It is then subjected to chromatography on DEAE-cellulose as in Step 2 of this preparation, and finally chromatographed on agarose exactly as outlined in Step 3 of the plasma factor XIII method.

TABLE II
PURIFICATION OF HUMAN FACTOR XIII

Step	Protein (mg)	Specific activity (units/mg)	Yield (%)
A. From blood plasma			
Crude plasma[a]	280,000[b]	0.009	
1. $BaCl_2$ treatment and glycine precipitation	11,100[b]	0.02	88
2. Heat treatment	160[b]	1.26	80
3. Agarose chromatography	40[c]	3.70	59
B. From blood platelets			
1. Supernatant fluid	5,200[b]	0.09	
2. DEAE-cellulose chromatography	110[b]	2.23	52
3. Hydroxyapatite chromatography	23[c]	7.18	35

[a] Crude plasma (as much as 0.02 ml) may be assayed directly for factor XIIIa without interference by the clotting of fibrinogen. It is necessary, however, to conduct the assay as follows: Factor XIII is activated by incubation for 20 min in a total volume of 0.425 ml containing 0.01 ml of thrombin solution and all components of the assay except [^{14}C]putrescine. Assay is commenced by addition of putrescine.

[b] Protein estimated as in Table I, footnote *a*.

[c] Protein estimated by the use of $E_{280}^{1\%}$, 13.8 (M. L. Schwartz, S. V. Pizzo, R. L. Hill, and P. A. McKee, *J. Biol. Chem.* **248,** 1395 (1973)].

2 m*M* in benzamidine and 1000 units of aprotinin is added for each 100 ml of volume. Finally, the clear supernatant is collected after removal of solids by centrifugation for 1 hr at 105,000 *g*.

Step 2. Chromatography on DEAE-Cellulose. The supernatant from Step 1 is applied to a column, 2.5 × 20 cm, of DEAE-cellulose equilibrated with 0.05 *M* Tris acetate buffer, pH 7.4, containing 1 m*M* EDTA. After application of the sample the column is washed with equilibration buffer until the effluent shows an absorbance of less than 0.05 at 280 nm in a 1 cm cell. At this point a 1.5-liter linear gradient of 0 to 0.3 *M* NaCl in the Tris-EDTA buffer is commenced. Elution of the zymogen occurs between 0.12 and 0.18 *M* NaCl. Fractions containing the zymogen are combined, ammonium sulfate (25 g/100 ml) is added, and the protein is collected by centrifugation.

Step 3. Chromatography on Hydroxyapatite. The protein is dissolved in the minimum volume of a 0.01 *M* solution of EDTA that had been adjusted to a pH of 7.0 and this solution is dialyzed against 2 changes of 40 volumes of 0.01 *M* sodium phosphate buffer, pH 6.8. Chromatography is carried out on a column, 1.5 × 9 cm, of hydroxyapatite (DNA-grade, BioGel HTP, Bio-Rad) equilibrated with the sodium phosphate buffer. Development is with a 400 ml linear gradient of 0.01 to 0.2 *M* sodium phosphate buffer, pH 6.8, and the zymogen elutes between 0.10 and 0.12 *M* buffer. Fractions that display the highest specific activity after thrombin activation are combined, dialyzed free of buffer salts with distilled water, and lyophilized.

The purification procedure is summarized in Table IIB.

Properties

Stability. Solutions of the plasma zymogen in the sodium citrate-NaCl buffer display no losses in potential catalytic activity over several months at 5°. Frozen solutions may be stored for over a year at −30° without noticable losses. Solution of this zymogen dialyzed free of salts may be lyophilied. The dried powders prepared in this way are stable for several years when stored at −30°. The platelet zymogen in the sodium phosphate chromatography buffer gradually loses potential activity over several weeks, even when stored at −30°. When solutions of this zymogen are dialyzed free of salts and lyophilized, the resulting powders may be stored for several months at −30° without significant losses in potential catalytic activity. The stability of factor XIIIa formed from the zymogens by thrombin activation has not been systematically studied. It is clear, however, that the enzyme is more stable in the absence of Ca^{2+} than in its presence.

Physical Properties. The plasma zymogen, M_r = ~300,000, is composed of two apparently identical catalytic subunits, *a* chains of M_r = ~75,000, and two apparently identical noncatalytic subunits, *b* chains of M_r = ~80,000.[37] The platelet zymogen, M_r = ~150,000, contains two *a* chains only.[31,37] Thrombin-catalyzed conversion of the zymogens to the enzymes occurs by limited proteolysis[42] and proceeds *in vitro* in the absence of Ca^{2+}.[43] The enzyme formed by thrombin from plasma factor XIII (a_2b_2) in the absence of Ca_{2+} is a tetramer ($a_2'b_2$, where a' is the proteolytically modified catalytic subunit, M_r = ~71,000.[32] In the presence of Ca^{2+} at a level needed for full catalytic activity, this tetrameric form of enzyme is reversibly dissociated to a' dimers and *b* dimers.[32] The platelet enzyme formed from its zymogen (a_2) by thrombin in the absence of Ca^{2+} has a dimeric structure (a_2'); addition of Ca^{2+} does not cause dissociation.[32] Thus it appears that factor XIIIa catalyzes its reactions in the form of a dimer of identical subunits. Only one of the six –SH groups in each a'

[42] T. Takagi and R. F. Doolittle, *Biochemistry* **13,**750 (1974).
[43] H. M. Tyler, *Biochim. Biophys. Acta* **222,** 397 (1970).

chain is essential for activity. The sequence of amino acids surrounding this cysteine is the same as that at the active site of the guinea pig liver enzyme.[44] Factor XIIIa displays strong cooperativity between subunits in the reactions of its active site –SH groups with several alkylating reagents.[32,45]

Specificity and Mechanism. A comparison of the specificities of factor XIIIa and guinea pig liver transglutaminase toward peptide-bound glutamine residues has revealed marked differences. Clearly secondary interactions of enzymes with amino acid residues surrounding glutamine have a direct influence on specificity.[2,3] On the other hand, the specificities of these two enzymes toward amines, including peptide-bound lysine residues, are very similar.[46] The details and considerations of these findings have led to a generalized theory on the mechanism and basis for specificity of transglutaminase-catalyzed ε-(γ-glutamyl)lysine bond formation.[47]

Human Epidermal Transglutaminase

Assay

The enzyme is assayed by measuring the incorporation of radiolabeled putrescine into casein.[48] The procedure is similar to that described above for factor XIIIa. Alternatively, the enzyme has been assayed using a fluorescently labeled amine in place of the radioactive amine.[49]

Procedure. Assays are carried out in 0.5 ml of 0.1 *M* Tris-acetate buffer, pH 7.5, containing 3 mg of α-casein (instead of Hammersten casein), 0.06 m*M* [^{14}C]putrescine (6.5 Ci/mol), 10 m*M* $CaCl_2$, 1 m*M* dithiothreitol, and 1 m*M* EDTA. Termination of the reactions is with equal volumes of 10% trichloroacetic acid after 30 min at 37° and the precipitates are collected, washed, and measured for radioactivity as outlined above in the factor XIIIa assay.

Definition of Unit and Specific Activity. A unit is defined as the amount of enzyme that catalyzes the incorporation of 1 nmol of putrescine into casein in 30 min. Specific activity is given as the units per mg of protein.

[44] J. J. Holbrook, R. D. Cooke, and I. B. Kingston, *Biochem. J.* **135,** 901 (1973).
[45] G. F. Seeling and J. E. Folk, *J. Biol. Chem.* **255,** 9589 (1980).
[46] J. Schrode and J. E. Folk, *J. Biol. Chem.* **254,** 653 (1979).
[47] J. E. Folk, *Adv. Enzymol.* **54,** 1 (1983).
[48] H. Ogawa and L. A. Goldsmith, *J. Biol. Chem.* **251,** 7281 (1976).
[49] M. M. Buxman and K. D. Wuepper, *Biochim. Biophys. Acta* **452,** 356 (1976).

Isolation of the Enzyme[48,50]

All operations are carried out at 0 to 4°.

Step 1. Preparation of Crude Extract. A 50 g portion of callus obtained from the hyperkeratotic human sole is finely minced and homogenized for 3 min in 500 ml of 0.01 *M* Tris-acetate buffer, pH 7.5, in a Tekmar homogenizer. The resulting homogenate is centrifuged for 30 min at 12,000 *g* and the precipitate is again homogenized with 500 ml of the Tris-acetate buffer and centrifuged. This extraction procedure is repeated four additional times. The extracts are finally combined and filtered through cheesecloth.

Step 2. Ultrafiltration. The extracts of Step 1 are concentrated to about 20 ml by ultrafiltration with the use of an Amicon PM-30 membrane. The concentrate is dialyzed against 5 m*M* Tris-acetate buffer, pH 7.5, containing 1 m*M* EDTA and the small precipitate formed is removed by centrifugation.

Step 3. Chromatography on DEAE-Cellulose. This clear solution is chromatographed on a column 1.2 × 15 cm, of DEAE-cellulose equilibrated with the Tris–EDTA buffer. After application of the sample the column is washed with 120 ml of the equilibration buffer and elution is carried out with a 200 ml linear gradient of 0 to 0.05 *M* NaCl in the same buffer. The enzyme activity elutes between 0.02 and 0.03 *M* NaCl. The eluate containing enyme is concentrated to ~ 2.5 ml by ultrafiltration with the use of an Amicon PM-10 membrane.

Step 4. Exclusion Chromatography on Sephadex G-75. Dialysis of this partially purified enzyme solution against 0.01 *M* Tris-acetate buffer, pH 7.5, containing 1 m*M* EDTA and 0.15 *M* NaCl is followed by chromatography on a column, 2.5 × 96 cm, of Sephadex G-75 equilibrated with the same buffer. The eluate fractions containing enzyme are concentrated to ~ 7.5 ml, adjusted to pH 6.0 with 0.1 *M* acetic acid, and dialyzed for 20 min against 6 liters of 5 m*M* Tris-acetate buffer, pH 6.0, containing 1 m*M* EDTA.

Step 5. Chromatography on CM-Cellulose. This material is applied to a column, 1.2 × 13 cm, of CM-cellulose equilibrated with the pH 6.0 buffer. After applicaton of the sample the column is washed with 100 ml of the same buffer and elution is carried out with a 200 ml linear gradient of 0–0.5 *M* NaCl in the same buffer. Enzyme activity is eluted between 0.1 and 0.2 *M* NaCl. The active fractions are combined and their volume is reduced to ~ 3.5 ml.

Step 6. Chromatography on Agarose. After dialysis against 0.01 *M*

[50] A similar purification has been carried out using as starting material glabrous cow snout.[49]

TABLE III
PURIFICATION OF TRANSGLUTAMINASE FROM HUMAN CALLUS

Step	Protein[a] (mg)	Specific activity (units/mg)	Yield (%)
1. Crude extract	8870	0.054	
2. Ultrafiltration	847	0.35	61
3. DEAE-cellulose chromatography	56.8	4.6	54
4. Sephadex G-75 chromatography	10.5	21.4	46
5. CM-cellulose chromatography	4.2	27.9	24
6. Agarose chromatography	1.8	39.4	15

[a] Protein estimated as in Table I, footnote *a*.

Tris-acetate buffer, pH 7.5, containing 1 mM EDTA and 0.15 M NaCl, the enzyme is chromatographed on a column, 1.5 × 72 cm, of 10% agarose (BioGel A-0.5m, Bio-Rad) equilibrated with the same buffer. Fractions showing specific activities near 40 are combined and stored at 4°.

Table III summarizes the purification procedure.

Properties

Stability. The purified enzyme is stable for at least 1 month at 4°.

Physical Properties. The enzyme shows an apparent M_r = ~55,000 both under native and denaturing conditions. It requires Ca^{2+} for enzymatic activity and enzymatic activity may be increased as much as 25-fold by heating at 56° in the presence of Ca^{2+}.[48] Treatment with trypsin, chaotropic salts, or certain organic solvents in the presence of Ca^{2+} also causes large increases in catalytic activity.[51] The basis for this enhancement in efficiency by these agents is not known at present.

Guinea Pig Hair Follicle Transglutaminase

Assay

Assay is by measurement of radiolabeled putrescine incorporation into casein.[52] The assay is carried out in a manner similar to that detailed above for factor XIIIa. Catalytic activity of the purified enzyme has been

[51] M. F. Plishker, J. M. Thrope, and L. A. Goldsmith, *Arch. Biochem. Biophys.* **191,** 49 (1978).

[52] S. I. Chung and J. E. Folk, *Proc. Natl. Acad. Sci. U.S.A.* **69,** 303 (1972).

determined by measuring radiolabeled methylamine incorporation into the acetylated B chain of oxidized insulin.[15]

Procedure. Assays are carried out at 37° in 0.5 ml of 0.1 *M* Tris-acetate buffer, pH 7.5, containing 3 mg of Hammersten casein, 0.0135 m*M* putrescine (7.41 Ci/mol), 10 m*M* $CaCl_2$, 1 m*M* dithiothreitol, and 1 m*M* EDTA. The reactions are stopped at appropriate times by addition of trichloroacetic acid to a final concentration of 5%. The precipitates are collected, washed, and measured for radioactivity as outlined above in the factor XIIIa assay.

Definition of Unit and Specific Activity. An enzyme unit is the amount of enzyme that catalyzes the incorporation of 1 nmol of [^{14}C]putrescine into casin per hour in the test. Specific activity is defined as enzyme units per mg of protein.

Isolation of the Enzyme[52]

Unless stated otherwise, all operations are conducted at 0–2°; 1 m*M* EDTA is a component of all solutions and used throughout the purification.

Step 1. Preparation of Crude Extract. The frozen skins from five 20- to 25- week-old guinea pigs are thawed and maintained at 0–4° on ice. Most of the hair is removed by clipping, leaving short segments of hair root shafts exposed. A beeswax-rosin (2 : 7 w/w) mixture[53] at 55° is spread over the skins and allowed to solidify. Upon removal of the skins the hair roots are exposed with the shafts embedded in the wax. The roots are harvested with the use of fine clippers and are homogenized in a Polytron homogenizer for 3 min at intermediate speed with 200 ml of 0.01 *M* Tris-acetate buffer, pH 7.5. This mixture is filtered through 3 layers of cheesecloth and is centrifuged for 60 min at 105,000 *g*. The clear supernatant is concentrated to ~6 ml by ultrafiltration with the use of a UM-10 membrane.

Step 2. Chromatography on 6% Agarose. This concentrate is placed on a column, 2.5 × 95 cm, of 6% agarose (BioGel A-5m, Bio-Rad) which was equilibrated with the Tris-acetate buffer containing 0.3 *M* NaCl. The fractions comprising the single peak of enzyme activity are combined and dialyzed against 5 m*M* Tris-acetate buffer, pH 7.5. A notable increase in total enzyme activity (~2-fold over the homogenate) is observed at this step (Table IV). The cause for this increased activity is not known at present.

Step 3. Treatment with DEAE-Cellulose. The solution of enzyme from Step 2 is percolated through a column, 1.5 × 25 cm, of DEAE-cellulose

[53] W. J. Ellis, *Nature (London)* **162,** 957 (1948).

TABLE IV
PURIFICATION OF TRANSGLUTAMINASE FROM GUINEA PIG HAIR FOLLICLE

Step	Protein (mg)	Specific activity (units/mg)	Yield (%)
1. Crude extract	360[a]	0.02	
2. Chromatography on 6% agarose	90[a]	0.17	95
3. DEAE-cellulose treatment	5.2[b]	2.5	84
4. Chromatography on CM-cellulose	1.5[b]	7.4	70
5. Chromatography on 10% agarose	0.88[b]	10.0	56

[a] Protein estimated by the Biuret method.
[b] Protein estimated by the use of an assumed $E_{280}^{1\%}$, 15.

that was equilibrated with the Tris buffer. This is followed by a 200 ml wash of the same buffer. About 90% of the total enzymatic activity is recovered in the effluent from this column. This is adjusted to pH 6.0 with 0.1 *M* acetic acid and used directly in the next step of purification.

Step 4. Chromatography on CM-Cellulose. Chromatography on a column, 1.5 × 15 cm, of CM-cellulose is carried out in 5 m*M* Tris-acetate buffer, pH 6.0. After application of the sample, the column is washed with 100 ml of equilibration buffer and development is with an 800-ml linear gradient of NaCl from 0 to 0.5 *M* in the same buffer. The enzymatic activity is eluted from the column almost immediately upon application of the gradient and coincident with the first peak of protein. The fractions containing enzyme are combined and the volume is reduced to ~ 5 ml by ultrafiltration.

Step 5. Chromatography on 10% Agarose. This solution is passed into a column, 2.5 × 93 cm, of 10% agarose (BioGel A-0.5m Bio-Rad) in 0.01 *M* Tris-acetate buffer, pH 7.5, containing 0.15 *M* NaCl. Elution is with this buffer-salt mixture. Fractions containing enzyme of specific activity about 10 are combined and concentrated by ultrafiltration to give a final concentration near 1 mg/ml.

In Table IV is summarized this purification procedure.

Properties

Stability. No loss in catalytic activity is observed during 3 months storage of the purified enzyme at −20°.

Physical Properties. The enzyme displays an apparent M_r = ~54,000 under native conditions and an apparent M_r = ~27,000 in sodium dodecyl sulfate. Based on these properties a dimeric structure of identical size subunits has been postulated.[52] The dimeric nature of the hair follicle enzyme and the monomeric structure of the epidermal enzyme (Section III) have been confirmed by direct comparison of the two enzymes from the rat.[54] This finding is in agreement with earlier indications of the individual and separate natures of the two enzymes.[49,55,56]

Mechanism. Comparison of the kinetic features of the hair follicle and liver transglutaminases suggests identical catalytic mechanisms for the two.[15] The constants for these enzymes with the acetylated B chain of oxidized insulin and methylamine are pronouncedly different, however, and thus provide compelling evidence for the individuality of the two enzymes.[15]

[54] L. L. Peterson and M. M. Buxman, *Biochim. Biophys. Acta* **657,** 286 (1981).
[55] H. Ogawa and L. A. Goldsmith, *J. Invest. Dermatol.* **68,** 32 (1977).
[56] M. M. Buxman and K. D. Wuepper, *J. Histochem. Cytochem.* **26,** 340 (1978).

Section III

Glutathione

[47] Glutathione Biosynthesis; γ-Glutamylcysteine Synthetase from Rat Kidney

By GAIL FOURÉ SEELIG and ALTON MEISTER

Glutathione is synthesized by the consecutive actions of γ-glutamylcysteine synthetase and glutathione synthetase. These enzymes catalyze, respectively, the following reactions[1-4]:

$$\text{L-Glutamate} + \text{L-cysteine} + \text{ATP} \rightleftharpoons \text{L-}\gamma\text{-glutamyl-L-cysteine} + \text{ADP} + P_i$$

$$\text{L-}\gamma\text{-Glutamyl-L-cysteine} + \text{glycine} + \text{ATP} \rightleftharpoons \text{glutathione} + \text{ADP} + P_i$$

γ-Glutamylcysteine synthetase has been purified from liver (hog,[1] toad,[5] rat,[5] human fetal[6]), kidney (rat[4,7-9]), lens (bovine[10]), erythrocytes (bovine,[11] human,[12] sheep,[13,14] rat[14]), wheat germ,[15] *E. coli,*[16] and *Proteus mirabilis*.[17] The most highly purified preparations of the enzyme have been obtained from rat kidney and rat and sheep erythrocytes. The isolation procedure described below is a modification[18] of the method of Sekura and Meister.[9]

[1] R. B. Johnston and K. Bloch, *J. Biol. Chem.* **179,** 493 (1949).
[2] A. Meister, *in* "The Enzymes" (P. D. Boyer, ed.), 3rd ed., Vol. 10, p. 671. Academic Press, New York, 1974.
[3] E. D. Mooz and A. Meister, this series, Vol. 17B, p. 483.
[4] M. Orlowski and A. Meister, this series, Vol. 17B, p. 495.
[5] J. S. Davis, J. B. Balinsky, J. S. Harington, and J. B. Shepherd, *Biochem. J.* **133,** 667 (1973).
[6] D. Rollins, A. Larsson, B. Steen, K. Krishnaswamy, and L. Hagenfeldt, *J. Pharmacol. Exp. Ther.* **217,** 697 (1981).
[7] M. Orlowski and A. Meister, *Biochemistry* **10,** 372 (1971).
[8] M. Orlowski and A. Meister, *J. Biol. Chem.* **246,** 7095 (1971).
[9] R. Sekura and A. Meister, *J. Biol. Chem.* **252,** 2599 (1977).
[10] W. B. Rathbun, *Arch. Biochem. Biophys.* **122,** 62 (1967).
[11] A. Wendel, *in* "Glutathione" (L. Flohe, H. C. Benohr, H. Sies, H. D. Waller, and A. Wendel, eds.), p. 69. Academic Press, New York, 1974.
[12] P. W. Majerus, M. J. Brauner, M. B. Smith, and V. Minnich, *J. Clin. Invest.* **50,** 1637 (1971).
[13] P. G. Board, J. E. Smith, K. Moore, and D. Ou, *Biochim. Biophys. Acta* **613,** 534 (1980).
[14] G. F. Seelig and A. Meister, *Anal. Biochem.* **141,** 510 (1984).
[15] G. C. Webster and J. E. Varner, *Arch. Biochem. Biophys.* **52,** 22 (1954).
[16] P. Apontoweil and W. Berends, *Biochim. Biophys. Acta* **399,** 1 (1975).
[17] H. Kumagai, R. Nakayama, and T. Tochikura, *Agric. Biol. Chem.* **46,** 1301 (1982).
[18] G. F. Seelig and A. Meister, *J. Biol. Chem.* **259,** 3534 (1984).

Copyright © 1985 by Academic Press, Inc.
All rights of reproduction in any form reserved.

Assay Method (ADP Formation)

Principle. Enzyme activity is measured in reaction mixtures containing L-glutamate, L-α-aminobutyrate, and ATP by a coupled enzyme procedure in which the rate of formation of ADP, in the presence of pyruvate kinase, lactate dehydrogenase, phosphoenolpyruvate, and NADH, is obtained from the decrease in the absorbance of NADH at 340 nm (20). L-α-Aminobutyrate is used in place of L-cysteine to avoid complications associated with oxidation of L-cysteine, which is not a substrate. About equal rates of reaction are found with saturating levels of L-cysteine and L-α-aminobutyrate.

Reagents

Tris–HCl buffer (pH 8.0 at 37°)
Potassium chloride
Phosphoenolpyruvate
Sodium L-glutamate
L-α-Aminobutyric acid
Magnesium chloride
Na_2EDTA
NADH
Na_2ATP
Pyruvate kinase
Lactate dehydrogenase

Procedure. Enzyme activity is determined at 37° in reaction mixtures (final volume, 1.0 ml) containing 0.1 *M* Tris–HCl buffer, 150 m*M* KCl, 5 m*M* Na_2ATP, 2 m*M* phosphoenolpyruvate, 10 m*M* L-glutamate, 10 m*M* L-α-aminobutyrate, 20 m*M* $MgCl_2$, 2 m*M* Na_2EDTA, 0.2 m*M* NADH, 17 μg of pyruvate kinase, and 17 μg of lactate dehydrogenase. The reaction is initiated by addition of γ-glutamylcysteine synthetase. The absorbance at 340 nm is monitored.

Assay Method (γ-Glutamyl-α-Aminobutyrate Formation, Determined as 5-Oxoproline)

Principle. Enzyme activity is measured in the presence of γ-glutamyl cyclotransferase, which catalyzes conversion of the γ-glutamyl-α-aminobutyrate formed to 5-oxoproline and α-aminobutyrate (7,8). γ-Glutamylcyclotransferase is considered in another chapter of this volume [52].

Reagents

L-[U-^{14}C]Glutamate, 100 m*M*, 7.2×10^5 cpm/mol
L-α-Aminobutyrate, 100 m*M*

$MgCl_2$, 200 m*M*
Na_2ATP, 50 m*M*
Na_2EDTA, 1 *M*, pH 8.2
Tris–HCl buffer (pH 8.0 at 37°), 1 *M*
γ-Glutamylcyclotransferase, 175 units/ml

Procedure. A reaction mixture containing 20 μl of each of the stock solutions listed above is brought to a final volume of 0.2 ml by adding a solution containing enzyme. The reaction is initiated by adding enzyme. After incubation for the desired time at 37°, the reaction mixture is terminated by adding 0.2 ml of 20% trichloroacetic acid. The protein is removed by centrifugation and a portion (0.25 ml) of the supernatant solution is applied to a column (about 1 ml) of Dowex 50 (H^+), and the column is eluted with 3.0 ml of water. The 5-oxo[^{14}C]proline in the water effluent is determined by liquid scintillation counting after correcting for blanks (e.g., the synthetase is omitted).

A similar assay procedure in which phosphocreatine and creatine kinase are added has been described.[4] The use of such an ATP-generating system may be desirable for assay of crude tissues homogenates. It should be noted, however, that an ATP-generating system consisting of pyruvate kinase and phosphoenolpyruvate may not be desirable because the pyruvate formed transaminates with glutamate leading to formation of α-keto[^{14}C]glutarate. This labeled compound appears in the effluent of the Dowex 50 (H^+) column together with labeled 5-oxoproline and thus gives erroneously high values of activity. During purification, it is desirable to assay the enzyme by the coupled procedure described above in which γ-glutamylcyclotransferase is added at least until step 2 in the purification procedure has been carried out. After step 2, activity may be conveniently followed by determinations of ADP or of inorganic phosphate.

Definitions of Enzyme Unit and Specific Activity. One unit of enzyme activity is defined as the amount that catalyzes formation of 1 μmol of product per hr at 37°. Specific activity is expressed as units per milligram of protein. Protein is determined by the method of Lowry *et al.*[19] using bovine serum albumin as standard.

Purification. Highly purified enzyme exhibiting a specific activity in the range of 1200–1500 units/mg can be obtained from rat kidney. All steps are carried out at 5° unless otherwise stated.

Step 1. Male Sprague–Dawley rate (225–250 g; about 30) are decapitated and exsanguinated. The kidneys (about 60 g) are removed, decapsulated, and placed at 0° in a solution consisting of 150 m*M* KCl, 1 m*M*

[19] O. H. Lowry, N. J. Rosebrough, A. L. Farr, and R. J. Randall, *J. Biol. Chem.* **193,** 265 (1951).

$MgCl_2$, and 5 m*M* 2-mercaptoethanol. The kidneys are minced and thoroughly homogenized in 5 volumes of the same solution with a Potter-Elvehjem homogenizer (at high speed). The homogenate is then centrifuged at 10,000 *g* for 15 min. The supernatant solution (about 300 ml) is saved.

Step 2. Solid ammonium sulfate (93.9 g/300 ml) is added slowly to the supernatant solution, which is stirred for 30 min and then centrifuged at 13,000 *g* for 15 min. The pellet is discarded. The supernatant solution is then treated with ammonium sulfate (9.4 g/100 ml). After stirring for 30 min, the suspension is centrifuged at 13,000 *g* for 15 min. The pellet is collected and dissolved in a solution of 50 m*M* Tris–HCl buffer (pH 7.4) containing 5 m*M* L-glutamate and 5 m*M* $MgCl_2$ to give a volume of about 30 ml. The enzyme is then dialyzed against 2 changes of 4 liters each of the same buffer.

Step 3. The dialyzed solution obtained in Step 2 is applied to a column (2.4 × 12 cm) of DE-52 cellulose which had previously been equilibrated with Tris–HCl buffer described in Step 2. The column is washed with 120 ml of the Tris–HCl buffer and the enzyme is eluted from the column with a linear gradient established between 400 ml of the Tris–HCl buffer and 400 ml of the Tris–HCl buffer containing 0.2 *M* NaCl. Fractions of 10 ml are collected at a rate of 50 to 60 ml/hr. Effluent fractions that contain specific activities of about 600 units/mg or greater are pooled to yield a total volume of about 150 ml with an enzyme activity recovery about 50%. This solution is concentrated to a volume of 10–15 ml with an Amicon ultrafiltration cell equipped with a YM-10 membrane. This solution is then dialyzed against 2 changes of 4 liters each of 50 m*M* sodium morpholinopropane sulfonate buffer (pH 7.0) containing 5 m*M* $MnCl_2$ and 5 m*M* L-glutamate (starting buffer).

Step 4. The dialyzed protein (about 8 mg in 4–5 ml) from Step 3 is placed on a 0.5 × 3.9 cm column of Agarose–hexane–adenosine–5′-triphosphate (AGATP) (P. L. Biochemicals) resin which has been previously equilibrated in the starting buffer. Protein is eluted from the column using 24 ml of the starting buffer containing 0.7 *M* NaCl and 10% glycerol followed by 24 ml of the starting buffer. Enzyme activity is eluted from the column using 50 m*M* morpholinopropane sulfonate buffer (pH 7.0) containing 5 m*M* L-glutamate, 5 m*M* $MgCl_2$, 0.2 *M* NaCl, and 1 m*M* Na_2ATP. Fractions of 2.0 ml are collected at a flow rate of 30 ml/hr. The fractions of highest enzyme activity are pooled and concentrated to about 6 ml on an Amicon ultrafiltration unit fitted with a YM-10 membrane and then dialyzed against 10 m*M* imidazole buffer (pH 8.4) containing 1 m*M* Na_2EDTA. The enzyme is stable for several months at 0–5° but rapidly loses activity upon freezing. The total recovery of enzyme activity by this

PURIFICATION OF γ-GLUTAMYLCYSTEINE SYNTHETASE FROM RAT KIDNEY[a]

Step	Volume (ml)	Total protein (mg)	Total activity (units)	Specific activity (unit/mg)	Yield (%)
1. Homogenate	297	2200	25,200	11.5	[100]
2. $(NH_4)_2SO_4$ fractionation	26.3	895	23,800	26.6	94
3. DE-52 column	19.0	50.4	16,600	329	66
4. ATP-Agarose column	10.5	5.9	9,300	1580	37

[a] From 60 g.

procedure is about 30% and the overall purification from rat kidney homogenate is about 140-fold. A representative purification scheme is given in the table.

Earlier studies in which the enzyme was purified from rat kidney led to preparations that exhibited specific activity values that are about 50% of those obtained by the procedure described above. Partial inactivation of the enzyme during purification could account for its markedly reduced activity. Other possible explanations for such "half of the sites" behavior of this enzyme have been discussed.[9]

Physical Properties, Molecular Weight, Subunit Structure, and Sulfhydryl Groups

The enzyme is homogeneous on gel electrophoresis under nondenaturing conditions. It exhibits a sedimentation coefficient of 5.5 S.[4] A solution of the enzyme containing 1 g/liter has an absorbance of 1.15 at 280 nm.

The molecular weight of the rat kidney enzyme is about 104,000.[22] The enzyme dissociates under denaturing conditions to yield two subunits of molecular weights about 73,000 and 27,700.[20,21] When the isolated enzyme is cross-linked with dimethylsuberimidate, the cross-linked species is found to be similar in molecular weight to that of the isolated enzyme.[20] Treatment of the enzyme with dithiothreitol followed by gel electrophoresis under nondenaturing conditions leads to partial dissociation into subunits.[21,22] These partially reassociate on removal of dithiothreitol.[22] The heavy subunit exhibits all of the catalytic activities of the isolated enzyme and, like the isolated enzyme, is feedback inhibited by glutathione.[21,22]

[20] R. Sekura and A. Meister, *J. Biol. Chem.* **252,** 2606 (1977).

[21] P. Richman, Doctoral Dissertation, Graduate School of Medical Sciences, Cornell University (1975).

[22] G. F. Seelig, R. Simondsen, and A. Meister, *J.Biol. Chem* **259,** 9345 (1984).

Treatment of the enzyme with 50 mM dithiothreitol for periods longer than about 1 hr at 4° leads to some loss of activity, which is partially reversible; further treatment leads to irreversible loss of activity. Such loss of activity is to some extent prevented by inclusion of 50 mM MgCl and 50 mM L-glutamate in the buffer solution. The amino acid content of the isolated enzyme and of its heavy subunit has been determined.[22] The light subunit of the enzyme has a significantly lower content of Trp, Phe, Tyr, Val, and Ala residues than the heavy subunit, while its content of Lys, His, Met, and Asx residues is higher. Although the enzyme is readily dissociated by treatment with dithiothreitol, a small amount of undissociated material is always found in the isolated enzyme. This may represent a precursor protein which, after specific cleavage, leads to formation of the 2 subunits of the enzyme. Thus, the enzyme may be synthesized as a polypeptide chain of 104,000 molecular weight, which is split to form 2 subunits which remain associated by disulfide bonds. This interpretation is consistent with studies in which a messenger RNA was obtained for the 104,000 M_r enzyme.[22a] It is possible that the light subunit of the enzyme contributes stability to the enzyme.

The enzyme has one sulfhydryl group that reacts with 5,5′-dithiobis-(2-nitrobenzoate).[18] This single exposed sulfhydryl group is not required for enzyme activity. As discussed below, the enzyme is potently inactivated by cystamine, which interacts with a sulfhydryl group at the active site to form a mixed disulfide. 5,5′-Dithiobis-(2-nitrobenzoate) does not react with the sulfhydryl group that reacts with cystamine. When the enzyme is 90% inactivated by reaction with cystamine, 3.4 mol of 5,5′-dithiobis-(2-nitrobenzoate) reacts per mole of enzyme suggesting that the binding of cystamine exposes sulfhydryl groups that may be buried or unreactive in the native enzyme.

Catalytic Properties

The pH optimum for the activity of the enzyme is in the range 8.0–8.4. The apparent K_m values for L-glutamate, ATP, L-α-aminobutyrate, and L-cysteine are 1.6, 0.2, 1.0, and 0.3 mM, respectively.[2]

L-Cysteine and L-α-aminobutyrate are the most active acceptor amino acid substrates of the enzyme. When hydroxylamine is substituted for the acceptor amino acid, a slow rate of γ-glutamyl hydroxamate formation is achieved. A number of other amino acids are also active as acceptor substrates including *S*-methyl-L-cysteine, DL-C-allylglycine, β-chloro-L-alanine, L-norvaline, L-homocysteine, L-threonine, L-alanine, and several

[22a] K. Murata, unpublished data (1984).

others. The acceptor amino acid specificity is somewhat broader when Mn^{2+} is substituted for Mg^{2+}.[7,8] Several analogs of L-glutamate are active including D-glutamate, β-aminoglutarate, *N*-methyl-L-glutamate, α-methylglutamate, and β-methylglutamate; γ-methylglutamate and L-aspartate are not active. Although D-glutamate binds to the enzyme (and reacts to form 5-oxoproline at a faster rate than observed with L-glutamate; see below), it is much less active than L-glutamate in dipeptide formation.[8] D-Glutamate is an effective inhibitor of the enzyme (apparent K_i, 0.8 mM).

Inhibition

γ-Glutamylcysteine synthetase is competitively inhibited by glutathione and it appears that such inhibition constitutes a physiologically significant feedback mechanism. The apparent K_i value for glutathione is about 2.3 mM, a value not far from that found for the level of glutathione in rat kidney.[23] Feedback inhibition by glutathione, which is not allosteric in nature, appears to involve binding of glutathione to the glutamate site of enzyme as well as to another enzyme site; this site may require a sulfhydryl group since ophthalmic acid (γ-glutamyl-α-aminobutyryl-glycine) is a weak inhibitor. That glutathione normally regulates its own synthesis in this way appears to explain observations on patients with the disease 5-oxoprolinuria.[24,25] These individuals have a marked deficiency of glutathione synthetase and consequently of glutathione. They overproduce γ-glutamylcysteine, which is converted (by the action of γ-glutamylcyclotransferase) to 5-oxoproline. The production of the latter compound exceeds the capacity of 5-oxoprolinase and therefore large amounts of 5-oxoproline accumulate and are excreted in the urine. γ-Glutamylcysteine synthetase from *Xenopus laevis* was reported not to be inhibited by glutathione.[5] Mutants of *Escherichia coli* have been selected whose γ-glutamylcysteine synthetase activity is not inhibited by glutathione.[26]

γ-Glutamylcysteine synthetase is inhibited by certain analogs of L-glutamate, such as D-glutamate, β-methylglutamate, and γ-methylgluta-

[23] P. G. Richman and A. Meister, *J. Biol. Chem.* **250,** 1422 (1975).

[24] V. P. Wellner, R. Sekura, A. Meister, and A. Larsson, *Proc. Natl. Acad. Sci. U.S.A.* **71,** 2505 (1974).

[25] A. Meister, *in* "The Metabolic Basis of Inherited Diseases" (J. B. Stanbury, J. B. Wyngaarden, D. S. Fredrickson, M. S. Brown, and J. L. Goldstein, eds.), 5th ed., Chapter 17, p. 348. McGraw Hill, New York, 1983.

[26] K. Murata and A. Kimura, *Appl. Environ. Microbiol.* **44,** 1444 (1982).

mate.[9] The enzyme is inactivated by cystamine,[27–30] L- and D-3-amino-1-chloro-2-pentanone,[30] and L-2-amino-4-oxo-5-chloropentanoate.[20]

Inactivation by L-2-amino-4-oxo-5-chloropentanoate requires very low concentrations (3–6 μM) of Mg^{2+} or of certain other divalent cations (e.g., Mn^{2+}, Ca^{2+}, Cd^{2+}, and Sr^{2+}).[20] L-Glutamate (but not D-glutamate) protects competitively against inactivation and such protection is increased in the presence of ATP or ADP. The rate of inactivation by this chloroketone is increased by L-α-aminobutyrate. Studies with ^{14}C-labeled L-2-amino-4-oxo-5-chloropentanoate show that this reagent binds stoichiometrically to the enzyme, and that it binds exclusively to its heavy subunit.[20]

The optical isomers of 3-amino-1-chloro-pentanone (the α-chloroketone analogs of L- and D-α-aminobutyrate) are also highly potent irreversible inactivators of the enzyme; these chloroketones are 20 to 30 times more active than L-2-amino-4-oxo-5-chloropentanoate.[30] Both L- and D-glutamate, in the presence of Mg^{2+} or Mn^{2+}, protect against inactivation. Increased protection is found when ATP is also present. Some protection is also afforded by the dipeptide product γ-glutamyl-α-aminobutyrate, especially in the presence of Mn^{2+} or Mg^{2+} and ATP. L-α-Aminobutyrate provides relatively little protection. Inhibition by the L- and D-isomers of 3-amino-1-chloro-2-pentanone, in contrast to the findings with L-2-amino-4-oxo-5-chloropentanoate, does not require divalent metal ions, and in fact Mn^{2+} decreases inhibition by the L-isomer of 3-amino-1-chloro-2-pentanone.

In contrast to inactivation produced by chloroketones, that produced by cystamine is completely reversible by dithiothreitol. Treatment of the enzyme with cystamine prevents interaction of the enzyme with the chloroketones.[30] Thus, when the enzyme is inhibited by cystamine and then treated with a chloroketone, it may be reactivated by dithiothreitol. The findings suggest that a sulfhydryl group at the active site of the enzyme interacts with the chloroketones and with cystamine, and also that the chloroketones and cystamine bind to the enzyme as glutamate analogs. The evidence also suggests that divalent metal ions play a role in the binding of certain of these compounds as well as in the binding of amino acid substrates.

[27] R. V. Lebo and N. M. Kredich, *J. Biol. Chem.* **253,** 2615 (1978).

[28] O. W. Griffith, A. Larsson, and A. Meister, *Biochem. Biophys. Res. Commun.* **79,** 919 (1977).

[29] G. F. Seelig and A. Meister, *J. Biol. Chem.* **257,** 5092 (1982).

[30] R. L. Beamer, O. W. Griffith, J. D. Gass, M. E. Anderson, and A. Meister, *J. Biol. Chem.* **255,** 11732 (1980).

The reaction of cystamine with the enzyme involves formation of a mixed disulfide between cysteamine and an enzyme sulfhydryl group. Although certain cystamine analogs and other disulfides do not inhibit the enzyme,[28] cystamine is not unique in its ability to form a mixed disulfide with the enzyme. Thus, the enzyme binds covalently to cystamine-Sepharose.[29] The interaction of the enzyme with cystamine-Sepharose is facilitated by ATP and decreased in the presence of L-glutamate plus magnesium ions. When the enzyme is applied to columns of cystamine-Sepharose a large fraction is bound and this may be released from by treatment with dithothreitol. Some of the enzyme applied to such columns is inactivated and is not covalently bound to the column. Interestingly, the enzyme does not bind to columns of *S*-(*S*-methyl)cysteamine-Sepharose, whereas free *S*-(*S*-methyl)cysteamine is a potent inhibitor. That the enzyme is partitioned on columns of cystamine-Sepharose between column-bound enzyme and nonbound inactivated enzyme suggests that the reactive sulfhydryl group forms a disulfide linkage with the sulfur atom at the immobilized end of cystamine to link the enzyme to the column and to liberate free cysteamine, and also that the enzyme interacts with the external cysteamine moiety of the bound cystamine. This may occur if the free cysteamine released is spontaneously oxidized to free cystamine followed by its inhibition of the enzyme, or if there is direct reaction between the enzyme-reactive sulfhydryl group and the sulfur atom of the external cysteamine moiety of cystamine-Sepharose. Binding of the enzyme to such columns may be useful as a purification step. Thus, enzyme preparations that are about 10% pure bind effectively to the column and protein impurities can be eluted. Subsequent addition of 50 m*M* dithiothreitol to the elution buffer leads to prompt release of about 60% of the enzyme applied to the column.

L-Glutamate (but not D-glutamate or L-α-aminobutyrate) protects against inactivation of the enzyme by cystamine. On the other hand, ATP enhances the rate of inactivation and the apparent K_m value for this effect is similar to that for ATP in the catalytic reaction.[18] Studies on the structural feature of cystamine that facilitate its interaction with the enzyme show that selenocystamine, monodansylcystamine, and *N*-[2((2-aminoethyl)-dithio)-ethyl]-4-azido-2-nitrobenzeneamine are also good inactivators. It can be demonstrated that the enzyme is almost completely inhibited by cystamine under conditions in which 0.5 mol of this compound is bound per mole of enzyme. This can occur because reaction of the enzyme sulfhydryl group with cystamine to form a mixed disulfide between the enzyme and cysteamine forms 1 equivalent of free cysteamine; spontaneous oxidation of free cysteamine yields cystamine thus accounting for inactivation of almost 2 mol of enzyme per mol of cystamine added initially.[30] The studies with cystamine-Sepharose show that the enzyme sulfhydryl group reacts to form a disulfide bound with the internal sulfur atom of cysteamine-Sepharose (i.e., to form enzyme–S–S–CH_2–CH_2–NH–Sepharose); however, reaction with the external sulfur atom may also occur.[18] *S*-(*S*-Methyl)cysteamine reacts almost entirely in a manner in which enzyme–S–S–CH_2–CH_2–NH_2 is formed.

Methionine sulfoximine is an effective irreversible inhibitor of the enzyme[31]; as in the case of glutamine synthetase,[32,33] this inhibitor serves as an analog of the transition state and is phosphorylated by ATP on the enzyme and is converted to methionine sulfoximine phosphate which binds tightly to the enzyme. Of the four stereoisomers of methionine sulfoximine only L-methionine-*S*-sulfoximine inhibits. This is the same stereoisomer that irreversibly inhibits glutamine synthetase and produces convulsions in animals.[34] Chemically prepared L-methionine sulfoximine phosphate inhibits the enzyme. Methionine sulfoximine phosphate binds less tightly to γ-glutamylcysteine synthetase than it does to glutamine synthetase and may be released from the inhibited enzyme leading to restoration of catalytic activity.[31] The studies with methionine sulfoximine are in accord with the intermediate formation of enzyme bound γ-glutamyl phosphate; thus, phosphorylation of the sulfoximine appears to reflect phosphorylation of glutamate in the normal catalytic reaction. Selective inhibitors of glutamine synthetase and of γ-glutamylcysteine synthetase have been prepared.[35–37] Thus, α-ethylmethionine sulfoximine induces convulsions and inhibits glutamine synthetase irreversibly, but does not inhibit γ-glutamylcysteine synthetase.[35] Analogs of methionine sulfoximine which inhibit γ-glutamylcysteine synthetase without affecting glutamine synthetase were prepared by replacing the S-methyl group of methionine sulfoximine with bulkier moieties, e.g., prothionine sulfoximine[36] and buthionine sulfoximine.[37] Buthionine sulfoximine inhibits the enzyme more effectively than prothionine sulfoximine and is much more effective than methionine sulfoximine. Presumably the S-alkyl moiety of the sulfoximine binds at the enzyme site that normally accepts L-cysteine. Higher homologs such as the penta, hexa, and hepathionine sulfoximines

[31] P. G. Richman, M. Orlowski, and A. Meister, *J. Biol. Chem.* **248,** 6684 (1973).

[32] R. Ronzio and A. Meister, *Proc. Natl. Acad. Sci. U.S.A.* **59,** 164 (1968).

[33] A. Meister, *in* "The Enzymes" (P. D. Boyer, ed.), 3rd ed., Vol. 10, p. 699. Academic Press, New York, 1974.

[34] W. B. Rowe and A. Meister, *Proc. Natl. Acad. U.S.A. Sci.* **66,** 500 (1970).

[35] O. W. Griffith and A. Meister, *J. Biol. Chem.* **253,** 2333 (1978).

[36] O. W. Griffith, M. E. Anderson, and A. Meister, *J. Biol. Chem.* **254,** 1205 (1979).

[37] O. W. Griffith and A. Meister, *J. Biol. Chem.* **254,** 7558 (1979).

also inhibit.[38] The hexa- and hepta-derivatives are toxic for reasons not yet known. Buthionine sulfoximine, like methionine sulfoximine,[31] is phosphorylated by ATP at the active site of γ-glutamylcysteine synthetase.[38]

Mechanism[2]

Highly purified preparations of the enzyme catalyze the slow hydrolysis of ATP, and there is evidence that this reaction is an inherent property of the synthetase. The enzyme does not catalyze ATP–ADP exchange at a significant rate nor does it catalyze exchange between ATP and inorganic phosphate. Interestingly, the enzyme can catalyze the phosphorylation of inorganic phosphate to form inorganic pyrophosphate. The synthesis of pyrophosphate is markedly inhibited by L-glutamate and by methionine sulfoximine, and is activated by the acceptor amino acid L-α-aminobutyrate.

The reaction catalyzed by γ-glutamylcysteine synthetase is reversible. When the enzyme is incubated with L-γ-glutamyl-L-α-aminobutyrate, ADP, Mg^{2+}, inorganic phosphate and L-α-[^{14}C]aminobutyrate, there is rapid incorporation of isotope into the dipeptide.[8] Such incorporation does not occur when either ADP or inorganic phosphate is omitted. Similarly, incorporation of L-[^{14}C]glutamate into the dipeptide also occurs, but at about 50% of the rate observed for incorporation of L-α-aminobutyrate.

Like glutamine synthetase, γ-glutamylcysteine synthetase can catalyze the formation of 5-oxoproline from L- and D-glutamate. The rate of formation of 5-oxoproline from D-glutamate is about 5 times greater than observed with L-glutamate. In contrast to glutamine synthetase, which catalyzes stoichiometric formation of 5-oxoproline from glutamate in the absence of ammonia,[33] γ-glutamylcysteine synthetase catalyzes formation of about 0.7 mol of 5-oxoproline per mole of inorganic phosphate formed suggesting that enzyme-bound γ-glutamyl phosphate can undergo both cyclization and hydrolysis. L-α-Aminobutyrate inhibits the formation of 5-oxoproline from L-glutamate, but has little effect on the formation of 5-oxoproline from D-glutamate.

That highly purified γ-glutamylcysteine synthetase can catalyze several partial reactions including the hydrolysis of ATP (inhibited by glutamate), phosphorylation of inorganic phosphate (inhibited by glutamate), the cyclization of glutamate in the presence of ATP to yield 5-oxoproline, ADP, and inorganic phosphate, and the phosphorylation of L-methionine-*S*-sulfoximine, a transition state inhibitor of the reaction, is in accord in

[38] O. W. Griffith, *J. Biol. Chem.* **257,** 13704 (1982).

the view that enzyme-bound γ-glutamyl phosphate is an intermediate in the reaction. It appears probable that an enzyme–ADP–γ–glutamyl phosphate complex is formed as in the case of glutamine synthetase. Kinetic studies on the rat kidney enzyme are consistent with the γ-glutamyl phosphate hypothesis and with the view that the enzyme reacts first with MgATP followed by reaction with glutamate and α-aminobutyrate.[39,40] However, the enzyme can evidentally react with L-glutamate in the absence of added ATP.[21] The mechanism of the reaction has been discussed in detail.[2]

[39] B. Yip and F. B. Rudolph, *J. Biol. Chem.* **251,** 3563 (1976).
[40] V. B. Shandle and F. B. Rudolph, *J. Biol. Chem.* **256,** 7590 (1981).

[48] γ-Glutamylcysteine Synthetase from Erythrocytes

By GAIL FOURÉ SEELIG and ALTON MEISTER

Interest in erythrocyte γ-glutamylcysteine synthetase has been stimulated by the discovery of humans[1] and sheep[2] that have a genetic deficiency in glutathione synthesis. Purified preparations of erythrocyte γ-glutamylcysteine synthetase have been reported to have relatively low specific activity, suggesting that the erythrocyte enzyme may be intrinsically less stable or less active than that isolated from rat kidney.[2-4] Since protein synthesis does not occur in the erythrocyte it is possible that the enzyme becomes gradually inactivated during the life span of the erythrocyte and that inactive enzyme molecules are present in the isolated preparations. The possibility of marked species variation in the activity of γ-glutamylcysteine synthetase needs also to be considered.

The procedure described below, which is that of Seelig and Meister,[5] provides a relatively rapid method of purification of γ-glutamylcysteine

[1] A. Meister, *in* "The Metabolic Basis of Inherited Diseases" (J. B. Stanbury, J. B. Wyngaarden, D. S. Fredrickson, M. S. Brown, and J. L. Goldstein, eds.), 5th ed., Chapter 17, p. 348. McGraw-Hill, New York, 1983.
[2] P. G. Board, J. E. Smith, K. Moore, and D. Ou, *Biochim. Biophys. Acta* **613,** 534 (1980).
[3] P. W. Majerus, M. J. Brauner, M. B. Smith, and V. Minnich, *J. Clin. Invest.* **50,** 1637 (1971).
[4] A. Wendel, *in* "Glutathione" (L. Flohé, H. C. Benohr, H. Sies, H. D. Waller, and A. Wendel, eds.), p. 69. Academic Press, New York, 1974.
[5] G. F. Seelig and A. Meister, *Anal. Biochem.* **141,** 510 (1984).

METHODS IN ENZYMOLOGY, VOL. 113
Copyright © 1985 by Academic Press, Inc.
All rights of reproduction in any form reserved.

synthetase from rat erythrocytes. This method has also been used for the isolation of the enzyme from sheep erythrocytes. Preparations of the enzyme from these sources have been found to exhibit specific enzymatic activities that are not far from that found for purified rat kidney γ-glutamylcysteine synthetase. The methods used for determination of enzyme activity are described above.

Step 1. Rats (225–250 g) are decapitated and the blood is collected with swirling in a beaker containing 200 mg of heparin. The heparinized blood (400 ml; obtained from 60 rats) is centrifuged at 4° at 1000 *g* for 30 min. The following procedures are carried out at 4°. The supernatant solution is discarded and the packed erythrocytes are washed by centrifugation with an equal volume of 50 m*M* Tris–HCl buffer (pH 8.0) containing 0.154 *M* NaCl. The packed cells are brought to ten times their volume by addition of cold (4°) water. After stirring vigorously with a magnetic stirrer for 15 mixtures, the mixture is centrifuged at 10,000 *g* for 15 min.

The supernatant solution is treated with solid ammonium sulfate (3.3 g/100 ml) and stirred for 30 min. After centrifugation at 13,000 *g* for 15 min, the supernatant solution is collected and treated with solid ammonium sulfate (9.4 g/100 ml). The solution is stirred for 15 min and then centrifuged at 13,000 *g* for 15 min. The precipitate is collected and dissolved in 50 m*M* Tris–HCl buffer (pH 7.4) containing 5 m*M* sodium L-glutamate and 5 m*M* $MgCl_2$. This solution is dialyzed against the same buffer. (Such solutions retain enzymatic activity for several weeks when stored at 4°.)

Step 2. The solution obtained in Step 1 is added to the top of a column (2.4 × 12 cm) of DE-52 cellulose and the column is washed with a portion of the same buffer equal in volume to the applied sample. The enzyme is eluted with a linear gradient established between 400 ml of the buffer and 400 ml of the buffer containing 0.2 *M* NaCl. The flow rate is 50–60 ml/hr and fractions of 10 ml each are collected. The active fractions are pooled and concentrated in an Amicon ultrafiltration cell equipped with a YM-10 membrane. The concentrated sample is then dialyzed against a solution containing 50 m*M* 3-[*N*-morpholino]-propanesulfonate (MOPS) buffer (pH 7.0) containing 5 m*M* $MgCl_2$ and 5 m*M* L-glutamate.

Step 3. After dialysis, a sample containing about 15 mg of protein is added to the top of a column (0.5 × 3.9 cm) of Agarose-ATP. The protein is eluted with 24 ml of the MOPS buffer containing 0.7 *M* NaCl and 10% glycerol followed by 24 ml of the MOPS buffer used initially. The enzyme activity is eluted with a MOPS buffer (pH 7.0) containing 5 m*M* L-glutamate, 5 m*M* $MgCl_2$, 0.2 *M* NaCl, and 1 m*M* Na_2ATP. The flow rate is 30 ml/hr and fractions of 2.0 ml each are collected. The fractions containing enzyme activity are pooled, concentrated, and dialyzed against a buffer

PURIFICATION OF γ-GLUTAMYLCYSTEINE SYNTHETASE FROM RAT BLOOD[a]

Step	Protein Total (mg)	Protein Conc. (mg/ml)	Activity Total (units)	Activity Conc. (mg/ml)	Volume (ml)	Yield (%)	Specific activity (units/mg)
1. $(NH_4)_2SO_4$ fractionation	1590	29.4	7820	145	54	[100]	4.91
2. DE-52 chromatography	60	7.5	2910	364	8	37	48.6
3. Agarose-ATP chromatography	1.3	0.10	1900	146	13	24	1460

[a] From 400 ml of whole blood; see the text.

consisting of 10 m*M* imidazole–HCl buffer (pH 8.4) containing 1 m*M* EDTA and 5 m*M* $MgCl_2$

A summary of the purification is given in the table. The isolated enzyme exhibits a specific activity of about 1400 U/mg of protein; the yield of activity is about 24%. The enzyme may be stored at 4° for 6–10 weeks without significant loss of activity.

The procedure described above was also applied to isolation of γ-glutamylcysteine synthetase from sheep erythrocytes. A highly purified preparation of the enzyme was obtained that had a specific enzymatic activity of 1100 Units per mg.

The purified enzyme migrates as a single band on gel electrophoresis carried out under nondenaturing conditions at pH 8.9 and also in the presence of 0.1% sodium dodecyl sulfate and 6 *M* urea at pH 7.15.[5] The molecular weight of the enzyme is about 100,000. Gel electrophoresis carried out under nondenaturing conditions in the presence of 50 m*M* dithiothreitol leads to appearance of 2 bands corresponding in molecular weight to 75,000 and 25,000, respectively. These properties of the rat erythrocyte enzyme are virtually the same as those exhibited by the rat kidney enzyme.[6] The catalytic properties of the erythrocyte enzyme are also similar to those found for the rat kidney enzyme. The rat erythrocyte enzyme reacts strongly with antiserum produced in rabbits injected with rat kidney γ-glutamylcysteine synthetase.[5]

[6] G. F. Seeling and A. Meister, this volume [47].

[49] Glutathione Synthetase from Rat Kidney

By ALTON MEISTER

This enzyme, first demonstrated in studies with pigeon liver extracts,[1,2] has subsequently been examined in yeast,[3-5] human,[6] and bovine[7] erythrocytes, ciliary epithelium,[8] wheat germ,[9] and *Escherichia coli*.[10] The most highly purified enzyme preparations thus far isolated have been obtained from bakers' yeast[3-5] and rat kidney.[11] A suitable procedure for the isolation of glutathione synthetase from yeast was previously described in this series.[5] The method described below for isolation of the enzyme from rat kidney is based on the work of Oppenheimer *et al.*[11]

Reagents

Tris–HCl buffer, pH 8.2 at 37°
Potassium phosphate buffer, pH 7.0
Potassium chloride
L-γ-Glutamyl-L-α-aminobutyric acid
Sodium ATP
Glycine
[^{14}C]Glycine
Magnesium chloride
Trichloroacetic acid
Sulfosalicylic acid
Lactate dehydrogenase

[1] J. E. Snoke and K. Bloch, *J. Biol. Chem.* **199,** 407 (1952).
[2] A. Meister, *in* "The Enzymes" (P. D. Boyer, ed.), 3rd ed., Vol. 10, p. 671. Academic Press, New York, 1974.
[3] J. E. Snoke, *J. Biol. Chem.* **213,** 813 (1955).
[4] E. D. Mooz and A. Meister, *Biochemistry* **6,** 1722 (1967).
[5] E. D. Mooz and A. Meister, this series, Vol. 17B, p. 483.
[6] P. W. Majerus, M. J. Brauner, M. B. Smith, and V. J. Minnich, *J. Clin. Invest.* **50,** 1637 (1971).
[7] A. Wendel, E. Schaich, V. Weber, and L. Flohe, *Hoppe-Zeyler's Z. Physiol. Chem.* **353,** 514 (1972).
[8] L. L. Ross, L. Barber, S. S. Tate, and A. Meister, *Proc. Natl. Acad. Sci. U.S.A.* **70,** 2211 (1973).
[9] G. C. Webster and J. E. Varner, *Arch. Biochem. Biophys.* **55,** 95 (1955).
[10] P. J. Samuels, *Biochem. J.* **55,** 441 (1953).
[11] L. Oppenheimer, V. P. Wellner, O. W. Griffith, and A. Meister, *J. Biol. Chem.* **254,** 5184 (1979).

Copyright © 1985 by Academic Press, Inc.
All rights of reproduction in any form reserved.

Creatine kinase
Phosphoenolpyruvate
Phosphocreatine
Acetic acid
Ammonium acetate
NADH

Procedure

Enzymatic activity may be determined by measuring the formation of inorganic phosphate, ADP, or [^{14}C]glycine-labeled tripeptide in reaction mixtures containing the enzyme and its substrates. Many workers have preferred to use L-γ-α-aminobutyrate in place of L-γ-glutamyl-L-cysteine, which readily oxidizes to form the corresponding disulfide, which is not a substrate of the enzyme. The formation of inorganic phosphate and ADP may be determined in reaction mixtures containing 100 m*M* Tris–HCl buffer (pH 8.2 at 37°), 50 m*M* potassium chloride, 5 m*M* L-γ-glutamyl-L-α-aminobutyrate, 10 m*M* ATP, 5 m*M* glycine, 20 m*M* magnesium chloride, 2 m*M* EDTA, and enzyme (added last) in a final volume of 0.1 ml. The assay mixture is incubated for 2.5–30 min at 37°. For determinations of inorganic phosphate, 0.4 ml of 10% trichloroacetic acid is added, and inorganic phosphate may be determined by the modification of the phosphomolybdate method described by Taussky and Shorr.[12] To determine ADP, the reaction mixtures are treated with 0.02 ml of 10% sulfosalicylic acid and 0.9 ml of a solution containing 0.5 m*M* phosphoenolpyruvate, 0.2 m*M* NADH, pyruvate kinase (1 unit), 40 m*M* magnesium chloride, 50 m*M* potassium chloride, and 250 m*M* potassium phosphate buffer (pH 7.0). The amount of ADP formed is calculated from the change in absorbance at 340 nm observed after addition of 0.1 ml (1 unit) of lactate dehydrogenase.

For determination of labeled tripeptide formation, [^{14}C]glycine (80,000 cpm/μmol), 5 m*M* phosphocreatine, and creatine kinase (1 unit) are added to the reaction mixture described above. The reactions are initiated by adding enzyme, and after incubation at 37°, the reactions are stopped by adding 0.9 ml of ice-cold 20 m*M* acetic acid. The assay solutions are added to small columns (0.5 × 7 cm) of Dowex-1-acetate. After washing with several column volumes of 20 m*M* acetic acid, the labeled product is eluted with 4 ml of 1.5 *M* ammonium acetate. The radioactivity present in the eluate is determined by scintillation counting. In the assay involving use of labeled glycine, it is preferable to use the creatine kinase system for

[12] H. H. Taussky and E. Shorr, *J. Biol. Chem.* **202,** 675 (1953).

regeneration of ATP in place of the pyruvate kinase system. Phosphoenolpyruvate inhibits the enzymatic reaction about 7–10% under the conditions given above. An ATP-regenerating system is desirable because in its absence the rates of reaction are linear with enzyme concentration to only 5% of completion as compared to about 50% of completion in the presence of the ATP-generating system.

A unit of glutathione synthetase activity is defined as the amount of enzyme that catalyzes the synthesis of 1 μmol of L-γ-glutamyl-L-α-aminobutyrylglycine in 1 hr under the conditions given above. Specific activity is expressed as units per mg of protein (determined by the method of Lowry *et al.*[13] using bovine serum albumin as standard).

During isolation of the enzyme, activity and protein may be conveniently determined as follows. Enzyme activity is determined by incubating 5 μl portions of each fraction in reaction mixtures (final volume, 0.5 ml) containing the components of the standard assay given above. After incubation at 37° for 20 min, 0.95 ml of diluted phosphate reagent (prepared by dissolving 2.0 g of $FeSO_4 \cdot 7H_2O$ in 90 ml of 10% trichloroacetic acid) is added, followed by addition of 4 ml of a solution containing 16% of ammonium molybdate in 5 *M* H_2SO_4. After standing for 5 min, the absorbance at 720 nm is determined. The protein concentration is monitored by determining the absorbance at 280 nm. The protein concentration in the effluents from the ATP affinity column (step 7 of the purification procedure) is conveniently determined by the Bradford procedure[14] using bovine serum albumin as standard. The absorbance at 595 nm is measured after adding 20 μl of each fraction to 1 ml of Bradford reagent (prepared by dissolving 100 mg of Coomassie Brilliant Blue G-250 in 50 ml of 95% ethanol; 100 ml of 85% phosphoric acid is then added, and the solution is diluted with water to a final volume of 1 liter).

Purification Method

All procedures are carried out at 0–5°. Enzyme activity and protein are monitored in the column effluents as stated above.

Step 1. Male rats weighing 400–600 g are killed by decapitation. After exsanguination, the kidneys are removed, decapsulated, and stored in ice-cold Buffer E (50 m*M* imidazole–HCl, 5 m*M* 5-oxo-L-proline, 10 m*M* magnesium chloride, pH 8.5). The kidneys (350 g) obtained from 150 rats are weighed, minced, and homogenized in a final volume of 2 liters of Buffer E. The homogenate is centrifuged at 16,300 *g* for 60 min. (5-Oxo-L-

[13] O. H. Lowry, N. J. Rosebrough, A. L. Farr, and R. J. Randall, *J. Biol. Chem.* **193,** 265 (1951).

[14] M. M. Bradford, *Anal. Biochem.* **72,** 248 (1976).

proline is included in Buffer E to protect the 5-oxoprolinase activity, which may be isolated from the first pellet obtained in Step 2.)

Step 2. The supernatant solution obtained in Step 1 is treated with solid ammonium sulfate (19.6 g/100 ml) to yield 33% of saturation. After stirring for 30 min, the mixture is centrifuged at 16,300 *g* for 20 min and the supernatant solution obtained (1.75 liters) is treated with solid ammonium sulfate (12.5 g/100 ml) to yield 53% saturation. After stirring for 30 min, the precipitate which forms is collected by centrifugation for 20 min. The pellet is dissolved in 200 ml of Buffer S (50 m*M* imidazole–HCl, 0.1 m*M* EDTA, pH 7.4).

Step 3. The solution is divided into 2 equal portions and placed in 500-ml Erlenmeyer flasks, which are placed in a boiling water bath and gently swirled. The temperature of the solutions is maintained at 58° for 5 min, and the suspensions are then rapidly cooled by immersion in ice. At this point, the two mixtures are combined and centrifuged at 31,000 *g* for 30 min. The pellet is washed twice by centrifugation with 50 ml portions of Buffer S and the supernatant solutions and washings are combined.

Step 4. The solution is treated with solid ammonium sulfate (47.2 g/100 ml) to yield 70% of saturation. After stirring for 30 min, the mixtures are centrifuged at 31,000 *g* for 20 min. The pellet is dissolved in the minimal amount (about 30 ml) of Buffer S. This solution is added to the top of a column (6 × 95 cm) on Sephadex G-100 previously equilibrated with Buffer S. The column is developed with Buffer S at a flow rate of 100 ml/hr; fractions of 23 ml are collected. Analysis of the fractions shows a peak of enzyme activity which elutes slightly later than the major protein peak. The enzymatically active fractions are combined.

Step 5. The pooled fractions are added to the top of a column (2.5 × 16 cm) of DEAE-cellulose previously equilibrated with Buffer S. The column is washed with 200 ml of Buffer S; elution is carried out with a linear gradient established between 750 ml of Buffer S and 750 ml of Buffer S containing 0.5 *M* sodium chloride. The flow rate is 130 ml/hr. Fractions of 23 ml are collected. Enzyme activity elutes near the middle of the salt gradient; the active fractions are pooled.

Step 6. The solution obtained in Step 5 is treated with solid ammonia sulfate (47.2 g/100 ml) to give 70% of saturation. The resulting suspension is stirred for 30 min and then centrifuged at 31,000 *g* for 25 min. The pellet is dissolved in the minimal amount (about 4 ml) of Buffer S. The solution is added to the bottom of a reverse flow column (3.5 × 122 cm) of Sephadex G-150. A solution [10 ml containing 20% glycerol in Buffer S (v/v)] is added to the column and elution is then carried out with Buffer S with a flow rate of 22 ml/hr. Fractions of 5 ml are collected, and the active fractions are combined.

Step 7. The pooled fractions are applied to the top of a column (1.5 × 15 cm) containing agarose-aminohexyl-N^6-ATP previously equilibrated with 200 ml of Buffer I (50 m*M* imidazole–HCl, 0.1 m*M* EDTA, and 1 m*M* $MnCl_2$, pH 7.4). The column is washed with 100 ml of Buffer I. Glutathione synthetase is eluted with a linear gradient established between 100 ml of Buffer I and 100 ml of Buffer I containing 5 m*M* ATP and 6 m*M* $MnCl_2$. The flow rate is 25 ml/hr and fractions of 5 ml are collected. The active fractions are pooled (about 44 ml). Ammonium sulfate (47.2 g/100 ml) is added and the resulting suspension is stirred for 30 min and then centrifuged to 31,000 *g* for 30 min. The pellet is dissolved in the minimal volume of Buffer S (final volume, about 2.5 ml). The enzyme solution is freed from manganese ions and ATP by gel filtration on a column (1 × 40 cm) of Sephadex G-25 using Buffer S for elution. The flow rate is 20 ml/hr and fractions of 2 ml are collected. The enzyme is eluted in about 7 ml and is well separated from ATP.

A summary of the purification is given in the table. The overall yield is about 40%. The enzyme is stable for more than 2 months when stored at 4°. It is notable that potassium chloride is not required for binding of the enzyme to the ATP affinity column. The enzyme does not bind to ATP-agarose columns in which the ATP is bound through the C^8 or ribose moiety. The capacity of N^6-aminohexyl-ATP-agarose for the enzyme is about 6 times greater in the presence of Mn^{2+} than with Mg^{2+}. However, the enzyme loses activity when exposed to Mn^{2+} for long periods, and

PURIFICATION OF GLUTATHIONE SYNTHETASE[a]

Step	Volume (ml)	Protein (total) (mg)	Activity: Total units	Activity: Specific units/mg	Yield (%)
1. Centrifuged homogenate	1,830	22,700	25,200	1.1	(100)
2. Ammonium sulfate fractionation	245	8,210	24,300	3.0	96
3. Heat denaturation of impurities	300	2,200	27,300	12.3	108
4. Gel filtration on Sephadex G-100	205	800	27,000	33.8	107
5. DEAE-cellulose chromatography	225	100	18,900	189	75
6. Gel filtration on Sephadex G-150	88	50	11,800	236	47
7. ATP-affinity chromatography	2.5	12	9,900	825	39

[a] From 350 g of rat kidney.

for this reason it is desirable to precipitate the enzyme with ammonium sulfate promptly, followed by gel filtration as described in Step 7. In Step 3, the enzyme is stable at 58° for 5 min in the absence of glutathione or other thiols.

Physical and Chemical Characterization

The enzyme exhibits maximal absorbance at 277 nm and the ratio of the absorbance at 280 nm to that at 260 nm is 2.6.[11] At 280 nm, the absorbance of any enzyme solution containing 1 mg/ml is 0.80. The purified enzyme exhibits a sharp single band on polyacrylamide gel electrophoresis in the presence of sodium dodecyl sulfate. Determination of molecular weight by gel filtration gives a value of 118,000 ± 4,000. Gel electrophoresis in the presence of sodium dodecyl sulfate indicates a subunit molecular weight of 59,000, indicating that the enzyme is composed of 2 apparently identical subunits. The purified enzyme contains neutral sugar, amino sugar, and sialic acid; the values for total carbohydrate vary from 1.7 to 2.04%. Amino acid analysis of this enzyme preparation has been reported.[11]

Catalytic Properties

The enzyme exhibits a pH optimum in the range 8.0–8.5. The apparent K_m values for glycine and ATP were reported to be 0.76 m*M* and 33 μ*M*, respectively, in the presence of 10 m*M* concentrations of the other substrates.[11] Attempts to determine the K_m value for L-γ-glutamyl-L-α-aminobutyrate gave atypical results which are consistent with an apparent K_m value between 20 and 200 μ*M*. The K_m value for L-γ-glutamyl-L-cysteine is probably less than 50 μ*M*. Although potassium ions are required for maximum activity, substantial activity has been found when 50 m*M* K^+ ions are replaced by equimolar concentrations of Na^+, Rb^+, Li^+, or NH_4^+. The enzyme is active when ATP is replaced by dATP and to a lesser extent when it is replaced by other nucleoside triphosphates. The enzyme is markedly inhibited by 10 m*M* *p*-chloromercuribenzoate; it is not inhibited by iodoacetamide, *N*-ethylmaleimide, or 5,5′-dithiobis-(2-nitrobenzoate).

When the enzyme is incubated in reaction mixtures containing hydroxylamine in place of glycine, the corresponding dipeptide hydroxamate is formed. The reaction catalyzed by glutathione synthetase is similar to those catalyzed by glutamine synthetase[15] and γ-glutamylcysteine synthetase[2] in that an intermediate enzyme-bound acyl phosphate is formed. Evidence for the intermediate formation of γ-glutamyl-α-

[15] A. Meister, *in* "The Enzymes" (P. D. Boyer, ed.), 3rd ed., Vol. 10, p. 699. Academic Press, New York, 1974.

aminobutyryl phosphate in the enzymatic synthesis of γ-glutamyl-α-aminobutyrylglycine (ophthalmic acid) was first obtained in studies[16,17] in which the pulse-labeling technique[18] was used. In this work the dipeptide phosphate was isolated.[16,17] When the enzyme was incubated with chemically synthesized γ-glutamyl-α-aminobutyryl phosphate, ADP, and magnesium ions, formation of ATP was observed. Incubation of the enzyme with γ-glutamyl-α-aminobutyryl phosphate, glycine, and MgADP led to formation of ophthalmic acid. A scheme for the mechanism of the reaction involving formation of a complex between enzyme, MgADP, and dipeptide phosphate has been proposed.[2] This scheme is consistent with the ability of the enzyme to catalyze several partial reactions:

$$\text{Tripeptide} + NH_2OH \rightleftharpoons \text{dipeptide hydroxamate} + \text{glycine}$$

$$\text{Tripeptide} + H_2O \xrightarrow{\text{arsenate}} \text{dipeptide} + \text{glycine}$$

$$\text{Dipeptide glycine*} + \text{glycine} \rightleftharpoons \text{dipeptide glycine} + \text{glycine*}$$

It is notable that these reactions require catalytic amounts of nucleotide, which seems to be involved in the binding and utilization of dipeptide phosphate. The mechanism of the reaction has been discussed.[2]

The substrate specificity of rat kidney glutathione synthetase has been examined in some detail.[11] In general, the findings indicate a high degree of specificity toward γ-glutamyl amino acids in which the amino acid moiety is sterically similar to cysteine (e.g., α-aminobutyrate, serine, alanine). Substitution of the L-α-aminobutyrate moiety by the corresponding D-isomer, β-amino-*n*-butyrate, or β-aminoisobutyrate leads to markedly reduced enzymatic susceptibility. None of the four possible monomethyl derivatives of the α-aminobutyrate moiety is significantly active. On the other hand, the L-γ-glutamyl moiety of L-γ-glutamyl-L-α-aminobutyrate can be modified considerably with retention of substrate activity. Thus, the D-γ-glutamyl derivative of L-α-aminobutyrate as well as the β-aminoglutaryl derivative, and all four possible monomethyl derivatives of the L-α-glutamyl moiety (*N*-methyl, α-methyl, β-methyl, and γ-methyl) are significantly active. It is interesting to note that both *N*-acetyl-L-α-aminobutyrate and *N*-acetyl-L-cysteine are also active. Studies on the mapping of the substrate binding sites thus indicate a high degree of specificity for the cysteinyl and glycine binding sites. The γ-glutamyl binding site of the enzyme seems to resemble glutamine synthetase[15] and γ-glutamyl transpeptidase[19,20] in that it accepts the D-γ-glutamyl and certain monomethyl-substituted L-γ-glutamyl moieties.

[16] J. S. Nishimura, E. A. Dodd, and A. Meister, *J. Biol. Chem.* **238,** 1179 (1963).
[17] J. S. Nishimura, E. A. Dodd, and A. Meister, *J. Biol. Chem.* **239,** 2553 (1964).
[18] P. R. Krishnaswamy, V. Pamiljans, and A. Meister, *J. Biol. Chem.* **237,** 2932 (1962).
[19] S. S. Tate and A. Meister, *Mol. Cell. Biochem.* **39,** 357 (1981).
[20] G. A. Thompson and A. Meister, *J. Biol. Chem.* **252,** 6792 (1977).

[50] γ-Glutamyl Transpeptidase from Kidney

By SURESH S. TATE and ALTON MEISTER

Introduction

The initial step in the degradation of glutathione and its derivatives is catalyzed by γ-glutamyl transpeptidase, an intrinsic membrane protein that transfers the γ-glutamyl (γ-glu) moiety of this tripeptide to an acceptor, which may be an amino acid, dipeptide, or glutathione itself, producing the respective γ-glutamyl derivatives. If the nucleophile is water, hydrolysis occurs resulting in the formation of glutamate. These three general types of reactions are illustrated in reactions (1)–(3).

$$\text{Glutathione} + \text{amino acid} \rightleftharpoons \gamma\text{-glu-amino acid} + \text{CYSH-GLY} \quad (1)$$
$$2\ \text{Glutathione} \rightleftharpoons \gamma\text{-glu-glutathione} + \text{CYSH-GLY} \quad (2)$$
$$\text{Glutathione} + H_2O \rightarrow \text{glutamate} + \text{CYSH-GLY} \quad (3)$$

The reaction catalyzed by γ-glutamyl transpeptidase is of major importance in the γ-glutamyl cycle, a metabolic pathway that accounts for the enzymatic synthesis and degradation of glutathione.[1-3] The evidence that the reactions of the γ-glutamyl cycle take place *in vivo* and related investigations on the metabolism and function of glutathione have been reviewed.[1-3] The activity of γ-glutamyl transpeptidase was first observed about 70 years ago and the enzyme has subsequently been studied by many investigators. Various aspects of this enzyme have been and are currently being explored. The research received strong impetus from the proposal that transpeptidase and other reactions of the γ-glutamyl cycle constitute one of the mechanisms for the transport of amino acids into certain cell types,[1-3] as well as from the recognition that transpeptidase plays an important role in such processes as the formation of mercapturic acids from S-conjugates of glutathione,[4] metabolism of leukotriene D,[5] one of a group of biological mediators that have important roles in inflammatory and allergic responses, and that the enzyme is a useful marker for

[1] A. Meister and S. S. Tate, *Annu. Rev. Biochem.* **45,** 559 (1976).
[2] A. Meister, *Curr. Top. Cell. Regul.* **18,** 21 (1981).
[3] A. Meister and M. E. Anderson, *Annu. Rev. Biochem.* **52,** 711 (1983).
[4] S. S. Tate, *in* "Enzymatic Basis of Detoxication" (W. B. Jakoby, ed.), Vol. 2, p. 95. Academic Press, New York, 1980.
[5] K. Bernström, L. Örning, and S. Hammarström, this series, Vol. 86, p. 38.

Copyright © 1985 by Academic Press, Inc.
All rights of reproduction in any form reserved.

preneoplastic changes in a number of tissues and of cell differentiation and cell aging.[3,6,7] The development of the various areas of research on γ-glutamyl transpeptidase has been reviewed.[1,3,5,8–10] Recent work has led to the isolation of highly purified preparations from a number of sources, and to elucidation of various aspects of the structure, catalytic properties, function, and biosynthesis of this enzyme. This chapter deals principally with relatively recent findings on mammalian γ-glutamyl transpeptidase, with emphasis on the rat kidney enzyme, the most extensively studied transpeptidase. Another chapter describes the specificity and kinetic studies on this enzyme.[11]

Homogenates of animal tissues exhibit a wide range of transpeptidase activity. Low transpeptidase activity in a tissue homogenate is often misleading since specific localization studies (see below) indicate that there are specific regions of intense enzyme activity in many tissues. In most mammals, the kidney exhibits by far the highest activity. The relative activities found in homogenates of adult rat tissues are as follows: kidney, 100; pancreas, 20; epididymis (caput), 27; epididymis (cauda), 4; seminal vesicle, 4; jejunal crypt cells, 0.8; jejunal villus tip cells, 3.3; liver, 0.2; spleen cells, 0.2.[4] Microdissection studies of rat nephron indicate that the proximal straight tubule exhibits about twice as much transpeptidase activity as the proximal convoluted tubule.[12] The activity found in both these regions is nevertheless significantly higher than those found in other rat tissues.

Introduction of histochemical methods for the localization of transpeptidase activity (see ref. 10 for the relevant literature) allowed precise localization of the enzyme in a variety of tissues and cells. These studies indicate that, in general, high enzyme activity is seen in cells which exhibit intense secretory or absorptive functions such as the epithelial cells of renal proximal tubules, jejunum, biliary tract, epididymis, seminal vesicles, choroid plexus, ciliary body, retinal pigment epithelium, bronchioles, thyroid follicles, canalicular region of hepatocytes, and the pancreatic acinar and ductule cells.[1,4,10] A variety of approaches (including histochemistry, subcellular fractionation, ultrastructural localization using cytochemistry, use of impermeable substrates and of proteinases to

[6] D. M. Goldberg, *CRC Crit. Rev. Clin. Lab. Sci.* **12,** 1 (1980).

[7] S. B. Rosalki, *Adv. Clin. Chem.* **17,** 53 (1975).

[8] S. S. Tate and A. Meister, *Mol. Cell. Biochem.* **39,** 357 (1981).

[9] A. Meister, S. S. Tate, and O. W. Griffith, this series, Vol. 77, p. 237.

[10] A. Meister, S. S. Tate, and L. L. Ross, *in* "The Enzymes of Biological Membranes" (A. Martinosi, ed.), Vol. 3, p. 315. Plenum Press, New York, 1976.

[11] R. D. Allison, this volume [51].

[12] H. Heinle, A. Wendel, and U. Schmidt, *FEBS Lett.* **73,** 220 (1977).

release the enzyme, and immunological techniques using specific antibodies) have all contributed to the conclusion that transpeptidase is predominantly membrane bound and is localized on the external surface of cells.[4,8] In epithelial cells, such as those of the jejunum and renal proximal tubule, which are characterized by the presence of microvilli on their lumenal surface, the primary site of transpeptidase activity is the outer surface of microvillus (brush border) membranes.[13,14] A recent immunocytochemical study indicates that some enzyme is also present on the basolateral (contralumenal) membranes of the renal proximal tubule cells.[15] The enzyme on the basolateral membranes appears to play a role, along with the glomerular filtration mechanism, in the clearance of plasma glutathione and its derivatives.[3]

As noted above, γ-glutamyl transpeptidase plays a key role in the γ-glutamyl cycle, a pathway for the synthesis and degradation of glutathione. Glutathione is synthesized from its constituent amino acids within the cells by the successive actions of γ-glutamylcysteine and glutathione synthetases. The reactions allow the storage of cysteine as glutathione and γ-glutamyl transpeptidase catalyzes the first step in the pathway that leads to release of the cysteine moiety from the tripeptide. Glutathione is translocated across cell membranes and thus serves as a substrate for cell-surface γ-glutamyl transpeptidase. γ-Glutamyl amino acids, formed at the cell surface, are transported into the cell. The role of the γ-glutamyl cycle and γ-glutamyl transpeptidase in this pathway of amino acid transport has been reviewed.[1-3]

Assay Methods

The broad specificity exhibited by γ-glutamyl transpeptidase toward γ-glutamyl compounds[8,11] has made it possible to devise a number of convenient spectrophotometric procedures for the determination of its activity. The most widely used substrate is L-γ-glutamyl-*p*-nitroanilide.[16] The usual assay consists of incubation of the enzyme with L-γ-glutamyl-*p*-nitroanilide and an acceptor such as glycylglycine; the *p*-nitroaniline released is readily determined from the increase in absorbance at 410 nm. In the absence of an added acceptor substrate, both hydrolysis and autotranspeptidation of L-γ-glutamyl-*p*-nitroanilide can occur. However, because this substrate is poorly bound as an acceptor, it can be used to determine

[13] G. V. Marathe, B. Nash, R. H. Haschemeyer, and S. S. Tate, *FEBS Lett.* **107,** 436 (1979).
[14] B. Tsao and N. P. Curthoys, *J. Biol. Chem.* **255,** 7708 (1980).
[15] H. W. Spater, M. S. Poruchynsky, N. Quintana, M. Inoue, and A. B. Novikoff, *Proc. Natl. Acad. Sci. U.S.A.* **79,** 3547 (1982).
[16] M. Orlowski and A. Meister, *Biochim. Biophys. Acta* **73,** 679 (1963).

the hydrolysis reaction provided its concentration is kept below 10 μM.[17] A more convenient substrate to study the hydrolytic reaction is D-γ-glutamyl-*p*-nitroanilide.[17,18] Since the enzyme exhibits strict L-stereospecificity toward acceptor substrates, the D isomer can not participate in autotranspeptidation and thus can be used for separate study of the hydrolytic reaction. The apparent K_m values for the L and D isomers of γ-glutamyl-*p*-nitroanilide (5 and 31 μM, respectively) are of the same order of magnitude as that of glutathione (and its S-derivatives) and the V_{max} values for hydrolysis of the two isomers are similar.[17] Details of the assay procedures using γ-glutamyl-*p*-nitroanilides have been described previously.[9] An outline of these procedures is given in the following sections.

Reagents

Buffer: 0.1 M Tris–HCl, pH 8.0
Glycylglycine, 0.1 M, pH 8.0
L-γ-Glutamyl-*p*-nitroanilide, 5 mM, pH 8.0
D-γ-Glutamyl-*p*-nitroanilide, 5 mM, pH 8.0: the D-isomer is synthesized as described[16]

Details of the preparation of these reagents have been given.[9]

Procedure

Transpeptidase Activity. Additions are made to a spectrophotometer cuvette (semimicro; 1 cm light path) as follows: 0.2 ml of L-γ-glutamyl-*p*-nitroanilide (1 mM, final concentration), 0.2 ml of glycylglycine (20 mM, final concentration), and 0.6 ml of Tris–HCl buffer. The volume of the buffer can be adjusted to allow for that of the enzyme solution added (final volume of the reaction mixture is 1.0 ml). The solution is allowed to reach 37° in a spectrophotometer equipped with a thermostatted cuvette holder. The reaction is initiated by adding a suitable amount of enzyme and the rate of release of *p*-nitroaniline is recorded at 410 nm (the ε for *p*-nitroaniline is 8800 M^{-1} cm^{-1}). The activity is expressed as micromoles of *p*-nitroaniline formed per minute (units) and the specific activity as the units per mg of protein. This procedure has been used to monitor transpeptidase activity during purification of the enzyme. The presence of glycylglycine (20 mM) suppresses both the hydrolysis of γ-glutamyl-*p*-nitroanilide and the autotranspeptidation reaction. The major reaction that occurs initially under these conditions is

$$\text{L-}\gamma\text{-Glutamyl-}p\text{-nitroanilide} + \text{Gly-Gly} \rightarrow \text{L-}\gamma\text{-Glu-Gly-Gly} + p\text{-nitroaniline} \quad (4)$$

[17] G. A. Thompson and A. Meister, *Biochem. Biophys. Res. Commun.* **71,** 32 (1976).
[18] G. A. Thompson and A. Meister, *J. Biol. Chem.* **254,** 2956 (1979).

When several assays are to be performed, a mixture containing L-γ-glutamyl-p-nitroanilide, Gly-Gly, and Tris–HCl buffer, 4:4:12 (v/v) is prepared; 1 ml of this solution is used per assay as described above. This mixture can be stored at −15° for 4 days and at 4° for 24 hr. When stored frozen, γ-glutamyl-p-nitroanilide tends to precipitate. The mixture should be defrosted and kept at 37° until clear.

Hydrolytic Activity. The reaction mixture (prepared in a cuvette) contains 0.1 ml of D-γ-glutamyl-p-nitroanilide (0.5 m*M*, final concentration) and 0.9 ml of Tris–HCl buffer. After equilibration at 37°, the reaction is initiated by adding the enzyme, and the p-nitroaniline release is recorded at 410 nm.

Separate determinations of the transpeptidase and hydrolytic activities of the enzyme have been useful in probing the mechanism of differential modulation of these two activities by compounds such as maleate, hippurate, bile acids, and related compounds.[18–21]

Other Assay Methods

A number of other procedures are available for the determination of γ-glutamyl transpeptidase activity.[9] The fluorogenic substrate, L-γ-glutamyl-7-amino-4-methylcoumarin (L-γ-glutamyl-AMC) provides a very sensitive assay.[22] A stock solution of the substrate (10 m*M*) is prepared in methoxyethanol. Prior to use, 0.2 ml of the stock L-γ-glutamyl-AMC is diluted with 9.8 ml of 0.1 *M* ammediol–HCl buffer (pH 8.5) containing 20 m*M* Gly-Gly and 0.1% (w/v) Triton X-100 (final concentration of L-γ-glutamyl-AMC = 0.2 m*M*). Enzyme activity is assayed by adding 0.1 ml of suitably diluted enzyme solution to 0.25 ml of the reaction mixture in 6-ml glass tubes and incubating at 37°. The reaction is terminated by the addition of 2.0 ml of ice-cold 0.05 *M* glycine buffer (pH 10.4). [It may be advantageous to include L-serine and sodium borate (5 m*M* each) in the glycine buffer to assure complete termination of transpeptidase activity (see below)]. The AMC produced is determined by reading its fluorescence at 440 nm with excitation at 370 nm. The fluorescence is compared with the appropriate AMC standard solutions. This method is useful for determining the activity in tissues containing relatively low levels of transpeptidase. The lower limit for detection of AMC is approximately 0.05 nmol/ml[22] whereas the lower limit of detection of p-nitroaniline is about 1 to 2 nmol/ml.

[19] G. A. Thompson and A. Meister, *J. Biol. Chem.* **255,** 2109 (1980).
[20] W. A. Abbott and A. Meister, *J. Biol. Chem.* **258,** 6193 (1983).
[21] S. J. Gardell and S. S. Tate, *J. Biol. Chem.* **258,** 6198 (1983).
[22] G. D. Smith, J. L. Ding, and T. J. Peters, *Anal. Biochem.* **100,** 136 (1979).

For references to the literature which describe the assay procedures that use glutathione and a number of its S-derivatives as substrates see Meister *et al.*[9] and Tate.[4]

Enzyme Purification

Kidneys from male Sprague–Dawley rats (250 to 350 g in weight) are used either fresh or after storage at −20° for as long as several months. γ-Glutamyl transpeptidase constitutes about 4% of the total proteins in rat renal brush borders. Since the enzyme is membrane-bound, it must be brought into a soluble form before attempting purification. The solubilization methods include treatment of membranes with either detergents or proteinases. Fortunately the enzyme is stable and thus withstands relatively harsh treatments. γ-Glutamyl transpeptidase is an amphipathic, intrinsic membrane protein that consists of a hydrophilic domain anchored to the microvillus membranes by a hydrophobic segment.[8] Detergents solubilize the entire molecule, whereas proteinases (e.g., papain) cleave the hydrophilic, active site-containing domain (representing about 90% of the total protein mass) from the membrane-bound anchor. The catalytic properties of the detergent- and proteinase-solubilized enzymes are identical but the two forms differ in certain physical characteristics. A relatively rapid procedure for the purification of the papain-solubilized rat renal γ-glutamyl transpeptidase has been described.[9,23] This procedure has been extended recently to allow the simultaneous purification of the three glutathione-degrading enzyme of the microvillus membranes: γ-glutamyl transpeptidase, dipeptidase (cysteinyl-glycine hydrolyzing), and aminopeptidase M (a broad specificity exopeptidase).[24,25] An outline of this method is given below. An immunoaffinity chromatographic technique is also described for the purification of the Triton-solubilized transpeptidase.

Purification of Papain-Solubilized Rat Kidney γ-Glutamyl Transpeptidase

Steps 1 to 4 are described in detail in this volume which also gives a flow-sheet outline of the purification procedure. Microvillus membranes are isolated (Step 1) essentially as described by Malathi *et al.*[26] except that

[23] E. M. Kozak and S. S. Tate, *FEBS Lett.* **122,** 175 (1980).
[24] E. M. Kozak and S. S. Tate, *J. Biol. Chem.* **257,** 6322 (1982).
[25] S. S. Tate, this volume [58].
[26] P. Malathi, H. Preiser, P. Fairclough, P. Mallett, and R. K. Crane, *Biochim. Biophys. Acta* **554,** 259 (1979).

Tris-mannitol buffer adjusted to pH 7.5 (instead of pH 7.0) is used for the homogenization of rat kidneys. This small change in pH of the homogenization buffer consistently gives higher yields (about 60 to 65%) of the microvillus membranes (as judged by the recovery of the marker enzymes).[24] The membranes are suspended in 5 m*M* Tris–HCl buffer (pH 7.4) containing 50 m*M* NaCl and treated with papain. The supernatant after papain treatment is subjected to $(NH_4)_2SO_4$ fractionation. Transpeptidase (as well as dipeptidase and aminopeptidase M) precipitates between 45 and 90% of $(NH_4)_2SO_4$ saturation. The precipitate, obtained by centrifugation, is dissolved in the minimal volume of 0.05 *M* Tris–HCl buffer (pH 8.0) (Step 2). This solution is then chromatographed on a Sephadex G-150 column equilibrated and eluted with 0.05 *M* Tris–HCl buffer (pH 8.0) (Step 3). As described previously[24] (see also ref. 25), the three enzyme activities separate on this column. There is, however, some overlap between dipeptidase and γ-glutamyl transpeptidase. Apparently homogeneous transpeptide can be obtained by pooling the appropriate fractions from the Sephadex G-150 column. To increase the yields of both the dipeptidase and transpeptidase, the Sephadex G-150 fractions containing both the dipeptidase and transpeptidase are pooled and dialyzed 18 hr against 30 volumes of 5 m*M* Tris–HCl buffer (pH 8.0). The solution is then lyophilized, the residue redissolved in a minimal volume of water (less than 5 ml), and dialyzed 18 hr against 100 volumes of 0.025 *M* Tris–HCl buffer (pH 8.0) containing 0.05 *M* NaCl. The dialyzed solution is applied to a DEAE-cellulose column (Whatman DE 52, microgranular, 1 × 21 cm), previously equilibrated with the Tris–NaCl buffer. The column is washed with about 4 column volumes of the buffer (Step 4). Transpeptidase activity emerges in the initial wash (see Fig. 2 in ref. 25). Fractions containing transpeptidase are pooled and treated with $(NH_4)_2SO_4$ to give 90% of saturation. The precipitate obtained by centrifugation is dissolved in 4 ml of 0.05 *M* Tris–HCl buffer (pH 8.0) and dialyzed for 18 hr against 200 volumes of the same buffer.

A summary of a typical purification of papain-solubilized γ-glutamyl transpeptidase is given in Table I. This procedure has consistently given between 50 and 55% yields of apparently homogeneous transpeptidase.

Purification of Triton-Solubilized Rat Kidney γ-Glutamyl Transpeptidase

A procedure for the purification of the Triton-solubilized rat renal γ-glutamyl transpeptidase has been published.[9] The method, based upon that used by Hughey and Curthoys,[27] is laborious and yields are generally

[27] R. P. Hughey and N. P. Curthoys, *J. Biol. Chem.* **251,** 7863 (1976).

TABLE I
PURIFICATION OF PAPAIN-SOLUBILIZED RAT KIDNEY γ-GLUTAMYL TRANSPEPTIDASE[a]

Step	Total protein (mg)	Transpeptidase activity[b] Total (units)	Specific (units/mg)	Yield (%)
1. Microvillus membranes[c]	2185	31,030	14.2	[100]
2. Papain treatment followed by $(NH_4)_2SO_4$ fractionation (45–90%)	272	25,130	92.4	81
3. Chromatography on Sephadex G-150	31.5	18,430	585	59
4. Chromatography on DEAE-cellulose	20.7	16,750	810	54

[a] Data taken from Kozak and Tate.[24]
[b] Activity determined with 1 m*M* L-γ-glutamyl-*p*-nitroanilide and 20 m*M* Gly-Gly.
[c] From 210 g of frozen rat kidney.

low (10 to 15%). An efficient method involving immunoaffinity chromatography has now been developed (S. S. Tate and L. Weissman, unpublished). The method makes use of antibodies prepared against the highly purified papain-solubilized γ-glutamyl transpeptidase.

Preparation of Antibodies Directed against γ-Glutamyl Transpeptidase and Their Covalent Attachment to Sepharose

Antibodies directed against the papain-solubilized γ-glutamyl transpeptidase[28] are obtained as follows: transpeptidase (0.5 ml containing 0.8 mg of the enzyme) is mixed with an equal volume of Freund's complete adjuvant and the mixture is injected intradermally into a male New Zealand white rabbit. A booster injection (0.5 mg of transpeptidase) is given after 6 weeks. The first bleeding is made 8 weeks after the initial immunization and additional bleedings are done very 2 weeks thereafter for a total of 6 bleedings. The γ-globulin fraction (anti-transpeptidase-IgG) is obtained from the antisera by $(NH_4)_2SO_4$ precipitation in borate–NaCl buffer (5 m*M* boric acid, 1.25 m*M* sodium tetraborate, 0.15 *M* NaCl, pH 8.4).[29] The anti-transpeptidase-IgG is covalently attached to CNBr-activated Sepharose CL-4B beads (Sigma) as described by Garvey *et al.*[29] (anti-transpeptidase-IgG-Sepharose).

[28] B. Nash and S. S. Tate, *J. Biol. Chem.* **257,** 585 (1982).
[29] J. S. Garvey, N. E. Cremer, and D. H. Sussdorf, "Methods in Immunology," 3rd ed., p. 215. Benjamin, Reading, Massachusetts, 1977.

Preimmune serum γ-globulin fraction is similarly prepared from bleedings taken from a rabbit which is not immunized. This γ-globulin fraction is also coupled to CNBr-activated Sepharose CL-4B as described above (preimmune γ-globulin-Sepharose).

Purification of Triton-Solubilized γ-Glutamyl Transpeptidase

Steps 1 and 2. The microvillus membranes (from 10 g of rat kidney) are isolated as described above.[25] The membranes are suspended in 5 ml of 0.01 *M* Tris–HCl buffer (pH 8.0) containing 0.15 *M* NaCl and 1% Triton X-100 (Tris–Triton buffer), and the mixture is homogenized with a Potter-Elvehjem homogenizer. After standing on ice for 30 min, the suspension is centrifuged at 43,000 *g* for 1 hr.

Step 3. Immunoaffinity Chromatography. The clear supernatant from the previous step is applied to the preimmune γ-globulin-Sepharose column (1 × 5 cm; equilibrated with Tris–Triton). The column is eluted with 10 ml of Tris–Triton. The eluates (which contain all of the transpeptidase activity) are combined and the solution is mixed with anti-transpeptidase-IgG-Sepharose (about 4 ml of settled slurry). The slurry is gently mixed at 25° for 60 min and then poured into a column (1 × 10 cm). The column is washed extensively with Tris–Triton. About 75% of the applied transpeptidase is bound to the anti-transpeptidase-IgG-Sepharose. The column is then eluted with 0.04 *M* NH_4OH containing 0.2% Triton X-100. The eluates (1 ml fractions) are collected in tubes containing 0.1 ml of 1 *M* Tris–HCl (pH 7.0) (to adjust the pH of the eluate to approximately 8 to 8.5). Fractions containing transpeptidase are pooled and the enzyme is precipitated by the addition of 10 volumes of acetone (precooled to −15°). The precipitate (collected by centrifugation at 8000 *g* for 15 min) is dissolved in 0.5 ml of 0.01 *M* Tris–HCl (pH 8.0) containing 0.2% Triton X-100.

A summary of a typical purification procedure for the Triton-solubilized γ-glutamyl transpeptidase is shown in Table II. The immunoaffinity column, after equilibration with Tris–Triton, can be used repeatedly without significant loss of binding capacity.

Properties of the Enzyme

Structural and Topological Features

Although the papain- and Triton-solubilized forms of rat kidney γ-glutamyl transpeptidase exhibit identical catalytic properties, the two forms differ in a number of physical and chemical characteristics. The papain-solubilized enzyme is soluble in aqueous buffers. In contrast, the Triton-solubilized form exhibits physical characteristics of a typical am-

TABLE II
PURIFICATION OF TRITON-SOLUBILIZED γ-GLUTAMYL TRANSPEPTIDASE FROM RAT KIDNEY[a]

Step	Total protein (mg)	Transpeptidase activity		
		Total (units)	Specific (units/mg)	Yield (%)
1. Microvillus membranes[b]	115	2010	17.5	[100]
2. Triton extract of microvillus membranes	93	1850	19.9	92
3. Immunoaffinity chromatography on anti-transpeptidase-IgG-Sepharose	1.3	960	738	48

[a] From S. S. Tate and L. Weissman, unpublished.
[b] From 10 g of rat kidney.

phipathic protein; thus it binds to detergent micelles, undergoes aggregation in the absence of detergent, and can associate with lecithin vesicles.[30] Some of the other properties of the two forms of rat kidney transpeptidase are compared in Table III.[31–34]

Both forms of the enzyme are heterodimeric glycoproteins. The light subunits (L) of the two forms of the enzyme exhibit identical molecular weights (M_r = 22,000), amino acid compositions, and N- and C-terminal residues.[31–33] In contrast, the heavy subunit (H) of the Triton-solubilized enzyme is larger than that of the papain-solubilized enzyme by about 52 amino acid residues (M_r of the H subunits are, respectively, about 51,000 and 46,000).[33] Furthermore, the two forms of the heavy subunit contain different N-terminal amino acid residues but identical C-terminal residues.[33] These and other data from studies in which the membrane associated domain of the enzyme was labeled using a membrane-soluble reagent[35] suggest that the N-terminal segment of the large subunit contains a papain-sensitive hydrophobic domain that serves to anchor the transpeptidase to the lumenal surface of the brush border membranes. The light subunit presumably does not interact directly with the lipid bilayer but is held by noncovalent interactions with the heavy subunit on the external

[30] N. P. Curthoys and R. P. Hughey, *Enzyme* **24,** 383 (1979).
[31] S. S. Tate and A. Meister, *Proc. Natl. Acad. Sci. U.S.A.* **73,** 2599 (1976).
[32] B. Nash and S. S. Tate, *J. Biol. Chem.* **259,** 678 (1984).
[33] Y. Matsuda, A. Tsuji, and N. Katunuma, *J. Biochem.* (*Tokyo*) **93,** 1427 (1983).
[34] K. Yamashita, Y. Tachibana, A. Hitoi, Y. Matsuda, A. Tsuji, N. Katunuma, and A. Kobata, *Arch. Biochem. Biophys.* **227,** 225 (1983).
[35] T. Frielle and N. P. Curthoys, *Ciba Found. Symp.* **95,** 73 (1983).

TABLE III
COMPARISON OF SOME STRUCTURAL FEATURES OF THE TRITON- AND PAPAIN-SOLUBILIZED RAT KIDNEY γ-GLUTAMYL TRANSPEPTIDASE

	Triton solubilized	Papain solubilized
M_r	73,000	68,000
Subunit M_r[a]		
H	51,000	46,000
L	22,000	22,000
N-Terminal residue[b]		
H	Methionine	Glycine
L	Threonine	Threonine
C-Terminal residue[b]		
H	Tyrosine	Tyrosine
L	Phenylalanine	Phenylalanine
Number of oligosaccharide chains (all N-linked) per mol[c]		
H	3	2
L	3	3

[a] The values for the subunit molecular weights are the best estimates from polyacrylamide gel electrophoresis in the presence of SDS.[31-33] The M_r values for the two forms of the enzyme are the sum of the two subunit M_r values. H and L represent the heavy and the light subunits of the enzyme, respectively.

[b] From Matsuda *et al.*[33]

[c] From Yamashita *et al.*[34]

surface of the membranes. Experiments, in which lactoperoxidase was positioned at either the internal surface of the brush border membrane vesicles or at the external surface of the sealed vesicles, thus enabling radioiodination of brush border proteins from either surface of the membrane, indicate that the heavy subunit is transmembranous, i.e., a portion of its anchor domain is exposed on the cytoplasmic surface of the membranes.[36]

The membrane-binding segment of the heavy subunit, released by papain treatment of the Triton-solubilized rat kidney transpeptidase, has been isolated. About 60% of its amino acid residues are hydrophobic in nature.[33] Amino acid sequence analysis of the N-terminal portion of the heavy subunit of Triton-solubilized enzyme (Met-Lys-Asn-Arg-Phe-Leu-

[36] B. Tsao and N. P. Curthoys, *Biochim. Biophys. Acta* **690,** 199 (1982).

Val-Leu-Gly-Leu-Val-Ala-Val-Val-Leu-Val-Phe-Val-Ile-Ile-Gly-Leu-) indicates that this region is highly hydrophobic.[33] The membrane-associated domain has also been shown to contain one N-linked oligosaccharide chain.[34,35]

The purified rat kidney transpeptidase exhibits considerable heterogeneity on polyacrylamide gels and is separable by isoelectric focusing into 12 enzymatically active isozymes ranging in pI from 5 to 8.[31] The isozymes are similar with respect to catalytic properties, amino acid composition, hexose and aminohexose content. However, their sialic acid content varies from 14 to 61 nmol/mg of protein. Neuraminidase treatment followed by isoelectric focusing indicates that the multiple forms are primarily related to different degrees of sialylation.[31] Subsequent studies by Matsuda *et al.*[37] have confirmed these findings. Recently, the carbohydrate moieties of rat kidney transpeptidase (all Asn-linked) have been subjected to detailed structural analysis and an extremely heterogeneous pattern has emerged.[38] The papain-solubilized enzyme contains 5 oligosaccharide chains per mol of enzyme, 3 of which are attached to the light subunit and 2 are found on the heavy subunit.[34] The enzyme has been shown to contain both neutral oligosaccharides of either the high mannose type or biantennary complex type chains, and acidic oligosaccharides of the bi-, tri-, and tetraantennary complex type. The significance of such heterogeneity in the oligosaccharide portion of the enzyme is not yet apparent at present.

Treatment of transpeptidase with dissociating agents such as urea and SDS at neutral pH values results in extensive proteolytic degradation of the heavy subunit.[39] The proteinase activity, which is not exhibited by the native dimeric enzyme, appears to be a catalytic function associated with the light subunit and involves the active center residue (probably a hydroxyl group) at which γ-glutamylation of the enzyme also occurs during the γ-glutamyl transfer reactions.[40,41] Intact subunits can be isolated upon urea treatment of the enzyme in 1 *M* acetic acid followed by gel filtration in the same medium.[42] The denatured subunits are inactive and renaturation of individual subunits, by dialysis against pH 6.8 buffers containing no urea, does not restore activity. Mixing of the renatured subunits also fails to restore activity. The latter finding may be ascribed in part to the

[37] Y. Matsuda, A. Tsuji, and N. Katunuma, *J. Biochem.* (*Tokyo*) **87,** 1243 (1980).

[38] K. Yamashita, A. Hitoi, Y. Matsuda, A. Tsuji, N. Katunuma, and A. Kobata, *J. Biol. Chem.* **258,** 1098 (1983).

[39] S. J. Gardell and S. S. Tate, *J. Biol. Chem.* **254,** 4942 (1979).

[40] S. S. Tate and A. Meister, *Proc. Natl. Acad. Sci. U.S.A.* **74,** 931 (1977).

[41] S. S. Tate and A. Meister, *Proc. Natl. Acad. Sci. U.S.A.* **75,** 4806 (1978).

[42] S. J. Gardell and S. S. Tate, *J. Biol. Chem.* **256,** 4799 (1981).

tendency of the denatured light subunit to form inactive polymers upon renaturation. Significant reconstitution of transpeptidase activity (up to 15% of the native enzyme) is achieved by prior mixing of denatured subunits followed by removal of urea by dialysis. The reconstituted active species has been shown to be similar to the native enzyme. These studies[42] provide no support for a report by Horiuchi *et al.*[43] that the isolated renatured heavy subunit possesses an active site which is not expressed in the native enzyme.

Highly purified preparations of renal transpeptidase have also been obtained from other mammalian species.[8] These enzymes, like the rat renal transpeptidase, are heterodimeric glycoproteins. The light subunits have molecular weights of about 22,000. However, the enzymes fall into two groups with respect to the size of the heavy subunit: those with heavy subunits exhibiting M_r between 46,000 and 50,000 (rat and rabbit), and those with heavy subunits of M_r in excess of 60,000 (cattle, sheep, dog, and human). The γ-glutamyl binding site is located on the light subunit in all species tested. The mode of association of the enzyme with membranes, in species other than the rat, is not known.

Biosynthesis and Processing of γ-Glutamyl Transpeptidase

The topological features of the rat kidney γ-glutamyl transpeptidase, a glycosylated heterodimer anchored to the external surface of the brush border membranes by a short N-terminal hydrophobic region of one of its subunits, generate a number of questions regarding its synthesis, assembly of its subunits, and the mode of achievement of its membrane orientation. In order to answer some of these questions, the biosynthesis and processing of γ-glutamyl transpeptidase were studied both in the tissue and by *in vitro* translation of its mRNA.[28,32] Rat renal slices were incubated with [^{35}S]methionine and transpeptidase-related proteins were isolated by immunoprecipitation with antibodies made against the purified enzyme as well as with antibodies directed against its two subunits.[28] The major species seen after relatively short pulse times was a M_r = 78,000 polypeptide containing the determinants for both the subunits. Kinetic and pulse-chase studies indicate that the M_r = 78,000 species is the common precursor of the two subunits of the enzyme. Like the dimeric enzyme, the single-chain precursor contains both core sugars and the peripheral sugar, fucose, on its oligosaccharide moieties. The brush border membranes were shown to contain primarily the labeled dimeric enzyme. Since fucose is a terminal sugar that is added to complex Asn-linked

[43] S. Horiuchi, M. Inoue, and Y. Morino, *Biochem. Biophys. Res. Commun.* **80,** 873 (1978).

chains in the distal portion of the Golgi complex, its presence on the M_r = 78,000 polypeptide indicates that the cleavage of the single-chain precursor takes place either in the trans Golgi, en route to brush border, or immediately upon its arrival at the brush border. Pulse-chase studies in the presence of pactamycin, an inhibitor of protein synthesis initiation, have shown that the heavy subunit of the enzyme is located at the N-terminal region of the M_r = 78,000 species.[32]

Whether or not the M_r = 78,000 species is catalytically active remains to be determined. The use of the various antibodies in these studies has revealed that a conformational change occurs upon cleavage of the M_r = 78,000 species to the two subunits which causes several determinants that are exposed in the precursor to be masked in the dimeric enzyme. Further studies are required to assess the role of the changes associated with the processing of the single-chain precursor in acquirement of the catalytic potential of the enzyme.

The initial events in the biosynthesis and processing of γ-glutamyl transpeptidase were investigated by translating the rat kidney mRNA in a heterologous cell-free system using rabbit reticulocyte lysate.[32] Such translation results in the synthesis of a M_r = 63,000 nonglycosylated polypeptide which has been shown immunologically to contain the domains of both subunits. The M_r = 63,000 species is processed to a M_r = 78,000 core-glycosylated (endo-β-*N*-acetylglucosaminidase H-sensitive) polypeptide when the translation of mRNA is carried out in the presence of dog pancreas microsomes. The M_r = 78,000 polypeptide is integrated into the cisternae of the microsomal membranes with an orientation that is analogous to that found on the brush border membranes. Glycosylation and membrane integration of the M_r = 78,000 species are cotranslational events, and the translocation and membrane integration of the primary translation product is not associated with the cleavage of an N-terminal leader sequence. Upon longer incubation, the M_r = 78,000 species sequestered within the microsomal vesicles is cleaved to species corresponding to the two subunits of the kidney enzyme.

These studies indicate that an uncleaved N-terminal hydrophobic sequence may function first as a cotranslational translocation signal sequence and later serve as the anchoring domain for γ-glutamyl transpeptidase either by reinsertion into the membrane, as proposed for the N-terminally anchored proteins by Kenny and Booth,[44] or by remaining in the membrane throughout the translocation process in accordance with the "helical hairpin" model of Engleman and Steitz.[45] In this regard, it is

[44] A. J. Kenny and A. G. Booth, *Essays Biochem.* **14,** 1 (1978).

[45] D. M. Engelman and T. A. Steitz, *Cell* **23,** 411 (1981).

interesting to note that the primary structure of the N-terminal anchor on the large subunit of γ-glutamyl transpeptidase (see above) resembles the amino acid sequences of the signal for cotranslational translocation of various secretory and membrane proteins across the rough endoplasmic reticulum of eukaryotic cells or the plasma membrane of prokaryotic cells.[46] This strategy may be a common one for other N-terminally anchored brush border proteins.

Following the first report of a common, single-chain precursor of the two subunits of γ-glutamyl transpeptidase,[28] at least two other laboratories have confirmed these findings,[47,48] although there is still some question regarding the intracellular site at which cleavage of the precursor occurs. The results taken together indicate that the cleavage may begin in the endoplasmic reticulum, before completion of the oligosaccharide chain processing in the Golgi, and is complete either before or soon after the enzyme reaches the brush border membranes. A number of questions regarding the intracellular route taken by the enzyme enroute to the brush border, and the factors involved in the sorting-out process remain to be addressed.

Catalytic Properties

Detailed discussion of the specificity, kinetics, and the present status of the mechanism of action of γ-glutamyl transpeptidase appears in the following chapter.[11] The pertinent findings are as follows. All of the mammalian γ-glutamyl transpeptidases studied thus far exhibit a broad optical and steric specificity towards γ-glutamyl compounds. Thus, glutathione, a large number of its S-derivatives, and other γ-substituted glutamyl derivatives including γ-glutamyl-*p*-nitroanilide and glutamine serve effectively as donor substrates.[4,8] The relative rates of transpeptidation for the rat kidney enzyme with L-methionine as the acceptor substrate are glutathione, 100; glutathione disulfide, 22; *S*-methylglutathione, 59; *S*-acetamidoglutathione, 172; *S*-acetophenoneglutathione, 185; *S*-acetylglutathione, 146; *S*-benzoylglutathione, 124; L-γ-glutamyl-L-glutamine, 68; and L-γ-glutamyl-*p*-nitroanilide, 218.[4] The relative V_{max} values for the hydrolysis of glutathione, glutathione disulfide, *S*-methylglutathione, and L-glutamine are, respectively, 100, 59, 78, and 14.[11,49]

[46] G. Blobel, P. Walter, C. N. Chang, B. N. Goldman, A. H. Erickson, and V. R. Lingappa, *Symp. Soc. Exp. Biol.* **33,** 9 (1979).

[47] M. A. Capraro and R. P. Hughey, *FEBS Lett.* **157,** 139 (1983).

[48] T. Kuno, Y. Matsuda, and N. Katunuma, *Biochem. Biophys. Res. Commun.* **114,** 889 (1983).

[49] T. M. McIntyre and N. P. Curthoys, *J. Biol. Chem.* **254,** 6499 (1979).

Specificity studies provide evidence for three separate subsites in the active center of transpeptidase.[8,21] The γ-glutamyl donor site (subsite 1) exhibits broad optical and steric specificity. Thus, L, D, and L-α-methyl γ-glutamyl compounds serve as donor substrates.[50] The acceptor site, consisting of subsites 2 and 3 for cysteinyl and glycyl moieties of glutathione, respectively, on the other hand, has restricted stereospecificity in that only L-amino acids and dipeptide acceptors in which both amino acids are of L-configuration serve as acceptors.[51,52] A large variety of dipeptide acceptors can bind to the Cys-Gly site and the amino acid acceptors interact with the cysteinyl subsite (subsite 2).[52] The preferred amino acid acceptors are the neutral amino acids such as the L-isomers of cystine, glutamine, methionine, alanine, and serine. Branched chain amino acids are relatively poor acceptors, whereas D-amino acids, L-proline, and α-substituted amino acids are inactive. The glycine subsite (subsite 3) shows preference for glycine; other amino acid substitutions reduce the acceptor activity considerably. Thus, the best dipeptide acceptors include L-Met-Gly, L-Gln-Gly, L-Ala-Gly, L-cystinyl-bis-Gly, L-Ser-Gly, Gly-Gly, and L-CysH-Gly (in decreasing order of efficiency).[52]

As indicated in reactions (1) to (3) (see above), the enzyme degrades γ-glutamyl compounds by transpeptidation as well as hydrolysis. With L-γ-glutamyl donors all three reactions occur in absence of an added acceptor. Autotranspeptidation can be minimized or eliminated by employing either L-γ-glutamyl substrates at concentrations well below their K_m values as acceptor substrates or by using a D-γ-glutamyl compound (e.g., D-γ-glutamyl-*p*-nitroanilide[17,18]) since the D-isomers do not serve as acceptors. The D-isomers have also proven useful in studying the hydrolytic reaction of the enzyme.

Kinetic studies are consistent with a mechanism involving a γ-glutamyl-enzyme intermediate.[51,52] An activated γ-glutamyl intermediate is in accord with the observation that the enzyme catalyzes γ-glutamyl transfer to hydroxylamine.[53] The possibility that a covalent γ-glutamyl-enzyme intermediate is formed in the course of the transfer reactions receives strong support from the findings that the γ-glutamyl (or glutamine) analogs, 6-diazo-5-oxo-L-norleucine (DON) and *O*-diazoacetyl-L-serine (L-azaserine) inactivate the enzyme by attaching covalently and stoichiometrically to the γ-glutamyl site.[40] Studies with DON-labeled enzyme indicate that the covalent attachment of the analog involves an O-ether

[50] O. W. Griffith and A. Meister, *Proc. Natl. Acad. Sci. U.S.A.* **74,** 3330 (1977).
[51] S. S. Tate and A. Meister, *J. Biol. Chem.* **249,** 7593 (1974).
[52] G. A. Thompson and A. Meister, *J. Biol. Chem.* **252,** 6792 (1977).
[53] S. S. Tate and A. Meister, *Proc. Natl. Acad. Sci. U.S.A.* **71,** 3329 (1974).

linkage to a hydroxyl group located in the γ-glutamyl binding subsite.[41] These results have led to the inference that an enzyme hydroxyl group (presumably a serine or a threonine residue) also participates in the formation of the covalent γ-glutamyl–enzyme intermediate. The site of attachment of DON and L-azaserine (and by inference the site involved in the γ-glutamyl–enzyme formation) is located on the light (L) subunit of the enzyme.[40]

Incubation of glutathione with γ-glutamyl transpeptidase results in its rapid oxidation to glutathione disulfide.[54,55] This apparent "GSH oxidase" activity has been ascribed to the transpeptidase-catalyzed production of CysH-Gly. This dipeptide oxidizes rapidly and nonenzymatically to form cystinyl-bis-glycine. The oxidation of glutathione takes place by nonenzymatic transhydrogenation between glutathione and cystinyl-bis-glycine and between glutathione and mixed disulfide of CysH-Gly and glutathione.[56] Only a very low concentration of CysH-Gly is needed to mediate glutathione oxidation. Although transpeptidase may initiate a nonenzymatic chain of reactions leading to the oxidation of glutathione, several tissues also contain proteins capable of catalyzing this reaction. Such sulfhydryl oxidases have been isolated from milk,[57] kidney,[58] and seminal vesicle secretions.[59]

Inhibition and Regulation

Both reversible and irreversible inhibitors as well as a number of modulators of γ-glutamyl transpeptidase have been described; these have proven useful in studies on the function and mechanism of action of this enzyme.

Reversible Inhibitors. Revel and Ball discovered that L-serine in the presence of borate strongly inhibits transpeptidase.[60] To date, transpeptidase from every source has been shown to be inhibited by this combination. Indeed, inhibition by L-serine plus borate is a valuable test for the enzyme since neither compound alone inhibits the enzyme. D-Serine is also an inhibitor in the presence of borate. Serine appears to bind to the γ-glutamyl binding subsite of the enzyme occupying the region that accepts

[54] S. S. Tate, E. M. Grau, and A. Meister, *Proc. Natl. Acad. Sci. U.S.A.* **76,** 2715 (1979).

[55] S. S. Tate and J. Orlando, *J. Biol. Chem.* **254,** 5573 (1979).

[56] O. W. Griffith and S. S. Tate, *J. Biol. Chem.* **255,** 5011 (1980).

[57] M. B. Sliwkowski, M. X. Sliwkowski, H. E. Swaisgood, and H. R. Horton, *Arch. Biochem. Biophys.* **211,** 731 (1981).

[58] C. H. Schmelzer, H. E. Swaisgood, and H. R. Horton, *Biochem. Biophys. Res. Commun.* **107,** 196 (1982).

[59] M. C. Ostrowski and W. S. Kistler, *Biochemistry* **19,** 2639 (1980).

[60] J. P. Revel and E. G. Ball, *J. Biol. Chem.* **234,** 577 (1959).

the α-amino and α-carboxyl groups of glutathione; the borate anion presumably forms a bridge-complex between the hydroxyl group of serine and an active site hydroxyl group.[4] Such a complex resembles the apparent transition-state analog. The K_i values for L- and D-serine measured in the presence of 10 mM borate are, respectively, 20 and 170 μM.[41]

γ-Glutamylhydrazones of a number of α-keto acids are competitive inhibitors with respect to the γ-glutamyl donor; the K_i values for the γ-glutamylhydrazones of pyruvate and α-ketoglutarate are about 400 μM.[51] Particularly potent inhibitors are the γ-glutamylphenylhydrazides.[9,61] The most potent of these are the L- and D-isomers of γ-glutamyl-(*O*-carboxy) phenylhydrazide (L-OC and D-OC; K_i values, 8 and 23 μM, respectively). Interestingly, L-γ-glutamyl-(*p*-carboxy)phenylhydrazide is much less effective (K_i = 800 μM). When administered to mice, both L-OC and D-OC induce marked glutathionuria.[61] These derivatives are, however, rather toxic; thus, doses greater than about 0.5 and 1.25 mmol/kg for the L- and D-isomers, respectively, are not well-tolerated by mice. The toxicity is probably due to slow release of *O*-carboxyphenylhydrazine, a potent convulsant, which is released by the slow hydrolysis of L-OC and D-OC, presumably catalyzed by transpeptidase.[9] Nevertheless, the *in vivo* use of these compounds has provided strong evidence that translocation of intracellular glutathione to membrane-bound transpeptidase is a discrete and necessary step in glutathione metabolism.[61]

The relative rates of the hydrolytic and transpeptidation reactions of the enzyme are affected by compounds such as maleate and hippurate.[18,19,53,62,63] These and related compounds reversibly stimulate the hydrolysis of glutathione and other γ-glutamyl compounds (a large stimulatory effect is seen on the hydrolysis of L-glutamine), but inhibit the formation of transpeptidation products. Data indicate that these modulators bind to the Cys-Gly binding site of the enzyme.[19] Evidence has been obtained indicating that hippurate, commonly considered only as a detoxification product of benzoate, is formed in liver from benzoate originating from the degradative metabolism of phenylalanine.[19] Normal concentrations of hippurate in serum are such that the activity of renal proximal tubule transpeptidase could be modulated by hippurate in the glomerular filtrate.

Recent evidence indicates that free bile acids and their glycine and taurine conjugates affect the enzyme in a manner similar to hippurate.[20,21] The glycine derivatives of bile acids, however, exhibit a much greater

[61] O. W. Griffith and A. Meister, *Proc. Natl. Acad. Sci. U.S.A.* **76,** 268 (1979).
[62] S. S. Tate and A. Meister, *J. Biol. Chem.* **250,** 4619 (1975).
[63] N. P. Curthoys and T. Kuhlenschmidt, *J. Biol. Chem.* **250,** 2099 (1975).

affinity for transpeptidase than hippurate (thus, the apparent dissociation constants, K_d, for glycocholate and hippurate are, respectively, 0.13 and 1.7 mM).[21] It is interesting to note that, although both cholate and glycocholate stimulate hydrolysis of *S*-methylglutathione (both inhibit transpeptidation between *S*-methylglutathione and L-methionine), only glycocholate accelerates glutamine hydrolysis. Similar effects are observed with benzoate and hippurate, emphasizing the significance of the binding of the glycine moiety of the glycine conjugates to the Gly subsite of the enzyme's active site.[21] Binding studies with [^{14}C]glycocholate indicate that close to 1 mol of the compounds binds per mol of the enzyme. The binding is enhanced by compounds that bind to the γ-glutamyl binding site of the enzyme. These and other studies point to cooperative influences in the interactions at the γ-glutamyl and cysteinylglycine binding domains of transpeptidase.[21] The effects of free and conjugated bile acids upon transpeptidase may be of significance in the regulation of its activity along the route of the enterohepatic circulation of these compounds, since histochemical and cytochemical techniques have localized transpeptidase to the outer surface of epithelial cells lining the route of this circulation.

Irreversible Inhibitors. Transpeptidase is effectively inactivated by incubation with the glutamine antagonists, L-azaserine,[40] DON,[40] or L-(αS,5S)-α-amino-3-chloro-4,5-dihydro-5-isoxazoleacetic acid (AT-125).[54,64–66] These compounds react covalently at the γ-glutamyl portion of the substrate binding site. The residue labeled by these compounds is localized on the light (L) subunits of the enzyme.[40,66] (It should be noted, however, that DON and L-azaserine appear to bind via a hydroxyamino acid residue in the active center, whereas the binding of AT-125 probably involves a different residue.[66]) With purified rat kidney transpeptidase, the relative rates of inactivation by L-azaserine, DON, and AT-125 are in the ratio 2 : 5 : 100.[9] The rates of inactivation by these compounds are decreased in the presence of compounds that bind to the γ-glutamyl site (e.g., glutathione, glutamine, serine plus borate); the inactivation rates are markedly accelerated by acceptor site-directed modulators (e.g., maleate, hippurate, glycocholate, etc.).

Administration of DON and AT-125 to mice leads to inactivation of tissue γ-glutamyl transpeptidase.[61,65] This inactivation is associated with the onset of a marked glutathionuria. AT-125 is particularly effective in this respect; the urine concentrations of glutathione found after the ad-

[64] L. Allen, R. Meck, and A. Yunis, *Res. Commun. Chem. Pathol. Pharmacol.* **27,** 175 (1980).

[65] O. W. Griffith and A. Meister, *Proc. Natl. Acad. Sci. U.S.A.* **77,** 3384 (1980).

[66] S. J. Gardell and S. S. Tate, *FEBS Lett.* **122,** 175 (1980).

ministration of AT-125 range between 15 to 30 m*M* (normal urine glutathione, 2–5 μ*M*). Although, *in vitro,* the inhibition of the enzyme by DON and AT-125 is effectively irreversible, *in vivo,* the inhibition appears to be slowly reversible as judged by the extent of glutathionuria.[65] It has not yet been established whether such reversal *in vivo* reflects *de novo* synthesis of enzyme or breakdown of the covalent enzyme–inhibitor complex, or both.

Phenylmethanesulfonylfluoride (PMSF), a reagent that inactivates serine-class peptidases and proteinases, inactivates rat kidney, γ-glutamyl transpeptidase.[67] The rate of inactivation is markedly accelerated by maleate. It seems likely that this reagent modifies the same active center hydroxyl group that is affinity labeled by compounds such as DON and L-azaserine. Thus, the binding of PMSF and DON is mutually exclusive (S. J. Gardell and S. S Tate, unpublished).

The irreversible inhibitors described above are all directed towards the γ-glutamyl site of the enzyme. The acceptor binding site (CysH-Gly site) can be modified by the diazonium derivatives of *p*-aminohippurate and *p*-aminobenzoate.[68] Thus, preincubation of transpeptidase with the diazonium derivatives results in rapid loss of transpeptidase activity. The modified enzyme can still hydrolyze γ-glutamyl substrates but the hydrolytic reactions are no longer modulated by the acceptor site-directed modulators such as hippurate and maleate. Treatment with the diazonium derivatives results in specific cross-linking of the two subunits of the enzyme and the extent of cross-linking parallels loss of transpeptidation activity; both inactivation and cross-linking of the subunits can be prevented by hippurate and maleate. These studies have provided evidence that the acceptor binding site is located in an intersubunit contact region and, thus, the active center of the enzyme may include domains contributed by both the subunits of transpeptidase.

[67] M. Inoue, S. Horiuchi, and Y. Morino, *Biochem. Biophys. Res. Commun.* **82,** 1183 (1978).
[68] S. J. Gardell and S. S. Tate, *Arch. Biochem. Biophys.* **216,** 719 (1982).

[51] γ-Glutamyl Transpeptidase: Kinetics and Mechanism

By R. DONALD ALLISON

Three catalytic activities are associated with γ-glutamyl transpeptidase. All of these activities involve the transfer of the γ-glutamyl moiety of glutathione (or of other γ-glutamyl compounds such as glutamine, glu-

Copyright © 1985 by Academic Press, Inc.
All rights of reproduction in any form reserved.

tathione disulfide, glutathione S-conjugates, and model substrates such as γ-glutamyl-*p*-nitroanilide) to an acceptor substrate. Transpeptidation occurs when the acceptor is an amino acid, dipeptide, or tripeptide, leading to formation of the corresponding γ-glutamyl amino acid or γ-glutamyl di- or tripeptide. There have been reports, particularly with the plant enzyme,[1] that amines may also function as acceptor substrates. This aspect of the enzyme specifically awaits further study. If the γ-glutamyl acceptor is water, then the overall result is hydrolysis of the γ-glutamyl bond. If the donor substrate also acts as an acceptor, autotranspeptidation occurs. Though the autotranspeptidation reaction is negligible at low donor concentrations (i.e., physiological conditions),[1a] it can be significant at millimolar levels of the γ-glutamyl substrate, conditions which are commonly used in clinical assays and in enzyme purifications.[2-4] This activity will cause alterations in the initial rate behavior of γ-glutamyl transpeptidase. Nordlie[5] has recently discussed certain approaches to enzymes with multiple reaction pathways, with particular emphasis on glucose 6-phosphatase. This chapter is concerned with the parameters, kinetics, and mechanistic studies peculiar to γ-glutamyl transpeptidase. The autotransfer pathway of this enzyme appears to be unique among those enzymes that have branch reaction pathways.

The activity of this enzyme was first observed by Dakin and Dudley[6] in 1913 and prior to the discovery that the enzyme catalyzes transpeptidation,[7,8] it was believed that the enzyme only functions as a glutathionase.[9,10] Recent efforts have been made to examine the relative contributions of the different pathways.[11-14] The formation of γ-glutamyl

[1] T. Kasai and P. O. Larsen, *Prog. Chem. Org. Nat. Prod.* **39,** 173 (1980).

[1a] G. A. Thompson and A. Meister, *Biochem. Biophys. Res. Commun.* **71,** 32 (1976).

[2] L. M. Shaw, J. W. London, and L. E. Petersen, *Clin. Chem.* (*Winston-Salem, N.C.*) **24,** 905 (1978).

[3] H. E. Solberg, L. Theodorsen, and J. H. Strømme, *Clin. Chem.* (*Winston-Salem, N.C.*) **27,** 303 (1981).

[4] L. M. Shaw and J. W. London, *in* "Gammaglutamyltransferases" (G. Siest and C. Heusghem, eds.), 3rd ser., p. 23. Masson, Paris, 1982.

[5] R. C. Nordlie, this series, Vol. 87, p. 319.

[6] H. D. Dakin and H. W. Dudley, *J. Biol. Chem.* **15,** 463 (1913).

[7] C. S. Hanes, F. J. R. Hird, and F. A. Isherwood, *Nature* (*London*) **166,** 288 (1950).

[8] C. S. Hanes, F. J. R. Hird, and F. A. Isherwood, *Biochem. J.* **51,** 25 (1952).

[9] E. F. Schroeder and G. E. Woodward, *J. Biol. Chem.* **120,** 209 (1937).

[10] F. Binkley and K. Nakamura, *J. Biol. Chem.* **173,** 411 (1948).

[11] J. S. Elce and B. Broxmeyer, *Biochem. J.* **153,** 223 (1976).

[12] T. M. McIntyre and N. P. Curthoys, *J. Biol. Chem.* **254,** 6499 (1979).

[13] A. M. Karkowsky, M. V. W. Bergamini, and M. Orlowski, *J. Biol. Chem.* **251,** 4736 (1976).

[14] R. D. Allison and A. Meister, *J. Biol. Chem.* **256,** 2988 (1981).

amino acids is crucial to a proposed amino acid transport system.[15–18] According to this hypothesis, the γ-glutamyl amino acids formed by γ-glutamyl transpeptidase are translocated into the cell where free amino acids are released by the next enzyme of the γ-glutamyl cycle, γ-glutamyl cyclotransferase. In addition to γ-glutamyl transpeptidase's role in the hydrolysis or transfer of the γ-glutamyl moiety of glutathione, glutamine, and other γ-glutamyl compounds, the enzyme has been proposed to participate in a number of other metabolic processes. It also functions in the processing of mercapturic acids[19,20] and in pathways of metabolism involving prostaglandins,[21,22] estrogens,[23–27] and leukotrienes.[28–30] This chapter concerns kinetics and mechanism of this multifunctional enzyme, whose other properties and assay protocols have been reviewed.[18,31–35]

Kinetic Mechanism

The kinetic mechanism of γ-glutamyl transpeptidase has been described by a number of investigators.[1–4,11,13,36–40] The first substrate to bind

[15] M. Orlowski and A. Meister, *Proc. Natl. Acad. Sci. U.S.A.* **67,** 1248 (1970).
[16] A. Meister, *Science* **180,** 33 (1973).
[17] A. Meister, *Curr. Top. Cell. Regul.* **18,** 21 (1981).
[18] A. Meister and M. E. Anderson, *Annu. Rev. Biochem.* **52,** 711 (1983).
[19] E. Boyland, *Proc. Int. Pharmacol. Meet., 1st, 1961* Vol. 6, p. 65 (1962).
[20] E. Boyland and L. F. Chesseaud, *Adv. Enzymol. Relat. Areas Mol. Biol.* **32,** 173 (1969).
[21] L. M. Cagen, H. M. Fales, and J. J. Pisano, *J. Biol. Chem.* **251,** 6550 (1976).
[22] L. M. Cagen and J. J. Pisano, *Biochim. Biophys. Acta* **573,** 547 (1979).
[23] E. Kuss, *Hoppe-Seyler's Z. Physiol. Chem.* **348,** 1707 (1967).
[24] E. Kuss, *Hoppe-Seyler's Z. Physiol. Chem.* **349,** 1234 (1968).
[25] E. Kuss, *Hoppe-Seyler's Z. Physiol. Chem.* **350,** 95 (1969).
[26] P. H. Jellinck, J. Lewis, and F. Boston, *Steroids* **10,** 329 (1967).
[27] J. S. Elce, *Biochem. J.* **116,** 913 (1970).
[28] L. Örning, S. Hammarström, and B. Samuelsson, *Proc. Natl. Acad. Sci. U.S.A.* **77,** 2014 (1980).
[29] S. Hammarström, B. Samuelsson, D. A. Clark, G. Groto, A. Marfat, C. Miowskowski, and E. J. Corey, *Biochem. Biophys. Res. Commun.* **92,** 946 (1980).
[30] M. E. Anderson, R. D. Allison, and A. Meister, *Proc. Natl. Acad. Sci. U.S.A.* **79,** 1088 (1982).
[31] A. Meister and S. S. Tate, *Annu. Rev. Biochem.* **45,** 559 (1976).
[32] A. Meister, S. S. Tate, and G. A. Thompson, *Ciba Found. Symp.* new ser., **50,** 123 (1977).
[33] A. Meister, S. S. Tate, and O. W. Griffith, this series, Vol. 77, p. 237.
[34] S. S. Tate and A. Meister, *Mol. Cell. Biochem.* **39,** 357 (1981).
[35] S. S. Tate, this volume [58].
[36] S. S. Tate and A. Meister, *J. Biol. Chem.* **249,** 7593 (1974).
[37] G. A. Thompson and A. Meister, *J. Biol. Chem.* **252,** 6792 (1977).
[38] A. M. Karkowsky and M. Orlowski, *J. Biol. Chem.* **253,** 1574 (1978).
[39] J. W. London, L. M. Shaw, D. Fetterolf, and D. Garfinkel, *Biochem. J.* **157,** 609 (1976).
[40] J. H. Strømme and L. Theodorsen, *Clin. Chem. (Winston-Salem, N.C.)* **22,** 417 (1976).

is the γ-glutamyl donor. A γ-glutamyl-enzyme intermediate forms and the first product is released (cyteinylglycine, if the donor was glutathione). The acceptor substrate then binds to the γ-glutamyl-enzyme, transferring the γ-glutamyl moiety to the α-amino group of the amino acid or peptide, to another molecule of donor substrate (thus, autotranspeptidation), or to water (thus, hydrolysis). The autotranspeptidation pathway causes difficulties in the kinetic analyses because squared terms of the donor concentration appear in the numerator of the initial rate kinetic expression. To avoid the difficulties of autotranspeptidation two separate routes of analysis may be followed. The first involves the use of a γ-glutamyl donor substrate that cannot act as an acceptor substrate (or, the true Michaelis constant of this γ-glutamyl compound acting as an acceptor is so large that the autotranspeptidation pathway is negligible). With the enzyme sources studied so far, D-γ-glutamyl derivatives act as donor substrates but not as acceptors.[1a] Likewise, L-γ-(α-methyl)glutamyl derivatives are not autotranspeptidated with the sheep kidney enzyme.[13,41] One may also avoid autotranspeptidation by using L-γ-glutamyl derivatives at concentrations well below the true K_m of the donor-acting-as-an-acceptor. In the cases examined so far, this K_m appears to be millimolar, whereas the true K_m of the donor-acting-as-the-donor is micromolar. With either method, using donors that cannot serve as acceptors or using donors at low concentrations, the autotranspeptidase activity is negligible and the kinetic mechanism for γ-glutamyl transpeptidase can be described as shown in Scheme 1.

$$E + A \underset{k_2}{\overset{k_1}{\rightleftharpoons}} [EA \rightarrow FP] \underset{k_4}{\overset{k_3}{\rightleftharpoons}} F + P$$

$$F + B \underset{k_6}{\overset{k_5}{\rightleftharpoons}} [FB \rightarrow EQ] \underset{k_8}{\overset{k_7}{\rightleftharpoons}} E + Q$$

$$F \xrightarrow{k_9} E + R$$

$$E + B \underset{k_{11}}{\overset{k_{10}}{\rightleftharpoons}} EB$$

SCHEME 1

In Scheme 1, A, B, P, Q, and R correspond to glutathione [or other γ-glutamyl donor (such as γ-glutamyl-*p*-nitroanilide)], acceptor (such as an amino acid), cysteinylglycine (or other leaving group), γ-glutamyl-acceptor, and glutamate, respectively. Free enzyme is designated by E and the γ-glutamyl enzyme is referred to as F. Experimentally, it has been observed that most acceptor substrates also act as inhibitors by competing with the donor substrate. Thus, an EB complex can also form.

[41] O. W. Griffith and A. Meister, *Proc. Natl. Acad. Sci. U.S.A.* **74,** 3330 (1977).

TABLE I
DEFINITION OF KINETIC PARAMETERS

V_h	= maximum velocity for hydrolysis	$= k_3k_9(E_{tot})/(k_3 + k_9)$
V_b	= maximum velocity for transfer	$= k_3k_7(E_{tot})/(k_3 + k_7)$
K_h	= Michaelis constant for donor	$= k_9(k_2 + k_3)/k_1(k_3 + k_9)$
K_i	= inhibition constant for acceptor	$= k_{11}/k_{10}$
K_b	= Michaelis constant for acceptor	$= (k_6 + k_7)(k_3 + k_9)/k_5(k_3 + k_7)$
K_{iab}	= partitioning constant	$= K_bV_h/V_b = k_9(k_6 + k_7)/k_5k_7$

The initial velocity equation of Scheme 1 for the formation of the common product, P, is:

$$v = \frac{V_h[A](1 + [B]/K_{iab})}{K_h(1 + [B]/K_i)(1 + [B]/K_{iab}) + [A](1 + [B]/K_b)} \tag{1}$$

The kinetic parameters are defined in Table I. Scheme 1 depicts a modified ping pong kinetic mechanism[5,42,43] in which a hydrolytic shunt releases the γ-glutamyl moiety from the modified enzyme in essentially an irreversible step. Scheme 1 is actually a special case of a ping pong mechanism with two alternative second-substrates[44–46] (C would correspond to water in Scheme 1):

$$E + A \underset{k_2}{\overset{k_1}{\rightleftharpoons}} [EA \rightleftharpoons FP] \underset{k_4}{\overset{k_3}{\rightleftharpoons}} F + P$$
$$F + B \underset{k_6}{\overset{k_5}{\rightleftharpoons}} [FB \rightleftharpoons EQ] \underset{k_8}{\overset{k_7}{\rightleftharpoons}} E + Q$$
$$F + C \underset{k_{10}}{\overset{k_9}{\rightleftharpoons}} [FC \rightleftharpoons ER] \underset{k_{12}}{\overset{k_{11}}{\rightleftharpoons}} E + R$$
$$E + B \underset{k_{14}}{\overset{k_{13}}{\rightleftharpoons}} EB$$

SCHEME 2

As can be seen from Eq. (1), straightforward double-reciprocal plots are linear when γ-glutamyl donor (A) is the varied substrate, but are curvilinear when the acceptor (B) is the varied substrate (the degree of nonlinearity being dependent upon the relative sizes of the $[B]/K_b$ and

[42] H. F. Fromm, "Initial Rate Enzyme Kinetics." Springer-Verlag, Berlin and New York, 1975.

[43] H. F. Fromm, this series, Vol. 63, p. 42.

[44] H. F. Fromm, *Biochim. Biophys. Acta* **81,** 413 (1964).

[45] C. Y. Huang, this series, Vol. 63, p. 486.

[46] The kinetic expression for the rate of formation of the common product P is $v = [V_h[A]([C]/K_c + [B]/K_{iab})]/[K_a(1 + [B]/K_{ib})([C]/K_c + [B]/K_{iab}) + [A](1 + [B]/K_b + [C]/K_c)]$.

$[B]/K_{iab}$ terms). In the absence of the acceptor substrate B, both V_h and K_h can be obtained from standard kinetic plots. As stated earlier, care has to be exercised that the third reaction pathway, autotranspeptidation, is not present or is negligible since, at millimolar levels of the donor substrate, the kinetic parameters culled from the plots would reflect this additional activity.

In the presence of an acceptor, the double-reciprocal form of Eq. (1) becomes

$$\frac{1}{v} = \frac{K_h(1 + [B]/K_i)}{V_h[A]} + \frac{(1 + [B]/K_b)}{V_h(1 + [B]/K_{iab})} \tag{2}$$

Note that replots of the slopes from double-reciprocal plots as a function of the acceptor concentration provide values for K_i, the inhibition constant for the acceptor. This parameter has been determined with the rat kidney enzyme for a number of common acceptors (Table II). Both the hydrolysis and transfer activities are competitively inhibited, implying that the acceptor is binding at the donor site. However, this hypothesis has not been rigorously tested.

A replot of the intercepts of Eq. (2) is nonlinear. Nevertheless, the partition coefficient K_{iab} can be determined. Let int_0 refer to the intercept in the absence of acceptor ($int_0 = V_h^{-1}$) and int_b to the intercept observed at a specific concentration of acceptor ($int_b = (1 + [B]/K_b)/[V_h(1 + [B]/K_{iab})]$). Plotting $(int_0–int_b)^{-1}$ as a function of $[B]^{-1}$, the abscissa intercept of this new replot would provide a value for K_{iab}. This is an important parameter, equal to $V_h(K_b/V_b)$. Since V_h is already known, the V_b/K_b ratio is readily determined. This ratio, the maximum velocity of the transfer activity divided by the true Michaelis constant for the acceptor, is a measure of relative efficiency of the substrate to serve as an acceptor. It is more accurately determined than either V_b or K_b. Nevertheless, since differences are plotted (in the form $int_0–int_b$), the errors associated with this parameter may be greater than typically encountered. Tables III and IV provide values for V_b/K_b for a number of acceptor substrates for the rat kidney enzyme (at pH 8.0 and 37°).[37]

The Michaelis constants for the acceptor can be obtained from computer fitting of the replot data or by an algebraic treatment. Note that the ordinate intercept of the replot is $(V_hK_b)/(K_b - K_{iab}) = b'$. Hence, $K_b = (b'K_{iab})/(b' - V_h)$. The determination of this parameter requires the calculation of a second difference, in this case $b' - V_h$. Hence, the error in this value can be severe. Unless one has accurately determined reaction velocities, a true Michaelis constant for the acceptor may have an unacceptable standard error associated with it. The true Michaelis constants for a number of acceptors have been measured for the rat kidney enzyme: e.g.,

TABLE II
INHIBITION BY AMINO ACIDS AND DIPEPTIDES OF RAT KIDNEY γ-GLUTAMYL TRANSPEPTIDASE[a]

Amino acid	K_i (mM)	Dipeptide	K_i (mM)
L-Glutamine	3.0 ± 0.5[b]	L-Trp-Gly	9.7 ± 2.7
Glycine	3.2 ± 0.8	L-Phe-Gly	16.0 ± 4.8
L-Alanine	4.6 ± 0.9	L-Tyr-Gly	18.6 ± 5.6
D-Glutamine	6.6 ± 1.2	L-His-Gly	37.8 ± 17.1
D-Alanine	8.8 ± 2.3	L-Ala-Gly	57.5 ± 18.7
L-Cysteine[c]	9.4 ± 1.5	L-Gln-Gly	59.8 ± 28.4
L-Histidine	10.2 ± 2.1	L-CysH-Gly[c]	63.4 ± 31.4
L-Serine	10.7 ± 1.8	L-Met-Gly	63.5 ± 28.0
L-Glutamate	14.1 ± 2.3	L-Ser-Gly	110.0 ± 69.1
L-Valine	15.0 ± 2.5	L-Leu-Gly	>150
L-Threonine	15.4 ± 3.7	L-Lys-Gly	>150
L-Typtophan	24.3 ± 6.6	Gly-L-Ala	>150
L-Aspartate	26.8 ± 5.6	Gly-Gly	>150
L-Methionine	26.9 ± 3.3		
L-Arginine	48.5 ± 12.0		
L-Isoleucine	64.5 ± 35.8		
L-Lysine	71.9 ± 22.4		
L-Asparagine	>100		
L-Leucine	>100		
L-Phenylalanine	>100		

[a] From Thompson and Meister.[37]
[b] SE.
[c] Studied in the presence of equimolar concentrations of dithiothreitol.

L-cystine (0.030 mM[1a]), L-methionylglycine (0.24 mM[37]), L-glutaminylglycine (0.23 mM[37]), L-cystinylbisglycine (0.17 mM[37]), L-glutamine (0.76 mM[47]), L-cysteinylglycine (1.09 mM[37]), glycylglycine (2.96 mM[37]), and L-methionine (4.7 mM[11]). It has been suggested that, since the reported Michaelis constants are often larger than physiological concentrations typically encountered, the transpeptidation reaction is not a physiologically important pathway. Such reasoning is questionable since the Michaelis constants for the substrates of many enzymes are much higher than the physiological concentrations of their substrates.[14] Of equal importance is the maximum velocity of transfer for a substrate. Of particular note is glycylglycine, which has a relatively weak Michaelis constant, but whose V_b is exceptionally large, relative to other acceptors. Thus, the

[47] R. D. Allison, unpublished observation.

TABLE III
AMINO ACID SPECIFICITY OF RAT KIDNEY γ-GLUTAMYL TRANSPEPTIDASE[a]

Amino acid	V_b/K_b
L-Cystine	250 ± 30[b]
L-Glutamine	23 ± 5
L-Methionine	12 ± 2
L-Alanine	6.8 ± 1.8
L-Glutamate	4.6 ± 0.8
L-Cysteine[c]	4.0 ± 0.8
L-Serine	3.3 ± 0.6
Glycine	2.6 ± 1.4
L-Asparagine	2.5 ± 0.3
L-Arginine	2.5 ± 0.3
L-Lysine	2.3 ± 0.3
L-Histidine	1.6 ± 0.5
L-Phenylalanine	1.1 ± 0.1
L-Tryptophan	1.0 ± 0.4
L-Leucine	0.6 ± 0.2
L-Threonine	<0.6
L-Isoleucine	<0.6
L-Valine	<0.6
L-Aspartate	<0.6

[a] From Thompson and Meister.[37] The donor substrate was D-γ-glutamyl-*p*-nitroanilide. Kinetic experiments were done at pH 8.0 and 37°. L-Tyrosine was not studied.

[b] SE.

[c] Studied in the presence of equimolar concentrations of dithiothreitol.

glycylglycine efficiency as a substrate, reflected in its V_b/K_b ratio, is excellent.

The Michaelis constant for the acceptor may be more easily determined by measuring the rate of formation of glutamate, the hydrolytic product (R), in the presence of the acceptor. Here, a replot of the intercept of the double-reciprocal plot as a function of acceptor concentration provides $-K_b$ as the abscissa intercept. The difficulty with this procedure is in accurate determination of low initial-rate levels of glutamic acid. Frère[47a] has also pointed out that linear plots are obtained when the reciprocal of the rate of glutamate formation is plotted as a function of [B]. This would only be true if the acceptor substrate does not act as a substrate

[47a] J.-M. Frère, *Biochem. J.* **135,** 469 (1973).

TABLE IV
DIPEPTIDE SPECIFICITY OF RAT KIDNEY γ-GLUTAMYL TRANSPEPTIDASE[a]

Peptide	V_b/K_b min^{-1}	Peptide[b]	V_b/K_b min^{-1}
L-Met-Gly	310 ± 70[c]	Gly-Gly	140 ± 20
L-Gln-Gly	270 ± 70	Gly-L-Ala	120 ± 10
L-Ala-Gly	240 ± 30	Gly-L-Pro	17 ± 2
(L-Cys-Gly)$_2$	183 ± 100	Gly-Ser	16 ± 2
L-Ser-Gly	140 ± 20	Gly-Gly-NH_2	14 ± 2
Gly-Gly	140 ± 20	Gly-dehydro-Ala	10 ± 1
L-CysH-Gly[d]	110 ± 20	Gly-L-Lys	6.3 ± 3.0
L-Lys-Gly	79 ± 11	Gly-L-Leu	6.1 ± 2.5
L-His-Gly	44 ± 7	Gly-L-Trp	5.6 ± 1.6
L-Trp-Gly	43 ± 15	Gly-L-His	5.3 ± 1.4
L-Tyr-Gly	43 ± 8	Gly-L-Met	4.8 ± 1.7
L-Phe-Gly	36 ± 7	Gly-L-Asn	3.6 ± 1.2
L-Leu-Gly	26 ± 4	Gly-L-Asp	3.2 ± 1.2
		Gly-L-Tyr	<3
		Gly-Gly-Gly	<3
		Gly-γ-aminobutyrate	<3
		Gly-L-Glu	<3
		Gly-L-Val	<3
		Gly-L-Ile	<3
		Gly-L-Thr	<3

[a] From Thompson and Meister.[68]
[b] No products were detected with L-Val-Gly, L-Pro-Gly, Gly-D-Phe, and (Gly)$_4$.
[c] SE.
[d] Studied in the presence of equimolar concentrations of dithiothreitol.

inhibitor. As pointed out earlier, many acceptors do display significant inhibition. If one measures the rate of formation of the γ-glutamyl acceptor, [Eq. (3)],

$$\frac{1}{v} = \frac{K_h(1 + [B]/K_i)(1 + [B]/K_b)}{V_h[A][B]/K_{iab}} + \frac{K_{iab}(1 + [B]/K_b)}{V_h[B]} \qquad (3)$$

the intercept replot (int vs $[B]^{-1}$) will provide values for both K_{iab} and K_b.

Specificity and Partitioning

The enzymes studied so far display a broad specificity for the leaving group of the γ-glutamyl donor (both D- and L-donors are active). A list of the donor substrates would include: glutathione, glutathione disulfide, glutamine, γ-glutamyl-*p*-nitroanilide,[1a,48,49] ophthalmic acid,[36,50,51] leukotriene C_4,[30] *S*-acetophenone-glutathione,[36,51,52] poly-γ-glutamyl derivatives,[49] L-γ-glutamyl-7-amino-4-methyl-coumarin,[53] etc. In addition to the hydrolysis of these peptides, the enzyme also displays an esterase activity; both γ-ethyl and β-benzyl glutamate are hydrolyzed by the hog kidney enzyme.[49] This esterase activity needs further study. One may suspect differences in the rate-determining step between the peptidase and esterase activities.

A ranking of different donors, based on their V_h/K_h ratio, has not been extensively done as was accomplished with the acceptor substrates. McIntyre and Curthoys[12] have provided the true kinetic parameters for a number of donor substrates. As one may expect, glutathione was the most active substrate (based on its V_h/K_h ratio) followed by *S*-methylglutathione, *S*-benzylglutathione, glutathione disulfide, *S*-sulfobromophthaleinglutathione, γ-glutamylalanine, γ-glutamylglutamine, and glutamine. It should be noted that glutamine is 530-fold less active as a donor than either glutathione or γ-glutamyl-*p*-nitroanilide. This apparent inefficiency is partly circumvented by the fact that physiological concentrations of glutamine exceed those concentrations of other γ-glutamyl compounds. It has been suggested that the utilization of glutamine by γ-glutamyl transpeptidase may be a significant pathway for glutamine metabolism and ammonia production in specific cases.[54–56] Tate and Meister[57] reported that the maleate-stimulated phosphate-independent glutaminase activity is identical to γ-glutamyl transpeptidase.[57–59] It should also be noted that L-γ-glutamyl-*p*-nitroanilide, the model substrate most commonly used, is reported to have the same maximum velocity and true Michaelis constant (5.6 μM) as glutathione for the rat kidney enzyme.[1,12,60]

[48] M. Orlowski and A. Meister, *Biochim. Biophys. Acta* **73,** 679 (1963).
[49] M. Orlowski and A. Meister, *J. Biol. Chem.* **240,** 338 (1965).
[50] E. E. Cliffe and S. G. Walez, *Biochem. J.* **79,** 118 (1961).
[51] S. S. Tate, *in* "Isozymes" (C. L. Markert, ed.), Vol. 2, p. 743. Academic Press, New York, 1975.
[52] S. S. Tate, *FEBS Lett.* **54,** 319 (1975).
[53] G. D. Smith, J. L. Ding, and T. J. Peters, *Anal. Biochem.* **100,** 136 (1979).
[54] T. C. Welbourne, D. Francoeur, G. Thornley-Brown, and C. J. Welbourne, *Biochim. Biophys. Acta* **444,** 644 (1976).
[55] T. C. Welbourne and P. D. Dass, *Life Sci.* **30,** 793 (1982).
[56] P. D. Dass, *Life Sci.* **33,** 1757 (1983).
[57] S. S. Tate and A. Meister, *Proc. Natl. Acad. Sci. U.S.A.* **71,** 3329 (1974).
[58] N. P. Curthoys and T. Kuhlenschmidt, *J. Biol. Chem.* **250,** 2099 (1975).
[59] S. S. Tate and A. Meister, *J. Biol. Chem.* **250,** 4619 (1975).
[60] G. A. Thompson and A. Meister, *J. Biol. Chem.* **254,** 2956 (1979).

Few reports have dealt with the specificity of the glutamyl portion of the donor substrate. The K_m for D-γ-glutamyl-*p*-nitroanilide is 31 μM, whereas the V_{max} is identical to the L-isomer.[1a,60] (However, the maximum velocity for hydrolysis of D-glutamine is almost 6-fold lower than that for L-glutamine. This suggests a change in the rate determining step when glutamine acts as a substrate than when either glutathione or γ-glutamyl-*p*-nitroanilide acts as a substrate.) A number of substrates were found not to act as donors[41,49,61–63]: e.g., L-asparagine, L-homoglutamine, β-aminoglutaryl-L-α-aminobutyric acid, L-γ-(β-methyl)glutamyl-L-α-aminobutyric acid, leucyl-β-naphthylamide, *N*-(α-L-glutamyl)-β-naphthylamide, L-γ-(*N*-methyl)glutamyl-L-α-aminobutyric acid. The interaction of the γ-glutamyl donor with the enzyme is further probed by pH rate studies on the true kinetic parameters. The Michaelis constant for the donor is relatively stable between pH 6.5 and 8.5, rising in both acid and alkaline solutions. Likewise, when V_{max} is plotted, the V_{max} for donor hydrolysis is relatively constant over a wide pH range, dropping only below pH 5.5. However, the effect of pH on the true V_{max}/K_m ratio is a typical bell-shaped curve, with p*K* values of about 6.1 and 9.0.[47] These three plots suggest that the p*K* values observed are due, in part, to the substrate, a free amino group apparently being needed on the γ-glutamyl moiety. This is substantiated by the earlier observation that L-γ-(*N*-methyl)glutamyl-L-α-aminobutyrate does not serve as a donor.[41] Also of interest are the recent reports of donor inhibition at high donor concentrations.[4,11,64–67] This effect is due in part to the presence of autotranspeptidation, but still requires further investigation.

The specificity of the acceptor substrate is provided in Tables III and IV[68] for the rat kidney enzyme at pH 8.0. In many systems studied, the amino acid specificities have been based on percent activities. Such tables have severe drawbacks: e.g., substrate inhibition by the free amino acid has been ignored, the amino acid concentration is constant, irrespective of the true K_m, autotranspeptidation is present and in competition with

[61] R. J. Bridges, O. W. Griffith, and A. Meister, *J. Biol. Chem.* **255,** 10787 (1980).

[62] A. Szewczuk and T. Baranowski, *Biochem. Z.* **338,** 317 (1963).

[63] G. G. Glenner, J. E. Folk, and P. J. McMillan, *J. Histochem. Cytochem.* **10,** 481 (1982).

[64] S. B. Rosalki and D. Tarlow, *Clin. Chem.* (*Winston-Salem, N.C.*) **20,** 1121 (1974).

[65] C. PetitClerc, F. Schiele, D. Bagrel, A. Mahassen, and G. Siest, *Clin. Chem.* (*Winston-Salem, N.C.*) **26,** 1688 (1980).

[66] D. Bagrel, C. PetitClerc, F. Schiele, and G. Siest, *Biochim. Biophys. Acta* **658,** 220 (1981).

[67] F. Schiele, Y. Artur, D. Bagrel, C. PetitClerc, and G. Siest, *Clin. Chim. Acta* **112,** 187 (1981).

[68] G. A. Thompson and A. Meister, *Proc. Natl. Acad. Sci. U.S.A.* **72,** 1985 (1975).

transpeptidation, etc. L-Cystine is the best acceptor of all the common amino acids.[37,38] As mentioned earlier, the efficiency of an acceptor is measured by the V_b/K_b ratio. In assessing the physiological role of hydrolysis relative to transpeptidation, it would be of value to determine the corresponding ratio for hydrolysis. This would require the ability to determine the Michaelis constant for water (not the Michaelis constant for the donor, as has been reported), with an appropriate correction for the protein's solvent shell. Though no such experiment has been reported, it should be noted that the rat kidney enzyme is remarkably stable in high concentrations of organic solvents.[47] In addition to those acceptors listed in Table III, many other compounds have been reported to act as acceptor substrates: e.g., hydroxylamine, homoarginine, leukotriene D_4, leukotriene E_4, L-isoglutamine, glycinamide, L-methionine methyl ester, meso-lanthionine, L-cystathionine, DL-allocystathionine, meso-α,ε-diaminopimelic acid, DL-ethionine, and *S*-methyl-L-cysteine. Interestingly enough, taurine and γ-aminobutyric acid have both been reported to act as substrates with the mouse cerebellar enzyme,[69] an effect not observed with the kidney enzyme. In general, imino acids, such as L-proline, α-substituted amino acids, and D-amino acids do not act as substrates. However, D-methionine has been reported to act as a substrate with the sheep kidney enzyme when L-γ-(α-methyl)glutamyl-α-aminobutyrate is used as the donor substrate.[13] This apparent change in stereospecificity needs further investigation. It has also been reported that the reactivity ratios (ratios of activities when two acceptors are present) change with different donors.[38] This suggests that the acceptor specificity is partially dependent on the donor substrate. Another report suggests that there is an apparent change of acceptor specificity when using glutathione as a donor as to when L-γ-glutamyl-*p*-nitroanilide is used as the donor.[70] Other peptides than those listed in Table IV have been demonstrated to act as acceptors: e.g., L-alanyl-L-alanine, L-methionyl-L-alanine, L-methionyl-L-methionine, L-methionyl-L-serine, L-glutamyl-L-alanine, L-alanylglycylglycine, and glycylglycyl-L-alanine.

As mentioned earlier, free amino acids also bind to the free enzyme, thus exhibiting substrate inhibition (Table II). However, important differences in the degree of this inhibition have been reported. The K_i for inhibition by L-methionine is 26.9 m*M* (Table II[37]), yet a tighter affinity has been reported (0.8 m*M*[12]). Likewise, the K_i for L-cysteinylglycine is 63.4 m*M* in one report[37] and 0.024 m*M* in another,[12] a 2600-fold difference. These modes of inhibition by free amino acids and dipeptides are important for the complete understanding of the physiological role of the pro-

[69] V. Lisý, G. R. Dutton, and D. N. Currie, *Life Sci.* **27,** 2615 (1980).

[70] P. Zelazo and M. Orlowski, *Eur. J. Biochem.* **61,** 147 (1976).

tein. The differences may be due to the stereospecificity of the donor substrate used: the values reported in Tables II to IV were obtained with D-γ-glutamyl-*p*-nitroanilide, whereas the smaller K_i values were obtained with glutathione. However, these K_i values are true dissociation constants for the free enzyme and are obtained from kinetic replots. Thus, the values should be independent of the donor substrate. Care also needs to be exercised that modulating substances, such as maleate, hippurate, or acetate, are not present. These compounds can alter the partitioning between hydrolysis, transpeptidation, and autotranspeptidation.

Osuji[71–74] discusses a second amino acid binding site in which an additional amino acid can bind to enzyme-γ-glutamylamino acid binary complex. At this time, there is no evidence for such a site.

If an amino acid acceptor contains two stereochemically distinct α-amino groups from which two separate γ-glutamyl products are formed, the experimentally determined kinetic parameters would be composite values reflecting both activities. L-Cystathionine is such a substrate for γ-glutamyl transpeptidase.[68] At L-γ-glutamyl donor concentrations less than 30 μM (autotranspeptidation thus being negligible), L-cystathionine exhibits a relative reactivity (V_b/K_b) comparable to that of L-cystine,[47] the most active acceptor of the common amino acids.[37] The apparent Michaelis constant is 440 μM;[47] cystathionine does not display substrate inhibition. The formation of the two γ-glutamyl products, however, is not equivalent.[75] The amino group of the alanine portion of L-cystathionine is γ-glutamylated in preference to the homocysteine portion. This would imply that the enzyme's active site is capable of partially distinguishing the location of the sulfide linkage is cystathionine. The initial rate expression for Scheme 1, in the presence of two acceptors and no substrate inhibition would be

$$\frac{d[\mathrm{P}]}{dt} = \frac{V_h[\mathrm{A}][1 + ([\mathrm{B}]/K_{iab}) + ([\mathrm{C}]/K_{iac})]}{K_h[1 + ([\mathrm{B}]/K_{iab}) + ([\mathrm{C}]/K_{iac})] + [\mathrm{A}][1 + ([\mathrm{B}]/K_b) + ([\mathrm{C}]/K_c)]} \tag{4}$$

Equation (4) simplifies when cystathionine serves as the acceptor substrate, since [B] = [C]. It is readily apparent that replots of the initial rate data will determine composite parameters of both transfer activities. The abscissa–intercept in the $(\mathrm{int}_0 - \mathrm{int}_B)^{-1}$ vs $[\mathrm{B}]^{-1}$ replot provides the composite partition function: $K_{iab,comp} = K_{iab}K_{iac}/(K_{iab} + K_{iac})$. This parameter

[71] G. O. Osuji, *FEBS Lett.* **108,** 240 (1979).
[72] G. O. Osuji, *FEBS Lett.* **110,** 192 (1980).
[73] G. O. Osuji, *FEBS Lett.* **139,** 167 (1982).
[74] G. O. Osuji, *Indian J. Biochem. Biophys.* **19,** 102 (1982).
[75] O. W. Griffith, personal communication.

was experimentally determined to be 140 μM for L-cystathionine. Likewise, the composite Michaelis constant, $K_{b,comp}$, for cystathionine would be $K_bK_c/(K_b + K_c)$.

Karkowsky and Orlowski[38] have presented an equation for γ-glutamyl transpeptidase with n acceptors. Since many acceptors also display strong substrate inhibition patterns, the initial rate equation has been extended to include this modulating activity:

$$v = \frac{V_h[A]\left[1 + \sum_{j=1}^{n} \frac{[B_j]}{K_{iaj}}\right]}{K_h\left[1 + \sum_{j=1}^{n} \frac{[B_j]}{K_{iaj}}\right]\left[1 + \sum_{l=1}^{m} \frac{[B_l]}{K_{I_l}}\right] + [A]\left[1 + \sum_{j=1}^{n} \frac{[B_j]}{K_j}\right]} \tag{5}$$

If a particular acceptor does not function as a substrate inhibitor, such as is the case with cystine or cystathionine, then the inhibition term is absent for that substrate. Likewise, if an inhibitor binds to the free enzyme, inhibiting both hydrolysis and transpeptidation, but does not function as a substrate, the K_{I_l} term is present, but both the K_{iaj} and K_j terms are absent.

Equation (5) can be used to calculate the theoretical degree of partitioning under initial rate conditions between the two reaction pathways in the presence of a pool of acceptor substrates. All the information that is needed is V_h, V_b/K_b for each acceptor, and the concentration of each acceptor. Equation (5) can be separated into its hydrolytic and transpeptidation functions and the partition function in the numerator can be replaced by Michaelis constants and maximum velocities (similar to those definitions found in Table I):

$$v = (V_h[A]/D) + \left([A] \sum_{j=1}^{n} (V_j[B_j]/K_j)/D\right)$$

D is the denominator in Eq. (5). Thus, the total utilization of glutathione (at micromolar concentrations) in the presence of a pool of acceptors can be represented. Transpeptidation will be the major activity for the enzyme when $\sum_{j=1}^{n} (V_j[B_j]/K_j) > V_h$. Knowing the values of V_j/K_j for the common amino acids determined at pH 8 by Thompson and Meister[37] (Table III) and typical plasma concentrations of each amino acid,[14] calculation of the total initial rate utilization of glutathione for transfer may be made. Under the same conditions used to determine V_j/K_j for each amino acid in Tables III and IV, Thompson and Meister[37] obtained a value of 15 units/mg for V_h. Thus, the percentage of glutathione directed toward transfer can be quickly determined: percent utilization = $[\Sigma(V_j[B_j]/K_j)]/[V_h + \Sigma(V_j[B_j]/K_j)] \times 100$. A simple calculation quickly discovers that at least 64% of the γ-glutamyl donor is directed toward transfer to acceptors rather than hydrolysis. Allison and Meister[14] experimentally obtained a value of 65.6% at these same concentrations, which compares well with the theoretical value. If neither cystine nor glutamine, the better amino acid acceptors, was present, then 34% of the donor would still be directed toward transpeptidation. Plasma concentrations of amino acids also increase postprandially. Thus, the degree of glutathione utilization can be larger than 64% with a physiologically relevant increase in acceptor concentrations. It should be pointed out, however, that the V_b/K_b values used in the above calculation were determined at pH 8.0.[37] One may suspect that these values may be different at a more physiological pH value, e.g., pH 7.4. The degree of the pH dependence of the V_b/K_b ratio appears to vary with each acceptor. For example, the rate of formation of γ-glutamylserine appears to be much more pH sensitive than does the rate of formation of γ-glutamylasparagine.[14] Nevertheless, at pH 7.4, approximately 50% of [^{35}S]glutathione was observed to be directed toward the transpeptidation pathway in a pool of amino acids[14] (in which the concentration of many amino acids were lower than levels often reported: e.g., the concentration of L-glutamine was only 0.4 m*M*).

At high donor concentrations, autotranspeptidation becomes a major reaction pathway. In fact, a number of reports[2-4,39,40,49,64-67,76,77] suggest that, in the presence of a strong acceptor and at millimolar levels of the donor, the hydrolytic activity may be negligible. Autotranspeptidation has been known for some time.[8] Tetra-γ-glutamyl products were detected with the hog kidney enzyme.[49] The apparent Michaelis constant for the donor substrate binding at the acceptor site appears to be millimolar. Since hydrolysis, step k_9 in Scheme 1, is a minor reaction pathway under these conditions, the apparent Michaelis constant for the donor substrate binding at the donor site will reflect this change in the partitioning among the reaction pathways, and this apparent Michaelis constant for the donor may be larger than the value experimentally determined at lower concentration ranges where autotranspeptidation is negligible.[78] It should also be noted that the donor substrate is competing with other amino acids for the acceptor site and the apparent Michaelis constant for amino acids may reflect this competition. Care should also be exercised that modulating effectors, such as maleate, hippurate, acetate, or bile acids, are not present since these acids will affect the partitioning among the three reaction pathways.

[76] L. M. Shaw, J. W. London, D. Fetterolf, and D. Garfinkel, *Clin. Chem.* (*Winston-Salem, N.C.*) **23,** 79 (1977).

[77] L. M. Shaw, L. Petersen-Archer, J. W. London, and E. Marsh, *Clin. Chem.* (*Winston-Salem, N.C.*) **26,** 1523 (1980).

[78] R. D. Allison, *Clin. Chem.* (*Winston-Salem, N.C.*) (submitted for publication).

A glutathione oxidase activity has also been associated with γ-glutamyl transpeptidase.[79-81] The cysteinylglycine produced by the enzyme undergoes rapid and nonenzymatic oxidation with glutathione to form the mixed disulfide or with another dipeptide product to form cystinylbisglycine. Either of these disulfides can undergo nonenzymatic transhydrogenation with free glutathione, regenerating cysteinylglycine and forming either glutathione disulfide or the mixed disulfide of cysteinylglycine and glutathione.[81] This transpeptidase-mediated pathway for glutathione oxidation may have significance in glutathione metabolism.[80-83] However, other sulfhydryl oxidases have been isolated.[84-87] The relative contribution to the formation of glutathione disulfide by these pathways awaits further investigation.

The kinetics of γ-glutamyl transpeptidase may also be affected by the presence of metal ions, though there is considerable variance in these effects with the enzyme obtained from different sources. Monovalent cations are reported to activate the enzyme when either γ-glutamyl-*p*-nitroanilide or γ-glutamylnaphthylamide serves as the donor.[88,89] The effect was not seen when glutathione was acting as the γ-glutamyl donor[50]; however, Zelazo and Orlowski[70] report activation by monovalent cations of the sheep kidney cortex enzyme with either glutathione or γ-glutamyl-*p*-nitroanilide. Autotranspeptidation is also affected. A variety of effects have been reported with divalent metal ions and further study is needed. Zinc ion appears to be a potent inhibitor.[90-94] Mercuric ion has also been reported to alter the activity of the enzyme.[53,62]

[79] S. S. Tate, E. M. Grau, and A. Meister, *Proc. Natl. Acad. Sci. U.S.A.* **76,** 2715 (1979).
[80] S. S. Tate and J. Orlando, *J. Biol. Chem.* **254,** 5573 (1979).
[81] O. W. Griffith and S. S. Tate, *J. Biol. Chem.* **255,** 5011 (1980).
[82] D. P. Jones, P. Moldeus, and S. Orrenius, *J. Biol. Chem.* **254,** 2787 (1979).
[83] M. E. Anderson, R. J. Bridges, and A. Meister, *Biochem. Biophys. Res. Commun.* **96,** 848 (1980).
[84] V. G. Janolino and H. E. Swaisgood, *J. Biol. Chem.* **250,** 2532 (1975).
[85] M. C. Ostrowski and W. S. Kistler, *Biochemistry* **19,** 2639 (1980).
[86] L. Flohé, W. A. Günzler, and R. Ladenstein, *in* "Glutathione: Metabolism and Function" (I. M. Arias and W. B. Jakoby, eds.), p. 115. Raven Press, New York, 1976.
[87] M. B. Sliwkowski, M. X. Sliwkowski, H. E. Swaisgood, and H. R. Horton, *Arch. Biochem. Biophys.* **211,** 731 (1981).
[88] M. Orlowski, P. O. Okonkwo, and J. P. Green, *FEBS Lett.* **31,** 237 (1973).
[89] J. S. Elce, J. Bryson, and L. G. McGirr, *Can. J. Biochem.* **52,** 33 (1974).
[90] M. Orlowski and A. Szewczuk, *Acta Biochem. Pol.* **8,** 189 (1961).
[91] J. L. Ding, G. D. Smith, and T. J. Peters, *Biochim. Biophys. Acta* **657,** 334 (1981).
[92] N. Taniguchi, *J. Biochem. (Tokyo)* **75,** 473 (1974).
[93] N. Taniguchi, K. Saito, and E. Takakuwa, *Biochim. Biophys. Acta* **391,** 265 (1975).
[94] P. Selvaraj, K. A. Balasubramanian, and P. G. Hill, *Enzyme* **26,** 57 (1981).

Inhibition

Both reversible and irreversible inhibitors of γ-glutamyl transpeptidase have been reported. These inhibitors have served as effective probes of the enzyme's mechanism and kinetics and of the role of γ-glutamyl transpeptidase in glutathione metabolism.

Both L- and D-serine strongly inhibit γ-glutamyl transpeptidase in the presence of borate ion.[95,96] The borate ion apparently is situated between the hydroxyl group of serine and the hydroxyl group of an active site amino acid residue. This tetrahedral complex appears to serve as a transition-state analog at the enzyme's active site.[96]

A particularly potent inhibitor synthesized is L-γ-glutamyl-(*o*-carboxy) phenylhydrazide (L-OC; also called anthglutin), which has an apparent K_i value of 8.2 μM.[97–99] This inhibitor has been isolated from a culture medium of *Penicillium oxalicum*.[98,99] Other phenylhydrazides, including the D-isomer of L-OC (with an apparent K_i of 22.5 μM), are also strong inhibitors.[97] Other inhibitors include certain γ-glutamyl hydrazones,[36] the sulfophathalein derivatives, sulfobromophthalein and bromocresol green,[100] and acetazolamide,[101] which appears to bind at the acceptor site. The apparent inhibition by bilirubin[102–104] is disputed.[105]

Of great interest are the effects of certain mono- and dicarboxylic acids. Maleate appears to stimulate the hydrolytic activity and inhibit the transfer activity by binding to the acceptor site.[57–60,106] Other compounds may also affect the partitioning,[36] notably several hippuric acid analogs.[106] Maleate and other compounds also stimulate the inactivation by certain irreversible inhibitors,[107–110] suggesting that maleate induces a conforma-

[95] J. P. Revel and E. G. Ball, *J. Biol. Chem.* **234,** 577 (1959).

[96] S. S. Tate and A. Meister, *Proc. Natl. Acad. Sci. U.S.A.* **75,** 4806 (1978).

[97] O. W. Griffith and A. Meister, *Proc. Natl. Acad. Sci. U.S.A.* **76,** 268 (1979).

[98] T. Kinoshita and S. Minato, *Bull. Chem. Soc. Jpn.* **51,** 3282 (1978).

[99] S. Minato, *Arch. Biochem. Biophys.* **192,** 235 (1979).

[100] F. Binkley, *J. Biol. Chem.* **236,** 1075 (1961).

[101] D. Thornley-Brown, P. D. Dass, and T. C. Welbourne, *Biochem. Pharmacol.* **31,** 3347 (1982).

[102] B. Combes, G. M. Shore, F. G. Cunningham, F. B. Walker, J. W. Shorey, and A. Ware, *Gastroenterology* **72,** 271 (1977).

[103] F. C. Sitzmann, K. Kellerer, and M. Bierchenk, *Med. Klin. (Munich)* **67,** 1183 (1972).

[104] A. D. Dickson and P. R. Beck, *Clin. Chem. (Winston-Salem, N.C.)* **24,** 723 (1978).

[105] J. H. Strømme and L. Theodorsen, *Clin. Biochem. (Ottawa)* **12,** 260 (1979).

[106] G. A. Thompson and A. Meister, *J. Biol. Chem.* **255,** 2109 (1980).

[107] S. S. Tate and A. Meister, *Proc. Natl. Acad. Sci. U.S.A.* **74,** 931 (1977).

[108] M. Inoue, S. Horiuchi, and Y. Morino, *Eur. J. Biochem.* **73,** 335 (1977).

[109] S. S. Tate and M. E. Ross, *J. Biol. Chem.* **252,** 6042 (1977).

[110] S. J. Gardell and S. S. Tate, *Arch. Biochem. Biophys.* **216,** 719 (1982).

tional change in the protein. Recently, conjugated bile acids were reported to inhibit both the hydrolytic and transpeptidation reactions[111] and stimulate the glutaminase activity.[112] Free bile acids stimulated both hydrolysis and transpeptidation[111] and had no effect on glutamine hydrolysis.[112]

Many workers have studied the effects of sulfhydryl reagents and a variety of results have been reported. These effects appear to be strongly species dependent. In general, γ-glutamyl transpeptidase is irreversibly inhibited by iodoacetamide, albeit slowly, but not by *N*-ethylmaleimide, DTNB, or *p*-chloromercuribenzoate. Phenobarbital and thiobarbituric acid have recently been reported to irreversibly inhibit the enzyme, though with relatively high apparent K_i values.[113] The enzyme is strongly inactivated by *O*-diazo-acetyl-L-serine (L-azaserine),[107] 6-diazo-5-oxo-L-norleucine (DON),[107] and by L-(αS,5S)-α-amino-3-chloro-4,5-dihydro-5-isoxazoleacetic acids (AT-125).[81,114,115] AT-125 is 20-fold more effective as an inhibitor than DON and the inactivation by each of these glutamine analogs is enhanced by maleate and hippurate. Phenylmethanesulfonylfluoride has also been found to inhibit the enzyme,[116] this inactivation being stimulated by maleate, and apparently binds at the same site as DON.[34]

Mechanism

Very little, other than a number of covalent modification studies, has been reported on the possible active-site chemistry of γ-glutamyl transpeptidase. Much more work is needed to characterize the enzyme mechanism and the interaction of modulators of the enzyme activity. The initial rate kinetic studies strongly suggest the formation of a covalent γ-glutamyl–enzyme intermediate. It should be recalled, however, that such an intermediate is not an absolute requirement for ping pong or modified ping pong mechanisms.[5,42,43] Though such a covalent bound species is assumed, this intermediate has not been isolated or characterized. The studies with glutamine analogs, such as DON mentioned earlier, strongly

[111] W. A. Abbott and A. Meister, *J. Biol. Chem.* **258,** 6193 (1983).

[112] S. J. Gardell and S. S. Tate, *J. Biol. Chem.* **258,** 6198 (1983).

[113] G. P. Sachdev, D. S. Leahy, and K. V. Chace, *Biochim. Biophys. Acta* **749,** 125 (1983).

[114] L. Allen, R. Meck, and A. Tunis, *Res. Commun. Chem. Pathol. Pharmacol.* **27,** 175 (1980).

[115] O. W. Griffith and A. Meister, *Proc. Natl. Acad. Sci. U.S.A.* **77,** 3384 (1980).

[116] J. S. Elce, *Biochem. J.* **185,** 473 (1980).

suggest the presence of a hydroxyl group at the active site. Elce[116,117] has presented some evidence for the participation of a free amino group and suggests that the glutamyl moiety is covalently attached to the ε-amino group of a lysyl residue. A carboxyl group has also been suggested to be present at the active site.[116–118]

Though many workers have studied the effects of pH on the reaction, there have been no published reports of the pH effects on the true kinetic parameters of transpeptidation, K_b, V_b, and the V_b/K_b ratio. It appears[47] that there is an amino acid residue at the active site with a pK_a of about 7. Such a pK_a agrees with the well-documented increase in the transpeptidation activity with an increase in pH and suggests that the responsible group is a histidyl residue. The presence of an imidazole group could facilitate the deprotonation of the α-amino group of the acceptor substrate. Recently, the presence of an arginine residue at the active site was strongly suggested by inactivation experiments with phenylglyoxal.[119] In this same study, it was reported that the methyl ester of AT-125 was inactive as an inhibitor. The authors suggest that the purported arginyl residue at the active site interacts electrostatically with the carboxyl group of AT-125 and of the γ-glutamyl donor substrate.[119]

As is apparent, considerable work is needed to fully characterize the kinetics and mechanism of this multifunctional enzyme. Purich[120] has provided a number of protocols for assaying the catalytic competence of enzyme–substrate intermediates. The relative stability of γ-glutamyl transpeptidase in organic solvents may allow the thermal trapping of the purported intermediate. In addition, temperature studies will characterize the purported differences in the rate-determining, or rate-contributing, steps between γ-glutamyl-*p*-nitroanilide and glutamine serving as the donor substrate.

[117] J. S. Elce, *in* "Gammaglutamyltransferases" (G. Siest and C. Heusghem, eds.), 3rd ser., p. 15. Masson, Paris, 1982.

[118] C. Huc, A. Olomucki, and F. Thome-Beau, *FEBS Lett.* **60,** 414 (1975).

[119] C. S. Schasteen, N. P. Curthoys, and D. J. Reed, *Biochem. Biophys. Res. Commun.* **112,** 564 (1983).

[120] D. L. Purich, this series, Vol. 87, p. 3.

[52] γ-Glutamylcyclotransferase from Rat Kidney

By ALTON MEISTER

γ-Glutamylcyclotransferase catalyzes the conversion of γ-glutamyl amino acids to 5-oxoproline and the corresponding free amino acid[1-3]:

L-γ-Glutamyl-L-amino acid → 5-oxo-L-proline + L-amino acid

The enzyme is widely distributed in mammalian tissues and is highly active toward the L-γ-glutamyl derivatives of glutamine, methionine, alanine, cysteine, cystine, and several other amino acids. The enzyme also acts on γ-glutamyl-γ-glutamyl amino acids; it is in fact much more active toward certain di-γ-glutamyl amino acids than toward the corresponding γ-glutamyl amino acids. The enzyme has been purified from human and sheep brain,[4] hog liver,[5] rat liver,[6] human erythrocytes,[7] and rat kidney.[8] Several forms of the enzyme which are separable by chromatography and electrophoresis have been reported. The enzyme is unstable during isolation and on storage; it undergoes substantial change in physical and catalytic properties. The most highly purified preparations of the enzyme have been obtained from rat kidney.[8] In one procedure for isolation of the enzyme, chromatography on thiol-Sepharose is used. In another procedure, the crude enzyme is treated with dithiothreitol and then with iodoacetamide. These procedures, which are described below, are based on the work of Taniguchi and Meister.[8]

Reagents

Tris–HCl buffer (pH 7.5 at 37°)
L-γ-[U-^{14}C]Glutamyl-L-α-aminobutyric acid
Trichloroacetic acid
Dowex 50 (H^+)

[1] G. E. Woodward and F. E. Reinhart, *J. Biol. Chem.* **145,** 471 (1942).
[2] G. E. Connell and C. S. Hanes, *Nature* (*London*) **177,** 377 (1956).
[3] A. Meister and M. E. Anderson, *Annu. Rev. Biochem.* **52,** 711 (1983).
[4] M. Orlowski, P. G. Richman, and A. Meister, *Biochemistry* **8,** 1048 (1969).
[5] E. D. Adamson, A. Szewczuk, and G. E. Connell, *Can. J. Biochem.* **49,** 218 (1971).
[6] M. Orlowski and A. Meister, *J. Biol. Chem.* **248,** 2836 (1973).
[7] P. G. Board, K. A. Moore, and J. E. Smith, *Biochem. J.* **173,** 427 (1978).
[8] N. Taniguchi and A. Meister, *J. Biol. Chem.* **253,** 1799 (1978).

Copyright © 1985 by Academic Press, Inc.
All rights of reproduction in any form reserved.

Determination of Enzyme Activity

The assay mixtures consist of 0.05 ml of 0.2 *M* Tris–HCl buffer (pH 7.5 at 37°), 0.01 ml of 0.1 *M* L-γ-[U-^{14}C]glutamyl-L-α-aminobutyrate (about 7900 cpm), and enzyme in a final volume of 0.1 ml. After incubation at 37° for 10–30 min, the reaction is stopped by adding 0.1 ml of 10% trichloroacetic acid. After centrifugation, a portion (0.1 ml) of the supernatant solution is added to the top of a small column (0.5 × 3 cm) of Dowex 50 (H^+; 200–400 mesh), and the column is washed with 3 ml of water. The radioactivity present in the effluent due to 5-oxo-[^{14}C]proline is determined by scintillation counting. At the concentration of substrate used, the rate of product formation is 67% of the maximum; the observed values are multiplied by a factor of 1.5. A unit of enzyme activity is defined as the amount that catalyzes the formation of 1 μmol of 5-oxoproline per minute. Protein is determined by the method of Lowry *et al.*[9] using bovine serum albumin as standard, or from the absorbance at 280 nm. The extinction coefficient (13.1 at 279 nm) of the enzyme is based on the amino acid composition of the enzyme and spectral measurements. Other assay procedures may be used, such as those involving determination of the formation of the L-amino acid, or of the model substrate γ-glu-γ-glu-*p*-nitroanilide.[4]

Purification of the Enzyme

Kidney preparations obtained by homogenizing the fresh tissue in 0.25 *M* sucrose in Tris–HCl buffer (pH 7.5 at 37°), followed by centrifugation and isoelectric focusing show a major peak of activity at pH 5.1, a somewhat smaller one at pH 6.1 and five smaller ones. When homogenate is prepared in the presence of dithiothreitol followed by treatment with 0.1 *M* iodoacetamide, the isoelectric focusing pattern shows a single predominant peak of activity at pH 4.6. The purification methods described below yield highly purified preparations of the p*I* 5.1 form, and of the reduced and alkylated form of the enzyme (p*I* 4.6 form). Unless otherwise stated, all steps were carried out at 4°.

Isolation of pI 5.1 Form: Procedure A

Step 1. Homogenate. Rats (250–350 g) are decapitated and exsanguinated, and the decapsulated kidneys (670 g) are homogenized in 5 volumes of a solution containing 0.15 *M* potassium chloride, 5 m*M* 2-mercaptoethanol, and 1 m*M* magnesium chloride with a Potter-Elvehjem

[9] O. H. Lowry, N. J. Rosebrough, A. L. Farr, and R. J. Randall, *J. Biol. Chem.* **193,** 265 (1951).

apparatus. After centrifugation at 0° for 15 min at 12,000 *g*, the supernatant solution is filtered through a layer of glass wool.

Step 2. Ammonium Sulfate Fractionation. Solid $(NH_4)_2SO_4$ is added to the enzyme solution to achieve 60% of saturation. After stirring for 40 min, the precipitate is removed by centrifugation at 12,000 *g* for 20 min. The solution is then brought to 90% of $(NH_4)_2SO_4$ saturation by adding solid $(NH_4)_2SO_4$, and the precipitate which forms is collected by centrifugation and then resuspended in 0.01 *M* Tris–HCl buffer (pH 8.0 at 4°) containing 5 m*M* 2-mercaptoethanol.

Step 3. Selective Heat Denaturation. The enzyme solution is placed in a flask which is immersed in a water bath at 80°. The solution is stirred continuously until its temperature reaches 56°; this temperature is maintained for 20 min after which the solution is cooled in ice. After centrifugation at 20,000 *g* for 30 min, the sedimented protein is resuspended in 20 ml of 0.01 *M* Tris–HCl buffer (pH 8.0) containing 5 m*M* 2-mercaptoethanol and then centrifuged; the supernatant solutions are combined.

Step 4. Chromatography on Sephadex G-75. Solid $(NH_4)_2SO_4$ is added to the enzyme solution to achieve 95% of saturation, and the precipitate which forms is collected by centrifugation and resuspended in 0.01 *M* Tris–HCl buffer (pH 8.0) to yield a final volume of 50 ml. This solution is added to the top of a column (5 × 100 cm) of Sephadex G-75 equilibrated with 0.01 *M* Tris–HCl buffer (pH 8.0). Elution is carried out with the same buffer and fractions of 20 ml are collected at a flow rate of 100 ml/hr. The major peak of activity emerges from the column immediately after appearance of the major protein peak. The activity appears in two peaks; the properties of the first and smaller of these (about 15 to 20% of the total activity) may represent enzyme dimer.[8] The fractions comprising the second peak were combined and processed as follows.

Step 5. Chromatography on DEAE-Cellulose. The enzyme solution is added to the top of a column (5 × 30 cm) of DEAE-cellulose equilibrated with 0.01 *M* Tris–HCl buffer (pH 8.0). The column is then washed with this buffer (flow rate, 120 ml/hr) until the absorbance at 280 nm is zero. Elution is carried out with a linear gradient established between 2 liters of 0.01 *M* Tris–HCl buffer (pH 8.0) and 2 liters of the same buffer containing 0.2 *M* NaCl; fractions of 20 ml are collected. In this chromatography, seven peaks of enzyme activity can be discerned. Aliquots of the pooled fractions of each peak of activity (and a sample of the material applied to the column) are subjected to isoelectric focusing. The solution applied to the column exhibits a pattern consisting of seven peaks of activity. Peak 4 contains mainly at p*I* 5.7 form; peaks 5 and 6 contain; respectively, p*I* 5.1 and p*I* 4.8 forms. The fractions comprising peak 5 (p*I* form) are pooled and further purified as follows.

Step 6. Chromatography on Thiol-Sepharose 4B. The pooled fractions containing peak 5 (Step 5) are placed in a dialysis sac which is buried in a beaker containing solid $(NH_4)_2SO_4$; this is placed at 4° for 14 hr. The solution [now saturated with $(NH_4)_2SO_4$] is centrifuged at 12,000 *g* for 30 min and the sediment is dissolved in 2 ml of 0.1 *M* Tris–HCl buffer (pH 8.0) containing 0.1 *M* dithiothreitol. The solution is allowed to stand under nitrogen for 1 hr and then added to the top of a column (1.5 × 5 cm) of Sephadex G-25 (PD 10), equilibrated with 0.1 *M* Tris–HCl buffer (pH 8.0) containing 2 m*M* EDTA and 0.3 *M* NaCl. The fractions containing the activity are pooled (volume, 3.5 ml) and this solution is then added to the top of a column (1 × 7 cm) of activated thiol-Sepharose 4B equilibrated with 0.1 *M* Tris–HCl buffer (pH 8.0) containing 1 m*M* EDTA and 0.3 *M* NaCl. The flow rate during loading is 5 ml/hr. The column is washed with the same buffer at a flow rate of 30 ml/hr until the absorbance at 280 nm of the effluent is zero. The enzyme elutes with a solution consisting of the equilibrated buffer containing 10 m*M* dithiothreitol. About 25% of the enzyme activity appears in the column; this fraction may contain enzyme dimer.[8] The fractions exhibiting high specific enzymatic activity are pooled, concentrated by precipitation with $(NH_4)_2SO_4$, and stored at 0°. As indicated in the table (A), a purification of more than 1000-fold is obtained; the overall yield is about 4% and the yield based on the activity present in peak 5 is about 30%. In contrast to the reduced and alkylated form of the enzyme (see below), the p*I* 5.1 form of the enzyme is relatively unstable and loses about 30% of its activity when stored for 3 weeks at 0°.

Purification of the Reduced and Alkylated Form of the Enzyme (pI 4.6 Form): Procedure B

Step 1. Homogenate. Rat kidneys (666 g) are homogenized essentially as described in procedure A in a buffer consisting of 0.05 *M* potassium phosphate, 5 *M* $MgCl_2$, 1 m*M* EDTA, and 5 m*M* dithiothreitol (pH 7.6).

Step 2. Ammonium Sulfate Fractionation and Alkylation. Ammonium sulfate fractionation is carried out as described in procedure A, and the enzyme solution obtained is brought to pH 8.0 and to a final concentration of 0.1 *M* iodoacetamide. The solution is then dialyzed for 18 hr against a large volume of 0.01 *M* Tris–HCl buffer (pH 8.0).

Step 3. Selective Heat Denaturation. The solution is maintained at 56° (see procedure A; Step 3) for 10 min and then cooled and centrifuged. The precipitate is washed with buffer and the combined supernatant solutions are dialyzed against 0.01 *M* potassium phosphate buffer (pH 6.5) for 18 hr.

Step 4. Chromatography on CM-Cellulose. The dialyzed solution is added to the top of a column (5 × 15 cm) of CM 52 equilibrated with

PURIFICATION OF γ-GLUTAMYLCYCLOTRANSFERASE BY TWO PROCEDURES

Step	Volume (ml)	Total protein (mg)	Activity: Total units	Activity: Specific units/mg	Yield (%)
A. p*I* 5.1 form					
1. Centrifuged homogenate	2,890	74,600	12,700	0.17	100
2. 60 to 90% $(NH_4)_2SO_4$	170	1,930	7,400	3.83	58
3. Selective heat denaturation	150	578	7,300	12.6	57
4. Sephadex G-75	440	233	3,990	17.1	31
5. DEAE-cellulose	380	11	1,470	134	12
6. Thiol-Sepharose 4B	20	2.6	470	181	4
B. p*I* form (reduced and alkylated)					
1. Centrifuged homogenate	3,600	94,300	12,300	0.13	100
2. 60 to 90% $(NH_4)_2SO_4$, iodoacetamide	160	1,510	4,880	3.23	40
3. Selective heat denaturation	180	920	4,500	4.89	37
4. CM-cellulose	1,450	390	3,200	8.21	26
5. Sephadex G-75	150	98	2,930	29.9	24
6. DEAE-cellulose	140	23	1,580	68.7	13
7. Isoelectric fractionation	10	6.4	640	100	5
8. Sephadex G-75	12	2.4	380	158	3

0.01 *M* potassium phosphate buffer (pH 6.5). The effluent fractions containing activity are combined.

Step 5. Chromatography on Sephadex G-75. The solution is concentrated by treatment with $(NH_4)_2SO_4$ (see procedure A; Step 6); the enzyme is recovered by centrifugation and then taken up in 0.01 *M* potassium phosphate buffer (pH 6.5) to a final volume of 30 ml. This solution is added to the top of a column (5 × 100 cm) of Sephadex G-75 equilibrated with 0.01 *M* potassium phosphate buffer (pH 6.5). Fractions of 10 ml are collected at a flow rate of 100 ml/hr. The major peak of activity, which emerges after the major protein peak, is collected.

Step 6. Chromatography on DEAE-Cellulose. The enzyme solution obtained in Step 5 is added to the top of a column (2.7 × 20 cm) of DE 52 equilibrated with 0.01 *M* potassium phosphate buffer (pH 6.5). The column is washed with this buffer (flow rate, 120 ml/hr) until the absorbance at 280 nm is zero. The enzyme is eluted with a linear gradient established between 300 ml of 0.01 *M* potassium phosphate buffer (pH

6.5) and 300 ml of the same buffer containing 0.2 *M* NaCl. The most active fractions emerge at about 0.1 *M* NaCl.

Step 7. Isoelectric Fractionation. The enzyme solution is concentrated by addition of $(NH_4)_2SO_4$ and the precipitated protein is dissolved in 2 ml of 0.01 *M* Tris–HCl buffer (pH 8.0). This solution is desalted using a column of Sephadex G-25 (PD 10) as described above (Step 6), and then subjected to isoelectric focusing (pH 4 to 6), which yields a peak of activity at pH 4.6. The active fractions are pooled.

Step 8. Chromatography on Sephadex G-75. The enzyme solution is concentrated as described in procedure A (Step 6) by precipitation with solid $(NH_4)_2SO_4$. The precipitated enzyme is taken up in 2 ml of 0.01 *M* Tris–HCl buffer (pH 8.0). The enzyme solution is added to the top of a column (2.7 × 100 cm) of Sephadex G-75 equilibrated with 0.01 *M* Tris–HCl buffer (pH 8.0). Elution is carried out with the same buffer. The fractions obtained within the major enzyme peak exhibit the same specific activities. The purification is more than 1200-fold. This form of the enzyme is stable on storage at 0° for at least 2 months.

Properties of the Purified Enzyme Preparations

Both the p*I* 5.1 and p*I* 4.6 forms give a single band on electrophoresis in 7% polyacrylamide gels.[8] It was demonstrated by determinations of enzyme activity that all of the activity coincided with the protein. Both enzyme preparations give single bands on polyacrylamide gel electrophoresis carried out in the presence of sodium dodecyl sulfate. Molecular weight determinations carried out by sodium dodecyl sulfate–polyacrylamide gel electrophoresis with appropriate standards give a value of 27,000 for the molecular weights of both the p*I* 4.6 and p*I* 5.1 forms of the enzyme.

Both forms of the enzyme exhibit identical ultraviolet absorption characteristics with a maximum at 279 nm. The extinction coefficient $E_{1\,\text{cm}}^{1\%}$ (279 nm) is 13.1. The ratio of the absorbance at 280 nm and at 260 nm is 1.57. The amino acid compositions of the two forms are substantially the same. Both forms have about 7 half-cystine residues per mole. Immunological studies carried out with antibodies produced in rabbits show that the two forms are immunologically identical.

When the p*I* 5.1 form of the enzyme is titrated with 5,5′-dithiobis(2-nitrobenzoate) and with 4-hydroxymercuribenzoate, values close to 4 and 5 mol of sulfhydryl groups per mole of enzyme are found in the absence and presence, respectively, of sodium dodecyl sulfate. When the titration is carried out immediately after treatment with dithiothreitol (followed by removal of excess dithiothreitol), the corresponding values are about 5 and 7. The latter values are in accord with those obtained on amino acid analysis, which are based on determinations of cysteic acid and *S*-carboxymethyl cysteine.

The purified p*I* 5.1 form of the enzyme can be converted to a form which appears to be identical on isoelectric focusing to the p*I* 4.6 form. This can be accomplished by treating the p*I* 5.1 form with dithiothreitol and then with iodoacetamide. When the purified p*I* 5.1 form is treated wth cystamine at pH 8 (after treatment with dithiothreitol), a heterogeneous pattern was found on isoelectric focusing which closely resembles that seen on similar treatment of kidney extracts.

Catalytic Properties

Optimal activity is observed in the pH range 7.5–8 in Tris–HCl buffers. The apparent K_m value for γ-glutamyl-α-aminobutyrate (p*I* 4.6 form) is 6.6 m*M*, a value close to that observed with freshly prepared kidney extracts. However, the corresponding K_m value determined with the p*I* 5.1 form of the enzyme increases during purification from a value of about 6 m*M* to values that are 4–10 times higher. These observations suggest that the active site of the enzyme undergoes substantial alteration during purification and storage. In contrast, the reduced and alkylated form of the enzyme (p*I* 4.6 form) is much more stable.

The substrate specificity of the p*I* 4.6 form indicates that the enzyme is highly active toward the γ-glutamyl derivatives of L-α-aminobutyrate, L-glutamine, L-alanine, and L-methionine. Somewhat less activity is observed toward the γ-glutamyl derivatives of L-glutamate, L-aspartate, and glycine. The enzyme acts very slowly on the γ-glutamyl derivatives of L-valine, L-tyrosine, L-isoleucine, L-leucine, L-proline, and L-phenylalanine. In contrast the di-L-γ-glutamyl derivatives of L-alanine, L-glutamine, glycine, L-valine, L-leucine, L-tyrosine, L-glutamate, L-phenylalanine, L-lysine, and L-proline exhibit substantial activity. The preferred substrates of this enzyme have the general structure γ-glu-γ-gluNHR, in which the nature of the R-moiety has relatively little effect on activity.

Although the L-γ-glutamyl derivatives of L-glutamine and L-α-aminobutyrate are good substrates of the enzyme, these compounds are also excellent substrates of γ-glutamyl transpeptidase. On the other hand, L-γ-(threo-β-methyl)glutamyl-L-α-aminobutyrate is not a substrate of γ-glutamyl transpeptidase; it is, however, about as active as L-γ-glutamyl-L-α-aminobutyrate as a substrate of γ-glutamylcyclotransferase.[10] This substrate is thus valuable for the determination of γ-glutamylcyclotrans-

[10] R. J. Bridges, O. W. Griffith, and A. Meister, *J. Biol. Chem.* **255,** 10,787 (1980).

ferase in tissue preparations that contain γ-glutamyl transpeptidase, and can thus be used for the determination of γ-glutamylcyclotransferase in tissue homogenates. An additional advantage of using this model substrate derives from the fact that it is converted by the enzyme to 3-methyl-5-oxoproline, which is not a very active substrate of 5-oxoprolinase. Therefore, the formation of 3-methyl-4-oxoproline can also be used as an indicator of γ-glutamylcyclotransferase activity. In another application of the model substrate, animals may be injected with L-γ-(threo-β-methyl)-glutamyl-L-α-amino[^{14}C]butyrate, which is rapidly metabolized to form respiratory CO_2, and since the release of L-α-amino[^{14}C]butyrate moiety from the substrates specifically requires the activity of γ-glutamylcyclotransferase, the formation of respiratory $^{14}CO_2$ after injection of the labeled model substrate provides a valid measure of *in vivo* γ-glutamylcyclotransferase activity.

β-Aminoglutaryl-L-α-aminobutyrate is a selective inhibitor of γ-glutamylcyclotransferase. The apparent K_i value for this inhibitor is about 0.46 m*M*. The inhibitor markedly inhibits the γ-glutamylcyclotransferase activity of many mouse tissues.[10] β-Aminoglutaryl-L-α-aminobutyrate is also effective *in vivo;* thus administration of β-aminoglutaryl-L-α-aminobutyrate to mice not only decreases the level of 5-oxoproline in the kidney of control mice but also of mice in which kidney 5-oxoproline levels are increased by administration of L-methionine. Administration of β-aminoglutaryl-L-α-aminobutyrate to mice decreases the *in vivo* metabolism of L-(threo-β-methyl)glutamyl-L-α-amino[^{14}C]butyrate as indicated by a marked decrease in the rate of respiratory rate of $^{14}CO_2$ formation. D-γ-Glutamyl-L-α-aminobutyrate also inhibits the action of γ-glutamylcyclotransferase toward L-γ-glutamyl-L-α-aminobutyrate.

[53] 5-Oxo-L-prolinase from Rat Kidney

By ALTON MEISTER, OWEN W. GRIFFITH,
and JOANNE M. WILLIAMSON

5-Oxoprolinase catalyzes the ATP-dependent cleavage of 5-oxoproline to glutamate:

$$\text{5-Oxo-L-proline} + \text{ATP} + 2\text{H}_2\text{O} \xrightarrow[\text{K}^+(\text{NH}_4{}^+)]{\text{Mg}^{2+}} \text{L-glutamate} + \text{ADP} + \text{P}_i$$

This reaction seems to be the only one presently known in which cleavage of ATP is required for the cleavage of a peptide bond. The requirement for

Copyright © 1985 by Academic Press, Inc.
All rights of reproduction in any form reserved.

ATP is mandated by the unusual stability of the internal peptide bond of 5-oxoproline. 5-Oxoprolinase has been found in mammalian tissues,[1-13] plants,[14] and microorganisms.[15,16] 5-Oxoproline arises from the action of γ-glutamylcyclotransferase on γ-glutamyl amino acids,[17,18] and may also be formed by other pathways[2,19] such as the enzymatic cleavage of amino terminal 5-oxoprolyl residues of peptides and proteins, and by the action of γ-glutamylcyclotransferase on ε-(γ-glutamyl)lysine derived from the degradation of proteins containing transglutaminase-generated cross-links.[19] The preparation and properties of γ-glutamylaminecyclotransferase have been described recently in another volume of this series.[20]

Reagents

Sodium Hepes buffer (pH 8.0)
5-Oxo-L-proline
5-Oxo-[U-^{14}C]L-proline
NaATP
$MgCl_2$
Phosphoenolpyruvate
KCl
Dithiothreitol
Pyruvate kinase

1 P. Van Der Werf, M. Orlowski, and A. Meister, *Proc. Natl. Acad. Sci. U.S.A.* **68,** 2892 (1971).
2 P. Van Der Werf and A. Meister, *Adv. Enzymol.* **43,** 519 (1975).
3 P. Van Der Werf, O. W. Griffith, and A. Meister, *J. Biol. Chem.* **250,** 6686 (1975).
4 O. W. Griffith and A. Meister, *Biochem. Biophys. Res. Commun.* **70,** 759 (1976).
5 O. W. Griffith and A. Meister, *J. Biol. Chem.* **257,** 4392 (1982).
6 O. W. Griffith and A. Meister, *J. Biol. Chem.* **256,** 9981 (1981).
7 J. Williamson and A. Meister, *J. Biol. Chem.* **257,** 9161 (1982).
8 J. M. Williamson and A. Meister, *J. Biol. Chem.* **257,** 12,039 (1982).
9 A. Wendel, U.-I. Flugge, and L. Flohe, *in* "Glutathione: Metabolism and Function" (I. M. Arias and W. B. Jakoby, eds.), p. 71. Raven Press, New York, 1975.
10 A. Wendel and U.-I. Flugge, *Hoppe-Seyler's Z. Physiol. Chem.* **356,** 873 (1975).
11 A. Wendel, U.-I. Flugge, and H.-S. Jenke, *Hoppe-Seyler's Z. Physiol. Chem.* **356,** 881 (1975).
12 S. S. Tate, L. L. Ross, and A. Meister, *Proc. Natl. Acad. Sci. U.S.A.* **70,** 1447 (1973).
13 L. L. Ross, L. Barber, S. S. Tate, and A. Meister, *Proc. Natl. Acad. Sci. U.S.A.* **70,** 2211 (1973).
14 M. Mazelis and R. K. Creveling, *Plant Physiol.* **62,** 798 (1978).
15 P. Van Der Werf and A. Meister, *Biochem. Biophys. Res. Commun.* **56,** 90 (1974).
16 E. D. Mooz and L. Wigglesworth, *Biochem. Biophys. Res. Commun.* **68,** 1066 (1976).
17 A. Meister, *Science* **180,** 33 (1973).
18 A. Meister and M. E. Anderson, *Annu. Rev. Biochem.* **52,** 711 (1983).
19 M. L. Fink, S. I. Chung, and J. E. Folk, *Proc. Natl. Acad. Sci. U.S.A.* **77,** 4554 (1980).
20 This series [94].

Tris
Dowex 50 (H^+)

Determination of Enzyme Activity

Enzyme activity is determined in reaction mixtures containing (final volume, 0.5 ml) 100 m*M* sodium Hepes buffer (pH 8.0), 2 m*M* 5-oxo-L-proline (containing 100 cpm/nmol of 5-oxo-L-[U-^{14}C]proline), 5 m*M* ATP, 8 m*M* MgCl, 2 m*M* phosphoenolpyruvate, 150 m*M* potassium chloride, 2 m*M* dithiothreitol, pyruvate kinase (5 units), and 5-oxoprolinase. The reaction mixture is incubated at 37° for 5–30 min. It is then treated with 0.1 volume of 1 *M* HCl and immediately placed at 0° for 5 min. An equivalent volume of 1 *M* Tris is then added and denatured protein is removed by centrifugation. Portions of the neutralized reaction mixtures are then analyzed for inorganic phosphate (by the method of Fiske and SubbaRow[21]), for [^{14}C]glutamate [by liquid scintillation counting after removal of unreacted labeled 5-oxoproline by passage through Dowex-50 (H^+),[3]] or for unlabeled glutamate by use of an amino acid analyzer. Protein is determined as described[22] using bovine serum albumin as standard.

Purification of the Enzyme

The procedure given below is that of Williamson and Meister.[7] All steps are carried out at 0–5°. Buffer A contains 50 m*M* Tris–HCl (pH 7.3 at 25°), 0.1 m*M* NaEDTA, 2 m*M* dithiothreitol, and 5 m*M* 5-oxo-L-proline. A summary of the purification is given in the table.

Step 1. Frozen rat kidneys (500; 515 g) are allowed to thaw in 1 liter of buffer A; this takes about 30 min. The kidneys are then homogenized in a glass Potter-Elvehjem apparatus equipped with a Teflon pestle. Additional buffer A is added to bring the total volume to 2 liters. Insoluble material is removed by centrifugation at 16,300 *g* for 90 min.

Step 2. The supernatant solution is filtered through a plug of glass wool and the volume of the filtered solution is adjusted to 2 liters by adding buffer A. Solid ammonium sulfate (20.9 g/100 ml) is added slowly with stirring over a period of 15 min. After addition is complete, the solution is allowed to stir for 30 min. The precipitated protein is then collected by centrifugation and dissolved in buffer A. The enzyme solution is dialyzed for 8 hr against 2 changes of 4 liters each of buffer A.

Step 3. The dialyzed ammonium sulfate fraction is applied to a column (5 × 25 cm) of DEAE-cellulose. The column is washed with 500 ml of

[21] C. H. Fiske and Y. SubbaRow, *J. Biol. Chem.* **66,** 375 (1925).
[22] O. H. Lowry, N. J. Rosebrough, A. L. Farr, and R. J. Randall, *J. Biol. Chem.* **193,** 265 (1951).

PURIFICATION OF 5-OXO-L-PROLINASE FROM RAT KIDNEY[a]

Step	Total volume (ml)	Total activity (units)[b]	Total protein (mg)	Specific activity (units/mg)	Recovery (%)	Relative specific activity
1. Crude supernatant solution	2,000	2,000	38,400	0.05	100	1.0
2. $(NH_4)_2SO_4$ fractionation	490	2,350	12,840	0.18	118[c]	3.7
3. Chromatography on DEAE-Cellulose	45	2,350	900	2.6	118[c]	52
4. Chromatography on Ultrogel AcA 34	210	1,580	167	9.4	79	185
5. Chromatography on phenyl-Sepharose	70	1,500	35.0	43	75	860
6. Chromatography on aminohexyl-Sepharose	4.9	1,170	14.7	80	59	1,590
7. Chromatography on Ultrogel AcA 34	4.1	1,000	11.5	87	50	1,740

[a] From 515 g.

[b] One unit of activity is the amount of enzyme needed for the production of 1 μmol of glutamate per hour under standard assay conditions.

[c] Recovery of activity is greater than 100% because of the stimulatory effect of residual ammonium ions.

buffer A, and then with a linear gradient established between 2 liters of buffer A and 2 liters of buffer A containing 0.3 *M* NaCl. The flow rate is 150 ml/hr; fractions (23 ml) are collected. The enzyme elutes in a single peak when about 67% of the gradient is collected. The fractions containing activity are combined (895 ml) and the protein present is precipitated by adding 217.4 g of ammonium sulfate (about 45% of ammonium sulfate saturation). The precipitate is recovered by centrifugation and dissolved in buffer A.

Step 4. The enzyme solution is applied to a column (5 × 116 cm) of Ultrogel AcA 34 equilibrated with buffer A. The column is developed with buffer A; fractions (12 ml) were collected at a rate of 100 ml/hr. The fractions containing enzyme activity, which were obtained when about 30% of the bed volume had been eluted, are combined.

Step 5. The solution obtained in Step 4 is applied to a column (1.5 × 28 cm) of phenyl-Sepharose equilibrated with buffer A. The column is washed with 250 ml of buffer A, and then eluted with 50% ethylene glycol (v/v) in buffer A. Fractions (3 ml) are collected at a rate of 25 ml/hr. The fractions containing enzyme activity are combined and dialyzed overnight against 2 liters of buffer A.

Step 6. The dialyzed enzyme solution is then added to the top of a column (1.3 × 28.5 cm) of aminohexyl-Sepharose equilibrated with buffer A. The column is washed with 50 ml of buffer A. Elution is carried out with a linear gradient established between 150 ml of buffer A and 150 ml of buffer A containing 0.9 *M* NaCl. Fractions (3.0 ml) are collected at a rate of 30 ml/hr. The fractions containing enzyme activity (which emerged near the end of the gradient) are combined and concentrated by vacuum dialysis against buffer A.

Step 7. The concentrated enzyme solution is then applied to a column (1.6 × 114 cm) of Ultrogel AcA-34 equilibrated with buffer A. The column is developed with buffer A and fractions (1.2 ml) are collected at a rate of 15 ml/hr. The fractions containing enzyme activity are combined and concentrated by vacuum dialysis against buffer A.

Comments on the Purification

Successful purification of 5-oxoprolinase requires that it be maintained constantly in the presence of 5-oxo-L-proline. Under these conditions, the enzyme is stable during purification. Storage of the enzyme in the absence of 5-oxo-L-proline lead to rapid loss of activity.

Molecular Weight, Subunit Structure, and Amino Acid Composition

The purified enzyme, which is homogenous on polyacrylamide gel electrophoresis, has a molecular weight of 325,000 as determined by gel filtration.[7] The enzyme contains two apparently identical subunits, Amino acid analysis of the enzyme has been carried out. The enzyme contains 6 ± 1 cysteine residues per subunit which are available for reaction with 5,5′-dithiobis(2-nitrobenzoic acid). Titration of the enzyme with this reagent in the presence of sodium dodecyl sulfate gives a value of 29 ± 0.8; this value is within experimental error, equivalent to the total cysteine residues of the protein indicating that the enzyme does not contain disulfide bonds. The enzyme can bind 5-oxo-L-proline in the absence of nucleoside triphosphate and it can bind nucleoside triphosphate in the absence of 5-oxoproline.

One of the sulfhydryl groups that can be titrated with 5,5′-dithiobis(2-nitrobenzoic acid) and with *N*-ethylmaleimide can be protected against modification by ATP or ITP.[7] It appears that at least one sulfhydryl group is at or close to the nucleoside triphosphate binding site and is involved in the cleavage of nucleoside triphosphate. 5′-*p*-Fluorosulfonylbenzoyl adenosine and 5′-*p*-fluorosulfonylbenzoyl inosine interact with the sulfhydryl group involved in cleavage of nucleoside triphosphate and also with another amino acid residue of the enzyme.[7]

Catalytic Properties

The enzyme exhibits a small amount of ATPase in the absence of 5-oxoproline; such activity appears to be an intrinsic property of the enzyme and is thus not due to contamination with other proteins. The enzyme also hydrolyzes ITP in the absence of 5-oxoproline at about the same rate that it hydrolyzes ATP. The apparent K_m value for 5-oxoproline is $8 \pm 2\ \mu M$. The K_m value for ATP is 0.17 mM. The K_m for ITP and for ATP (measured in the absence of 5-oxoproline) is 0.13 mM.[7]

The enzyme can catalyze cleavage of several nucleoside triphosphates in the absence of 5-oxoproline; when 5-oxoproline is added to the enzyme during catalysis of nucleoside triphosphate cleavage, there is a slow time-dependent decrease in the V_{max} and an increase in the value of K_m for the nucleoside triphosphate.[4,7] It has been concluded that 5-oxoprolinase is a hysteretic enzyme, and it has been postulated that in the normal reaction, the binding of both ATP and 5-oxoproline induces conformational changes that brings the substrates into a juxtaposition that facilitates the enzymatic reaction.

Substrate Specificity

The enzyme interacts with several nucleoside triphosphates,[4,7] and also with a number of analogs of 5-oxo-L-proline.[3,8] Coupling between the cleavage of the internal peptide bond of 5-oxoproline and that of ATP is efficient, i.e., the ratio of ATP cleavage to peptide bond cleavage is unity. A similar result is obtained in the reaction with the 5-oxoproline analog L-2-oxothiazolidine-4-carboxylate.[8,23,24] However, the reactions with other 5-oxoproline analogs are only partially coupled or uncoupled. For example, 2-imidazolidone-4-carboxylate, a 5-oxoproline analog in which the 4-methylene moiety is replaced by NH, stimulates rapid ATP cleavage, but is not hydrolyzed itself.[3,8] Other 5-oxoproline analogs, such 2-piperidone-6-carboxylate, stimulate ATP cleavage that exceeds net peptide bond cleavage severalfold.[3] Examination of a number of 5-oxoproline analogs in this reaction[3,8] indicates that the binding of the imino acid substrate requires a carbonyl (or = NH) group at position 5. Although the ring

[23] J. M. Williamson and A. Meister, *Proc. Natl. Acad. Sci. U.S.A.* **78,** 936 (1981).

[24] J. M. Williamson, B. Boettcher, and A. Meister, *Proc. Natl. Acad. Sci. U.S.A.* **79,** 6246 (1981).

nitrogen can be replaced by oxygen (with low activity), the enzyme cannot accommodate an N-methyl group. The substrate appears to interact with the enzyme in a manner such that its α-hydrogen atom is directed toward the enzyme, but the presence of an α-hydrogen atom is not an absolute requirement for binding. It appears that the 2-carboxyl group is required and that substantial modification of the 5-oxo-L-proline molecule in the region of C-3 and C-4 is possible with retention of binding properties. Thus, 2-piperidone-6-carboxylate, 3- and 4-oxy-5-oxoproline, 3- and 4-methyl-5-oxoproline, as well as compounds that contain sulfur, oxygen, or nitrogen atoms in the 4-position of the ring exhibit significant interaction with the active site.

The two hydrolytic activities of 5-oxoprolinase may also be uncoupled by substitution of ATP by certain other nucleoside triphosphates.[4,7] Uncoupling also may arise from substitution of Mg^{2+} by Mn^{2+} in excess, Ca^{2+} or Co^{2+}.[7]

Studies on the Mechanism

The enzyme does not catalyze ATP–ADP exchange reactions. Studies with ^{18}O-labeled substrates show that (1) all 3 oxygen atoms of 5-oxoproline are recovered in glutamate; and (2) the water molecules used in the reaction contribute one oxygen atom to inorganic phosphate and one to the γ-carboxyl group.[6] The enzyme also catalyzes the intrinsically exergonic hydrolysis of α-hydroxyglutarate lactone, a reaction that is ATP dependent.[6] Although conversion of 5-oxo-L-proline to L-glutamate is essentially irreversible, very slow reversal can be shown by measurement of ATP formation in the presence of high glutamate concentrations.[6] A plausible mechanism for the reaction involves phosphorylation of 5-oxoproline by ATP, perhaps on the amide carbonyl oxygen atom, followed by hydrolysis of the resulting intermediate to yield γ-glutamyl phosphate. The latter would then be hydrolyzed to glutamate and inorganic phosphate.

[54] 5-Oxo-L-prolinase from *Pseudomonas putida*

By ANDREW P. SEDDON, LUYUAN LI, and ALTON MEISTER

Extracts of *Pseudomonas putida*, an organism isolated from soil by enrichment culture, contain about 50 times more 5-oxo-L-prolinase activity than do homogenates of rat kidney. However, attempts to purify the

Copyright © 1985 by Academic Press, Inc.
All rights of reproduction in any form reserved.

enzyme from extracts of this organism were repeatedly unsuccessful and it was found that virtually all the activity disappeared when the bacterial extracts were chromatographed on DEAE-Sepharose or DEAE-cellulose. As described in detail below (and elsewhere[1]), the explanation for the apparent disappearance of enzymatic activity arose from the discovery that combination of two of the protein fractions obtained in the chromatography led to complete restoration of 5-oxoprolinase activity. Analysis of the effluent fractions obtained from chromatography on DEAE-Sepharose revealed 3 separate ATPase activities (designated α, β, and γ), none of which exhibited 5-oxoprolinase activity. The elution profile was subsequently divided into 8 protein-containing fractions. By examining all possible combinations of the ATPase, α, β, and γ (fractions 1, 5 and 7, respectively), with fractions that did not contain ATPase activity, it was established that only a mixture of one of the ATPase fractions (β^5) and fraction 8 (F^8) leads to restoration of virtually all of the 5-oxoprolinase activity present before DEAE-chromatography. The separation and purification of the two protein components required for catalysis are described below. One of these (Component β^5) catalyzes 5-oxo-L-proline-dependent ATPase activity. The other (Component F^8), when added to Component β^5, couples the hydrolysis of ATP with cleavage of the internal peptide bond of 5-oxo-L-proline to yield L-glutamate.

Determination of Enzyme Activity

Principles

The activity of Component β^5, which catalyzes the 5-oxo-L-proline-dependent hydrolysis of ATP, may be determined from the rate of formation of inorganic phosphate or that of ADP. 5-Oxoprolinase activity, which is catalyzed by mixtures of Components β^5 and F^8, may be determined from the rate of formation of [^{14}C]glutamate in reaction mixtures containing 5-oxo-L-[U-^{14}C]proline and from the rates of formation of inorganic phosphate or ADP.

Assay Procedures[2]

5-Oxo-L-proline-dependent ATPase is determined in reaction mixtures containing the following components in a final volume of 0.25 ml: 100 m*M*

[1] A. P. Seddon, L. Li, and A. Meister, *J. Biol. Chem.* **259,** 8091 (1984). In this publication the terms β^5 and F^8 were changed to the nondescriptive terms Component "A" and Component "B", respectively.

[2] In principle, these are the same as given above for the rat kidney enzyme; a few details are different.

Tris–HCl (pH 8.2), 2 m*M* 5-oxo-L-proline, 5 m*M* ATP, 10 m*M* $MgCl_2$, 2 m*M* dithiothreitol, 80 m*M* KCl, and enzyme. After incubation at 37° for 5–30 min, the inorganic phosphate formed is determined. This is carried out as follows. The reaction mixture is treated with 0.8 ml of ice-cold phosphate reagent[3] and the solution is allowed to stand at room temperature for 10 min. The absorbance at 720 nm is read against a blank containing all of the components listed above except the enzyme. The amount of inorganic phosphate present is determined from a standard curve (range 0.05–0.25 μmol). Appropriate controls such as one in which 5-oxoproline is omitted should also be carried out.

The formation of ADP in the 5-oxoproline-dependent ATPase reaction may be carried out in reaction mixtures identical to that described above except that the final volume is 1.0 ml and 1.6 m*M* phosphoenolpyruvate, 0.24 m*M* NADH, pyruvate kinase (15 units) and lactate dehydrogenase (15 units) are added. The reaction is carried out in a quartz cuvette (1 cm light path), which is placed in a thermostated holder in a spectrophotometer at 37°. The reaction is initiated by adding enzyme. The rate of NADH oxidation, recorded at 340 nm ($E = 6220\ M^{-1}\ cm^{-1}$), is equivalent to that of ADP formation.

5-Oxo-L-prolinase activity may be determined in reaction mixtures (final volume, 0.5 ml) containing 2 m*M* 5-oxo-L-proline [containing 5-oxo-L-[U-^{14}C]proline; (50 cpm/nmol)], 5 m*M* ATP, 10 m*M* $MgCl_2$, 80 m*M* KCl, 5 m*M* L-methionine-*SR*-sulfoximine, 2 m*M* phosphoenolpyruvate, pyruvate kinase (5 units), 2 m*M* dithiothreitol, 100 m*M* Tris–HCl buffer (pH 8.2), and enzyme. Methionine sulfoximine is added when assaying crude preparations to inhibit any glutamine synthetase that may be present. After incubation at 37°, for 10–60 min, the mixtures are placed at 100° for 2 min and then cooled in ice. After removal of labeled 5-oxo-L-proline by passage of the reaction mixture through a small column (1 ml in a Pasteur pipette) of Dowex-50 [H^+], and by washing the column with 10 ml of water, the column is treated with 3 ml of 3 *M* ammonium hydroxide to elute [^{14}C]glutamate; radioactivity is determined by liquid scintillation counting.

Protein is determined as described.[4] The activities of components β[5] and F[8] are expressed as micromoles of product formed per hour (units). Specific activities are expressed as units/mg of protein.

Growth of Pseudomonas putida

Pseudomonas putida (strain ALA), an organism isolated by enrichment culture from soil, is grown at 26° with aeration in a medium contain-

[3] C. H. Fiske and Y. SubbaRow, *J. Biol. Chem.* **66,** 375 (1925).
[4] M. M. Bradford, *Anal. Biochem.* **72,** 248 (1976).

ing 40 mM 5-oxo-L-proline, 5.5 mM KH_2PO_4, 3.5 mM Na_2HPO_4, 0.80 mM $MgSO_4$, 0.20 mM $CaCl_2$, 15 μM $FeCl_3$, 15 μM $MnCl_2$, and 4.0 μM Na_2MoO_4; adjusted to pH 7.0. Batch production of cells is carried out to the end of log phase. The cells are harvested by centrifugation and washed with 0.154 M NaCl. The cells may be stored at −20°. The cultures are maintained on solid medium containing 1.8% Agar at 4° under mineral oil.

Separation and Partial Purification of Components β^5 and F^8

All steps were conducted at 4° unless otherwise stated. Buffer A contains Tris–HCl (50 mM; pH 7.8 at 4°), 5-oxo-L-proline (5 mM), dithiothreitol (2 mM, added just prior to use) and K_2EDTA (0.1 mM).

Step 1. The frozen cells (250 g) are thawed in Buffer A and thoroughly mixed. The cells are broken by two passages through a Gaulin homogenizer (Model 15 M). The homogenate is centrifuged at 16,000 g for 60 min, and the supernatant solution obtained (2000 ml) is filtered through a layer of glass wool and applied to a column (5 × 27 cm) of DEAE-Sepharose Cl-6B equilibrated with Buffer A. The column is washed with 500 ml of Buffer A and then with 500 ml of Buffer A containing 0.1 M KCl. The remaining column-bound material is eluted with a linear gradient established between 2 liters of Buffer A containing 0.1 M KCl and 2 liters of Buffer A containing 0.4 M KCl. Fractions of 20 ml are collected at flow rate at 190 ml/hr. The elution profile has 3 peaks of ATPase activity; these are designated α, β, and γ. Assay for 5-oxoprolinase activity in the effluent fractions does not reveal any activity. The fraction corresponding to the β-ATPase contains Component β^5. Component F^8 is eluted between 1500 and 2000 ml of the gradient, which correspond, respectively, to 0.18 and 0.24 M KCl. The fractions containing Component β^5 and those containing Component F^8 are separately pooled, and the protein present in each pool is precipitated by addition of 30 g of ammonium sulfate per 100 ml. The precipitated proteins are collected by centrifugation and dissolved in the minimum volumes of Buffer A. Component β^5 exhibits 5-oxoproline-dependent ATPase activity but does exhibit 5-oxoprolinase activity unless Component F^8 is added. Component F^8 does not catalyze cleavage of ATP nor does it catalyze 5-oxoprolinase activity in the absence of Component β^5.

Step 2. The solution containing Component β^5 is applied to a column (5 × 105 cm) of Ultrogel ACA-34 equilibrated with buffer A. The column is developed with Buffer A and fraction of 12 ml are collected at a rate of 90 ml/hr.

Step 3. The fractions containing Component β^5 activity obtained in Step 2 (225 ml) are applied to a column (0.9 × 25 cm) of aminohexyl-

Sepharose equilibrated with Buffer A. The column is developed with 50 ml of Buffer A followed by 50 ml of Buffer A containing 0.1 *M* KCl. Component β^5 is eluted with a linear gradient established between 250 ml of Buffer A containing 0.1 *M* KCl and 250 ml of Buffer A containing 0.8 *M* KCl. The flow rate is 30 ml/hr and fractions of 3 ml are collected. Component β^5 elutes at a KCl concentration of 0.5 *M*.

Step 4. This step is carried out on Component F^8 exactly as described above for the purification of Component β^5 (Step 2).

Step 5. The pooled fractions containing Component F^8 (450 ml) obtained in Step 4 are applied to a column (0.9 × 25 cm) of aminohexyl-Sepharose equilibrated with Buffer A. The column is treated as described in Step 3; Component F^8 elutes with a KCl concentration of 0.3 *M*.

Results of this purification procedure are summarized in the table.

Stability of Component β^5

Successful purification of Component β^5 requires the presence of 5-oxo-L-proline, dithiothreitol, and potassium ions; these are also required for storage of the activity at 4°. When 5-oxo-L-proline is omitted during

SUMMARY OF PURIFICATION OF 5-OXOPROLINASE FROM *Pseudomonas putida*

Step	Total volume (ml)	Total activity[a] (units)	Total protein (mg)	Specific activity (units/mg)	Recovery (%)
1. Crude supernatant	2,000	15,400[b]	10,400	1.48	
	Purification of β^5				
1a. DEAE-Sepharose	620	23,300	1,830	12.7	100
2. Ultrogel ACA-34	225	12,250	180	68.0	53
3. Amino-hexyl Sepharose 4B	57	11,900	49	243	51
	Purification of F^8[c]				
1b. DEAE Sepharose	880	15,000	2,200	3.10	100
4. Ultrogel ACA-34	450	14,300	360	44.0	95
5. Amino-hexyl Sepharose 4B	73	9,900	80	75.0	66

[a] Determined as 5-oxo-L-proline-dependent ATPase activity.

[b] Determined as 5-oxoprolinase activity; this value may not reflect the total amount of enzyme present. Activity may be reduced by the presence of inhibitors in the crude supernatant solution. Note that the activity obtained after step 2 is somewhat greater.

[c] Determined as 5-oxoprolinase activity. In these assays, a constant amount of component β^5 was present in the assay mixtures and activity was determined in the presence of saturating amounts of component F^8. The values of specific activity given in the table are based on the total protein present (i.e., sum of components β^5 and F^8).

purification or storage, irreversible inactivation results. Separate omission of dithiothreitol leads to loss of activity which can be fully restored by addition of 2 mM dithiothreitol. When enzyme solutions are frozen, even in the presence of 5-oxo-L-proline, dithiothreitol, and potassium ions, there is irreversible loss of activity.

Stability of Component F[8]

In contrast to Component β[5], Component F[8] may be purified successfully in the absence of 5-oxo-L-proline, dithiothreitol, and potassium ions. These compounds are not required for successful storage of the enzyme at 4°.

Catalytic Properties

When 5-oxoproline is removed from solutions of Component β[5] by gel filtration using 50 mM Tris–HCl buffer (pH 8.2) containing 2 mM dithiothreitol, a preparation is obtained that exhibits no ATPase activity. However, when 5-oxo-L-proline is added, the initial level of ATPase activity is restored. The maximal activity of Component β[5] as an ATPase requires the presence of potassium ions; optimal range, 80–300 mM. Potassium ions may be replaced by ammonium ions (ammonium chloride). However, no ATPase activity is observed when potassium ions are replaced by sodium ions.

The pH optimum for Component β[5]-catalyzed 5-oxoproline-dependent cleavage of ATP is about 8.2. In the presence of Component F[8], the pH optimum for the ATP-dependent conversion of 5-oxoproline to glutamate is about 9.5. For the 5-oxoproline-dependent hydrolysis of ATP catalyzed by Component β[5], the optimal ratio of ATP to Mg^{2+} is 1 : 2. The apparent K_m value for MgATP is about 150 μM, and that for 5-oxo-L-proline is about 43 μM. Similar apparent K_m values are obtained for the 5-oxoprolinase reaction which is catalyzed by mixtures of Component β[5] and Component F[8]. No activity was observed when 5-oxo-L-proline is replaced with 5-oxo-D-proline.

When ATP is replaced by CTP, GTP, ATP, UTP, and XTP in the ATPase reaction catalyzed by Component β[5], less than 5% of the activity observed with ATP is found. However, when ATP is replaced by dATP appreciable activity is found.

Component β[5] exhibits a progressive decrease in activity when stored in the absence of dithiothreitol or 2-mercaptoethanol, and appears to require sulfhydryl group(s) for ATPase activity. Iodoacetamide (2×10^{-2}

M) inhibits the ATPase substantially as does incubation with *p*-hydroxymercuribenzoate (10^{-6} *M*) and with 5×10^{-4} *M* *N*-ethylmaleimide.

Coupling of ATP Hydrolysis to L-Glutamate Synthesis

Although Component β^5 catalyzes the 5-oxoproline-dependent hydrolysis of ATP, it does not catalyze cleavage of 5-oxoproline. When Component F^8 is added to Component, β^5 ATP hydrolysis and glutamate formation are coupled. When the amount of Component β^5 is kept constant and that of Component F^8 varied, apparent saturation of Component β^5 is observed; at saturation there is stoichiometry between ATP hydrolysis and glutamate formation. In an analogous experiment, the concentration of Component F^8 was kept constant and that of Component β^5 was varied. At low ratios of Component β^5 to Component F^8, the formation of inorganic phosphate exceeded that of glutamate, but as the ratio of Component β^5 to Component F^8 was increased, the ratio approached unity. In the presence of excess Component β^5, the coupling ratio departs from unity, with ATP hydrolysis exceeding that of glutamate formation.

Polyacrylamide Gel Electrophoresis of Component β^5 and Component F^8

Gel electrophoresis of Component β^5 in the presence of sodium dodecyl sulfate yields two major protein bands (M_r 50,000 and 64,000); under native conditions, gel electrophoresis shows a single protein band. When the gels obtained under native conditions are cut into slices and the slices then assayed for 5-oxoproline-dependent ATPase, the mobility of the protein-stained band is identical to that of the position of the ATPase activity.

Gel electrophoresis of Component F^8 under native conditions shows a single major band, and when these gels are cut into slices and the slices analyzed for 5-oxoprolinase activity in the presence of added Component β^5, the respective mobilities of the major protein-stained band and that which exhibits activity are identical. Gel electrophoresis of Component F^8 in the presence of sodium dodecyl sulfate gives a major band with apparent molecular weight 82,000.

As discussed above,[5] rat kidney 5-oxoprolinase requires K^+ for optimal activity and this enzyme has a number of sulfhydryl groups, at least one of which is required for the cleavage of ATP. The 5-oxoprolinase of *Pseudomonas putida* is composed of two functionally different proteins,

[5] See this volume [53].

neither of which alone catalyzes conversion of 5-oxoproline to glutamate. The 5-oxoproline-dependent ATPase exhibited by Component β^5 requires K^+ for optimal activity, and this enzyme also requires sulfhydryl groups. The purification of Component β^5 and its storage require the presence of potassium ions, dithiothreitol, and 5-oxo-L-proline; these compounds are not required for purification or stability of Component F^8. The findings suggest that Component β^5 interacts with both 5-oxo-L-proline and ATP, and are thus consistent with the view that Component β^5 may catalyze an initial step involving these substrates such as phosphorylation of 5-oxoproline. Component F^8 appears to be involved in the reaction in a different way. Component F^8 may alter the conformation of Component β^5 so as to facilitate the coupled reaction, or more likely, Component F^8 may be a catalyst that converts a phosphorylated form of 5-oxoproline to glutamate. If this interpretation is correct, it may be postulated that the phosphorylated form of 5-oxoproline is formed on Component β^5 and that it breaks down to yield 5-oxoproline and inorganic phosphate in the absence of Component F^8. In the presence of Component F^8, the intermediate is converted to glutamate.

[55] Synthesis of L-2-Oxothiazolidine-4-carboxylic Acid

By BRIAN BOETTCHER and ALTON MEISTER

L-2-Oxothiazolidine-4-carboxylic acid (compound **1**), an analog of 5-oxo-L-proline, is a good substrate of 5-oxoprolinase (apparent K_m, 2 μM; 5-oxo-L-proline, 8 μM).[1,2] This thiazolidine is the only analog of 5-oxo-L-proline that behaves in a fully coupled manner, i.e., the formation of ADP

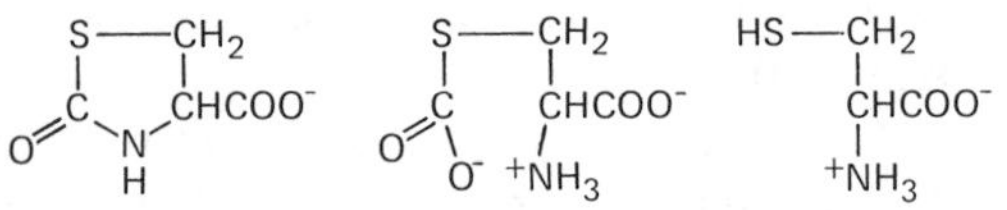

S—CH₂ / O=C, CHCOO⁻ / N–H (ring) **1**

S—CH₂ / O=C–O⁻, CHCOO⁻ / ⁺NH₃ **2**

HS—CH₂ / CHCOO⁻ / ⁺NH₃ **3**

[1] J. M. Williamson and A. Meister, *Proc. Natl. Acad. Sci. U.S.A.* **78,** 936 (1981).
[2] J. M. Williamson and A. Meister, *J. Biol. Chem.* **257,** 12,039 (1982).

Copyright © 1985 by Academic Press, Inc.
All rights of reproduction in any form reserved.

(and of inorganic phosphate) is stoichiometric with peptide bond cleavage. L-2-Oxothiazolidine-4-carboxylate is useful in enzyme studies and also because it functions in an intracellular delivery system for L-cysteine.[1,3,4] This thiazolidine can protect cells against toxicity produced by a variety of compounds including oxygen. The biochemical basis of such protection is that L-2-oxothiazolidine-4-carboxylate is readily transported into many cells where it is converted by intracellular 5-oxoprolinase to *S*-carboxyl-L-cysteine (compound **2**). The latter compound decarboxylates spontaneously to yield L-cysteine (compound **3**), which is effectively utilized for the synthesis of glutathione. Administration of L-cysteine itself is much less effective in stimulating glutathione synthesis; in addition, cysteine produces toxicity.[5-8] L-2-Oxothiazolidine-4-carboxylate may also be useful as a component of amino acid mixtures used in diets and in solutions employed for parenteral administration. A cysteine delivery system of this type is appropriate for cells and organisms that cannot use methionine sulfur for the synthesis of cysteine.

L-2-Oxothiazolidine-4-carboxylate was first prepared by reaction of L-cysteine with phosgene[9]; the phosgene procedure has been successfully used by several investigators.[1,3,10] Since phosgene is highly toxic, its use in the laboratory is hazardous and requires a number of precautions. In the procedure described here, the relatively safe and convenient synthesis of L-2-oxothiazolidine-4-carboxylic acid is described using phenyl chloroformate in place of phosgene. Phenyl chloroformate is a corrosive liquid which should be used with due caution, but it is much safer to handle than is phosgene. The procedure described below also has another advantage in that it gives somewhat higher yields of product (70% as compared to 49%) than obtained by the phosgene procedure. The method has been used for the convenient preparation of ^{35}S-labeled L-2-oxothiazolidine-4-carboxylic acid from L-[^{35}S]cysteine.[11] The procedure described below is that of Boettcher and Meister.[12]

[3] J. M. Williamson, B. Boettcher, and A. Meister, *Proc. Natl. Acad. Sci. U.S.A.* **79,** 6246 (1982).

[4] A. Meister, *Science* **220,** 472 (1983).

[5] J. W. Olney, O. L. Ho, and V. Rhee, *Exp. Brain Res.* **14,** 61 (1971).

[6] R. L. Karlsen, I. Grofova, D. Malthe-Sorensen, and F. Fonnum, *Brain Res.* **208,** 167 (1981).

[7] S. M. Birnbaum, M. Winitz, and J. P. Greenstein, *Arch. Biochem. Biophys.* **72,** 428 (1957).

[8] Y. Nishiuch, M. Sasaki, M. Nakayasu, and A. Oikawa, *In Vitro* **12,** 635 (1976).

[9] T. Kaneko, T. Shimokobe, Y. Ota, E. Toyokawa, T. Unui, and T. Shiba, *Bull. Chem. Soc. Jpn.* **37,** 242 (1964).

[10] H. Shah, S. P. Hartman, and S. Weinhouse, *Cancer Res.* **39,** 3942 (1979).

[11] M. E. Anderson, O. Hwang, and A. Meister, unpublished (1983).

[12] B. Boettcher and A. Meister, *Anal. Biochem.* **138,** 449 (1984).

Reagents

Potassium hydroxide
Na_2EDTA
L-Cysteine (free base)
Phenyl chloroformate
Toluene
Concentrated HCl
Diethyl ether
Ethyl acetate
Sodium sulfate, anhydrous

A mixture consisting of KOH (0.4 mol), Na_2EDTA (0.18 g; 0.5 mmol) in 350 ml of deionized water is placed in a 3-neck 1-liter round-bottom flask. The solution is cooled to 5° in an ice-salt bath, and argon (or nitrogen) is bubbled into the solution for 10–15 min. L-Cysteine (free base; 24.2 g; 0.2 mol) is then added with magnetic stirring. After the cysteine dissolves, 100 ml of 4 *M* KOH and 100 ml of a 1 : 1 mixture (v/v) of phenyl chloroformate and toluene are added dropwise by means of separate dropping funnels. (These solutions are flushed with nitrogen for 10 min prior to addition.) The solutions are added simultaneously and at the same rate over a period of 1 hr, and the mixture is stirred for an additional hour after the additions are complete. The solution is then filtered through a sintered glass Buchner funnel to remove a small white precipitate which forms. The toluene layer is removed and the aqueous layer is brought to about pH 7 by addition of concentrated HCl. The acidified solution is extracted three times with 200 ml portions each of diethyl ether. The aqueous layer is then adjusted to pH 1.5 by adding concentrated HCl, and extracted with 4 portions of 240 ml each of ethyl acetate. The ethyl acetate fractions are combined, dried over anhydrous sodium sulfate, and evaporated to dryness under vacuum. The product is crystallized from hot water to give about 21 g (70%) of L-2-oxothiazolidine-4-carboxylic acid. The melting point is 171–173°.

[56] Formation of Respiratory $^{14}CO_2$ from Radiolabeled Substrates

By OWEN W. GRIFFITH

Introduction

The ability of an intact animal to catabolize a radiolabeled metabolite or metabolite analog is conveniently determined by monitoring the formation of respiratory $^{14}CO_2$ in a metabolic chamber. Under favorable circumstances it is possible to investigate *in vivo* the activity of a specific enzyme or to quantitate the metabolic flux through a particular pathway. Similarly, the activity of enzyme inhibitors can often be demonstrated *in vivo* as a diminution in the rate or extent of $^{14}CO_2$ formation from radiolabeled substrates requiring the inhibited enzyme for their catabolism. This chapter discusses the construction and use of metabolic chambers suitable for mice or rats.

Apparatus

Metabolic chambers designed for studies with whole animals have been described in the literature[1,2] and are available from several commercial suppliers. The apparatus shown in Fig. 1 has proven useful in our laboratory[3–7] and is easily constructed with simple glass-blowing techniques. Although intended primarily for the collection of respiratory CO_2, urine samples can be collected with relatively little contamination by feces. The lower portion of the apparatus is constructed on a $ 71/60 female joint which is tapered to a funnel at the lower end and attached to a $ 14/20 male joint. Several indents are made around the circumference at a level just above the top of the funnel portion and these are used to support a circular screen cut from any suitable wire mesh (e.g., window screen). If it is necessary to collect the urine completely free of feces, it is often useful to place a second screen just above the $ 14/20 male joint.

[1] S. Weinhouse and B. Friedman, *J. Biol. Chem.* **191,** 707 (1951).
[2] M. Baggiolini and M. H. Bickel, *Anal. Biochem.* **14,** 290 (1966).
[3] O. W. Griffith and A. Meister, *Proc. Natl. Acad. Sci. U.S.A.* **76,** 268 (1979).
[4] O. W. Griffith and A. Meister, *Proc. Natl. Acad. Sci. U.S.A.* **76,** 5606 (1979).
[5] R. J. Bridges, O. W. Griffith, and A. Meister, *J. Biol. Chem.* **255,** 10787 (1980).
[6] O. W. Griffith, *J. Biol. Chem.* **257,** 13704 (1982).
[7] O. W. Griffith, *J. Biol. Chem.* **258,** 1591 (1983).

Copyright © 1985 by Academic Press, Inc.
All rights of reproduction in any form reserved.

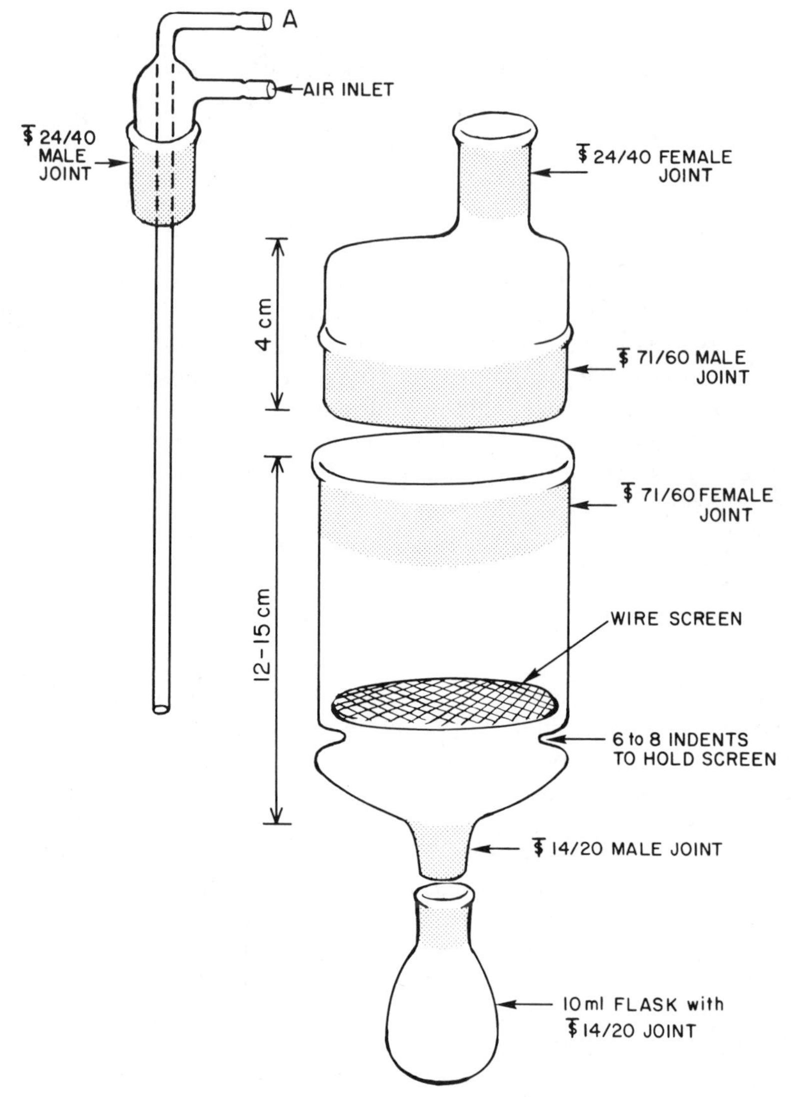

FIG. 1. Metabolic chamber.

The upper portion of the chamber is constructed from a $ 71/60 male joint which is closed and sealed to a $ 24/40 female joint as shown. The adapter shown to the left of the chamber is a commercially available product generally referred to as a long-stem vacuum takeoff adapter. The length of the stem should be adjusted so that when it is placed in the upper portion of the assembled chamber the stem extends about 1 cm through a hole in

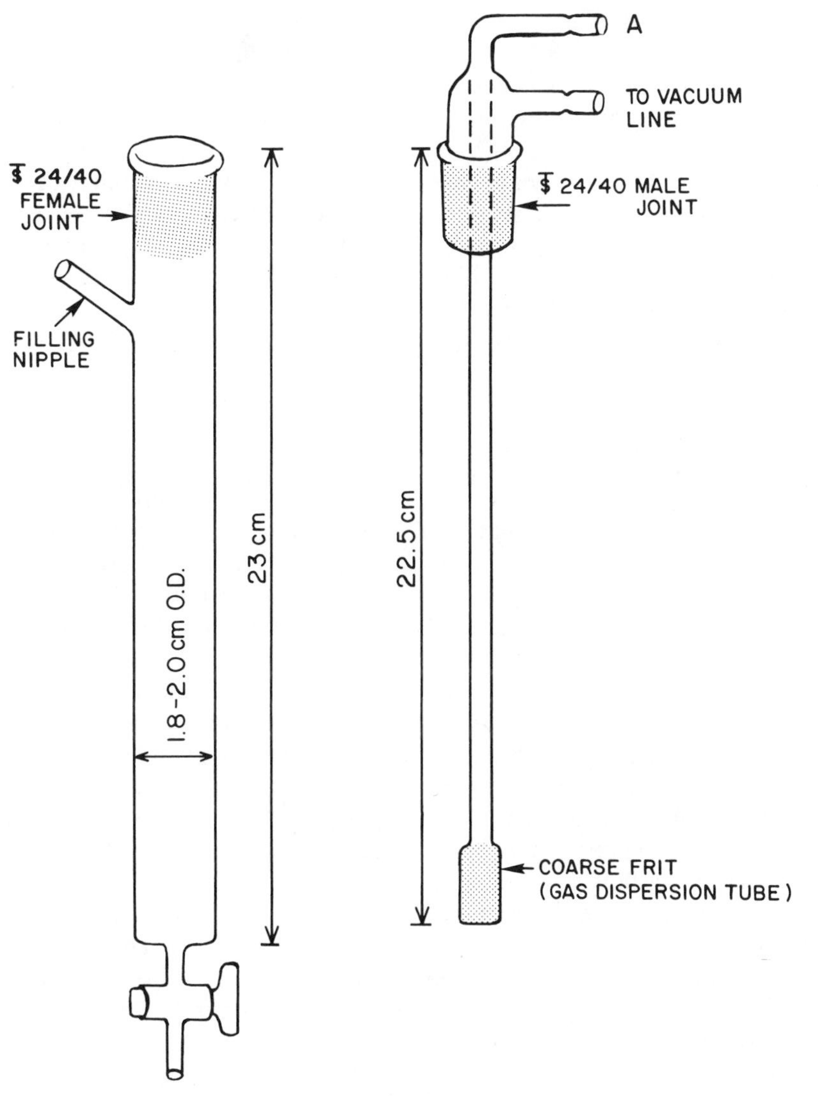

FIG. 2. Gas bubbler.

the screen. The chamber described is intended for mice but is usable with rats weighing up to about 150 g. Rats larger than 150 g are more conveniently studied with a larger chamber.

Respiratory $^{14}CO_2$ is trapped in a gas bubbler constructed as shown in Fig. 2. The exact dimensions are not critical but it is necessary that the coarse frit which serves as bubbler be near the bottom of the outer chamber when the apparatus is assembled. The filling nipple should be fitted with a short length of tubing and a screw clamp which allows that port to be opened and closed as necessary during the course of the experiment. In the assembled apparatus port A of the metabolic chamber adapter is connected to port A of the gas bubbler by a length of tubing; the remaining port of the gas bubbler is connected to the vacuum line by similar tubing. It is prudent to interpose a second gas trap between the latter port and the vacuum line to trap any $^{14}CO_2$ which may escape the first gas bubbler. Since this second trap need not be emptied during the course of the experiment it can be simply constructed from a 250-ml vacuum flask fitted with a stopper carrying a gas dispersion tube.

Operating Procedure

To place the chamber in operation the gas bubbler and second $^{14}CO_2$ trap (if used) must be charged with a suitable alkaline solution. An ideal $^{14}CO_2$ trapping solution would (1) have a large capacity, (2) form small bubbles with a large surface to volume ratio, and (3) not interfere with radioactivity determinations by liquid scintillation counting. Suitable solutions are available from many suppliers of liquid scintillation cocktails; they are generally expensive but are formulated to offer minimal interference with liquid scintillation counting. An inexpensive and useful trapping solution can be made by mixing ethanolamine and methanol in a 20 : 80 ratio.[2] Using a syringe, 20 ml of this solution (or other suitable solution) is placed in the gas bubbler and approximately 75 ml is placed in the secondary $^{14}CO_2$ trap (if used). Air is then slowly drawn through the metabolic chamber and gas bubblers by opening the vacuum line (it is convenient to modulate the air flow by placing a screw clamp on the tubing leading from the gas bubbler to the vacuum line). The air flow rate should be approximately 1 liter per minute which is sufficient to fill the chamber of the gas bubbler with a multitude of very small bubbles. If the flow rate is too high, large bubbles are formed and trapping of $^{14}CO_2$ is inefficient. If gas flow is too low, $^{14}CO_2$ is not efficiently drawn from the metabolic chamber into the trapping solution.

Prior to placing the experimental animal in the chamber it is necessary to measure the background level of radioactivity. Air is therefore drawn through the chamber for 15–30 min, and then the screw clamp on the filling nipple is opened. Since the filling nipple is above the level of solution in the bubbling chamber, bubbling immediately stops when that port is opened and air flow through the metabolic chamber ceases. A 25- or 50-ml graduated cylinder (preferably with a ground glass stopper) is placed under the stop cock of the bubbling chamber and the solution is quickly drained from the chamber. The collected volume should be 20.0 ml. If it is not, and the bubbling chamber including the center tube has been completely drained, then the volume is made up to 20 ml using fresh solution.[8] The graduated cylinder is then stoppered and briefly agitated. Five milliliters of the solution is removed and placed in a liquid scintillation counting vial, and 5 ml of fresh solution is added to the remaining 15 ml. Using a syringe the resulting 20 ml is returned to the bubbling chamber through the filling nipple and that port is again closed. Air flow through the metabolic chamber and bubbling is automatically and immediately resumed. The entire sequence of steps can be completed in 60–90 sec. The chamber should be used with an experimental animal immediately after the background level of radioactivity is determined since continued flow of air through the bubbling solution will eventually exhaust its capacity for trapping $^{14}CO_2$.

To begin an experiment, a mouse or rat is administered a ^{14}C-radiolabeled substance (either orally or by injection) and is immediately placed in the metabolic chamber. The dose should contain 10^5 to 10^6 cpm of radiolabeled substrate. At appropriate intervals 5 ml samples are removed from the gas bubbler using the same procedure employed to determine the background level of radioactivity. In each case, 5 ml of fresh solution is used to replace that removed. For many rapidly metabolized substances it is convenient to take samples every 15–30 min. If intervals longer than 30 min are employed, it may be necessary to replace the entire 20 ml of CO_2 trapping solution at each interval in order to avoid exhausting its trapping capacity.

Each 5 ml sample is placed in a liquid scintillation counting vial and mixed with 10 ml of suitable cocktail. Contained radioactivity is determined by liquid scintillation counting. Because the CO_2 trapping solution is alkaline, quenching and chemoluminescence are potential problems; they can be minimized by using premium grades of liquid scintillation cocktail and by counting in a refrigerated instrument. Since some quenching appears inevitable, it is convenient to determine the amount of radioactivity administered to the experimental animal by counting a small aliquot of the administered solution in admixture with 5 ml of the CO_2 trapping solution. If this is done, the efficiency of counting is sufficiently uniform that no additional corrections or exact determination of efficiency are necessary.

[8] Volumes slightly less than 20.0 ml are usually attributable to evaporative losses of methanol on warm days. Larger losses may indicate that the air flow rate is too high and that the trapping solution is being carried over into the second trap.

Calculation of Results

If the entire CO_2 trapping solution is replaced at each time point, the procedure for calculating total formation of respiratory $^{14}CO_2$ is straightforward. For each time point the observed radioactivity is multiplied by 4 to reflect the fact that only 5 ml of 20 ml was taken for liquid scintillation counting; total formation of respiratory $^{14}CO_2$ at any time point is calculated by summing the total radioactivity collected at that time point with that collected at all previous time points. If the alternative procedure is used in which only 5 ml of fresh trapping solution is added at each time point, then total formation of respiratory $^{14}CO_2$ (cpm_t) in the period leading up to and including that time point is calculated as shown in Eq. 1.[9]

$$cpm_t = (cpm_{obs} \times 4) + cpm_r \quad (1)$$

where cpm_{obs} is cpm observed in the 5 ml sample removed at the time point of interest and cpm_r is radioactivity removed in all previous determinations (i.e., the sum of cpm_{obs} for all previous time points). It is often convenient to express total respiratory $^{14}CO_2$ as a percentage of the total ^{14}C administered to the animal. A graph showing "percentage of administered ^{14}C compound recovered as $^{14}CO_2$" as a function of time is then a continuously increasing curve. Compounds which are fairly rapidly catabolized yield curves which generally plateau within 6 hr; percentage recoveries often reach 80–90% of the administered dose.

Cautionary Notes and Suggestions

The apparatus and procedures described permit convenient, accurate, and reproducible assessment of $^{14}CO_2$ formation in intact animals. Poorly reproducible results or unusually shaped $^{14}CO_2$ formation curves can generally be attributed to one or more of the causes listed below.

Radiolabeled compounds can be administered to mice or rats orally or by subcutaneous, intraperitoneal, intramuscular, or intravenous injection. Only the last, if done without extravasation, assures essentially instantaneous distribution of the injected material throughout the extracellular space. Some of the other routes of administration are technically simple but absorption of the radiolabeled material into the blood stream can require several minutes to hours. To minimize difficulties, injected volumes should be large enough to measure accurately but small enough not to distort or overfill the space injected into. For mice volumes of 0.1 to 0.3 ml are easily given subcutaneously; volumes of 0.1 to 1 ml can be

[9] Hand calculators are easily programmed to store cpm_r data and to then perform this repetitive calculation.

given intraperitoneally. For convenience and reproducibility, these are preferred routes of administration. Intramuscular injections to mice are difficult; volumes should be small. Oral doses up to 0.5 ml are possible but have little to recommend them unless oral pharmacokinetics are of particular interest; rates of absorption are often slow and variable. It should be remembered that compounds given orally or intraperitoneally will be delivered into the portal system and liver preferentially. With all injections it is important to verify that the solution administered has not leaked out at the needle puncture. In our experience poor injections are the most common cause of irreproducible results; both the rate of $^{14}CO_2$ formation and the final percent recovery can be affected.

The rate and constancy of air flow through the metabolic chamber and gas bubblers can substantially affect the apparent rate of $^{14}CO_2$ formation. As noted, the air flow must be sufficiently rapid to effectively sweep CO_2 from the chamber but still be slow enough to be efficiently scrubbed by the gas bubbler. If the rate of air flow drops (e.g., through operator inattention) and is later returned to normal, the $^{14}CO_2$ formation curve generally shows an abrupt jump upward representing the sudden trapping of $^{14}CO_2$ accumulated in the chamber. With large chambers such as those used for large rats, air flow must be quite high to assure adequate $^{14}CO_2$ transport. Larger gas bubblers are then required to accommodate the high flow rates. The volumes of CO_2 trapping solution must also be increased to prevent saturation. It is noted that the second gas trap should not accumulate any $^{14}CO_2$; if it does, then either the air flow is too high or the trapping solution in the first bubbler has been saturated. The adequacy of air flow can be assessed approximately by placing a lighted cigarette in the metabolic chamber and determining that the smoke does not accumulate but exits quickly. The entire system can be tested by injecting $NaH^{14}CO_3$ (1 mmol/kg body weight) into a mouse and determining that $>90\%$ of the administered dose is recovered as $^{14}CO_2$ within 1 hr.

Finally it should be remembered that administered compounds will be catabolized to CO_2 in competition with one or more endogenous substrates which use the same enzyme systems. Nutritional status of the animals and the quantity of compound given can thus greatly affect the rate of $^{14}CO_2$ formation. For determinations of an animal's maximum catabolic capacity for a particular compound it is generally best to use fasted animals and to give a relatively large molar dose of radiolabeled compound (doses of 1–5 mmol/kg body weight are appropriate for several amino acids). In general, it is necessary to test several doses to determine the amount necessary to saturate the pathway of interest; when multiple catabolic pathways are involved, it may be possible to probe them individually by using radiolabeled substrate analogs which are metabolized only by the pathway of interest.[3-5]

Acknowledgments

Irma T. Hirschl Career Scientist. Work in the author's laboratory was supported by Grants AM26912 and GM32907 from the National Institutes of Health; studies in collaboration with Dr. A. Meister were also supported by Grant AM12034.

[57] Metabolism *in Vivo* of 5-Oxo-L-proline and Its Inhibition by Analogs of 5-Oxo-L-proline

By TINGSEN HSU and ALTON MEISTER

Introduction

5-Oxo-L-proline is rapidly metabolized *in vivo*. This can be shown by injecting 5-oxo-L-[U-^{14}C]proline into mice and collecting the respiratory $^{14}CO_2$.[1] In such experiments, about 50–70% of the injected radioactivity can usually be recovered in the respiratory carbon dioxide within about 2 hr. It appears that the limiting step in the conversion of 5-oxoproline carbon to carbon dioxide is that catalyzed by 5-oxoprolinase, which converts 5-oxo-L-proline to L-glutamate. The rate of appearance of $^{14}CO_2$ after administration of L-[U-^{14}C]glutamate is significantly more rapid than found after giving 5-oxo-L-[U-^{14}C]proline. The effects of various inhibitors of 5-oxoprolinase can be estimated by injecting them into mice together with 5-oxo-L-[U-^{14}C]proline and measuring the rate of $^{14}CO_2$ formation. It is necessary as a control in such studies to also determine the effect of the inhibitors on the metabolism of glutamate. The 5-oxoprolinase inhibitors thus far studied have little effect on the rate of $^{14}CO_2$ formation from L-[U-^{14}C]glutamate, but they have significant effects on the rate of $^{14}CO_2$ formation from 5-oxo-L-[U-^{14}C]proline. The compounds known to inhibit 5-oxoprolinase *in vitro* include L-2-imidazolidone-4-carboxylate,[1] L-2-oxothiazolidine-4-carboxylate,[2] and 3-methyl-5-oxo-L-proline.[3]

[1] P. Van Der Werf, R. A. Stephani, M. Orlowski, and A. Meister, *Proc. Natl. Acad. Sci. U.S.A.* **70,** 759 (1973).

[2] J. M. Williamson and A. Meister, *Proc. Natl. Acad. Sci. U.S.A.* **78,** 936 (1981).

[3] O. W. Griffith and A. Meister, *Proc. Natl. Acad. Sci. U.S.A.* **74,** 3330 (1977).

Copyright © 1985 by Academic Press, Inc.
All rights of reproduction in any form reserved.

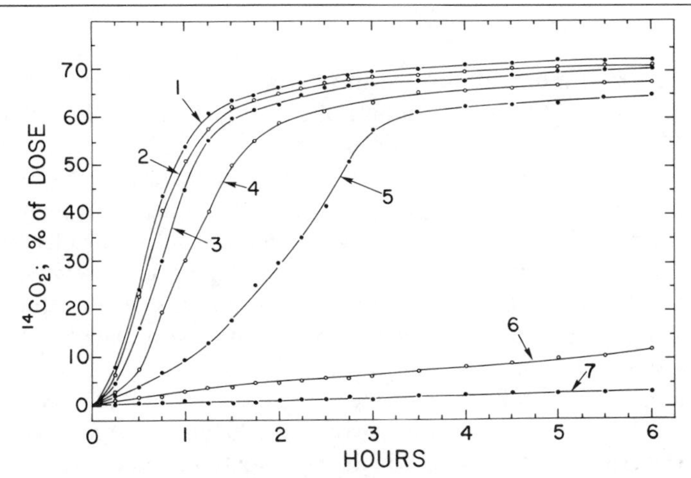

FIG. 1. Metabolism of 5-oxo-L-proline. The mice (fasted overnight) were injected subcutaneously with 5-oxo-L-[U-^{14}C]proline (Curve 1, 2 μmol; Curve 2, 10 μmol; Curve 3, 20 μmol; Curve 4, 40 μmol; Curve 5, 80 μmol; Curve 6, 200 μmol; Curve 7, 2 mmol), and the formation of respiratory $^{14}CO_2$ was determined.

Materials

Mice, NCS strain, male, 6–7 weeks old; 30 g
5-Oxo-L-[U-^{14}C]proline
L-[U-^{14}C]Glutamic acid
L-2-Oxothiazolidine-4-carboxylic acid
L-2-Imidazolidone-4-carboxylic acid
β-Methylglutamic acid (mixture of 4 isomers)
3-Methyl-5-oxoproline (mixture of 4 isomers)
Note: All solutions are adjusted to pH 7.4 by addition of NaOH

Methods

The mice are fasted overnight and injected subcutaneously with a solution in 0.15 *M* sodium chloride containing 5-oxo-L-[U-^{14}C]proline (1.2×10^6 cpm) and other compounds as indicated. Each mouse is placed in a metabolic chamber and the respiratory $^{14}CO_2$ is collected by the procedure given elsewhere in this volume.[4] In the experiments described below, the curves are based on the means of closely agreeing data obtained in experiments on 2 or 3 mice.

[4] O. W. Griffith, this volume [56].

Results and Comments

In the experiments described in Fig. 1, the mice were injected with different amounts (2 μmol–2 mmol) of 5-oxo-L-[U-^{14}C]proline containing 1.2×10^6 cpm of ^{14}C. In the studies in which doses of 2, 10, 20, 40, and 80 μmol were given, the recovery of $^{14}CO_2$ was equivalent to 58–70% of the administered dose after 3 hr, and 65–72% after 6 hr. With higher doses (0.2 and 2 mmol), the recovery of the administered radioactivity in the respiratory carbon dioxide was much lower. However, a relatively large amount of 5-oxoproline was metabolized during this period; thus after a dose of 2 mmol, about 240 μmol of 5-oxoproline was metabolized within 6 hr. These studies indicate that the mouse has substantial capacity for utilization of 5-oxoproline. Since the percent conversion of 5-oxo[^{14}C]proline to $^{14}CO_2$ was similar after doses of 2, 10, 20, and 40 μmol, a dose of 10 μmol was used in studies in which the effects of inhibitors were measured.

Figure 2 gives the results of experiments in which the mice were given subcutaneous injections of 5-oxo-L-[^{14}C]proline (10 μmol; 1.2×10^6 cpm) and an inhibitor of 5-oxoprolinase. The findings indicate substantial inhibition of respiratory $^{14}CO_2$ formation by L-2-imidazolidone-4-carboxylate, L-2-oxothiazolidine-4-carboxylate, and 3-methyl-5-oxoproline. In similar studies it was found that these compounds (and also β-methylglutamate)

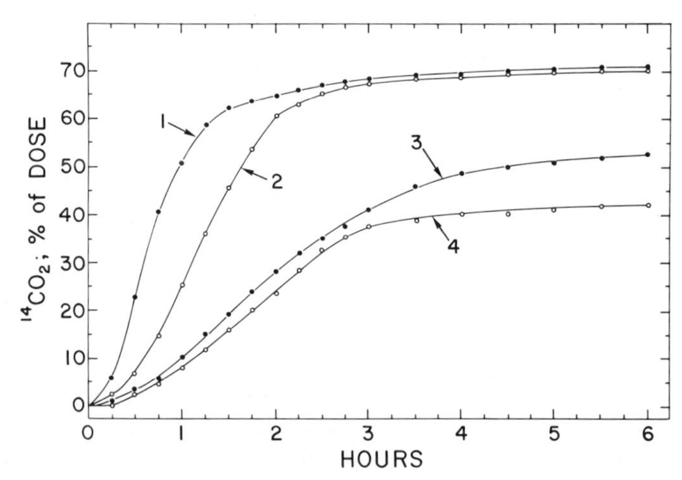

FIG. 2. Inhibition of 5-oxoproline metabolism. The mice were injected with 5-oxo-L-[U-^{14}C]proline (10 μmol) 10 min after subcutaneous injection of saline (Curve 1), L-2-imidazolidone-4-carboxylate (60 μmol) (Curve 2), L-2-oxothiazolidine-4-carboxylate (60 μmol) (Curve 3), and DL-3-methyl 5-oxoproline (mixture of 4 isomers: 240 μmol) (Curve 4). The formation of respiratory $^{14}CO_2$ was determined.

did not inhibit formation of respiratory $^{14}CO_2$ from L-[U-^{14}C]glutamate to greater than about 10–15%.

The inhibitors used in the experiments described in Fig. 2 have been shown in *in vitro* studies with the purified rat kidney enzyme[5] to be effective competitive inhibitors of 5-oxo-L-prolinase. It is of interest that L-2-imidazolidone-4-carboxylate is not cleaved by the action of the enzyme although its presence is accompanied by substantial cleavage of ATP to ADP and P_i.[1,6,7] On the other hand, L-2-oxothiazolidine-4-carboxylate is an excellent substrate of the enzyme and is cleaved, presumably via *S*-carboxycysteine to L-cysteine.[2,7] The threo and erythro isomers of 3-methyl-5-oxo-DL-proline are relatively poor substrates; the *erythro* isomer is a more potent inhibitor than the *threo* isomer.[7]

[5] A. Meister, O. W. Griffith, and J. M. Williamson, this volume [53].
[6] P. Van Der Werf, O. Griffith, and A. Meister, *J. Biol. Chem.* **250,** 6686 (1975).
[7] J. M. Williamson and A. Meister, *J. Biol. Chem.* **257,** 12,039 (1982).

[58] Microvillus Membrane Peptidases That Catalyze Hydrolysis of Cysteinylglycine and Its Derivatives

By Suresh S. Tate

Introduction

The metabolism of glutathione in animal cells involves intracellular as well as extracellular components.[1,2] Its biosynthesis is mediated by enzymes present in the cytosol, whereas its degradation is initiated by γ-glutamyl transpeptidase, an enzyme associated with the external surface of cell membranes[1-4] (see also Tate and Meister, this volume [50]). Locations of high transpeptidase activity include epithelial cells, such as those of the renal proximal tubule, jejunum, and epididymis which are characterized by the presence of numerous, closely packed microvilli on their apical (lumenal) surface.

The role of transpeptidase in catalyzing the initial step (the removal of γ-glutamyl moiety) in the degradation of glutathione and its S-derivatives

[1] A. Meister, *Curr. Top. Cell. Regul.* **18,** 21 (1981).
[2] A. Meister and M. E. Anderson, *Annu. Rev. Biochem.* **52,** 711 (1983).
[3] S. S. Tate, *in* "Enzymatic Basis of Detoxication" (W. B. Jakoby, ed.), Vol. 2, p. 95. Academic Press, New York, 1980.
[4] S. S. Tate and A. Meister, *Mol. Cell. Biochem.* **39,** 357 (1981).

Copyright © 1985 by Academic Press, Inc.
All rights of reproduction in any form reserved.

is well established. The cysteinylglycine and its S-derivatives produced during glutathione catabolism must also be hydrolyzed to their component amino acids. A number of reports appeared assigning this role to aminopeptidase M,[5-7] an enzyme concentrated in the renal and intestinal microvilli.[8] This enzyme displays a broad substrate specificity hydrolyzing a variety of dipeptides as well as larger peptides. Our studies on the utilization of glutathione S-derivatives by microvillus membranes from rat kidney, jejunum, and epididymis indicated that another dipeptide-hydrolyzing activity, not ascribable to aminopeptidase M, might play a far greater role in the hydrolysis of cysteinylglycine and its S-derivatives.[9,10] Thus, the epididymal membranes, which lack aminopeptidase M, nevertheless, effectively converted glutathione S-derivatives to its constituent amino acids. This chapter describes the purification of the dipeptidase and aminopeptidase M from rat renal microvillus membranes (brush border membranes) and compares the two enzymes especially with regard to their ability to hydrolyze cysteinylglycine and its S-derivatives.

Assay Methods

Dipeptidase

The ninhydrin-cyanide procedure of Matheson and Tattrie[11] as modified by Patterson[12] is used to determine the amino acids released from most dipeptides (including L-Cys-Gly and its derivatives). This method, however, cannot be used when the substrates are dipeptides containing NH_2-terminal glycine or serine. The difference in absorbance at 220 to 235 nm between the dipeptides and their constituent amino acids can be utilized to follow the hydrolysis of such peptides.[12] Other procedures that have been used include an assay using [^{3}H]glycine-labeled L-Cys-Gly and mercurial-agarose to separate the substrate from the labeled product, and the use of L-amino acid oxidase to determine the released amino acids.[13]

[5] R. P. Hughey, B. B. Rankin, J. S. Elce, and N. P. Curthoys, *Arch. Biochem. Biophys.* **186,** 211 (1978).

[6] B. B. Rankin, T. M. McIntyre, and N. P. Curthoys, *Biochem. Biophys. Res. Commun.* **96,** 991 (1980).

[7] K. Okajima, M. Inoue, and Y. Morino, *Biochim. Biophys. Acta* **675,** 379 (1981).

[8] A. J. Kenny and A. G. Booth, *Essays Biochem.* **14,** 1 (1978).

[9] E. M. Kozak and S. S. Tate, *J. Biol. Chem.* **257,** 6322 (1982).

[10] E. M. Kozak, Ph.D. Thesis, Cornell University, Ithaca, New York (1982).

[11] A. T. Matheson and B. L. Tattrie, *Can. J. Biochem.* **42,** 95 (1964).

[12] E. K. Patterson, this series, Vol. 45B, p. 377.

[13] T. McIntyre and N. P. Curthoys, *J. Biol. Chem.* **257,** 11915 (1982).

The Ninhydrin-Cyanide Method

Reagents

Buffer: 0.1 *M* Tris–HCl, pH 8.0

Dipeptide substrate, 0.01 *M* in Tris–HCl buffer. The solution can be stored frozen at −15° for several weeks. L-Ala-Gly was used to follow dipeptidase activity during purification of the enzymes

HCl, 0.1 *N* in 95% ethanol

Sodium citrate buffer, 0.2 *M*, pH 5.7

Ninhydrin-cyanide reagent: The following stock solutions are made monthly and stored at 4°: 5 g of ninhydrin dissolved in 100 ml of methyl cellosolve (ethylene glycol monomethyl ether) and 0.01 *M* KCN in water. The reagent is prepared by first diluting 2 ml of 0.01 *M* KCN to 100 ml with methyl cellosolve and then adding to it 20 ml of the stock ninhydrin solution. A red color initially develops which then fades to a yellow color. The reagent is stable for about 2 weeks when stored at 4°.

Procedure. Additions are made to the bottom of test tubes as follows: 25 μl of the dipeptide solution (final concentration, 2.5 m*M*) and 75 μl of Tris–HCl buffer (the volume is adjusted to allow for that of the enzyme solution added; the final volume of the reaction mixture is 100 μl). The solution is brought to 37° and the reaction is initiated by adding a suitable amount of enzyme. The following controls are run in parallel: (1) 100 μl of Tris–HCl buffer, and (2) the complete reaction mixture to which the enzyme solution is added after stopping the reaction. After appropriate incubation at 37°, the reaction is terminated by the addition of 50 μl of 0.1 *N* HCl in ethanol, and in the following order are added: 1 ml of H_2O, 0.5 ml of citrate buffer, and 1 ml of ninhydrin-cyanide reagent. The solutions are mixed thoroughly, the tubes are capped with glass marbles, and heated in a boiling water bath for 7.5 min. After cooling in an ice bath, 2.5 ml of 60% ethanol is added, the contents mixed thoroughly, and the absorbance at 570 nm read in a spectrophotometer. A standard curve is obtained by making mixtures of 0.01 *M* dipeptide and 0.01 *M* solutions of its constituent amino acids to simulate various degrees of hydrolysis. Under the conditions described here, linearity in the enzymic reaction was obtained when the extent of hydrolysis of the dipeptide did not exceed 40%.

Other Methods

The hydrolysis of dipeptides (as well as larger peptides) can also be followed by determining the amino acids produced with an amino acid analyzer (we have routinely used a Durrum model D500 analyzer). For this, the reaction mixtures were made up as described in the ninhydrin-cyanide procedure. After incubation at 37°, the reactions were terminated by the addition of 10 μl of 50% sulfosalicylic acid. An aliquot of the protein-free supernatant was taken for amino acid analysis.

A spectrophotometric method that depends on the difference in absorbance in the low ultraviolet between dipeptides and their constituent amino acids can be used when the concentration of enzyme is low such that its absorbance does not interfere with the absorbancy of the dipeptide. This method is useful for kinetic studies with purified enzymes. In this method, the additions are made to a spectrophotometer cuvette (semimicro; 1 cm light path) as follows: 0.25 ml of 0.01 *M* dipeptide (final concentration, 2.5 m*M*), 0.5 ml of 0.1 *M* Tris–HCl buffer (pH 8.0), and 0.25 ml of water (the volume is adjusted to allow for that of the enzyme solution; the final volume of the reaction mixture is 1.0 ml). The solution is brought to 37° in a spectrophotometer equipped with a thermostatted cuvette holder [we routinely use a Cary Model 219 spectrophotometer equipped with a range selector with eight full-scale absorbance options (ranging from 0.01 to 2.0 absorbance units)]. Reaction is initiated by adding a suitable amount of enzyme and the decrease in absorbance at 220 nm or 235 nm is recorded. The change in absorbance per minute is converted to the rate of hydrolysis from a standard curve prepared by reading the absorbance of mixtures which contain the dipeptide and its constituent amino acids to simulate various degrees of hydrolysis.

In all methods, the activity is expressed as micromoles of substrate utilized or products formed per minute (units) and the specific activity as the units per milligram of protein.

Aminopeptidase M

The aminopeptidase M activity is most conveniently followed by spectrophotometrically determining the release of *p*-nitroaniline from the *p*-nitroanilide derivatives of neutral amino acids.[9]

Reagents

Buffer: 0.1 *M* Tris–HCl, pH 8.0

L-Leucyl-*p*-nitroanilide: 5 m*M* in methyl cellosolve. This solution can be stored at −15° for up to 3 months

Procedure. Additions are made to a spectrophotometer cuvette as follows: 0.1 ml of L-leucyl-*p*-nitroanilide (final concentration, 0.5 m*M*) and 0.9 ml of Tris-HCl buffer (the volume is adjusted to allow for that of the enzyme solution added; final volume, 1.0 ml). The solution is brought to 37° in a spectrophotometer equipped with a thermostatted cuvette holder and the reaction is initiated by adding an appropriate amount of enzyme. The release of *p*-nitroaniline is recorded at 410 nm ($E = 8800\ M^{-1}\ cm^{-1}$). When several assays are to be performed, a mixture containing L-leucyl-*p*-nitroanilide and Tris–HCl buffer, 1 : 9 (v/v) is prepared and 1 ml of this solution is used per assay. This mixture can be stored at 4° for 5 days.

Enzyme Purification. Rat kidney is the tissue of choice for the purification of glutathione-degrading enzymes because of the high concentration of these enzymes in this organ.[3,9] Aminopeptidase M and γ-glutamyl transpeptidase, together, represent approximately 10% of the total proteins in rat renal brush borders. Kidneys from male Sprague–Dawley rats of 250 to 350 g in weight have been used either fresh or after storage at −20° for as long as several months. Since both the dipeptidase discussed here and aminopeptidase M (as well as γ-glutamyl transpeptidase) are bound to the brush border membranes, the enzymes must be rendered into soluble forms. The solubilization methods include treatment of membranes with either detergents or proteinases. The hydrolases of the brush border membrane are amphipathic, intrinsic membrane proteins which consist of a hydrophilic moiety anchored to the membrane by a hydrophobic segment.[14] Detergents solubilize the entire molecule, whereas proteinases cleave the hydrophilic portion containing the active site off the membrane-bound anchor. The catalytic properties of the detergent- and the proteinase-solubilized enzymes have been, in most cases, shown to be similar. The purification procedure described below is for the simultaneous purification of the papain-solubilized forms of the dipeptidase and aminopeptidase M. The method also yields highly purified γ-glutamyl transpeptidase (see Fig. 1).

Step 1. Microvillus Membranes. These are isolated essentially as described by Malathi *et al.*[15] Fresh or frozen kidneys are homogenized in 15 volumes (v/w) of 10 m*M* Tris–HCl (pH 7.5) containing 5 m*M* D-mannitol. For homogenization, a Potter-Elvehjem homogenizer equipped with a Teflon pestle (10 strokes) is used followed by homogenization in an all-glass Dounce homogenizer with a loose-fitting pestle (10 strokes). All procedures are carried out at 4°. To the homogenate, 1 *M* $CaCl_2$ solution is added to achieve a final $CaCl_2$ concentration of 10 m*M*, and the mixture is placed on ice with gentle stirring for 10 min. The homogenate is then centrifuged at 3000 *g* for 20 min. The supernatant fluid is carefully de-

[14] A. J. Kenny and S. Maroux, *Physiol. Rev.* **62,** 91 (1982).

[15] P. Malathi, H. Preiser, P. Fairclough, P. Mallett, and R. K. Crane, *Biochim. Biophys. Acta* **554,** 259 (1979).

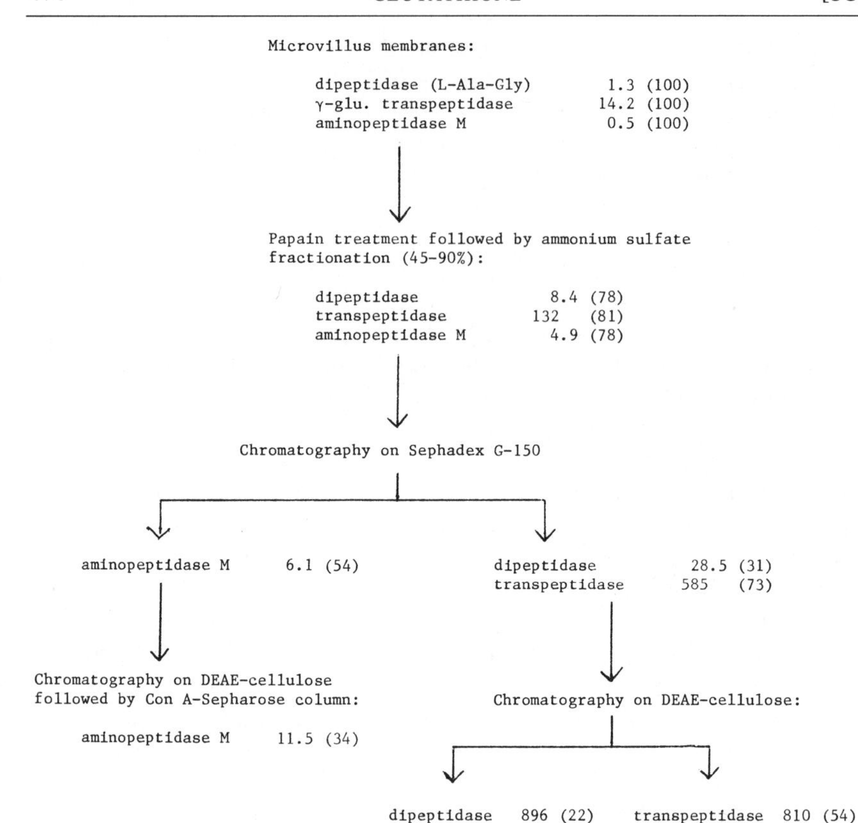

FIG. 1. Outline of the protocol used for the purification of dipeptidase, γ-glutamyl transpeptidase, and aminopeptidase M from rat kidney microvillus membranes. The renal microvillus membranes (total protein, 2185 mg) were isolated from 210 g of rat kidney. The dipeptidase, aminopeptidase M, and transpeptidase activities were assayed using L-Ala-Gly, L-Leu-*p*-nitroanilide, and L-γ-Glu-*p*-nitroanilide (in presence of Gly-Gly), respectively.[9] The specific activities of the peptidases (units/mg of protein) are given next to the enzyme and the values in parentheses represent the yield (%) at each step. Adapted from Kozak.[10]

canted and centrifuged at 39,000 *g* for 20 min. The pellet thus obtained is suspended in the Tris-mannitol buffer (one-half the volume of the original homogenate) using the Dounce homogenizer with a loose pestle. The suspension is again centrifuged at 39,000 *g* for 20 min and the pellet is suspended (using the Dounce homogenizer) in 5 m*M* Tris–HCl buffer (pH 7.4) containing 50 m*M* NaCl [the volume used is about equal to the weight of the kidneys used (v/w)].

Step 2. Papain Treatment and Ammonium Sulfate Fractionation. 2-Mercaptoethanol is added to the suspension to a final concentration of 10 m*M*. Papain (28 units/mg; Sigma) is dissolved in Tris-NaCl buffer containing 20 m*M* 2-mercaptoethanol to give a solution containing 10 mg papain/ml and then incubated at 25° for 30 min. The papain solution is added to the membrane suspension (final concentration, 1 mg papain/10 mg of total membrane proteins) and the mixture is stirred gently at 37° for 2 hr. The suspension is centrifuged at 39,000 *g* for 20 min. The supernatant (which contains greater than 90% of the membrane peptidases) is treated with $(NH_4)_2SO_4$ to give 45% of saturation (27.7 g/100 ml) and the precipitate removed by centrifugation (18,000 *g* for 30 min). The supernatant is treated again with $(NH_4)_2SO_4$ to give 90% of saturation (34 g/100 ml) and the precipitate obtained by centrifugation is dissolved in the minimal volume [about one-twentieth the weight of kidneys (v/w)] of 0.05 *M* Tris–HCl buffer (pH 8.0).

Step 3. Chromatography on Sephadex G-150. The 45–90% $(NH_4)_2SO_4$ fraction is chromatographed on a Sephadex G-150 column (2.5 × 100 cm colume for a preparation from 50 to 200 g of kidneys), equilibrated and eluted with 0.05 *M* Tris–HCl buffer (pH 8.0). As shown in a typical experiment depicted in Fig. 2A, aminopeptidase M, dipeptidase, and γ-glutamyl transpeptidase activities separate on the Sephadex column. Two peaks of L-Ala-Gly-hydrolyzing activity are obtained. The peak emerging near the excluded volume of the column (peak 1) coincides with the aminopeptidase M activity. Peak 2, associated with a protein of M_r approximately 105,000, represents the dipeptidase.

Step 4. Chromatography on DEAE-Cellulose. The Sephadex G-150 fractions containing dipeptidase (peak 2) and γ-glutamyl transpeptidase are pooled and dialyzed 18 hr against 30 volumes of 5 m*M* Tris–HCl buffer (pH 8.0). The solution is then lyophilized, the residue redissolved in a minimal volume of water (less than 5 ml), and dialyzed 18 hr against 100 volumes of 0.025 *M* Tris–HCl buffer (pH 8.0) containing 0.05 *M* NaCl. The dialyzed solution is applied to a DEAE-cellulose column (Whatman DE 52, microgranular, 1 × 21 cm), previously equilibrated with the Tris-NaCl buffer. The column is washed with about 4 column volumes of the buffer. Transpeptidase activity emerges in the initial wash (specific activity, 800 units/mg of protein) (Fig. 3). The column is then developed with a linear gradient established between 200 ml each of 0.025 *M* Tris–HCl buffer containing 0.05 *M* NaCl and 0.025 *M* Tris–HCl containing 0.2 *M* NaCl. Dipeptidase activity elutes between 0.13 and 0.15 *M* NaCl (Fig. 3). The active fractions are pooled, dialyzed against 25 volumes of 2 m*M* Tris–HCl (pH 8.0), and then lyophilized. The residue is dissolved in 1 ml of water and dialyzed against 100 volumes of 0.01 *M* Tris–HCl buffer (pH 8.0). (A similar procedure is followed for the fractions containing transpeptidase.) A summary of a typical purification of renal dipeptidase is

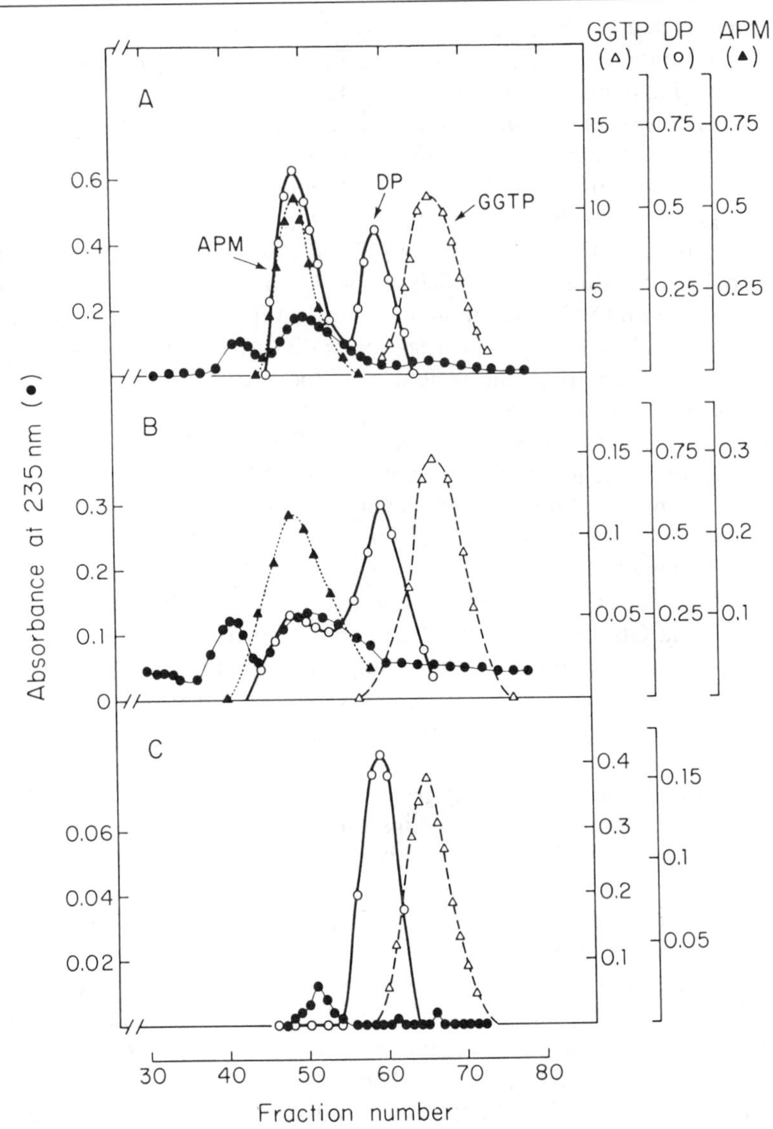

FIG. 2. Gel filtration profiles of microvillus membrane peptidases. Membranes from rat kidney (A), jejunum (B), and epididymis (C) were treated with papain and the solubilized proteins chromatographed on a Sephadex G-150 column (1.5 × 90 cm; 1.4 ml/fraction), equilibrated and eluted with 0.05 *M* Tris–HCl buffer (pH 8.0). GGTP (△), γ-glutamyl transpeptidase; DP (○), dipeptidase activity determined with L-Ala-Gly; and APM (▲), aminopeptidase M (L-Leu-*p*-nitroanilide hydrolyzing). From Kozak and Tate.[9]

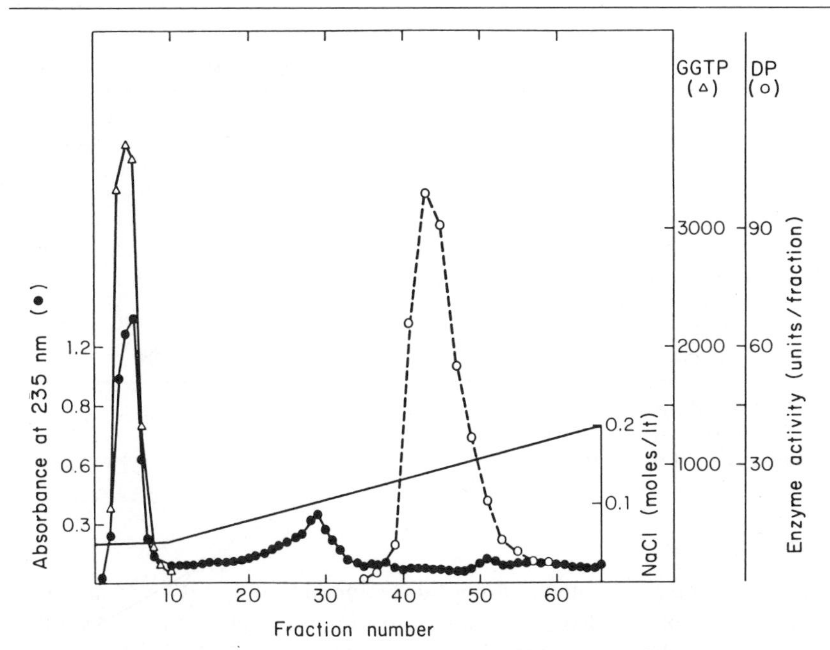

FIG. 3. DEAE-Cellulose chromatography of rat kidney dipeptidase and γ-glutamyl transpeptidase (Table I, step 4). DP, Dipeptidase; GGTP, γ-glutamyl transpeptidase. From Kozak.[10]

given in Table I. The flow diagram in Fig. 1 gives the purification and yield of γ-glutamyl transpeptidase.

Purification of Aminopeptidase M. The Sephadex G-150 fractions from Step 3 above, containing the aminopeptidase M activity (L-Leu-*p*-nitroanilide hydrolyzing) are pooled and diluted with an equal volume of water. The solution is then applied to a DEAE-cellulose column (1 × 21 cm) previously equilibrated with 0.025 *M* Tris–HCl buffer (pH 8.0). The column is washed with 100 ml of the same buffer and then developed with a linear gradient established between 150 ml of the Tris–HCl buffer and 150 ml of the buffer containing 0.25 *M* NaCl. The aminopeptidase M activity elutes between 0.1 and 0.13 *M* NaCl. The active fractions are pooled and applied to a Con A-Sepharose column (concanavalin A covalently bound to Sepharose 4B, Pharmacia; 1 × 15 cm), previously equilibrated with 0.025 *M* Tris–HCl buffer (pH 8.0) containing 0.1 *M* NaCl. The bound enzyme is eluted with a linear gradient established between 60 ml of the Tris–NaCl buffer and 60 ml of the buffer containing α-methylmannoside (25 mg/ml). The active fractions were pooled and dialyzed against 100 volumes of 0.01 *M* Tris–HCl buffer (pH 8.0). Figure 1 summarizes the purification and yield of aminopeptidase M.

TABLE I
PURIFICATION OF RAT RENAL MICROVILLUS MEMBRANE DIPEPTIDASE[a]

Step	Total protein (mg)	Dipeptidase activity[b]		
		Total (units)	Specific (units/mg)	Yield (%)
1. Microvillus membranes[c]	2185	2932	1.3	[100]
2. Papain treatment followed by $(NH_4)_2SO_4$ fractionation (45–90%)	272	2288	8.4	78
3. Chromatography on Sephadex G-150	31.5	897	28.5	31
4. Chromatography on DEAE-cellulose	0.73	650	896	22

[a] From Kozak and Tate.[9]
[b] Dipeptidase activity determined with L-Ala-Gly.
[c] From 210 g of frozen rat kidney.

Molecular Properties

Some of the properties of dipeptidase and aminopeptidase M are summarized in Table II. Both enzymes are dimers, each composed of two apparently similar glycopeptide subunits. The subunit M_r of the papain-solubilized dipeptidase and aminopeptidase M are approximately 50,000 and 130,000, respectively. The detergent (Triton X-100)-solubilized form of aminopeptidase M from rat kidney has also been purified[16] and the subunit M_r of this has been reported to be about 150,000. The difference in the subunit M_r of the enzyme purified following the two solubilization techniques presumably reflects the size of the membrane anchoring domain of the native enzyme. The detergent-solubilized form of the dipeptidase has not been characterized as yet.

Both peptidases are zinc metalloproteins.[9,17] Thus, incubation of the two enzymes with *o*-phenanthroline results in loss of activity which can be restored by zinc. Incubation of dipeptidase in phosphate buffers also results in a slow loss of activity.[10] Full protection against this loss is afforded by $ZnSo_4$. Phosphate has virtually no effect on aminopeptidase M activity. Aminopeptidase M contains 2 mol of Zn/mol of enzyme, whereas the dipeptidase contains 4 mol of Zn/mol of enzyme. It is likely that zinc in dipeptidase plays a structural as well as a catalytic role. The pH optima for the dipeptidase- and aminopeptidase M-catalyzed hydrolysis of L-Ala-Gly are, respectively, pH 8.0–8.5 and 7.5–8.0.

[16] T. M. McIntyre and N. P. Curthoys, *Int. J. Biochem.* **12,** 545 (1980).
[17] A. J. Kenny, *in* "Proteinases in Mammalian Cells and Tissues" (A. J. Barrett, ed.), p. 311. Am. Elsevier, New York, 1977.

TABLE II
PROPERTIES OF PAPAIN-SOLUBILIZED RAT RENAL PEPTIDASES[a]

	Dipeptidase	Aminopeptidase M
M_r (approximate)	105,000	260,000
Subunit M_r (approximate)	50,000	130,000
Zn (mol/mol of enzyme)	4	2
Carbohydrate	+	+
Substrates		
Dipeptides	+	+
Oligopeptides	−	+
Inhibitors		
o-Phenanthroline	+	+
Bestatin	−	+
Thiol compounds	+	−
Phosphate	+	−

[a] Adapted from Kozak.[10]

A number of techniques (immunocytochemical, use of impermeant substrates, selective solubilization from the external surface of microvillus membranes using proteinases, etc.) indicate that, like γ-glutamyl transpeptidase, aminopeptidase M and dipeptidase are also located on the external (lumenal) surface of the microvillus membranes.[7,9,18,19]

A hog kidney participate dipeptidase has been characterized.[20] Although its subcellular localization is not clear, the molecular and catalytic properties suggest similarities to the rat kidney enzyme. McIntyre and Curthoys[13] have also characterized the rat renal dipeptidase and studied its role in glutathione catabolism. Their results and conclusions are essentially the same as reported earlier by us.[9]

Aminopeptidase M from a variety of tissues and species has been extensively studied. Thus, the renal and intestinal enzyme has been purified from the hog, rabbit, and rat (see literature cited in references 10, 14, and 17). The molecular and catalytic properties of the rat kidney preparation described in this chapter are in accord with published reports.

Catalytic Properties

The specific activity of dipeptidase with L-Ala-Gly (2.5 mM) as the substrate is about 70-fold higher than that of aminopeptidase M. The activity of the dipeptidase and aminopeptidase M toward a variety of

[18] B. Tsao and N. P. Curthoys, *J. Biol. Chem.* **255,** 7708 (1980).
[19] W. Haase, A. Schäfer, H. Murer, and R. Kinne, *Biochem. J.* **172,** 57 (1978).
[20] B. J. Campbell, this series, Vol. 19, p. 722.

TABLE III
SUBSTRATE SPECIFICITY OF RENAL MICROVILLUS MEMBRANE DIPEPTIDASE AND AMINOPEPTIDASE M[a]

Substrate	Relative activity[b]	
	Dipeptidase	Aminopeptidase M
L-Ala-Gly (1.0 m*M*)	100 (690)	100 (7.6)
L-α-Aminobutyryl-Gly	58	61
L-Cys-Gly	0[c]	26
L-Cystinyl-bis-Gly	38	1
S-Methyl-L-Cys-Gly	82	28
L-Ala-Gly (0.08 m*M*)	100 (117)	100 (0.8)
Leukotriene D_4	14	<0.1
L-Ala-Gly (2.5 m*M*)	100 (896)	100 (12.9)
L-Ala-L-Ala	74	160
L-Ala-L-Lys	11	262
Gly-Gly	58	27
Gly-L-Ala	30	88
Gly-D-Ala	198	0
L-Leu-Gly	72	48
L-Met-Gly	92	104
L-Phe-Gly	25	125
L-Met-L-Met	61	176
L-Met-L-Met-L-Met	0	324
L-Ala-L-Ala-L-Ala	0	911
L-Leu-*p*-nitroanilide	0	89
S-Benzyl-L-Cys-*p*-nitroanilide	0	9

[a] From Kozak and Tate.[9]

[b] Activities are expressed relative to L-Ala-Gly. The numbers in parentheses represent the specific activities (μmol of L-Ala-Gly hydrolyzed/min/mg of enzyme) of dipeptidase and aminopeptidase M at the concentration of L-Ala-Gly shown. Within each group of peptides, the activities are compared at the concentration shown next to L-Ala-Gly. The hydrolysis of *p*-nitroanilide derivatives was measured at 0.5 m*M* substrate concentration.

[c] L-Cys-Gly is a strong inhibitor of dipeptidase at these concentrations.

peptides (relative to L-Ala-Gly) is shown in Table III. Dipeptidase, in contrast to aminopeptidase M, exhibits no activity toward tripeptides and the *p*-nitroanilide derivatives of amino acids. Replacement of the Gly residue in L-Ala-Gly with larger amino acids markedly reduces the effectiveness of the peptide as a substrate for dipeptidase. On the other hand, its activity is relatively less affected when the NH_2-terminal L-Ala is replaced by amino acids such as Gly, L-Leu, L-Met, and L-Lys.

In general, the introduction of bulkier amino acid residues in position one or two of L-Ala-Gly has less effect on their ability to serve as substrates for aminopeptidase M. Indeed, dipeptides such as L-Ala-L-Ala, L-Ala-L-Lys, L-Met-Gly, L-Met-L-Met, and L-Phe-Gly were better substrates than L-Ala-Gly. Both the dipeptidase and aminopeptidase M exhibit virtually no activity toward D-Ala-Gly, L-Val-Gly, L-Asp-Gly, Pro-containing peptides, and glutathione. Interestingly, however, Gly-D-Ala is an excellent substrate for dipeptidase but is not cleaved by aminopeptidase M.

Significant differences emerge in the ability of dipeptidase and aminopeptidase M to hydrolyze L-Cys-Gly and its derivatives. Thus, the dipeptidase is several hundred-fold more efficient than aminopeptidase M in the hydrolysis of peptides such as *S*-methyl-L-Cys-Gly, a mercapturic acid precursor, L-cystinyl-bis-Gly, product of the action of γ-glutamyl transpeptidase on glutathione disulfide (L-Cys-Gly is also rapidly oxidized in solution to the disulfide[21]), and leukotriene D_4 (5-hydroxy-6*S*-cysteinylglycyl-7,9-*trans*-11,14-*cis*-eicosatetraenoic acid), a spasmogenic and vasoactive Cys-Gly derivative.[22,23] Thus, dipeptidase is about 270-, 2,600-, and 16,000-fold more active in the hydrolysis of *S*-methyl-L-Cys-Gly, L-cystinyl-bis-Gly, and leukotriene D_4, respectively, than aminopeptidase M. Of interest is the finding that L-Cys-Gly above 50 μM severely inhibits dipeptidase.[9] The peptide, however, serves a good substrate for dipeptidase at lower concentrations.[13] L-Cys-Gly, in contrast to L-cystinyl-bis-Gly, is hydrolyzed efficiently by aminopeptidase M.

The inhibition of dipeptidase by L-Cys-Gly reflects the ability of thiol compounds to inhibit this enzyme. Thus, in addition to L-Cys-Gly, thiols such as dithiothreitol, L and D isomers of pencillamine are potent inhibitors of dipeptidase.[9] Interestingly, glutathione is much less effective. The inhibition by the thiol compounds is reversible upon dilution or dialysis. Aminopeptidase M, in contrast to dipeptidase, is not significantly affected by the thiol compounds. On the other hand, bestatin ([(2*S*,3*R*)-3-amino-2-hydroxy-4-phenylbutanoyl]-L-leucine), a strong inhibitor of aminopeptidase M,[13] had no effect on dipeptidase activity.

Dipeptidase and Aminopeptidase M in Other Tissues

Subcellular fractionation studies show that, like other microvillus membrane markers, γ-glutamyl transpeptidase and aminopeptidase M are enriched in the rat renal and jejunal microvilli.[9] However, the epididymal

[21] O. W. Griffith and S. S. Tate, *J. Biol. Chem.* **255,** 5011 (1980).

[22] R. A. Lewis and K. F. Austen, *Nature* (*London*) **293,** 103 (1981).

[23] K. Bernström, L. Örning, and S. Hammarström, this series, Vol. 86, p. 38.

homogenates and the stereocilia (epididymal microvilli) exhibit virtually no aminopeptidase M activity. In contrast, the dipeptidase activity (L-Ala-Gly-hydrolyzing) in the microvillus membrane preparations from all three tissues in approximately the same and is enriched (except in the jejunum where significant dipeptidase activity is also present in the cytosol) to the same extent as the other microvillus membrane markers. Figure 2 shows the gel filtration profiles of the papain-solubilized peptidases from rat renal, jejunal, and epididymal microvillus membranes. Note the absence of aminopeptidase M in the epididymal membranes. The properties of γ-glutamyl transpeptidase and dipeptidase from the three tissues are similar. Also, renal and jujunal aminopeptidases M exhibit similarities.

It should be noted that virtually all of the L-Ala-Gly-hydrolyzing capacity of the microvillus membranes from the three rat tissues is accounted for by the dipeptidase and aminopeptidase M. Selective inhibition of dipeptidase and aminopeptidase M (by thiols and bestatin, respectively) allows one to estimate the relative contribution of the two enzymes in the hydrolysis of peptides such as L-Ala-Gly and L-Cys-Gly derivatives.[9,13] Such studies show that about 65, 50, and 95% of L-Ala-Gly- and *S*-methyl-L-Cys-Gly-hydrolyzing activities of the renal, jejunal, and epididymal membranes, respectively, are due to the dipeptidase.[9] Furthermore, dipeptidase is the major activity responsible for the hydrolysis of L-cystinyl-bis-Gly and leukotriene D_4.

[59] Glutathione Reductase

By INGER CARLBERG and BENGT MANNERVIK

Glutathione reductase (EC 1.6.4.2) is a flavoprotein catalyzing the NADPH-dependent reduction of glutathione disulfide (GSSG) to glutathione (GSH)[1]:

$$\text{NADPH} + \text{H}^+ + \text{GSSG} \rightarrow \text{NADP}^+ + 2\text{GSH}$$

The reaction is essential for the maintenance of glutathione levels. Glutathione has a major role as a reductant in oxidation–reduction processes, and also serves in detoxication and several other cellular functions of great importance.[2]

[1] E. Racker, *J. Biol. Chem.* **217,** 855 (1955).

[2] C. H. Williams, Jr., *in* "The Enzymes" (P. D. Boyer, ed.), 3rd ed., Vol. 13, p. 89. Academic Press, New York, 1976.

Copyright © 1985 by Academic Press, Inc.
All rights of reproduction in any form reserved.

Glutathione reductase has been isolated from a number of different organisms.

A purification method is described in this chapter which yields glutathione reductase free from the closely related flavoenzyme thioredoxin reductase.

Assay Method

Principle. The oxidation of NADPH is followed spectrophotometrically at 340 nm.[1]

Reagents

Potassium phosphate buffer 0.2 *M*, pH 7.0, containing 2 m*M* EDTA
NADPH, 2 m*M*, in 10 m*M* Tris–HCl (pH 7.0)
GSSG, 20 m*M*, in water

Procedure. To a 1-ml cuvette, 0.5 ml phosphate buffer at 30°, 50 μl NADPH, 50 μl GSSG, and a volume of deionized water giving a final total volume of 1 ml are added. The reaction is initiated by the addition of enzyme to the cuvette and the decrease in absorbance at 340 nm is followed at 30°.

Definition of Unit and Specific Activity

A unit of glutathione reductase activity is defined as the amount of enzyme that catalyzes the reduction of 1 μmol of NADPH per minute. Specific activity is expressed in units per milligram of protein.

Purification Procedures[3–5]

Procedures are given here for the purification of glutathione reductase from calf liver[4] and rat liver.[5] Similar methods have been used for the purification of the enzyme from yeast,[6] porcine,[7] and human[8] erythrocytes.

All steps are carried out at about 5°.

[3] I. Carlberg and B. Mannervik, *J. Biol. Chem.* **250,** 5475 (1975).
[4] I. Carlberg and B. Mannervik, *Anal. Biochem.* **116,** 531 (1981).
[5] I. Carlberg, B. Altmejd, and B. Mannervik, *Biochim. Biophys. Acta* **677,** 146 (1981).
[6] I. Carlberg and B. Mannervik, *Biochim. Biophys. Acta* **484,** 268 (1977).
[7] V. Boggaram, T. Brobjer, K. Larson, and B. Mannervik, *Anal. Biochem.* **98,** 335 (1979).
[8] G. Krohne-Erich, R. H. Schirmer, and R. Untucht-Grau, *Eur. J. Biochem.* **80,** 65 (1977).

TABLE I
PURIFICATION OF CALF LIVER GLUTATHIONE REDUCTASE[a]

Step	Vol (ml)	Activity (units)	Protein (mg)	Specific activity (units/mg)
1. Cytosol fraction	7,100	10,000	79,800	0.125
2. DEAE-Sephadex	1,320	9,500	25,300	0.373
3. 2.26 M $(NH_4)_2SO_4$	238	9,470	19,300	0.49
4. 2′,5′-ADP Sepharose I	232	7,470	70	107
5. 2′,5′-ADP Sepharose II	22	6,810	31	219[b]
6. Sephadex G-100	16	5,730	33	174
7. Hydroxyapatite	122	5,100	25	204

[a] Data from Carlberg and Mannervik.[8]
[b] Protein determined by the Coomassie Brilliant Blue method of M. M. Bradford. [*Anal. Biochem.* **72,** 248 (1976).]

Calf Liver

Fresh calf liver is obtained from a local slaughterhouse. Table I summarizes the results of a typical purification.

Step 1. Preparation of Cytosol Fraction. Calf liver (1200 g) is cut into small pieces and homogenized in ice-cold 0.25 M sucrose (6 to 7 ml/g liver) in a Turmix blender. EDTA (1 mM), phenylmethanesulfonylfluoride (PMSF; 50 μM), dithioerythritol (DTE; 0.1 mM), and FAD (6 μM) are added to the homogenate which is then centrifuged for 60 min at 7000 g. The supernatant fraction is adjusted to pH 5.3 by addition of ice-cold 0.2 M acetic acid and centrifuged at 15,000 g for 40 min. The pH of the supernatant is adjusted to pH 7.4 by addition of 1 M Tris base.

Step 2. Chromatography on DEAE-Sephadex. The cytosol fraction is diluted with deionized water to a total volume of 10 liters. Sedimented DEAE-Sephadex (600 ml) equilibrated with 10 mM Tris–HCl, pH 7.4, containing 50 μM PMSF, 0.1 mM DTE and 1 mM EDTA is added to the cytosol fraction. The suspension is gently stirred for 90 min at 5°, whereafter the DEAE-Sephadex is allowed to sediment. After decantation the gel is poored into a 9-cm column, in which it is again allowed to settle and then washed with equilibration buffer. Elution of glutathione reductase is performed with about 3 liters of 50 mM Tris–HCl, pH 7.4, containing 0.3 M KCl, 0.1 mM DTE, and 1 mM EDTA.

All subsequent buffers should contain 0.1 mM DTE and 1 mM EDTA.

Step 3. Precipitation with Ammonium Sulfate. Solid ammonium sulfate (350 g/liter) is added with stirring to the pooled fractions from step 2.

After 1 hr the precipitate is collected by centrifugation, dissolved in 50 m*M* potassium phosphate, pH 7.5, and dialyzed over night against the same buffer.

Step 4. Chromatography on 2′,5′-ADP Sepharose I. The dialyzed enzyme fraction is applied to a 2′,5′-ADP Sepharose column (3.2 × 5 cm) equilibrated with 50 m*M* potassium phosphate, pH 7.5. After sample application the column is washed with 0.5 *M* potassium phosphate, pH 7.5 (500 ml). Glutathione reductase is eluted with 500 ml of equilibration buffer fortified with 0.5 *M* KCl.

Step 5. Chromatography on 2′,5′-ADP Sepharose II. The glutathione reductase containing effluent is dialyzed against 50 m*M* potassium phosphate, pH 7.5, and applied to a second 2′,5′-ADP Sepharose column (2 × 3 cm) equilibrated with the same buffer. The column is washed with 0.4 *M* potassium phosphate, pH 7.5, and the enzyme eluted with a linear gradient of NADPH (0–0.5 m*M*) in 50 m*M* potassium phosphate, pH 7.5 (100 ml total volume).

Step 6. Chromatography on Sephadex G-200. The fraction obtained in step 5 is concentrated to about 3 ml by ultrafiltration and applied to a Sephadex G-200 column (1.5 × 90 cm), equilibrated with 50 m*M* potassium phosphate, pH 6.7, containing 0.1 *M* KCl, and eluted with the same buffer.

Step 7. Chromatography on Hydroxyapatite. The material obtained after gel filtration is diluted with deionized water to 10 m*M* phosphate concentration, and applied to a hydroxyapatite column (2 × 4.3 cm) equilibrated with 10 m*M* potassium phosphate, pH 6.7. Glutathione reductase is eluted with a linear gradient of potassium phosphate, pH 6.7 (10–350 m*M*, total volume 300 ml). The enzyme containing fractions are concentrated to a protein concentration of 5–10 mg/ml and dialyzed against 50 m*M* Tris–HCl, pH 7.5.

Rat Liver

In our laboratory purification of glutathione reductase from rat liver has usually been combined with the preparation of glutathione transferases,[10] thioltransferase,[11] and glyoxalase I.[12]

Table II[13] summarizes the results of a representative purification. All buffers contain 1 m*M* EDTA.

[10] P. Askelöf, C. Guthenberg, I. Jakobson, and B. Mannervik, *Biochem. J.* **147,** 513 (1975).
[11] K. Axelsson, S. Eriksson, and B. Mannervik, *Biochemistry* **17,** 2978 (1978).
[12] E. Marmstål and B. Mannervik, *Biochim. Biophys. Acta* **566,** 362 (1979).

TABLE II
PURIFICATION OF RAT LIVER GLUTATHIONE REDUCTASE

Step	Vol (ml)	Activity (units)	Protein (mg)[a]	Specific activity (units/mg)
1. Cytosol fraction[b]	4,000	4,700	88,680	0.053 (0.037–0.069)
2. CM-cellulose[b]	3,000	4,150	7,280	0.57 (0.42–0.68)
3. Sephadex G-75[b]	350	2,920	2,160	1.35 (0.95–2.1)
4. 2′,5′-ADP Sepharose I[c]	126	1,822	13.4	136
5. Sephacryl S-200[c]	12	2,045	11.3	181
6. 2′,5′-ADP-Sepharose II[c]	14	1,425	5.3	269

[a] Protein determined from the absorbance at 260 and 280 nm [H. M. Kalckar, *J. Biol. Chem.* **167,** 461 (1947)].
[b] Data presented are a summary of results from several different purifications. Values in parenthesis show the range in specific activities obtained from at least 6 preparations.
[c] Data from Carlberg *et al.*[5]

Step 1. Preparation of Cytosol Fraction. The cytosol fraction from 100 livers (male Sprague–Dawley rats) is prepared essentially as described for calf liver (see above). A 20% (w/v) homogenate in 0.25 *M* sucrose should be used as the starting material for the further purification steps. PMSF or FAD are usually not added.

Step 2. Chromatography on CM-Cellulose. The supernatant fraction, after passage through Sephadex G-25 (9 × 126 cm; 10 m*M* phosphate, pH 6.1), is applied to a CM-cellulose column (9 × 14 cm) in the same buffer. Elution is performed with a linear concentration gradient of NaCl (0 to 0.2 *M*) in 10 m*M* phosphate, pH 6.1 (a total of 4 liters).

Step 3. Chromatography on Sephadex G-75. The pooled material from step 2 is concentrated by ultrafiltration to <30 ml and applied to a Sephadex G-75 column (12 × 125 cm) equilibrated with 10 m*M* sodium phosphate, pH 6.7.

This gel filtration separates glutathione reductase from the other glutathione-dependent enzymes mentioned above. At this stage glutathione reductase can be frozen and stored for several months before the following steps are carried out.

Step 4. Chromatography on 2′,5′-ADP Sepharose I. Before application to the affinity column the pooled material from step 3 is concentrated by addition of solid ammonium sulfate (400 g/liter), centrifugation and dissolution in about 60 ml 50 m*M* potassium phosphate, pH 7.5 containing 0.1 m*M* DTE. The dissolved precipitate is applied to 2′,5′-ADP

Sepharose (2 × 5 cm) equilibrated with 50 m*M* potassium phosphate, pH 7.5 containing 0.1 m*M* DTE. The column is washed and eluted in the same manner as in the calf liver preparation.

Step 5. Chromatography on Sephacryl S-200. The fraction from step 4, after concentration, is run through a Sephacryl S-200 column (3 × 40 cm) in 50 m*M* phosphate, pH 7.5 containing 0.1 *M* KCl.

Step 6. Chromatography on 2',5'-ADP Sepharose II. The pooled material from step 5 is applied to a second 2',5'-ADP Sepharose column (1 × 5 cm), which is again equilibrated, washed and eluted in the same manner as in the calf liver preparation. Total gradient volume in this case is 40 ml.

Comments

Washing the 2',5'-ADP Sepharose affinity column with 0.4 *M* phosphate before elution of glutathione reductase increases the purification in this step by a factor of 2–5. Furthermore, a great advantage is that 0.4 *M* phosphate elutes the closely related flavoenzyme thioredoxin reductase from the affinity column, while leaving glutathione reductase bound.

TABLE III
MOLECULAR AND KINETIC PROPERTIES OF GLUTATHIONE REDUCTASE

Property	Calf liver[a]	Rat liver[b]
Molecular weight	100,000	100,000
Number of subunits (probably identical)	2	2
Prosthetic group (one per subunit)	FAD	FAD
Isoelectric point	6.1	6.4[c]
Apparent K_m for GSSG	101 ± 7 μM	26.3 ± 5.7 μM
Apparent K_m for NADPH	21 ± 1 μM	8.2 ± 0.8 μM
Molecular activity (μmol/min/μmol FAD)	17,100 ± 600	13,500 ± 400

[a] Data from Carlberg and Mannervik.[4]

[b] Data from Carlberg *et al.* [ref. 5, and I. Carlberg, J. W. DePierre, and B. Mannervik, *Biochim. Biophys. Acta* **677**, 140 (1981)].

[c] Value from Eriksson *et al.* [S. Eriksson, P. Askelöf, K. Axelsson, I. Carlberg, C. Gutenberg, and B. Mannervik, *Acta Chem. Scand. Ser. B* **28**, 922 (1974)].

Properties

The purified enzyme from both sources is homogeneous in several electrophoretic and chromatographic systems. Pure glutathione reductase can be stored in the refrigerator for very long periods of time (without apparent change of its properties) provided that the enzyme is concentrated to 2 mg/ml (or more) and dialyzed against neutral Tris buffer. A crystal of thymol is added as an antibacterial agent. In the pure state glutathione reductase should preferably not be frozen, since freezing often leads to denaturation and loss of activity.

Some of the molecular and kinetic properties are summarized in Table III. Glutathione reductase shows a broad pH optimum centered at about pH 7.0.

Acknowledgment

The work in our laboratory was supported by grants (to B.M.) from the Swedish Natural Science Research Council.

[60] Glutathione Peroxidase

By BENGT MANNERVIK

Reactions Catalyzed

Glutathione peroxidases[1-3] catalyze the reduction of hydroperoxides (ROOH) by glutathione (GSH):

$$ROOH + 2\ GSH \rightarrow ROH + H_2O + GSSG$$

R may be an aliphatic or aromatic organic group or, simply, hydrogen. The products are H_2O, an alcohol (ROH) (or a second H_2O when H_2O_2 serves as substrate) and glutathione disulfide (GSSG). Regeneration of GSH from GSSG in the cell is effected by the enzyme glutathione reductase.

Assays of glutathione peroxidase activity are based on measurement of ROOH or GSH consumption.[4,5] Alternatively, GSSG production is

[1] G. C. Mills, *J. Biol. Chem.* **229,** 189 (1957).

[2] A. Wendel, *in* "Enzymatic Basis of Detoxication" (W. B. Jakoby, ed.), Vol. 1, p. 333. Academic Press, New York, 1980.

[3] L. Flohé, *in* "Free Radicals in Biology" (W. A. Pryor, ed.), Vol. 5, p. 223. Academic Press, New York, 1982.

[4] L. Flohé and W. A. Günzler, this series, Vol. 105, p. 114.

[5] A. Wendel, this series, Vol. 77, p. 325.

Copyright © 1985 by Academic Press, Inc.
All rights of reproduction in any form reserved.

monitored by coupling to the reaction catalyzed by glutathione reductase:

$$GSSG + NADPH + H^+ \rightarrow 2\ GSH + NADP^+$$

Oxidation of NADPH is recorded spectrophotometrically or fluorometrically. A critical survey of assay methods has recently been published.[4]

Enzyme Proteins

Two major types of glutathione peroxidase have been found. One type is distinguished by containing selenium in the form of covalently bound selenocysteine in its active site.[6,7] This selenium-dependent enzyme is active with both organic hydroperoxides and H_2O_2. The enzyme from bovine erythrocytes is a tetrameric protein of $M_r \sim 80{,}000$.[8] Its 3D structure has been determined by X-ray diffraction analysis,[9] and its amino acid sequence has been elucidated by conventional techniques of protein chemistry.[10]

The second type of glutathione peroxidase consists of proteins that do not depend on selenium for catalysis and have negligible activity with H_2O_2.[11–13] This class is constituted by glutathione transferases,[14] first described as proteins catalyzing the conjugation of GSH with electrophilic compounds such as aryl halides.[15,16] These enzymes are dimeric proteins that often occur in multiple forms in the same organ, and some of the isoenzymes have distinctly higher glutathione peroxidase activities than others.[17–20] None of the transferases is yet characterized to the same

[6] J. T. Rotruck, A. L. Pope, H. E. Ganther, A. B. Swanson, D. G. Hafeman, and W. G. Hoekstra, *Science* **179,** 588 (1973).

[7] L. Flohé, W. A. Günzler, and H. H. Schock, *FEBS Lett.* **32,** 132 (1973).

[8] L. Flohé, B. Eisele, and A. Wendel, *Hoppe-Seyler's Z. Physiol. Chem.* **352,** 151 (1971).

[9] O. Epp, R. Ladenstein, and A. Wendel, *Eur. J. Biochem.* **133,** 51 (1983).

[10] L. Flohé, G. J. Steffens, W. A. Günzler, S.-M. A. Kim, F. Ötting, A. Grossman, and A. Wendel, *Life Chem. Rep., Suppl.* **2,** *Oxidative Damage Relat. Enzymes,* 358 (1984).

[11] R. A. Lawrence and R. F. Burk, *Biochem. Biophys. Res. Commun.* **71,** 952 (1976).

[12] J. R. Prohaska and H. E. Ganther, *J. Neurochem.* **27,** 1379 (1976).

[13] F. E. Hunter, Jr., F. Posadas del Rio, and A. A. Painter, *Fed. Proc., Fed. Am. Soc. Exp. Biol.* **35,** 1529 (1976).

[14] J. R. Prohaska and H. E. Ganther, *Biochem. Biophys. Res. Commun.* **76,** 437 (1977).

[15] B. Combes and G. S. Stakelum, *J. Clin. Invest.* **40,** 981 (1961).

[16] J. Booth, E. Boyland, and P. Sims, *Biochem. J.* **79,** 516 (1961).

[17] J. R. Prohaska, *Biochim. Biophys. Acta* **611,** 87 (1980).

[18] B. Mannervik, C. Guthenberg, and K. Åkerfeldt, *in* "Microsomes, Drug Oxidations and Chemical Carcinogenesis" (M. J. Coon, A. H. Conney, R. W. Estabrook, H. V. Gelboin, J. R. Gillette, and P. J. O'Brien, eds.), Vol. 2, p. 663. Academic Press, New York, 1980.

[19] B. Mannervik and H. Jensson, *J. Biol. Chem.* **257,** 9909 (1982).

[20] B. Mannervik, C. Guthenberg, H. Jensson, M. Warholm, and P. Ålin, *in* "Functions of Glutathione: Biochemical, Physiological, Toxicological, and Clinical Aspects" (A. Lars-

extent with respect to molecular structure as the selenoprotein. It is currently not believed that the peroxidase activity of the glutathione transferases represents their major biological function, even though its significance is well documented.

It is the purpose of the present chapter to delineate similarities and differences in the properties of the two types of protein exhibiting glutathione peroxidase activity. Detailed descriptions of the preparation and characteristics of the selenium-dependent glutathione peroxidase[5,21] and various glutathione transferases glutathione peroxidase activity have been published in this series.[22–26]

Occurrence

Glutathione peroxidase activity has been demonstrated in all mammalian tissues examined.[27] In most animals the selenoprotein is responsible for a substantial fraction of the activity, but in the guinea pig (liver) the selenoprotein is absent or present in a very small amount.[28] The non-selenium-dependent activity is, in all case investigated in depth, ascribable to the glutathione transferases. The ratio between the selenium and non-selenium-dependent activities may vary not only between animal species but also from tissue to tissue in the same species. In the rat, the contribution of the glutathione transferases to the total peroxidase activity is especially high in testis.[28]

In selenium deficiency the relative importance of the peroxidase activity of the glutathione transferase increases.

Subcellular Localization

The distribution of glutathione peroxidase has been most extensively studied in the rat. In hepatocytes selenium-dependent glutathione peroxidase is localized primarily in the cytosol and in the matrix of mitochondria.[29] Glutathione transferases are present in the same compartments,

son, S. Orrenius, A. Holmgren, and B. Mannervik, eds.), p. 75. Raven Press, New York, 1983.

[21] A. L. Tappel, this series, Vol. 52, p. 506.

[22] W. H. Habig and W. B. Jakoby, this series, Vol. 77, p. 218.

[23] B. Mannervik and C. Guthenberg, this series, Vol. 77, p. 231; P. C. Simons and D. L. Vander Jagt, *ibid.* p. 235.

[24] M. Warholm, C. Guthenberg, C. von Bahr, and B. Mannervik, this volume [62].

[25] H. Jensson, P. Ålin, and B. Mannervik, this volume [63].

[26] C. Guthenberg, P. Ålin, and B. Mannervik, this volume [64].

[27] L. Flohé, W. A. Günzler, and G. Loschen, *in* "Trace Metals in Health and Disease" (N. Kharasch, ed.), p. 263. Raven Press, New York, 1979.

[28] R. A. Lawrence and R. F. Burk, *J. Nutr.* **108,** 211 (1978).

but also in membrane-containing subcellular fractions.[30–32] The six major glutathione transferase isoenzymes in rat cytosol all exhibit peroxidase activity.[19] Transferase 2-2 (earlier named transferase AA or B_2[32a]) has the highest specific activity,[17,19] but transferase 1-2, in view of its high relative concentration, carries more peroxidase activity than any of the other isoenzymes. Judging from specific activities and relative abundance, transferase 1-1, 3-4, and 3-3 also contribute substantially to the total activity (measured with cumene hydroperoxide); transferase 4-4 and the isoenzymes with lower isoelectric points[19] have significant specific activities as well.

The specific peroxidase activities of the three groups of human transferases characterized differ more than those of the rat isoenzymes. The glutathione transferases with high isoelectric points have the highest peroxidase activity.[20,33]

The microsomal glutathione transferase in rat liver, which is distinct from the cytosolic isoenzymes, also has glutathione peroxidase activity with cumene hydroperoxide.[34,35] This peroxidase activity, like the transferase activity with 1-chloro-2,4-dinitrobenzene, is activatable severalfold by pretreatment of the microsomal enzyme with *N*-ethylmaleimide.[34,35] The microsomes also contain the major "cytosolic" isoenzymes and quantitative determinations of the relative amounts of the different transferases in this subcellular fraction have been made.[36] Judging from the specific activities with cumene hydroperoxide, it can be calculated that the peroxidase activity of the "cytosolic" isoenzymes in the microsomes are quantitatively at least as important as the "microsomal" isoenzyme, even after activation with *N*-ethylmaleimide.

The "microsomal" glutathione transferase is also present in the outer mitochondrial membrane in which it constitutes 5% of the total protein content (the value for microsomes is 3%).[37] Less is known about the

[29] L. Flohé and W. Schlegel, *Hoppe-Seyler's Z. Physiol. Chem.* **352,** 1401 (1971).

[30] P. Kraus, *in* "Conjugation Reactions in Drug Biotransformation" (A. Aitio, ed.), p. 503. Elsevier/North-Holland, Amsterdam, 1978.

[31] A. Wahlländer, S. Soboll, and H. Sies, *FEBS Lett.* **97,** 138 (1979).

[32] T. Friedberg, P. Bentley, P. Stasiecki, H. R. Glatt, D. Raphael, and F. Oesch, *J. Biol. Chem.* **254,** 12028 (1979).

[32a] W. B. Jakoby, B. Ketterer, and B. Mannervik, *Biochem. Pharmacol.* **33,** 2539 (1984).

[33] M. Warholm, C. Guthenberg, and B. Mannervik, *Biochemistry* **22,** 3610 (1983).

[34] C. C. Reddy, C.-P. D. Tu, J. R. Burgess, C.-Y. Ho, R. W. Scholz, and E. J. Massaro, *Biochem. Biophys. Res. Commun.* **101,** 970 (1981).

[35] R. Morgenstern and J. W. DePierre, *Eur. J. Biochem.* **134,** 591 (1983).

[36] R. Morgenstern, C. Guthenberg, B. Mannervik, and J. W. DePierre, *FEBS Lett.* **160,** 264 (1983).

[37] R. Morgenstern, Ph.D. dissertation, Univ. of Stockholm, 1983.

nature and amounts of the other transferases in the different subcellular fractions.

Importance

Numerous chemical processes in aerobic cells lead to the production of peroxides by activated forms of oxygen.[38–40] The peroxides may cause oxidative damage in biological tissues as well as decompose to generate free radicals and other reactive chemical species. The simplest hydroperoxide, H_2O_2, can be detoxified by the selenium-dependent glutathione peroxidase. Catalase, which also decomposes H_2O_2, is primarily localized in peroxisomes of the hepatocytes, whereas the selenoprotein is found in the cytosol and the mitochondrial matrix. Thus, these two enzymes appear to have complementary intracellular localizations as well as complementary catalytic activities.[2]

Organic hydroperoxides derived from polyunsaturated fatty acids such as linoleic and linolenic acid or certain prostaglandins can occur *in vivo*. Cholesterol 7β-hydroperoxide as well as derivatives of some steroid hormones and vitamin K are also relevant in this context. These hydroperoxides, like thymine hydroperoxide and "peroxidized DNA," have all been reported to be reduced by glutathione under the influence of glutathione peroxidase (see refs. 3 and 39 for original references). It appears as if most organic hydroperoxides are substrates both for the selenium-dependent and the non-selenium-dependent enzymes, even though not all compounds have been tested with both types of enzyme. However, it should be noted that among the glutathione transferases the relative specific peroxidase activities between different isoenzymes may depend on the nature of the hydroperoxide. For example, linoleic acid hydroperoxide has been reported as a better substrate for a fraction of partially purified cytosolic kidney transferases than for corresponding liver transferases, in spite of the fact that cumene and *t*-butyl hydroperoxides are better substrates for the liver enzymes.[41] The explanation is that the sets of isoenzymes are different in the two tissues.

A more complex chemical process of biological interest is lipid peroxidation.[39] This process has been studied in microsomal subcellular fractions and leads to degradation of the microsomal membranes. In this

[38] B. Chance, H. Sies, and A. Boveris, *Physiol. Rev.* **59,** 527 (1979).

[39] H. Sies, A. Wendel, and W. Bors, *in* "Metabolic Basis of Detoxication" (W. B. Jakoby, J. R. Bend, and J. Caldwell, eds.), p. 307. Academic Press, New York, 1982.

[40] K. Yagi, ed., "Lipid Peroxides in Biology and Medicine." Academic Press, New York, 1983.

[41] C.-P. D. Tu, M. J. Weiss, N. Li, and C. C. Reddy, *J. Biol. Chem.* **258,** 4659 (1983).

experimental system, lipid peroxidation is inhibited by certain glutathione transferases in the presence of glutathione.[42–44] Very significant differences in inhibitory activity of the isoenzymes are apparent. Also other cellular membranes appear to be protected by the glutathione peroxidase activity.[3]

That the glutathione peroxidase activity of the selenoprotein as well as of the glutathione transferases operates *in vivo* can be concluded from liver perfusion experiments.[45–47] In normal rats glutathione disulfide is released into bile upon infusion of both H_2O_2 and *t*-butyl hydroperoxide. In selenium-deficient animals only *t*-butyl hydroperoxide is effective (owing to the lack of the H_2O_2-specific selenoprotein).[48] Thus, it appears evident that the glutathione peroxidase activity of both types of enzyme is important in cellular defense against a wide variety of hydroperoxides.

Acknowledgments

Work from the author's laboratory was supported by the Swedish Cancer Society, the Swedish Council for Planning and Coordination of Research, and the Swedish Natural Science Research Council.

[42] R. F. Burk, M. J. Trumble, and R. A. Lawrence, *Biochim. Biophys. Acta* **618,** 35 (1980).
[43] R. Morgenstern, J. W. DePierre, C. Lind, C. Guthenberg, B. Mannervik, and L. Ernster, *Biochem. Biophys. Res. Commun.* **99,** 682 (1981).
[44] K. H. Tan, D. J. Meyer, and B. Ketterer, *Biochem. Soc. Trans.* **11,** 308 (1983).
[45] H. Sies, C. Gerstenecker, H. Menzel, and L. Flohé, *FEBS Lett.* **27,** 171 (1972).
[46] N. Oshino and B. Chance, *Biochem. J.* **162,** 509 (1977).
[47] H. Sies, R. Brigelius, and T. P. M. Akerboom, *in* "Functions of Glutathione: Biochemical, Physiological, Toxicological, and Clinical Aspects" (A. Larsson, S. Orrenius, A. Holmgren, and B. Mannervik, eds.), p. 51. Raven Press, New York, 1983.
[48] R. F. Burk and R. A. Lawrence, *in* "Functions of Glutathione in Liver and Kidney" (H. Sies and A. Wendel, eds.), p. 114. Springer-Verlag, Berlin and New York, 1978.

[61] Glutathione Transferases: An Overview

By WILLIAM B. JAKOBY

The glutathione transferases (EC 2.5.1.18) are among the catalysts that participate in the process of detoxication,[1] the means by which those compounds without nutritional value are eliminated, usually after metabolic processing. The glutathione transferases are normally present in

[1] W. B. Jakoby, ed., "Enzymatic Basis of Detoxication," Vols. 1 and 2. Academic Press, New York, 1980.

Copyright © 1985 by Academic Press, Inc.
All rights of reproduction in any form reserved.

large quantities, representing about 10% of the extractable protein of rat liver[2] but can be induced to greater than 20%.[3] Most of the work has been carried out with enzymes from rat and human liver[2,4–6] but human erythrocytes[7] and placenta,[8] as well as sheep[9] and mouse[10] liver have also been sources for homogeneous preparations. The enzymes have been found in all mammalian tissue tested as well as in insects, protozoa, algae, fungi, and bacteria.[2,4] Included in this volume [62–64], and elsewhere in this series,[11–13] are detailed procedures for isolation of homogeneous transferases from human and rat tissues; methods for assay of the glutathione transferases have also been made available.[14,15]

The large number of glutathione transferase isoenzymes that are found in each of several species has caused considerable confusion when, as with the enzymes from rat, different means of naming them were used. A consistent system of nomenclature has now been developed for the rat[16] and may serve as a model for other species. The nomenclature system adopted, shown in the table together with the large number of synonyms used in the past, is based on the subunit composition of the transferases. Each different subunit is denoted by an Arabic numeral. The particular isoenzymes shown in the table are presented in the order of generally decreasing isoelectric points.

The major outlines of the reactions catalyzed by these enzymes, whatever their source, are clear. The glutathione transferases may be considered as catalysts of all reactions in which glutathione, as the thiolate anion, can participate as a nucleophile, providing only that a compound

[2] W. B. Jakoby, *Adv. Enzymol.* **46,** 383 (1978).

[3] I. M. Arias, G. Fleischner, R. Kirsch, S. Mishkin, and Z. Gatmaitan, *Arch. Biochem. Biophys.* **188,** 287 (1978).

[4] W. B. Jakoby and W. H. Habig, *in* "Enzymatic Basis of Detoxication" (W. B. Jakoby, ed.), Vol. 2, p. 63. Academic Press, New York, 1980.

[5] B. Mannervik and H. Jansson, *J. Biol. Chem.* **257,** 235 (1982).

[6] K. Kamisaka, W. H. Habig, J. N. Ketley, I. M. Arias, and W. B. Jakoby, *Eur. J. Biochem.* **60,** 153 (1975).

[7] C. J. Marcus, W. H. Habig, and W. B. Jakoby, *Arch. Biochem. Biophys.* **188,** 287 (1978).

[8] C. Guthenberg and B. Mannervik, *Biochim. Biophys. Acta* **661,** 255 (1981).

[9] C. C. Reddy, J. R. Burgess, Z. Z. Gang, E. J. Massaro, and C.-P. D. Tu, *Arch. Biochem. Biophys.* **224,** 87 (1983).

[10] C.-Y. Lee, L. Johnson, R. H. Cox, J. D. McKinney, and S.-M. Lee, *J. Biol. Chem.* **256,** 8110 (1981).

[11] W. H. Habig and W. B. Jakoby, this series, Vol. 77, p. 218.

[12] B. Mannervik and C. Guthenberg, this series, Vol. 77, p. 231.

[13] P. C. Simons and D. L. VanderJagt, this series, Vol. 77, p. 235.

[14] W. H. Habig and W. B. Jakoby, this series, Vol. 77, p. 298.

[15] W. H. Habig, M. J. Pabst, and W. B. Jakoby, *J. Biol. Chem.* **249,** 7130 (1974).

[16] W. B. Jakoby, B. Ketterer, and B. Mannervik, *Biochem. Pharmacol.* **33,** 2539 (1984).

NOMENCLATURE OF THE RAT GLUTATHIONE TRANSFERASES

New nomenclature[a]	Previous systems of nomenclature					
	c	*d*	*e*	*f, g, e*	*h*	*i, j, k*
Glutathione transferase 1-1	B^b	Ligandin	B_1	Y_aY_a	L_2	
Glutathione transferase 1-2		B	B_2	Y_aY_c	BL	
Glutathione transferase 2-2	AA		AA	Y_cY_c	B_2	
Glutathione transferase 3-3	A		A	$Y_b{}^1Y_b{}^1$	A_2	
Glutathione transferase 3-4	C		C	$Y_b{}^1Y_b{}^2$	AC	
Glutathione transferase 4-4	D		"D"	$Y_b{}^2Y_b{}^2$	C_2	
Glutathione transferase 5-5	E		E	—	—	
Glutathione transferase 6-6						M_T
Glutathione transferase 7-7[l–n]						

[a] This systems is based entirely on the nature of the subunits, each of which are identified by an Arabic numeral. The table is adapted from W. B. Jakoby, B. Ketterer, and B. Mannervik, *Biochem. Pharmacol.* **33,** 2539 (1984).

[b] Both species have been referred to as ligandin and as glutathione transferase B.

[c] W. B. Jakoby, *Adv. Enzymol.* **46,** (1978).

[d] J. D. Hayes, R. C. Strange, and I. W. Percy-Robb, *Biochem. J.* **197,** 491 (1981).

[e] B. Ketterer, D. Beale, J. B. Taylor, and D. J. Meyer, *Biochem. Soc. Trans.* **11,** 466 (1983).

[f] N. M. Bass, R. E. Kirsch, S. A. Taff, I. Marks, and S. J. Saunders, *Biochim. Biophys. Acta* **492,** 163 (1977).

[g] N. C. Scully and T. J. Mantle, *Biochem. Soc. Trans.* **8,** 45 (1980).

[h] B. Mannervik and H. Jennson, *J. Biol. Chem.* **257,** 9909 (1982).

[i] P. J. Diericks and J. O. DeBeer, *Biochem. Int.* **3,** 565 (1981).

[j] C. Gutenberg, P. Ålin, I.M. Åstrand, S. Jalçin, and B. Mannervik, *in* "Extrahepatic Drug Metabolism and Chemical Carcinogenesis" (J. Rydström, J. Montelius, and M. Bengtsson, eds.), p. 171. Elsevier, Amsterdam, 1983.

[k] D. J. Meyer, L. G. Christodonlides, D. Nyan, B. R. Schuster, and B. Ketterer, *ibid,* p. 189.

[l] I. G. C. Robertson, H. Jensson, C. Gutenberg, M. K. Tahir, B. Jernstrom, and B. Mannervik, *Biochem. Biophys. Res. Comm.* **127,** 80 (1985).

[m] D. J. Meyer, D. Beale, K. H. Tan, B. Coles, and B. Ketterer, *FEBS Lett.,* in press (1985).

[n] C. Gutenberg, H. Jensson, N. Nyström, E. Österlund, M. K. Tahir, and B. Mannervik, *Biochem. J.,* submitted (1985).

with a sufficiently electrophilic group binds to the enzyme.[17,18] Leaving the caveat of binding for subsequent discussion, it is obvious that the above definition is very broad. It proposes that the glutathione transferases can utilize any ligand with a sufficiently electronegative atom, whether C, S, N, or O, as the electrophile. The enzyme participates, for example, in the initial step of mercapturic acid synthesis[15] in which a thioether is formed between GSH and the *carbon* of a large variety of

[17] J. H. Keen, W. H. Habig, and W. B. Jakoby, *J. Biol. Chem.* **251,** 6183 (1976).

[18] J. H. Keen and W. B. Jakoby, *J. Biol. Chem.* **253,** 5654 (1978).

$$GSH + CH_3I \rightarrow GSCH_3 + HI \quad (1)$$

$$GSH + Cl\text{-}C_6H_3(NO_2)\text{-}NO_2 \rightarrow GS\text{-}C_6H_3(NO_2)\text{-}NO_2 + HCl \quad (2)$$

$$GSH + C_6H_5\text{-}CH{=}CHCOCH_3 \rightarrow C_6H_5\text{-}CH_2CH(SG)COCH_3 \quad (3)$$

$$GSH + RCH_2NO_2 \rightarrow RCH_2OH + (GSNO)_2 \xrightarrow{GSH} GSSG + HNO_2 \quad (4)$$

$$GSH + RSSR' \rightarrow GSSR + R'SH \quad (5)$$

$$GSH + RSCH \rightarrow RSSG + HCN \quad (6)$$

$$GSH + CH_3(CH_3)C(OOH)\text{-}C_6H_5 \rightarrow CH_3(CH_3)C(OH)\text{-}C_6H_5 + GSSG + H_2O \quad (7)$$

electrophiles as exemplified by Reactions (1)–(3).[2,4] *Sulfur* is attacked in thiocyanates [Reaction (4)] to yield the appropriate mixed disulfide and HCN,[17] or, is attacked in a mixed disulfide to result in disulfide interchange [Reaction (5)].[18] In Reaction (6), a nitrate *nitrogen* of trinitroglycerol forms an S-nitration product which undergoes reaction with a second mole of GSH to form nitrous acid.[17] The transferases also act as glutathione peroxidases (cf. this volume [60]), i.e., they attacked an *oxygen,* as shown for cumene hydroperoxide [Reaction (7)], but differ from the selenium-containing peroxidase by their inability to utilize hydrogen peroxide.[19] To this list of reactions must be added the capability of acting as isomerases as in the conversion of Δ^5-androstene-3,17-dione to Δ^4-androstene-3,17-dione,[20] or of maleylacetoacetate to fumarylacetoacetate,[18] reactions in which GSH is required only in catalytic quantities. The transferases also catalyze thiolysis, e.g., with nitrophenyl acetate to form acetylCoA and the appropriate phenol.[18]

Specificity for the thiol is limited to GSH and to its β-alanine analog.[4] A report that 2-propylthiouracil can replace GSH[21] could not be confirmed.[22]

[19] J. R. Prohaska and H. E. Ganther, *Biochem. Biophys. Res. Commun.* **71,** 952 (1977).

[20] A. M. Benson, P. Talalay, J. H. Keen, and W. B. Jakoby, *Proc. Natl. Acad. Sci. U.S.A.* **74,** 158 (1977).

[21] T. Yamada and N. Kaplowitz, *J. Biol. Chem.* **255,** 3508 (1980).

[22] W. H. Habig, W. B. Jakoby, C. Guthenberg, B. Mannervik, and D. L. Vander Jagt, *J. Biol. Chem.* **259,** 7409 (1984).

This variety of type reactions is intrinsic to the glutathione transferases, a group of enzymes that display a qualitatively overlapping pattern of activity even when several species of the enzyme are found in an individual animal.[2,4,23] In addition to catalysis, these enzymes serve a storage function[24] in that they act in much the same manner, and with as great a range of ligands within cells, as does albumin in the circulation. The transferases were originally described as binding proteins under the name of ligandin,[25] emphasizing the broad range of affinity that seems to include most molecules with a lipophilic aspect. Indeed, the substrates noted in Reactions 1 though 7 testify to such versatility of binding, which is equally effective for those ligands that are insufficiently reactive electrophiles and, therefore, are not substrates.[26] Also illustrated by Reactions (1) through (7) is the enormous catalytic versatility that serves as an efficient means of coping with our exposure to the products of both Nature and the chemical industry.

[23] W. H. Habig, M. J. Pabst, and W. B. Jakoby, *Arch. Biochem. Biophys.* **175,** 710 (1976).
[24] A. W. Wolkoff, R. A. Weisiger, and W. B. Jakoby, *Prog. Liver Dis.* **6,** 213 (1979).
[25] G. Litwack, B. Ketterer, and I. M. Arias, *Nature (London)* **234,** 466 (1971).
[26] J. N. Ketley, W. H. Habig, and W. B. Jakoby, *J. Biol. Chem.* **250,** 8670 (1975).

[62] Glutathione Transferases from Human Liver

By MARGARETA WARHOLM, CLAES GUTHENBERG, CHRISTER VON BAHR, and BENGT MANNERVIK

The glutathione transferases are a group of related enzymes that catalyze the conjugation of glutathione with a variety of hydrophobic compounds bearing an electrophilic center.[1] The proteins also act as intracellular binding proteins for a large number of lipophilic substances, including bilirubin.[2] Human glutathione transferases have been purified from liver,[3-6] erythrocytes,[7] placenta,[8,9] and lung.[10]

[1] W. B. Jakoby and W. H. Habig, *in* "Enzymatic Basis of Detoxication" (W. B. Jakoby, ed.), Vol. 2, p. 63. Academic Press, New York, 1980.
[2] G. J. Smith and G. Litwack, *Rev. Biochem. Toxicol.* **2,** 1 (1980).
[3] K. Kamisaka, W. H. Habig, J. N. Ketley, I. M. Arias, and W. B. Jakoby, *Eur. J. Biochem.* **60,** 153 (1975).
[4] P. C. Simons and D. L. Vander Jagt, *Anal. Biochem.* **82,** 334 (1977).
[5] Y. C. Awasthi, D. D. Dao, and R. P. Saneto, *Biochem. J.* **191,** 1 (1980).
[6] M. Warholm, C. Guthenberg, B. Mannervik, and C. von Bahr, *Biochem. Biophys. Res. Commun.* **98,** 512 (1981).

Copyright © 1985 by Academic Press, Inc.
All rights of reproduction in any form reserved.

The different human isoenzymes have been denoted by Greek letters; transferases α, β, γ, δ, and ε are basic proteins with very similar properties,[3] μ is a protein with an isoelectric point at pH 6.6,[6,11] and ρ[7] or π[12] (probably identical) is an acidic protein.

We here describe a simple and rapid procedure for the purification of basic (α–ε) and neutral (μ) glutathione transferases from human liver cytosol. A procedure for preparation of the acidic glutathione transferase (π) from human placenta has previously been published in this series.[13]

Assay Method

Principle. Enzyme activity during purification is determined spectrophotometrically at 340 nm by measuring the formation of the conjugate of glutathione (GSH) and 1-chloro-2,4-dinitrobenzene (CDNB).[14]

Reagents

Sodium phosphate buffer, 0.1 *M*, pH 6.5, containing 1 m*M* EDTA
GSH, 20 m*M*, in deionized water
CDNB, 20 m*M*, in 95% ethanol

Procedure. To a 1-ml cuvette are added 850–895 μl of buffer, 50 μl of 20 m*M* GSH, and 50 μl of 20 m*M* CDNB. The reaction, which is carried out at 30°, is started by addition of a suitable amount of enzyme (5–50 μl). The increase in absorbance at 340 nm (ε_{340} = 9.6 mM^{-1} cm^{-1}) is monitored. Correction for the spontaneous reaction is made by measuring and subtracting the rate in the absence of enzyme.

Definition of Units and Specific Activity. A unit of enzyme activity is defined as the amount of enzyme that catalyzes the formation of 1 μmol of *S*-2,4-dinitrophenylglutathione per minute at 30° using 1 m*M* concentrations of GSH and CDNB. Specific activity is defined as units per mg of protein; protein concentrations are based on absorbance at 260 and 280 nm[15] during the purification procedure. For solutions of pure enzyme,

[7] C. J. Marcus, W. H. Habig, and W. B. Jakoby, *Arch. Biochem. Biophys.* **188,** 287 (1978).
[8] C. Guthenberg, K. Åkerfeldt, and B. Mannervik, *Acta Chem. Scand., Ser. B* **33,** 595 (1979).
[9] G. Polidoro, C. Di Ilio, A. Arduini, and G. Federici, *Biochem. Med.* **25,** 247 (1981).
[10] K. Koskelo, E. Valmet, and R. Tenhunen, *Scand. J. Clin. Lab. Invest.* **41,** 683 (1981).
[11] M. Warholm, C. Guthenberg, and B. Mannervik, *Biochemistry* **22,** 3610 (1983).
[12] C. Guthenberg and B. Mannervik, *Biochim. Biophys. Acta* **661,** 255 (1981).
[13] B. Mannervik and C. Guthenberg, this series, Vol. 77, p. 231.
[14] W. H. Habig, M. J. Pabst, and W. B. Jakoby, *J. Biol. Chem.* **249,** 7130 (1974).
[15] H. M. Kalckar, *J. Biol. Chem.* **167,** 461 (1947).

protein concentration is determined as described by Lowry *et al.*[16] with bovine serum albumin as a standard.

Purification Procedure

The enzyme should be maintained at 5° during the entire purification procedure.

Step 1. Preparation of Cytosol Fraction. In our studies human liver tissue was obtained from kidney donors.[17] The donors did not suffer from any liver disease and the livers were judged to be normal (by morphological criteria). The liver tissue is collected immediately after circulatory arrest in connection with removal of the kidneys for transplantation. Postmortem changes are thus expected to be minimal.

The supernatant fraction is prepared immediately or after thawing of intact liver tissue stored at −80°. The liver tissue is homogenized in 0.3 *M* sucrose by standard procedures. The homogenate is diluted with sucrose to give a concentration of 20% (w/v) and then filtered through 2 layers of gauze to remove connective tissue. The filtrate is rehomogenized with 2 strokes in a Potter-Elvehjem homogenizer set at 200 rpm. The homogenate is centrifuged at 9000 *g* for 15 min. The resulting supernatant is then centrifuged at 100,000 *g* for 1 hr. The cytosol fraction thus obtained can be stored at −80° prior to further purification. Any frozen material is centrifuged at 14,000 rpm for 20 min after thawing.

Step 2. Chromatography on Sephadex G-25. The 100,000 *g* supernatant (100 ml) is passed through a Sephadex G-25 column (4 × 48 cm) packed in 10 m*M* Tris–HCl (pH 7.8, measured at room temperature) containing 1 m*M* EDTA (Buffer A). The active fractions are pooled.

Step 3. Chromatography on DEAE-Cellulose. The pooled fractions from Step 2 are applied to a column (4 × 9 cm) of DEAE-cellulose (Whatman DE-52) equilibrated with Buffer A. The column is washed with the same buffer until glutathione transferase activity ceases to appear in the effluent. The active fractions are pooled (DEAE pool I). About 25% of the activity is retained on the column. This activity is eluted with a 1200 ml linear gradient of NaCl (0–0.1 *M*) in Buffer A.[18] Fractions showing transferase activity are pooled (DEAE-pool II).

[16] O. H. Lowry, N. J. Rosebrough, A. L. Farr, and R. J. Randall, *J. Biol. Chem.* **193,** 265 (1951).

[17] C. von Bahr, C.-G. Groth, H. Jansson, G. Lundgren, M. Lind, and H. Glaumann, *Clin. Pharmacol. Ther.* **27,** 711 (1980).

[18] In earlier studies[6] no significant amount of glutathione transferase was retained on this column. In this case the NaCl gradient can be omitted.

Step 4. Affinity Chromatography on S-Hexylglutathione Coupled to Epoxy-Activated Sepharose 6B. Both pools from the chromatography on DEAE-cellulose are applied on an affinity column (2 × 10 cm) prepared as earlier described[13] and equilibrated with Buffer A. Nonspecifically adsorbed protein is eluted with 150 ml of 0.2 *M* NaCl in Buffer A. Glutathione transferases are eluted with 5 m*M* *S*-hexylglutathione dissolved in Buffer A containing 0.2 *M* NaCl. The active fractions are pooled. The protein concentration cannot be measured by Kalckar's method,[15] owing to the high absorbance of *S*-hexylglutathione.

Step 5. Chromatography on Sephadex G-25. The active pool from the affinity chromatography step is desalted on a Sephadex G-25 column (4 × 48 cm), packed in 10 m*M* potassium phosphate (pH 6.7) containing 1 m*M* EDTA (Buffer B).

Step 6. Hydroxyapatite. The pooled fractions of Step 5 are adsorbed on a column of hydroxyapatite (2 × 10 cm) equilibrated with Buffer B. After washing with two volumes of starting buffer the glutathione transferases are eluted by a linear concentration gradient (10–350 m*M*, total volume 400 ml) of potassium phosphate (pH 6.7) containing 1 m*M* EDTA. Two peaks of transferase activity, clearly separated, are obtained. The first peak contains apparently homogeneous glutathione transferase μ (if present, cf. refs. 6, 11) and the second peak a mixture of basic forms (if several are present). The ratio of the two peaks varies considerably between preparations from livers of different individuals. Transferase μ is present in approximately 60% of the human livers tested. The purification procedure is summarized in the table.

PURIFICATION OF GLUTATHIONE TRANSFERASES FROM HUMAN LIVER

Step	Volume (ml)	Total protein (mg)	Total activity (units)	Specific activity (units/mg)
1. Cytosol fraction	100	1720	3550	2.1
2. Sephadex G-25	180	1630	3895	2.4
3. DEAE-cellulose				
Pool I	308	363	2750	7.6
Pool II	257	105	731	7.0
4–5. Affinity chromatography + Sephadex G-25	134	51	2700	52.9
6. Hydroxyapatite				
Peak I	30	5.3	421	79.4 (140)[a]
Peak II	60	39	2204	56.5 (70)[a]

[a] Protein determined by the method of Lowry *et al.*[16]

Separation of Different Basic Forms of Glutathione Transferase. Approximately 300 units of activity of peak II (containing a mixture of basic forms) eluted from the hydroxyapatite column are passed through a column of Sephadex G-25 (2 × 15 cm) packed in 10 m*M* sodium phosphate (pH 6.0) containing 1 m*M* EDTA (Buffer C). The active fractions are pooled and applied to a CM-Sepharose column (1.5 × 16 cm) equilibrated with Buffer C. After washing the column with the same buffer, the basic transferases are eluted with a linear gradient formed by mixing 550 ml Buffer C and 550 ml 50 m*M* sodium phosphate (pH 6.0) containing 1 m*M* EDTA and 0.1 *M* NaCl. The glutathione transferase activity is resolved into several peaks. The specific activities of the different basic isozymes isolated so far are similar (50 units/mg). Which of the transferases α, β, γ, δ, or ε, described by Kamisaka *et al.*,[3] the various peaks correspond to, is not known.

Properties

Purity of the Enzymes. Purified glutathione transferase μ (peak I from the elution of the hydroxyapatite column) displays only one protein band when analyzed by disc gel electrophoresis; this band shows glutathione transferase activity. Sodium dodecyl sulfate (SDS)–polyacrylamide slab gel electrophoresis confirms the homogeneity and indicates that the enzyme is composed of two identical subunits. Likewise, SDS–gel electrophoresis of peak II from the hydroxyapatite chromatography only shows one protein band. Disc gel electrophoresis of peak II usually shows 2–3 protein bands, all having glutathione transferase activity.

Molecular Properties. Glutathione transferase μ has a M_r of 53,000 and is composed of two very similar or identical subunits.[11] The basic forms are slightly smaller (M_r = 51,000; subunit M_r = 25,500).[3,11] The two groups of human liver transferases have different isoelectric points, at pH 6.6 (μ) and >pH 7.8 (α–ε), respectively, different amino acid compositions, and different circular dichroism spectra in the near-ultraviolet region.[11] They do not cross-react immunologically.[11]

Kinetic Properties. Transferase μ has high activity with CDNB, *trans*-4-phenyl-3-butene-2-one, benzo[*a*]pyrene 4,5-oxide, and styrene 7,8-oxide.[11] The basic transferases have much lower activity with the last three compounds. In contrast, the basic forms are much more active with cumene hydroperoxide and Δ^5-androstene-3,7-dione.[19]

[19] B. Mannervik, C. Guthenberg, H. Jensson, M. Warholm, and P. Ålin, *in* "Functions of Glutathione—Biochemical, Physiological, Toxicological and Clinical Aspects" (A. Larsson, S. Orrenius, A. Holmgren, and B. Mannervik, eds.), p. 75. Raven Press, New York, 1983.

Both transferase μ[11] and the basic isozymes (α–ε)[3] have been shown to bind nonsubstrate ligands, e.g., bilirubin.

Acknowledgments

This work was supported by grants from the Swedish Natural Science Research Council, the Swedish Cancer Society (to B. M.), and the Swedish Medical Research Council (to C. v. B.).

[63] Glutathione Transferase Isoenzymes from Rat Liver Cytosol

By HELGI JENSSON, PER ÅLIN, and BENGT MANNERVIK

Purification of different forms of glutathione transferase in rat liver cytosol has previously been described in this series.[1] However, advances in the understanding of the nature of different isoenzymes[2] and in methods of purification[3] warrant the publication of a new procedure. This chapter describes the preparation of seven major glutathione transferases in rat liver cytosol. Six of these isoenzymes are formed as binary combinations of four different protein subunits (denoted by Arabic numerals) which result in different catalytic activities.[2] A seventh isoenzyme is not yet fully characterized in terms of subunit composition. The glutathione transferases have now been named on the basis of their constituent subunits;[4] an outline of the nomenclature is presented in this volume [61]. Thus, glutathione transferase 1-2, 2-2, 3-3, and 3-4 correspond to B, AA, A, and C, respectively, in the previous nomenclature.[1,4]

Assay Method

Principle. The enzymatic activity is determined spectrophotometrically by measuring formation of the conjugate of the GSH and the CDNB at 340 nm.[1]

[1] W. H. Habig and W. B. Jakoby, this series, Vol. 77, p. 218.
[2] B. Mannervik and H. Jensson, *J. Biol. Chem.* **257,** 9909 (1982).
[3] H. Jensson, P. Ålin, and B. Mannervik, *Acta Chem. Scand., Ser. B* **36,** 205 (1982).
[4] W. B. Jakoby, B. Ketterer, and B. Mannervik, *Biochem. Pharmacol.* **33,** 2539 (1984).

Copyright © 1985 by Academic Press, Inc.
All rights of reproduction in any form reserved.

Reagents

Glutathione (GSH), 20 mM in deionized water
1,2-Chloro-2,4-dinitrobenzene (CDNB), 20 mM in 95% ethanol
Buffer A: 0.11 M sodium phosphate, pH 6.5
Buffer B: 10 mM Tris–HCl, 2 mM dithioerythritol (DTE), pH 7.8
Buffer C: 10 mM Tris–HCl, 0.2 M NaCl, 2 mM DTE, pH 7.8
Buffer D: 5 mM Tris–HCl, 2 mM DTE, pH 8.0
Buffer E: 25 mM triethylamine-HCl, pH 10.8 in degassed water
Buffer F: 6.25 ml Pharmalyte (pH 10.5–8), diluted with degassed water and adjusted to pH 8.0 with HCl to give a final volume of 500 ml

Note that the pH of the Tris–HCl buffers (Buffers A–D) refers to room temperature, but the pH of the others (E and F) at 5°.

Procedure. To a 1-ml cuvette are added suitable amount of Buffer A, 50 μl of 20 mM GSH and 50 μl of 20 mM CDNB. The reaction which is carried out at 30° is started by addition of enzyme (5–50 μl); the final volume is 1 ml. The reaction is monitored spectrophotometrically by the increase in absorbance at 340 nm ($\varepsilon = 9.6 \text{ m}M^{-1} \text{ cm}^{-1}$). Correction for the spontaneous activity is made by subtracting the rate in absence of enzyme.

Definition of Unit and Specific Activity. A unit of enzymatic activity is defined as the amount of enzyme that catalyzes formation of 1 μmol of *S*-2,4-dinitrophenylglutathione per minute at 30° using 1 mM concentration of GSH and CDNB. Specific activities are units per milligram of protein. Protein concentration is measured by the method of Lowry *et al.*[5] after precipitation with trichloroacetic acid.

Purification Procedure

The entire purification should be performed at 5°. The procedure described here is dimensioned for one rat liver (7–10 g). It can readily be scaled up severalfold.

Step 1. Preparation of Microsome-Free Cytosol Fraction. A male Sprague–Dawley rat is decapitated and the liver placed in ice-cold 0.25 M sucrose and homogenized in a Potter-Elvehjem homogenizer (three strokes at 440 rpm). The homogenate is made 20% (w/v) and centrifuged at 16,000 g for 20 min. The pellet is discarded and the supernatant is centrifuged at 105,000 g for 60 min.

Step 2. Chromatography in Sephadex G-25 (I). The supernatant from

[5] O. H. Lowry, N. J. Rosebrough, A. L. Farr, and R. J. Randall, *J. Biol. Chem.* **193,** 265 (1951).

PURIFICATION OF GLUTATHIONE TRANSFERASES FROM RAT LIVER CYTOSOL

Step	Volume (ml)	Total protein (mg)[a]	Total activity (units)	Specific activity (units/mg)	Yield (%)
1. Cytosol fraction	20	290	280	0.97	100
2. Sephadex G-25 (I)	45	243	243	1.00	86.8
3. Affinity chromatography	18	n.d.[b]	n.d.[b]	n.d.	n.d.
4. Sephadex G-25 (II)	30	9	240	27	85.7
5. Chromatofocusing					
Peak I (transferase 1-1)	8	0.30	14.9	50	39.5
Peak II (transferase 1-2)	12	0.85	21.3	25	
Peak III (transferase 2-2)	12	0.37	6.3	17	
Peak IV (transferase 3-3)	10	0.29	17.0	58	
Peak V (transferase 3-4)	10	0.84	38.0	45	
Peak VI (transferase 3-?)	14	0.11	7.0	64	
Peak VII (transferase 4-4)	16	0.35	6.0	17	

[a] Protein determined by the method of Lowry *et al.*[5] after precipitation with trichloroacetic acid.

[b] n.d., Not determined; *S*-hexylglutathione inhibits activity and interferes with protein measurements.

Step 1 is passed through a column (4 × 25 cm) of Sephadex G-25 Coarse equilibrated and eluted with Buffer B.

Step 3. Affinity Chromatography on S-Hexylglutathione Coupled to Sepharose 6B. The pooled material from Step 2 is applied to an affinity column (2 × 12 cm), prepared as described.[6] The gel is equilibrated with Buffer B. After application of the sample, the column is washed with 250 ml of Buffer C and then eluted with Buffer C fortified with 5 m*M* *S*-hexylglutathione. Glutathione transferase 5-5[1,4] is recovered in the effluent from the affinity column which is collected during application of the sample and washing.

Step 4. Chromatography on Sephadex G-25 (II). The pool from Step 3 is desalted on a Sephadex G-25 Coarse column (4 × 25 cm) equilibrated and eluted with Buffer D. The active fractions are pooled and concentrated to 4.5 ml by ultrafiltration using an Amicon PM-10 membrane.

Step 5. Chromatofocusing. The chromatofocusing gel PBE 118 is equilibrated with buffer E according to the instructions of the manufacturer. Before applying the concentrated pool from Step 4, 5 ml of Buffer F is pumped onto the column. The column is eluted with 500 ml of Buffer F. The flow rate should be 20 ml/hr and 2-ml fractions are collected. The

[6] B. Mannervik and C. Guthenberg, this series, Vol. 77, p. 231.

completion of the elution is checked by measuring the pH in the last fraction eluted. The pH value should have decreased to 8.0. Finally, the column is washed with 1 *M* NaCl to elute the less basic transferase isoenzymes.[2,3]

The table summarizes the results of a representative purification.

Properties

Some of the molecular and catalytic characteristics of the six basic isoenzymes of glutathione transferase in rat liver cytosol have been published.[2,7] The seventh isoenzyme, denoted 3-?, appears related to transferase 3-3. The properties of the dimeric proteins reflect the properties of their constituent subunits. The different subunits have distinct substrate specificities[2] and differential sensitivities to inhibitors.[7]

Acknowledgments

This work was supported by the Swedish Natural Science Research Council and the Swedish Cancer Society.

[7] S. Yalçin, H. Jensson, and B. Mannervik, *Biochem. Biophys. Res. Commun.* **114,** 829 (1983).

[64] Glutathione Transferase from Rat Testis

By Claes Guthenberg, Per Ålin, and Bengt Mannervik

Glutathione transferases play an important role in the biotransformation and detoxication of electrophilic xenobiotics.[1] The occurrence of glutathione transferase in animal species is widespread. Several transferase isoenzymes have been isolated from rat liver.[2] Six basic transferase isoenzymes in rat hepatic cytosol have been characterized as binary combinations of four protein subunits[3] designated by Arabic numerals.[4] The presence of glutathione transferase is not restricted to the liver but also has been demonstrated in extrahepatic organs. However, in comparison with the liver most other organs show considerably lower activity. One

[1] L. F. Chasseaud, *Adv. Cancer Res.* **29,** 175 (1979).
[2] W. B. Jakoby and W. H. Habig, *in* "Enzymatic Basis of Detoxication" (W. B. Jakoby, ed.), Vol. 2, p. 63. Academic Press, New York, 1980.
[3] B. Mannervik and H. Jensson, *J. Biol. Chem.* **257,** 9909 (1982).
[4] W. B. Jakoby, B. Ketterer, and B. Mannervik, *Biochem. Pharmacol.* **33,** 2539 (1984).

Copyright © 1985 by Academic Press, Inc.
All rights of reproduction in any form reserved.

exception is rat testis which also show high transferase activity. In contrast with liver, and most other organs, a major part of the glutathione transferase activity in rat testis is borne by an isoenzyme, glutathione transferase 6-6, with an acidic isoelectric point.[5,6] The purification of this enzyme, which accounts for approximately 50% of the cytosolic glutathione transferase activity, is described in this chapter.

In addition to this major acidic isoenzyme, smaller amounts of the basic species, glutathione transferases 2-2, 3-3, 3-4, and 4-4 have been identified in testis.[5]

Assay Method

Enzyme activity during purification is monitored spectrophotometrically at 340 nm by measuring the formation of the conjugate of glutathione and 1-chloro-2,4-dinitrobenzene.[7,8] The reagents, assay procedures, and the definition of unit of enzyme activity are described in the accompanying chapter [62].

Purification Procedure

The enzyme should be maintained at about 5° during the entire purification procedure.

Step 1. Preparation of Cytosol Fraction. Twenty Sprague–Dawley rats (180–200 g) are killed by decapitation. The testes are removed, rinsed with ice-cold 0.25 *M* sucrose, and trimmed free of adjoining tissues. The testes are cut into small pieces with scissors and homogenized (20% w/v) in ice-cold 0.25 *M* sucrose with a Potter-Elvehjem homogenizer equipped with a motor-driven Teflon pestle. The homogenate is centrifuged at 100,000 *g* for 90 min. The supernatant fraction is filtered through two layers of gauze. The proteins in the supernatant fraction are separated from low-molecular-weight compounds by chromatography on a column (4 × 40 cm) of Sephadex G-25 Coarse equilibrated with 10 m*M* Tris–HCl at pH 7.8. The active fractions are pooled.

Step 2. Affinity Chromatography on S-Hexylglutathione Coupled to Epoxy-Activated Sepharose 6B. The pooled fractions from Step 2 are applied to an affinity column (2 × 14 cm), packed in 10 m*M* Tris–HCl (pH 7.8). Nonspecifically adsorbed protein is eluted with 200 ml of 0.2 *M* NaCl

[5] C. Guthenberg, I.-M. Åstrand, P. Ålin, and B. Mannervik, *Acta Chem. Scand., Ser. B* **37,** 261 (1983).

[6] P. J. Dierickx and J. O. De Beer, *Biochem. Int.* **3,** 565 (1981).

[7] W. H. Habig, M. J. Pabst, and W. B. Jakoby, *J. Biol. Chem.* **249,** 7130 (1974).

[8] W. H. Habig and W. B. Jakoby, this series, Vol. 77, p. 218.

PURIFICATION OF THE MAJOR ACIDIC FORM OF GLUTATHIONE TRANSFERASE IN RAT TESTIS CYTOSOL

Step	Volume (ml)	Total protein (mg)[a]	Total activity (units)	Specific activity (units/mg)	Yield (%)
1. Supernatant fraction	116	840	1220	1.5	100
2. Affinity chromatography	27	—	(970)[b]		(80)[b]
3. Chromatofocusing	28	5.0	650	130	53

[a] Protein was estimated from the absorbance at 260 and 280 nm,[9] except in the last step where the method of Lowry *et al.*[10] was used.
[b] The inhibitor *S*-hexylglutathione present causes underestimation of the activity.

in the starting buffer. Glutathione transferase is eluted with 100 ml of 5 m*M* *S*-hexylglutathione in the buffer fortified with 0.2 *M* NaCl.

The active fractions from the affinity chromatography are separated from the inhibiting *S*-hexylglutathione and desalted on a column (4 × 20 cm) of Sephadex G-25 Coarse, equilibrated with 25 m*M* Tris-HAc (pH 8.0). The active fractions of effluent are pooled and concentrated to 5 ml with an Amicon PM-10 membrane.

Step 3. Chromatofocusing on gel PBE 94. Chromatofocusing is performed according to the instructions of the manufacturer (Pharmacia Fine Chemicals). The concentrated pool from Step 2 is applied to a column of chromatofocusing gel PBE 94, equilibrated with 25 m*M* Tris-HAc (pH 8.0). The column is washed with 40 ml of the starting buffer. The adsorbed transferases are eluted with 350 ml of a mixture of Polybuffer 96 (30%) and Polybuffer 74 (70%) diluted 1 : 10 and adjusted to pH 5.0 with acetic acid. The flow rate during the elution procedure is 15 ml/hr and the effluent is collected in 2-ml fractions. One major peak (transferase 6-6) and several minor peak are resolved. The most active fractions of the major peak are pooled. A summary of the results of a representative purification is given in the table.[9,10]

Preparation of Affinity Gel

The ligand, *S*-hexylglutathione, was synthesized according to Method A of Vince *et al.*[11,12] The ligand is linked to epoxy-activated Sepharose 6B

[9] H. M. Kalckar, *J. Biol. Chem.* **167,** 461 (1947).
[10] O. H. Lowry, N. J. Rosebrough, A. L. Farr, and R. J. Randall, *J. Biol. Chem.* **193,** 265 (1951).
[11] R. Vince, S. Daluge, and W. B. Wadd, *J. Med. Chem.* **14,** 402 (1971).
[12] B. Mannervik and C. Guthenberg, this series, Vol. 77, p. 231.

according to the instructions of the manufacturer (Pharmacia Fine Chemicals). Details are given by Mannervik and Guthenberg.[12]

Properties

Purity and Yield of the Enzyme. The glutathione transferase obtained is >90% pure as judged by sodium dodecyl sulfate slab gel electrophoresis. The yield of this major testicular isoenzyme is almost quantitative; almost all activity not recovered as transferase 6-6 in the purification is associated with other isoenzymes.

Some Molecular and Kinetic Properties. The enzyme has an isoelectric point at pH 5.8. The subunit molecular weight was determined as 26,000 by use of sodium dodecyl sulfate slab gel electrophoresis. The purified enzyme does not give a precipitin reaction with antibodies raised against rat liver transferases 1-1, 2-2, 3-3, or 4-4.

The enzyme has high activity with the substrate 1-chloro-2,4-dinitrobenzene (130 μmol/min/mg protein) but low or very low activity with all other glutathione transferase substrates tested. The substrates used include 1,2-dichloro-4-nitrobenzene, ethacrynic acid, *trans*-4-phenyl-3-buten-2-one, bromosulfophthalein, 1,2-epoxy-3-(*p*-nitrophenoxy) propane, and *p*-nitrobenzylchloride.

Acknowledgment

This work was supported by grants (to B.M.) from the Swedish Council for Planning and Coordination of Research.

[65] Cysteine Conjugate β-Lyase

By James L. Stevens and William B. Jakoby

$$RSCH_2\underset{\displaystyle NH_2}{\underset{|}{C}}HCOOH + H_2O \rightarrow RSH + CH_3COCOOH + NH_3 \quad (1)$$

The enzyme was originally described as active only with thioethers of cysteine[1,2] but is now known to also be effective with derivatives of

[1] M. Tateishi, S. Suzaki, and H. Shimizu, *J. Biol. Chem.* **253,** 8854 (1978); see also this series, Vol. 77, p. 253.

[2] M. Tateishi and H. Shimizu, *in* "Enzymatic Basis of Detoxication" (W. B. Jakoby, ed.), Vol. 2, p. 313. Academic Press, New York, 1980.

Copyright © 1985 by Academic Press, Inc.
All rights of reproduction in any form reserved.

alanine bearing a good leaving group at the β-carbon.[3–5] The β-lyase represents the initial enzyme in a shunt pathway from mercapturic acid synthesis,[4–6] that also includes thiol S-methyltransferase.[7]

Assay Methods

Principle. Both recommended methods allow continuous spectrophotometric estimation of the reaction with time. In Method A, the formation of 2-mercaptobenzothiazole [Eq. (2)] at 316 nm provides a sensitive assay

$$\text{(benzothiazol-2-yl)}-SCH_2CHNH_2COOH + H_2O \rightarrow \text{(benzothiazoline-2-)}{=}S + CH_3COCOOH + NH_3 \quad (2)$$

in both crude and purified systems; the disadvantage of the method is the need to synthesize the substrate.[8] Method B is presented for β-chloroalanine, a substrate that is commercially available, but the assay can be adapted for use with any of the substrates of the enzyme that produce pyruvate. The procedure depends on coupling the reaction with that of lactate dehydrogenase so as to follow spectrophotometrically the conversion of pyruvate to lactate. Method B is ineffective for crude extracts because of the presence of interfering activities.

Reagents

10 m*M* NADH
0.5 *M* potassium phosphate, pH 7.0
0.1 m*M* *S*-2-benzothiazolyl-L-cysteine (abbreviated CBZ for cystenylbenzothiazole)
100 m*M* β-chloroalanine (freshly prepared solution)

Method A.[3] To a cuvette (1 cm light path) containing a total volume of 1.0 ml, is added 0.1 ml of the buffer and 0.5 ml of CBZ. An appropriate

[3] J. L. Stevens and W. B. Jakoby, *Mol. Pharmacol.* **23,** 761 (1983).
[4] W. B. Jakoby and J. L. Stevens, *Biochem. Soc. Trans.* **12,** 33 (1984).
[5] W. B. Jakoby, J. Stevens, M. W. Duffel, and R. A. Weisiger, *Rev. Biochem. Toxicol.* **6,** 97 (1984).
[6] M. Tateishi and H. Shimizu, *Xenobiotica* **6,** 431 (1976).
[7] R. A. Weisiger and W. B. Jakoby, *Arch. Biochem. Biophys.* **196,** 631 (1979).
[8] D. R. Dohn and M. W. Anders, *Anal. Biochem.* **120,** 379 (1982).

amount of enzyme is used to initiate the reaction, the course of which is measured at 316 nm and 30° ($\Delta\varepsilon_{316}$ = 19,600 at pH 7.0). Under these conditions, the rate of the reaction appears as essentially linear for 3 min with respect to both time and protein concentration when less than 0.3 nmol of product has been formed. It should be noted that *each* of the substrates of the enzyme acts as a suicide inactivator[3] and, therefore, results in deviation from linearity that becomes obvious at high pH or when larger amounts of product are formed.

Method B.[3] To a cuvette containing a total volume of 1 ml, is added 0.1 ml of the buffer, 10 μl NADH solution, 0.1 unit of muscle lactate dehydrogenase, 0.1 ml β-chloroalanine, and an appropriate amount of enzyme. The course of the reaction is measured at 340 nm and 30° ($\Delta\varepsilon_{340}$ = 6200). Apparent linearity is achieved under these conditions subject only to the same reservations as expressed for Method A.

Units. A unit of enzyme activity is defined as the amount catalyzing the formation of 1 nmol of product per min under the conditions presented for Methods A and B. With CBZ as substrate, both methods result in the same value for activity. Specific activity is defined as units per mg of protein. Protein is determined with bovine serum albumin (Armour) as standard.[9]

Other Methods of Assay. Enzyme activity was originally measured after chromatography of a reaction mixture in which 2,4-[^{32}S]dinitrophenyl-*S*-cysteine served as substrate.[1] An endpoint assay has been introduced recently in which CBZ serves as substrate.[8] The enzyme can also be assayed with either kynurenin or 3-hydroxykynurenin as substrates by sensitive fluorometric means.[10]

Enzyme Purification[3]

The method makes uses of livers, obtained from male Sprague–Dawley rats weighing 200–250 g, that are stored at −70° for several months. Each preparation is based on 400 g of liver and all steps are carried out at 4°.

Step 1. Extract. The livers are allowed to thaw partially and are rinsed in water. Thawing is completed in 800 ml of 0.25 *M* sucrose containing 10 m*M* Tris–HCl at pH 7.5. Homogenization in a Waring blender is carried out for 30 sec using two batches of liver of 200 g each. After filtration through glass wool, the suspension is centrifuged at 100,000 *g* for 1 hr and the residue discarded.

[9] M. M. Bradford, *Anal. Biochem.* **72,** 248 (1975).

[10] W. B. Jakoby and D. M. Bonner, *J. Biol. Chem.* **205,** 699 (1953).

Step 2. DEAE-Cellulose I. The supernatant fluid is charged onto a column of DEAE cellulose (5 × 50 cm) that had been equilibrated with Buffer A (10 m*M* Tris–HCl at pH 7.5, containing 50 m*M* potassium chloride). The column is washed with 1 liter of Buffer A and a 3-liter gradient is applied in which KCl, in Buffer A, is increased from 50 to 350 m*M*. Active fractions, eluting at approximately 6 mmho, are pooled and concentrated 10-fold (Amicon with a PM 30 filter); all subsequent concentration steps are performed with the same ultrafiltration method.

Step 3. DEAE-Cellulose II. The concentrate is diluted 2-fold with 10 m*M* Tris chloride at pH 7.5 and charged onto a column of DEAE-cellulose (2 × 20 cm) that has been equilibrated with the same buffer. After washing with 100 ml of the buffer, the enzyme is eluted with a 1-liter gradient of the buffer increasing from 0 to 0.3 *M* potassium chloride. Fractions of 15 ml are collected in tubes in which 50 μl of 15 m*M* pyridoxal phosphate (pH 6.5) is present. Active fractions are concentrated 10-fold.

Step 4. Hydroxylapatite I. The protein solution is applied to hydroxylapatite (2 × 20 cm) that has been equilibrated with Buffer B (10 m*M* potassium phosphate at pH 7.0 containing 50 μM pyridoxal phosphate). After washing with 200 ml of Buffer B, elution is begun with an 800 ml gradient of Buffer B in which the potassium phosphate concentration rises linearly from 10 to 200 m*M*. The enzyme is eluted at about 7 mmho and is concentrated 10-fold.

Step 5. Hydroxylapatite II. After dilution of the protein solution with an equal volume of Buffer B, Step 4 is repeated using a 2 × 10 cm column of hydroxylapatite and a 400 ml gradient of Buffer B that ranges from 10 to 300 m*M* in potassium phosphate. Active fractions are again concentration 10-fold and are dialyzed against 1 liter of 10 m*M* potassium phosphate at pH 7.0 containing 50 μM pyridoxal phosphate.

Step 6. Chromatofocusing. A 0.5 × 20 cm column of Polybuffer Exchanger Gel (Pharmacia PBE94) is washed and equilibrated with 25 m*M* histidine–HCl at pH 6.0. A 10-fold dilution of Polybuffer 96 concentrate (Pharmacia) is adjusted to pH 5.0 and 5 ml is applied to the column. The dialysate from Step 5 is added and eluted with the diluted Polybuffer 96 until the column effluent reaches pH 5.0. Active fractions appear at pH 5.3, a pH at which the enzyme is unstable. For this reason, 0.5 ml fractions are collected in tubes containing 0.5 ml of 0.5 *M* potassium phosphate (pH 7.0) and 50 μl of 50 μM pyridoxal phosphate. Active fractions are pooled and concentrated to approximately 2 ml.

Step 7. Sepharose 6B. The concentrate is charged onto a column of Sepharose 6B (1.5 × 90 cm) equilibrated with 0.1 *M* sodium phosphate at pH 7.5 containing 50 μM pyridoxal phosphate, and washed with the same buffer. After collection of 2 ml fractions, the active fractions are pooled and concentrated to 1.5 ml.

Step 8. AH-Sepharose. To the concentrated protein from Step 7 is added 3 ml of water and the solution is applied to a column of AH-Sepharose (0.5 × 5 cm) equilibrated with 10 m*M* sodium phosphate at pH 7.8. After washing with 20 ml of the same buffer, enzyme is eluted with a 10 ml gradient of sodium phosphate rising linearly from 10 to 200 m*M*. Sufficient pyridoxal phosphate is added to each collecting tube to yield 50 μM in the final volume of 2 ml. Active fractions are pooled and concentrated to 2 ml.

Properties of the Enzyme[3]

Purification of about 1500-fold (see the table) led to a single, concomitant band of protein and β-lyase activity upon polyacrylamide gel electrophoresis. SDS-gel electrophoresis, however, revealed two major protein bands (M_r of 37,000 and 43,000) and two trace contaminants of somewhat higher molecular weight. Cross-linking of the subunits led to a band at M_r = 94,000 upon SDS-gel electrophoresis. Antibody obtained to the enzyme reacts (Ouchterlony immunodiffusion) with enzyme preparations from rat liver but not from kidney.

Substrates. Thioethers of cysteine, and derivatives of alanine, each bearing a good leaving group at the β-carbon, serve as substrates. Among them are *S*-2-benzothiazolyl-L-cysteine (K_m = 0.01 m*M*), *S*-1,2-dichlorovinyl-L-cysteine (K_m = 100 μM), *S*-2,4-dinitrophenyl-L-cysteine[1] (K_m = 0.08 m*M*), and β-chloroalanine (K_m = 2.6 m*M*). Maximal activity

SUMMARY OF PURIFICATION OF CYSTEINE CONJUGATE β-LYASE FROM RAT LIVER (400 g)[a]

Step	Volume (ml)	Total protein (mg)	Total activity (nmol min^{-1})	Specific activity (nmol min^{-1} mg^{-1})
1. Extract	800	25,600	5,120	0.2
2. DEAE-cellulose I	202	1,480	5,590	3.8
3. DEAE-cellulose II	190	266	5,080	19
4. Hydroxylapatite I	80	72	3,380	47
5. Hydroxylapatite II	88	62	3,280	53
6. Chromatofocusing	1.85	7.0	2,090	297
7. Sepharose 6B	1.35	3.5	1,100	314
8. AH-Sepharose	1.4	2.1	668	318

[a] From Stevens and Jakoby.[3]

for these compounds varied between 1.5 and 2.0 μmol min^{-1} mg^{-1}. *S*-2,4-Dinitrophenyl-L-homocysteine, *S*-4-nitrobenzyl-L-cysteine, and *S*-4-bromophenyl-L-cysteine have also been shown to act as substrates by using a chromatographic assay.[1,2] Kynurenine and 3-hydroxykynurenine are substrates but pyruvate is not a product of these reactions.[4,5] Indeed, purification of rat hepatic kynureninase[11] yields a preparation with both kynureninase and cysteine conjugate β-lyase activities.[12] All of the substrates that were measured spectrophotometrically behaved as suicide inactivators.[3] The partition ratio, i.e., the rate of product formed per inactivation event, is essentially the same, about 600, despite greater *rates* of inactivation at higher pH values. The following were inactive as substrates: cystathionie; the 4-nitrophenylethyl, benzyl, methyl, 2-chloroethyl, and butyl thioethers of L-cysteine; L-aspartate; L-serine; L-tryptophan; and L-*O*-phosphoserine.

pH Optimum. Estimation of initial rates discloses a pH optimum at 8.8 with half of maximal activity at pH 7.5 and 9.5 in the CBZ assay with Tris–HCl. However, in order to approximate liner rates, it was necessary to use potassium phosphate at pH 7.0 in the standard assay system. Comparison of 50 m*M* buffers at pH 7.5 in the otherwise standard CBZ assay revealed the following relative rates: potassium phosphate, 1.0; sodium pyrophosphate, 1.8; Tris hydrochloride, 0.8; and sodium borate, 0.6. EDTA was without effect. Tris, added to the normal phosphate buffer of the same pH, results in inhibition (50% at 0.3 *M*).

Pyridoxal Phosphate. Direct evidence for the participation of this coenzyme is now available. All of the findings agree with such a relationship based on analogy with the reaction catalyzed by cystathionase, a pyridoxal phosphate enzyme, and the requirement for pyridoxal phosphate in order to overcome the instability of cysteine conjugate β-lyase. Inhibition of the β-lyase by carbonyl group reagents (hydroxylamine, I_{50} = 50 μ*M*; potassium cyanide, I_{50} = 3 m*M*) and by L-cysteine (I_{50} = 7 m*M*) is in accord with a role for pyridoxal phosphate. L-Cysteine appears to be noncompetitive with the substrates.

[11] F. Takeuchi, H. Otsuka, and T. Shibata, *J. Biol. Chem. Tokyo*, **88**, 97 (1980).
[12] J. L. Stevens, *J. Biol. Chem.*, in press.

[66] Cysteine S-Conjugate N-Acetyltransferase

By MICHAEL W. DUFFEL and WILLIAM B. JAKOBY

$$\begin{array}{l} \text{COOH} \\ | \\ \text{CHNH}_2 \\ | \\ \text{CH}_2 \\ | \\ \text{S—R} \end{array} + \text{CoASAc} \rightarrow \begin{array}{l} \text{COOH} \\ | \\ \text{CHNHAc} \\ | \\ \text{CH}_2 \\ | \\ \text{S—R} \end{array} + \text{CoASH}$$

Cysteine *S*-conjugate *N*-acetyltransferase from rat kidney microsomes catalyzes the final reaction in mercapturic acid biosynthesis, the acetylation of cysteine thioethers.[1] This enzyme is not active with substrates of the cytosolic amine *N*-acetyltransferase that had been described previously.[2]

Assay Method[1]

Principle. The transfer of a radiolabeled acetyl group from CoASAc to a cysteine conjugate is used as the basis for measurement of the enzyme.[1,3] After acidification, the mercapturic acid is extracted into cyclohexanone for determination of radioactivity.

Reagents

- 0.2 *M* potassium phosphate, pH 7.0
- 8 m*M* [1-^{14}C]CoASAc. The commercial radioactive compound is adjusted to a specific activity of 0.25 μCi/μmol with unlabeled sodium CoASAc and water (pH 5.0 to 6.0)
- 1.33 *M* acetic acid
- Cyclohexanone
- 10 m*M* *S*-benzyl-L-cysteine

Procedure. In a total volume of 200 μl are added the following reagents in the indicated sequence: 125 μl of the phosphate buffer, 25 μl *S*-benzyl-L-cysteine, 25 μl CoASAc, and 50 μl water. After 2 min of incubation at 37°, the reaction is initiated by addition of enzyme in a volume of

[1] M. W. Duffel and W. B. Jakoby, *Mol. Pharmacol.* **21,** 444 (1982).
[2] H. Tabor, A. H. Mehler, and E. R. Stadtman, *J. Biol. Chem.* **204,** 127 (1953).
[3] R. M. Green and J. S. Elce, *Biochem. J.* **147,** 283 (1975).

METHODS IN ENZYMOLOGY, VOL. 113
Copyright © 1985 by Academic Press, Inc.
All rights of reproduction in any form reserved.

25 μl. Reaction mixtures are incubated at 37° for 4 min, and stopped by addition of 0.75 ml of the acetic acid solution. Immediately after addition of the acid, 2 ml of cyclohexanone is added and the mixture is subjected to vortex mixing for 10 sec. The phases are separated by centrifugation for 3 min at 1500 rpm, and a 1-ml aliquot of the cyclohexanone layer is added to 10 ml of Hydrofluor (National Diagnostics, Sommerville, NJ) for determination of radioactivity. Control incubations are carried out in the absence of *S*-benzyl-L-cysteine and the values obtained are subtracted from values for complete assays. Control values less than 5% of experimental values are considered acceptable.

Under the assay conditions described, product formation is linear for at least 8 min with 0.5 mg or less of rat kidney microsomal protein. For the solubilized enzyme, activity is a linear function of protein concentration when less than 15 units of enzyme is used per standard assay. Extraction efficiency for 25 nmol of *N*-acetyl-*S*-benzyl-cysteine is 101 ± 1.5%.[1] The lower volatility and superior extraction properties of cyclohexanone present distinct advantages over previous procedures using ethyl acetate.[3]

Definition of Units. A unit of activity is the amount of enzyme required for formation of 1 nmol of product per min in the standard assay. Specific activity is expressed in terms of units of activity per mg of protein. Protein is determined colorimetrically[4] with crystalline bovine serum albumin (Armour) as a standard.

Purification Procedure[1]

Solubilization and purification of cysteine *S*-conjugate *N*-acetyltransferase is carried out with kidneys from male Sprague–Dawley rats of 175–200 g (ARS Sprague-Dawley, Madison, WI). Kidneys are stored at −70° prior to use. All purification steps are conducted at 4° unless otherwise noted. The results of the purification procedure are summarized in Table I.

Step 1. Preparation of Microsomes. Rat kidneys, about 75 g, are thawed and homogenized in 300 ml of 0.25 *M* sucrose containing 1 m*M* dithiothreitol. Homogenization is conducted with three 10-sec pulses of a Waring Blender, followed by four passes with a Teflon-glass homogenizer. The suspension is centrifuged at 12,000 *g* for 10 min and the supernatant fluid treated at 100,000 *g* for 1 hr.[5] The resultant residue, the microsomal fraction, is suspended in a solution of 0.25 *M* sucrose, 1 m*M* dithiothreitol, and 0.2 *M* potassium chloride. After centrifugation at

[4] M. M. Bradford, *Anal. Biochem.* **72,** 248 (1976).

[5] P. Siekevitz, this series, Vol. 5, p. 61.

TABLE I
SUMMARY OF PURIFICATION OF CYSTEINE *S*-CONJUGATE *N*-ACETYLTRANSFERASE

Step	Volume (ml)	Total activity (units)	Total protein (mg)	Specific activity (units/mg)
1. Microsomes	86	126,000	969	130
2. Solubilization	167	131,000	935	140
3. Polyethylene glycol	159	92,000	715	129
4. DEAE-cellulose	43	21,000	90	232
5. Hydroxylapatite	10	11,000	28	394

100,000 *g* for 80 min, the pellet is suspended to a protein concentration of between 10 and 15 mg/ml in fresh 0.25 *M* sucrose, containing 1 m*M* dithiothreitol and 0.2 *M* potassium chloride.

Step 2. Solubilization. The microsomal suspension is treated with an equal volume of a mixture containing 0.25 *M* sucrose, 20 m*M* potassium phosphate, pH 7.0, 1 m*M* dithiothreitol, 0.2 *M* KCl, and 20 mg of the DGDC[6] detergent per ml. The ratio of DGDC to microsomal protein is approximately 2 : 1 (w/w). After stirring for 10 min, the mixture is homogenized in a Teflon-glass tissue grinder and subjected to centrifugation at 100,000 *g* for 80 min. The supernatant liquid is retained.

Step 3. Polyethylene Glycol. A 50% (w/w) solution of polyethylene glycol (PEG-6000) is added to the supernatant fluid to yield a final polymer concentration of 20% (w/v). The mixture is stirred for 20 min and the precipitate collected by centrifugation for 25 min at 18,000 *g*. The pellet is suspended in 150 ml of Buffer A [10 m*M* Tris–HCl, pH 7.8 at 25°, 30% (v/v) glycerol, 1 m*M* dithiothreitol, and 5 mg DGDC per ml].

Step 4. DEAE-Cellulose. A column of DEAE-cellulose (Whatman DE-52, 4 × 22 cm) is equilibrated with Buffer A, and the enzyme preparation is applied at a flow rate of 1.2 ml/min. The column is washed with 200 ml Buffer A and eluted with a linear gradient formed between 700 ml Buffer A and 700 ml Buffer A supplemented to 0.3 *M* KCl. Fractions containing *N*-acetyltransferase activity are combined and concentrated to approximately 5 mg protein per ml in an Amicon apparatus with PM10 membrane. Salt is removed by precipitation of the enzyme with 20% (w/v) PEG-6000, followed by centrifugation at 25,000 *g* for 30 min. The residue

[6] DGDC serves as an abbreviation for *N*,*N*-bis(3-D-gluconamidopropyl)-3α,12α-dihydroxy-5β-cholan-24-amide. The compound was prepared by the method of L. M. Hjelmeland, *Proc. Nat. Acad. Sci. USA* **77,** 6368 (1980).

TABLE II
SUBSTRATE SPECIFICITY OF CYSTEINE S-CONJUGATE N-ACETYLTRANSFERASE[a]

Acetyl acceptor	Relative rate[b]	K_m[c] (μM)	V_m[c] protein)
S-Benzyl-L-cysteine	100	140	440
S-Butyl-L-cysteine	69	63	320
S-Propyl-L-cysteine	46	670	330
S-Ethyl-L-cysteine	11	7100	360
S-Methyl-L-cysteine	0.7		
L-Cysteine	<0.1		
S-Benzyl-D-cysteine	1.1		
O-Benzyl-L-serine	16	2600	210
S-Carboxymethyl-L-cysteine	<0.1		
L-Tryptophan	0.3		

[a] From Duffel and Jakoby.[1]

[b] Determined under standard assay conditions at an acetyl acceptor concentration of 1 m*M*.

[c] Apparent K_m and V_m at 0.8 m*M* CoASAc under standard assay conditions.

is suspended in Buffer B [10 m*M* potassium phosphate, pH 7.0 at 25°, 30% (v/v) glycerol, 1 m*M* dithiothreitol, and 5 mg DGDC per ml].

Step 5. Hydroxylapatite. The product of Step 4 is applied at a rate of 1.2 ml/min to a column of spheroidal hydroxylapatite (Gallard-Schlesinger)[7] (2.5 × 29 cm) which has been equilibrated with Buffer B. The column is washed with 280 ml of Buffer B and eluted with a linear gradient formed between 225 ml of Buffer B and 225 ml of Buffer B containing 0.3 *M* potassium phosphate (pH 7.0 at 25°). After this gradient, 250 ml of Buffer B, containing 0.3 *M* potassium phosphate, is applied and fractions with *N*-acetyltransferase activity are pooled, supplemented with sodium azide to 3 m*M*, and concentrated with an Amicon PM10 membrane.

Properties[1]

Purity and Stability. Although the enzyme obtained by this procedure is not purified to high specific activity, it is soluble and looses only 30% of its activity after storage at 4° for 3 months. Hydrolytic enzymes that are active in the scision of CoASAc and of *N*-acetyl-*S*-benzyl-L-cysteine,

[7] The usual hydroxylapatite preparations, either commercial or made in the laboratory by the method of O. Levin (this series, Vol. 5, p. 27), were ineffective for this separation.

both present prior to Step 4, have been removed from the transferase preparation.

Substrate Specificity. The purified *N*-acetyltransferase catalyzes acetylation of lipophilic thioethers of L-cysteine (Table II). Neither L-cysteine itself nor polar thioether conjugates such as *S*-carboxymethyl-L-cysteine serve as acetyl acceptors. The efficiency of enzyme catalysis, as measured by V_m/K_m, increases with the lipophilicity of the substituent (butyl > benzyl > ethyl). The V_m/K_m values correlate well with Hansch π constants for ethyl, propyl, butyl, and benzyl substituents.[8]

S-Acetyl coenzyme A serves as acetyl donor with an apparent K_m of 26 μM (V_m = 420 nmol min^{-1}) in an otherwise standard assay system.

Substitution of oxygen for sulfur, as in *O*-benzyl-L-serine, greatly increases the apparent K_m, with only slight decrease in the maximal velocity of the reaction. The configuration of the amino acid α-carbon is also important; rates for D-cysteine and D-serine derivatives are either too low or are negligible when limits of stereochemical purity are considered. The following do not serve as substrates: aniline, *p*-aminobenzoate, D-cysteine, L-methionine, L-leucine, L-phenylalanine, and L-glutamate.

pH Optimum. The optimum pH for acetylation of *S*-benzyl-L-cysteine catalyzed by the partially purified *N*-acetyltransferase is pH 6.8–7.0. This pH optimum for the soluble enzyme is the same in both Tris–HCl and potassium phosphate buffers, in contrast to a previously reported difference in pH optima between the two buffers when microsomes were used.[3]

[8] C. Hansch and A. Leo, "Substiuant Constants for Correlation Analysis in Chemistry and Biology." Wiley, New York, 1979.

[67] Thioltransferase from Human Placenta

By KERSTIN LARSON, VÉRONIQUE ERIKSSON, and BENGT MANNERVIK

Thioltransferase catalyzes the reversible thiol-disulfide interchange reactions

$$RSH + R'SSR' \rightleftharpoons RSSR' + R'SH$$

where RSH is a thiol and R'SSR' a disulfide. The enzyme has a major role in maintaining intracellular thiols in the reduced state and functions in this capacity by coupling to glutathione and glutathione reductase.[1] In addi-

[1] B. Mannervik, *in* "Enzymatic Basis of Detoxication" (W. B. Jakoby, ed.), Vol. 2, p. 229. Academic Press, New York, 1980.

Copyright © 1985 by Academic Press, Inc.
All rights of reproduction in any form reserved.

tion, there is evidence that thioltransferase has a role in cellular regulation by catalyzing the reversible modification of proteins by thiol-disulfide interchange.[2] Purification of thioltransferase from rat liver cytosol has been described earlier.[3,4]

This chapter describes the preparation of a cytosolic thioltransferase from human placenta.

Assay Method

Principle. Thioltransferase activity is monitored spectrophotometrically at 340 nm by coupling of the thioltransferase-catalyzed reaction, with glutathione as the thiol, to the reduction of glutathione disulfide catalyzed by glutathione reductase:

$$\mathrm{GSSG + NADPH + H^+ \rightarrow 2GSH + NADP^+}$$

The velocity of a blank reaction, run in the absence of thioltransferase, must be subtracted from the velocity determined in the presence of thioltransferase, because thiol-disulfide interchange takes place spontaneously.

We recommend the use of the thiosulfate ester *S*-sulfocysteine, $CySSO_3^-$, as a substrate, because it gives a lower spontaneous reaction rate than many alternative disulfide substrates. The synthesis of $CySSO_3^-$ is described in ref. 5. Thioltransferase catalyzes the two consecutive reactions:

$$\mathrm{GSH + CySSO_3^- \rightarrow GSSCy + HSO_3^-}$$
$$\mathrm{GSH + GSSCy \rightarrow GSSG + CySH}$$

The GSSG formed is reduced by glutathione reductase.

Reagents

Potassium phosphate buffer, 0.2 *M*, pH 7.6 containing 2 m*M* EDTA
NADPH, 2 m*M*, dissolved in 50 m*M* Tris–HCl, pH 7.0
GSH, 10 m*M*, dissolved in deionized water
Glutathione reductase, 200 units/ml (commercially available)
$CySSO_3^-$, 60 m*M*, dissolved in deionized water

[2] B. Mannervik and K. Axelsson, *Biochem. J.* **190,** 125 (1980).
[3] K. Axelsson, S. Eriksson, and B. Mannervik, *Biochemistry* **17,** 2978 (1978).
[4] B. Mannervik, K. Axelsson, and K. Larson, this series, Vol. 77, p. 281.
[5] I. H. Segel and M. J. Johnson, *Anal. Biochem.* **5,** 330 (1963).

Procedure

To a 1-ml cuvette are added at 30°: 500 μl of buffer, 50 μl of NADPH, 50 μl of GSH, 10 μl of glutathione reductase, and deionized water to give a final volume of 1 ml in the cuvette, after addition of the enzyme solution and 50 μl of $CySSO_3^-$.

The reaction is initiated by addition of $CySSO_3^-$.

The spontaneous reaction is measured separately in the absence of thioltransferase.

Definition of Unit and Specific Activity

A unit of thioltransferase activity is defined as the amount of enzyme that catalyzes the formation of 1 μmol of GSSG per min at 30° under steady-state conditions (which should be reached within 2 min).

It is essential that the coupling enzyme, glutathione reductase, is present at a concentration high enough to prevent the coupling reaction from being rate limiting.

Specific activity is expressed as units per milligram of protein.

Purification Procedure

The enzyme is kept at about 5° during the entire purification procedure. All buffers contain 1 m*M* EDTA. The use of plastic tubes when the enzyme is relatively pure seems to increase the yield (from step 5 on).

Step 1. Preparation of Cytosol Fraction. One full-time placenta, between 2 and 3 hr postpartum, is obtained from a nearby hospital. We use only one placenta at a time to avoid possible genetic differences.

The placenta is cut free from connective tissue and membranes. Thereafter, it is cut in 1-cm^3 pieces. A 30% (w/v) homogenate is made up in 0.25 *M* ice-cold sucrose by homogenization for 2 × 1 min in a blender. The homogenate is centrifuged for 60 min at 11,000 *g*. The supernatant fraction is filtered through cheese cloth to remove a layer of lipid at the surface of the liquid.

Step 2. Chromatography on Sephadex G-25 (*1*). The enzyme from Step 1 is run through a column (12.5 × 80 cm) of Sephadex G-25 Coarse, equilibrated with 10 m*M* Tris–HCl (pH 7.8). The protein peak, about 1500 ml, emerges after 3 liters of effluent.

Step 3. Chromatography on DEAE-Cellulose. The pooled effluent from Step 2 is applied to a column (9 × 15 cm) of DEAE-cellulose (Whatman DE 52) equilibrated with 10 m*M* Tris–HCl (pH 7.8). This step removes hemoglobin while thioltransferase passes unretarded. The effluent from the column is collected after about 500 ml. Thioltransferase is collected in about 2 liters.

Step 4. Chromatography on SP-Sephadex C-25 (I). The pool from Step 3 is adjusted to pH 5.7 with ice-cold 0.2 *M* acetic acid and loaded on an SP-Sephadex C-25 column (4 × 30 cm) equilibrated with 10 m*M* sodium phosphate (pH 5.7), containing 0.1 m*M* dithioerythritol (DTE). After washing the gel with 3–5 bed volumes of starting buffer, thioltransferase is eluted with the same buffer containing 0.5 *M* NaCl.

Step 5. Chromatography on Thiopropyl-Sepharose 6B (Pharmacia). The pooled active fractions from the SP-Sephadex chromatography (total volume of about 200 μl) is applied to a Thiopropyl-Sepharose 6B column (2 × 4 cm) equilibrated with 10 m*M* sodium phosphate (pH 7.0) containing 0.5 *M* NaCl. The pH of the sample is adjusted to 7.0 with ice-cold 0.2 *M* NaOH before application. The column is washed with starting buffer until the effluent is colorless. Thioltransferase is eluted by 20 m*M* DTE in the sodium phosphate buffer. NaCl is excluded from the buffer during elution. The colored fractions are pooled (approximately 25 ml). Determination of activity or protein in this effluent is interfered with by 2-thiopyridine liberated from the column. This compound has high absorbance at 340 nm.

Step 6. Chromatography on Sephadex G-25 (II). The pool of the colored fractions from Step 5 is chromatographed on a column (4 × 25 cm) of Sephadex G-25 (Coarse) equilibrated with 10 m*M* potassium phosphate (pH 7.0).

Step 7. Chromatography on Hydroxyapatite. The pool from the Sephadex G-25 column (about 55 ml) is passed through a column (2 × 6 cm) of hydroxyapatite (Bio-Rad Laboratories), equilibrated with 10 m*M* potassium phosphate (pH 7.0). Thioltransferase passes unretarded through the column.

Step 8. Chromatography on SP-Sephadex C-25 (II). The effluent from the hydroxyapatite, adjusted to pH 5.7 with ice-cold acetic acid, is applied to a second SP-Sephadex C-25 column (1 × 3 cm), equilibrated with 10 m*M* potassium phosphate (pH 5.7). Thioltransferase is eluted with a 20-ml pulse of 0.5 *M* NaCl in the starting buffer after washing the column with 3–5 bed volumes of starting buffer.

The specific activity of thioltransferase in the peak fractions after Step 8 is approximately 200 μmol/min/mg protein. The overall recovery is somewhat less than 10% (see the table).

Properties

Stability. The purified thioltransferase from placenta is stored frozen in plastic tubes. It is stable for several weeks at −20°.

Molecular Properties. The molecular weight is 6000, as determined independently by gel filtration and sodium dodecyl sulfate–polyacryl-

PURIFICATION OF THIOLTRANSFERASE FROM HUMAN PLACENTA

Step	Volume (ml)	Total protein (mg)	Total activity (units)	Specific activity (units/mg)
1. Cytosol fraction	920	41,400	867	0.021
2. Sephadex G-25 I	1,500	20,400	783	0.029
3. DEAE cellulose	2,350	893	540	0.60
4. SP-Sephadex A-25 I	208	254	334	1.31
5. Thiopropyl-Sepharose	25	—	—	
6. Sephadex G-25 II	54	30.2	88	2.9
7. Hydroxyapatite	54	1.84	83.9	45.6
8. SP-Sephadex II	12	0.5	70	140

amide gel electrophoresis. Amino acid analysis demonstrates the presence of 2 cysteine residues, which presumably participate in the redox process of the catalytic mechanism. In the absence of reducing agents, isoelectric focusing of the purified thioltransferase shows 2 components separated by approximately 0.6 pH unit. Pretreatment of the enzyme with 20 m*M* DTE for 1 hr eliminated the form with the higher pH value with a corresponding increase of the component focusing at about pH 7.

Substrate Specificity. Low-molecular-weight disulfides such as cystine, homocystine, and cystamine, as well as the peptides vasopressin, oxytocin, and insulin, previously studied with the cytosolic rat enzyme,[3] serve as substrates also for the placental thioltransferase. Fructose-1,6-bisphosphatase from rabbit liver is a substrate as shown by the catalytic effect of thioltransferase on the activation of the enzyme with cystamine (cf. ref. 6). This finding indicates that, like the rat enzyme, thioltransferase from placenta may have a regulatory role in modulating the biological activity of proteins.

Acknowledgment

This work was supported by grant (to B.M.) from the Swedish Natural Science Research Council.

[6] S. Pontremoli, S. Traniello, M. Enser, S. Shapiro, and B. L. Horecker, *Proc. Natl. Acad. Sci. U.S.A.* **58,** 286 (1967).

[68] Glutaredoxin from *Escherichia coli* and Calf Thymus

By ARNE HOLMGREN

Introduction

The reduction of ribonucleotides to the corresponding deoxyribonucleotides, required for DNA synthesis, is catalyzed by the enzyme ribonucleoside diphosphate reductase.[1] Two different hydrogen donor systems for this enzyme have been identified. One is the well studied NADPH-dependent thioredoxin system (1) and the other is a glutathione-dependent system where the protein glutaredoxin[2–5] couples the oxidation of GSH to the reduction of ribonucleotides. The glutathione disulfide is reduced by glutathione reductase and NADPH [Reaction (2)].

$$\mathrm{NDP + 2GSH \xrightarrow[\text{Glutaredoxin}]{\text{Ribonucleotide reductase}} dNDP + GSSG + H_2O} \quad (1)$$

$$\mathrm{GSSG + NADPH + H^+ \xrightarrow{\text{Glutathione reductase}} 2GSH + NADP^+} \quad (2)$$

Glutaredoxin was originally discovered[2] in a mutant of *Escherichia coli* (*tsnC* 7004)[6] lacking detectable thioredoxin[7] but with a fully active NADPH-dependent ribonucleotide reduction.[2,7] Glutaredoxin has been purified from wild type *E. coli* B cells[4] and characterized as a small (M_r 10,000) acidic protein, containing a single disulfide bond, which could be reduced to a dithiol by GSH, glutathione reductase and NADPH, but not by thioredoxin reductase and NADPH.[4]

The first evidence of a mammalian glutaredoxin was obtained in calf thymus,[3] using the homologous calf thymus ribonucleotide reductase.[8] Glutaredoxin has subsequently been purified to homogeneity from calf thymus.[9]

E. coli and calf thymus glutaredoxin, as well as phage T4 thioredoxin,[10] have inherent GSH-disulfide oxidoreductase (transhydrogenase)

[1] A. Holmgren, *Curr. Top. Cell. Regul.* **19,** 47 (1981).
[2] A. Holmgren, *Proc. Natl. Acad. Sci. U.S.A.* **73,** 2275 (1976).
[3] M. Luthman, S. Eriksson, A. Holmgren, and L. Thelander, *Proc. Natl. Acad. Sci. U.S.A.* **76,** 2158 (1979).
[4] A. Holmgren, *J. Biol. Chem.* **254,** 3664 (1979).
[5] A. Holmgren, *J. Biol. Chem.* **254,** 3672 (1979).
[6] M. Chamberlin, *J. Virol.* **14,** 509 (1974).
[7] A. Holmgren, I. Ohlsson, and M.-L. Grankvist, *J. Biol. Chem.* **253,** 430 (1978).
[8] Y. Engström, S. Eriksson, L. Thelander, and M. Åkerman, *Biochemistry* **18,** 2941 (1979).
[9] M. Luthman and A. Holmgren, *J. Biol. Chem.* **257,** 6686 (1982).
[10] A. Holmgren, *J. Biol. Chem.* **253,** 7424 (1978).

Copyright © 1985 by Academic Press, Inc.
All rights of reproduction in any form reserved.

activity. This allows a simple and rapid spectrophotometric assay, independent of ribonucleotide reductase.

Glutaredoxin from *E. coli* B and K12

Ribonucleotide Reductase Assay

Principle. Glutaredoxin is required for GSH-dependent reduction of ribonucleotides to deoxyribonucleotides by *E. coli* ribonucleotide reductase. The purification of ribonucleotide reductase from the overproducing strain *E. coli* KK 546 that contains a defective lambda carrying the genes for this enzyme has been covered in this series previously.[11] Glutaredoxin activity is determined by measuring the formation of [^{3}H]dCDP from [^{3}H]CDP.[1,4,5]

Procedure. Glutaredoxin is determined in a coupled assay with ribonucleotide reductase.[4] The incubation mixture contains 200 nmol of ATP, 90 nmol of [^{3}H]CDP (5000 cpm/nmol), 1.5 μmol of $MgCl_2$, 80 nmol of NADPH, 5 μmol of *N*-2-hydroxyethylpiperazine-*N'*-2-ethanesulfonic acid buffer, pH 7.6, 0.5 mg/ml of bovine serum albumin, 4 m*M* GSH, and 7 μg/ml of yeast glutathione reductase. Glutaredoxin-containing fractions (less than 50 μl) are added and the reaction is started by adding 3 μl of *E. coli* ribonucleotide reductase (20 mg/ml).[11] The enzyme is a mixture of protein B1 and B2 in a 1 : 1 ratio and is stored frozen at −70° in 50 m*M* Tris–Cl, pH 7.6–15 m*M* $MgCl_2$–10 m*M* dithiothreitol (DTT)–20% glycerol in 15-μl aliquots which are thawed once.[11] The enzyme is diluted 4-fold with 50 m*M* Tris–Cl, pH 7.5 immediately before use. The final assay volume is 125 μl. After incubations for 20 min at 37° the reaction is stopped by the addition of 1 ml of 1 *M* $HClO_4$. The amount of [^{3}H]dCDP formed is determined after hydrolysis to [^{3}H]dCMP by chromatography on 3 ml Dowex-50 columns eluted with 55 and 25 ml of 0.2 *M* acetic acid, respectively.[11] The amount of [^{3}H]dCMP is determined[11] by liquid scintillation counting of 1 ml of the dCMP eluate (25 ml) in 10 ml of Instagel (Packard).

Definition of Unit and Specific Activity. One unit of glutaredoxin corresponds to the net formation of 1 nmol of dCDP/min at 37°. The assays for glutaredoxin may be standardized for variations in the activity of ribonucleotide reductase by two controls with 70 and 140 pmol of thioredoxin plus excess NADPH and thioredoxin reductase (3 μg).[4] Thioredoxin, 70 pmol, gave 0.60 nmol of dCDP/min. The assay for glutaredoxin in extracts, which also contained the thioredoxin system, was done after

[11] L. Thelander, B.-M. Sjöberg, and S. Eriksson, this series, Vol. 51, p. 227.

addition of 20 μl of rabbit antithioredoxin γ-globulin.[2] Specific activity is defined as units of enzyme per mg of protein.

Comments. Assays of glutaredoxin by ribonucleotide reductase may also be performed spectrophotometrically by following the oxidation of NADPH [Reactions (1) and (2)].[4] However, this assay can only be used at later stages of purification due to its lower sensitivity[4] and interference with other NADPH-consuming reactions in the first steps of the preparation.

Hydroxyethyldisulfide (HED) Assay

Principle. Glutaredoxin catalyses reduction of certain disulfides by GSH[4] and thus has GSH-disulfide-transhydrogenase[12,13] or oxidoreductase activity (EC 1.8.4.4 and 1.8.4.1). When the reaction is coupled to NADPH and glutathione reductase, the overall reaction may be characterized as an NADPH-dependent disulfide reduction. As seen from Reaction (3)–(5), below, the reaction rate may be followed spectrophotometrically at 340 nm from the disappearance of NADPH:

$$2GSH + X\text{-}S\text{-}S\text{-}X \xrightarrow{\text{Glutaredoxin}} GSSG + 2X\text{-}SH \qquad (3)$$

$$GSSG + NADPH + H^+ \xrightarrow{\text{Glutathione reductase}} 2GSH + NADP^+ \qquad (4)$$

$$\text{Net: } NADPH + H^+ + X\text{-}S\text{-}S\text{-}X \rightarrow NADP_+ + 2XSH \qquad (5)$$

Procedure

Reagents

NADPH 40 mM in H_2O, stored at $-20°$

HED (hydroxyethyldisulfide) (Aldrich 98%), 15 mM solution is made in 50 mM Tris–Cl, pH 8.0; stored frozen at $-20°$

Tris–Cl, 1.00 M, pH 8.0

EDTA, 0.2 M neutralized to pH 7.45 with NaOH

Glutathione reductase 2.0 mg/ml from yeast (Boehringer or Sigma, 200 U/mg) dialyzed against 50 mM Tris–Cl–1 mM EDTA and concentrated by ultrafiltration to 2.0 mg/ml (2.7 A_{280} nm); stored at $-20°$

Bovine serum albumin (British Drug House), 10 mg/ml in H_2O; stored at $-20°$

GSH 100 mM. The solution of GSH is made from the solid substance that is dissolved in H_2O and pH adjusted to 5.0 with 1 M NaOH; stored at $-20°$.

[12] E. Racher, *J. Biol. Chem.* **217,** 867 (1955).

[13] S. Nagai and S. Black, *J. Biol. Chem.* **243,** 1942 (1968).

Preparation of Mixture. The following mixture (10 ml) is prepared in an icebath, fresh before use: 0.1 *M* Tris–Cl, pH 8.0, 2 m*M* EDTA, 0.4 m*M* NADPH, 0.1 mg/ml bovine serum albumin, 1 m*M* GSH, and 6 μg/ml glutathione reductase.

Spectrophotometric Assay. A Zeiss PM Q 3 spectrophotometric system (Carl Zeiss) consisting of an automatic reference value unit and a programmable cuvette changer and a Servogor S recorder is used. Two quartz semimicro cuvettes with 1 cm light path contained 500 μl of mixture at 25°. To both cuvettes were added 25 μl of 15 m*M* HED (0.7 m*M* final concentration) and the absorbance at 340 nm was recorded for 2 min with 0.5 min intervals to ensure that both cuvettes were balanced with respect to the nonenzymatic spontaneous reaction between GSH and HED. (This amounts to ~a ΔA_{340} of $0.020 \times \text{min}^{-1}$.) Glutaredoxin, 1–20 μl, was added to the sample cuvette and an equal volume of buffer was added to the reference cuvette. The decrease in absorbance was then recorded for 3 min with 0.5 min intervals. The result is calculated as $\Delta A_{340} \times \text{min}^{-1}$ by using the change between the first and second min. The assay shows a linear dependence on glutaredoxin up to ΔA_{340} values of $0.100 \times \text{min}^{-1}$. If required the glutaredoxin fractions are diluted to achieve this activity.

Definition of Unit and Specific Activity. One unit of activity is defined as 1 μmol of NADPH oxidized per min and is calculated from the expression $(\Delta A_{340} \times \text{min}^{-1} \times V \times 10^3)/6200$ where $\Delta A_{340} \times \text{min}^{-1}$ is the net decrease in absorbance at 340 nm (due to oxidation of NADPH), V is the volume in ml, and 6200 is the molar extinction coefficient for NADPH. Specific activity is defined as units of enzyme per mg of protein.

Comments. Other disulfides such as L-cystine are also substrates for glutaredoxin (see below). However, the limited solubility of L-cystine and the considerably higher spontaneous thiol-disulfide interchange reaction between GSH and L-cystine makes HED the most useful substrate.

Purification Procedure

Glutaredoxin from E. coli B

Reagents. *E. coli* B cells grown aerobically on a minimal medium with glycerol[4,14] commercially available from the Microbiological Research Establishment, Porton, England. *E. coli* B cells grown on glucose and harvested in late log phase are also possible to use.[2]

[14] R. Elsworth, G. A. Miller, A. R. Whitaker, D. Kitching, and P. D. Sayer, *J. Appl. Chem.* **17**, 157 (1968).

Tris–Cl 1.0 *M*, pH 8.0. EDTA 0.2 *M*, pH 7.45. Streptomycin sulfate (Glaxo). Spectrapor membrane tubing (No 3) for dialysis (Spectrum Medical Industries, Ltd., Los Angeles, CA). DEAE-cellulose (Whatman DE-32). Sephadex G-50 fine (Pharmacia). CM-cellulose (Whatman CM52).

All operations were performed at 0–4°. All centrifugations were done in a Sorvall RC2B centrifuge at 10,000 *g* for different length of times.

Disintegration of Bacteria. Frozen cells (1.0 kg) were disintegrated by using a bacterial X-press[15] at −20 to −25°.

Crude Extract. The cells were homogenized in 5000 ml of 50 m*M* Tris–Cl, pH 7.6–5 m*M* EDTA and centrifuged for 60 min. The supernatant fraction, fraction I, was saved.

Streptomycin Sulfate Precipitation. Streptomycin sulfate was slowly added to fraction I to give a final concentration of 0.8%. After slowly stirring for 15 min, the precipitate was removed by centrifugation for 15 min, and the supernatant (fraction II) was saved.

Acid Treatment. Acetic acid, 1.0 *M* was used to adjust the pH of fraction II to 5.0 (takes about 40 ml/liter). The resulting precipitate was removed by centrifugation and the supernatant fraction was neutralized to pH 7.5 by addition of 1 *M* NH_4OH (fraction III). (The precipitate, fraction IV, may be saved and used for preparation of thioredoxin reductase.)

Heat Treatment. To fraction III was added 25 ml of 0.2 *M* EDTA, pH 7.5 per 1000 ml and the mixture was rapidly heated to 75° followed by rapid cooling to +4°. The precipitated protein was removed by centrifugation, and the supernatant solution (fraction V) was saved.

Ammonium Sulfate Precipitation. Solid ammonium sulfate (612 g/1000 ml, 85% saturation) was added slowly to fraction V. The mixture was stirred for 2 hr; the precipitate was collected by centrifugation.

The ammonium sulfate precipitate was dissolved in 50 ml of 50 m*M* Tris–Cl, pH 7.6, 5 m*M* EDTA, by stirring for 1 hr. Insoluble material was removed by centrifugation and reextracted three times using 15 ml of buffer. The combined supernatant fractions were dialyzed extensively over a 20-hr period against 0.02 *M* potassium phosphate, pH 7.0, 2 m*M* EDTA, with several changes of buffer until the solution was free of ammonium sulfate. This yielded 150 ml of solution with an $A_{260}:A_{280}$ ratio of 0.65 (fraction VI). Fraction VI contains both glutaredoxin and thioredoxin activity when assayed with ribonucleotide reductase.

First DEAE-Cellulose Chromatography. Fraction VI (150 ml) was applied to a column (28 cm^2 × 18 cm) of DEAE-cellulose equilibrated with 0.02 *M* potassium phosphate, pH 7.0, 2 m*M* EDTA. The column was eluted with a linear gradient of 3000 ml of 0.02 *M* potassium phosphate,

[15] L. Edebo, *J. Biochem. Microbiol. Technol. Eng.* p. 151 (1961).

pH 7.0, 2 mM EDTA, and 3000 ml of 0.20 M potassium phosphate, pH 7.0, 2 mM EDTA. Glutaredoxin separated completely from thioredoxin under these conditions. Thioredoxin showed a peak at 0.08 M phosphate, whereas glutaredoxin appeared as a more acidic protein with a peak at 0.17 M phosphate. The glutaredoxin peak was pooled and concentrated to about 100 ml by flash evaporation at +15°. The material was then dialyzed extensively against 0.06 M NH_4HCO_3, pH 7.8, and further concentrated by flash evaporation to a volume of about 10 ml, to give fraction VII.

Sephadex G-50 Chromatography. A column (12 cm^2 × 130 cm) of Sephadex G-50 fine was equilibrated with 0.06 M NH_4HCO_3, pH 7.8. Fraction VII was applied to the Sephadex G-50 column, which was eluted with 0.06 M NH_4HCO_3, pH 7.8. The peak of glutaredoxin eluted with a K_{av} value of 0.25. The fractions containing glutaredoxin activity (110 ml) were pooled (fraction VIII).

Second DEAE-Cellulose Chromatography. Fraction VIII was lyophilized twice and dissolved in 3.0 ml of 0.05 M Tris–Cl, pH 7.5, 1 mM EDTA, followed by dialysis against the same buffer. This sample was applied to a column (1 cm^2 × 10 cm) of DE32 cellulose equilibrated with 0.05 M Tris–Cl, pH 7.5, 1 mM EDTA. The column was eluted with a linear gradient consisting of 50 ml of 0.05 M Tris–Cl, pH 7.5, 1 mM EDTA, 0.10 M sodium acetate, and 50 ml of 0.05 M Tris–Cl, pH 7.5, 1 mM EDTA, 0.40 M sodium acetate. The active fractions were combined and concentrated by flash evaporation to about 3 ml to give fraction IX. This fraction had a high absorbance at 310 nm and an $A_{260}:A_{280}$ ratio of 0.93. This was caused by an impurity with a mobility on gel electrophoresis[4] identical with the bromphenol blue marker.[16]

CM-Cellulose Chromatography. Fraction IX was dialyzed against 2 changes of 500 ml of 10 mM sodium acetate, pH 5.0. The material after dialysis, 7.0 ml, was applied to a column (0.9 cm^2 × 2 cm) of CM52 cellulose equilibrated with 10 mM sodium acetate, pH 5.0. The glass column and test tubes used to collect fractions were siliconized before use. The column was eluted with a linear gradient of 10 ml of 0.01 M sodium acetate, pH 5.0, and 10 ml of 0.10 M sodium acetate. Inactive material, which had a high $A_{260}:A_{280}$ ratio, passed through the column and glutaredoxin activity was eluted as a symmetrical peak of protein and activity. The fractions containing glutaredoxin were combined and the pH was adjusted to 7.5 by addition of 1.8 M Tris–Cl, pH 8.0 and concentrated by flash evaporation. This pool was stored frozen at −20° (fraction X).

Yield and Purity. The preparation is summarized in Table I. The final yield of glutaredoxin was around 3%. The material of fraction X was

[16] L. Ornstein, *Ann. N.Y. Acad. Sci.* **121,** 321 (1964).

TABLE I

PURIFICATION OF GLUTAREDOXIN FROM 1.0 kg OF *E. coli* B (WILD TYPE)

Fraction	Step	Volume (ml)	Protein (mg)	Glutaredoxin activity (nmol dCDP/min)	Specific activity (nmol dCDP/mg protein)	Yield glutaredoxin (%)	Purification (fold)	GSH-disulfide oxidoreductase activity (units)[a]
I	Crude extract	5,500	80,000	10,200[b]	0.13	100	1	40,150[c]
VI	Ammonium sulfate precipitation	150	2,210	1,530	0.69	15	5.3	1,335
VII	First DEAE-cellulose chromatography	10	133	1,070	8.1	11	62	33
VIII	Sephadex G-50 chromatography	110	1.9	790	416	8	3,200	18
IX	Second DEAE-cellulose chromatography	3.0	0.23	460	2,040	5	15,700	10.5
X	CM-cellulose chromatography	0.5	0.06	306	5,100	3	39,230	6.5

[a] One unit corresponds to 1 μmol of NADPH oxidized/min.

[b] Approximate value. The activity determination in the crude extract is complicated by the presence of thioredoxin and inhibitors. This applies through Steps II to V.

[c] Total activity. Only 1–2% correspond to glutaredoxin at this stage.

homogenous when analyzed on both native[16] and SDS–polyacrylamide gel electrophoresis.[17]

Comments. The first steps in the present method were designed to purify both glutaredoxin and thioredoxin. The yield of thioredoxin is better than the apparent yield of glutaredoxin in fraction VI (See Table I). Some considerable but variable losses of glutaredoxin may occur in the streptomycin sulfate precipitation step. Yet this step is important for obtaining a pure final protein. Thioredoxin separated completely from glutaredoxin in the first DEAE-cellulose chromatography step.

Physical Properties

Stability. The glutaredoxin activity of fraction X was stable for months when stored frozen at −20°. The protein may easily be lost on glass surfaces when dilute solutions are handled.

Molecular Weight and Composition. Both native[16] and SDS–gel electrophoresis[17] as well as the elution position of glutaredoxin on a column of Sephadex G-50 indicated a molecular weight of 11,000.[4] The protein is acidic. Amino acid analysis showed a content of about 89 residues[4] with two half-cystines forming a single disulfide bond in the oxidized form.[4] Glutaredoxin from *E. coli* B contains all common amino acids except methionine. No evidence of carbohydrate or metals was found.

Spectrum and Molar Extinction Coefficient. The spectrum of glutaredoxin showed only contribution from amino acid residues.[4] From the spectrum and the concentration as determined by amino acid analysis, a molar extinction coefficient at 280 nm of 11,600 was calculated.[4] This is equivalent to a value of $E^{1\%}_{280\ \mathrm{nm}} = 10.0$.

NH_2-Terminal Residue. Reduced and carboxymethylated glutaredoxin from *E. coli* B showed dansyl-glutamic acid as the only N-terminal amino acid. This indicates that the N-terminal is glutamic acid or glutamine.[4]

Presence of a Cystine Disulfide Bridge in Glutaredoxin. Reduction of glutaredoxin by excess dithiothreitol followed by carboxymethylation with [^{14}C]iodoacetic acid resulted in incorporation of 176 nmol of carboxymethylcysteine/mg of protein. This is consistent with 2.0 mol of half-cystine residues/11,600 g of protein.[4] The amino acid analysis and carboxymethylation experiments suggested that glutaredoxin contains a functional disulfide bridge made up from 2 half-cystine residues in the protein. The reduction of this disulfide bond occurred rapidly in the presence of 1 m*M* GSH, NADPH, and glutathione reductase. No reaction was seen with NADPH and thioredoxin reductase, nor did glutaredoxin ex-

[17] P. H. O'Farrel, *J. Biol. Chem.* **250,** 4007 (1975).

hibit thioredoxin activity. Eighty-five nanomoles of NADPH was oxidized/1.00 mg of glutaredoxin. The result is consistent with Reactions (6) and (7) below:

$$\text{Glutaredoxin-S}_2 + 2\ \text{GSH} \rightarrow \text{glutaredoxin-(SH)}_2 + \text{GSSG} \quad (6)$$

$$\text{GSSG} + \text{NADPH} + \text{H}^+ \xrightarrow{\text{glutathione reductase}} 2\ \text{GSH} + \text{NADP}^+ \quad (7)$$

where glutaredoxin-S_2 is an oxidized form of glutaredoxin. From the stoichiometry, it is apparent that complete reduction of a disulfide in glutaredoxin by NADPH through the glutathione system is obtained. The stoichiometry also provides independent evidence for the homogeneity of the glutaredoxin preparation.

Catalytic Properties

Activity with Ribonucleotide Reductase. The assay for glutaredoxin measures its ability to catalyze the formation of [^{3}H]dCDP from [^{3}H]CDP and GSH (4 mM) by ribonucleotide reductase. The dependence of the reaction rate on glutaredoxin concentration in this assay shows saturation kinetics. This allows determination of the apparent K_m value for glutaredoxin. A Lineweaver–Burk plot gave a value of 1.3×10^{-7} M.[5] In these experiments, glutaredoxin was present in concentrations similar to that of ribonucleotide reductase.

The activity of glutaredoxin with ribonucleotide reductase was compared with the activity of the thioredoxin system (thioredoxin, NADPH, and thioredoxin reductase).[5] Glutaredoxin was about 10-fold more active on a molar basis. Apparent K_m values for thioredoxin of 1.25×10^{-6} M and for glutaredoxin of 0.13×10^{-6} M were obtained. The V_{max} values for ribonucleotide reductase were essentially identical (within $\pm 10\%$) with the two systems. The molecular activities of glutaredoxin and thioredoxin were calculated and compared with those of subunits B1 and B2 of ribonucleotide reductase (Table II). It is apparent from this calculation that glutaredoxin has a turnover number which is similar to proteins B1 and B2.

Reduction of Glutaredoxin with Dithiothreitol. A low concentration of dithiothreitol (5 mM) stabilizes *E. coli* ribonucleotide reductase, but only gave a low activity as hydrogen donor for the enzyme. With glutaredoxin and dithiothreitol (5 mM) in the assay as ultimate hydrogen donor full ribonucleotide reductase activity was obtained. The K_m for glutaredoxin was 0.15×10^{-6} M.[5] The V_{max} in this experiment appeared to be somewhat higher than in the presence of 5 mM dithiothreitol and thioredoxin.[5]

Assay for Glutaredoxin by NADPH Oxidation. Because the equilibrium of the glutathione reductase reaction at pH 7.5 in the presence of

TABLE II
MOLECULAR ACTIVITY OF GLUTAREDOXIN AND THIOREDOXIN IN RIBONUCLEOTIDE REDUCTASE REACTION[a]

Protein	Turnover number[b]
Glutaredoxin	110–150
B1 subunit	40–80
B2 subunit	110–240
Thioredoxin	13–15

[a] Calculated from experiments using the linear part of saturation curve up to 10% V_{max}.
[b] Molecules of substrate reduced/protein molecule/min.

excess NADPH is shifted toward reduction of all GSSG, the assays of glutaredoxin may also be performed spectrophotometrically by following the oxidation of NADPH at 340 nm.

Thioredoxin from *E. coli* gave no activity in the glutaredoxin assay system which is an indirect demonstration of the inability of GSH to reduce thioredoxin-S_2[10] in catalytic reactions. Assays for glutaredoxin may thus be performed in the presence of thioredoxin provided that all thioredoxin reductase has been inactivated (i.e., by heating to 100°).

Apparent K_m Value for GSH. An apparent K_m value of 0.4 m*M* was determined for GSH by using an assay mixture at saturating CDP with proteins B1 and B2 and glutaredoxin.[5] No activity was observed in the absence of GSH, thereby excluding a direct reduction of glutaredoxin by NADPH and glutathione reductase.[5]

Thiol Specificity of Glutaredoxin. To study the thiol specificity in Reaction (1), GSH was exchanged for 2-mercaptoethanol or dithiothreitol. The activity was determined as a function of time and compared to that of GSH. No activity was obtained with mercaptoethanol. As observed previously,[2] the activity with dithiothreitol as substrate is high. The rate of the reaction with GSH is strongly influenced by the presence of NADPH and glutathione reductase to keep all GSSG as GSH.

Effect of GSSG on the Reaction. The effect of GSSG in the overall process of the glutaredoxin reaction was studied.[5] Even in carefully prepared mixtures of GSH under argon, some GSSG will be present. The removal of glutathione reductase and NADPH from the assays resulted in a large decrease in the rate of synthesis of dCDP (26% remaining). The addition of small defined quantities of GSSG caused further losses of activity. At a 1 : 1 ratio of GSH to GSSG the glutaredoxin-catalyzed ribonucleotide reductase activity was totally inhibited.

Glutaredoxin from E. coli K12

Reagents

Strain: *E. coli* C-10-17.[18] K12 strain derived from KK 1006 with *metB, upp, ndK, thyA* (ts) *trxB* (thioredoxin reductase)

Culture medium (in grams/liter): Bacto-tryptone (10), yeast extract (5), NaCl (10), glucose (10); pH is adjusted to 7.0 before autoclaving. This medium is supplemented with 50 mg/liter of thymine

Buffer A: 50 m*M* Tris–Cl pH 7.5–3 m*M* EDTA

Antiglutaredoxin immunoadsorbent column: containing antiglutaredoxin covalently bound to Sepharose 4B (Pharmacia) by CNBr activation. The glutaredoxin antiserum was obtained by immunization of a rabbit with 100 μg doses of pure glutaredoxin from *E. coli* B.[18] The column is prepared by linking the γ-globulin fraction of a rabbit antiserum against glutaredoxin to Sepharose. The coupling of the γ-globulin fraction is made at pH 6.5 in 0.2 *M* sodium citrate.[19]

Buffer B: Tris–Cl 1.0 *M*, pH 8.0

Sephadex G-50 (Pharmacia Fine Chemicals)

Spectrapor membrane tubing (No 3) for dialysis (Spectrum Medical Industries Ltd, Los Angeles, CA)

Growth and Disintegration of Bacteria. E. coli C10-17 cells were grown at 37° in 700 liters of medium with 1% glucose supplemented with 50 μg/ml of thymine in a 1000-liter fermentor.[7] The inoculum was prepared by taking cells from an agar plate and transfer to a culture flask that was shaken overnight at 37°. It was essential not to aerate this inoculum by blowing air through the medium since the C10-17 cells appeared to be sensitive to oxygen.

Crude Extract. This was prepared by homogenization of x-pressed cells with five times their weight of 50 m*M* Tris–Cl, 3 m*M* EDTA, pH 7.5, followed by centrifugation at 10,000 *g* for 90 min. The supernatant fraction (crude extract) was carefully decanted.

Antiglutaredoxin Sepharose Column. The crude extract was applied to antiglutaredoxin Sepharose. The column (100 ml) was then washed

[18] J.-O. Höög, H. Jörnvall, A. Holmgren, M. Carlquist, and M. Persson, *Eur. J. Biochem.* **136** 223 (1983).

[19] B.-M. Sjöberg and A. Holmgren, *Biochim. Biophys. Acta* **315,** 176 (1973).

TABLE III
PURIFICATION OF GLUTAREDOXIN (FROM 200 g WET WEIGHT C10-17 CELLS)

Step	Volume (ml)	Protein (mg)	Glutaredoxin activity[a] (units)
Crude extract	1,000	16,000	210[b]
Antiglutaredoxin-Sepharose pool	80	10	125
Sephadex G-50 pool	10	1.0	110

[a] Determined by HED assay (see above); 1 unit = 1 μmol NADPH oxidized × min^{-1}.

[b] This value was estimated from the content of glutaredoxin determined by assay with ribonucleotide reductase. The HED assay gave a much higher apparent content due to the presence of other GSH-disulfide transhydrogenases (see Table I).

with several column volumes of 1.0 *M* Tris–Cl, 3 m*M* EDTA, pH 8.0 and finally with one volume of 50 m*M* Tris–Cl, 3 m*M* EDTA, pH 7.5. Bound glutaredoxin was eluted with 0.1 *M* acetic acid–0.5 *M* formic acid pH 2.1 and neutralized with Buffer B. The pooled fractions were concentrated by lyophilization followed by dialysis against Buffer A and repeated lyophilization.

Sephadex G-50 Column. The antiglutaredoxin pool was dissolved in 0.8 ml of 0.06 *M* ammonium bicarbonate and taken to the last step. This consisted of a column of Sephadex G-50 (1 × 120 cm) in 0.06 *M* ammonium bicarbonate. Glutaredoxin activity was well separated from the bulk of protein and eluted with a K_{av} value of 0.25. The pooled material was stored frozen at −20°. It could also be lyophilized and dissolved in the appropriate buffer.

Yield and Purity. The preparation is summarized in Table III. The final yield of glutaredoxin was 52%. The final material was homogeneous as analyzed on both native[16] and SDS–polyacrylamide gel electrophoresis.[17]

Comments. The *E. coli* K12 strain C10-17 gives a considerable (3–10X) overproduction of glutaredoxin when compared with wild type *E. coli* cells.[20] No thioredoxin is bound on the antiglutaredoxin column, demonstrating a lack of antigenic crossreactivity with thioredoxin.

Molecular Properties. The glutaredoxin from *E. coli* K12 is very similar in amino acid composition and molecular properties to glutaredoxin

[20] A. Holmgren, in preparation.

from *E. coli* B.[19] One important difference concerns the N-terminal residue which in the K12 glutaredoxin is Met whereas it is Gln or Glu in the *E. coli* B glutaredoxin. This may represent a difference in N-terminal processing. The other differences in total composition are small and may not be significant (see Table V).

Amino Acid Sequence. The complete amino acid sequence of the 85 residues of *E. coli* K12 glutaredoxin has been determined.[19] The active center of the molecule in its oxidized form consists of a 14-membered disulfide ring with the structure

```
 ┌──S-S────┐
-Cys-Pro-Tyr-Cys-
```

Glutaredoxin shows a clear homology to phage T4 thioredoxin[19] and a remote similarity to *E. coli* thioredoxin. This means thioredoxins and glutaredoxins together constitute a superfamily of small proteins with a catalytically active cystine disulfide bridge.[19] A three-dimensional structure for glutaredoxin has been proposed based on the homology with T4 thioredoxin.[21]

Glutaredoxin from Calf Thymus

Reagents. Thymuses from 2- to 4-month-old calves were obtained from local slaughterhouses. Dithiothreitol was purchased from Sigma, Streptomycin sulfate from Glaxo, DEAE (DE 52) from Whatman, CM-Sepharose and Sephadex G-50 from Pharmacia Fine Chemicals.

All steps were carried out at +4°. Prior to all chromatographic steps, extracts were incubated with 2 m*M* dithiothreitol at +4° for 1 hr. Glutaredoxin was assayed with the HED assay (see above).

Preparation of Crude Extract. Frozen thymuses (1500 g) were split into pieces which were thawed in 4400 ml of 50 m*M* Tris–Cl, pH 7.6, for 60 min at 4°. This half-thawed tissue was minced and extracted in a homogenizer (CUT-O-MAT, type H 10, Kneubühler & Co., Luzern) for 5 min, and the slurry was centrifuged for 30 min at 20,000 *g* to remove insoluble debris. The supernatant solution was filtered through glass wool to remove flotating lipids, giving 3900 ml of opalescent crude extract (fraction I).

Precipitation with Streptomycin. A 6.5% (w/v) solution of streptomycin sulfate (390 ml) was added during 30 min to 3900 ml of the crude extract with vigorous stirring. The solution was centrifuged for 20 min at 20,000 *g*, the precipitate was discarded and the supernatant (4200 ml) fraction II, was used in the next step.

[21] H. Eklund, C. Cambillau, B.-M. Sjöberg, A. Holmgren, H. Jörnvall, J.-O. Höög, and C.-I. Brändén, *EMBO J.* **3,** 1443 (1984).

Precipitation with Ammonium Sulfate. Solid ammonium sulfate (0.243 g/ml, corresponding to 40% saturation) was slowly added to fraction II with stirring and, after 30 min the suspension was centrifuged 20 min at 20,000 *g*. The supernatant solution was saved, and the precipitate taken to the ribonucleotide reductase preparation.[22]

Ammonium Sulfate Precipitation. The supernatant after 0 to 40% saturated ammonium sulfate precipitation was brought to 90% of saturation by the addition of solid ammonium sulfate. The precipitate obtained after centrifugation (8000 *g*) was dissolved in a small volume of 50 m*M* Tris–Cl, pH 7.5, and dialyzed extensively against 10 m*M* Tris–Cl, pH 7.5, 1 m*M* EDTA until the extract reached the same specific conductivity as the buffer.

DEAE-Cellulose Chromatography. The dialyzed ammonium sulfate precipitate was adjusted to pH 9.0 by the addition of 1.0 *M* Tris base, and adsorbed on a column of DEAE-cellulose (78.5 cm^2 × 28 cm) equilibrated with 10 m*M* Tris–Cl, pH 9.0. The column was eluted with a linear gradient of sodium chloride from 0 to 0.05 *M* (5 liters of each in 10 m*M* Tris–Cl, pH 9.0). Glutaredoxin was fully separated from thioredoxin and was purified more than 100-fold with good recovery. Fractions containing glutaredoxin activity were pooled, adjusted to pH 7.5 by the addition of 1.0 *M* sodium acetate, pH 5.5, and concentrated by lyophilization.

Sephadex G-50 Chromatography. A column of Sephadex G-50 (7.0 cm^2 × 150 cm) was equilibrated with 50 m*M* Tris–Cl, pH 7.5, 1 m*M* EDTA. The DEAE-pool (10 ml) was applied to the column and eluted with the above buffer. The peak of glutaredoxin appeared at a K_{av} value of 0.25 (same as *E. coli* glutaredoxin as well as thioredoxin from both species). Glutaredoxin-containing fractions were pooled and concentrated by lyophilization.

CM-Sepharose Chromatography. The pool from Sephadex G-50 (1 ml) was dialyzed overnight against 10 m*M* sodium acetate, pH 5.52, 0.05 m*M* EDTA, and applied to a column of CM-Sepharose (0.8 cm^2 × 12 cm) equilibrated with 10 m*M* sodium acetate, 0.05 m*M* EDTA, pH 5.52. The column was eluted with a linear gradient formed from 50 ml of 10 m*M* sodium acetate, pH 5.52, and 50 ml of 50 m*M* sodium acetate. Glutaredoxin activity was eluted at 30 m*M* salt as a symmetrical peak of protein and activity. The active fractions were pooled, adjusted to pH 7.5 with 2 *M* Tris–Cl, pH 8.0, and concentrated to 0.5 ml. When stored frozen at −20° the enzyme was stable for at least 12 months.

Summary of the Purification. The purification of glutaredoxin from 1.5 kg of calf thymus was 3200-fold with a final yield of 10% (Table IV).

[22] Y. Engström, S. Eriksson, L. Thelander, and M. Åkerman, *Biochemistry* **18,** 2941 (1969).

TABLE IV
PURIFICATION OF GLUTAREDOXIN FROM 1.5 kg OF CALF THYMUS

Fraction	Protein (mg)	Glutaredoxin activity[a] (units)	Specific activity (units/mg)
I. Crude extract	82,000	11,000	0.1
IV. Ammonium sulfate precipitate	18,000	5,000	0.3
V. DEAE-cellulose	100	3,650	36
VI. Sephadex G-50	9.9	1,930	195
VII. CM-Sepharose	2.55	1,050	412

[a] One unit corresponds to 1 μmol of NADPH oxidized/min (HED assay).

Comment. Glutaredoxin from calf thymus behaves as a small basic protein with GSH-disulfide-transhydrogenase activity in the HED assay[9] (see above). In contrast to *E. coli,* a crude extract of calf thymus seems to contain only glutaredoxin as the main component with HED activity.[9] Note that it is absolutely essential to incubate with dithiothreitol prior to all chromatographic steps to obtain one molecular species.[9]

Physical Properties

Purity. The final material was pure by the criteria of gel electrophoresis under reducing conditions (preincubation with 2 mM dithiothreitol)[9] and high-performance liquid chromatography.[9]

Molecular Weight. This was 11,000 as determined by gel chromatography.

Amino Acid Sequence. The total composition of glutaredoxin with 101 amino acid residues is given in Table V. The amino acid sequence of calf thymus glutaredoxin[23] shows that the active center structure Cys-Pro-Tyr-Cys is identical to that of *E. coli* glutaredoxin.[19] Calf thymus glutaredoxin is extended at both ends and has 31% overall residue identities with the corresponding *E. coli* protein.[23] One important difference is the presence of extra half-cystines in calf thymus glutaredoxin apart from the two required for the catalytic activity of the protein.[5]

Catalytic Properties

GSH-Disulfide-Transhydrogenase Activity. Glutaredoxin in fraction VII had a transhydrogenase activity of 412 μmol of NADPH oxidized/

[23] I.-M. Klintrot, J.-O. Höög, H. Jörnvall, A. Holmgren, and M. Luthman, *Eur. J. Biochem.* **144,** 417 (1984).

TABLE V
AMINO ACID COMPOSITION OF GLUTAREDOXINS

Amino acid	*E. coli* K12[a]		*E. coli* B	Calf thymus[b]
Cys	1.5	2	1.6	4
Asx	10.0	10	10.3	6
Thr	4.6	4	4.8	6
Ser	3.1	2	3.1	3
Glx	14.1	14	14.0	17
Pro	3.0	3	2.8	4
Gly	6.9	6	7.3	9
Ala	7.6	7	7.4	6
Val	7.1	7	7.9	7
Met	0.9	1	0.1	2
Ile	4.7	5	4.8	7
Leu	4.9	4	4.6	12
Tyr	3.1	4	3.9	2
Phe	3.5	4	4.0	5
Trp	1.0	1	0.9	0
Lys	5.4	6	6.5	6
His	0.9	1	1.0	1
Arg	3.6	4	3.9	4

[a] First column from acid hydrolysis; second column from complete amino acid sequence.[19]

[b] Amino acid composition from complete amino acid sequence.[23]

min/mg of protein. Hydroxyethyldisulfide, L-cystine, *S*-sulfocysteine, and *S*-sulfoglutatione were also disulfide substrates for *E. coli* and calf thymus glutaredoxin.[9] Insulin was a very poor substrate because.[9]

Activity with Ribonucleotide Reductase. The activity of calf thymus glutaredoxin with the homologous calf thymus ribonucleotide reductase gave an apparent K_m of 6×10^{-7} *M*. The V_{max}, 2.0 nmol of dCDP formed/30 min, was the same as that obtained with 10 m*M* dithiothreitol.[9]

In the heterologous system with calf thymus glutaredoxin as hydrogen donor for *E. coli* ribonucleotide reductase only about 10% of the activity of *E. coli* glutaredoxin was obtained. The apparent K_m for calf thymus glutaredoxin was 1.9×10^{-6} *M* in this system. This should be compared to a K_m value of 1.3×10^{-7} *M* for *E. coli* glutaredoxin demonstrating species specificity.

Acknowledgments

This investigation was supported by grants from the Swedish Medical Research Council 13X-3529 and the Swedish Cancer Society (961). The excellent secretarial assistance of Mrs. Agneta Sjövall is gratefully acknowledged.

[69] Thiol : Protein Disulfide Exchange Enzymes[1]

By JOHN E. MORIN and JACK E. DIXON

The enzymatic catalysis of thiol : protein disulfide exchange reactions has been the subject of numerous investigations for more than two decades. The enzymes were initially defined by the activity assays used in their purification. Tomizawa's[1–3] bovine liver preparation produced TCA-soluble insulin by reducing its disulfide bonds with GSH and was referred to as glutathione:insulin transhydrogenase. The purification procedure was adopted by other investigators who assayed for the product, GSSG, spectrophotometrically with glutathione reductase.[4]

Anfinsen and Haber[5] were concurrently studying the formation of disulfide bonds in proteins and chose as a model the reactivation of scrambled ribonuclease through thiol:protein disulfide exchange reactions which restored the correct disulfide bonds and ribonuclease activity. They succeeded in purifying an enzyme from bovine liver with an active cysteine residue which catalyzed reactivation of scrambled ribonuclease in the presence of GSH and became known as protein disulphide-isomerase.[6–9] Venetianer and Straub[10] also purified a similar activity from pigeon pancreas, however this enzyme was not extensively characterized.

The essential similarity of the two thiol : protein disulfide exchange reactions was noted by several investigators who reported that the substrate specificity of glutathione : insulin transhydrogenase was rather broad.[11,12] The question of identity between these enzymes was also

[1] The enzyme described herein has conventionally been named for the catalysis observed. In the earliest studies, it was referred to by different groups as protein disulphide-isomerase or as glutathione : insulin transhydrogenase. More recently the name thiol : protein disulfide oxidoreductase (TPOR) has come into use.

[1a] H. H. Tomizawa and Y. D. Halsey, *J. Biol. Chem.* **234,** 307 (1959).

[2] H. H. Tomizawa, *J. Biol. Chem.* **237,** 428 (1962).

[3] H. H. Tomizawa, *J. Biol. Chem.* **237,** 3393 (1962).

[4] H. M. Katzen and D. Stetten, *Diabetes* **11,** 271 (1962).

[5] C. B. Anfinsen and E. Haber, *J. Biol. Chem.* **236,** 1361 (1961).

[6] R. F. Goldberger, C. J. Epstein, and C. B. Anfinsen, *J. Biol. Chem.* **238,** 628 (1963).

[7] F. DeLorenzo, R. F. Goldberger, E. Steers, D. Givol, and C. B. Anfinsen, *J. Biol. Chem.* **241,** 1562 (1966).

[8] F. DeLorenzo, S. Fuchs, and C. B. Anfinsen, *Biochemistry* **5,** 3961 (1966).

[9] S. Fuchs, F. DeLorenzo, and C. B. Anfinsen, *J. Biol. Chem.* **242,** 398 (1967).

[10] P. Venetianer and F. B. Straub, *Biochim. Biophys. Acta* **89,** 189 (1964).

[11] H. M. Katzen and F. Tietze, *J. Biol. Chem.* **241,** 3561 (1966).

[12] P. T. Varandani, *Biochim. Biophys. Acta* **371,** 577 (1974).

Copyright © 1985 by Academic Press, Inc.
All rights of reproduction in any form reserved.

raised when Ansorge *et al.*[13,14] demonstrated both glutathione:insulin transhydrogenase activity and protein disulphide-isomerase activity in a highly purified preparation from rat liver.

Varandani has published prolifically in support of the thesis that the action of glutathione:insulin transhydrogenase is the initial step in the degradation of insulin. Varandani and Nafz[15] demonstrated that reductive cleavage of insulin precedes extensive proteolysis in liver homogenates buffered to neutrality and treated with EDTA. The marked heterogeneity of the bovine pancrease preparation[16,17] makes it difficult to interpret its broad substrate specificity.[18,19]

Freedman and co-workers have attempted to identify the enzyme(s) responsible for thiol:protein disulfide exchange by determining the ratio of glutathione:insulin transhydrogenase activity to protein disulphide-isomerase activity throughout their purification. The two activities copurify, but these ratio shifts from fraction to fraction of their bovine liver preparation from which they infer the existence of multiple polypeptide moieties with different but overlapping specificities. Unfortunately, they were unable to isolate either activity from the other and their purified preparation was not homogeneous.[20,21]

This laboratory has concentrated effort on obtaining a single enzyme with thiol:protein disulfide exchange activity from a single source and to thoroughly characterize the enzyme. Carmichael *et al.*[22] reported the purification and characterization of a thiol:protein disulfide oxidoreductase from bovine liver. The purification procedure afforded enzyme in good yield and avoided many undesirable steps which had been used in earlier purifications. This new procedure has been of significant practical importance in that others have used it to obtain TPOR activity from several tissue sources. The purity and yield of TPOR purified by this procedure have permitted biochemical and immunological analyses which have helped to resolve the question of identity of glutathione:insulin transhydrogenase and protein disulphide-isomerase. The procedure of Carmi-

[13] S. Ansorge, P. Bohley, H. Kirschke, J. Langner, I. Marquardt, B. Wiederanders, and H. Hanson, *FEBS Lett.* **37,** 238 (1973).

[14] S. Ansorge, P. Bohley, H. Kirschke, J. Langner, B. Wiederanders, and H. Hanson, *Eur. J. Biochem.* **32,** 27 (1973).

[15] P. T. Varandani and M. A. Nafz, *Diabetes* **25,** 173 (1976).

[16] P. T. Varandani, *Biochim. Biophys. Acta* **371,** 577 (1974).

[17] P. T. Varandani and M. A. Nafz, *Int. J. Biochem.* **1,** 313 (1970).

[18] M. L. Chandler and P. T. Varandani, *Biochemistry* **14,** 2107 (1975).

[19] P. T. Varandani, M. A. Nafz, and M. L. Chandler, *Biochemistry* **14,** 2115 (1975).

[20] A. L. Ibbetson and R. B. Freedman, *Biochem. J.* **159,** 377 (1976).

[21] H. C. Hawkins and R. B. Freedman, *Biochem. J.* **159,** 385 (1976).

[22] D. F. Carmichael, J. E. Morin, and J. E. Dixon, *J. Biol. Chem.* **252,** 7163 (1977).

chael *et al.*[22] has been used by Roth and Koshland[23] to isolate and characterize a TPOR from mouse liver and plasmacytoma. They demonstrated that the enzyme catalyzed the formation of interchain disulfides in the assembly of monomer immunoglobulin *in vitro*. Recently Lambert and Freedman[24] have modified the purification procedure of Carmichael *et al.*[22] and used it in the purification of an enzyme which will catalyze isomerization of the disulfide bonds in "scrambled" ribonuclease. The results of Lambert and Freedman[24] suggest that a single enzyme catalyzes both activities, a point which was emphasized by Morin *et al.*[25] in his characterization of the bovine liver enzyme. Recently, Svein Bjelland[26] has also shown that enzyme preparations from Dixon's, Freedman's, and Varandani's laboratories all share common antigenic determinants.

Physicochemical Parameters of TPOR. The table is a summary of the physical characteristics of enzymes purified in this laboratory[22] and by Lambert and Freedman,[24] Tomizawa,[1] Anfinsen and colleagues,[6–9] Varandani and colleagues,[15,16] Ansorge *et al.*,[13,14] and Roth and Koshland.[23] The source materials vary but the enzymes are roughly similar in molecular weight, were isolated from the particulate fraction and several have acidic isoelectric points. Taken alone, the physical characteristics are not strongly suggestive of homology, but when the similarity of their activities and their shared susceptibility to alkylation by iodoacetamide only in the presence of thiols is considered, some degree of homology seems likely. It should be recalled that Roth and Koshland[23] used the purification protocol described by Carmichael *et al.*[22] They report contamination of the purified preparation with trace amounts of proteolytic enzymes which were evident if the enzyme was stored above $-70°$. Some of the heterogeneity apparent from the results presented in the table may be due to proteolysis. It should also be noted that carbohydrate content of (12%) reported by Carmichael *et al.*[22] has been reexamined in several preparations of the enzyme and appears to be between 1 and 2%. This value is consistent with that reported by Varandani[16] and by Lambert and Freedman.[24]

Physiological Functions of TPOR. The physiological function of TPOR is difficult to deduce from the properties of the isolated enzyme. Varandani's[27] homologous glutathione:insulin transhydrogenase is ubiquitous in its tissue distribution. Examination of rat hepatocytes by immunoferritin and electron microscopy revealed its presence primarily on smooth vesicles just inside the plasma membrane and on smooth endo-

[23] R. A. Roth and M. E. Koshland, *Biochemistry* **20**, 6594 (1981).
[24] N. Lambert and R. Freedman, *Biochem. J.* (in press).
[25] J. E. Morin, D. F. Carmichael, and J. E. Dixon, *Arch. Biochem. Biophys.* **189**, 354 (1978).
[26] S. Bjelland, personal communication.
[27] P. T. Varandani, *Biochim. Biophys. Acta* **295**, 630 (1973).

PHYSICOCHEMICAL PARAMETERS OF THIOL : PROTEIN DISULFIDE OXIDOREDUCTASES

	Dixon beef liver[23]	Freedman beef liver[24]	Anfinsen beef liver[7]	Katzen beef liver[11]	Tomizawa beef liver[1a]	Varandani beef pancreas[15,16]	Ansorge rat liver[13,14]	Roth mouse liver[23]
Electrophoresis								
Paper				1 band	1 band			
Starch gel			3 bands					
Polyacrylamide gel disc	1 band	1 band	3 bands[a]				2 bands[b]	1 band
Sodium dodecyl sulfate polyacrylamide gel	1 band	1 band				3 bands	2 bands	1 band
Isoelectric focusing polyacrylamide gel	1 major band, pH 4.10 with a shoulder	4.2 microheterogeneity						2 bands pH 4.0, 4.5
Sedimentation velocity centrifugation		3.5	3.27					
Molecular weight estimates								
Sedimentation equilibrium centrifugation	60,000		42,000			54,000		
Sodium dodecyl sulfate–polyacrylamide gel electrophoresis	60,000	57,000					60,000, 121,000 (faint)	
Molecular sieve chromatography	92,000	107,000				58,000	50,000–55,000	60,000
N-terminal amino acid	Ala	His				Lys,Leu,Val		
Carbohydrate content (percent by weight)	1–2	1				1.6		

[a] The two major bands migrated as one band following reduction and alkylation.

[b] The major band was apparently converted to the slower band despite storage at −20°.

plasmic reticulum, with some labeling on external microvilli and the inner surface of the plasma membrane and on rough endoplasmic reticulum.[28] Varandani could have missed significant amounts of enzyme, since Williams *et al.*[29] report that removing the ribosomes unmasks protein disulphide-isomerase activity in rough endoplasmic reticulum. Thus the activity is associated with the sites of both protein synthesis and degradation and may interact with plasma membrane constituents, as well.

Freedman and co-workers determined the protein disulphide-isomerase activity in homogenates of chick embryo tissues. They found the highest activity in tissues active in procollagen synthesis and this activity peaked in parallel with collagen hydroxylase activities at the developmental stage of maximal procollagen synthesis.[30] Roth and Koshland[23] found that the pure mouse liver or plasmacytoma TPOR catalyzes the formation of interchain disulfides in the assembly of monomer immunoglobulin M, *in vitro*. Radioimmunoassay indicated that the amount of TPOR within cells actively secreting immunoglobulin was one to two orders of magnitude greater than that in unstimulated B cells or non-immunoglobulin producing T cells.[23] These results are cogent support for the hypothesis that TPOR is involved in protein synthesis.

Varandani[15] has long maintained that glutathione:insulin transhydrogenase has a primary role in insulin degradation, that reduction of the disulfide bonds precedes significant proteolysis. Anti-bovine liver TPOR antibodies will clear bovine and rat liver homogenates of glutathione:insulin transhydrogenase activity, but the assay is performed at pH 7.5 in the presence of EDTA, conditions which inhibit many lysosomal enzymes. Examination of the iodinated insulin substrate indicates that considerable proteolysis, including the production of desoctapeptide insulin, might occur without prior reduction of substrate disulfides. This proteolysis might not produce radioactive TCA-soluble fragments of insulin since the radiolabel is bound to an internal tyrosine residue. The latest evidence, using hepatocytes in culture, suggests that insulin is degraded only subsequent to binding with its plasma membrane receptor and internalization, in a lysosomal system inhibited by chloroquine.[31] While TPOR may accelerate proteolysis of disulfide containing proteins in lysosomal systems[32] there is no compelling evidence for Varandani's contention at this time.

[28] P. T. Varandani, D. Raveed, and M. A. Nafz, *Biochim. Biophys. Acta* **538,** 343 (1978).
[29] D. J. Williams, D. Gurari, and B. R. Rabin, *FEBS Lett.* **2,** 133 (1968).
[30] B. E. Brockway, S. J. Forster, and R. B. Freedman, *Biochem. J.* **191,** 873 (1980).
[31] M. N. Krupp and M. D. Lane, *J. Biol. Chem.* **257,** 1372 (1982).
[32] S. Grisolia and R. Wallace, *Biochem. Biophys. Res. Commun.* **70,** 22 (1976).

Moss and co-workers[33] have recently shown that bovine liver TPOR purified in this laboratory will catalyze the activation of choleragen through reductive cleavage of the disulfide linking the A_1 and A_2 peptides of the toxin, *in vitro*. While there is no evidence that TPOR is necessary for this activation *in vivo,* its presence in the plasma membrane invites speculation as to its activities there. Phelps and Varandani[34] have published a fascinating account of an inhibitory effect of antibodies to glutathione:insulin transhydrogenase on the activation of phosphodiesterase by insulin in isolated rat adipocytes. The situation was complicated by the observation that even dialyzed normal and antisera stimulated phosphodiesterase activity above basal levels, but a further insulin-dependent stimulation was evident with normal but not with antiserum. Czech and Massague[35] have found that the insulin receptor is a high-molecular-weight complex of disulfide bonded subunits, and Schweitzer *et al.*[36] have demonstrated that reduction of these disulfides alters insulin binding in rat adipocytes. It is conceivable that the oxidation/reduction state of hormone receptor disulfides plays a role in ligand binding and it may be part of the mechanism by which an occupied receptor effects changes in metabolism. TPOR may be involved, perhaps even loosely associated with the receptor complex. Though not to be construed as direct support for these speculations, the observations of Pillion *et al.*[37] are worth noting here. They found that the insulinomimetic activity of antibodies raised against adipocyte intrinsic plasma membrane proteins was not the result of binding directly to either the insulin receptor or the glucose transport system.

Axelsson *et al.*[38] have purified a cytoplasmic thiol:protein disulfide exchange activity from rat liver to apparent homogeneity. The intracellular location, molecular weight of 11,000, and isoelectric point of 9.6 all distinguish their thioltransferase from TPOR and homologs, although it displays a similarly broad substrate specificity. Mannervik and Axelsson[39] have recently demonstrated that thioltranserase will catalyze the reactivation of the mixed disulfide of GSH and rat liver pyruvate kinase, *in vitro*. They suggest that such modification of enzyme thiol groups by the forma-

[33] J. Moss, S. J. Stanley, J. E. Morin, and J. E. Dixon, *J. Biol. Chem.* **255,** 11085 (1980).

[34] B. H. Phelps and P. T. Varandani, *Biochem. Biophys. Res. Commun.* **75,** 302 (1977).

[35] M. P. Czech and J. Massague, *Diabetes* **29,** 945 (1980).

[36] J. B. Schweitzer, R. M. Smith, and L. Jarett, *Proc. Natl. Acad. Sci. U.S.A.* **77,** 4692 (1980).

[37] D. J. Pillion, C. A. Carter-Su, P. F. Pilch, and M. P. Czech, *J. Biol. Chem.* **255,** 9168 (1980).

[38] K. Axelsson, S. Eriksson, and B. Mannervik, *Biochemistry* **17,** 2978 (1978).

[39] B. Mannervik and K. Axelsson, *Biochem. J.* **190,** 125 (1980).

tion of mixed disulfides may be an important regulatory mechanism linking the activity of metabolic networks to the oxidation/reduction status of the cell and that this regulation is effected by thioltransferase. As regards TPOR, Francis and Ballard[40] have recently reported the reversible inactivation of glucose-6-phosphate dehydrogenase by formation of an internal disulfide, catalyzed by a membrane bound enzyme. The microsomal enzyme was not purified or characterized save that its inactivation by iodoacetamide required prior treatment with thiol reagent, a characteristic of TPOR.

In summary, most evidence indicates that protein disulphide-isomerase, glutathione:insulin transhydrogenase, and thiol:protein disulfide oxidoreductase are probably the same enzyme. Other thiol protein disulfide exchange activities are also present within the cell.[38] While the function of TPOR is not clear, it is present in most tissues and is prominent in those active in protein metabolism and secretion, such as liver and pancreas.[27] It is known that in diabetic rat liver,[41] and in fibroblasts cultured in minimal media,[42] exogenous insulin is necessary to maintain normal levels of TPOR activity. The large increase in TPOR activity associated with the synthesis of procollagen[30] and of immunoglobulin[23] suggests an involvement in protein synthesis which should be further investigated.

Acknowledgments

This work was supported in part by a grant from the National Institutes of Health AM 18024. J. E. M. was a N. I. H. predoctoral trainee supported by Grant GM 07211. This is Journal Paper number 9438 from the Purdue University Agricultural Experiment Station.

[40] G. L. Francis and F. J. Ballard, *Biochem. J.* **186,** 581 (1980).

[41] E. P. Hern and P. T. Varandani, *J. Biol. Chem.* **255,** 697 (1980).

[42] J. E. Morin, J. E. Dixon, P. P. Chang, and J. Moss, *Biochem. Biophys. Res. Commun.* **111,** 872 (1983).

[70] Determination of Glutathione and Glutathione Disulfide in Biological Samples

By MARY E. ANDERSON

There are a number of procedures (e.g., chemical, enzymatic, chromatographic) for the determination of glutathione (GSH) and/or glutathione disulfide (GSSG) in biological samples.[1-4] Enzymatic and chromatographic methods for the determination of glutathione in biological samples are described here.

Preparation of Biological Samples

Although both GSH and GSSG occur in tissues, GSH is by far the predominant form. Greater than 99.5% of tissue "total glutathione" (i.e., GSH + GSSG, in GSH equivalents) is in the form of GSH. Since GSH readily oxidizes nonenzymatically and because it is a good substrate of γ-glutamyl transpeptidase (as is also GSSG), biological samples must be acidified quickly to reduce oxidation of GSH to GSSG and to mixed disulfides, and also to inactivate γ-glutamyl transpeptidase. Glutathione oxidizes rapidly at pH values greater than 7. Acid treatment inactivates γ-glutamyl transpeptidase which catalyzes the following reactions that decrease the levels of both GSH and GSSG.

$$\text{GSH} + \text{amino acid} \rightleftharpoons \gamma\text{-Glu-amino acid} + \text{CysH-Gly} \quad \text{(transpeptidation)}$$

$$\text{GSH} + \text{H}_2\text{O} \rightarrow \text{Glu} + \text{CysH-Gly} \quad \text{(hydrolysis)}$$

$$\text{GSH} + \text{GSH} \rightleftharpoons \gamma\text{-Glu-GSH} + \text{CysH-Gly} \quad \text{(autotranspeptidation)}$$

Tissues that have high levels of γ-glutamyl transpeptidase include kidney, pancreas, and the epithelia of small intestine, choroid plexus, and biliary ductules.

Biological samples must be obtained and processed rapidly. The optimum method for treating biological samples depends upon the tissue and

[1] A. Meister and M. E. Anderson, *Annu. Rev. Biochem.* **52,** 711 (1983).

[2] A. Larsson, S. Orrenius, A. Holmgren, and B. Mannervik, eds.), "Functions of Glutathione—Biochemical, Physiological and Toxicological Aspects (Nobel Conference)." Raven Press, New York, 1983.

[3] A. Meister, *Science* **220,** 472 (1983).

[4] M. E. Anderson, "Handbook for Oxy Radical Research." CRC Press, Cleveland, Ohio, 1985 (in press).

Copyright © 1985 by Academic Press, Inc.
All rights of reproduction in any form reserved.

the experimental system. Cells grown in tissue culture, lymphocytes, and erythrocytes may be lysed in dilute hydrochloric acid by freezing and thawing; the protein may be removed by treatment with 5-sulfosalicylic acid. Perfused tissues may be obtained by the freeze stop method.[5] Blood plasma must be obtained rapidly and immediately acidified. When rat[6] or mouse plasma is obtained rapidly (<4 min) the total glutathione levels are about 20–25 μM; however, if the plasma is allowed to stand, GSH disappears rapidly.[7] Tissues with high levels of γ-glutamyl transpeptidase must be homogenized rapidly, while those with very little transpeptidase, such as the liver, brain, spleen, and heart, may be frozen and stored at $-20°$ for 10–20 hr. The kidney, pancreas, and other transpeptidase-rich tissues should not be frozen and then thawed because transpeptidase cleaves GSH as the tissue is warmed.

Various acids, i.e., 5-sulfosalicylic, picric, metaphosphoric, trichloroacetic, and perchloric, have been used for deproteinization. Perchloric, trichloroacetic, and metaphosphoric acids may not maintain the GSH to GSSG ratio for all biological samples[8,9]; thus, picric acid (1%, w/v) and 5-sulfosalicylic acid (usually 5%, w/v) are preferable.

Tissue Treatment

Typically, tissues from mice or rats may be treated as follows. After decapitation, blood is collected for about 20 sec in a 150-ml beaker containing 50 μl of Na_4-EDTA (500 mM; pH 7.0) for 1 to 5 ml of mouse blood and 100 μl Na_4-EDTA for 5 to 10 ml of rat blood. The EDTA-treated blood is carefully transferred to 400-μl plastic tubes and centrifuged in a microfuge (Beakman; 10,000 g) for 1.5 min. The top of the tube is cut off with a razor blade and the plasma (100 μl) is quickly pipetted into a clean 400-μl tube containing 50 μl of 5-sulfosalicylic acid (10%, w/v) and agitated (preferably with a mechanical device). The sulfosalicylic acid supernatant solution is obtained after centrifuging (Beckman microfuge; 10,000 g) for 5 min.

Whole blood (20 μl) is added to hydrochloric acid (10 mM; 100 μl) in a 400-μl plastic tube. The red blood cells are lysed by freezing in a Dry-Ice acetone bath and then thawed; this process is repeated three times. After centrifuging in a microfuge (Beckman; 10,000 g) for 5 min, the superna-

[5] T. P. M. Akerboom and H. Sies, this volume, Vol. 77, p. 373.

[6] O. W. Griffith and A. Meister, *Proc. Natl. Acad. Sci. U.S.A.* **76,** 268 (1979).

[7] M. E. Anderson and A. Meister, *J. Biol. Chem.* **255,** 9530 (1980).

[8] D. J. Reed, J. R. Babson, P. W. Beatty, A. E. Brodie, W. W. Ellis, and D. W. Potter, *Anal. Biochem.* **106,** 55 (1980).

[9] S. K. Srivastava and E. Beutler, *Anal. Biochem.* **25,** 70 (1968).

tant solution (100 μl) is deproteinized with 5-sulfosalicylic acid (10%; 50 μl) and the protein-free supernatant solution is obtained after centrifuging for 5 min. Tissue culture cells are treated in a similar manner. Kidneys and the pancreas are rapidly removed, rinsed in water, blotted dry, and homogenized in 5 vol/g of wet tissue weight of 5% 5-sulfosalicylic acid. For smaller organs such as the pancreas and spleen, 10 volumes of 5% 5-sulfosalicylic acid may be used. The whole head (the brain is removed from the skull while frozen), muscle (hind quarter), spleen, heart, liver [minus gall bladder (mice)], and lung are frozen in individual beakers containing ethanol and Dry-Ice. Later, these tissues are completely defrosted, rinsed in water, and homogenized in 5% 5-sulfosalicylic acid as described above. The tissue homogenates are centrifuged in 1.5-ml plastic vials in a microfuge for 5 min. The supernatant solutions are stored at 4° until assayed. The analysis should be performed soon; however, storage overnight at 4° usually has minimal effects on the assay for total GSH.

Enzymatic Methods

DTNB-GSSG Reductase Recycling Assay for GSH and GSSG

Principle. Many workers have used the DTNB–GSSG reductase recycling procedure first reported by Owens and Belcher[10] and later modified by Tietze.[11] The modification described here is basically that of Griffith.[12] The recycling assay for total GSH (GSH + GSSG, in GSH equivalents) is a sensitive and specific enzymatic procedure. As indicated in reaction (1), GSH is oxidized by 5,5′-dithiobis(2-nitrobenzoic acid) (DTNB) to give GSSG with stoichiometric formation of 5-thio-2-nitrobenzoic acid (TNB). GSSG is reduced to GSH by the action of the highly specific glutathione reductase (GSSG reductase) and NADPH. The rate of TNB formation is followed at 412 nm (or 405 nm) and is proportional to the sum of GSH and GSSG present. The assay may be monitored at 340 nm (NADPH) or fluorometrically (NADPH).[9,13]

$$2\ \mathrm{GSH} + \mathrm{DTNB} \rightarrow \mathrm{GSSG} + \mathrm{TNB} \tag{1}$$

$$\mathrm{GSSG} + \mathrm{NADPH} + \mathrm{H}^{+} \xrightarrow[\text{Reductase}]{\text{GSSG}} 2\ \mathrm{GSH} + \mathrm{NADP}^{+} \tag{2}$$

[10] C. W. I. Owens, and R. V. Belcher, *Biochem. J.* **94,** 705 (1965).
[11] F. Tietze, *Anal. Biochem.* **27,** 502 (1969).
[12] O. W. Griffith, *Anal. Biochem.* **106,** 207 (1980).
[13] J. E. Brehe and H. B. Burch, *Anal. Biochem.* **74,** 189 (1976).

Reagents

Stock buffer: sodium phosphate (143 m*M*) and Na_4-EDTA (6.3 m*M*); final pH 7.5

Daily buffer: NADPH (0.248 mg/ml) in stock buffer is stored at 4° and prepared daily

DTNB: DTNB (6 m*M*) is prepared in stock buffer (store frozen)

GSSG reductase: The yeast enzyme is diluted to 266 U/ml from commercially available stock enzyme with stock buffer and stored at 4°

GSH standards: Standards are diluted daily in 5-sulfosalicylic acid from frozen stock solution (100 m*M* GSH or 50 m*M* GSSG) that is prepared weekly.

Procedure. Daily buffer (700 μl), DTNB solution (100 μl), and water (200 μl minus the sample volume, i.e., if the sample volume is 25 μl, then use 175 μl of water) are pipetted into each cuvette. The cuvettes are warmed at 30° in a jacketed spectrophotometer compartment or in a water bath for 12 to 15 min. The sample is added with mixing. GSSG reductase (5–10 μl) is added with mixing to initiate the assay. The formation of TNB is followed continuously with a recorder for a total of 2 or 2.5 absorbance units (full scale, 0.5 absorbance unit). The linear portion of the curve is usually between 1 and 2 absorbance units. For most biological samples, 1 to 25 μl of sample supernatant is assayed (e.g., 1 μl of liver, 2 μl of kidney, 10 μl of spleen, and 25 μl of plasma or cells grown in tissue culture). The amount of GSH is determined from a standard curve in which the GSH equivalents present (usually 1, 2, 3 and 4 nmol) is plotted against the rate of change of absorbance at 412 nm. A standard curve for each sample volume and enzyme volume is prepared. It is advantageous to use a total sample supernatant volume of 25 μl, i.e., 25 μl = sulfosalicylic acid volume plus sample volume. A sample blank lacking GSH is used to determine the background rate. Values are usually reported in GSH equivalents, e.g., μmol per gram for tissues, nmol per 10^6 cells, μ*M* for plasma.

Comments. The DTNB–GSSG reductase recycling assay for total glutathione is a specific, sensitive, rapid, and reliable procedure. However, since the method depends on an accurate standard curve, appropriate standards containing the protein precipitating agent are essential.

DTNB-GSSG Reductase Recycling Assay for GSSG

Principle. The determination of GSSG in biological samples is often difficult because GSSG is normally present at very low levels as compared to GSH (usually >99% GSH). Therefore, oxidation of GSH must be minimized; rapid sample preparation is important. GSSG may be determined using GSSG reductase by monitoring NADPH spectrophotometrically or fluorometrically.[9,13] The DTNB–GSSG reductase recycling assay described above for total GSH may be used for GSSG after treating GSH with *N*-ethylmaleimide (NEM)[5] or 2-vinylpyridine.[14] Both NEM and 2-vinylpyridine react with GSH, but NEM reacts more rapidly with 2-vinylpyridine. However, since as little as 10 μM NEM inhibits the GSSG reductase recycling assay by 30% NEM must be removed by extraction or chromatography prior to the enzymatic assay or erroneously low GSSG values will be obtained.[12] 2-Vinylpyridine at the concentrations given, however, does not inhibit the reductase; therefore, it need not be removed prior to the assay. The procedure below is that of Griffith.[12]

Reagents

2-Vinylpyridine: 2-vinylpyridine is used undiluted and stored at −20° (replace if the compound becomes brown or viscous)

Standard solutions: GSSG (50 mM stock solution) is diluted daily in 5-sulfosalicylic acid

The other reagents are as described above for the DTNB–GSSG reductase recycling assay for GSH and GSSG.

Procedure. The 2-vinylpyridine derivatizations should be carried out in a hood; 2-vinylpyridine has a low vapor pressure and constant exposure may be irritating.

The 5-sulfosalicylic acid supernatant solution (100 μl) is placed in a 1.5-ml plastic tube and 2-vinylpyridine (2 μl) is added with mixing. Triethanolamine (6 μl) is added to the side of the tube and the solution is vigorously mixed. The final pH, which should be carefully checked with pH paper, should be between 6 and 7. If the pH inadvertently exceeds 7, derivatization should be redone on a new sample using less triethanolamine. After 60 min, the derivatized samples are assayed as described above in the DTNB–GSSG reductase recycling assay. Assay volumes are typically 25 μl and the reductase volume is usually at least 10 μl. GSSG standards and sample blanks containing only 2-vinylpyridine are also run.

Comments. Other assays for GSSG are available[5]; however, the 2-vinylpyridine derivatization procedure is also useful for the chromatographic method for determining GSH and GSSG described below.

Chromatographic Procedures for the Determination of GSH

A number of sensitive and specific chromatographic methods are available for the determination of GSH, but they are somewhat

[14] O. W. Griffith and A. Meister, *Proc. Natl. Acad. Sci. U.S.A.* **77,** 3384 (1980).

lengthy.[15–18] Recent advances in high-performance liquid chromatography (HPLC) and the variety of HPLC columns now available have led to development of several rapid chromatographic methods.[4] Derivatization of GSH with 2-vinylpyridine followed by ion-exchange chromatography and ninhydrin detection and the derivatization of GSH with Kosower's[19] monobromobimane reagent followed by reverse-phase HPLC and fluorescent detection are described below.

Ion-Exchange Chromatography after 2-Vinylpyridine Derivatization

Principle. GSH and other thiols react rapidly at pH ~7 with 2-vinylpyridine to form a conjugate, which is then determined by ion-exchange amino acid analysis.

Procedure. Samples are derivatized as described above for the enzymatic determination of GSSG. Total glutathione (GSH + GSSG, in GSH equivalents) may also be determined after sample reduction with dithiothreitol (50 m*M*; pH 7, 1 hr) and derivatization with 2-vinylpyridine as described above.

Chromatography is carried out on an amino acid analyzer (such as Durrum Model 500) using sodium or lithium citrate buffers. The 2-vinylpyridine derivatives of GSH, cysteine, and γ-glutamylcysteine elute at ~78, 165, 72 min, respectively (sodium buffer).[14] The amount of GSH present is quantitated by comparison with standards.

High-Performance Liquid Chromatography (HPLC) Determination of GSH

Principle. GSH and other thiols, such as cysteine, react with Kosower's monobromobimane[19] (3,7-dimethyl-4-bromomethyl-6-methyl-1,5-diazobicyclo[3.3.0]octa-3,6-diene-2,8-dione) (mBBr). The GSH-bimane derivative is then quantitated after HPLC using fluorescence detection. The method described below is an isocratic modification of Newton, Dorian, and Fahey.[20,21]

Reagents

MBBr solution: mBBr (0.1 *M*) is prepared in acetonitrile (mBBr is available as Thiolyte MB from Calbiochem-Mannheim). This solu-

[15] C. W. Tabor and H. Tabor, *Anal. Biochem.* **78,** 543 (1977).

[16] S. Moore and W. H. Stein, *J. Biol. Chem.* **211,** 893 (1954).

[17] J. W. Purdie and D. E. Hanafi, *J. Chromatogr.* **59,** 181 (1971).

[18] M. Hsiung, Y. Y. Yeo, K. Itiaba, and J. C. Crawhill, *Biochem. Med.* **19,** 305 (1978).

[19] E. M. Kosower, B. Pazhenchevsky, and E. Hershkowitz, *J. Am. Chem. Soc.* **100,** 6516 (1978).

[20] G. L. Newton, R. Dorian, and R. C. Fahey, *Anal. Biochem.* **114,** 383 (1981).

[21] M. E. Anderson and A. Meister, *Proc. Natl. Acad. Sci. U.S.A.* **80,** 707 (1983).

tion is stored at −20° in a brown bottle. Although this light-sensitive solution is stable for several weeks, only small (2 to 4 ml) quantities are usually prepared.

N-Ethylmorpholine: *N*-ethylmorpholine (1 *M*) is prepared daily from redistilled *N*-ethylmorpholine (7.8 *M*; store in a brown bottle).

Thiol standard solution: GSH (1 m*M*) in 5-sulfosalicylic acid is prepared daily from a 100 m*M* stock solution that is stored frozen and prepared weekly. Cysteine standard solutions are prepared in 5-sulfosalicylic acid immediately before use.

Derivatization Procedure. The derivatization is carried out in reduced light to reduce the formation of mBBr degradation products. The proportions given are for mouse or rat kidney and liver (5-sulfosalicylic acid supernatant solutions; 5% 5-sulfosalicylic acid is used to homogenize kidney and liver at 5 vol/g of wet tissue weight). For tissues with low GSH levels, the volumes of supernatant solution and *N*-ethylmorpholine used are greater and the volume of water is decreased (to allow for adjustment of the pH to ~7.6).

Water (340 μl) and 5-sulfosalicylic acid supernatant solution (120 μl) are pipetted into a 1.5-ml plastic tube. *N*-Ethylmorpholine (100 μl, 1 *M*) is added to the wall of the tube. mBBr (20 μl, 0.1 *M*) is immediately added with mixing. The reaction is allowed to proceed for 20 min in the dark; then glacial acetic acid (20 μl) is added to decrease the pH. The samples may be stored at −20° in the dark until analyzed.

HPLC Procedure

Apparatus. A Waters Associates' HPLC consisting of two M6000A pumps, 721 system controller, a 710B WISP autosampler, a 420 fluorescent detector with *o*-phthalaldehyde filters (or a Gilson fluorometer), and a 720 data module (or a Nelson Analytical data system) are used; however, only a manual isocratic HPLC system with solvent switching is required. For the solvent system described a 10 μm C_{18} (4.6 × 250 mm) μBondpak column (Waters Associates) is used; an Ultrasphere ODS (3.9 × 250 mm, 5 μm) column (Beckman) or other C_{18} columns may be used with only a slight modification of the percentage methanol in Solvent A. The retention times of the thiol conjugates vary with temperature; typically 22° is used.

Reagents

Solvent A: 14.2% methanol (v/v) and 0.25% glacial acetic acid (v/v) adjusted to pH 3.9 with sodium hydroxide (50%, w/v)

Solvent B: 90% methanol (v/v) and 0.25% glacial acetic acid (v/v) adjusted to pH 3.9 with sodium hydroxide (50%, w/v)

Procedure. The column is equilibrated for 12 min with Solvent A (1 ml/min). The sample is injected. For a given series of tissue samples, the injection volume is constant, e.g., for kidney and liver samples the injection volumes are 200 μl and 100 μl, respectively. Elution with Solvent A is carried out for 30 min. Sulfosalicylic acid elutes with the void volume (4 min), while the mBBr conjugates of cysteine, cysteinylglycine, γ-glutamylcysteine, and GSH elute at 10, 13, 15, and 18 min, respectively. The column is cleaned for 8 min with Solvent B (1 ml/min) and reequilibrated as described above. GSH and other thiols are quantitated by using a standard curve.

Comments. This method for determination of GSH is readily automated. It may be used in conjunction with determinations of radioactivity in GSH and its thiol metabolites after radioactive GSH precursors are administered to laboratory animals.[21]

[71] Preparation of γ-Glutamyl Amino Acids by Chemical and Enzymatic Methods

By MARY E. ANDERSON and ALTON MEISTER

Several chemical methods are available for the synthesis of γ-glutamyl amino acids.[1-11] The method described below involves reaction of *N*-phthaloyl-L-glutamic anhydride with an amino acid in glacial acetic acid to form the corresponding *N*-phthaloyl-L-glutamyl amino acid.[5,6] The phthaloyl moiety is removed by treatment with hydrazine and the product is separated by column chromatography. This procedure is applicable to the preparation of a large number of γ-glutamyl amino acids and of certain other related compounds such as γ-glutamyl *p*-nitroanilide.[7-11] It is not

[1] M. E. Anderson and A. Meister, *Proc. Natl. Acad. Sci. U.S.A.* **80,** 707 (1983).
[2] W. J. LeQuesne and G. T. Young, *J. Chem. Soc.* p. 1959 (1950).
[3] A. Meister, S. S. Tate, and O. W. Griffith, this series, Vol. 77, p. 237.
[4] H. Sachs and E. Brand, *J. Am. Chem. Soc.* **75,** 4608 (1953).
[5] F. E. King and D. A. A. Kidd, *J. Chem. Soc.* p. 3315 (1949).
[6] F. E. King, J. W. Clark-Lewis, and R. Wade, *J. Chem. Soc.* p. 886 (1957).
[7] M. Orlowski and A. Meister, *J. Biol. Chem.* **240,** 338 (1965).
[8] M. Orlowski and A. Meister, *J. Biol. Chem.* **246,** 7095 (1971).
[9] M. Orlowski and A. Meister, *J. Biol. Chem.* **248,** 2836 (1973).
[10] O. W. Griffith and A. Meister, *Proc. Natl. Acad. Sci. U.S.A.* **74,** 3330 (1977).
[11] O. W. Griffith, R. J. Bridges, and A. Meister, *Proc. Natl. Acad. Sci. U.S.A.* **76,** 6319 (1979).

Copyright © 1985 by Academic Press, Inc.
All rights of reproduction in any form reserved.

useful for preparation of γ-glutamylcystine because of the insolubility of cystine in glacial acetic acid.

Synthesis of L-*γ-Glutamyl*-L-*α-Aminobutyric Acid*[8]

N-Phthaloyl-L-glutamic anhydride (0.05 mol; 12.96 g)[3] and L-α-aminobutyric acid (0.05 mol; 5.16 g) are placed in a 500-ml round bottom flask equipped with a calcium chloride drying tube. The flask is evacuated for 15 min to remove moisture. Glacial acetic acid (150 ml) is added with swirling. The flask is swirled for 20 min at 110–115° in an oil bath as the reagents dissolve. (Note: The temperature should not exceed this range because of the possibility of racemization.[5]) The flask is allowed to cool to room temperature; a white precipitate forms in a golden solution. The solvent is removed by flash evaporation (bath temperature 35°) and the residue is washed successively with four portions (100 ml) of ethanol to remove acetic acid. The remaining yellow syrup is dissolved in methanol (100 ml) and hydrazine hydrate (10 ml) is added in a hood with swirling. Triethylamine (10 ml) is added with mixing. [The temperature increases and cooling in ice may be necessary.] The mixture is allowed to stand at room temperature in a hood in a flask equipped with a drying tube for 2 days, during which time a large precipitate forms. The solvent is removed by flash evaporation, and the residue is washed successively with four portions (100 ml) of methanol. Water (100 ml) is added to dissolve the residue. The pH is adjusted to 3 by adding concentrated hydrochloric acid and the flask is allowed to stand overnight at 4°. The precipitate which forms is removed by filtration under reduced pressure through a sintered glass funnel. The pH of the filtrate is adjusted to 6 by adding sodium hydroxide (10 *M*). It is then applied to a Dowex-1-acetate column (4.1 × 30 cm). The column is washed with 1 liter of water, and the column is then treated with a linear gradient established between 4 liters of water and 4 liters of 1.5 *M* acetic acid. Fractions of 25 ml are collected; the fractions containing product (numbers 100–150) are combined and lyophilized. The yield is 70%; the melting point is 183–815°.

Synthesis of L-*γ-Glutamyl*-L-*Tyrosine*[12]

N-Phthaloyl-L-glutamic anhydride (0.1 mol; 25.9 g) and L-tyrosine (0.1 mol; 18.1 g) are treated as described above for the synthesis of L-γ-glutamyl-L-α-aminobutyric acid, but the volumes and flask size are in-

[12] M. E. Anderson, Doctoral Dissertation, Cornell University Medical College, Ithaca, New York (1983).

creased. The pH of the filtrate is adjusted to 6 and the solution is applied to a Dowex-1-acetate column (4.1 × 30 cm). The column is washed successively with 1.2 liters each of water, 0.1 *M* acetic acid (tyrosine elutes), 1.8 *M* acetic acid (glutamic acid elutes), and 4 *M* acetic acid (γ-glutamyltyrosine elutes). The fractions containing the product are combined and lyophilized. The yield is 30%. The melting point is 243–246°.

Preparation of γ-Glutamylcystine

The procedure given below for the synthesis of γ-glutamylcystine also facilitates the preparation of several related γ-glutamyl compounds such as γ-glutamylcysteine disulfide and γ-glutamylcysteine.[1,12] The general procedure for mixed disulfide synthesis is limited by the ability to separate the product from the starting materials and byproducts. This method has been used to make the mixed disulfides between CoA and glutathione, cysteine and glutathione, panteheine and glutathione, and others.[12–14] The following steps are involved: preparation of glutathione disulfide, enzymatic formation of γ-glutamylcysteine disulfide, chemical synthesis of cystine thiosulfonate, reduction of γ-glutamylcysteine disulfide, and reaction of γ-glutamylcysteine with cystine thiosulfonate to yield γ-glutamylcystine.

Preparation of Glutathione Disulfide (GSSG) from Glutathione

Although glutathione disulfide is commercially available, its preparation is less costly and leads to a product that is free of ethanol. The preparation may be scaled-up easily.

Glutathione (5 g) is dissolved in water (100 ml) in an Erlenmeyer flask. The pH is adjusted to 8 by adding concentrated ammonium hydroxide. The flask is placed at 0° and oxygen is bubbled into it for 24 to 48 hr until the solution gives a negative test for thiol. A small portion (e.g., 10 μl) of the reaction mixture is added to 1 ml of 5,5′-dithiobis(2-nitrobenzoic acid) (DTNB) solution (6 m*M* DTNB in 143 m*M* sodium phosphate, 6.3 m*M* sodium EDTA, pH 7.5) in a cuvette; the absorbance at 412 nm is read against a blank containing DTNB (ε_M = 13,600 at 412 nm[16]). The product is obtained by lyophilization of the solution; high yields (>98%) of a glassy product are obtained.

[13] B. Eriksson (B. Mannervik), *Acta Chem. Scand.* **20,** 1178 (1966).
[14] B. Eriksson and S. A. Eriksson, *Acta Chem. Scand.* **21,** 1304 (1967).
[15] B. Mannervik and G. Nise, *Arch. Biochem. Biophys.* **134,** 90 (1969).
[16] G. L. Ellman, *Arch. Biochem. Biophys.* **82,** 70 (1959).

Preparation of γ-Glutamylcysteine Disulfide $[(GC)_2]$[17]

The solution described above prior to lyophilization may be used. Alternatively, GSSG (5 g) is suspended in 100 ml of water and the pH is adjusted to 8 by adding ammonium hydroxide. Carboxypeptidase A (12,500 U) is added with stirring. After 16–20 hr at 20–37°, additional carboxypeptidase A (500 U) is added and the enzymatic reaction is allowed to continue for 10 to 12 hr. The course of the reaction is monitored by following the production of glycine on a Durrum 500 amino acid analyzer. The reaction mixture is filtered and the filtrate is applied to a Dowex-1-formate column (12.1 × 20 cm). The column is washed with 1 liter of water, and then with 1 liter of 0.15 *M* formic acid. The product, $(GC)_2$, is eluted with 0.27 *M* formic acid (300 ml). $(GC)_2$ may be obtained by lyophilization. Since any residual GSSG will also elute with $(GC)_2$, it may be necessary to dissolve the product in water (25 ml), adjust to pH 8 by adding ammonium hydroxide, and incubate again with carboxypeptidase A. A useful modification[12] involves cycling the solution through a column (1.1 × 20 cm) of Agarose-carboxypeptidase A for 12–16 hr at 20° using a small peristaltic pump. The product [$(GC)_2$ free of GSSG] is obtained by chromatography on Dowex-1-formate as described above. The yield of GSSG-free $(GC)_2$ is about 95%.

Cystine Thiosulfonate[18]

L-Cystine (9.6 g; 0.04 mol) is placed in a 500-ml round bottom flask equipped with a magnetic stirrer. Formic acid (88%; 222 ml) is added with swirling and the cystine is dissolved. The mixture is placed in an ice bath to maintain the temperature of the solution at 20°. Concentrated hydrochloric acid (8 ml) is then slowly added. Hydrogen peroxide (30%; 10 ml) is added slowly with stirring. After stirring for 2.5 hr at 20°, the solvent is carefully removed under reduced pressure at less than 20°. *Caution:* Do not allow the reaction mixture to become dry since there is a high concentration of performic acid present. The procedure should be done behind a protective Plexiglas shield. The solvent is evaporated until a thick syrup is obtained. Water (100 ml) is added with mixing, followed by addition of ammonium hydroxide (concentrated, 10 ml) to adjust the pH to between 3 and 4. The flask is allowed to stand overnight at 4°. The precipitate is collected by filtration under reduced pressure and rinsed with cold water (30 ml). It is dried under reduced pressure over phosphorus pentoxide. The yield is 85%. The compound decomposes (172-5°

[17] D. Strumeyer and K. Bloch, *Biochem. Prep.* **9,** 52 (1962).
[18] R. Emiliozzi and L. Pichat, *Bull. Soc. Chim. Fr.* **25,** 1887 (1959).

uncorr.). Although this procedure has been carried out successfully many times without difficulty, it is probably not advisable to increase the scale of the preparation because of the relatively large amount of performic acid involved.

Reduction of γ-Glutamylcysteine Disulfide to γ-Glutamylcysteine

This procedure for the reduction of γ-glutamylcysteine disulfide [$(GC)_2$] to γ-glutamylcysteine (GCH) is a slightly modified version of the method of Strumeyer and Bloch.[17] Caution: Hydrogen sulfide is toxic; therefore, use a fume hood.

$(GC)_2$ (5 g; 0.01 mol) is dissolved in 4 *M* formic acid (250 ml) in a 1-liter Erlenmeyer flask equipped with a magnetic stirrer. This solution is stirred at 35–50° while oxygen-free nitrogen is bubbled into it. Zinc dust (20 g) is added in 5 portions over a 1-hr period. The solution is decanted into a 2-liter Erlenmeyer side-arm flask equipped with a magnetic stirrer. The zinc residue is washed 3 times with 4 *M* formic acid (nitrogen saturated) (~100 ml). The washings are added to the filtrate. Hydrogen sulfide is bubbled gently into the filtrate under slight pressure and the solution is vigorously stirred. After 1 hr the solution is rapidly filtered under reduced pressure (Whatman 50). The filtration is repeated if necessary until the filtrate is clear. The zinc sulfide is washed with two 25 ml portions of hydrogen sulfide-saturated formic acid. The wash solutions are added to the filtrate. Nitrogen is bubbled into the solution at 0–4° until the odor of hydrogen sulfide cannot be detected (usually about 12–16 hr). The solution is concentrated under reduced pressure and then lyophilized to give a white powder. The product may be stored in a desiccator over Drierite at −10°.

Synthesis of γ-Glutamylcystine (GCC)

Cystine thiosulfonate (16 g; 0.06 mol) and γ-glutamylcysteine (5 g; 0.02 mmol) are dissolved in 500 ml of 10 m*M* formic acid in a 1-liter round bottom flask. The solution is stirred at 20° for 2 hr and filtered under reduced pressure. The filtrate is applied to a Dowex-1-formate column (2.1 × 25 cm). Cystine and cystine thiosulfonate by-products are removed from the column by elution with water (1 liter). GCC elutes with 0.15 *M* formic acid (250 ml). The product is obtained by lyophilization. The yield is about 70% based on γ-glutamylcysteine.

Enzymatic Synthesis of L-*γ-Glutamyl*-L-*Amino Acids*

L-γ-Glutamyl-L-amino acids may be synthesized enzymatically by use of γ-glutamylcysteine synthetase or of γ-glutamyl transpeptidase. Enzymatic synthesis of γ-glutamyl amino acids is often the preferred method for preparation of radiolabeled compounds. γ-Glutamyl transpeptidase has been used for the synthesis of L-γ-glutamyl-L-glutamine,[19] L-γ-glutamyl-L-serine,[20] and L-γ-glutamyl-L-cystine.[12] γ-Glutamylcysteine synthetase has been used to prepare L-γ-glutamyl-L-α-aminobutyric acid,[21] L-γ-glutamyl-L-cysteine (and its disulfide),[12] L-γ-(β-methyl)–glutamyl-L-α-aminobutyric acid,[10] and β-aminoglutaryl-L-α-aminobutyric acid.[10] γ-Glutamylcysteine synthetase may be used for the synthesis of the L-γ-glutamyl derivatives of L-alanine, L-valine, L-serine, L-methionine, L-homocysteine, L-threonine, L-homoserine, and glycine.[20,22]

The enzymatic preparation of L-γ-glutamyl-L-α-aminobutyric acid, L-γ-glutamyl-L-cysteine disulfide, L-γ-glutamyl-L-cystine, and L-γ-glutamyl-L-glutamine are described below.

Enzymatic Synthesis of L-γ-[*U*-^{14}*C*]*Glutamyl*-L-α-*Aminobutyric Acid*[21]

γ-Glutamylcysteine synthetase[22] (2 units) is incubated at 37° in a solution containing L-[U-^{14}C]glutamate (25 μCi; 0.954 μmol), Tris–HCl (300 μl mol; pH 8.2), Na_2-EDTA (0.25 μmol), phosphoenolpyruvate (4.69 μmol), Na_2ATP (4 μmol), L-α-aminobutyric acid (2 μmol), $MgCl_2$ (10 μmol), NADH (357 μmol), pyruvate kinase (20 units), and lactate dehydrogenase (10 units) in a final volume of 3.036 ml. After 40 min, the reaction is cooled to 0° and unlabeled L-γ-glutamyl-L-α-aminobutyric acid (100 mg) is added. The solution is applied to a Dowex-50-H^+ column (2.2 $\times$ 6 cm) (W $\times$ 8, 200–400 mesh). The column is washed with water (60 ml) to elute the nucleotides and lactate. The column is then treated with 0.4 *M* pyridine (adjusted to pH 7.0 by adding acetic acid). Fractions of 5 ml are collected; the fractions containing radioactivity are combined and lyophilized. The lyophilized residue is dissolved in 10 ml of water and applied to a Dowex-1-acetate column (1 $\times$ 30 cm; $\times$8; 200–400 mesh). The column is eluted with a linear gradient established between 450 ml of water and 450 ml of 0.5 *M* acetic acid. Fractions of 10 ml are collected. The product elutes in fraction 21 to 23 and is obtained in dry form by lyophilization of these fractions (final specific activity 0.381 μCi/μmol).

Enzymatic Synthesis of L-γ-*Glutamyl*-L-[^{35}*S*]*Cysteine Disulfide* [(γ-*Glu-Cy*^{35}*S*)$_2$][12]

The reaction mixture contains Tris–HCl (0.1 *M*; pH 8.0), ATP (5 m*M*), L-glutamate (10 m*M*), $MgCl_2$ (20 m*M*), EDTA (2 m*M*), KCl (150

[19] M. Orlowski, P. G. Richman, and A. Meister, *Biochemistry* **8,** 1048 (1969).
[20] M. Orlowski and A. Meister, *Biochemistry* **10,** 372 (1971).
[21] N. Taniguchi and A. Meister, *J. Biol. Chem.* **253,** 1799 (1978).
[22] R. Sekura and A. Meister, *J. Biol. Chem.* **252,** 2599 (1977).

m*M*), L-[^{35}S]cysteine (0.125 m*M*; 5 mCi), dithiothreitol (1.58 m*M*) in a final volume of 4 ml. The solution is protected from air after blowing a stream of dry nitrogen over it, and after warming to 37° for 5 min, the reaction is initiated by addition of 22 μl of γ-glutamylcysteine synthetase (232 U/ml). After 2 hr, additional enzyme (22 μl) is added, and the reaction is allowed to proceed for 2 hr. Hydrochloric acid (concentrated, 75 μl) is added to stop the reaction. The mixture is extracted six times with 1 ml of ethyl acetate (water-saturated), and the aqueous solution is freed from traces of ethyl acetate by use of a stream of nitrogen at 30°. Unlabeled γ-glutamylcysteine (5 μmol) is added and the pH is adjusted to 8 by adding sodium hydroxide (10 *M*; 100 μl). Oxygen is bubbled into the solution for 18 hr. The pH is adjusted to 5 by addition of acetic acid, and the solution is then applied to a Dowex-1-formate column (1.2 × 15.7 cm). The column is washed with 50 ml of water. Elution is carried out with a linear gradient established between 250 ml of water and 250 ml of 0.4 *M* formic acid (flow rate, 0.5 ml/min). The fractions (5 ml) containing the product (numbers 52–64; corresponding to 0.15–0.25 *M* formic acid) are combined and lyophilized. The yield is about 88%. [Note: (γ-Glu-Cy^{35}S)$_2$ is prepared because γ-glu-cysteine is unstable and oxidizes readily.]

Synthesis of the Mixed Disulfide between L-*γ-Glutamyl*-L-[^{35}*S*]*Cysteine and Cysteine,* L-*γ-Glu*-[^{35}*S*]*Cys Cys ("Internally-Labeled" γ-Glutamylcystine*)[1,12]

This mixed disulfide may be prepared by the thiosulfonate method described above after reduction of γ-glutamyl-L-[^{35}S]cysteine disulfide. The method described above for the reduction of (GC)$_2$ is more suitable for larger scale preparations than for small scale preparations (e.g., those involving labeled compounds). For the latter, reduction may be carried out with dithiothreitol as follows.[23]

L-γ-Glutamyl-L-[^{35}S]cysteine disulfide (1 μmol; 0.8 mCi) is dissolved in 2 ml of ammonium bicarbonate (10 m*M*, pH 8.0) containing 2 m*M* Na-EDTA (pH 7). Dithiothreitol (0.1 mmol) is added with mixing until it is dissolved. After standing for 60 min at 25°, formic acid (88%) is added to the solution to achieve a pH value of 2. The solution is extracted five times with 6 ml portions of water-saturated ethyl acetate. The residue ethyl acetate is removed from the aqueous solution by blowing a stream of nitrogen over the solution at 30° for 20 min. The solvent is removed carefully under reduced pressure. The solid L-γ-glutamyl-L-[^{35}S]cysteine is then dissolved in 50 ml of 10 m*M* formic acid and the mixed disulfide between L-γ-glutamyl-L-[^{35}S]cysteine and L-cysteine is synthesized by the thiosulfonate method described above.

[23] J. Butler, S. P. Spielberg, and J. D. Schulman, *Anal. Biochem.* **75**, 674 (1976).

An enzymatic method for the synthesis of L-γ-glutamyl-L-[^{35}S]cystine and for the mixed disulfide between L-γ-glutamyl-L-[^{35}S]cysteine and cysteine using γ-glutamyl transpeptidase is given below.

Synthesis of L-γ-Glutamyl-L-[^{35}S]Cystine and the Mixed Disulfide between γ-Glutamyl-L-[^{35}S]Cysteine and Cysteine[1,12]

L-[^{35}]Cystine (30 μmol; 3 mCi), L-γ-glutamyl-*p*-nitroanilide (15 μmol), and sodium phosphate (300 μmol, pH 8.3) in a final volume of 7.5 ml are placed at 37° for 5 min. The reaction is initiated by addition of 0.038 unit of rat kidney γ-glutamyl transpeptidase. The reaction is terminated after 15 min by addition of 3.75 ml of 70 m*M* acetic acid. The γ-glutamyl[^{35}S]cystine formed is purified by chromatography on a Dowex-1-acetate column (1.9 × 12.6 cm). After washing with water and 0.1 *M* acetic acid, the product is eluted with 1.5 *M* ammonium acetate (pH 7). Unreacted [^{35}S]cystine may be recovered in the washings. L-γ-Glutamyl-L-[^{35}S]cystine (4 μmol) is further purified by chromatography on a Dowex 50 (H^+) column (3.1 × 25.5 cm) and lyophilized. The dry residue from above is dissolved in 10 ml of 10 m*M* NH_4HCO_3 (pH 8) containing 2 m*M* Na-EDTA (pH 7) and 0.2 ml of 2-mercaptoethanol. After standing at 25° for 50 min, the solution is acidified by addition of 1 ml of 4 *M* formic acid, and is then lyophilized. The residue is dissolved in 24 ml of 10 m*M* formic acid and 2 mmol of L-cystine thiosulfonate is added. This solution is stirred for 14 hr at 25°. The L-γ-glu-L-[^{35}S]cys L-cys is filtered and purified by Dowex-1 formate chromatography as described above and has a specific radioactivity of 22,200 cpm/nmol.

Synthesis of the Mixed Disulfide between L-γ-Glutamyl-L-Cysteine and L-[^{35}S]Cysteine, L-γ-Glu-Cys [^{35}S]Cys ("Externally-Labeled" γ-Glutamyl-[^{35}S]Cysteine)[1,12]

L-[^{35}]Cystine (20 μmol; 2 mCi) and 0.2 mmol of L-γ-glutamylcysteine disulfide are dissolved in 1 ml of 100 m*M* NH_4HCO_3 (pH 8.0) containing 2.1 m*M* EDTA. Dithiothreitol (2 mmol) is dissolved in the solution by warming it at 55° for 10 min. After standing for 60 min at 25°, 0.4 ml of 88% formic acid is added (pH 2), and the solution is extracted five times with 3 ml portions of water-saturated ethyl acetate. The residual ethyl acetate is removed from the aqueous solution by blowing a stream of dry nitrogen over it at 30° for 10 min. The pH is then adjusted to 8 by adding NH_4OH; a crystal of $FeSO_4$ (<1 mg) is added and oxygen is bubbled into the mixture. After 60 min, formic acid (88%, 0.75 ml) is added and the solution is carefully evaporated with an evapo-mix (Buchler). The residue is dissolved in 10 ml of water and chromatographed on a Dowex-1 formate column (2.1 × 11 cm) as described above for the chemical synthesis of γ-glutamylcystine. The product, which elutes with 0.15 *M* formic acid, has a specific radioactivity of 22,200 cpm/mol.

Enzymatic Synthesis of L-*γ-Glutamyl*-L-*Glutamine*[19]

Glutathione (3.07 g; 10 mmol) and L-glutamine (4.39 g; 30 mmol) are dissolved in water (200 ml, final volume), and the solution is adjusted to pH 8.7 by adding sodium hydroxide. γ-Glutamyl transpeptidase (50 units) is added to the solution, which is placed at 37° for 1 hr. The solution is then chilled to 0°, and applied to a Dowex-1-acetate column (×8, 200–400 mesh, 4 × 75 cm; 5°). The column is washed with 8 liters of water. The column is then treated with 4 liters of each of the following acetic acid solutions: 0.05, 0.10, 0.15, and 0.30 *M*. γ-Glutamylglutamine elutes with 0.30 *M* acetic acid. The fractions containing the product are combined and evaporated to a low volume under reduced pressure (below 40°) and then lyophilized. The product is crystallized from ethanol–water. The yield is 40%; mp 194–195°.

Characterization of γ-Glutamyl Amino Acids

γ-Glutamyl amino acid preparations obtained by organic synthesis may contain variable amounts of the corresponding α-glutamyl amino acids.[4,5,24] Racemization of the γ-glutamyl moiety may occur at least to some extent during preparation of phthaloyl-L-glutamic acid and in the reaction between this compound and an amino acid.[5] γ-Glutamyl amino acids have been characterized by the ninhydrin-carbon dioxide procedure.[4,25] Thus, treatment of γ-glutamyl amino acids with ninhydrin yields stoichiometric amounts of carbon dioxide, whereas α-glutamyl amino acids do not yield carbon dioxide under these conditions. Chromatographic procedures may also distinguish between α- and γ-glutamyl amino acids. NMR spectra have been used to distinguish α- and γ-glutamyl peptides.[26,27] Many, but not all L-γ-glutamyl-L-amino acids are substrates of γ-glutamylcyclotransferase (see also this volume [52]), and therefore exhaustive treatment of certain L-γ-glutamyl-L-amino acids with this enzyme yields stoichiometric amounts of 5-oxo-L-proline and the corresponding L-amino acid.[9,21] γ-Glutamyl transpeptidase exhibits a broad specificity for the L- and D-γ-glutamyl amino acids and this enzyme there-

[24] D. A. Kidd and F. E. King, *Nature (London)* **162,** 776 (1948).
[25] T. T. Otani and A. Meister, *J. Biol. Chem.* **224,** 137 (1957).
[26] T. Kasai and S. Sakamura, *Agric. Biol. Chem.* **37,** 2155 (1973).
[27] T. Kasai and S. Sakamura, *Agric. Biol. Chem.* **37,** 685 (1973).

fore has usefulness in the characterization of these compounds (see this volume [50]). The extent of racemization of the γ-glutamyl amino acid residue of γ-glutamyl amino acids may be determined by careful examination of the product after acid hydrolysis (6 *M* HCl, 100°, 12–14 hr). After hydrolysis, glutamate and the other amino acid present may be separated by ion-exchange chromatography and the fractions containing glutamic acid may be combined and lyophilized. The amount of L-glutamate present may then be determined by assay with L-glutamate dehydrogenase,[28] and the amount of D-glutamate may be determined with D-glutamate cyclase (see this volume [9]). L- and D-Glutamic acid may also be quantitatively determined after derivatization with L-leucine according to the procedure of Manning and Moore[29]; this column chromatographic method is capable of separating L-leucyl-L-amino acids from L-leucyl-D-amino acids.[30] A recently described method for distinguishing D- and L-amino acids involves high-performance liquid chromatography using a chiral solvent.[31,32]

[28] H. J. Strecker, this series, Vol. 2, p. 220.
[29] J. M. Manning and S. Moore, *J. Biol. Chem.* **243,** 5591 (1968).
[30] T. Takaya, Y. Kishida, and S. Sakakibara, *J. Chromatogr.* **215,** 279 (1981).
[31] E. Gil-Av, A. Tishbee, and P. E. Hare, *J. Am. Chem. Soc.* **102,** 5115 (1980).
[32] S. Weinstein, M. H. Engel, and P. E. Hare, *Anal. Biochem.* **121,** 370 (1982).

[72] Separation of γ-Glutamyl Amino Acids by Ion-Exchange Chromatography

By Daniel Wellner

Introduction

γ-Glutamyl amino acids are widely distributed in nature. The formation of γ-glutamylcysteine and the γ-glutamyl derivatives of certain other amino acids is catalyzed by γ-glutamylcysteine synthetase and many γ-glutamyl amino acids are formed by transpeptidation.[1] γ-Glutamyl amino acids have been found in brain,[2,3] urine,[4–6] and blood.[7] More than 70

[1] A. Meister and S. S. Tate, *Annu. Rev. Biochem.* **45,** 559 (1976).
[2] I. Sano, *Int. Rev. Neurobiol.* **12,** 235 (1970).
[3] K. L. Reichelt, *J. Neurochem.* **17,** 19 (1970).
[4] D. L. Buchanan, E. E. Haley, and R. T. Markiw, *Biochemistry* **1,** 612 (1962).
[5] M. F. Lou, *Biochemistry* **14,** 3503 (1975).
[6] H. Peck and R. J. Pollitt, *Clin. Chim. Acta* **94,** 237 (1979).
[7] L. Hagenfeldt, A. Arvidsson, and A. Larsson, *Clin. Chim. Acta* **85,** 167 (1978).

Copyright © 1985 by Academic Press, Inc.
All rights of reproduction in any form reserved.

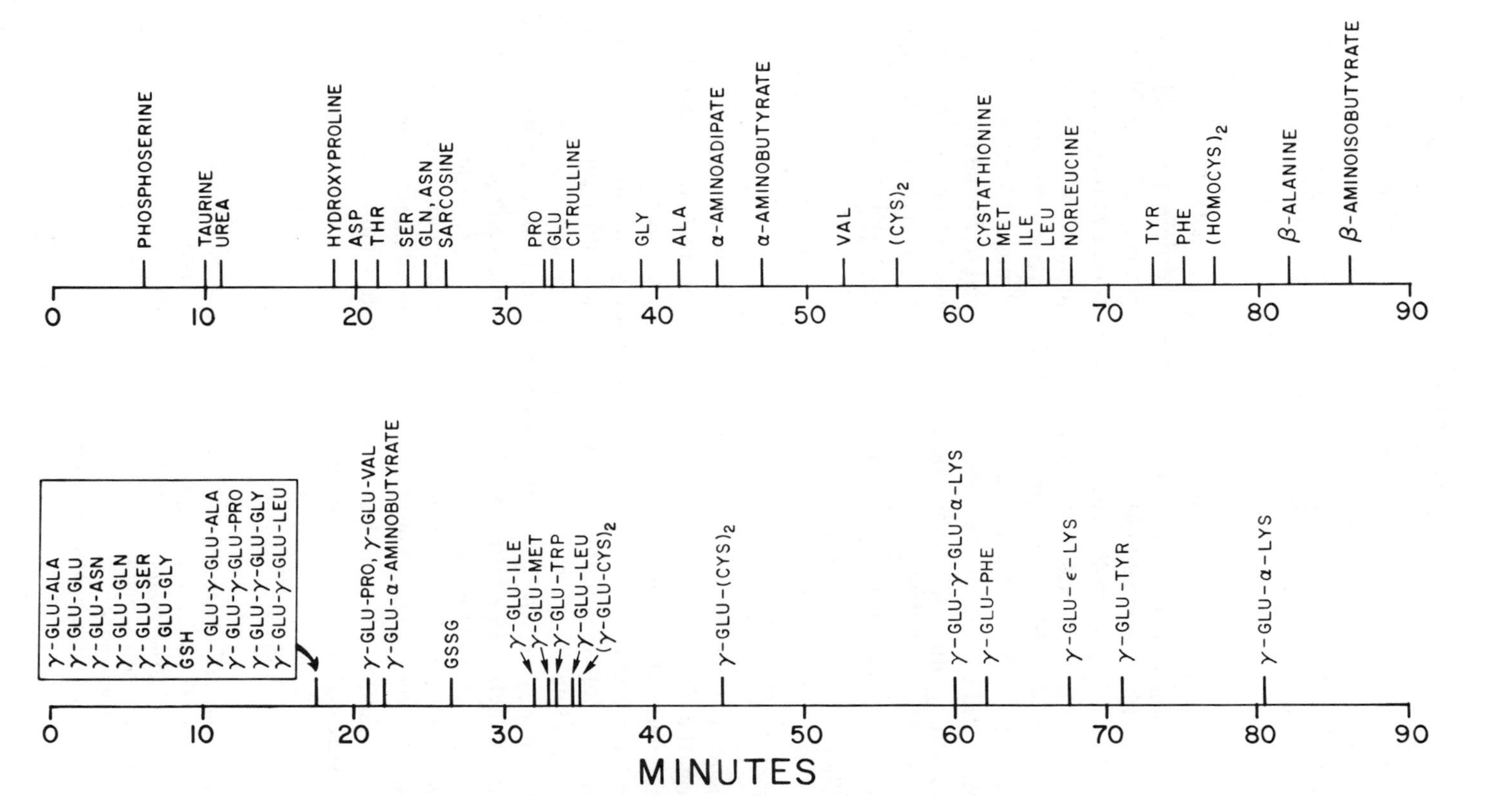

FIG. 1. Bottom: Elution time of γ-L-glutamyl derivatives of L-amino acids. Diastereoisomeric pairs, such as γ-L-glutamyl-L-methionine and γ-L-glutamyl-D-methionine, are not separated in this system. Top: Elution times of standard amino acids under the same conditions.

γ-glutamyl derivatives have been isolated from plants.[8] The separation of some γ-glutamyl amino acids by ion-exchange chromatography has been described.[9,10]

Method

The separation may be carried out on a Durrum D-500 amino acid analyzer using a single-column sodium citrate procedure designed for the analysis of physiological fluids.[11] A sample containing 10 nmol of each amino acid or peptide in 40 μl of pH 2.2 sodium citrate buffer (0.2 M Na^+), is applied to a 0.175 × 48 cm column of a sulfonated polystyrene resin (Durrum DC-4A). Elution is carried out at 30° with a sodium citrate buffer of pH 3.25 (0.172 M Na^+) at a flow rate of 7 ml/hr. After elution of sarcosine and cystine, the temperature is changed to 50 and 61°, respectively. After elution of cystine, the buffer is changed to a sodium citrate buffer of pH 4.25 (0.2 M Na^+). The column effluent is treated with ninhydrin reagent[12] at 130° for 1 min and analyzed spectrophotometrically at 590 nm.

Comments

Figure 1 shows the elution times of 26 γ-glutamyl peptides. For comparison, the elution times of a standard mixture of amino acids chromatographed under the same conditions are also shown. The γ-glutamyl derivatives of alanine, glutamic acid, asparagine, glutamine, serine, and glycine, as well as glutathione and several γ-glutamyl-γ-glutamyl amino acids, all elute together before aspartic acid. Human urine exhibits a prominent peak (Peak W) in this position. This peak contains several peptides which can be further fractionated by ion-exchange chromatography. One of these is β-aspartyl-glycine.[4] In addition to aspartic acid and glycine, acid hydrolysis of Peak W yields glutamic acid, serine, alanine, and other amino acids. Thus, Peak W probably contains a number of γ-glutamyl peptides. When plasma is subjected to amino acid analysis, a small peak is also observed in this position.

[8] T. Kasai and P. O. Larsen, *Fortschr. Chem. Org. Naturst.* **39,** 173 (1980).
[9] R. M. Zacharius and E. A. Talley, *Anal. Chem.* **34,** 1551 (1962).
[10] T. Kasai and S. Sakamura, *J. Chromatogr.* **103,** 189 (1975).
[11] P. L. Y. Lee, Durrum Application Notes No. 10. Dionex Corporation, 1228 Titan Way, Sunnyvale, California.
[12] S. Moore, *J. Biol. Chem.* **243,** 6281 (1968).

[73] Enzymatic Synthesis of Isotopically Labeled Glutathione by Means of an Immobilized Cell Matrix

By RICHARD J. BRIDGES, KOUSAKU MURATA, WILLIAM ABBOTT, and ALTON MEISTER

Studies on the metabolism of glutathione often require glutathione preparations that are labeled with isotope in specific positions of the molecule. Certain types of isotopically labeled glutathiones are currently obtainable from commercial sources, but availability as well as purity is variable. The most efficient procedures for preparation of isotopically labeled glutathiones are based on enzymatic methods, which requires preparations of the two synthetases.[1,2] In the procedure described here, which is applied to the preparation of [U-^{14}C-Glu]glutathione, use is made of an immobilized cell matrix[3] containing cells of a strain of *E. coli* B that is enriched in its content of the two synthetases. This strain was constructed by recombinant DNA techniques.[4] The genes for the synthetases were isolated from *E. coli* B (RC912), inserted into a vector (pBR325), and used to transform the original strain with the hybrid plasmid. The cells obtained, which overproduce the synthetases, provide an ideal "bioreactor" system.

Preparation of *E. coli* Cells

Reagents. Growth medium: 2% glucose, 2% peptone, 2.5% yeast extract, 0.5% KH_2PO_4, and 0.1% $MgSO_4 \cdot 7H_2O$. The pH of the solution is adjusted to 7.2 by adding 10 *M* NaOH. The media is then sterilized in an autoclave (120°, 30 min). Chloramphenicol (20 μg/ml) is added to the medium after sterilization and just prior to inoculation with cells. Since the transforming vector also carries chloramphenicol resistance, inclusion of the antibiotic selects for transformed cells.

Procedure

Primary cultures, inoculated from an agar culture, are each grown in 50 ml of medium that is shaken at 37° for 24 hr. These cultures are each

[1] G. F. Seelig and A. Meister, this volume [48].
[2] A. Meister, this volume [49].
[3] K. Murata, K. Tani, J. Kato, and I. Chibata, *Biochimie* **62,** 347 (1980).
[4] H. Gushima, T. Miya, K. Murata, and A. Kimura, *J. Appl. Biochem.* **5,** 43 (1983).

Copyright © 1985 by Academic Press, Inc.
All rights of reproduction in any form reserved.

transferred to flasks containing 1 liter of medium; these are then shaken at 37° for an additional 24 hr. One liter of media will usually produce about 5 g of wet cells.

Preparation of the Immobilized Cell Matrix

Step 1. Cell Preparation. The 1 liter cultures are cooled to 4° and the cells are collected by centrifugation (7000 *g*, 5 min). A typical preparation containing 9 liters of culture medium produces about 50 g of wet cells. The pelleted cells are pooled, suspended in ice-cold 0.85% NaCl, and centrifuged again (7000 *g*, 5 min). The washed cells are then resuspended (1 : 1, wet weight : volume) in 0.85% NaCl that is carefully maintained at 40°.

Step 2. Carrageenan Suspension. Two volumes of 3.1% kappa carrageenan (Sigma Type III) solution, maintained at 40°, is added to the cell suspension. The mixture is thoroughly stirred and carefully kept at 40°.

Step 3a. Preparation of Matrix Beads. To prepare the immobilized cells in a bead shaped matrix, the carrageenan cell suspension is taken up in a syringe, passed through an 18-gauge needle, and added dropwise to an ice-cold solution (~1 liter) of 2% KCl. The syringe requires frequent warming to prevent solidification in the needle. This technique produces spherical beads that are about 2 mm in diameter. The beads are stirred gently in ice-cold 2% KCl for 1 hr.

Step 3b. Preparation of Matrix Cubes. Alternatively, the immobilized cell matrix can be prepared in the shape of cubes. This is achieved by allowing the carrageenan-cell suspension to cool to room temperature at the bottom of a beaker. The beaker size should be chosen so that the suspension forms a sheet about 3 mm thick. The mixture is allowed to cool and ice-cold 2% KCl (about 1 liter) is carefully poured into the beaker. After standing for 1 hr, the KCl solution is decanted and the sheet is diced into 3 mm cubes with a scalpel.

Step 4. Matrix Crosslinking. The 2% KCl solution is decanted and replaced with an ice-cold solution containing 2% KCl, 80 m*M* 1,6-hexanediamine, and 0.3 *M* potassium phosphate (pH 7.0). After gentle stirring for 10 min, 25 ml of 25% glutaraldehyde is added per 600 ml of solution (final glutaraldehyde concentration, 0.1 *M*). The crosslinking reaction mixture is then gently stirred in an ice-bath for 1 hr.

Step 5. Toluene Treatment. The crosslinking solution is decanted and replaced with a 5 m*M* Tris–HCl solution (pH 7.0) containing 0.5 m*M* L-cysteine and 8% (v : v) toluene (about 1 liter). (The toluene renders the cell walls permeable to the substrates.) The immobilized cells are shaken in the toluene solution for 1 hr at 25°.

Step 6. Collection and Storage. The immobilized cells are collected by filtering through cheesecloth. While still suspended in the cheesecloth, the cells are rinsed with about 500 ml of ice-cold 0.85% NaCl. The matrix is stored at 4° in about 1 liter of 5 m*M* Tris–HCl (pH 7.0) containing 0.5 m*M* L-cysteine. The final immobilized cell matrix, which contains about 0.25 g of wet cells per g of immobilized material, retains most of its activity for at least 6 months.

Preparative Synthesis of Glutathione

The immobilized cell matrix is first tested for activity in a preparative scale synthesis of glutathione (400 μmol) in which unlabeled amino acids are used.

Reagents. Reaction solution: 200 m*M* L-glutamate, 80 m*M* L-cysteine, 80 m*M* glycine, 5 m*M* ATP, 50 m*M* $MgCl_2$, 80 m*M* acetylphosphate, and 50 m*M* potassium phosphate. The pH of the reaction solution is adjusted to 7.0 by adding 10 *M* NaOH.

Synthesis of Glutathione. The synthesis is initiated by combining 5 ml of the reaction solution with 5 g of immobilized cells in a 25-ml Erlenmeyer flask. The flask is stoppered and immediately placed in a 37° water bath, where it is shaken gently. Powdered acetylphosphate (200 μmol) is added hourly to the mixture during the first 4 hr of the synthesis. The acetylphosphate is used in the regeneration of ATP; the immobilized cells contain sufficient amounts of acetyl kinase. Powdered glycine (200 μmol) is also added to the mixture 4 hr after the synthesis is started. (The required replenishment of amino acids suggests they are also consumed in side reactions.) Aliquots of the reaction mixture are removed at regular intervals and assayed for glutathione using the glutathione reductase–DTNB recycling method.[5] The yield, based on the initial cysteine concentration, reaches about 50% in 6 hr. Continual additions (200 μmol each) of acetylphosphate (every 2 hr), glycine (every 4 hr), and L-glutamate (every 6 hr) lead to a final yield of glutathione of about 80% in 12 hr.

Preparation of [^{14}C-Glu]Glutathione

To minimize dilution of the isotope, the synthesis is carried out on a smaller scale (3.06 μmol) than that used in the preparative synthesis.

Reagents. Reaction solution: 1.53 m*M* L-glutamate (including 0.9 mCi of L-[U-^{14}C]glutamate), 10 m*M* L-cysteine, 10 m*M* glycine, 1 m*M* ATP, 10 m*M* $MgCl_2$, 5 m*M* acetylphosphate, 1.25 m*M* NaEDTA, and 25 m*M* potassium phosphate. The pH of the solution is adjusted to 7.0 by adding 10 *M* NaOH.

[5] M. E. Anderson, this volume [70].

Procedure

Synthesis of [^{14}C-Glu]Glutathione. The synthesis is initiated by combining 2 ml of reaction solution with 2 g of minced immobilized cell matrix in a 5 ml reacti-vial (Pierce). The vial is capped and stirred continuously at 37°. Additions of powdered L-cysteine (10 μmol), glycine (10 μmol), and acetylphosphate (20 μmol) are made to the reaction mixture every hour. To minimize oxidation of L-cysteine, N_2 is carefully bubbled into the reaction mixture immediately after the reaction is initiated and also after each addition. Aliquots are removed prior to each addition and assayed for glutathione. The time course of the reaction reaches a yield, based on the initial [^{14}C]glutamate concentration, of about 90% after 4 hr.

Purification of [^{14}C-Glu]Glutathione. After 4 hr of incubation, the entire reaction mixture is removed from the reacti-vial and homogenized (Potter-Elvehjem) with 1 volume of 5% picric acid in ethanol. The homogenate is centrifuged (Beckman Model B Microfuge, 3 min) and the supernatant solution is applied to a column (0.6 × 2 cm) of Dowex-1 (acetate). After washing with 10 ml of water, the glutathione is then eluted with a stepwise gradient of acetic acid (2 ml each of 0.2, 0.4, 0.6, 0.8, 1.0, 1.2, 1.4, 1.6, and 1.8 *M*). The fractions containing product are identified by determinations of radioactivity and assayed for glutathione by means of the glutathione reductase–DTNB recycling assay. To ensure purity, the glutathione containing fractions are pooled, evaporated under reduced pressure, and rechromatographed on Dowex-1 (acetate). The purity of the [^{14}C-glu]glutathione prepared in this manner is greater than 95%, when characterized by amino acid analysis of the 2-vinylpyridine-treated product.[6]

Comment

The method is readily adaptable to the preparation of glutathione containing isotopes in various positions, and is also useful for the preparation of certain analogs of glutathione.[7]

[6] O. W. Griffith, *Anal. Biochem.* **106,** 207 (1980).

[7] R. J. Bridges, W. A. Abbott, K. Murata, and A. Meister, unpublished.

[74] Methods for the Selective Modification of Glutathione Metabolism and Study of Glutathione Transport

By ALTON MEISTER

Selective Modification of Glutathione Metabolism

Introduction

Glutathione,[1] found in almost all cells,[2] functions in metabolism, transport, and cellular protection.[3] It participates in the reduction of the disulfide linkages of proteins and of other molecules. It functions in the synthesis of the deoxyribonucleotide precursors of DNA. Glutathione also protects cells against the effects of free-radicals and of reactive oxygen intermediates such as peroxides that are formed in metabolism. Glutathione is transported out of many cells; this process seems to be connected with a transport system for γ-glutamyl amino acids, reactions that involve the cell membrane and its immediate environment, and the interorgan transport of amino acid sulfur. Glutathione plays a role in the inactivation of a number of drugs and in the metabolic processing of certain endogenous compounds such as estrogens, prostaglandins, and leukotrienes. Glutathione is a coenzyme for a number of enzymes. The concentration of glutathione within cells is generally much higher than that of cysteine and it appears that glutathione serves as a storage form as well as a transport form of cysteine.

Much information about the functions of glutathione has come from studies in which glutathione metabolism was modified, and many of the currently available procedures for the modification of glutathione metabolism have developed from studies on particular enzymes that are involved in glutathione synthesis and utilization. For example, it is possible to

[1] Glutathione is L-γ-glutamyl-L-cysteinylglycine (abbreviated, GSH). The disulfide derived from glutathione is glutathione disulfide (abbreviated, GSSG). The term "oxidized glutathione" is potentially misleading since there are several possible forms of oxidized glutathione. The term "total glutathione" has been used to mean the sum of GSH and GSSG in GSH equivalents. The term "reduced glutathione" is redundant.

[2] It has been reported that certain anaerobic bacteria [R. C. Fahey, W. C. Brown, W. B. Adams, M. B. Worham, *J. Bacteriol.* **133,** 1126 (1978)] and amoeba [R. C. Fahey, G. L. Newton, B. Arrick, T. Ovendank-Bogart, and S. B. Aley, *Science* **224,** 70 (1984)] do not contain glutathione.

[3] Recent reviews include: "Functions of GSH" (A. Larssen, S. Orrenius, A. Holmgren, and B. Mannervik, eds.). Raven Press, New York, 1983; A. Meister and M. E. Anderson, *Annu. Rev. Biochem.* **52,** 722 (1983).

Copyright © 1985 by Academic Press, Inc.
All rights of reproduction in any form reserved.

decrease tissue concentrations of glutathione by inhibition of its synthesis by use of selective inhibitors of γ-glutamylcysteine synthetase. Several methods now available for increasing cellular levels of glutathione involve the actions of enzymes. Reactions involved in the metabolism of glutathione can be inhibited *in vivo* by selective inhibitors which were developed from studies on enzymes. Selective modification of glutathione metabolism may have applications in therapy.[4,5] Some of the approaches and experimental procedures that have been used are reviewed below. Glutathione is synthesized intracellularly and many types of cells have been found to exhibit γ-glutamylcysteine synthetase and glutathione synthetase activities. Studies on glutathione transport have thus far shown that the intact tripeptide is not taken up by cells. The apparent uptake of glutathione by some cells has been shown to be due to extracellular breakdown of glutathione, uptake of the products formed, and intracellular resynthesis of glutathione.[6,7] Nevertheless, one cannot exclude, without appropriate experimental study, the possibility that intact glutathione is transported into some cells. In contrast, the transport of glutathione out of cells seems to be a property of many, perhaps most, cells.[3,8,9] Experimental approaches which have been used in studies on glutathione transport are reviewed below.

Methods for Decreasing Cellular Glutathione Levels

Diazene carboxylic acid derivatives such as methylphenyldiazene carboxylate (azo ester) and diazene dicarboxylic acid bis(*N*,*N*-dimethylamide) (diamide) were introduced by Kosower *et al.*[10–19] These com-

[4] A. Meister and O. W. Griffith, *Cancer Treat. Rep.* **63,** 1115 (1979).

[5] A. Meister, *Science* **220,** 471 (1983).

[6] G. L. Jensen and A. Meister, *Proc. Natl. Acad. Sci. U.S.A.* **80,** 4714 (1983).

[7] M. E. Anderson and A. Meister, *Proc. Natl. Acad. Sci. U.S.A.* **80,** 707 (1983).

[8] A. Meister, O. W. Griffith, A. Novogrodsky, and S. S. Tate, *Ciba Found. Symp.* **72** (new ser.), 135 (1980).

[9] A. Meister, *Curr. Top. Cell. Regul.* **18,** 21 (1981).

[10] E. M. Kosower and N. S. Kosower, *Nature (London)* **224,** 117 (1969).

[11] N. S. Kosower, E. M. Kosower, and B. Wertheim, *Biochem. Biophys. Res. Commun.* **37,** 593 (1969).

[12] N. S. Kosower, K.-R. Song, and E. M. Kosower, *Biochim. Biophys. Acta* **192,** 1 (1969).

[13] N. S. Kosower, K.-R. Song, E. M. Kosower, and W. Correa, *Biochim. Biophys. Acta* **192,** 8 (1969).

[14] N. S. Kosower, K.-R. Song, and E. M. Kosower, *Biochim. Biophys. Acta* **192,** 15 (1969).

[15] T. Zehavi-Willner, N. S. Kosower, T. Hunt, and E. M. Kosower, *Biochem. Biophys. Res. Commun.* **40,** 37 (1970).

[16] T. Zehavi-Willner, E. M. Kosower, T. Hunt, and N. S. Kosower, *Biochim. Biophys. Acta* **228,** 245 (1971).

pounds oxidize glutathione, and are effective in decreasing glutathione levels in many systems. The use of these compounds and other oxidants has been reviewed in a previous volume of this series.[20] Although use of such agents has led to new and interesting information, especially in *in vitro* studies, these compounds also interact with compounds other than glutathione and such lack of specificity diminishes their value. Furthermore, their use leads to high levels of glutathione disulfide, which is toxic. The oxidation achieved by diamide and related compounds is often short-lived and this also limits the usefulness of this approach.

As discussed in a previous volume of this series,[20] cellular glutathione may be depleted by administration of a variety of compounds that interact with glutathione to form adducts. Thus, many electrophilic compounds decrease cellular glutathione levels; interactions of these compounds with glutathione may be promoted by the action of glutathione *S*-transferases. A widely used agent for glutathione depletion is diethyl maleate, which is very effective. However, its use is associated with several problems. Thus, after treatment with diethyl maleate, the rate of glutathione synthesis in some tissues (e.g., liver) is increased, a phenomenon related to the fact that glutathione regulates its own biosynthesis by feedback inhibition.[21] Commercial diethyl maleate often contains impurities; both these and diethyl maleate itself have a number of biological effects in addition to reacting with glutathione. It should be noted that diethyl maleate is rapidly hydrolyzed by tissue esterase to form maleate, which exhibits other biological effects, including for example, on the activity of γ-glutamyl transpeptidase.[22]

Other compounds that have been used for the depletion of cellular glutathione include diisopropylidene acetone (phorone)[23] and 2-cyclohexene-1-one.[24] Although these compounds are possibly less toxic than diethyl maleate, their effects on glutathione levels are also subject to rapid reversal due to increased glutathione synthesis.

Glutathione may be decreased by interfering with the supply of the necessary amino acids, especially cysteine (and methionine). Systems in

[17] N. S. Kosower, G. A. Vanderhoff, B. Benerofe, and T. Hunt, *Biochem. Biophys. Res. Commun.* **45,** 816 (1971).

[18] N. S. Kosower, G. A. Vanderhoff, and E. M. Kosower, *Biochim. Biophys. Acta* **272,** 623 (1972).

[19] E. M. Kosower, W. Correa, B. J. Kinon, and N. S. Kosower, *Biochim. Biophys. Acta* **264,** 39 (1972).

[20] J. L. Plummer, B. R. Smith, H. Sies, and J. R. Bend, this series, Vol. 77, p. 50.

[21] P. Richman and A. Meister, *J. Biol. Chem.* **250,** 1422 (1975).

[22] G. A. Thompson and A. Meister, *J. Biol. Chem.* **254,** 2956 (1979).

[23] M. Younes and C.-P. Siegers, *Chem.-Biol. Interact.* **34,** 257 (1981).

[24] E. Boyland and L. F. Chasseaud, *Biochem. J.* **104,** 95 (1967).

which there is depletion of the sulfur containing amino acids are also subject to other effects including those related to decreased synthesis of proteins.

Depletion of glutathione by inhibition of its biosynthesis seems to offer the most useful approach. Selective inhibition of glutathione synthetase may be considered. However, in such an approach to glutathione depletion, one would expect marked accumulation of 5-oxoproline and consequent acidosis.[25,26] In patients who have a marked genetic deficiency of glutathione synthetase, 5-oxoproline accumulates in the blood plasma, cerebrospinal fluid, and the tissues. Since glutathione feedback inhibits γ-glutamylcysteine synthetase,[21] when there is a severe deficiency of glutathione γ-glutamylcysteine is overproduced and converted by the action of γ-glutamylcyclotransferase to 5-oxoproline. The formation of 5-oxoproline exceeds the capacity of 5-oxoprolinase so that 5-oxoproline accumulates. It would be expected that the use of a selective inhibitor of glutathione synthetase for depletion of cellular glutathione would be complicated by the occurrence of similar metabolic phenomena.

A more satisfactory approach to the inhibition of glutathione synthesis is inhibition of γ-glutamylcysteine synthetase. This enzyme is inhibited by methionine sulfoximine,[27,27a] but this agent also inhibits glutamine synthetase and its effects on cerebral glutamine synthetase lead to convulsions. Buthionine sulfoximine,[28] which has been considered elsewhere in this volume,[29] does not inhibit glutamine synthetase, and selectively interacts with γ-glutamylcysteine synthetase inhibiting it irreversibly. Like methionine sulfoximine, buthionine sulfoximine is a transition state inhibitor and thus mimics the transient intermediate or transition state formed in the reaction between enzyme-bound γ-glutamyl phosphate and the amino group of L-cysteine. Buthionine sulfoximine, like methionine sulfoximine, is phosphorylated on the sulfoximine nitrogen atom. Phosphorylated buthionine sulfoximine binds tightly to the active site. Although the binding of buthionine sulfoximine is competitive with L-glutamate for the active site of the enzyme, once bound and phosphory-

[25] V. P. Wellner, R. Sekura, A. Meister, and A. Larsson, *Proc. Natl. Acad. Sci. U.S.A.* **71,** 2505 (1974).

[26] A. Meister, *in* "The Metabolic Basis of Inherited Diseases" (J. B. Stanbury, J. B. Wyngaarden, and D. S. Fredrickson, eds.), 5th ed., Chapter 17, p. 348. McGraw-Hill, New York, 1983.

[27] P. G. Richman, M. Orlowski, and A. Meister, *Proc. Natl. Acad. Sci. U.S.A.* **70,** 2717 (1973).

[27a] A. G. Palekar, S. S. Tate, and A. Meister, *Biochem. Biophys. Res. Commun.* **62,** 651 (1975).

[28] O. W. Griffith and A. Meister, *J. Biol. Chem.* **254,** 7558 (1979).

[29] This volume [47].

lated, the intermediate is tightly bound. In *in vitro* studies, inhibition is essentially irreversible. When buthionine sulfoximine is used *in vivo* or in systems consisting of cell suspensions, some apparent reversal of inhibition has been observed. The nature of this effect in isolated cell systems and *in vivo* is not yet known. It may be associated with an increase in enzyme synthesis. Another possibility is that the phosphorylated inhibitor is acted upon by tissue phosphatases so as to reduce the amount of bound inhibitor. It is known that free methionine sulfoximine phosphate is a substrate of several phosphatases.[30-32]

When experimental animals are treated with buthionine sulfoximine, they exhibit a rapid decline in the glutathione levels of the kidney, liver, pancreas, muscle, and blood plasma.[33] When repeated doses of buthionine sulfoximine are given, or when buthionine sulfoximine is added to the drinking water, markedly decreased levels of glutathione occur in other tissues as well. The rate at which the glutathione level of tissues declines after administration of buthionine sulfoximine reflects the rate of utilization of glutathione, which is about equivalent to the rate of export of glutathione from the cells. The rapid decrease in blood plasma glutathione levels reflects the substantial interorgan transport of glutathione (see below). The levels of glutathione in cells grown in tissue culture also decline when buthionine sulfoximine is added to the medium, and such decline is associated with export of glutathione to the medium.[34] The idea that depletion of cellular glutathione by treatment with sulfoximine inhibitors of γ-glutamylcysteine synthetase might make tumor cells more susceptible to radiation and certain chemotherapeutic agents[35] has been followed by a variety of investigations.[36] In this approach, the destructive effects of chemical agents, reactive oxygen compounds such as superoxide and peroxides, and free-radicals are used advantageously. Selectivity may be achieved in situations in which the tumor cells and the normal cells have different quantitative requirements for glutathione. It appears that many normal cells probably have a considerable excess of glutathione whereas certain tumors and parasites may have levels of glutathione that are close to those minimally required for cell survival. Selectivity may also be achieved because many tumors, in contrast to normal

[30] R. Ronzio and A. Meister, *Proc. Natl. Acad. Sci. U.S.A.* **59,** 164 (1968).
[31] R. A. Ronzio, W. B. Rowe, and A. Meister, *Biochemistry* **8,** 1066 (1979).
[32] W. B. Rowe, R. A. Ronzio, and A. Meister, *Biochemistry* **8,** 2674 (1969).
[33] O. W. Griffith and A. Meister, *Proc. Natl. Acad. Sci. U.S.A.* **76,** 5606 (1979).
[34] J. K. Dethmers and A. Meister, *Proc. Natl. Acad. Sci. U.S.A.* **78,** 7492 (1981).
[35] A. Meister and O. W. Griffith, *Cancer Treat. Rep.* **63,** 1115 (1979).
[36] A. Meister, *Science* **220,** 471 (1983).

tissues, have markedly low levels of superoxide dismutase and catalase.[37] The development of resistance to a chemotherapeutic agent may be associated with increased glutathione synthesis; treatment with buthionine sulfoximine may be expected to restore sensitivity to the chemotherapeutic agent. Thus, mouse L1210 leukemia cells resistant to treatment with L-phenylalanine mustard have been found to have glutathione levels that are about twice those of sensitive L1210 cells.[38–41] Treatment of the resistant cells *in vitro* and *in vivo* wtih buthionine sulfoximine led to sensitization of the tumor cells to phenylalanine mustard.

Studies on several human lymphoid cell lines that were depleted of glutathione by incubation in media containing buthionine sulfoximine were much more sensitive to γ-irradiation than the controls.[34] Similar results were obtained on cultured human lung carcinoma and other tumor cells.[42,43] In studies on V79 cells, glutathione depletion produced by suspending the cells in buthionine sulfoximine led to selective sensitization to irradiation under hypoxic conditions and to a decrease of the oxygen enhancement ratio.[42,44] Investigations on oxidative cytolysis of several tumor cell lines by glucose oxidase and by activated macrophages and granulocytes in the presence of phorbol myristate acetate showed that depletion of glutathione by incubation of the cells in media containing buthionine sulfoximine led to increased cytolysis.[45] The recovery of tumor cells resistance to peroxide was found to be closely associated with the resynthesis of cellular glutathione. Buthionine sulfoximine depletion of cellular glutathione has also been found to augment the lysis of mouse tumor cells in the presence of sulfhydryl-reactive antineoplastic drugs.[46]

Cells that have markedly reduced levels of catalase would be expected to be destroyed by severe depletion of glutathione. The parasite *Trypanosoma brucei brucei* contains no catalase and has a very high intracellular

[37] L. W. Oberley, "Superoxide Dismutase," Vol. 2, Chapter 6, p. 127. CRC Press, Boca Raton, Florida, 1982.

[38] K. Suzkak, B. J. Petro, and D. T. Vistica, *Biochem. Pharmacol.* **31,** 121 (1982).

[39] K. Suzkake, B. J. Petro, and D. T. Vistica, *Biochem. Pharmacol.* **32,** 165 (1983).

[40] D. T. Vistica, S. Somfai-Relle, K. Suzuke, and B. J. Petro, *J. Cell. Biochem. Suppl.* **6,** 375 (1982).

[41] D. T. Vistica, *Pharmacol. Ther.* **22,** 379 (1983).

[42] J. E. Biaglow and M. E. Varnes, *Radiat. Res.* **95,** 437 (1983).

[43] J. B. Mitchell, A. Russo, J. E. Biaglow, S. McPherson, *Radiat. Res.* **96,** 422 (1983).

[44] M. Guichard, G. Jensen, A. Meister, and E. P. Malaise, *Proc. Radiat. Res. Soc.* Abstr. Dc-10 (1983).

[45] B. A. Arrick, C. F. Nathan, O. W. Griffith, and Z. A. Cohn, *J. Biol. Chem.* **257,** 1231 (1982).

[46] B. A. Arrick, C. F. Nathan, and Z. A. Cohn, *J. Clin. Invest.* **71,** 258 (1983).

level of hydrogen peroxide; the level is about 70 μM which is estimated to be at least 100 times higher than the concentration found in mammalian cells. It would be expected that depletion of glutathione in mice infected with this protozoan parasite would be more damaging to the parasite than to the host. Preliminary studies suggest that this is the case.[47] Depletion of glutathione by treatment with buthionine sulfoximine may also be useful in treating other types of infections and in preventing the induction of thermotolerance, a phenomenon associated with increased cellular glutathione levels.[48,49] Other effects of the administration of buthionine sulfoximine have been discussed elsewhere.[5]

Methods of Increasing Cellular GSH Levels

Glutathione synthesis may be increased by procedures that increase the levels of γ-glutamylcysteine synthetase and glutathione synthetase. Little is now known about the factors that control the synthesis of the two enzymes involved in glutathione biosynthesis. Administration of certain compounds has been reported to increase glutathione levels, and these may act by increasing the levels of the synthetases. Such compounds include cobaltous ions,[50] butylated hydroxyanisole, butylated hydroxytoluene,[51,52] and certain carcinogens.[53]

The first step in glutathione synthesis is controlled by feedback inhibition by glutathione, and this mechanism seems to normally determine the upper level of cellular glutathione. The level of glutathione in a given cell may also depend upon the availability of substrates for the enzyme. Cells generally contain substantial amounts of glutamate and glycine, but cysteine is often the limiting amino acid component. In certain cells, the formation of cysteine may depend upon transulfuration via the cystathionine pathway from methionine. Cysteine may also be derived from dietary protein and from the breakdown of tissue proteins. The synthesis of glutathione can be increased, under certain conditions, by increasing the supply of substrates to the two synthetases. Administration of cysteine is

[47] B. A. Arrick, O. W. Griffith, and A. Cerami, *J. Exp. Med.* **153,** 720 (1981).

[48] J. B. Mitchell, A. Russo, T. J. Kinsella, and E. Glatstein, *Cancer Res.* **43,** 987 (1983).

[49] J. B. Mitchell and A. Russo, *Radiat. Res.* **95,** 471 (1983).

[50] H. A. Sasame and M. R. Boyd, *J. Pharmacol. Exp. Ther.* **205,** 718 (1978).

[51] A. M. Benson, R. P. Batzinger, S.-Y. L. Ou, E. Bueding, Y.-N. Cha, and P. Talalay, *Cancer Res.* **38,** 4486 (1979).

[52] L. K. T. Lam, V. L. Sparnins, J. B. Hochalter, and L. W. Wattenberg, *Cancer Res.* **41,** 3940 (1981).

[53] A. Meister, *in* "Metabolism of Sulfur Compounds" (D. M. Greenberg, ed.), Vol. 7, p. 101. Academic Press, New York, 1975.

often not an ideal way to increase glutathione levels because this amino acid is rapidly metabolized. In addition, cysteine is toxic apparently at least in part by virtue of its extracellular effects.

A useful precursor of cysteine that seems to be nontoxic and to be efficiently transported into cells is L-2-oxothiazolidine-4-carboxylate.[54–56] This compound is an analog of 5-oxo-L-proline in which the 4-methylene moiety is replaced by a sulfur atom. The thiazolidine is a good substrate of 5-oxoprolinase, which catalyzes the conversion of the thiazolidine, presumably via *S*-carboxy-L-cysteine, which breaks down spontaneously to yield L-cysteine. When the thiazolidine is given to mice, the level of glutathione in the liver increases substantially. Thus, L-2-oxothiazolidine-4-carboxylate is an effective intracellular cysteine delivery agent. It is notable that the increase in liver glutathione found after administration of the thiazolidine is inhibited by administration of buthionine sulfoximine, thus supporting the interpretations given above. Administration of L-2-oxothiazolidine-4-carboxylate was found to protect mice against acetaminophen toxicity; the thiazolidine was found to be more effective in increasing glutathione levels of mice treated with acetaminophen than *N*-acetyl-L-cysteine. *N*-Acetyl-L-cysteine is also an effective intracellular cysteine delivery agent which has been successfully used as a therapy for acetaminophen toxicity.[57–61] Presumably *N*-acetyl cysteine is transported into a cell where it is deacetylated. Other compounds that might increase intracellular cysteine levels have been suggested[56]; these include thiazolidine-4-carboxylate[62] and its 2-methyl derivative.[63] Compounds that are transported into cells and which are converted intracellularly by an en-

[54] J. M. Williamson and A. Meister, *Proc. Natl. Acad. Sci. U.S.A.* **78,** 936 (1981).

[55] J. M. Williamson and A. Meister, *J. Biol. Chem.* **257,** 12,039 (1982).

[56] J. M. Williamson, B. Boettcher, and A. Meister, *Proc. Natl. Acad. Sci. U.S.A.* **79,** 6246 (1982).

[57] J. R. Mitchell, D. J. Jollow, W. Z. Potter, D. C. Davis, J. R. Gillette, and B. B. Brodie, *J. Pharmacol. Exp. Ther.* **187,** 185 (1973).

[58] D. J. Jollow, J. R. Mitchell, W. Z. Potter, D. C. Davis, J. R. Gillette, and B. B. Brodie, *J. Pharmacol. Exp. Ther.* **187,** 195 (1973).

[59] W. Z. Potter, D. C. Davis, J. R. Mitchell, D. J. Jollow, J. R. Gillette, and B. B. Brodie, *J. Pharmacol. Exp. Ther.* **187,** 202 (1973).

[60] J. R. Mitchell, D. J. Jollow, W. Z. Potter, J. R. Gillette, and B. B. Brodie, *J. Pharmacol. Exp. Ther.* **187,** 211 (1973).

[61] D. P. Jones, A. H. Stead, P. Moldeus, and S. Orrenius, *in* "Functions of Glutathione in Liver and Kidney" (H. Sies and A. Wendel, eds.), p. 194. Springer-Verlag, Berlin and New York, 1978.

[62] H. J. Debby, J. B. MacKenzie, and C. G. MacKenzie, *J. Nutr.* **66,** 607 (1958).

[63] H. T. Nagasawa, D. J. W. Goon, R. T. Zera, and D. L. Yuzon, *J. Med. Chem.* **25,** 489 (1982).

zyme-based mechanism to cysteine would probably be more effective than those that undergo nonenzymatic conversion to cysteine.

Tissue glutathione levels may also be increased by supplying substrate for glutathione synthetase. Thus, administration of γ-glutamylcysteine (or its disulfide form) or of γ-glutamylcystine leads to increased levels of GSH in the kidney. Since glutathione synthetase is not subject to feedback regulation in the manner that controls γ-glutamylcysteine synthetase, administration of γ-glutamylcysteine or of related compounds leads to levels of kidney glutathione that are substantially higher than the feedback regulated levels.[7] This approach is especially useful for the kidney, which has an active transport system for γ-glutamyl amino acids.[64]

Although administered glutathione does not seem to enter most cells to an appreciable extent, there is considerable evidence that many cells export glutathione (see below). An ideal system for delivering glutathione into a cell would be based on a derivative of glutathione that is much more effectively transported into cells than is glutathione itself, and which furthermore is readily converted to glutathione intracellularly after transport. Several derivatives of glutathione have been examined for their ability to increase intracellular glutathione levels. The monoethyl and monomethyl esters of glutathione (molecules in which the glycine carboxyl group is esterified) have properties consistent with their use in such a glutathione delivery system. Thus, administration of the monoethyl (or methyl) ester of glutathione to mice leads to substantial increases in the glutathione levels of the liver and kidney. The data indicate that these compounds are effectively transported and then hydrolyzed intracellularly. The esters are also effective in raising glutathione levels after pretreatment of the animals with buthionine sulfoximine, thus excluding the possibility that the increased intracellular levels of glutathione found are due to intracellular synthesis of glutathione from amino acids.[65]

Although glutathione is not effectively transported into human lymphoid cells,[66] normal human skin fibroblasts, and fibroblasts from patients with genetic deficiencies of γ-glutamylcysteine synthetase or glutathione synthetase,[67] the monoethyl ester of glutathione is readily transported into these cells and is hydrolyzed intracellularly.[67] In these studies, levels of cellular glutathione were found that far exceed those found normally.

[64] O. W. Griffith, R. J. Bridges, and A. Meister, *Proc. Natl. Acad. Sci. U.S.A.* **76,** 6319 (1979).

[65] R. N. Puri and A. Meister, *Proc. Natl. Acad. Sci. U.S.A.* **80,** 5258 (1983).

[66] G. L. Jensen and A. Meister, *Proc. Natl. Acad. Sci U.S.A.* **80,** 4714 (1983).

[67] V. P. Wellner, M. E. Anderson, R. N. Puri, G. L. Jensen, and A. Meister, *Proc. Natl. Acad. Sci. U.S.A.* **81,** 4732 (1984).

In animal studies, administration of glutathione ester effectively protected against acetaminophen toxicity. In studies on human lymphoid cells, glutathione ester protected against the lethal effects of irradiation. Evidence was obtained that glutathione may function also in repair processes that take place after irradiation.[67] The use of esters of glutathione is attractive because many cells appear to have high levels of esterase that can hydrolyze such compounds after transport. Cellular glutathione delivery by use of glutathione esters does not require intracellular energy as is required for glutathione synthesis from amino acids. Since the enzymes required for glutathione synthesis are not involved, delivery by the ester system is not subject to feedback inhibition.

In Vivo Inhibition of 5-Oxo-L-prolinase

This enzyme is competitively inhibited by several 5-oxoproline analogs.[55,68] Replacement of the 4-methylene moiety of 5-oxoproline by NH or S yields analogs which are highly effective competitive inhibitors which are active *in vitro* and *in vivo*. For example, administration of L-2-imidazolidone-4-carboxylate decreases the *in vivo* metabolism of 5-oxoproline.[69,70] Similar results were later obtained with L-2-oxothiazolidine-4-carboxylate[54] and 3-methyl-5-oxoproline.[70] Animals treated with inhibitors of 5-oxoprolinase accumulate 5-oxoproline in their tissues and excrete this compound in their urine.[69]

In Vivo Inhibition of γ-Glutamylcyclotransferase

The reaction catalyzed by this enzyme is of major significance in the *in vivo* formation of 5-oxo-L-proline. Effective competitive inhibitors of γ-glutamylcyclotransferase have been found.[71] When β-aminoglutaryl-*N*-α-aminobutyrate was administered to mice there was a substantial decrease in the level of 5-oxoproline in the kidney.[72] Administration of this inhibitor to mice in which kidney 5-oxoproline levels were increased by inhibition of 5-oxoprolinase also showed a substantial decrease in kidney levels of 5-oxoproline. L-γ-(Threo-β-methyl)glutamyl-L-α-amino[^{14}C]butyrate is a useful model substrate for indicating *in vivo* γ-glutamylcyclotransferase activity.[72] The model compound is not a substrate for γ-glutamyl transpeptidase, but is effectively acted upon by γ-glutamylcyclotransferase.

[68] See this volume [53].

[69] P. Van Der Werf, R. A. Staphani, and A. Meister, *Proc. Natl. Acad. Sci. U.S.A.* **71,** 1026 (1974).

[70] See this volume [57].

[71] See this volume [52].

[72] R. J. Bridges, O. W. Griffith, and A. Meister, *J. Biol. Chem.* **255,** 10.787 (1980).

When the model compound is administered to mice, it is converted to 3-methyl-5-oxoproline and labeled L-α-aminobutyrate; the latter compound is rapidly metabolized leading to formation of respiratory $^{14}CO_2$. Since 3-methyl-5-oxoproline is a relatively poor substrate of 5-oxoprolinase, the reaction may also be followed *in vivo* by determination of 3-methyl-5-oxoproline in various tissues and in the urine. Administration of β-aminoglutaryl-L-α-aminobutyrate to animals that had been given the model compound showed a marked decrease in the rate of respiratory $^{14}CO_2$ formation.

In Vivo Inhibition of γ-Glutamyl Transpeptidase

Several inhibitors of this enzyme are known including the L- and D-isomers of γ-glutamyl (*o*-carboxy)phenylhydrazide, the combination of L-serine plus borate, and the glutamine analogs L-azaserine, 6-diazo-5-oxo-L-norleucine, and L-[α*S*,5*S*]-α-amino-3-chloro-4,5-dihydro-5-isoxazole acetic acid.[73] The γ-glutamyl (*o*-carboxy) phenylhydrazides are competitive inhibitors; the L-isomer of this compound exhibits an apparent K_i value of about 8 μm. Inhibition *in vivo* is apparently selective; however, these hydrazides are split to a slight extent leading to the formation of *o*-carboxy phenylhydrazine, which is toxic. Doses of about 2 mmol/kg may be lethal to mice. There is also evidence that these hydrazides interact with glutathione to form conjugates which may also complicate their use in studies on glutathione metabolism. The combination of L-serine plus borate may be used *in vivo* but relatively high doses are needed to achieve *in vivo* effects. It is of importance to administer L-serine and sodium borate in separate experiments as controls. L-Serine plus borate inhibits the enzyme by forming an analog of the transition state involving a complex between serine, borate, and the group on the enzyme.[74] The glutamine analogs that inhibit γ-glutamyl transpeptidase activity are potent irreversible inhibitors. However, they are nonspecific. Presumably they bind by forming a covalent linkage with the group (possibly a hydroxyl group) on the enzyme.

A useful model substrate for following *in vivo* γ-glutamyl transpeptidase activity is D-γ-glutamyl-L-α-amino[^{14}C]butyrate.[75] This compound is not a substrate of γ-glutamylcyclotransferase, which is L-specific. Therefore, only transpeptidase can split the γ-peptide bond of D-γ-glutamyl-L-α-amino[^{14}C]butyrate. L-α-Aminobutyrate is rapidly metabolized leading to the formation of labeled respiratory carbon dioxide. When mice are

[73] See this volume [50].

[74] S. S. Tate and A. Meister, *Proc. Natl. Acad. Sci. U.S.A.* **75,** 4806 (1978).

[75] O. W. Griffith and A. Meister, *Proc. Natl. Acad. Sci. U.S.A.* **76,** 268 (1979).

given the labeled model compound, they excrete as $^{14}CO_2$ about 70–80% of the injected radioactivity within 5 hr. When such mice are treated with inhibitors of γ-glutamyl transpeptidase, the rate of respiratory CO_2 formation and the extent of formation are substantially decreased. In studies in which several inhibitors were examined in this way, significant effects were observed *in vivo* with all of the inhibitors, but complete inhibition was not observed except perhaps for 10 or 15 min after injection of D-γ-glutamyl-(*o*-carboxy)phenylhydrazide (dose, 1.25 mmol/kg body weight). Somewhat greater inhibition was observed in studies with L-(α*S*, 5*S*)-α-amino-3-chloro-4,5-dihydro-5-isoxazole acetic acid. However, even in these experiments complete inhibition of transpeptidase was not achieved *in vivo*. It is a general experience that although some of these inhibitors can produce virtually complete inhibition *in vitro* their effects *in vivo* are much less marked.

Studies on Glutathione Transport

Studies on dogs, in which arterial, portal, and hepatic venous blood glutathione and hepatic blood flow were measured, led to the suggestion that glutathione synthesized in liver and gut is transported to the periphery in red blood cells.[76] In this work, the changes in total blood glutathione were as much as 2-fold on passage through gut and liver. Additional studies on the apparent transport of glutathione from gut and liver via red blood cells have apparently not been done, and an explanation of these interesting findings requires further study.

When red blood cells are exposed to oxidant stresses they export glutathione disulfide from the cell to the suspending medium.[77] Such transport appears to be an energy-requiring process and is thought to account for most or perhaps all of the turnover of glutathione in red blood cells. Glutathione disulfide transport has been studied in resealed red cell membranes and in inside-out vesicles. Although it might be expected that glutathione would be transported from red cells and possibly appear as glutathione disulfide externally, this does not seem to be supported by studies on inside-out red cell vesicles, which do not transport glutathione. The export of glutathione disulfide has also been examined in lens[78] which, like red blood cells, has a high level of glutathione and a low level of glutathione disulfide. Export of glutathione disulfide from lens seems to resemble that observed in red cells.

In studies in which isolated rat liver preparations were perfused with

[76] D. H. Elwyn, H. C. Parikh, and W. C. Shoemaker, *Am. J. Physiol.* **215,** 1260 (1968).

[77] E. Beutler, *in* "Functions of Glutathione" (A. Larsson, S. Orrenius, A. Holmgren, and B. Mannervik, eds.), p. 65. Raven Press, New York, 1983.

[78] S. K. Srivastava and E. Beutler, *Proc. Soc. Exp. Biol. Med.* **127,** 512 (1968).

hemoglobin-free solutions, it was found that addition of hydrogen peroxide or of other peroxides to the perfusate led to oxidation of glutathione and to its release into the total perfusate as glutathione disulfide.[79,80] This phenomenon was ascribed to an effect of glutathione peroxidase. Further studies on this system showed that glutathione disulfide production also occurred after addition of certain drugs to the system, and that the extent of efflux of glutathione disulfide reflected the intracellular level of this compound.[81–83] These and related studies led to the conclusion that glutathione disulfide is released into the bile. It was also concluded that glutathione is released into the caval perfusate in this system. The steady state rates of release of glutathione and glutathione disulfide from the liver were found to be, respectively, about 10 and 1 nmol/g/min of perfused liver.[81,84] The conclusion that glutathione disulfide is preferentially released into the bile, whereas glutathione is released to the hepatic venous out flow, was based on experiments in which the isolated perfused rat liver system was used. Little or no glutathione is found in the bile in the perfused system, whereas bile normally contains 1–6 m*M* glutathione and much lower levels of glutathione disulfide (see refs. 85 and 86 for additional literature citations). It is now clear that liver glutathione is transported into both bile and hepatic venous plasma.

The normal occurrence of transport of glutathione across cell membranes became apparent from studies with selective inhibitors of γ-glutamyl transpeptidase and of γ-glutamylcysteine synthetase. The reactions of the γ-glutamyl cycle account for the synthesis and degradation of glutathione; glutathione is synthesized intracellularly by the sequential actions of γ-glutamyl cysteine synthetase and glutathione synthetase. The initial step in the breakdown of glutathione is catalyzed by γ-glutamyl transpeptidase, which is largely bound to the surfaces of cell membranes. Nevertheless, there are many data indicating that intracellular glutathione is the major substrate of the transpeptidase. The finding of an enzyme and its substrate on opposite sides of the membrane led to the hypothesis that

[79] H. Sies, C. Gerstenecker, K. H. Summer, H. Menzel, and L. Flohé, *in* "Glutathione (L. Flohé, H. C. Benohr, H. Sies, H. D. Waller, and A. Wendel, eds.), p. 261. Academic Press, New York, 1974.

[80] H. Sies and K.-H. Summer, *Eur. J. Biochem.* **57,** 503 (1975).

[81] H. Sies, G. M. Bartoli, R. F. Burk, and C. Waydhas, *Eur. J. Biochem.* **89,** 113 (1978).

[82] N. Oshino and B. Chance, *Biochem. J.* **162,** 509 (1977).

[83] H. Sies, A. Wahllander, C. Waydhas, S. Soboll, and D. Haberle, *Adv. Enzyme Regul.* **18,** 303 (1980).

[84] G. M. Bartoli and H. Sies, *FEBS Lett.* **86,** 89 (1978).

[85] A. Meister and M. E. Anderson, *Annu. Rev. Biochem.* **52,** 711 (1983).

[86] H. Sies, R. Brigelius, and T. P. M. Akerboom, *in* "Functions of Glutathione" (A. Larssen, S. Orrenius, A. Holmgren, and B. Mannervik, eds.), p. 51. Raven Press, New York, 1983.

intracellular glutathione is transported to the membrane-bound transpeptidase.[75,87] A patient was found who exhibited marked deficiency of γ-glutamyl transpeptidase and who exhibited glutathionuria and glutathionemia, and it was concluded that this patient's glutathionuria is secondary to glutathionemia and that glutathione is secreted by or leaks from renal cells into the glomerular filtrate.[88] It was proposed that the transport of intracellular glutathione to the plasma in this patient is part of a process that occurs normally which provides substrates to membrane-bound transpeptidase.[75] Therefore, in the absence of significant transpeptidase activity, substantial amounts of glutathione appear extracellularly.

Treatment of experimental animals with inhibitors of transpeptidase leads to marked glutathionuria and glutathionemia.[33,75] This provides an animal model for the human disease, and also illuminates a significant aspect of glutathione metabolism. When inhibitors of glutathione synthesis are given to experimental animals, the plasma level of glutathione decreases rapidly and substantially. The findings indicate that there is normally an appreciable flow of glutathione from tissues, mainly the liver, into the plasma, and that plasma glutathione is used by transpeptidase, predominantly located in the kidney. Studies on anephric animals treated with transpeptidase inhibitors showed that about two-thirds of the plasma glutathione is used by the kidney and the remainder by extrarenal transpeptidase.[33] Thus, glutathione is normally translocated to the membrane-bound transpeptidase as a discrete step in the γ-glutamyl cycle.[75]

Independent support for such interorgan glutathione transport was obtained by measurements of the levels of plasma glutathione in various blood vessels of the rat.[89] Hepatic vein plasma exhibits a much higher glutathione level than does arterial blood plasma. On the other hand, renal vein plasma has a much lower level of glutathione than found in arterial plasma. Since glomerular filtration can remove only about 25–30% of plasma glutathione, the data indicate that kidney has a mechanism in addition to glomerular filtration for removal on the basolateral circulation of glutathione.[89–91] The nature of this mechanism was investigated and shown to involve the activities of basolateral γ-glutamyl transpeptidase

[87] O. W. Griffith, R. J. Bridges, and A. Meister, *Proc. Natl. Acad. Sci. U.S.A.* **75,** 5405 (1978).

[88] J. D. Schulman, S. I. Goodman, J. W. Mace, A. D. Patrick, F. Tietze, and E. J. Butler, *Biochem. Biophys. Res. Commun.* **65,** 68 (1975).

[89] M. E. Anderson, R. J. Bridges, and A. Meister, *Biochem. Biophys. Res. Commun.* **96,** 848 (1980).

[90] Haberle, D., Wahllander, A., and Sies, H., *FEBS Lett.* **108,** 335 (1979). The values of the glutathione levels given in this paper were subsequently shown to be about 10% of the correct values (see refs. 89 and 91).

[91] M. E. Anderson and A. Meister, *J. Biol. Chem.* **255,** 9530 (1980).

and dipeptidase.[92] Thus, the decrease in plasma glutathione that occurs over the kidney is due to glutathione metabolism in both the tubule and the basolateral circulation by membrane-bound γ-glutamyl transpeptidase and dipeptidase. The turnover rate of kidney glutathione indicates an approximate flow of about 4.1 μmol/hr from renal cells to tubule (on the basis of a calculation for a 30 g mouse). Similarly, it may be estimated that the amount of glutathione entering the tubule by glomerular filtration of the plasma is about 0.8 μmol/hr, and that about 2.4 μmol/hr of glutathione reaches the basolateral circulation. The data indicate that the kidney transports much more glutathione to the tubule than does the liver to bile and blood plasma. Export of glutathione from renal cells does not reach the plasma but is completely utilized within the tubule, and the products formed are taken up by renal cells.

About 99.5% of intracellular total glutathione is in the glutathione form. Therefore it is not surprising that the major transport form is normally glutathione rather than glutathione disulfide. In some earlier investigations, it was reported that the predominant form present in blood plasma and bile is glutathione disulfide. However, later studies in which improved methods of assay were used[91] showed that rat blood plasma (from aorta, hepatic vein, renal vein, and vena cava) as well as rat bile contain total glutathione levels that consist of at least 90% glutathione. About 70% of mouse urinary total glutathione was found to be in the glutathione form after potent inhibition of transpeptidase. As discussed above, under conditions of marked toxicity or oxidative stress, intracellular glutathione disulfide levels increase appreciably, and this leads to substantial export of glutathione disulfide.

Transport of glutathione from mammalian cells grown in tissue culture has been demonstrated in studies carried out in the presence of buthionine sulfoximine.[34] Under these conditions, cells have little or no capacity to synthesize glutathione and the intracellular level of glutathione decreases rapidly. In studies on three human lymphoid cell lines cultured in the presence of buthionine sulfoximine, the intracellular level of glutathione decreased to about 50% of that of the untreated control after about 5 hr and to about 3% of the control after 30 hr. The disappearance of cellular glutathione under such conditions is due to its transport out of the cells. Cells that have substantial amounts of γ-glutamyl transpeptidase, however, show little extracellular glutathione. With lymphoid cells of the CEM line, about 50% of the transported glutathione could be recovered in the medium. Studies on cultured mouse resident peritoneal macrophages also showed export of glutathione.[93]

[92] W. Abbott, R. J. Bridges, and A. Meister, *J. Biol. Chem.* **259,** 15,393 (1984).

[93] C. A. Rouzer, W. A. Scott, O. W. Griffith, A. L. Hamill, and Cohn, Z. A., *Proc. Natl. Acad. Sci. U.S.A.* **79,** 1621 (1982).

Section IV

Aspartate and Asparagine

[75] L-Aspartate α-Decarboxylase

By JOANNE M. WILLIAMSON

L-Aspartate α-decarboxylase catalyzes the release of C-1 of aspartate as CO_2 to yield β-alanine.[1] β-Alanine is required in microorganisms for the enzymatic synthesis of pantothenic acid,[2,3] which is then converted to coenzyme A, its metabolically active form.[4] That aspartate α-decarboxylase is the enzyme responsible for the formation of β-alanine needed for the biosynthesis of pantothenic acid is shown by the fact that it is missing in a mutant (99-2) of *Escherichia coli* that requires either β-alanine or pantothenate as a nutritional factor and is present in a prototrophic revertent of the mutant strain.[1] Cronan has independently discovered this enzyme and shown that decarboxylase activity is associated with the genetic locus *pan D*.[5] The preparation and properties of aspartate α-decarboxylase are the subject of this chapter.

Assay Method

Principle. Aspartate α-decarboxylase activity can be determined by analysis for either of its two products, CO_2 or β-alanine, with the use of the appropriate radioactive aspartate as substrate. For the measurement of CO_2 release, [1-^{14}C]aspartate is used and the $^{14}CO_2$ released is trapped in base. For the measurement of β-alanine synthesis, either [4-^{14}C]- or [U^{14}C]aspartate is used and the unreacted substrate is separated from product by paper chromatography. One unit of enzyme is defined as the amount required for the formation of 1 nmol of product per mi at 42°. Protein is determined by the method of Lowry *et al.*[6] with bovine serum albumin as standard.

Reagents

Potassium phosphate buffer, 1.0 *M*, pH 7.5
EDTA (potassium salt, pH 7.0), 100 m*M*

[1] J. M. Williamson and G. M. Brown, *J. Biol. Chem.* **254,** 8074 (1979).
[2] W. K. Maas, *J. Biol. Chem.* **198,** 23 (1952).
[3] T. Wieland, A. Kreiling, W. Buck, and G. Pfleiderer, *Biochem. Z.* **333,** 311 (1960).
[4] G. M. Brown, *J. Biol. Chem.* **234,** 370 (1959).
[5] J. E. Cronan, *J. Bacteriol.* **141,** 1291 (1980).
[6] O. H. Lowry, N. J. Rosebrough, A. L. Farr, and R. J. Randall, *J. Biol. Chem.* **193,** 265 (1951).

Copyright © 1985 by Academic Press, Inc.
All rights of reproduction in any form reserved.

L-Aspartate (potassium salt; with 0.17 Ci/mol DL-[1-^{14}C]aspartate; or 2.8 Ci/mol DL-[4-^{14}C]aspartate or 1.4 Ci/mol L-[U-^{14}C]aspartate), 50 m*M*

Hyamine hydroxide (used as commercially supplied)

Trichloroacetic acid, 50% (w/v H_2O)

Procedure.[1] *CO_2 Release.* This procedure is patterned after that of Vanderbilt *et al.*[7] Reaction mixtures are prepared in polyallomer tubes (5/8″ × 2.5″; Beckman) to contain, per 0.5 ml total volume, the following: potassium phosphate buffer (0.05 ml), EDTA (0.05 ml), L-aspartate (0.01 ml; with [l-^{14}C]-aspartate), and decarboxylase (0.08–1.0 unit). After reaction is initiated by addition of either substrate or enzyme, the tube is fitted with a rubber stopper that is pierced with a glass rod; the tapered end of this rod is dipped in hyamine hydroxide just before it is placed in the tube. The reaction mixtures are then incubated for 20 min at 42°. The reaction is quenched by addition of tricholoroacetic acid (TCA) to a final concentration of 5%. After the TCA is added, the rod is replaced in the tube for an additional 20 min at 42° to allow collection of CO_2 liberated by addition of the acid. The hyamine is then washed off the tip of the rod into a scintillation vial with H_2O (1.5 ml). Scintillant is added and the amount of $^{14}CO_2$ trapped by the base is determined by liquid scintillation counting. The efficiency of capture and detection of $^{14}CO_2$ is about 80%. This assay is used routinely in the preparation of the enzyme.

β-Alanine Synthesis. Reaction mixtures are prepared to contain, per 0.2 ml total volume, the following: potassium phosphate buffer (0.02 ml), EDTA (0.02 ml), L-aspartate (0.004 ml), and decarboxylase (0.05–0.5 units). The reaction mixtures are incubated for 20 min at 42°. Reaction is quenched by the addition of TCA (0.02 ml). If the enzyme used is a crude preparation, the solution is filtered to remove insoluble material by being drawn up into a Pasteur pipette through a small ball of cotton added to the mixture. The residue is then washed with water (0.2 ml) filtered as before. The filtrates are combined. With more purified preparations of enzyme the filtration step is unnecessary. Each reaction mixture is then spotted onto a 3-cm strip of Whatman 3 MM chromatography paper. The chromatograms are developed with 1-butanol : acetic acid : H_2O (12 : 3 : 5) in a descending fashion overnight. The papers are then dried and the positions of the peaks of radioactivity determined with a Packard Model 7200 Radiochromatogram Scanner. A control reaction mixture containing a known amount of β-[1^{14}C]alanine and with TCA added initially is always included to estimate recovery of the product. Recoveries range from 69 to

[7] A. S. Vanderbilt, N. S. Gabey, and V. W. Rodwell, *J. Biol. Chem.* **250,** 5322 (1975).

80% depending on the purity of the enzyme preparation used. This assay method should be used with crude preparations of enzyme to obtain a more accurate estimate of total decarboxylase activity. Only about 70% of the CO_2 release from [1-^{14}C]aspartate with crude extracts is due to the activity of aspartate α-decarboxylase.

Procedure for Purification of Enzyme[1]

Preparation of Extracts of E. coli. Escherichia coli B is grown in a medium[8] containing (per liter): KH_2PO_4, 13.6 g; $(NH_4)_2SO_4$, 2.0 g; $MgSO_4 \cdot 7H_2O$, 0.2 g; $FeSO_4 \cdot 7H_2O$, 0.5 mg; glucose, 5.0; yeast extract, 3.0 g; and Bacto-peptone, 5.0 g. Cells are grown at 37° with vigorous aeration in 12-liter quantities in 16-liter carboys and harvested at late exponential growth phase with a Sharples centrifuge; aeration is maintained during the harvesting procedure. The cell paste is kept frozen until used. (Note: It was found during the course of these studies that aspartate α-decarboxylase is not stable to freezing. Cells stored at −20° lose decarboxylase activity at a rate of ~6%/week of storage. It is therefore best to use freshly grown cells for enzyme purification.) The frozen cells (400 g, moist weight) are ruptured in a Hughes Press and the resulting frozen material is suspended in 1.0 liter of Buffer A (50 m*M* potassium phosphate, 5 m*M* EDTA, 50 μ*M* dithiothreitol, pH 7.0). After the material is thawed, the suspension is homogenized in a Waring Blender and then incubated with DNase (50 mg) for 20 min at room temperature. The "crude extract" is obtained by removal of insoluble material at 24,000 *g* for 2 hr. Cells grown on this medium or on minimal medium (minus yeast extract and Bacto-peptone) contain the same amount of decarboxylase activity in the crude extract.

Heat Treatment. The crude extract (993 ml) is heated (in 200 ml portions) to 55° and maintained at that temperature for 3 min, after which each portion is cooled rapidly to 4°. The resulting precipitate is removed by centrifugation and discarded. All subsequent steps are carried out at 4° unless otherwise specified.

Fractionation with Ammonium Sulfate. The heat-treated extract (960 ml) is treated with solid ammonium sulfate to yield a fraction precipitating between 40 and 60% saturation. The precipitate is recovered by centrifugation and dissolved in 150 ml of Buffer A. The resulting solution is dialyzed for 8 hr each against two changes of Buffer A (6-liter portions).

Chromatography on DEAE-Sephadex A-50. The dialyzed ammonium

[8] A. D. Pardee, F. Jacob, and J. Monod, *J. Mol. Biol.* **1,** 165 (1959).

sulfate fraction (180 ml, 11 g of protein) is applied to a column (8.1 × 40 cm) of DEAE-Sephadex A-50 that is equilibrated with Buffer A. The colume is developed with 2 liters of the same buffer followed by a linear gradient (8 liters total) of KCl (0–0.3 *M*) in Buffer A. Fractions (25 ml each) are collected at a rate of 160 ml/hr. A portion of the enzyme activity is eluted (Fractions 11 to 35) in the absence of KCl in the developing buffer and a second much larger peak of activity (Fractions 275 to 390; peak fraction at 0.1 *M* KCl) appears during the development in the presence of KCl. The fractions from the two peaks are combined (3100 ml) and the protein is precipitated by addition of solid ammonium sulfate to 80% of saturation (1577 g). The precipitate is recovered by centrifugation and dissolved in Buffer A (40 ml). The resulting solution is dialyzed against two 6-liter portions (8 hr each) of the same buffer.

Chromatography on QAE-Sephadex Q-50. The dialyzed solution (58 ml, 1.1 g of protein) from the previous step is applied to a column (5.8 × 55 cm) of QAE-Sephadex Q-50 equilibrated with Buffer A. The column is developed with 1.5 liters of Buffer A, followed by a total of 3 liters of a linear gradient (0–0.3 *M*) of KCl in the same buffer. Fractions of 13.2 ml each are collected at a rate of 60 ml/hr. Two peaks of activity are present; one (Peak I, 15% of the total activity) appears in Fractions 21 to 36 (without KCl), and the remainder of the activity appears (Peak II) in Fractions 300 to 370. The fractions of Peak I are combined and concentrated (by ultrafiltration in an Amicon stirred cell equipped with a PM-10 membrane) to 6.6 ml; this enzyme activity is not further fractionated. Peak II fractions are combined and concentrated by ultrafiltration (as before) to 20 ml and dialyzed for 12 hr against 6 liters of Buffer A.

Chromatography on Ultrogel AcA 44. The dialyzed solution (20 ml, 696 mg of protein) from the previous step is applied to a column (5.4 × 110 cm) of Ultrogel AcA 44 that has been equilibrated with Buffer A. The column is developed with the same buffer, and fractions (11.2 ml each) are collected at a rate of 45 ml/hr. Fractions (94 to 106) containing enzyme activity are combined and concentrated by ultrafiltration to 9.0 ml. The solution is then dialyzed against two 6-liter portions (8 hr each) of Buffer B (5 m*M* potassium phosphate, 5 m*M* EDTA, and 50 μM dithiothreitol, pH 7.0).

Chromatography on Hydroxylapatite. The preparation (9.0 ml, 146 mg protein) from the previous step is applied to a column (3.7 × 27 cm) of hydroxylapatite which has been equilibrated with Buffer B. The column is developed with the same buffer, and fractions of 6.6 ml each are collected at a rate of 50 ml/hr. Fractions (9 to 27) containing enzyme activity are combined and concentrated by ultrafiltration to 9.6 ml.

Preparative Scale Isoelectric Focusing. This step is carried out according to the general procedure of Vesterberg.[9] The preparation (9.6 ml, 23 mg of protein) from the previous step is subjected to isoelectric focusing in an LKB 8101 (110 ml capacity) column. The sucrose gradient is formed manually and contains 6% ampholytes (pH 3.5 to 5.0). After the pH gradient forms (39.5hr at 200 to 400 V; 10°), the column is emptied at a rate of 1.0 ml/min and 1.0 ml fractions are collected. Fractions containing enzyme activity (32 to 58) are combined and used to form a second gradient without the addition of more ampholytes. The focusing procedure is repeated for 40.3hr (400 to 450 V; 10°), after which the column is emptied at a rate of 0.5 ml/min. Fractions (1.0 ml each) containing enzyme activity (50 to 79) are combined and filtered through an Ultrogel AcA 44 column (5.4 × 110 cm) to remove the ampholytes. Fractions (11.2 ml each) containing enzyme activity are combined and concentrated by ultrafiltration to 1.8 ml. The volume of the enzyme solution is further reduced to 0.5 ml by dialysis against a solution of 50% sucrose (w/v) in Buffer A.

Preparative Gel Electrophoresis. Polyacrylamide gels (10% acrylamide, 1.4 × 12.6 cm) are prepared by the procedure of Davis.[10] The buffer (pH 8.3) used in the electrode chambers contains (per liter): Tris (10.8 g), Na_2 EDTA (0.9 g), and boric acid (5.5 g).[11] To the enzyme solution (0.5 ml; 4 mg of protein) from the previous step is added 0.1 ml of 0.05% bromophenol blue (dye marker), and 0.2 ml of the preparation is added to each of three gels. Electrophoresis is carried out at 4° at 10 mA/gel until the dye marker is 1.5 cm from the end of the gel. Each gel is cut into slices (0.5 cm thick) and the corresponding slices from the three gels are combined and washed with four 3-ml portions of Buffer A at 4° over a 36-hr period. The four washes from each set are combined, concentrated to 2.0 ml, and assayed for enzyme activity. Fractions of highest specific activity are combined and filtered through a column (1.2 × 118 cm) of Ultrogel AcA 54 (equilibrated and developed with Buffer A) to remove traces of acrylamide. Fractions (1.0 ml each) containing enzyme activity (67 to 83) are combined and concentrated to 3.5 ml (1.1 mg of protein). This preparation can be stored for at least 6 months at 4° without loss of activity but activity is lost gradually if the enzyme is stored frozen.

A summary of the purification at each stage of the purification scheme described above is given in the table. Fractionation of hydroxylapatite appears to remove an inhibitor as the amount of activity recovered after this step (3144 units) significantly exceeds the amount applied to the column (2422 units).

[9] O. Vesterberg, this series, Vol. 22, p. 389.

[10] B. J. Davis, *Ann. N.Y. Acad. Sci.* **121,** 404 (1964).

[11] A. C. Peacock, S. L. Bunting, and K. G. Queen, *Science* **147,** 1451 (1965).

SUMMARY OF PURIFICATION PROCEDURE FOR ASPARTATE α-DECARBOXYLASE

Step	Total activity (Units)	Specific activity (nmol min^{-1} mg^{-1})	Relative specific activity
1. Crude extract	6700	0.17	1.0
2. Heat treatment	6940	0.24	1.4
3. Ammonium sulfate fractionation	6270	0.57	3.4
4. DEAE-Sephadex	5920	5.6	33
5. QAE-Sephadex			
Peak I	680	23	137
Peak II	3970	5.7	34
6. Ultrogel AcA 44	2420	17	98
7. Hydroxylapatite	3140	140	800
8. Isoelectric focusing	1380	350	2000
9. Preparative electrophoresis	710	650	3800

Properties[1]

Homogeneity. Enzyme obtained from step 9 of the purification procedure exhibits a single band of protein that contains all of the decarboxylase activity when subjected to polyacrylamide gel electrophoresis at pH 9.5 or 7.0.[12]

Physical Properties. The molecular weight of the decarboxylase is 58,000 as estimated by filtration through a calibrated column of Ultrogel AcA 54 according to the method of Leach and O'Shea.[13] Electrophoresis in the presence of SDS[14] gives complicated results which are a function both of the temperature at which denaturation is carried out and the composition of the denaturant used. If denaturation is carried out at 37° for 1.5hr in the presence of SDS (2%) and dithiothreitol (20 m*M*), a single protein band is observed after electrophoresis. The molecular weight of this species is estimated as 54,000 by comparison with the migration of standard proteins. If denaturation is carried out at 100° for 20 min in the presence of SDS, with or without dithiothreitol, the high-molecular-weight band disappears completely and is replaced by three protein bands of molecular weights, 11,800, 9,800, and 6,400. This conversion is stimulated by, but does not absolutely require, the presence of SDS in the preliminary denaturation mixture. As yet no definitive explanation is

[12] Gelman Instrument Co., Technical Bulletin No. 24, p. 21 (1975).
[13] A. A. Leach and P. C. O'Shea, *J. Chromatogr.* **17,** 245 (1965).
[14] K. Weber and M. Osborn, *J. Biol. Chem.* **244,** 4406 (1969).

available for the formation of these low-molecular-weight polypeptides when the purified enzyme is heated with SDS. The isoelectric point of the enzyme is 4.67.

Identification of the Prosthetic Group. The decarboxylase is inactivated by treatment with carbonyl reagents such as phenylhydrazine, sodium borohydride, and hydroxylamine. The compound containing the carbonyl group has been identified as covalently bound pyruvate.[1] Approximately 1 mol of pyruvate is found per mol of enzyme (58,000 molecular weight).

Catalytic Properties. The formation of product is linear with time for at least 80 min at 42°. The temperature optimum of the enzyme is 55°, with half maxima at 26 and 78°. The pH optimum is 6.5–7.5, with half maxima at pH 5.3 and 8.6

The activity of the enzyme is unaffected by monovalent cations such as Li^+, Na^+, NH_4^+, and K^+ at 10 m*M*; however, KCl at 200 m*M* is inhibitory (63%). The activity of the enzyme is unaffected by the following divalent cations at concentrations of 10 m*M*: Mn^{2+}, Mg^{2+}, Ca^{2+}, Fe^{2+}, and Co^{2+}. Activity is inhibited 88% by 10 m*M* $CuCl_2$.

The K_m for L-aspartate is 0.16 m*M*. D-Aspartate is neither a substrate nor a competitive inhibitor. The following compounds are competitive inhibitors (K_i, m*M*): L-glutamate (0.76); succinate (0.73); oxaloacetate (0.81); L-serine (0.73); L-cysteic acid (0.08); β-hydroxy-DL-aspartate (0.13); and D-serine (0.16). There is no evidence for any regulation of the activity of the decarboxylase by feedback inhibition by metabolic end products such as coenzyme A and pantothenic acid.

Properties of Enzyme in Peak I from QAE-Sephadex Column. Unlike the Peak II enzyme, the Peak I decarboxylase elutes from anion exchange columns such as DEAE- or QAE-Sephadex without the addition of KCl. The molecular weight of this form of the enzyme is 250,000 as estimated by the procedure described above. The temperature and pH optima of the two enzyme preparations are identical but the K_m for aspartate for the Peak I form is 0.30 m*M*, somewhat higher than that (0.16 m*M*) for the purified enzyme.

Both the Peak I and Peak II forms of the decarboxylase are missing in the mutant of *E. coli* ((99-2) that requires β-alanine for growth and are present in a prototophic revertent of the mutant strain. When *E. coli* is grown on a minimal medium,[8] a signficantly lower proportion of the total enzyme activity is present as this high-molecular-weight form than that found in extracts of cells grown on a rich medium. The relationship between these two forms of the decarboxylase is not yet clear.

[76] Aspartate Kinases I, II, and III from *Escherichia coli*[1]

By GEORGES N. COHEN

$$\begin{array}{c} COOH \\ | \\ H_2N-CH \\ | \\ CH_2 \\ | \\ COOH \end{array} + ATP \rightleftharpoons \begin{array}{c} COOH \\ | \\ H_2N-CH \\ | \\ CH_2 \\ | \\ O=COPO_3H_2 \end{array} + ADP$$

L-Aspartic acid L-β-Aspartyl phosphate

Aspartate Kinase Assay Method

Methods A and B, measuring respectively (1) the amount of aspartohydroxamate formed by incubation of the enzyme with the substrates and hydroxylamine, or (2) the amount of NADP formed from L-β-aspartylphosphate and NADPH in the presence of aspartate semialdehyde dehydrogenase, have been described previously.[2] A more convenient method is described below, originally proposed by Wampler and Westhead.[3]

Method C

Reagents

Tris–HCl buffer, pH 7.4 (0.5 *M* final) 61 g; adjust with concentrated HCl: $MgSO_4 \cdot 7H_2O$ (60 m*M* final), 15 g; KCl (1 *M* final), 75 g; H_2O, complete to 1 liter (A)
Lactate dehydrogenase (LDH), Sigma, rabbit muscle type II (400–900 units/mg protein; 10 mg/ml)
Pyruvate kinase (PK), Sigma, rabbit skeletal muscle, type II (350–500 units/mg protein; 10 mg/ml)
Phosphoenolpyruvate, cyclohexylamine salt (PEP)
NADH
ATP, sodium salt, 0.1 *M*, pH 7.4
Potassium L-aspartate 0.5 *M*, pH 7.4

[1] EC 2.7.2.4; L-aspartate-4-phosphotransferase; EC 1.1.1.3: L-homoserine : NADP oxidoreductase.
[2] P. Truffa-Bachi and G. N. Cohen, this series, Vol. 17A, p. 694.
[3] J. Janin and G. N. Cohen, *Eur. J. Biochem.* **11,** 520 (1969).

Copyright © 1985 by Academic Press, Inc.
All rights of reproduction in any form reserved.

Reaction Mixture. To 25 ml of solution A, add 25 ml of H_2O, 13 mg of NADH, 8 mg of PEP, 0.15 ml each of LDH and PK.

Procedure. In a quartz cuvette of 1 cm light path add 0.7 ml of the above mixture, 50 μl ATP and 20 μl L-aspartate. Shake. Read at 340 nm until stabilization. Then add the aspartokinase solution and H_2O to 1 ml. Shake and read at 340 nm. One unit is defined as the amount of enzyme that catalyzes the oxidation of 1 μmol of NADH under the conditions described.

Homoserine Dehydrogenase Assay Method. NADPH oxidation in the presence of aspartate semialdehyde is followed spectrophotometrically. The reagents and the procedure have been described in an earlier volume of this series.[2]

Purification of Aspartokinase I–Homoserine Dehydrogenase I. It has been described previously.[2] The main changes are the following[3]: (1) We use a strain derived from *E. coli* K12 (Tir 8) which is genetically derepressed for aspartokinase I–homoserine dehydrogenase I: it produces routinely 1.5–2 g of this protein per kg of frozen bacterial paste. (2) We grow the strain ourselves in 300 liter fermenters with strong aeration.

Purification of Aspartokinase II–Homoserine Dehydrogenase II. We have not modified the method described previously.[4]

Purification of Aspartokinase III[5]

Extraction

One kilogram of frozen cells was thawed and homogenized in 2 liters (final volume) of buffer (20 m*M* potassium phosphate buffer, pH 7.2, 2 m*M* magnesium Titriplex, 10 m*M* mercaptoethanol, 150 m*M* KCl, and 0.5 m*M* L-lysine) by mechanical stirring. After a homogeneous suspension had been obtained, the cells were subjected by 40 ml portions to sonic disruption in an MSE 100 W ultrasonic desintegrator for 5 min. This suspension of broken cells was centrifuged first at 23,000 *g* for 15 min, then at 78,000 *g* in a Spinco preparative ultracentrifuge for 60 min (rotor No. 30). The supernatant (total volume: 1100 ml) was called Fraction I.

Precipitation of the Nucleic Acids

The bulk of the nucleic acids were precipitated with streptomycin sulfate (3% w/v). The precipitate was allowed to accumulate for 16 hr and was eliminated by centrifugation at 23,000 *g* for 15 min. The supernatant (total volume: 1000 ml) was called Fraction II.

[4] See this series, Vol. 17A, p. 699.

[5] C. Lafuma, C. Gros, and J.-C. Patte, *Eur. J. Biochem.* **15**, 111 (1970).

Ammonium Sulfate Precipitation

Ammonium sulfate was added slowly to the enzyme solution in order to bring the salt to 45% saturation. After 6 hr, the precipitate was collected by centrifugation at 23,000 *g* for 15 min and resuspended in such a volume of buffer that the protein concentration was approximately 40 mg/ml. This fraction (Fraction III) was centrifuged at 78,000 *g* for 60 min in order to get a clear solution (Fraction IV).

A second precipitation was performed by adding ammonium sulfate until 35% saturation was reached. After 16 hr, the precipitate was resuspended in buffer to a volume of 200 ml (Fraction V).

DEAE-Sephadex Chromatography

Fraction V was dialyzed 3 times against 2 liters of buffer containing 0.3 *M* KCl. It was then adsorbed onto a DEAE-Sephadex column (25×28 cm^2) previously equilibrated with this buffer. After adsorption, the proteins were eluted with a linear gradient established with 3 liters of buffer containing 0.3 *M* KCl (first chamber) and 3 liters of buffer containing 0.6 *M* KCl (second chamber). The enzyme activity was eluted around 0.45 *M* KCl. All fractions containing enzyme activity were pooled (total volume: 1500 ml) and this eluate was brought to 50% saturation by addition of ammonium sulfate. After 16 hr, the precipitate was centrifuged (23,000 *g* for 15 min) and resuspended in 30 ml (total volume) of buffer (Fraction VI).

Sephadex Chromatography

Fraction VI was divided in two parts. Each one was filtered through a Sephadex G-200 column (90 cm $\times$ 4.9 cm^2) equilibrated with buffer. All fractions containing enzyme activity were pooled (total volume: 100 ml); this pool was called Fraction VII.

Hydroxyapatite Chromatography

The proteins were precipitated with ammonium sulfate to 50% saturation. The precipitate was dissolved in 7 m*M* potassium phosphate buffer pH 7.2 containing the same additions as normal buffer except magnesium Titriplex. This solution was adsorbed onto a hydroxyapatite column (6 cm $\times$ 12.5 cm^2) preequilibrated with the same buffer. The elution was performed with 800 ml of a linear gradient of phosphate buffer pH 7.2 from 7 to 100 m*M*. The activity was eluted at a concentration of phosphate of 50 m*M*. The tubes were pooled (Fraction VIII). Magnesium Titriplex was added at a final concentration of 2 m*M* and enzyme was kept in 50%

PURIFICATION OF ASPARTOKINASE III[a]

Fraction	Proteins (mg)	Aspartokinase III activity (units)	
		(units)	(units/mg)
I	62000	3070	0.05
II	62000	3070	0.05
III	35000	3070	0.08
IV	35000	3070	0.08
V	10000	2670	0.26
VI	1300	1990	1.63
VII	380	1900	5.00
VIII	100	1840	18.40

[a] Units are defined as micromoles of aspartolydroxamate formed per minute.

ammonium sulfate at +4°. A summary of the purification procedure is given in the table.

Properties

Some properties of aspartokinase I–homoserine dehydrogenase I and aspartokinase II–homoserine, dehydrogenase II have been reviewed previously in this series.[2,4] Some values have been corrected however: the extinction coefficient of aspartokinase I–homoserine dehydrogenase I is 0.63 OD units cm^2/mg at 278 nm. The protein contains 48 cysteine residues (12 per monomer). A detailed description of the structural, kinetic, and regulatory properties of these two proteins has recently appeared.[6]

The genes *thrA* and *metL,* corresponding respectively to aspartokinase I–homoserine dehydrogenase I and to aspartokinase II–homoserine dehydrogenase II, have been cloned on plasmid vectors and sequenced.[7,8]

Aspartokinase III, prepared as described above, is a homogeneous protein. Its extinction coefficient is 0.50 cm^2/mg. Its sedimentation coefficient is very sensitive to the ionic strength and to the presence of lysine, and is dependent on the protein concentration. The enzyme in the presence of lysine and 0.15 M KCl is a dimer (M_r = 105,000) of identical subunits.[9]

[6] G. N. Cohen and A. Dautry-Varsat, *in* "Multifunctional Proteins" (H. Bisswanger and E. Schminke-Ott, eds.), p. 49. Wiley, New York, 1980.

[7] M. Katinka, P. Cossart, L. Sibilli, I. Saint-Girons, M.-A. Chalvignac, G. Le Bras, G. N. Cohen, and M. Yaniv, *Proc. Natl. Acad. Sci. U.S.A.* **77,** 5730 (1980).

[8] M. M. Zakin, N. Duchange, P. Ferrara, and G. N. Cohen, *J. Biol. Chem.* **258,** 3028 (1983).

[9] C. Richaud, J.-P. Mazat, C. Gros, and J.-C. Patte, *Eur. J. Biochem.* **40,** 619 (1973).

[77] Aspartate-semialdehyde Dehydrogenase[1] from *Escherichia coli*

By GEORGES N. COHEN

$$\text{NADP}^+ + \text{P}_i + \text{H}_2\text{N-CH(COOH)-CH}_2\text{-CHO} \rightleftharpoons \text{H}_2\text{N-CH(COOH)-CH}_2\text{-C(=O)OPO}_3\text{H}_2 + \text{NADPH} + \text{H}^+$$

L-aspartate β-semialdehyde — Aspartyl phosphate

Assay Method. As described previously.[2] The triethanolamine buffer is 0.02 *M*, pH 8 instead of 0.03 *M*, pH 9.[2]

Buffers

Buffer 1 contained imidazole (12.25 g/liter), barbital (free acid, 3.78 g/liter), EDTA disodium salt (0.372 g/liter) and dithioerythritol (1.5 g/liter). The pH was adjusted to 8.0.

Buffer 2 contained 0.01 *M* potassium phosphate, pH 7.2, 0.1 m*M* EDTA, and 1 m*M* dithioerythritol.

Buffer 3 contained 0.01 *M* potassium phosphate, pH 7.2, 0.5 *M* potassium chloride, 0.1 m*M* EDTA, and 1 m*M* dithioerythritol.

Preparation of the Enzyme.[3] From a genetically derepressed strain (OE 73) of *E. coli* K12. Cell-free extracts in 30 m*M* potassium phosphate buffer, pH 7.2, were prepared by treatment in a Gifford-Wood homogenizer (Hudson, NY) in the presence of glass beads (0.1 mm diameter) followed by centrifugation at 15,000 *g* (1 hr).

The DEAE-cellulose (200 g) was equilibrated in 30 m*M* phosphate buffer, pH 6.8. The filtered DEAE-cellulose was added to the protein solution. After filtration, the gel was suspended in the same buffer and filtered. The enzymatic activity was eluted with 350 m*M* phosphate buffer, pH 6.8. The enzyme was dialyzed against buffer 1, 10 times diluted and containing 0.15 *M* potassium chloride, with several changes.

The DEAE-Sephadex column (3 × 30 cm) was equilibrated with the

[1] EC 1.2.1.11; L-aspartate-β-semialdehyde oxidoreductase (phosphorylating).

[2] See also this series, Vol. 17A, p. 708.

[3] J.-F. Biellmann, P. Eid, C. Hirth, and J. Jörnvall, *Eur. J. Biochem.* **104,** 53 (1980).

Copyright © 1985 by Academic Press, Inc.
All rights of reproduction in any form reserved.

same buffer. A linear gradient (2 × 1 liter) of potassium chloride (0.15–0.5 *M*) was applied. The enzyme was eluted at 0.35 *M*. The protein was dialyzed against buffer 2 (without EDTA) and the hydroxyapatite column (3 × 20 cm) was equilibrated and eluted with the same buffer (without EDTA). The enzyme was eluted after the void volume.

After ammonium sulfate precipitation (65% saturation), the concentrated enzyme solution, 20 mg/ml in buffer 2, was layered on an Ultrogel ACA-44 column (4 × 80 cm). The column was equilibrated and eluted with buffer 2. The enzyme (2 mg/ml in buffer 2) was adsorbed on a Sepharose C_3 column (3 × 35 cm) equilibrated with buffer 3 containing 0.5 *M* ammonium sulfate. A linear gradient from 0.5 *M* ammonium sulfate in buffer 3 (0.5 liter) to buffer 3 (0.5 liter) was applied. The enzyme was eluted at 0.2 *M* ammonium sulfate. The enzyme (10 mg/ml in buffer 2.50% glycerol) was stored at −20° in sealed tubes under argon.

Properties

The highest specific activity previously described for the enzyme was 120 U/mg determined at pH 9.0. The values obtained in the present activity test are not directly comparable to those observed previously since the instability of the substrate at high pH led to work at pH 8.0. The previous activity test gave for pure preparation an activity of 150 U/mg.

The enzyme was stable for several days in buffer 2, at 4°. It was stored in buffer 2 in the presence of 50% glycerol at −20°. Under these conditions, no loss of activity was detected within 2 months. A final specific activity of 100–110 U/mg was routinely obtained.

The enzyme proved to be homogeneous by electrophoresis, ultracentrifugation, electrofocusing, and N-terminal sequence determination: dansylation in 8 *M* urea revealed that the N-terminus is methionine. Similar determinations after successive Edman degradations showed the N-terminal sequence of the protein to be Met-Lys-Asx-Val-Gly-.

The molecular weight of the native enzyme determined by ultracentrifugation is 77,500 ± 1,500. The subunit molecular weight on sodium dodecyl sulfate gel electrophoresis is 38,000 ± 2,000. This is in agreement with the results obtained earlier from an impure preparation by centrifugation of the active enzyme. Therefore, the protein is dimeric and composed of two probably identical subunits.

The absorption spectrum ($A_{1\text{ cm } 280\text{ nm}}^{1\text{ mg/ml}} = A_{280}/A_{260} = 1.8$) agrees with the absence of any bound nucleotide. The isoelectric point is low (pH 4.3) in agreement with the behavior of the enzyme during isolation.

An intermediate acyl-enzyme has been detected and the stereochemistry of the hydrogen transfer with $NADP^+$ has been determined.[3]

The substrate binding site of the enzyme as well as the coenzyme binding site have been studied by affinity labeling.[4,5]

The gene coding in *E. coli* for aspartate-semialdehyde dehydrogenase has been cloned and sequenced.[6]

[4] J.-F. Biellmann, P. Eid, C. Hirth, and H. Jörnvall, *Eur. J. Biochem.* **104,** 59 (1980).
[5] J.-F. Biellmann, P. Eid, C. Hirth, and H. Jörnvall, *Eur. J. Biochem.* **104,** 65 (1980).
[6] C. Haziza, P. Stragier, and J.-C. Patte, *EMBO J.* **1,** 379 (1982).

[78] Asparagine Transaminase from Rat Liver

By ARTHUR J. L. COOPER and ALTON MEISTER

Introduction

Greenstein and co-workers showed that rat liver possesses two asparaginases, i.e., "asparaginase I" which is not enhanced by pyruvate and "asparaginase II" which is enhanced by pyruvate.[1-5] The two activities were distinguished by their relative susceptibility to heat and acid denaturation.[3,5] It was later shown that "asparaginase II" is composed of two enzymes: an asparagine transaminase (EC 2.6.1.14) that catalyzes the reversible transfer of the α-amino group of asparagine to a suitable α-keto acid acceptor [Eq. (1)] and ω-amidase (ω-amidodicarboxylate amidohydrolyase, EC 3.5.1.3) that catalyzes the deamidation of α-ketosuccinamate [Eq. (2)].[6-8]

$$\text{L-Asparagine} + \alpha\text{-keto acid} \rightleftharpoons \alpha\text{-ketosuccinamate} + \text{L-amino acid} \quad (1)$$
$$\alpha\text{-Ketosuccinamate} + H_2O \rightarrow \text{oxaloacetate} + \text{ammonia} \quad (2)$$

Asparagine transaminase activity is widespread.[9] The metabolic function of this enzyme in mammals may be to convert glyoxylate to glycine.[9]

[1] V. E. Price and J. P. Greenstein, *J. Natl. Cancer Inst.* (*U.S.*) **7,** 275 (1947).
[2] M. Errera and J. P. Greenstein, *J. Natl. Cancer Inst.* (*U.S.*) **7,** 285 (1947).
[3] M. Errera and J. P. Greenstein, *J. Natl. Cancer Inst.* (*U.S.*) **7,** 437 (1947).
[4] M. Errera and J. P. Greenstein, *Arch. Biochem. Biophys.* **15,** 449 (1947).
[5] J. P. Greenstein and V. E. Price, *J. Biol. Chem.* **178,** 695 (1949).
[6] A. Meister, H. A. Sober, S. V. Tice, and P. E. Fraser, *J. Biol. Chem.* **197,** 319 (1952).
[7] A. Meister, *J. Biol. Chem.* **200,** 571 (1953).
[8] A. Meister and P. E. Fraser, *J. Biol. Chem.* **210,** 37 (1954).
[9] See discussion by A. J. L. Cooper and A. Meister, *in* "The Transaminases" (D. E. Metzler and P. Christen, eds.), p. 397. Wiley, 1985.

Copyright © 1985 by Academic Press, Inc.
All rights of reproduction in any form reserved.

Assay Methods

Principles. The assay procedure first used required coupling of the transaminase and ω-amidase.[6-8] Ammonia liberated from amide cleavage of α-ketosuccinamate was measured by Nesslerization after diffusion into acid.[6] More recently, a radiochemical procedure has been employed; α-keto-[U-^{14}C]succinamate formed by transamination of L-[U-^{14}C]asparagine is decarboxylated with H_2O_2 and the labeled CO_2 is trapped in base.[10] A more convenient assay involves measurement of α-ketosuccinamate directly by a spectrophotometric technique following addition of dilute NaOH to the reaction mixture.[11] In 0.1 *M* NaOH, α-ketosuccinamate exhibits a strong absorbance at 290 nm.[7] The absorbance of most α-keto acids under these conditions is negligible (phenylpyruvate and *p*-hydroxyphenylpyruvate are exceptions).[11]

The reverse reaction, i.e., L-amino acid + α-ketosuccinamate → α-keto acid + L-asparagine, may be assayed by determining the rate of disappearance of α-ketosuccinamate.[11]

Preparation of Sodium α-Ketosuccinamate. L-Asparagine is oxidized with L-amino acid oxidase in the presence of catalase.[7,12] Following removal of the enzymes by dialysis or by ultrafiltration, ammonia and unreacted amino acid are removed on a Dowex 50 (Na^+) column. The effluent is decolorized with charcoal and concentrated by flash evaporation, or preferably by lyophilization, and the sodium salt of α-ketosuccinamic acid dimer is precipitated with acetone.[7,13] The dimer is now known to be 4,6-dihydroxy-2-oxopiperidine-5-carboxamide-4,6-dicarboxylic acid.[14] In practice, the dimer readily yields the monomer (α-ketosuccinamate) in dilute base.[7,15,16]

[10] D. A. Cooney, H. N. Jayaram, S. G. Swengros, S. C. Alter, and M. Levine, *Int. J. Biochem.* **11,** 69 (1980).

[11] A. J. L. Cooper, *J. Biol. Chem.* **252,** 2032 (1977).

[12] A. Meister, this series, Vol. 3, p. 404.

[13] T. T. Otani and A. Meister, *J. Biol. Chem.* **224,** 137 (1957).

[14] R. A. Stephani and A. Meister, *J. Biol. Chem.* **246,** 7115 (1971).

[15] The monomer has also been made by a nonenzymatic procedure [M. I. Lerman and S. R. Mardashev, *Biokhimiya* **25,** 701 (1960)].

[16] Solutions of α-ketosuccinamic acid or its dimer are unstable. Therefore, for best results it is important to carry out the oxidation of asparagine as rapidly as possible and to use lyophilization for concentrating solutions. We have observed that stored solutions of α-ketosuccinamic acid or the sodium salt of the dimer heated at 100° for several hours yield at least 10 UV absorbing products which have been tentatively identified as various derivatives of pyridine, quinolinimide, and cinchomeronimide.

Asparagine–α-Keto Acid Transaminase Assay[11]

Reagents

1 *M* glycylglycine–KOH, pH 8.4 (filtered through activated charcoal)
1 *M* sodium borate, pH 8.4
0.05 *M* L-asparagine
0.1 *M* EDTA
0.1 *M* sodium pyruvate; freshly prepared
0.1 *M* NaOH

Procedure. The reaction mixture (0.1 ml) contains 200 m*M* glycylglycine-buffer [or 100 m*M* sodium borate buffer (pH 8.4)], 20 m*M* L-asparagine, 5 m*M* EDTA, 20 m*M* α-keto acid, and enzyme (5–15 μl). After incubation at 37° for 1 hr, 0.9 ml of 0.1 *M* NaOH is added and the absorbance at 290 nm is determined. The blank contains the assay mixture to which enzyme is added just prior to addition of base. Under these conditions the absorbance due to α-ketosuccinamic acid is high ($\varepsilon = 10^4$).[17] For crude preparations containing ω-amidase, 200 m*M* glycylglycine is employed as a buffer. This compound is a competitive inhibitor of ω-amidase (apparent K_i for glycylglycine = 5 m*M*[11]; K_m for α-ketoglutaramate = 3 m*M*[18]; K_m for α-ketosuccinamate = 3.8 m*M*[7]). Thus, conversion of α-ketosuccinamate to oxaloacetate in preparations containing ω-amidase is minimized in the presence of 200 m*M* glycylglycine. In preparations free of ω-amidase, the borate buffer may be used in place of glycylglycine buffer. There is no difference in activity with the two buffers.

L-*Amino Acid–α-Ketosuccinamate Transaminase Activity*[11]

Reagents

1 *M* glycylglycine–KOH, pH 8.5
1 *M* sodium borate, pH 8.5
0.1 *M* EDTA
0.1 *M* sodium α-ketosuccinamate dimer; freshly prepared
0.1 *M* NaOH
100 m*M* L-Amino acid

[17] In studies of α-keto acid specificity, a large number of α-keto acids can substitute for pyruvate in the standard assay mixture. In the case of phenylpyruvate (or *p*-hydroxyphenylpyruvate) transamination can be estimated from the decrease in absorbance at 322 nm following addition 3.3 *M* KOH to 5-μl aliquots of the reaction mixture. The aromatic α-keto acids absorb much more strongly than α-ketosuccinamic acid under these conditions.

[18] L. B. Hersh, *Biochemistry* **10,** 2884 (1971).

Procedure. The reaction mixture contains 10 m*M* L-amino acid (except in the case of phenylalanine or tyrosine), 5 m*M* EDTA, 2 m*M* α-ketosuccinamate dimer, 200 m*M* glycylglycine buffer (or 100 m*M* borate buffer), and enzyme in a final volume of 0.05 ml.[19] After incubating at 37° for 1 hr, 0.95 ml of 0.1 *M* NaOH is added and the decrease in absorbance at 290 nm is compared to a blank in which enzyme is added just prior to addition of base. To measure activity with L-phenylalanine, the assay mixture contains 20 m*M* L-phenylalanine, 20 m*M* pyruvate, 100 m*M* sodium borate buffer, and enzyme in a final volume of 0.05 ml. After incubation at 37° for 1 hr, 0.95 ml of 3 *M* KOH is added and the increase in absorbance at 322 nm due to phenylpyruvate-enol is determined.[11]

Purification Procedure[20]

All steps are carried out at 0–4° unless otherwise noted, and all centrifugations are carried out at 40,000 *g* for 30 min.

Step 1. Ten male Sprague–Dawley rats, each weighing approximately 350 g, are decapitated and exsanguinated. Livers are homogenized for 30 sec at full speed in Waring Blender in 3 volumes of 25 m*M* potassium phosphate buffer, pH 7.2.

Step 2. To the supernate from Step 1 is added solid sodium pyruvate to a final concentration of 40 m*M*. The solution is then heated at 60–62° for 20 min, cooled, and centrifuged.

Step 3. Solid ammonium sulfate is slowly added to the supernate from Step 2 until the solution is 60% saturated. After stirring for 1 hr the suspension is centrifuged. The precipitate is dissolved in the minimal amount of potassium phosphate buffer, pH 7.2, and dialyzed against 10 liter of the same buffer for 48 hr. An inactive precipitate which forms during the dialysis is removed by centrifugation.

Step 4. The clear solution is added to the top of a DE-52 column (2 × 10 cm) previously equilibrated with 2.5 m*M* potassium phosphate buffer, pH 7.2. The column is eluted with 500 ml of 2.5 m*M* potassium phosphate buffer, pH 7.2, followed by 1 liter of 15 m*M* potassium phosphate buffer, pH 7.2. The enzyme is slowly eluted from the column with this latter buffer.

[19] Under the slightly basic conditions of the assay, α-ketosuccinamate dimer is rapidly converted to open-chain monomer.

[20] A unit of enzyme activity is defined as the amount that catalyzes the formation of 1 μmol of α-ketosuccinamate/hr at 37° in the standard L-asparagine–pyruvate assay mixture. Specific activity is expressed as units/mg of protein. Protein was measured by the method of Lowry *et al*. [O. H. Lowry, N. J. Rosebrough, A. L. Farr, and R. J. Randall, *J. Biol. Chem*. **193,** 265 (1951)].

Step 5. Distilled water is added to the active fractions to yield a final phosphate concentration of 10 m*M*. This solution is then added directly to a hydroxylapatite column (2×5 cm) equilibrated with 10 m*M* potassium phosphate buffer, pH 7.2. After eluting the column with 500 ml of this same buffer, the column is further eluted with a linear gradient established between 1 liter of 10 m*M* potassium phosphate and 1 liter of 100 m*M* potassium phosphate buffers, pH 7.2. A single peak of enzyme activity is eluted at about 75 m*M* phosphate. The enzyme is concentrated by dialysis against 5 m*M* potassium phosphate buffer, pH 7.2, addition to a hydroxylapatite column (0.6×5 cm) equilibrated with the same buffer, followed by elution with 2 ml of 100 m*M* potassium phosphate, pH 7.2.

The purification scheme is summarized in the table.

Comments on the Purification Scheme

At no time during the purification procedure does addition of pyridoxal 5′-phosphate result in activation. The enzyme loses 50% of its activity in 3 days when stored in 100 m*M* potassium phosphate buffer, pH 7.2, at 4°. However, in the same buffer system containing 20% glycerol, the enzyme is stable for at least 3 months at 4°. The isolation procedure yields an enzyme preparation that is devoid of ω-amidase, glutamate-aspartate transaminase, and glutamate-alanine transaminase activities. The preparation possesses some glutamine transaminase activity but this is a property of the purified enzyme.

Comments on the Specificity

Altogether more than 20 α-keto acids are known to be substrates.[6,11] Although the V_{max} increases slightly in the order glyoxylate < pyruvate < α-ketobutyrate, the affinity of the enzyme for substrate is in the order glyoxylate >> pyruvate >> α-ketobutyrate. The apparent K_m of glyoxylate of 0.077 m*M* is notably low, especially since in aqueous solution glyoxylate is 95% hydrated to the *gem*-diol whereas pyruvate is only 8% hydrated.[21] The binding characteristics cannot be ascribed solely to increasing chain length, because α-ketosuccinamate, phenylpyruvate, and α-keto-γ-methiolbutyrate bind effectively. The binding specificity may be due to the fact that the active site possesses a region that binds the $-C(O)CO_2^-$ moiety and a second region that interacts with large nonpolar groups ($-C_6H_5$, $-C(O)NH_2$, $-S-CH_3$).

[21] A. J. L. Cooper and A. G. Redfield, *J. Biol. Chem.* **250,** 527 (1976), and references quoted therein.

PURIFICATION OF RAT LIVER ASPARAGINE TRANSAMINASE

Step	Volume (ml)	Protein (mg)	Units	Specific activity	Yield (%)	Purification (fold)
1. Crude extract from 180 g liver	580	12,800	980	0.077	100	1
2. After heat treatment	480	3,840	844	0.220	86	2.9
3. After ammonium sulfate fractionation and dialysis	35	840	350	0.416	36	5.4
4. DE-52 chromatography	900	100	105	1.05	11	13.6
5. Hydroxylapatite chromatography and concentration	2	5.4	81	15.00	8	195

The amino acid specificity of asparagine transaminase is also broad, with at least 13 amino acids acting as substrates.[2,11] Glycine inhibits the asparagine–pyruvate transaminase reaction but it is not a detectable substrate.[11] It is interesting to note that two of the best amino acid substrates are asparagine and *S*-methylcysteine. Similarly, glutamine and methionine are among the best substrates of the glutamine transaminases. The $-C(O)NH_2$ and $-SCH_3$ groups may interact with the active sites of these transaminases in a similar fashion.

Unfortunately, due to its very wide substrate specificity, asparagine transaminase has probably been purified unknowingly by several groups of workers each using a different substrate pair.[22]

[22] Thus Noguchi *et al.* consider the following enzymes, described in the literature as probably identical to asparagine transaminase: histidine-pyruvate isozyme I, pyruvate (glyoxylate), serine-pyruvate, phenylalanine-pyruvate isozyme I, alanine-glyoxylate isozyme I and phenylalanine (leucine)-glyoxylate [T. Noguchi, Y. Takada, and Y. Oota, *Hoppe-Seyler's Z. Physiol. Chem.* **360,** 919 (1979), and references quoted therein].

[79] Asparaginase[1]

By JOHN C. WRISTON, JR.

$$\underset{\text{L-Asparagine}}{\mathrm{HOOC{-}\underset{NH_2}{\underset{|}{CH}}{-}CH_2{-}CONH_2}} + H_2O \rightarrow \underset{\text{L-Aspartic acid}}{\mathrm{HOOC{-}\underset{NH_2}{\underset{|}{CH}}{-}CH_2{-}COOH}} + NH_3$$

Introduction

Methods for the purification of four asparaginases (guinea pig serum, *E. coli B, Serratia marcescens,* and *Proteus vulgaris*) were described in earlier volumes of this series.[2,3] Interest in this enzyme has remained high because certain asparaginases (those from *E. coli B* and *Erwinia carotovora* in particular) are effective against acute lymphocytic leukemia. This has led to a search for other effective asparaginases with lower K_m values, longer circulation times, etc., or for asparaginases of clinical advantage for other reasons (e.g., lack of immunological cross-reactivity with the *E. coli B* or *Erwinia carotovora* enzymes).

Work on asparaginase has been reviewed from several points of view.[4–8] Topics discussed here are as follows: (1) assay methods; (2) purification of additional asparaginases; (3) methodological developments in connection with *E. coli* asparaginase, and properties of that enzyme; (4) mammalian asparaginases.

[1] EC 3.5.1.1; L-asparagine amidohydrolase; see also this volume, p. 602.

[2] J. C. Wriston, this series, Vol. 17A [98].

[3] I. Chibata, T. Tosa, T. Sato, R. Sano, K. Yamamoto, and Y. Matuo, this series, Vol. 34 [44].

[4] J. C. Wriston and T. O. Yellin, *Adv. Enzymol.* **39,** 185 (1973).

[5] R. L. Capizzi and Y.-C. Cheng, *in* "Enzymes as Drugs" (J. S. Holcenberg and J. Roberts, eds.), p. 1. Wiley, New York, 1981.

[6] H. E. Wade, *in* "Microorganisms and Nitrogen Sources" (J. W. Payne, ed.), p. 563. Wiley, New York, 1980.

[7] T. M. S. Change, this series, Vol. 44 [46]; W. H. Vieth and K. Venkatasubramanian, *ibid.* [19]; G. B. Broun, *ibid.* [20]; W. E. Hornby and L. Goldstein, *ibid.* [9]; G. Gregoriadis, *ibid.* [47].

[8] E. D. S. Chong and T. M. S. Chang, *Biomed. Appl. Immobilized Enzymes Proteins* **1,** 105 (1977).

Copyright © 1985 by Academic Press, Inc.
All rights of reproduction in any form reserved.

Assay Method[2,4]

Principle. Asparaginase activity can be measured by determination of ammonia or aspartic acid, or by following the disappearance of asparagine or alternate substrates (e.g., 5-diazo-4-oxo-L-norvaline, DONV[9] or β-aspartyl hydroxamic acid[10]). Direct nesslerization, the method most commonly used, is suitable for most samples at concentrations of enzyme as low as 0.1 IU/ml. Greater sensitivity (10^{-2} to 10^{-3} IU/ml) can be achieved using radiolabeled asparagine and determining aspartic acid after separating it from asparagine by paper electrophoresis,[11] chromatography on ion-exchange paper,[12] or ion-exchange chromatography.[13] There is a two-step radiometric method sensitive to 10^{-5} IU/ml; the first incubation leads to formation of aspartic acid from L-[U-^{14}C]asparagine; α-ketoglutarate and L-glutamate oxaloacetate transaminase are present in the second incubation, as well as zinc ion, leading to release of $^{14}CO_2$ from oxaloacetate.[14] Sensitive coupled assays have been described,[15] and there is a micromethod based on diffusion of ammonia and its determination by the Berthelot reaction.[16] Automated modifications of the Nessler and Berthelot procedures have also been described.[17] The direct nesslerization procedure described below appeared in an earlier volume of this series.[2]

Reagents

L-Asparagine monohydrate, 0.04 *M*, in sodium borate buffer, 0.1 *M*, pH 8.5 (replaced by 0.1 *M* Tris–HCl buffer, 0.1 *M*, pH 8.5, by several investigators)
Trichloroacetic acid, 15%
Nessler's reagent (Fisher Scientific, No. So-N-24)

[9] R. C. Jackson and R. E. Handschumacher, *Biochemistry* **9,** 3585 (1970).

[10] M. Ehrman, H. Cedar, and J. H. Schwartz, *J. Biol. Chem.* **246,** 88 (1971); Y. Z. Frohwein, M. Friedman, J. Reizer, and N. Grossowicz, *Nature* (*London*), *New Biol.* **230,** 158 (1971).

[11] J. D. Broome, *Br. J. Cancer* **22,** 595 (1968); P. P. K. Ho and L. Jones, *Biochim. Biophys. Acta* **177,** 172 (1969).

[12] J. H. Schwartz, J. Y. Reeves, and J. D. Broome, *Proc. Natl. Acad. Sci. U.S.A.* **56,** 1516 (1966).

[13] S. Prusiner and L. Milner, *Anal. Biochem.* **37,** 429 (1970).

[14] D. A. Cooney, H. A. Milman, and B. Taylor, *Biochem. Med.* **15,** 190 (1976).

[15] D. A. Cooney and R. E. Handschumacher, *Proc. Am. Assoc. Cancer Res.* **9,** 15 (1968); J. W. Boyd and A. W. Phillips, *J. Bacteriol.* **106,** 578 (1971); Y. Kojima and W. E. C. Wacker, *J. Lab. Clin. Med.* **74,** 521 (1969).

[16] M. A. Ramadan and D. M. Greenberg, *Anal. Biochem.* **6,** 144 (1963); A. Kaplan, *in* "Methods of Biochemical Analysis" (D. Glick, ed.), Vol. 17, p. 311. Wiley, New York, 1969.

[17] H. E. Wade and B. P. Phillips, *Anal. Biochem.* **44,** 189 (1971); M. K. Schwartz, E. D. Lash, H. F. Oettgen, and F. A. Tomao, *Cancer* **25,** 244 (1970); M. K. Schwartz, this series, Vol. 17B [261B].

Procedure. Enzyme solution (10–100 μl) is added to buffer to give a volume of 1.5 ml. The reaction is started with 0.5 ml of substrate solution and allowed to proceed for 5–60 min at 37° (30 min in routine assays). The incubation is stopped with 0.5 ml of 15% trichloroacetic acid and centrifuged if necessary in a clinical centrifuge. The supernatant is transferred quantitatively to 10-ml graduated test tubes and diluted to 9 ml with water. One milliliter of Nessler's reagent is added and the mixture allowed to stand for 15 min at room temperature. Ammonia is estimated by determining optical density at 500 nm (for 1–6 μmol of NH_3) or 425 nm (0.2–2 μmol) with a Spectronic 20 colorimeter. Enzyme and substrate blanks are included in all assays and a standard curve prepared with ammonium sulfate is used. Bovine serum albumin is added in assays involving highly purified asparaginase in dilute solution. Asparagine solutions should be stored in the cold and made fresh at least every 2 weeks. A unit of activity is defined as that amount of enzyme which will catalyze the formation of 1 μmol of ammonia/minute under the conditions of the assay.

Purification of Asparaginases

Asparaginase from Vibrio succinogenes. Distasio *et al.*[18] described in 1976 a 4-step purification of L-asparaginase from crude extracts of *Vibrio succinogenes,* a gram-negative anaerobic bacterium isolated from the bovine rumen. An overall yield of 40–45% and a final specific activity of 200 ± 2 IU/mg were obtained. More recently Abuchowski *et al.*[19] have described a procedure involving only two chromatographic steps with an overall yield of 69% and a specific activity of 233 IU/mg. This procedure is described below.

Step 1. Crude Extract. Vibrio succinogenes (ATCC 29543; Virginia Polytechnic Institute Anaerobe Laboratory LO659) was grown in VSF medium[20] modified by the addition of cysteine–HCl to 0.1% before autoclaving and the addition of $MgCl_2 \cdot 6H_2O$ to 0.01% afterward. Packed frozen cells (22 g) were thawed in an equal volume of 0.01 *M* Tris–HCl, pH 6.8, containing 1 mg deoxyribonuclease, and the suspension sonicated 4 times, 30 sec each, at 90 W with a Branson sonifier with 30 sec cooling in an ice-bath between sonications. Cell debris was removed at 40,000 *g* for 30 min at 4°, the pellet resuspended in the same buffer, and sonication and centrifugation repeated. The pellet was washed with buffer and the com-

[18] J. A. Distasio, R. A. Niederman, D. Kafkewitz, and D. Goodman, *J. Biol. Chem.* **251,** 6929 (1976).

[19] A. Abuchowski, D. Kafkewitz, and F. F. Davis, *Prep. Biochem.* **9,** 205 (1979).

[20] D. Kafkewitz, *Appl. Microbiol.* **29,** 121 (1975).

bined supernatants and wash centrifuged at 150,000 *g* for 90 min. The supernatant (190 ml, 19,200 units, 1940 mg protein) was dialyzed against 2 changes of 0.01 *M* Tris–HCl, pH 6.8.

Step 2. CM-BioGel A Chromatography. The sample was applied to a CM-BioGel A column (2.6 × 30 cm; 33 ml/hr) equilibrated with the same Tris buffer, washed with 200 ml of buffer, and eluted with a linear gradient of NaCl in the same buffer (1000 ml, 0–1.0 *M* NaCl). Fractions (5 ml) were tested for asparaginase activity qualitatively by the L-aspartyl-β-hydroxamate method of Frohwein *et al.*[10] Enzyme eluted at a salt concentration of approximately 0.25 *M*. Assay of pooled active fractions showed an 11-fold purification with a 73% yield.

Step 3. DEAE-BioGel A Chromatography. The above sample was dialyzed against 5 volumes of 0.01 *M* Tris–HCl, pH 9.0, using an Amicon Thin-channel Ultrafilter (PM 30 membrane) and applied to a DEAE-BioGel A column (2.6 × 30 cm, 33 ml/hr, equilibrated with 0.01 *M* Tris–HCl, pH 9.0). After a 200-ml buffer wash, enzyme was eluted with a linear gradient (1000 ml, 0–0.3 *M* NaCl in the same buffer). Active fractions were pooled, dialyzed, and concentrated in 0.05 *M* sodium phosphate, pH 7.3. A yield of 69% was obtained, with 23.5-fold purification over the initial extract. The product gave a single band on SDS polyacrylamide gel electrophoresis and had a specific activity of 233 IU/mg. Some of its properties are listed in Table I.

Other Asparaginases. Approximately 20 asparaginases (exclusive of amidases with equal or greater activity toward L-glutamine, and of a stereospecific D-asparaginase from *Thermus aquaticus,* an extreme thermophile[21]) have been purified to essential homogeneity, as shown in Table I. This list includes plant enzymes and an asparaginase from a marine algae in addition to mammalian, bacterial, fungal, and yeast sources.

The purification of several bacterial asparaginases has been facilitated by affinity chromatography. Kristiansen *et al.*[22] prepared a Sepharose 6B column with L-asparaginyl residues coupled via a hexamethylenediamine spacer. A similar procedure involving L-aspartyl residues, applied to the purification of L-asparaginase from *Proteus vulgaris,* has been described in an earlier volume of this series by Chibata *et al.*[3] Two other applications of this procedure are listed in Table I.[23–42]

[21] G. R. Guy and R. M. Daniel, *Biochem. J.* **203,** 787 (1982).

[22] T. Kristiansen, M. Einarson, L. Sundberg, and J. Porath, *FEBS Lett.* **7,** 294 (1970).

[23] T. O. Yellin and J. C. Wriston, *Biochemistry* **5,** 1605 (1966).

[24] L. T. Mashburn and J. C. Wriston, *Arch. Biochem. Biophys.* **105,** 450 (1964); H. Whelan and J. C. Wriston, *Biochemistry* **8,** 2386 (1969).

[25] P. P. K. Ho, E. B. Milikin, J. L. Bobbitt, E. L. Grinnan, P. J. Burck, B. H. Frank, L. D. Boeck, and R. W. Squires, *J. Biol. Chem.* **245,** 3708 (1970).

TABLE I
L-ASPARAGINASES PURIFIED TO HOMOGENEITY

Enzyme	MW	K_m	p*I*	pH opt.	Glnase act. (%)	Antitumor act.[a]	Reference
Guinea pig serum[b]	138,000	7.2×10^{-5}	3.6–4.5[c]	7.5–8.5	–	+	2, 23
E. coli B[b] (EC-2)	130,000	1.25×10^{-5}	4.8–5.2	Broad, pH 8	2	+	2, 22, 24–27
Serratia marcescens[b,d]	147,000	1×10^{-4} 1.2×10^{-5}	5.1, 5.85	Broad, 6.9	4	+	2, 28
Fusarium tricinctum[b]	165,000	5×10^{-4}	5.18	7.5–8.7	–	–	29
Proteus vulgaris[d]	120,000	2.6×10^{-5}	5.08	7–8	2	+	3, 30
Erwinia carotovora[b,d]	135,000	1×10^{-5}	8.6	ND	15	+	31
Acinetobacter calcoaceticus[e]	105,000	2×10^{-3}	5.2	8.6	10	–	32
Azotobacter vinelandii[e]	84,000	1.1×10^{-4}	ND	8.6	ND	+	33
Klebsiella aerogenes	141,000	$>10^{-5}$	ND	7.0–8.2	5	ND	34
Vibrio succinogenes[b,d]	146,000	4.78×10^{-5}	8.74	7.3	–	+	18, 19
Citrobacter freundii[b,f]	140,000	2.9×10^{-5}	5.75	Broad, 7.5	3	+	35

Lupinus polyphyllus	72,000	1.22×10^{-2}	ND	ND	–	ND	36
Lupinus arboreus[g]	75,000	6.6×10^{-3}	ND	8.0	–	ND	37
L. angustifolius	75,000	7.0×10^{-3}	ND	8.5	ND	ND	37
Pisum sativum[h]	68,300	$3.2–3.7 \times 10^{-3}$	ND	ND	ND	ND	38
Saccharomyces cerevisiae[i]							
I	400,000	7.4×10^{-4}	ND	8.5	–	ND[j]	39
II	800,000	3.5×10^{-4}	ND	6.8	–	ND[j]	
Pseudomonas geniculata							
A	96,000	1×10^{-3}	3.80	Broad, 9.0	–	–	40
AG[b]	135,000	1.5×10^{-5}	6.97	Broad, 9.5	9	+	
Chlamydomonas[k]	275,000	1.34×10^{-4}	ND	6.9–9.5	–	Limited	41

[a] Determined in most cases against 6C3HED tumor in C3H mice.
[b] Amino acid composition reported.
[c] α_2-Globulin behavior.
[d] Immunologically distinct from *E. coli*.
[e] Affinity chromatography used in purification.[3,22]
[f] There are 3 asparaginases in *Citrobacter freundii* extracts: AG, with substantial glutaminase activity, is very labile, sensitive to mercurials, not protected by dithiothreitol (DTT); B, also sensitive to mercurials, but protected by DTT, $K_m = 1.7 \times 10^{-3}$ *M*; and A (described here), insensitive to mercurials, stable on storage, high affinity for asparagine. Reaction of partial identity to *E. coli B* by Oucherlony double diffusion.
[g] From developing seeds; enzyme present in other tissues.
[h] Active only in the presence of K^+ ion.
[i] I is internal, constitutive; II is secreted in response to nitrogen starvation, biochemically and genetically distinct. II contains mannose, I no carbohydrate.
[j] Earlier negative tests suggested rapid clearance.[42]
[k] A marine microalgae.

E. coli Asparaginase

Other Purifications. The procedure described in an earlier volume of this series for the purification of L-asparaginase from *E. coli B* involved 5 steps from the crude extract, four of them chromatographic or electrophoretic, and gave homogeneous enzyme with a 15% recovery.[2] Other purifications of this enzyme have been described. Arens *et al.*[26] described a simpler procedure suitable for large-scale purification (applied to *E. coli* ATCC 9637 cells from a 3000-liter culture). Their procedure involved heat treatment, 2 solvent fractionations, and a gel filtration step, and gave homogeneous (crystallizable) enzyme with an overall yield of 4.5%. The Lilly group (Ho *et al.*)[25] has also described a 4-step procedure applicable to large-scale purifications, involving $(NH_4)_2SO_4$ and ethanol fractionations, which gives a homogeneous crystalline product with 25% overall yield. Nakamura *et al.*[27] have described a 5-step procedure involving 2 chromatographic steps that yields crystalline enzyme with a 16% yield. Enzyme used clinically in this country is produced by Merck ("Elspar") by a process involving a series of methanol precipitations, two chromatographic steps (DEAE 50 and CM Sephadex), and a final alcohol precipitation.[43] The specific activity of all these preparations is about the same.

[26] A. Arens, E. Rauenbusch, E. Irion, O. Wagner, K. Bauer, and W. Kaufmann, *Hoppe-Seyler's Z. Physiol. Chem.* **351,** 197 (1970).

[27] N. Nakamura, Y. Morikawa, T. Fujio, and M. Tanaka, *Agric. Biol. Chem.* **35,** 219 (1971).

[28] B. Rowley and J. C. Wriston, *Biochem. Biophys. Res. Commun.* **28,** 160 (1967).

[29] R. W. Scheetz, H. A. Whelan, and J. C. Wriston, *Arch. Biochem. Biophys.* **142,** 184 (1971).

[30] T. Tosa, T. Sato, R. Sano, K. Yamamoto, Y. Matuo, and I. Chibata, *Biochim. Biophys. Acta* **334,** 1 (1974); T. Tosa, R. Sano, K. Yamamoto, M. Nakamura and I. Chibata, *Biochemistry* **11,** 217 (1972).

[31] K. A. Cammack, D. I. Marlborough, and D. S. Miller, *Biochem. J.* **126,** 361 (1972).

[32] P. E. Joner, T. Kristianssen, and M. Einarsson, *Biochim. Biophys. Acta* **327,** 146 (1973).

[33] S. A. Gaffar and Y. I. Shethna, *Appl. Environ. Microbiol.* **33,** 508 (1977).

[34] A. D. Resnick and B. Magasanik, *J. Biol. Chem.* **251,** 2772 (1976).

[35] L. Davidson, M. Burkom, S. Ahn, L.-C. Chang, and B. Kitto, *Biochim. Biophys. Acta* **480,** 282 (1977).

[36] P. J. Lea, L. Fowden, and B. J. Miflin, *Phytochemistry* **17,** 217 (1978).

[37] K. S. Chang and K. J. F. Farnden, *Arch. Biochem. Biophys.* **208,** 49 (1981).

[38] L. Sodek, P. J. Lea, and B. J. Miflin, *Plant Physiol.* **65,** 22 (1980).

[39] P. C. Dunlop, G. M. Meyer, D. Ban, and R. J. Roon, *J. Biol. Chem.* **253,** 1297 (1978).

[40] G. B. Kitto, G. Smith, T.-Q. Thiet, M. Mason, and L. Davidson, *J. Bacteriol.* **137,** 204 (1979).

[41] J. H. Paul and K. E. Cooksey, *Can. J. Microbiol.* **25,** 1443 (1979); J. H. Paul, *Biochem. J.* **203,** 109 (1982).

[42] J. D. Broome, *J. Natl. Cancer Inst. (U.S.)* **35,** 967 (1965).

[43] R. F. Hirschmann, personal communication.

Properties. Properties of this enzyme were reviewed earlier[2,4]; only the most salient points from the early work, together with certain recent results, are presented here.

Asparaginase content varies widely in *E. coli* strains, some having no activity at all. There are two asparaginases with very different properties in *E. coli B* and certain other *E. coli* strains.[12,24,44] Only one of these, the periplasmic EC-2 enzyme, has antilymphoma activity. It is this enzyme (and its counterpart from other strains with high levels of asparaginase) that has been used clinically and for most biochemical studies.

The enzyme is a tetramer with identical subunits. Amino acid compositions of several different *E. coli* asparaginases have been reported (see Table I for references) and one enzyme (from *E. coli* A-1-3) completely sequenced.[45] Based on studies with [^{14}C]DONV (which is an inhibitor or an alternate substrate, depending on reaction conditions) the sequence of an active site decapeptide of the *E. coli B* enzyme has been established.[46] Chang *et al.*[47] have studied model compounds relevant to this active site-directed inactivation. There are sequence homologies between *E. coli* asparaginase and glutaminase-asparaginase enzymes from *Acinetobacter* and *Pseudomonas 7A,* but the site labeled by [^{14}C]diazooxonorleucine (DON) in the mixed amidases is different from that labeled by [^{14}C]DONV in *E. coli* asparaginase.[48]

The enzyme readily aggregates to multiples of the native tetramer.[49] It dissociates reversibly in 8 *M* urea, but the question of whether it dissociates to dimers or even monomers in very low concentrations is not resolved.[4,50] The catalytic activity of *E. coli* asparaginase depends on the integrity of its tetrameric structure; the protective effect of α_2-macroglobulin may be due to the fact that it inhibits dissociation of asparaginase in dilute solutions.[51] The enzyme has no free sulfhydryl groups and is not sensitive to pCMB, *N*-ethylmaleimide, and iodoacetic acid[2,4] nor to maleimide.[52] *E. coli* asparaginase is also inactivated by 2,3-butanedione with

[44] H. A. Campbell, L. T. Mashburn, E. A. Boyse, and L. J. Old, *Biochemistry* **6,** 721 (1967).

[45] T. Maita, K. Morokuma, and G. Matsuda, *J. Biochem.* (*Tokyo*) **76,** 1351 (1974); T. Maita and G. Matsuda, *Hoppe-Seyler's Z. Physiol. Chem.* **361,** 105 (1980).

[46] R. G. Peterson, F. F. Richards, and R. E. Handschumacher, *J. Biol. Chem.* **252,** 2072 (1977); R. E. Handschumacher, this series, Vol. 46 [47].

[47] P. K. Chang, L. B. Lachman, and R. E. Handschumacher, *Int. J. Pep. Protein Res.* **14,** 27 (1979).

[48] J. S. Holcenberg, L. Ericsson, and J. Roberts, *Biochemistry* **17,** 411 (1978).

[49] P. P. K. Ho and E. B. Milikin, *Biochim. Biophys. Acta* **206,** 196 (1970).

[50] N. S. Murthy and J. R. Knox, *J. Mol. Biol.* **105,** 567 (1976).

[51] E. Soru, *Mol. Cell. Biochem.* **23,** 185 (1979).

[52] H. A. Milman, D. A. Cooney, and R. G. Cable, *Biochem. Pharmacol.* **27,** 824 (1978).

TABLE II
MODIFICATIONS OF BACTERIAL ASPARAGINASES[a]

Modification	Microorganism	Reference
1. Succinylation	*E. coli*	59, 60
	Erwinia carotovora	61
	E. coli, Erwinia	62
	Acinetobacter[b]	63
2. Glutaraldehyde coupling of glycopeptides	*Acinetobacter*[b]	63
3. Glutaraldehyde cross-linking to mouse albumin	*E. coli*	64
4. Modification with poly-D, L-alanyl peptides	*E coli* *Erwinia*	65
5. Coupling to polyethylene glycols (cyanuric chloride)	*E. coli*	66
6. Glycosylation with sodium cyanoborohydride		67
With lactose, *N*-acetyl-neuraminic acid	*E. coli*	67
With glucuronic acid	*E. coli*	60
7. Coupling to dextran	*Erwinia*	68, 69
8. Modification with imidates	*Erwinia*	69
	E. coli, Erwinia	62
	E. coli	70
9. Miscellaneous		
Carbodiimide coupling of *N,N*-dimethyl propane-diamine	*E. coli*	60
O-Methylisourea	*E. coli, Erwinia*	62
$HCHO/NaBH_4$	*E. coli, Erwinia*	62
Woodward's reagent K	*Erwinia*	69
Maleyl anhydride	*E. coli*	70

[a] See refs. 2, 4, and 58 for reviews.
[b] A glutaminase-asparaginase enzyme.

alteration of one arginine and one histidine residue per subunit.[53] It hydrolyzes L-glutamine and D-asparagine at 4 and 5%, respectively, of the rate at which it hydrolyzes L-asparagine, but is not inhibited by L- or D-aspartic acids nor by L- or D-glutamic acids. β-Cyano-L-alanine is an alternate substrate for both the guinea pig serum and *E. coli* enzymes. A series of

[53] D. Petz, H.-G. Loffler, and F. Schneider, *Z. Naturforsch. C: Biosci.* **34C,** 742 (1979).

X-ray crystallographic studies (with the enzyme from *E. coli* A-1-3 most recently) has shown that the molecule has 222 point group symmetry.[54]

Both of the anti-leukemic asparaginases (*E. coli* and *Erwinia*) are immunosuppressive, and it has been suggested that this may be due to their glutaminase activity (2 and 15%, respectively). Support for this suggestion was provided recently by the observation that the *Vibro succinogenes* enzyme, devoid of glutaminase activity, is not immunosuppressive.[55]

With the exception of a study of the pH dependence of the kinetic parameters of *E. coli* asparaginase,[56] and a detailed study of the rather atypical cell wall asparaginase II of *Saccharomyces cerevisiae,* little work has been done on mechanism since this subject was last reviewed.[57]

Numerous modifications and coupling studies have been carried out, for a variety of purposes: to study factors influencing clearance from the circulation, to modify antigenicity, and to provide clinically more useful forms. Some of these studies were cited in earlier reviews[2,4] of asparaginase and in a review of modification of proteins by glycosylation.[58] Recent papers dealing with modification are listed in Table II.[59–70] Several articles on immobilization of asparaginase appeared in an earlier volume of this series.[7]

[54] A. Itai, M. Yonei, Y. Mitsui, and Y. Iitaka, *J. Mol. Biol.* **105,** 321 (1976); M. Yonei, Y. Mitsui, and Y. Iitaka, *ibid.* **110,** 179 (1977).

[55] D. L. Durden and J. A. Diastasio, *Cancer Res.* **40,** 1125 (1980).

[56] M. H. O'Leary and S. L. Mattes, *Biochim. Biophys. Acta* **522,** 238 (1978).

[57] P. C. Dunlop, G. M. Mayer, and R. J. Roon, *J. Biol. Chem.* **255,** 1542 (1980).

[58] J. D. Aplin and J. C. Wriston, *CRC Crit. Rev. Biochem.* **10,** 259 (1981).

[59] S. Shifrin and B. J. Grochowski, *J. Biol. Chem.* **247,** 1048 (1972).

[60] E. C. Nickle, R. D. Solomon, T. E. Torchia, and J. C. Wriston, *Biochim. Biophys. Acta* **704,** 345 (1982).

[61] S. Shifrin, B. G. Solis, and I. M. Chaiken, *J. Biol. Chem.* **248,** 3464 (1973).

[62] R. Blazek and J. E. Benbough, *Biochim. Biophys. Acta* **677,** 220 (1981).

[63] J. S. Holcenberg, G. Schmer, D. C. Teller, and J. Roberts, *J. Biol. Chem.* **250,** 4165 (1975).

[64] T. Yagura, Y. Kamisaki, H. Wada, and Y. Yamamura, *Int. Arch. Allergy Appl. Immunol.* **64,** 11 (1981).

[65] J. R. Uren and R. C. Ragin, *Cancer Res.* **39,** 1927 (1979).

[66] Y. Ashihara, T. Kono, S. Yamazaki, and Y. Inada, *Biochem. Biophys. Res. Commun.* **83,** 385 12978); Y. Kamisaki, H. Wada, T. Yagura, A. Matsushima, and Y. Inada, *J. Pharmacol. Exp. Ther.* **216,** 410 (1981); A. Matsushima, H. Nishimura, Y. Ashihara, Y. Yokota, and Y. Inada, *Chem. Lett.* **7,** 773 (1980).

[67] J. W. Marsh, J. Denis, and J. C. Wriston, *J. Biol. Chem.* **252,** 7678 (1977).

[68] R. L. Foster and T. Wileman, *J. Pharm. Pharmacol., Suppl.* 37P (1979).

[69] J. E. Benbough, C. N. Wiblin, T. N. A. Rafter, and J. Lee, *Biochem. Pharmacol.* **28,** 833 (1979).

[70] L. E. Hare and R. E. Handschumacher, *Mol. Pharmacol.* **9,** 531 (1973).

Mammalian Asparaginases

The only mammalian asparaginase purified to homogeneity is that from guinea pig serum; its purification and properties were described in an earlier volume of this series,[2] and little work has been done in this area since. A total of 45 species have been examined, but serum asparaginase is found in more than trace amounts only in the guinea pig and other members of the superfamily Cavioidea, and in New World monkeys.[71] A partial purification of asparaginase from agouti serum, which contains several times as many units/ml as the guinea pig, has been reported[72]; but this enzyme has resisted further purification attempts due to its apparent instability.[73]

Rat liver extracts appear to contain a labile, phosphate-dependent asparaginase,[4] but the only liver enzyme studied to any extent is that from guinea pig liver. Suld and Herbut[74] reported that papain digestion of a partially purified guinea pig liver asparaginase converted it to a species which behaved like the serum enzyme on ion-exchange chromatography and polyacrylamide gel electrophoresis; the modified liver enzyme also had tumor inhibitory activity, and gave a reaction of identity in immunodiffusion experiments with rabbit antisera to guinea pig serum asparaginase.

[71] J. C. Wriston, *J. Mol. Evol.* **17,** 1 (1981); *Experientia* **39,** 383 (1983).
[72] V. L. G. Calich and R. Guimaraes, *Rev. Inst. Med. Trop. Sao Paulo* **18** (4), 239 (1976).
[73] H. N. Jayaram, personal communication.
[74] H. M. Suld and P. A. Herbut, *J. Biol. Chem.* **245,** 2797 (1970).

[80] Aspartate Ammonia-lyase

By MASANOBU TOKUSHIGE

Escherichia coli

$$\underset{\text{L-Aspartic acid}}{\mathrm{HOOC{-}\underset{\underset{NH_2}{|}}{CH}{-}CH_2{-}COOH}} \rightleftharpoons \underset{\text{Fumaric acid}}{\mathrm{HOOC{-}CH{=}CH{-}COOH}} + NH_3$$

Aspartate ammonia-lyase (aspartase, EC 4.3.1.1) catalyzes the reversible conversion of L-aspartic acid to fumarate and ammonia. Enzyme synthesis in *Escherichia coli* W cells is subject to catabolite repression by glucose and is suppressed under aerobic conditions.[1]

[1] S. Suzuki, J. Yamaguchi, and M. Tokushige, *Biochim. Biophys. Acta* **321,** 369 (1973).

Copyright © 1985 by Academic Press, Inc.
All rights of reproduction in any form reserved.

Assay Method

Principle. The method routinely employed is based on the spectrophotometric determination of fumarate formed.[1]

Reagents

Tris–HCl buffer, 0.5 *M*, pH 8.8[2]
Sodium L-aspartate, 0.5 *M*
$MgCl_2$, 0.1 *M*
Enzyme dissolved in Buffer A (50 m*M* potassium phosphate buffer, pH 6.8 containing 0.1 *M* KCl, 1 m*M* EDTA, and 5 m*M* 2-mercaptoethanol)

Procedure. The reaction mixture, in a total volume of 1.0 ml, contains 0.2 ml of Tris–HCl buffer, 0.2 ml of sodium L-aspartate, and 0.02 ml of $MgCl_2$. The reaction is initiated at 30° by the addition of the enzyme and the increase in the absorption at 240 nm is determined. The molar extinction coefficient of fumarate, 2530 M^{-1} cm^{-1} at 240 nm reported by Emery[3] is used. One unit of enzyme is defined as the amount producing 1 μmol of fumarate per min under the standard conditions. Specific activity is expressed as units per mg of protein. The protein concentration is determined either by the procedure of Lowry *et al.*[4] or by using an $E_{1\,cm}^{1\%}$ value of 5.9 at 280 nm.[5,6]

Alternative Methods. (1) Ammonium ions released by the deamination reaction of L-aspartate are determined colorimetrically using the Nessler's reagent.[7] (2) L-Aspartate formed is determined by the bioassay method of Henderson and Snell using *Leuconostoc mesenteroides*.[8] (3) Fumarate consumption is determined spectrophotometrically by essentially the same method as the standard procedure, except that a longer wavelength.[1,9] is chosen in order to facilitate the use of high concentrations of fumarate.[10]

[2] Although the activity was assayed at pH 7.4 until recently, pH 8.8 is recommended, since the latter pH is optimum for the enzyme activity.

[3] T. F. Emery, *Biochemistry* **2,** 1041 (1963).

[4] O. H. Lowry, N. J. Rosebrough, A. L. Farr, and R. J. Randall, *J. Biol. Chem.* **193,** 265 (1951).

[5] M. Tokushige, G. Eguchi, and F. Hirata, *Biochim. Biophys. Acta* **480,** 479 (1977).

[6] The method of Lowry *et al.* is used for crude and partially purified enzyme preparations and that using the $E_{1\,cm}^{1\%}$ value for homogeneous enzyme preparations.

[7] The Nessler's method is used for the activity assay in crude preparations. The Conway technique (see ref. 7a) is useful for this purpose in order to minimize the high blank.

[7a] V. R. Williams and D. J. Lartigue, this series, Vol. 13, p. 354.

[8] L. M. Henderson and E. E. Snell, *J. Biol. Chem.* **172,** 15 (1948).

[9] N. Ida and M. Tokushige, *J. Biochem (Tokyo)* **98,** 1985, in press.

[10] The wavelengths of 270 and 293 nm are used in our previous works.

Purification Procedure[11]

Growth of Organism. Escherichia coli W is grown at 30° for 15 hr without aeration in a medium (4–5 liters) containing 0.5% K_2HPO_4, 1% polypeptone (Takeda Chemical Industries, Osaka), 1% yeast extract, and 1% sodium L-aspartate (neutralized) in 5-liter Erlenmeyer flasks.[12] Cells are then harvested by centrifugation. The yield of wet packed cells is about 1.5 g/liter of the medium. All subsequent procedures are carried out at 0–4°, unless otherwise specified.

Step 1. Preparation of Crude Extract. Frozen cells (50 g in wet weight) are suspended in 150 ml of Buffer A and disrupted in 80-ml portions in a sonic disruptor (20 kHz) for 20 min. Cell debris is removed by centrifugation at 10,000 *g* for 20 min.

Step 2. Streptomycin Treatment. To the crude extract (188 ml) is added an equal volume of Buffer A, and 188 ml of 5% (w/v) streptomycin sulfate (pH 7.0) is slowly added while stirring. After about 1 hr of stirring, the precipitate is removed by centrifugation at 10,000 *g* for 20 min. The protein in the supernatant solution (555 ml) is concentrated by adding solid ammonium sulfate (390 g/liter) to attain 60% saturation. After 1 hr of stirring, protein components are collected by centrifugation at 10,000 *g* for 20 min.

Step 3. Heat Treatment. The precipitate is dissolved in 250 ml of Buffer A containing 10% glycerol. The solution is divided into 80-ml portions and heated at 50° for 7 min with gentle shaking. After rapid cooling to 4°, the precipitate formed is removed by centrifugation at 10,000 *g* for 10 min.

Step 4. Ammonium Sulfate Fractionation. Solid ammonium sulfate (196 g/liter) is added while stirring to the supernatant solution from Step 3 to reach 33% saturation. After 1 hr of stirring, resulting precipitate is removed by centrifugation at 10,000 *g* for 15 min. Solid ammonium sulfate (107 g/liter) is then added to this solution to reach 50% saturation. After 1 hr of stirring, the precipitate is collected by centrifugation and dissolved in a minimal volume of Buffer A.

Step 5. Calcium Phosphate Gel Treatment. The solution from Step 4 (27 ml) is dialyzed overnight against 3 liters of 10 m*M* potassium phosphate buffer, pH 6.8 containing 5 m*M* 2-mercaptoethanol. The dialyzed solution is diluted to the protein concentration of 10 mg/ml with the same buffer. The gel is then slowly added to the solution to reach the gel–

[11] An improved method of Suzuki *et al.*[1]

[12] This system is essentially the same as that of the deep-grown culture of *Escherichia coli* used for biodegradative threonine deaminase (Y. Shizuta and M. Tokushige, this series, Vol. 17B, p. 575).

protein ratio of 1.8. The mixture is stirred for 15 min, and the gel is removed by centrifugation. The enzyme in the supernatant solution (192 ml) is precipitated by adding solid ammonium sulfate (390 g/liter) to attain 60% saturation. The precipitate is collected by centrifugation at 10,000 *g* for 15 min and dissolved in a minimal volume of Buffer A.

Step 6. DEAE-Sephadex Column Chromatography. The enzyme solution from Step 5 (10.5 ml, 939 mg of protein) is dialyzed overnight against 2 liters of 50 m*M* Tris–HCl buffer, pH 7.0 containing 5 m*M* 2-mercaptoethanol, 10 m*M* $MgCl_2$, and 100 m*M* KCl, and applied to a column of DEAE-Sephadex A-50 (3.0 × 40 cm), which has previously been equilibrated with the same buffer. After washing the column with 1 liter of the same buffer, elution is carried out with a linear concentration gradient of $MgCl_2$ from 10 to 80 m*M*, 1 liter of each solution. The main peak of the enzyme activity is eluted at a $MgCl_2$ concentration of 35 m*M*. The active fractions (146 ml) are pooled and the protein components are precipitated by adding solid ammonium sulfate (390 g/liter) in the presence of 1 m*M* dithiothreitol to attain 60% saturation. The precipitate is collected by centrifugation at 10,000 *g* for 15 min.

Step 7. Hydroxylapatite Column Chromatography. The precipitate from Step 6 is dissolved in a minimal volume of 10 m*M* potassium phosphate buffer, pH 6.0 containing 5 m*M* 2-mercaptoethanol (7.5 ml, 141 mg of protein) and is dialyzed overnight against 2 liters of the same buffer. Insoluble materials in the dialyzed enzyme solution are removed by centrifugation and the supernatant solution is applied to a hydroxylapatite column (1.6 × 30 cm), which has been equilibrated with the same buffer. After washing the column with 200 ml of the same buffer, elution is carried out with a linear concentration gradient of potassium phosphate from 10 (pH 6.0) to 200 m*M* (pH 6.8) containing 5 m*M* 2-mercaptoethanol, 200 ml of each solution. At this step, most of the colored materials are removed. The main peak of enzyme activity is eluted at a phosphate concentration of 170 m*M*. The active fractions (52 ml) are collected and the protein components are concentrated to 2.5 ml (62 mg of protein) with a membrane filter.

Step 8. Gel Permeation Chromatography. The solution from Step 7 is applied to a Sepharose 6B[13] column (2.0 × 110 cm). Fractionation is carried out with Buffer A. The flow rate is 4 ml/hr. The active fractions (30 ml) with a specific activity of 68.5[14] are pooled and the protein components are concentrated to 1.6 ml (49.0 mg of protein) with a membrane

[13] Recently, Sephacryl S-200 is more effectively used.

[14] The specific activity at pH 8.8 is approximately a little over 200 units/mg of protein, which is about 3 times higher than the value determined at pH 7.4.

TABLE I
PURIFICATION OF ASPARTASE FROM *Escherichia coli* W

Steps	Volume (ml)	Total protein (mg)	Total activity[a] (units)	Specific activity[a] (units/mg protein)	Yield (%)
Crude extract	188.0	7,820	8,700	1.1	100
Streptomycin	37.0	4,240	11,200	2.6	129
Heat treatment	252.0	3,430	9,900	2.9	114
Ammonium sulfate	27.0	1,620	8,000	4.9	92.0
Ca phosphate gel	10.5	939	7,350	7.8	84.5
DEAE-Sephadex	7.5	141	6,520	46.2	74.9
Hydroxylapatite	2.5	62.0	3,900	63.0	44.8
Sepharose 6B	1.6	49.0	3,360	68.5	38.6

[a] The enzyme activity in this table is assayed at pH 7.4.

filter. The enzyme solution thus obtained is stored in an ice bath and used throughout all experiments, unless otherwise specified. A summary of typical purification procedures is shown in Table I.[15] The enzyme obtained is homogeneous as judged from ultracentrifugation and gel electrophoresis.

Properties

Substrate Specificity.[1] The substrate specificity of the enzyme is quite strict and the following structural analogs of L-aspartate at 40 m*M* do not serve as the substrate, when the activity for deamination is examined by the Nessler's reagent: D-aspartate, α-methyl-DL-aspartate, β-methyl-DL-aspartate, DL-*threo*-β-hydroxyaspartate, DL-*erythro*-β-hydroxyaspartate, L-cysteate, L-α-aminobutyrate, L-asparagine, L-alanine, and L-glutamate. D-Aspartate acts as a competitive inhibitor. For the reverse reaction, NH_2OH and hydrazine can replace NH_3. Methylamine does not serve as an amino donor. Mesaconate and maleate do not replace fumarate.

K_m Values.[1,16] Due to the fact that the substrate saturation profiles exhibit complex kinetics including positive or negative cooperativity as a function of pH and other factors, exact K_m values are hardly estimated. The half saturation concentrations ($S_{0.5}$) of L-aspartate at pH 7.4 and 8.5 are 1.0 and 6.0 m*M*, respectively, in the presence of 2 m*M* $MgCl_2$.

[15] Recent results give about a half quantity of the purest enzyme preparation per g cells in wet weight, as compared with those in this table.

[16] N. Yumoto and M. Tokushige, *Biochim. Biophys. Acta* **749,** 101 (1983).

Activators and Inhibitors. Divalent metal ions, such as Mg^{2+} and Mn^{2+} activate the enzyme at the alkaline pH. In the neutral and slightly acidic pH ranges, the divalent metal ions do not affect the activity to an appreciable extent, and the inhibitory effect of EDTA is only partial.[1] Various sulfhydryl reagents such as *p*-hydroxymercuribenzoate and 5,5′-dithiobis(2-nitrobenzoic acid) inactivate the enzyme.[1,17] The enzyme is inactivated by diethylpyrocarbonate due to the modification of the histidyl residues and the inactivated enzyme is reactivated by hydroxylamine.[17a] $NaBH_4$ and carbonyl reagents do not affect the enzyme activity to an appreciable extent. KCl and NaCl affect the enzyme activity in complex manners as a function of pH and the substrate concentration altering not only the activity, but also the cooperativity of the enzyme.[1] Partially or severely inactivated enzyme preparations due to various causes involving denaturation can be reactivated by extensive unfolding and renaturation treatments.[18]

Effect of pH. The enzyme exhibits the maximal activity at pH 8.8 in the presence of 2 m*M* $MgCl_2$, while it shifts to about pH 7.8 in the absence of added metal ions.

Stability. The enzyme is fairly stable in the presence of rather high concentrations of inorganic salts, such as ammonium sulfate, potassium phosphate, and KCl. The optimum pH for the enzyme stability is rather broad (pH 6–9). Thiol compounds, such as 2-mercaptoethanol and dithiothreitol, protect the enzyme against inactivation during storage. Glycerol and other polyols stabilize the enzyme against heat inactivation.[1,19]

Other Characteristics. The results of the sedimentation analysis indicate that the enzyme has a molecular weight of 193,000 and is composed of four subunits of seemingly identical molecular weight (M_r = 48,500).[1] Our recent study on the cloning and DNA sequencing of the aspartase gene *aspA* of *E. coli* W shows that the enzyme contains 477 amino acid residues and has a molecular weight of 52,224.[19a] The amino- and carboxyl-termini of the subunits are serine and glutamine, respectively.[20] $s^0_{20,w}$ is 9.3 S. The molecular ellipticity of the enzyme protein at 222 nm is

[17] K. Mizuta and M. Tokushige, *Biochim. Biophys. Acta* **403,** 221 (1975).

[17a] N. Ida and M. Tokushige, *J. Biochem. (Tokyo)* **96,** 1315 (1984).

[18] M. Tokushige and G. Eguchi, *Biochim. Biophys. Acta* **522,** 243 (1978).

[19] Glycerol and some other polyols have also an activating effect [M. Tokushige and K. Mizuta, *Biochem. Biophys. Res. Commun.* **68,** 1082 (1976)].

[19a] J. S. Takagi, N. Ida, M. Tokushige, H. Sakamoto, and Y. Shimura, in preparation.

[20] N. Yumoto, K. Mizuta, M. Tokushige, and R. Hayashi, *Physiol. Chem. Phys.* **14,** 391 (1982).

[21] N. Yumoto, M. Tokushige, and R. Hayashi, *Biochim. Biophys. Acta* **616,** 319 (1980).

−29,300.[21] The tetrameric structure of the enzyme consists of two pairs of dimers, in which the two pairs of rod-shaped subunits meet perpendicularly, being typical of D_2 symmetry, as revealed by negative-stain electron microscopy and the simulation analysis of the cross-linking experiments using bifunctional reagents.[22] Aspartic β-semialdehyde inactivates the enzyme as an active-site directed (affinity labeling) agent.[23] The enzyme is 3- to 5-fold activated upon limited proteolysis with trypsin, subtilisin BPN′, and several other proteases concomitant with a release of COOH-terminal peptides.[20,21,24,25] The enzyme is also 2-fold activated by acetylation with acetic anhydride or *N*-hydroxysuccinimide acetate.[16] L-Aspartate acts not only as a substrate, but also as an activator, when the reaction proceeds using fumarate and ammonia as the substrates. A similar activation by L-aspartate is observed, when fumarate and hydroxylamine are used as the substrates also. Besides L-aspartate, α-methyl-DL-aspartate acts as an activator without serving as the substrate.[9]

Pseudomonas fluorescens

Aspartase is highly purified and crystallized from the organism as follows.[26,27]

Assay Method

The enzyme activity is determined by essentially the same method as that for the *E. coli* enzyme, except that the standard assay mixture contains 3 m*M* $MgCl_2$ and 0.1 *M* Tris–HCl buffer, pH 9.0.

Purification Procedure

Growth of Organism. Pseudomonas fluorescens (IFO 3081) cells are grown for 24 hr at 27° with shaking in 1 liter of the medium containing 0.5% sodium L-glutamate, 0.5% sodium L-aspartate, 0.2% yeast extract, 0.1% K_2HPO_4, 0.25% NaCl, and 0.025% $MgSO_4 \cdot 7H_2O$ in 5-liter Erlenmeyer flasks. Cells are harvested by centrifugation at room temperature.

[22] Y. Watanabe, M. Iwakura, M. Tokushige, and G. Eguchi, *Biochim. Biophys. Acta* **661,** 261 (1981).

[23] N. Yumoto, M. Okada, and M. Tokushige, *Biochem. Biophys. Res. Commun.* **104,** 859 (1982).

[24] K. Mizuta and M. Tokushige, *Biochem. Biophys. Res. Commun.* **67,** 741 (1975).

[25] K. Mizuta and M. Tokushige, *Biochim. Biophys. Acta* **452,** 253 (1976).

[26] M. Tokushige, K. Miyamoto, and H. Katsuki, *J. Biochem. (Tokyo)* **85,** 1415 (1979).

[27] J. S. Takagi, R. Fukunaga, M. Tokushige, and H. Katsuki, *J. Biochem. (Tokyo)* **96,** 545 (1984).

The yield of wet packed cells is about 5 g/liter of the medium. The harvested cells are washed with chilled 0.1 *M* potassium phosphate buffer, pH 7.2, and kept frozen until use.[26] All subsequent procedures are carried out at 0–4°, unless otherwise specified.

Step 1. Preparation of Crude Extract. Frozen cells (92.5 g in wet weight) are suspended in 278 ml of Buffer A and disrupted in 100-ml portions in a sonic disruptor (20 kHz) for 10 min. Cell debris is removed by centrifugation at 12,000 *g* for 30 min and the supernatant solution is stored. The precipitate is again suspended in 92.5 ml of the buffer, disrupted as above and centrifuged at 20,000 *g* for 30 min. The supernatant solutions from the first and second centrifugations are combined.

Step 2. Streptomycin Treatment. To the crude extract (400 ml) is slowly added an equal volume of 5% (w/v) streptomycin sulfate (pH 7.0) while stirring. After 1 hr of stirring, the precipitate is removed by centrifugation at 10,000 *g* for 20 min. To the resulting supernatant solution (790 ml) is added 247 g of solid ammonium sulfate (50% saturation). After 1 hr of stirring, the precipitate is collected by centrifugation at 10,000 *g* for 20 min.

Step 3. Heat Treatment. The precipitate from Step 2 is dissolved in 400 ml of Buffer A. The solution is divided into 50-ml portions and heated at 55° in 500-ml Erlenmeyer flasks for 10 min with gentle shaking. After rapid cooling to 4°, the precipitate formed is removed by centrifugation at 35,000 *g* for 30 min.

Step 4. Ammonium Sulfate Fractionation. Solid ammonium sulfate (209 g/liter) is added with a constant stirring to the supernatant solution from Step 3 to attain 35% saturation. After stirring for 1 hr, the resulting precipitate is collected by centrifugation at 10,000 *g* for 20 min and is dissolved in 100 ml of Buffer B (50 m*M* potassium phosphate buffer, pH 7.2 containing 1 m*M* EDTA and 5 m*M* 2-mercaptoethanol).

Step 5. DEAE-Sephadex Column Chromatography. The enzyme solution from Step 4 is dialyzed overnight against 3 liters of Buffer B. After the dialysis, solid materials in the sample solution are removed by centrifugation and the enzyme solution obtained is applied to a column of DEAE-Sephadex A-50 (3.2 × 55 cm), which has been equilibrated with Buffer B. After washing the column with 600 ml of the same buffer, elution is carried out with a linear concentration gradient of potassium phosphate from 50 to 400 m*M*, 1 liter of each solution. The active enzyme is eluted at a phosphate concentration of 200 m*M*. The active fractions (175 ml) are pooled and concentrated to 4 ml with a membrane filter.

Step 6. Sephacryl S-200 Column Chromatography. The enzyme solution from Step 5 is applied to a column of Sephacryl S-200 (2.0 × 120 cm), which has been equilibrated with Buffer A. Elution is carried out with the

TABLE II
PURIFICATION OF ASPARTASE FROM *Pseudonomas fluorescens*

Steps	Volume (ml)	Total protein (mg)	Total activity (units)	Specific activity (units/mg protein)	Yield (%)
Crude extract	400	7,800	20,364	2.6	100
Streptomycin	400	4,520	23,020	5.1	113
Heat treatment	345	1,794	18,573	10.4	91.2
Ammonium sulfate	125	897.5	16,551	18.4	81.3
DEAE-Sephadex	175	101.5	14,000	138	68.7
Sephacryl S-200	18.5	41.6	9,517	229	46.7

same buffer. The active fractions (18.5 ml) with a constant specific activity of 229 are pooled and concentrated with a membrane filter. The enzyme thus obtained is homogeneous as judged from ultracentrifugation and gel electrophoresis.

Step 7. Crystallization. The following procedures are based on the method of Jakoby.[28] Powdered ammonium sulfate (243 g/liter) is added to the enzyme solution in Buffer A (protein concentration, 1–10 mg/ml) to attain 40% saturation and the suspension is allowed to stand for 1 hr. The enzyme is then collected by centrifugation at 16,000 *g* for 15 min. The precipitated enzyme is dissolved in Buffer A containing ammonium sulfate (20% saturation) to attain a protein concentration of 10 mg/ml. This solution is left overnight at room temperature and small needle-shaped crystals appear. The yield of the crystals is about 44%. The specific activity of the enzyme remains constant after repeated crystallization procedures. So far as the pH of the ammonium sulfate solution is kept at 6.8, crystallization is readily reproducible without addition of preformed crystals as nuclei. A summary of typical purification procedures is shown in Table II.

Properties

Kinetic Properties. The K_m value for L-aspartate at pH 9.0 is 2.4 m*M* in the presence of 3 m*M* $MgCl_2$. The enzyme exhibits cooperative kinetics in its substrate saturation profiles depending upon pH, divalent metal concentration, and some other salt concentrations. Similarly to the *E. coli* enzyme, pH dependency of the activity and divalent metal ion-dependent activation of the enzyme activity are observed.

[28] W. B. Jakoby, *Anal. Biochem.* **26,** 295 (1968).

Other Characteristics. The purified enzyme sediments as a monodisperse entity upon ultracentrifugation with an $s^0_{20,w}$ value of 8.6 S. Upon polyacrylamide gel electrophoresis, the enzyme migrates as a single band. The enzyme has a molecular weight of 173,000, as determined by sedimentation equilibrium analysis and that of the subunit is 50,000, as determined by SDS-gel electrophoresis. Cross-linking experiments with dimethyl suberimidate indicate that the enzyme is composed of four subunits. Unlike aspartase of *E. coli* W, the enzyme of *P. fluorescens* is not activated by a limited proteolysis.

Other Organisms

Homogeneous enzymes are obtained from *Escherichia coli* B by Rudolph and Fromm[29] and also by Masaki *et al.*[30] According to Rudolph and Fromm, the enzyme requires divalent metal ions for the activity. The molecular weight of the native enzyme is 170,000, as determined by Sephadex gel permeation chromatography and that of the subunit is 45,000. The *s* value is 8.3 S. Aspartases are partially purified from *Bacterium cadaveris* and from *Pasteurella pestis* by Williams and Lartigue[7a,31] and by Korobeinik and Domaradskii,[32] respectively. The reaction mechanism of the enzyme from *Hafnia alvei* is studied in detail and a two-step mechanism with an intermediate carbanion in which C–N bond cleavage limits the overall rate is proposed.[33]

[29] F. B. Rudolph and H. J. Fromm *Arch. Biochem. Biophys.* **147,** 92 (1971).
[30] Y. Masaki, C. Sato, and T. Tosa, *J. Agric. Chem. Soc. Jpn.* **57,** 1293 (1983).
[31] V. R. Williams and D. J. Lartigue, *J. Biol. Chem.* **242,** 2973 (1967).
[32] N. V. Korobeinik and I. V. Domaradskii, *Biokhimiya* **33,** 1128 (1968).
[33] I. I. Nuiry, J. D. Hermes, P. M. Weiss, C-Y Chen, and P. F. Cook, *Biochemistry* **23,** 5168 (1984).

[81] Aspartate Transcarbamylase from *Bacillus subtilis*[1]

By JOHN S. BRABSON, MICHAEL R. MAURIZI, and ROBERT L. SWITZER

L-Aspartate + carbamyl phosphate → *N*-carbamyl-L-aspartate + P_i + H^+

The aspartate transcarbamylases (ATCases) of various bacterial species differ markedly in their subunit structures and patterns of allosteric

[1] This research was supported by U.S. Public Health Service Grant AI 11121 from the National Institute of Allergy and Infectious Diseases. We acknowledge Simon Rosenzweig for valuable technical assistance.

Copyright © 1985 by Academic Press, Inc.
All rights of reproduction in any form reserved.

regulation by nucleotides.[2] ATCase from *B. subtilis* is an example of those ATCases that contain only catalytic subunits and are insensitive to inhibition by pyrimidine nucleotides.[3] This ATCase is also of special interest because it is subject to selective, energy-dependent degradation in starving *B. subtilis* cells.[4,5]

Assay Methods

For most purposes the enzyme was assayed by the colorimetric determination of carbamyl aspartate using the procedure of Prescott and Jones,[6] as modified by Shindler and Prescott.[7]

Reagents. The standard assay mixture contained 0.2 *M* Tris-acetate (pH 8.2), 50 m*M* L-aspartate (pH 8.2), 10 m*M* carbamyl phosphate (freshly dissolved), and enzyme (0.1 to 1 μg of purified enzyme) to yield a final volume of 1 ml. The color development reagents consisted of 4 g/liter of antipyrine in 40% (v/v) H_2SO_4 and a freshly prepared solution of 0.5 g diacetylmonoxime in 100 ml of 5% (v/v) acetic acid.

Procedure. ATCase assays were incubated at 30° for a fixed time, usually 20 min; reaction was initiated with either enzyme or carbamyl phosphate. Reactions were terminated by addition of 1.0 ml of 5% (v/v) $HClO_4$ and placing the tubes on ice for 15 min or longer. Any precipitate that formed was removed by centrifugation in a clinical centrifuge. Appropriate samples of the assay solutions (usually 200 μl containing 20 to 150 nmol of carbamyl aspartate) were diluted to 1.0 ml with water and mixed with 1.0 ml antipyrine reagent and 0.5 ml diacetylmonoxime solution. A series of standards containing 10 to 150 nmol of carbamyl aspartate in 1.0 ml water were assayed in the same way. Color development was at 60° for 2 hr under yellow light.[7]

Comments and Alternative Procedures. The procedure described includes the refinements of Shindler and Prescott.[7] The Prescott-Jones[6] procedure reported earlier is also quite satisfactory. Acetate and other carboxylic acids activate *B. subtilis* ATCase.[3] The activity obtained with 0.2 *M* Tris–HCl buffer (pH 8.2) will be about 40% of that obtained with Tris acetate. Kinetic studies with very dilute solutions of the purified enzyme included 2 mg bovine serum albumin per ml.

[2] M. R. Bethell and M. E. Jones, *Arch. Biochem. Biophys.* **104,** 438 (1969).

[3] J. S. Brabson and R. L. Switzer, *J. Biol. Chem.* **250,** 8664 (1975).

[4] L. M. Waindle and R. L. Switzer, *J. Bacteriol.* **114,** 517 (1973).

[5] M. R. Maurizi, J. S. Brabson, and R. L. Switzer, *J. Biol. Chem.* **253,** 5585 (1978).

[6] L. M. Prescott and M. E. Jones, *Anal. Biochem.* **32,** 408 (1969).

[7] D. B. Shindler and L. M. Prescott, *Anal. Biochem.* **97,** 421 (1979).

For studies with crude extracts that contained or developed significant yellow color in the colorimetric assay and for kinetic studies with subsaturating concentrations of substrates more reliable assays were obtained using the [^{14}C]aspartate assay of Porter *et al.*[8] or the [^{14}C]carbamyl phosphate assay described by Davies *et al.*[9]

Purification Procedure[3,10]

All procedures were conducted at 2–4° unless stated otherwise.

Growth of Cells. ATCase was purified from a uracil-requiring mutant of *B. subtilis* 168 (also Trp^-) grown in a 200 liter fermenter on supplemented nutrient broth containing limiting pyrimidines to derepress the enzyme. The cells were grown at 37° with vigorous aeration. The growth medium contained the following per liter: 9 g nutrient broth, 5 g glucose (autoclaved separately), 7 g K_2HPO_4, 3 g KH_2PO_4, 3 mg $FeSO_4 \cdot 7H_2O$, 4 mg $MnCl_2 \cdot 4H_2O$, 250 mg $MgSO_4 \cdot 7H_2O$, 147 mg $CaCl_2 \cdot 2H_2O$, 7 mg uracil, and 40 mg L-tryptophan. A less expensive medium substituting 4.5 g nutrient broth, 10 mg uracil, and 10 mg L-tryptophan per liter also gave satisfactory results. The cells were harvested about 3.5 hr after the end of exponential growth and stored frozen.

Cell Extraction. Frozen cell paste (1200 g) was suspended in 2 liters of warm (37°) 0.1 *M* Tris–HCl buffer (pH 8.6) containing 1 m*M* EDTA by blending in three 400 g batches in a Waring blender. The suspension was made 1 m*M* in phenylmethylsulfonylfluoride (PMSF) by adding a freshly prepared solution of 530 mg PMSF in 30 ml absolute ethanol. Lysozyme [300 mg dissolved in 20 ml 0.1 *M* Tris–HCl (pH 8.6)] and DNase I (about 2 mg) were stirred into the suspension, which was then incubated at 37° for 30 min. The suspension was cooled to $\leq$10° and cell rupture was completed by two passages through a Manton-Gualin mill at 6000 to 8000 psi. The solution was cooled after each passage by allowing it to flow through an ice-chilled 24 ft long × 3/8 in. stainless-steel cooling coil. (The incubation at 37° with lysozyme and DNase has been omitted without seriously reducing cell rupture, but it is necessary to add DNase to such suspensions to reduce their viscosity.)

Protamine Sulfate Precipitation. Protamine sulfate was added to the crude extract at a final concentration of 0.4% (w/v) by allowing a 2% solution [15 g dissolved in 300 ml 0.1 *M* Tris–HCl (pH 8.6)] to drip into the cold solution with stirring over 20 min. Stirring was continued for another

[8] R. W. Porter, M. O. Modebe, and G. R. Stark, *J. Biol. Chem.* **244,** 1846 (1969).
[9] G. E. Davies, T. C. Vanaman, and G. R. Stark, *J. Biol. Chem.* **245,** 1175 (1970).
[10] M. R. Maurizi, Ph.D. Thesis, University of Illinois, Urbana (1978).

10 min, then the voluminous tan precipitate was removed by centrifuging for 1 hr at 23,000 *g*.

Heat Treatment. The supernatant fluid from the protamine sulfate precipitation was pooled and brought to 0.1 *M* in K_3 citrate by dissolving 32.5 g of the solid salt per liter of fluid. This solution was heated to 65° by placing 500 ml lots in 1-liter flasks in a 70° water bath. The solution was incubated with occasional swirling at 65° for 30 min. Then it was cooled by pouring into stainless-steel centrifuge cans in an ice bath. A substantial precipitate was removed by centrifugation for 1 hr at 23,000 *g* at 4°.

Ammonium Sulfate Fractionation. The supernatant fluid from the heat treatment was cooled in an ice bath and brought to 45% saturation with $(NH_4)_2SO_4$ (277 g/liter). The $(NH_4)_2SO_4$ was added over 2 hr with continuous slow stirring. The solution was then removed from the ice bath and allowed to stand at room temperature for 30 min. The precipitate was removed by centrifugation at 23,000 *g* for 1 hr. The precipitate was discarded and the $(NH_4)_2SO_4$ concentration of the supernatant fraction was brought to 70% saturation by addition of 171 g/liter of $(NH_4)_2SO_4$ as described above. The precipitate was collected by centrifugation as above and dissolved in a minimal volume (100 to 150 ml) of 50 m*M* Tris–HCl (pH 8.1).

Sephadex G-150 Chromatography. The solution from the previous step was allowed to flow slowly onto a large (10 × 75 cm) column of Sephadex G-150, which had been equilibrated with 50 m*M* Tris–HCl (pH 8.1) containing 0.2 *M* KCl. The column was developed with the same buffer, maintaining a hydrostatic head of about 20 cm of fluid and a flow rate of about 120 ml/hr. Fractions of 15 ml each were collected. Fractions containing ATCase activity were located by assay, and the central fractions containing about 85% of the total ATCase activity (approximately fraction numbers 110 to 140) were pooled.

DEAE-Sephadex Chromatography. The pooled ATCase fractions from Sephadex chromatography (about 300 ml) were made 10 m*M* in 2-mercaptoethanol by addition of 71 μl/100 ml solution. This solution was pumped onto a 5 × 30 cm column of DEAE-Sephadex A-50, which had been previously equilibrated with 50 m*M* Tris–HCl (pH 8.1). The conductivity of the sample loaded and the equilibrating buffer should be 10 to 12 mmho at 4°. The column was then developed with 300 ml of 50 m*M* Tris–HCl (pH 8.1) containing 0.25 *M* KCl and 10 m*M* 2-mercaptoethanol, followed by a linear gradient formed from 300 ml of 50 m*M* Tris–HCl (pH 8.1) containing 0.25 *M* KCl and 10 m*M* 2-mercaptoethanol and 300 ml of the same buffer containing 0.35 *M* KCl. Fractions of 3 ml each were collected and the central fractions containing about 85% of the total ATCase activity (usually fraction numbers 120 to 150) were pooled.

Second Heat Treatment. K_3 citrate was added to the pooled DEAE-cellulose fractions to 0.1 *M* (3.5 g/100 ml), and the solution was brought to 65° in a hot water bath. After 30 min at 65°, the solution was cooled on ice and centrifuged at 30,000 *g* for 20 min. The supernatant fluid was concentrated by pressure dialysis in an Amicon pressure cell (PM-10 filter) to a final volume of about 10 ml. Any precipitate that formed was removed by centrifugation.

Sephadex G-150 Superfine Chromatography. The concentrated solution from the previous step was pumped onto a 3.5 cm × 25 cm column of Sephadex G-150 (Superfine). The equilibrating and developing buffer for this column was 50 m*M* Tris–HCl (pH 8.1). Fractions of 4 to 5 ml each were collected at a flow rate of about 20 ml/hr. All the fractions containing 10% or more of the peak ATCase activity were pooled (normally fraction numbers 50 to 70).

DEAE-Cellulose Chromatography. The pooled ATCase fractions from the previous step were pumped onto a 2.5 × 30 cm column of DEAE-cellulose (Whatman DE-52, microgranular, which gave more consistent purification than other kinds of DEAE-cellulose or DEAE-Sephadex). This column had been equilibrated with 50 m*M* Tris–HCl (pH 8.1). The column was washed with 150 ml of the same buffer containing 0.1 *M* KCl, then eluted with a linear gradient formed from 200 ml each of the same buffer containing 0.1 *M* KCl and 0.3 *M* KCl. After the gradient development was completed, the column was washed with 200 ml of buffer containing 0.3 *M* KCl. A major purpose of this step was to separate ATCase from a persistent carbohydrate contaminant, which eluted from DEAE-cellulose after the ATCase. Fractions of 4 to 5 ml each containing ATCase, but no carbohydrate (usually numbers 70 to 90) were pooled and dialyzed against a 50 m*M* Tris–HCl–50 m*M* K_3 citrate (pH 7.5) buffer. Fractions containing carbohydrate were identified with an anthrone test.[11]

DEAE-Sephadex Chromatography with Citrate Gradient Elution. For most purposes, ATCase of suitable purity was obtained by pooling the central, carbohydrate-free fraction from the previous chromatographic purification step. However, some minor contaminants could be removed by the following procedure. The dialyzed pooled fractions from the previous step were applied to a 2 × 7 cm column of DEAE-Sephadex A-50, which had been equilibrated with 50 m*M* Tris–HCl–50 m*M* K_3 citrate (pH 7.5). The column was washed with 50 ml of 50 m*M* Tris–HCl–0.1 *M* K_3 citrate (pH 7.5), then developed with a linear gradient made from 150 ml each of 50 m*M* Tris–HCl (pH 7.5) containing 0.1 and 0.2 *M* K_3 citrate. Fractions of 3 ml each were collected at a flow rate of 20 ml/hr. All

[11] R. G. Spiro, this series, Vol. 8, p. 4.

PURIFICATION OF *B. subtilis* ASPARTATE TRANSCARBAMYLASE

Fraction	Volume (ml)	Protein[a] (mg/ml)	Total activity (μmol/min)	Specific activity (μmol/min/mg)	Recovery (%)
Crude extract (from 1200 g cells)	3300	52	200,000	1.1	100
Protamine sulfate	2900	44	175,000	1.4	88
Heat treatment	2350	16	130,000	3.4	65
$(NH_4)_2SO_4$ fractionation	200	65	110,000	8.5	55
Sephadex G-150	500	4	100,000	50	50
DEAE-Sephadex	180	4.5	65,000	80	33
Second heat treatment	175	2.4	70,000	165	35
Sephadex G-150	76	2.8	67,000	310	34
DEAE-cellulose[b]	40	1.4	32,000	560	16

[a] Determined by the method of O. H. Lowry, N. J. Rosebrough, A. L. Farr, and R. J. Randall, *J. Biol. Chem.* **193,** 265 (1951) using bovine serum albumin as a protein standard.

[b] The apparent yield on this step was low because only selected fractions were pooled in order to maximize the purity of the final preparation.

operations during this step were conducted at room temperature. The ATCase emerged between fraction 15 and fraction 40, after the gradient elution was begun. ATCase-containing fractions were pooled and could be concentrated by pressure dialysis and dialyzed for use as desired.

Crystallization. ATCase from either of the last two purification steps could be crystallized, precipitating about 4 mg of enzyme in 50 m*M* Tris–HCl (pH 8.1) with $(NH_4)_2SO_4$ at 70% of saturation. The precipitate was extracted with 1 ml of 53% saturated $(NH_4)_2SO_4$ (350 mg/ml) at 0° with constant gentle stirring for 5 min. Undissolved ATCase was removed by centrifugation at 5000 *g* for 10 min at 4°. The supernatant solution was transferred to a clean tube, covered, and left at room temperature for 12 to 24 hr. The crystals[10] were thin plates, measuring up to 150 μm long. The specific activity or electrophoretic purity of crystallized ATCase was not significantly different from material that had not been crystallized.

The results of a typical purification of *B. subtilis* ATCase are shown in the table.

Properties

Stability and Storage. Except in very dilute solutions and at pH values below 6, ATCase is quite stable at 0 to 25°. The enzyme is stabilized to thermal denaturation at 65 to 70° by carboxylic acids (acetate, citrate) and high ionic strength (0.2 *M* KCl) and is protected from proteolysis in the presence of carbamyl phosphate. The enzyme can be stored in frozen solutions at −10 to −20° for many months without losses of activity.

Molecular Properties.[3] *B. subtilis* ATCase is a trimeric protein (M_r = 100,000) assembled from identical subunits (M_r = 33,500 ± 1000). The enzyme contains no regulatory subunits and shows no evidence of interaction with isolated regulatory subunits from the *Escherichia coli* ATCase. The sedimentation coefficient for the native enzyme ($s_{20,w}$) is 5.4 S, and the diffusion coefficient under the same conditions is 4.7×10^{-7} cm^2/sec. An extinction coefficient of 5.7 at 280 nm for a 1.0% solution was determined from dry weight measurements. A partial specific volume of 0.729 ml/g was calculated from the amino acid composition. The amino acid composition reveals some similarity to that of the *E. coli* catalytic subunit, but the *B. subtilis* enzyme has a higher content of acidic residues (with fewer glycines and leucines). The isoelectric point is 5.0. Amino acid analysis of ATCase hydrolyzed in the presence of dimethyl sulfoxide yielded three cysteic acids per subunit. Since titration of the completely denatured ATCase with dithio*bis*nitrobenzoic acid indicated only one free sulfhydryl group per subunit, the cysteine residues of the enzyme appear to be in the form of one disulfide and one free sulfhydryl group.

Microheterogeneity.[10] Many preparations of ATCase were heterogeneous when they were analyzed by electrophoresis on sodium dodecyl sulfate-containing polyacrylamide slab gels poured to form a linear gradient of 15% acrylamide monomer (at the bottom) to 7.5% acrylamide monomer[12] (Fig. 1). Two very closely spaced protein bands were resolved under these conditions but not by other electrophoretic procedures. This microheterogeneity is the consequence of limited proteolytic cleavage of the native ATCase during purification, which releases a fragment or fragments comprising some 10 to 15 amino acid residues, probably from the carboxyl terminus. Evidence for this conclusion comes from several observations (Fig. 1). (1) Different preparations of ATCase with similar specific activities contained very different proportions of the two bands. (2) Both bands were precipitable by anti-ATCase antibody. (3) When cells were extracted under conditions that minimize proteolysis, only a single, larger band was present. (4) The relative amount of the smaller band tended to increase during purification. (5) The ATCase could be converted completely to material migrating as the smaller band by incubation with cell-free culture fluid from stationary phase cells (i.e., by an extracellular peptidase).

The purification procedure described in this article was modified to

[12] U. K. Laemmli, *Nature (London)* **227,** 680 (1970).

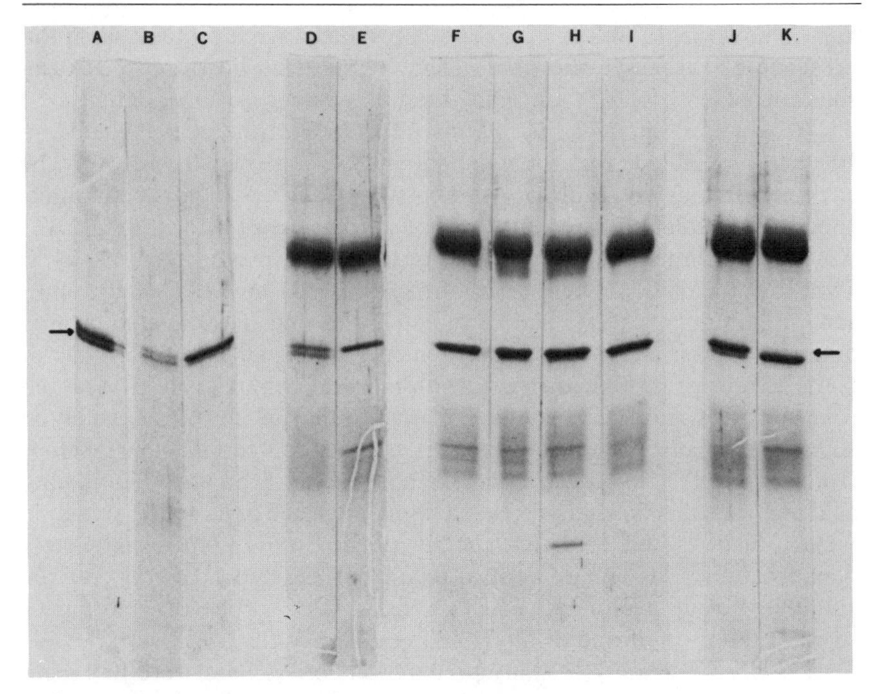

FIG. 1. Microheterogeneity of *B. subtilis* ATCase. All samples were analyzed by sodium dodecyl sulfate polyacrylamide slab gel electrophoresis on a 7.5 to 15% acrylamide gradient using the buffer system of Laemmli.[12] (A–C) Three different preparations of highly purified ATCase, showing variable extent of microheterogeneity. Lanes D–K were analyzed after immunoprecipitation with specific anti-ATCase antibodies. The broad bands above and below the sharp ATCase bands (arrows) are the heavy and light chains of IgG, respectively. (D) Immunoprecipitate of purified ATCase. (E) Immunoprecipitate of ATCase from a crude *B. subtilis* extract treated to prevent proteolysis. (F–I) Immunoprecipitates of ATCase from the crude extract, heat treatment, first Sephadex G-150, and second heat treatment steps of the purification procedure. Microheterogeneity appears to increase during purification. (J and K) Purified ATCase before (J) and after (K) incubation at 37° for 2 hr with cell-free culture fluid from stationary phase *B. subtilis*.

minimize the conditions that lead to microheterogeneity of the enzyme, but its occurrence in a given preparation cannot be ruled out without appropriate electrophoretic analysis.

ATCase which had been converted to the "nicked" form by treatment with extracellular fluid had the same ATCase specific activity as the native enzyme and reacted equally well with anti-ATCase antibody. The amino terminus of both the "nicked" and native ATCase was methionine. Native and "nicked" ATCase have the same susceptibility to proteolysis *in vitro*. No evidence has been found that "nicking" occurs *in vivo* or is involved in the intracellular degradation of ATCase in *B. subtilis*.[10]

Catalytic Properties.[3] The pH optimum for *B. subtilis* ATCase activity is 8.5. Steady state kinetic analysis indicates a sequential mechanism. The Michealis constants for L-aspartate and carbamyl phosphate are 7.0 ± 0.6 and 0.11 ± 0.01 m*M*, respectively. The enzyme is activated 2- to 6-fold by various anions, including acetate, phosphate, citrate, and lactate. Even at subsaturating substrate concentrations the activity of *B. subtilis* ATCase is not affected by common pyrimidine or purine nucleotides (tested at 5 m*M*).

B. subtilis ATCase is strongly inhibited by *N*-(phosphonoacetyl)-L-aspartate (PALA).[13] Inhibition is competitive with respect to carbamyl phosphate ($K_i = 6 \times 10^{-9}$ *M*) and noncompetitive with respect to aspartate.[14] Binding of PALA to the enzyme induces an ultraviolet difference spectrum with maxima at 277, 287, and 294 nm.[14] These interactions of PALA with *B. subtilis* ATCase are very similar to those described for its interaction with the isolated catalytic subunits of the *E. coli* ATCase[13] and suggest strong mechanistic similarity between the two enzymes.

[13] K. D. Collins and G. R. Stark, *J. Biol. Chem.* **246,** 6599 (1971).
[14] J. D. Brofman, B.S. Thesis, University of Illinois, Urbana (1979).

Section V

α-Aminoadipate

[82] α-Aminoadipate

By RICHARD A. HARTLINE

α-Aminoadipate (2-aminohexandioate) (**I**) is the six carbon homolog of glutamate. It was first isolated from a biological system, *Cholera vibrio,* in

$$^{-}OOC-CH_2-CH_2-CH_2-CHN^{+}H_3-COO^{-}$$

α - Aminoadipate

[2 - Aminohexanedioate]

$C_6H_{10}O_4N$, mol. wt. 160.16

[I]

1946[1] and is found in the free state in many plants,[2,3] in a variety of frog embryos,[4] and as a precursor to the pyridone ring of mimosine.[5] The L-isomer is an intermediate in the metabolism of lysine[6] and L-hydroxylysine[7] and a component of the tripeptide precursor to the penicillin and cephalosporin natural β-lactam antibiotics.[8] D-α-Aminoadipate is a constituent of some natural β-lactam antibiotics[8] and is an antogonist of L-glutamate and L-aspartate mediated synaptic transmissions.[9] α-Aminoadipate is normally found in human urine, 5–13 mg/24 hr for males and 0–13 mg/24 hr for females,[10] but α-aminoadipic acidurias have been reported.[11–14] α-Aminoadipate accumulates in the rat following injection of

[1] J. Blass and M. Macheboeuf, *Helv. Chim. Acta* **29,** 1315 (1946).

[2] A.-M. Berg, S. Kari, M. Alfthan, and A. I. Vitanen, *Acta. Chem. Scand.* **8,** 358 (1954).

[3] K. Takanari, U. Mariko, S. Sadao, and S. Kazukibo, *Nippon Nogei Kagaku Kaishi* **47,** 583 (1973).

[4] P. S. Chen, *Int. J. Biochem.* **1,** 669 (1970).

[5] E. P. Tiuari, W. R. Penrose, and I. D. Spenser, *Phytochemistry* **6,** 1245 (1967).

[6] V. W. Rodwell, *in* "Metabolic Pathways" (D. M. Greenberg, ed.), 3rd ed., p. 223.

[7] R. A. Hiles, K. Triebwasser, C. K. Triebwasser, and L. M. Henderson, *Biochem. Biophys. Res. Commun.* **41,** 667 (1970).

[8] S. W. Queener and N. Neuss, *in* "The Chemistry and Biology of β-Lactam Antibiotics" (R. B. Morin, and M. Gorman, eds.), Vol. 3, p. 1. Academic Press, New York, 1982.

[9] H. McLennan and J. G. Hall, *Brain Res.* **149,** 541 (1978).

[10] R. G. Westall, *in* "Amino Acid Pools, Distribution, Formation and Function of Free Amino Acids" (J. T. Holden, ed.), Chapter VI, Am. Elsevier, New York, 1962.

[11] A. J. Manders, C. G. von Oostrom, J. M. Trijbels, F. J. Rutten, and W. J. Kleijer, *Eur. J. Pediatr.* **136,** 51 (1981).

[12] H. Przyrembel, D. Bachmann, and I. Lombeck, *Clin. Chim. Acta* **58,** 257 (1975).

[13] R. W. Wilson, G. M. Wilson, S. C. Gates, and J. V. Higgins, *Pediatr. Res.* **9,** 522 (1975).

[14] R. E. Casey, W. A. Zaleski, M. Philp, I. S. Mendelson, and S. L. Mackenzie, *J. Inherited Metab. Dis.* **1,** 129 (1978).

Copyright © 1985 by Academic Press, Inc.
All rights of reproduction in any form reserved.

the anti-vitamin B_6 compound 1-aminoproline,[15] after subtotal nephrectomization,[16] or following administration of hydrazine.[17] In neuroophthalmological research DL-α-aminoadipate is used as a tool to selectively destroy Muller cells by intravitreal injection.[18]

Properties

Solubility. α-Aminoadipate forms a monohydrate if crystallized from water below 20°. One gram of the acidic form dissolves in 450 ml of water but is sparingly soluble in alcohol or ether. The salt form at pH 7 is more soluble in water and solutions to at least 0.6 *M* are possible.

Melting Point. Melting point values of 185–186, 197–198, 200–202, and 206° have been reported for α-aminoadipate. Waalkes *et al.*[19] found that the melting point varied from 165 to 202° depending on the rate of heating and the bath temperature at which the sample is introduced.

Ionization Constants and Optical Rotations. Ionization constants of α-aminoadipate are $pk_1 = 2.14$, $pk_2 = 4.21$, and $pk_3 = 9.77$. Optical rotations reported for the L-isomer are $[M]_D^{25} = +5.2°$ (H_2O) and +40.3° (5 *N* HCl), c = 2[20], $[\alpha]_D^{25} = +23.5°$ (5 *N* HCl), c = 2[21], $[\alpha]_D^{25} = +25$ (5 *N* HCl), c = 2.[22] For the D-isomer an $[\alpha]_D^{25}$ of −25° (5 *N* HCl), c = 2, has been reported.[22]

Derivatives. *N*-Chloroacetyl-DL-α-aminoadipic acid, MP 129[22] and 124–126[23]; carbobenzoxy-DL-aminoadipic diamide, MP 189°; *N*-benzoyl-α-aminoadipic acid, MP 184–185°; α-pthalimidoadipic acid, MP 227°[24]; diethyl-α-aminoadipic acid, bp_{13} 155–156°.[25]

Reaction with Ninhydrin. α-Aminoadipate reacts with ninhydrin to form a purple complex that can be quantitated at 570 nm.[20,26] Ninhydrin dissolved in a volatile solvent (0.2% in acetone or butanol) and sprayed on

[15] K. Sasaoka, T. Ogawa, K. Moritoki, and M. Kimoto, *Biochim. Biophys. Acta* **428,** 396 (1976).

[16] A. Mannan, L. I. Wiebe, A. A. Noujaim, and D. C. Secord, *Clin. Biochem.* (*Ottawa*) **8,** 303 (1975).

[17] T. L. Perry, S. J. Kish, S. Hansen, J. M. Wright, R. A. Wall, W. L. Dunn, and G. D. Bellward, *J. Neurochem.* **37,** 32 (1981).

[18] O. O. Pedersen and R. L. Karlsen, *Exp. Eye Res.* **28,** 569 (1979).

[19] T. P. Waalkes, W. S. Fones, and J. White, *J. Am. Chem. Soc.* **72,** 5760 (1950).

[20] J. P. Greenstien and M. Winitz, "Chemistry of the Amino Acids," 3 vols. Wiley, New York, 1961.

[21] K. Ramsamy, R. K. Olsen, and T. Emery, *Synthesis* No. 1, p. 42 (1982).

[22] J. P. Greenstien, S. M. Birnbaum, and M. C. Otey, *J. Am. Chem. Soc.* **75,** 1994 (1953).

[23] T. G. Wood and R. A. Hartline, *Anal. Biochem.* **43,** 282 (1971).

[24] R. Gaudry, *Can. J. Res., Sect. B* **37,** 21 (1949).

[25] "The Merk Index," 9th ed., p. 56. Merck & Co., Inc., Rahway, New Jersey, 1976.

[26] H. Rosen, *Arch. Biochem. Biophys.* **67,** 10 (1957).

a thin layer or paper chromatogram reacts with α-aminoadipate to form a blue colored complex. Heating the chromatogram hastens color development.

The Lactam. Heating solid α-aminoadipate **(I)** at its melting point or refluxing an aqueous solution for 2 hr promotes dehydration and ring closure resulting in the formation of its lactam, piperidonecarboxylate (6-oxo-piperidine-2-carboxylate) **(II)** (MP 177–178°).[20] The lactam also forms on a thin layer or paper chromatogram during drying at room temperature following application of an α-aminoadipate sample.[27] Reflux of an aqueous acidic or basic solution of the lactam converts it back to α-aminoadipate.

Δ, $-H_2O$

[I]

α-Aminoadipate

[II]

Piperidonecarboxylate
(6-Oxo-piperideine-α-carboxylate)

The lactam can be separated from α-aminoadipate on a Dowex 1 (formate form) column by a stepwise gradient of ammonium formate or by paper chromatography in a variety of solvents.[28] Piperiodonecarboxylate cannot be detected on a paper or thin layer chromatogram with ninhydrin. A sensitive color detection of the lactam is possible by exposing the thoroughly dried chromatogram to chlorine vapors for 10 min, placing it in an air hood for several hours, and spraying it with a 1% starch–1% KI solution.[29] The lactam appears as a blue-black spot on a faint blue background. This detection method can be used following a ninhydrin spray.

Racemic piperidonecarboxylate is found in the broth from fermentation by *Penicillum chrysogenum*[30] and its D-isomer is formed from α-aminoadipate in resting yeast cells.[28] There is no evidence for lactam utilization by these systems. Whole cells or cell extracts of *Pseudomonas*

[27] P. Pekala and R. A. Hartline, *Anal. Biochem.* **55,** 411 (1973).

[28] M. H. Kuo, P. P. Saunders, and H. P. Broquist, *J. Biol. Chem.* **239,** 508 (1964).

[29] H. N. Ryden and P. W. G. Smith, *Nature (London)* **169,** 922 (1952); See also this series, Vol. 17A, pp. 949–950.

[30] S. P. Brundidge, F. C. Gaeta, D. J. Hook, C. Sapino, Jr., R. P. Elander, and R. B. Morin, *J. Antibiot.* **33,** 1348 (1980).

putida (ATCC 25571) grown on DL-α-aminoadipate as the sole source of carbon and nitrogen do not oxidize radioactive DL-piperidonecarboxylate.[31]

Miscellaneous. The taste of the individual isomers, the solid state infrared absorption spectra of L-α-aminoadipic acid, the rotary dispersion of α-aminoadipate, the rates of oxidation of the isomers by amino acid oxidases, and the rates of hydrolysis of the isomers of *N*-chloroacetyl-α-aminoadipate by renal acylase I have been reported.[20,32] Partially purified glutamate decarboxylases from squash, cucumber, yeast, and calf brain are unusual compared to bacterial decarboxylases in that their specificities are broad enough to include α-aminoadipate.[33]

Separation and Identification

Being an acidic amino acid, α-aminoadipate is easily separated and identified in the presence of neutral or basic amino acids by thin-layer, paper, or ion-exchange chromatography or paper electrophoresis. Most texts on chromatography[34] list solvent systems for chromatography of α-aminoadipate on thin layer or paper. Four solvent systems are given in the table that separate it from related compounds.

As the six carbon homolog of glutamate, α-aminoadipate cannot generally be separated from its C_5 homolog in many chromatography solvent systems or ion exchange conditions. The two can be separated and identified by two-dimensional thin layer or paper chromatography using 1-butanol/acetic acid/water (4 : 1 : 1) followed by 71% phenol,[35] by high voltage electrophoresis[36] at 1500–2000 V at pH 3.5 and 6.4 in pyridine/glacial acetic acid/water buffers in ratios of 1 : 10 : 89 and 100 : 4 : 900, respectively, and by ion-exchange chromatography with Dowex 1 (acetate form; 100 to 200 mesh) by means of a nonlinear acetic acid gradient.[37] Separation of α-aminoadipate in a mixture containing aspartate, glutamate, and all the basic and neutral amino acids can be conveniently accomplished with a Dowex 1 (acetate form; 100–200 mesh) ion-exchange column. The neutral and basic compounds are eluted with water followed by a nonlinear acetic acid gradient to separate the dicarboxylic amino acids.[37]

[31] S. Khustoss and R. A. Hartline, unpublished data.

[32] A. Meister, "Biochemistry of the Amino Acids." Academic Press, New York, 1965.

[33] P. Lahdesmaki, H. Korhonen, and L. Anneli, *Finn. Chem. Lett.* **5–6,** 75 (1981).

[34] I. Smith, *in* "Chromatographic and Electrophoretic Techniques" (I. Smith, ed.), Vol. I, Chapter 5, p. 82. Wiley (Interscience), New York, 1960.

[35] R. Perfetti, R. J. Campbell, J. Titus, and R. A. Hartline, *J. Biol. Chem.* **247,** 4089 (1972).

[36] L. V. Basso, D. R. Rao, and V. W. Rodwell, *J. Biol. Chem.* **237,** 2239 (1962).

[37] D. R. Rao and V. W. Rodwell, *J. Biol. Chem.* **237,** 2232 (1962).

PAPER CHROMATOGRAPHY OF α-AMINOADIPATE AND RELATED COMPOUNDS[a]

Amino acid	R_f[b] A	B	C	D
α-Aminoadipic-δ-semialdehyde	0.44	0.85	0.54	0.62
α-Aminoadipic acid	0.29	0.26	0.58	0.31
2-Amino-6-heptenoic acid	0.70	0.81	0.85	0.79
2-Aminoheptanoic acid	0.75	0.84	0.89	0.82
2-Amino-6-chloroheptanoic acid	0.78	0.81	0.85	0.78
α-Amino-ε-hydroxycaproic acid	0.36	0.55	0.63	0.56
Lysine	0.08	0.23	0.30	0.08
α-Keto-ε-aminocaproic acid	0.43	0.46	0.62	0.67
Pipecolic acid	0.45	0.89	0.75	0.64
ε-*N*-Acetyllysine	0.40	0.83	0.71	0.56
α-*N*-Acetyllysine	0.42	0.76	0.69	0.49

[a] Data from A. J. Aspen and A. Meister, *Biochemistry* **1,** 600 (1962).
[b] Solvent systems: (A) butanol-1/acetic acid/water; 4 : 1 : 1; (B) phenol/0.1 *M* potassium phosphate buffer, pH 7.2, 80 : 20; (C) *t*-butyl alcohol/formic acid/water, 70 : 15 : 15; (D) ethanol/water 77 : 23.

α-Aminoadipate, 145 ninhydrin positive compounds, penicillin precursors, and penicillins can be separated with a cation exchange column by elutions with lithium buffers.[38] Separation and quantitation of α-aminoadipate, 18 nonprotein amino acids, and 20 protein amino acids can be accomplished by gas–liquid chromatography of their *N*-triflouroacetyl-*n*-butyl esters.[39] By means of an Ostion high-resolution cation exchange resin, LGKSO803 (Li), it is possible to separate 24 neutral and acidic amino acids including α-aminoadipate.[40]

Synthesis and Resolution

The synthesis of α-aminoadipate was first accomplished in 1894 by Diekman. This and other synthetic routes are described elsewhere[20] but, because they are mainly of historical interest, will not be included here. Recently several new synthetic methods starting with lysine,[41] tosyl-L-

[38] P. Adriaens, B. Meesschaert, W. Wutys, H. Vanderhaeghe, and H. Eyssen, *J. Chromatogr.* **140,** 103 (1977).
[39] F. Raulin, D. Shapshak, and B. N. Khare, *J. Chromatogr.* **73,** 35 (1972).
[40] P. Vratny and J. Ouhrabkora, *J. Chromatogr.* **152,** 214 (1978).
[41] A. I. Scott and T. J. Wilkinson, *Synth. Commun.* **10,** 127 (1980).

glutamate,[42] 3-chlorobutyronitrile,[43] or aspartate[21] have been reported. While these procedures are sound, they all involve a relatively large number of steps. A simple and efficient method of α-aminoadipate synthesis involves bromination of monoethyl adipic acid (commercially available from Eastman Organic Chemicals), amination with NH_4OH, and hydrolysis of the ester.[23]

The classical method of resolution by the action of renal acylase I on *N*-chloroacetyl-DL-α-aminoadipate has been described.[20,22] More recently, a method has been reported that takes advantage of preferential isomer uptake and metabolism by bacteria to obtain each enantiomer from racemic α-[^{14}C]aminoadipate..[27] The essence of this method is the preferential uptake and utilization of D-α-aminoadipate over the L-isomer by *Alcaligenes dentrificans* (ATCC 15173), and the utilization of only the L-isomer by *P. putida* (ATCC 25571). Each isomer is then purified by ion exchange chromatography.[37] While the disadvantage of this method is that to procure one isomer the other is lost, the advantages are that it is simple and exceedingly small quantities of each isomer of radioactive α-aminoadipate with high specific activity can be isolated. That this technique can be successfully used with relatively large quantities of DL-α-aminoadipate (2.5 g) has been shown by Chang and Massey.[44]

Cellular Uptake[45]

Bacterial. A. denitrificans (ATCC 15173) growing on DL-α-aminoadipate exhibits a diauxic growth in which the D-isomer is preferentially used in the first phase of growth.[46] Since at 10-fold concentrations neither isomer inhibits uptake of the opposite enantiomer, inhibition of L-α-aminoadipate uptake by the D-isomer is not the origin of preferential uptake.[47] Instead, it results from rapid conversion of D-α-aminoadipate to an intracellular L-isomer pool which prevents uptake of extracellular L-α-aminoadipate.

The K_m and V for uptake of L-α-aminoadipate are 1.13 μM and 3.57 nmol/min/mg dry cell wt; the K_m and V for D-isomer uptake are 0.27 μM and 4.12 nmol/min/mg dry cell wt. Uptake of both isomers is inducible and is inhibited by sulfhydryl group and electron transport chain inhibi-

[42] J. Rudinger and H. Forkasora, *Collect. Czech. Chem. Commun.* **28,** 2941 (1963).

[43] S. Wolfe and M. G. Johinen, *Can. J. Chem.* **57,** 1388 (1979).

[44] Y. F. Chang and S. C. Massey, *Biochem. Prep.* **10,** 215 (1980).

[45] Uptake is the process of translocation in the presence of subsequent metabolism as opposed to transport where the translocation occurs without further metabolism. See O. K. Ghei and W. W. Kay, *J. Bacteriol.* **114,** 65 (1973).

[46] T. G. Wood and R. A. Hartline, *Biochim. Biophys. Acta* **236,** 446 (1971).

[47] P. H. Pekala, T. Perfetti, and R. A. Hartline, *Biochim. Biophys. Acta* **394,** 65 (1975).

tors suggesting protein carriers and an ATP or $\Delta\mu H^+$ requirement. L-α-Aminoadipate, but not the D-isomer, is taken up against a concentration gradient. The lack of concentration of D-α-aminoadipate could be due to its rapid conversion to the L-isomer. However, it is possible that in a cell blocked in the as yet uncharacterized conversion of D- to L-α-aminoadipate D-isomer concentration may occur.

Mammalian. D- and L-α-Aminoadipate uptake by rat cerebral cortical slices is stereospecific favoring the L-configuration and is Na^+ and temperature dependent.[48] Uptake of the L-isomer is inhibited by L-glutamate and L-aspartate. L-Isomer uptake gives rise to a tissue-to-medium ratio of 250 : 1 whereas uptake of D-α-aminoadipate attained a ratio of only 33 : 1.

There is a saturatable component of uptake for each isomer at low substrate concentrations (high-affinity uptake), and there may be nonsaturatable diffusion components at high amino acid concentrations (low-affinity uptake). The K_m values for the high-affinity uptake are D-isomer 70 μM; L-isomer 160 μM. The K_m values for low-affinity uptake are D-isomer 1.5 mM; L-isomer 1.4 mM. The V values for the high-affinity uptake of the *D*- and L-enantiomers are 21 and 67 nmol/min/g of tissue, respectively. These values for low-affinity uptake are 87 and 517 nmol/min/g of tissue for the D- and L-isomers, respectively.

Release of L-α-aminoadipate from cortical slices is stimulated by K^+ in the presence of Ca^{2+} in perfusion buffer. During 40 min of perfusion following uptake, 90% of each isomer is released indicating less than 10% metabolism.

Metabolism

Lysine biosynthesis. α-Aminoadipate is an intermediate in the biosynthesis of L-lysine by the homocitrate-aminoadipate pathway in higher fungi, Euglenids, and some classes of lower fungi (Fig. 1).[49,50] Bacteria, cyanobacteria, green algae, higher plants, and some lower fungi synthesize lysine by the diaminopimelic acid pathway in which α-aminoadipate is not an intermediate. Mammals do not synthesize lysine.

As early as 1948 experiments with α-[^{14}C]aminoadipate utilization in a *Neurospora* mutant blocked in lysine biosynthesis provided evidence for α-aminoadipate as an intermediate. Shortly thereafter it was also shown to be an intermediate in lysine anabolism in yeast. Studies with isotopic tracers revealed that the initial reaction of the pathway is a condensation between acetyl-CoA and α-ketoglutarate to yield homocitrate. Homoci-

[48] A. K. Charles and Y. F. Chang, *J. Neurochem.* **36,** 1127 (1981).
[49] H. J. Vogel, *Proc. Int. Congr. Biochem., 5th, 1961* Vol. 3, p. 341 (1963).
[50] H. J. Vogel, *Am. Nat.* **98,** 435 (1964).

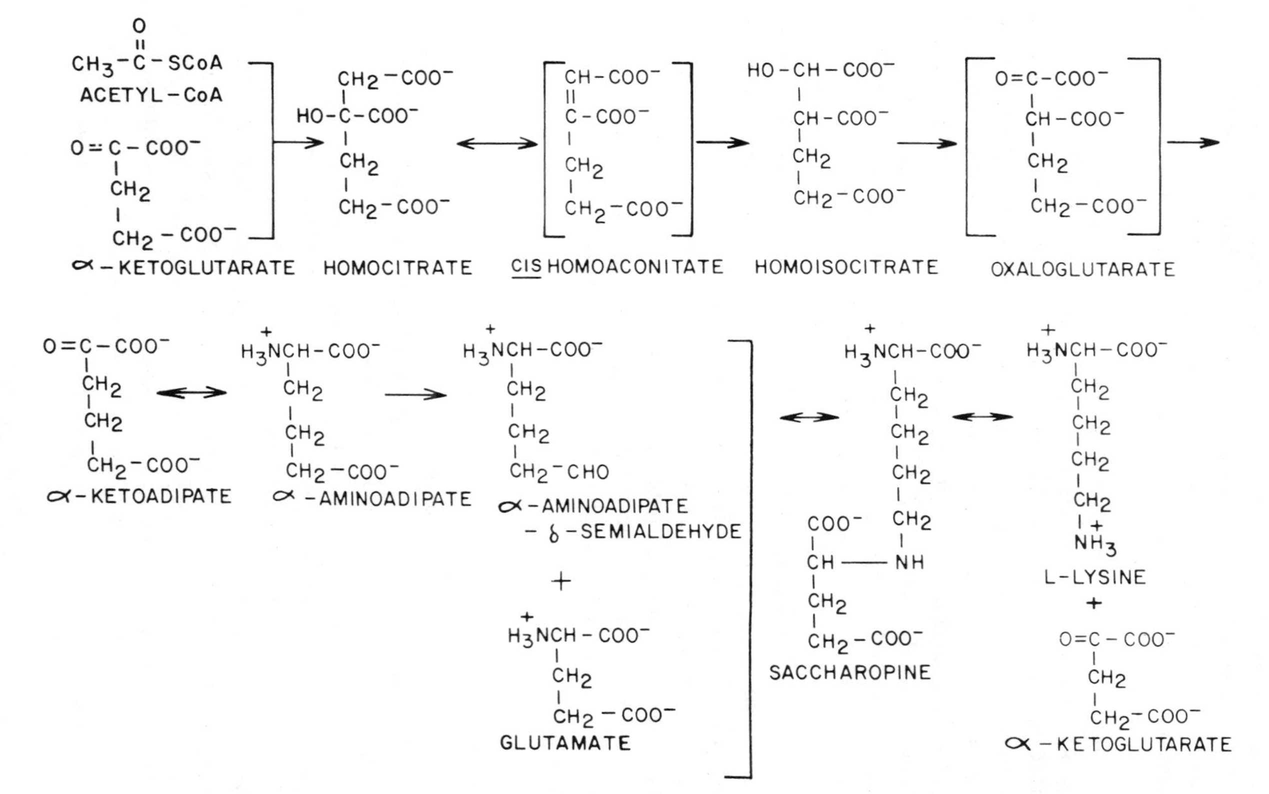

FIG. 1. Homocitrate-aminoadipate pathway for lysine biosynthesis in higher fungi, Euglenids, and some classes of lower fungi. Intermediates in brackets are proposed and may be enzyme bound.

trate gives rise to α-ketoadipate via homoisocitrate. This portion of the pathway has been covered in detail and the reader is referred to the reviews by Rodwell[6] and Meister[32] for literature references.

α-Ketoadipate formed by oxidative decarboxylation of homoisocitrate is transaminated to α-aminoadipate by an α-ketoadipate : glutamate aminotransferase.[51] Reduction of α-aminoadipate to α-aminoadipate-δ-semialdehyde as an obligatory step in lysine biosynthesis was first predicted in 1959.[52] However, evidence for the semialdehyde as an intermediate was largely inferential[53–55] until its formation in a *Sacchromyces cerevisiae* lysine auxotroph was detected by its characterization as the *o*-aminobenzaldehyde adduct.[56] Reduction of α-aminoadipate by α-aminoadipate reductase[51] consists of three steps[57,58] (Fig. 2): (1) an ATP and Mg^+ requiring activation of the δ-carboxyl of α-aminoadipate to form an adenylate with AMP; (2) an NADPH and Mg^{2+} requiring reduction of the ε-carbon of the amino acid in the adenylate; (3) cleavage of the adenylate to form α-aminoadipate-δ-semialdehyde and AMP. The semialdehyde is converted to saccharopine [ε-*N*-(glutaryl-2-)-lysine] by a saccharopine reductase (α-aminoadipate-δ-semialdehyde : glutamate reductase) requiring NADPH and glutamate.[59,60] Saccharopine is oxidized to lysine plus α-ketoglutarate in the presence of NAD^+ by saccharopine dehydrogenase.[61] The combined action of the reductase and dehydrogenase enzymes results in amination of the semialdehyde by glutamate to form lysine. Indirect evidence that the L-isomer of α-aminoadipate is the precursor to saccharopine has been obtained from experiments with *S. cerevisiae*.[28] For details on the assays, purifications, and properties of the enzymes of the homocitrate-aminoadipate pathway see the review by Broquist.[51]

D-*Lysine Degradation in Bacteria*. Pseudomonad bacteria degrade D-lysine by a pathway in which α-aminoadipate is an intermediate (Fig. 3).[62,63] α-Aminoadipate is not a catabolite of L-lysine degradation by bacteria or yeast.

[51] H. P. Broquist, this series, Vol. 17B, p. 119.
[52] S. Sagisaka and K. Shimura, *Nature* (*London*) **184,** 1710 (1959).
[53] S. Sagisaka and K. Shimura, *J. Biochem.* (*Tokyo*) **51,** 398 (1962).
[54] S. Sagisaka and K. Shimura, *Nature* (*London*) **188,** 1189 (1960).
[55] R. L. Larson, W. D. Sandine, and H. P. Broquist, *J. Biol. Chem.* **238,** 275 (1963).
[56] E. E. Jones and H. P. Broquist, *J. Biol. Chem.* **240,** 2531 (1965).
[57] J. R. Mattoon, T. A. Mashier, and T. H. Kreiser, *Biochim. Biophys. Acta* **51,** 615 (1961).
[58] A. K. Sinha and J. K. Bhattacharjee, *Biochem. J.* **125,** 743 (1971).
[59] E. E. Jones and H. P. Broquist, *J. Biol. Chem.* **241,** 3430 (1966).
[60] J. S. Turpin and H. P. Broquist, *J. Biol. Chem.* **240,** 2524 (1965).
[61] P. P. Saunders and H. P. Broquist, *J. Biol. Chem.* **241,** 3435 (1966).
[62] D. L. Miller and V. W. Rodwell, *J. Biol. Chem.* **246,** 2758 (1970).
[63] Y. Chang and E. Adams, *J. Bacteriol.* **117,** 753 (1974).

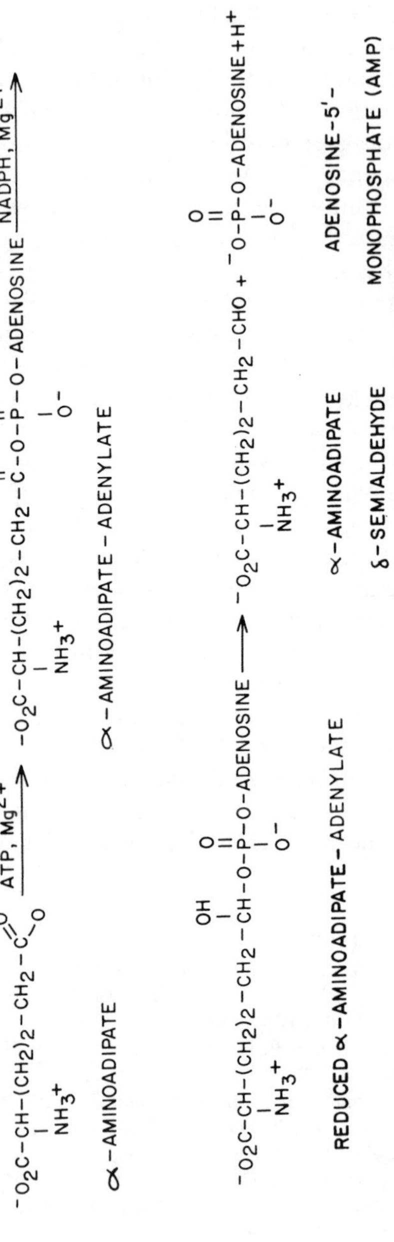

$$^{-}O_2C{-}CH(NH_3^+){-}(CH_2)_2{-}CH_2{-}CO_2^- \xrightarrow{ATP,\ Mg^{2+}} {}^{-}O_2C{-}CH(NH_3^+){-}(CH_2)_2{-}CH_2{-}C(=O){-}O{-}P(=O)(O^-){-}O{-}ADENOSINE$$

α-AMINOADIPATE → α-AMINOADIPATE-ADENYLATE

$$\xrightarrow{NADPH,\ Mg^{2+}} {}^{-}O_2C{-}CH(NH_3^+){-}(CH_2)_2{-}CH_2{-}CH(OH){-}O{-}P(=O)(O^-){-}O{-}ADENOSINE$$

REDUCED α-AMINOADIPATE-ADENYLATE

$$\longrightarrow {}^{-}O_2C{-}CH(NH_3^+){-}(CH_2)_2{-}CH_2{-}CHO + {}^{-}O{-}P(=O)(O^-){-}O{-}ADENOSINE + H^+$$

α-AMINOADIPATE δ-SEMIALDEHYDE + ADENOSINE-5'-MONOPHOSPHATE (AMP)

FIG. 2. Steps in the reduction of α-aminoadipate to α-aminoadipate-δ-semialdehyde.

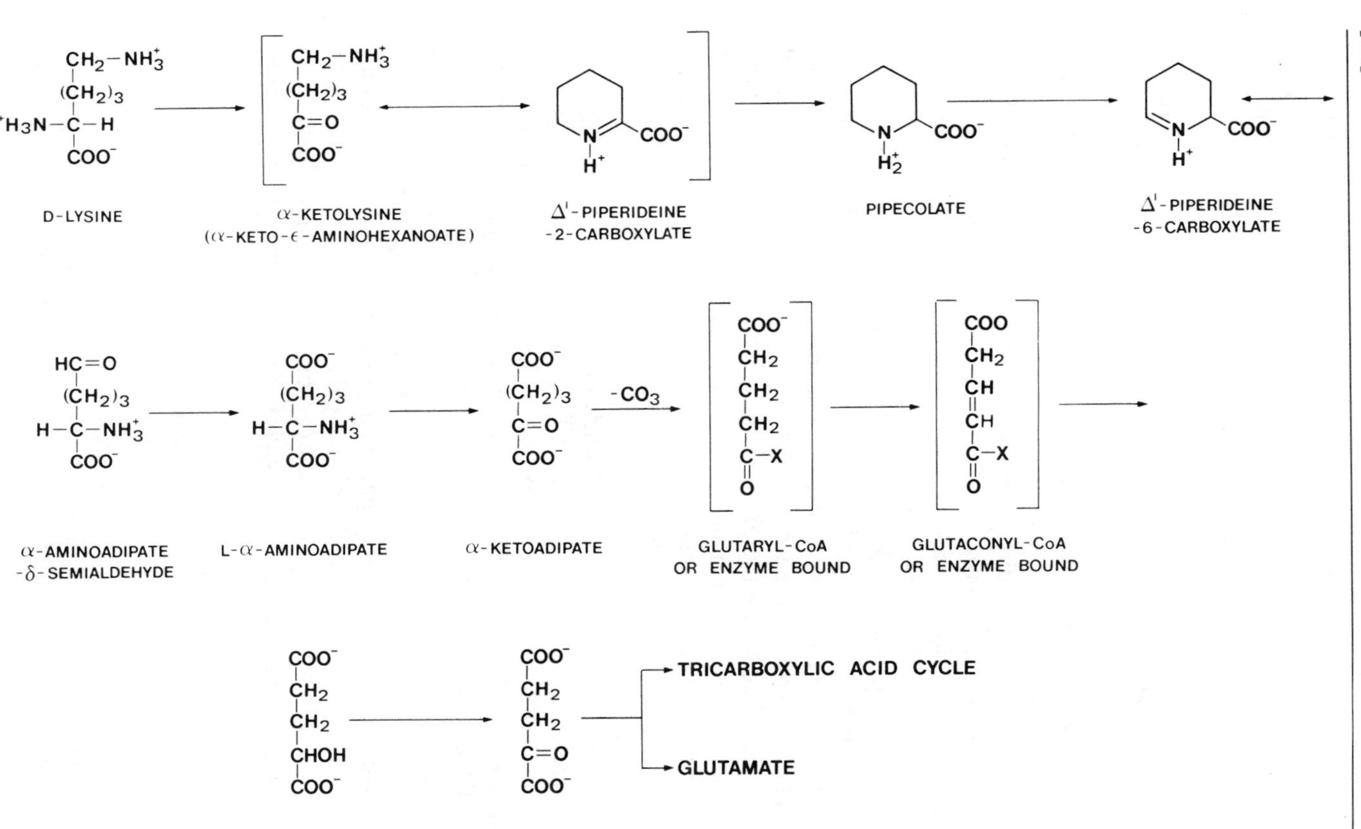

FIG. 3. D-Lysine catabolism in Pseudomonads. Intermediates in brackets are proposed.

Metabolism of D-[1-^{14}C]lysine in pseudomonads yields [^{14}C]pipecolate.[62] Incubation of whole cells with DL-[α-^{15}N]lysine, D-[ε-^{15}N]lysine, or DL-[α,ε-^{15}N]lysine gives rise to pipecolate retaining only the ε-nitrogen. Loss of the α-nitrogen of lysine is consistent with α-ketolysine (α-keto-ε-aminohexanoate) as an intermediate. The presumed α-ketolysine catabolite would be in chemical equilibrium with its cyclic form, Δ′-piperideine-2-carboxylate,[64] which on reduction would yield pipecolate. Δ′-Piperideine-2-carboxylate formation from L-lysine has been shown to occur by the action of a partially purified L-lysine:α-ketoglutarate aminotransferase from *Flavobacterium fuscum*, *F. flavescens*, and *Achromobacter liquidium*,[65,66] but direct evidence for its formation from D-lysine in pseudomonads has not been established. However, an inducible pipecolate:NADP$^+$ 2-oxidoreductase from *P. putida* that reduces Δ′-piperideine-2-carboxylate to pipecolate has recently been reported.[67]

α-Aminoadipate-δ-semialdehyde was shown to be an intermediate when it was trapped as the bisulfite adduct during incubation of whole cells of *P. putida* with pipecolate.[36] Eventually, a membrane bound L-pipecolate dehydrogenase was partially purified that catalyzed the oxidation of pipecolate to Δ′-piperideine-6-carboxylate in chemical equilibrium with α-aminoadipate-δ-semialdehyde.[68] The semialdehyde is reduced by an L-α-aminoadipate-δ-semialdehyde : NAD oxidoreductase to L-α-aminoadipate[69] which is transmitted to α-ketoadipate by an inducible transaminase.[70]

Resting cell suspensions of *P. putida* incubated with DL-[l^{14}C]pipecolate excretes L-α-[l-^{14}C]aminoadipate.[37] If the incubation is carried out with DL-[6-^{14}C]pipecolate cells excrete radioactive α-aminoadipate and glutamate.[70] Radioactive glutamate is excreted by cells incubated with DL-α-[6-^{14}C]aminoadipate.These observations served to verify L-aminoadipate as an intermediate and prompted investigations to determine whether the glutamate was formed from the intact α-aminoadipate molecule. Isotope distributions of excreted glutamate isolated from an incubation of resting cells with DL-α-[6^{14}C]aminoadipate or DL-[2-^{14}C]pipecolate are consistent with a pathway in which carbon 1 of α-aminoadipate is oxidized to CO_2 while carbons 2 through 6 are converted, intact, to gluta-

[64] A. Meister, *J. Biol. Chem.* **206,** 577 (1954).
[65] K. Soda, H. Misono, and T. Yamamoto, *Biochemistry* **7,** 4102 (1968).
[66] K. Soda and H. Misono, *Biochemistry* **7,** 4110 (1968).
[67] C. W. Pay and Y. Chang, *J. Bacteriol.* **149,** 864 (1982).
[68] V. W. Rodwell, this series, Vol. 17B, p. 174.
[69] A. F. Calvert and V. W. Rodwell, *J. Biol. Chem.* **241,** 409 (1966).
[70] R. A. Hartline and V. W. Rodwell, *Arch. Biochem. Biophys.* **142,** 32 (1971).

mate.[35] The glutamate labeling patterns also exclude free glutarate as an intermediate.[35,70] Since *P. putida* does not utilize D-α-aminoadipate,[27] incubations with racemic α-aminoadipate represent metabolism of the L-isomer.

Consideration of α-hydroxyglutarate as a potential intermediate gave rise to experiments providing evidence that in *P. putida* the particulate cell fraction contains an inducible ability to convert the hydroxyacid to α-ketoglutarate.[70] If the soluble cell fraction and aspartate are added, the hydroxyacid gives rise to glutamate. Subsequently, Reitz and Rodwell[71] identified an inducible particulate α-hydroxyglutarate-specific oxidoreductase from *P. putida* which catalyzes formation of α-ketoglutarate from the hydroxyacid. Direct evidence for α-hydroxyglutarate as an intermediate was the isolation of α-[6-^{14}C]hydroxyglutarate following incubation of a cell extract with DL-α-[6-^{14}C]aminoadipate.[72] A maximum conversion of 63% of available L-α-aminoadipate requires α-ketoglutarate for transamination of the aminoadipate, glutathione, and CoASH. In cells, α-ketoglutarate formed from oxidation of α-hydroxyglutarate can presumably be oxidized via the tricarboxylic acid cycle and transaminated to glutamate.

Enzyme-bound or CoASH derivatives of glutarate and glutaconate have been proposed as intermediates in the formation of α-hydroxyglutarate from α-ketoadipate.[35,70] Incubation of a cell extract with radioactive α-ketoadipate and hydroxylamine to trap CoA or sulfur-linked enzyme-bound intermediates produced radioactive compounds which on paper chromatography were tentatively identified as the monohydroxamates of glutarate and glutaconate.[73]

L-*Lysine Degradation in Mammals*. Two pathways which converge at α-aminoadipate-δ-semialdehyde have been reported for the initial steps in mammalian L-lysine catabolism (Fig. 4). The first pathway, in which saccharopine is an intermediate, is the reverse of the final two steps of lysine synthesis in fungi and yeast (Fig. 1). The other pathway is identical to the three initial steps of D-lysine catabolism by bacteria (Fig. 3) and contains pipecolate as an intermediate.

[^{14}C]Saccharopine and α-[^{14}C]aminoadipate have been isolated from mouse liver following injection of L-[U^{14}C]lysine.[74] Saccharopine is also formed from L-lysine and α-ketoglutarate in the presence of rat liver

[71] M. S. Reitz and V. W. Rodwell, *J. Bacteriol.* **100,** 708 (1969).

[72] J. J. Kopchick and R. A. Hartline, *J. Biol. Chem.* **254,** 3259 (1979).

[73] J. Huettner and R. A. Hartline, unpublished data.

[74] K. Hagashino, M. Fujioka, and Y. Yamamura, *Arch. Biochem. Biophys.* **142,** 606 (1971).

α-KETOGLUTARATE

SACCHAROPINE

L-LYSINE

α-KETOLYSINE
(α-KETO-ε-AMINOHEXANOATE)

α-AMINOADIPATE
-δ-SEMIALDEHYDE

Δ[1]-PIPERIDEINE
-6-CARBOXYLATE

PIPECOLATE

Δ[1]-PIPERIDEINE
-2-CARBOXYLATE

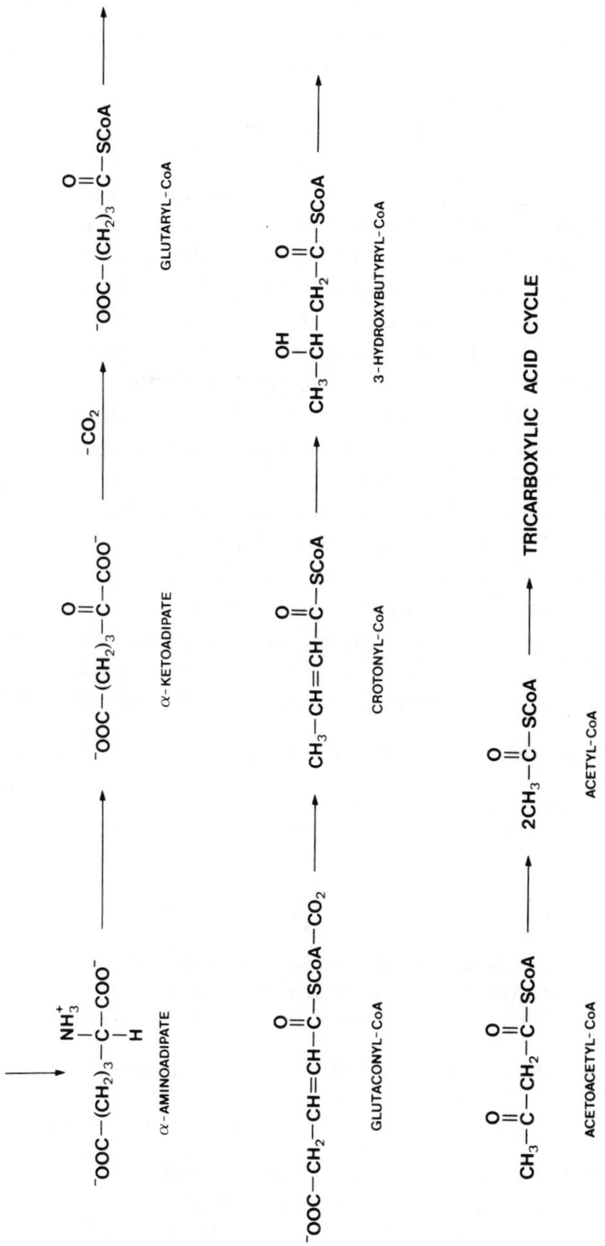

$$\rightarrow {}^{-}OOC-(CH_2)_3-\overset{NH_3^+}{\underset{H}{C}}-COO^- \longrightarrow {}^{-}OOC-(CH_2)_3-\overset{O}{\overset{\|}{C}}-COO^- \xrightarrow{-CO_2} {}^{-}OOC-(CH_2)_3-\overset{O}{\overset{\|}{C}}-SCoA \longrightarrow$$

α-AMINOADIPATE; α-KETOADIPATE; GLUTARYL-CoA

$$^{-}OOC-CH_2-CH{=}CH-\overset{O}{\overset{\|}{C}}-SCoA-CO_2 \longrightarrow CH_3-CH{=}CH-\overset{O}{\overset{\|}{C}}-SCoA \longrightarrow CH_3-\overset{OH}{\overset{|}{CH}}-CH_2-\overset{O}{\overset{\|}{C}}-SCoA \longrightarrow$$

GLUTACONYL-CoA; CROTONYL-CoA; 3-HYDROXYBUTYRYL-CoA

$$CH_3-\overset{O}{\overset{\|}{C}}-CH_2-\overset{O}{\overset{\|}{C}}-SCoA \longrightarrow 2CH_3-\overset{O}{\overset{\|}{C}}-SCoA \longrightarrow \text{TRICARBOXYLIC ACID CYCLE}$$

ACETOACETYL-CoA; ACETYL-CoA

FIG. 4. L-Lysine catabolism in mammals.

mitochondria,[75] a soluble enzyme from human liver,[76] mouse,[74] and dog liver[77] homogenates, and rabbit liver mitochondria.[78] It is converted to α-aminoadipate by extracts of mouse liver,[74] rat,[79,80] and human liver mitochondria.[81] In the intact rate[82] and extracts of rat liver,[83] L-[α^{15}N]- and [ε-^{15}N]lysine give rise to α-aminoadipate retaining only the α-nitrogen, a pattern consistent with a lysine degradation to α-aminoadipate through saccharopine.

While formation of saccharopine from lysine was not examined in kidney, this organ and heart and skin tissue of humans are reported to contain lysine, α-ketoglutarate reductase (saccharopine forming), and saccharopine dehydrogenase (α-aminoadipate-δ-semialdehyde forming) activities, but at lower levels than liver.[84] These enzyme activities were not detected above trace amounts in muscle, brain, spleen, pancrease, adrenal, lung, and gut tissue or in leukocytes or erythrocytes.

Experimental findings between 1948 and 1960 prompting investigators to propose the pipecolate pathway include: formation of α-aminoadipate from lysine by guinea pig liver homogenates[85] and rat liver slices,[86] pipecolate formation from DL-lysine in the intact rat[87] with retention of only the ε-amino group,[88] reduction of Δ′-piperideine-2-carboxylate to pipecolate in many mammalian tissues,[89] oxidation of pipecolate by the intact rat and rat liver mitochondria,[90] and conversion of pipecolate to α-aminoadipate by beef liver mitochondria.[91] At the time, the pipecolate pathway was generally considered to be the route for mammalian lysine degradation. However, three facts that seemed inconsistent with the pipecolate path-

[75] K. Higashino, K. Tsukada, and J. Lieberman, *Biochem. Biophys. Res. Commun.* **20,** 285 (1965).

[76] J. Hutzler and J. Dancis, *Biochim. Biophys. Acta* **158,** 62 (1968).

[77] H. Ghadimi, and W. S. Chou, and L. Kesner, *Biochem. Med.* **5,** 56 (1971).

[78] J. A. Grove, F. Young, H. G. Roghair, and P. Schipke, *Arch. Biochem. Biophys.* **151,** 464 (1972).

[79] K. Higashino, M. Fujioka, T. Aoki, and Y. Yamamura, *Biochem. Biophys. Res. Commun.* **29,** 95 (1957).

[80] J. Grove and L. M. Henderson, *Biochim. Biophys. Acta* **165,** 113 (1968).

[81] J. Hutzler and J. Dancis, *Biochim. Biophys. Acta* **206,** 205 (1970).

[82] J. A. Grove, T. G. Linn, C. J. Willett, and L. M. Henderson, *Biochim. Biophys. Acta* **215,** 191 (1970).

[83] P. Boulanger and J. Mizon, *Biochim. Biophys. Acta* **451,** 161 (1976).

[84] J. Hutzler and J. Dancis, *Biochim. Biophys. Acta* **377,** 42 (1975).

[85] H. Borsook, C. L. Deasy, A. J. Haagen-Smit, G. Keighly, and P. H. Lowry, *J. Biol. Chem.* **176,** 1383 (1948).

[86] J. W. Dubnoff and H. Borsook, *J. Biol. Chem.* **173,** 425 (1948).

[87] S. Lindstedt and G. Lindstedt, *Arch. Biochem. Biophys.* **85,** 565 (1959).

[88] M. Rothstein and L. L. Miller, *J. Biol. Chem.* **211,** 851 (1954).

[89] A. Meister, A. N. Radhakrishnan, and S. D. Buckley, *J. Biol. Chem.* **229,** 789 (1957).

[90] M. Rothstein and D. M. Greenberg, *J. Biol. Chem.* **235,** 714 (1960).

[91] M. Rothstein, K. E. Cooksey, and D. M. Greenberg, *J. Biol. Chem.* **237,** 2828 (1962).

way were: (1) the apparent inability of mammalian enzyme preparations to substantially catalyze α-deamination of L-lysine, a requirement for Δ′-piperideine-2-carboxylate formation[6,32,74]; (2) the metabolic inertness of pipecolate in intact "germ-free" rats,[92] perfused rat liver, or rat kidney and liver mitochondria[80,93,94]; (3) the retention of only the α-amino group in the metabolism of L-lysine by the intact rat[82] and rat liver homogenates.[83]

Paik and Benoiton[95] have pointed out that lysine α-deamination would not be detected if the substrate was ε-acylated lysine and not lysine. The basis for this proposal was the discovery of an enzyme from beef liver that catalyzes the formation of ε-*N*-acetyllysine from L-lysine and acetyl phosphate. An ε-lysine hog kidney acylase that hydrolyzes α-keto-ε-acetoamidocaproic acid (the ε-*N*-acetyl derivative of α-ketolysine) to α-ketolysine (Δ′-piperideine-2-carboxylate) has also been reported.[96] Experiments designed to test this concept have not provided data for its support. Hagashino *et al.*[74] were unable to show disappearance of ε-*N*-acetyllysine or its conversion to Δ′-piperdeine-2-carboxylate in mouse liver homogenates. In rat kidney and brain homogenates L-[U-^{14}C]lysine but not ε-*N*-acetyl-L-[U-^{14}C]lysine gives rise to [^{14}C]pipecolate.[97]

Since early investigations were carried out with DL-lysine rather than L-lysine, it has been suggested that pipecolate was formed, and retained the ε-amino group of lysine, because of D-lysine rather than L-isomer metabolism. The basis for this suggestion is, despite an earlier report to the contrary,[98] the observation that 96% of the D-lysine administered to rats is found in the urine as pipecolate.[80] However, it should be noted that patients with hyperlysinemia who have a decreased capacity to degrade L-lysine by the saccharopine pathway excrete pipecolate.[99] It has, therefore, been proposed that the pipecolate pathway for L-lysine degradation in mammals is a minor alternate pathway.

Recently, metabolism of L-[U-^{14}C)lysine has been studied in human autopsy tissues and in intact monkeys by intracerebroventricular and intravenous injection.[100] The human tissues were reported to be more

[92] P. Boulanger, R. Osteux, E. Sacquet, and H. Charlier, *Biochim. Biophys. Acta* **184,** 338 (1969).

[93] J. Grove and L. M. Henderson, *Fed. Proc., Fed. Am. Soc. Exp. Biol.* **27,** 614 (1968).

[94] J. A. Grove, T. J. Gilbertson, R. H. Hammerstedt, and L. M. Henderson, *Biochim. Biophys. Acta* **184,** 329 (1969).

[95] W. K. Paik and L. Benoiton, *Can. J. Biochem. Physiol.* **41,** 1643 (1963).

[96] W. K. Paik, *Biochim. Biophys. Acta* **65,** 518 (1962).

[97] M. F. Hernandez and Y. F. Chang, *Biochem. Biophys. Res. Commun.* **93,** 762 (1980).

[98] M. Rothstein, C. G. Bly, and L. L. Miller, *Arch. Biochem. Biophys.* **50,** 252 (1954).

[99] N. C. Woody and M. B. Pupene, *Pediatr. Res.* **4,** 89 (1970).

[100] Y. F. Chang, *Neurochem. Res.* **7,** 577 (1982).

active in lysine to pipecolate conversion than previously described for rat tissues with brain and kidney being higher than liver. In addition to radioactive pipecolate, traces of α-[^{14}C]aminoadipate and [^{14}C]saccharopine, but not labeled ε-*N*-acetyllysine, were detected in the incubation with liver. Twenty-four hours after intraventricular injection of the monkeys, substantial amounts of labeled pipecolate and α-aminoadipate were detected in brain and spinal cord. Much less was found in kidney, liver, and plasma. Compared with intraventricular administration, intravenous injection showed greatly reduced levels of metabolites in all organs and plasma. Conversion of L-[^{14}C]lysine to [^{14}C]pipecolate and α-[^{14}C]aminoadipate and of DL-[^{3}H]pipecolate to radioactive α-aminoadipate has also been shown to occur in mouse brain[101,102] and rat kidney[97] and brain.[103,104] Attempts to detect labeled saccharopine in one of the experiments with rat brain were unsuccessful.[103] This is consistent with earlier reports that mammalian brain appears inactive in terms of lysine metabolism by the saccharopine pathway.[75,76,84]

In summary, the saccharopine pathway appears to be the major route for mammalian metabolism of L-lysine to α-aminoadipate in liver. While the pipecolate pathway, negative data notwithstanding, might make some contribution to L-lysine catabolism in some organs, in brain it appears to be the only pathway. Kidney may contain both pathways at significant levels. Finally, while enzymes that acylate L-lysine and catalyze a reaction with acylated lysine are known, direct evidence that these reactions or an enzyme-bound ε-*N*-substituted L-lysine play a significant role in mammalian lysine degradation has yet to be established.

No matter which pathway ultimately generates α-aminoadipate, its subsequent catabolism should be the same. α-Ketoadipate has been detected in urine after feeding rats lysine[105] and α-aminoadipate : α-ketoglutarate aminotransferase activity in rat liver[106] and kidney[106,107] has been demonstrated. Thus, transamination to α-ketoadipate represents the first reaction in α-aminoadipate metabolism.

In spite of an earlier report that free glutarate is a lysine catabolite,[108] the isotope distributions in glutamate isolated from rat and dog tissues following administration of DL-[2-^{14}C]- and DL-[6-^{14}C]lysine were consistent

[101] T. Schmidt-Glenewinkel, Y. Nomura, and E. Giacobini, *Neurochem. Res.* **2,** 619 (1977).
[102] H. Nishio, E. Giacobini, and J. Ortiz, *Neurochem. Res.* **7,** 373 (1982).
[103] Y. F. Chang. *Biochem. Biophys. Res. Commun.* **69,** 174 (1976).
[104] Y. F. Chang, *J. Neurochem.* **30,** 347 (1978).
[105] D. Cavallini and B. Mondovi, *Arch. Sci. Biol. (Bologna)* **35,** 468 (1952).
[106] M. C. Tobes and M. Mason, *Biochem. Biophys. Res. Commun.* **62,** 390 (1975).
[107] M. C. Tobes and M. Mason, *J. Biol. Chem.* **252,** 4591 (1977).
[108] M. Rothstein and L. L. Miller, *J. Biol. Chem.* **206,** 243 (1954).

with glutaryl-CoA rather than glutarate as an intermediate.[109] Therefore, glutaryl-CoA and not glutarate is formed from the oxidative decarboxylation of α-ketoadipate. The proposed product of glutaryl-CoA oxidation is glutaconyl-CoA, but evidence for its formation is only inferential.[110,111]

Further insight into the degradation of glutaryl-CoA was obtained from data on mammalian glutarate metabolism. Early investigations centered on glutarate because of its putative role as an intermediate in lysine catabolism.[12,113] When it became apparent that the intermediate was the CoA form of glutarate,[109] compounds identified as products of glutarate degradation were considered catabolites of glutaryl-CoA[32,110,113]; an apparently valid assumption since glutarate is metabolized by mammalian tissus as its CoA derivative.[110,114] Accordingly, it was concluded that glutaryl-CoA is degraded to acetyl-CoA via crotonyl-CoA, β-hydroxybutyryl-CoA, and acetoacetyl-CoA. The complete scheme for mammalian catabolism of glutaryl-CoA is illustrated in Fig. 3.

Hydroxylysine Degradation. Degradation of 5-hydroxy-L-lysine by rat liver homogenates proceeds by way of *O*-phosphohydroxylysine and α-aminoadipate-δ-semialdehyde to yield α-aminoadipate.[7] The overall conversion requires ATP or GTP, pyridoxal-5′-phosphate, and NAD^+ or $NADP^+$. 5-Hydroxy-D-lysine is catabolized by *P. fluorescence* via 2-amino-5-hydroxyadipate (5-hydroxy-α-aminoadipate).[115]

The β-Lactam Antibiotics

Because of their importance as antibiotics the literature on cephalosporins, cephamycins, and penicillins, the β-lactam antibiotics, is extensive. Therefore, this discussion will center on the role of α-aminoadipate in these compounds. For those interested in a more comprehensive coverage, relatively recent reviews on their chemistry,[8,116–118] biol-

[109] S. K. Meghal, H. S. Cheung, R. M. O'Neal, and R. E. Koeppe, *J. Biol. Chem.* **241,** 2622 (1966).

[110] A. Besrat, C. E. Polan, and L. M. Henderson, *J. Biol. Chem.* **244,** 1461 (1969).

[111] E. R. Tustanoff and J. R. Stern, *Biochem. Biophys. Res. Commun.* **3,** 81 (1960).

[112] D. C. Hobbs and R. E. Koeppe, *J. Biol. Chem.* **230,** 655 (1958).

[113] S. P. Bagchi, I. K. Mushhwar, T. Chang, R. E. Koeppe, and G. A. Mourkides, *J. Biol. Chem.* **235,** 370 (1961).

[114] G. K. Menon, D. L. Friedman, and J. R. Stern, *Biochim. Biophys. Acta* **44,** 375 (1960).

[115] J. D. Friede and L. M. Henderson, *J. Bacteriol.* **127,** 1239 (1976).

[116] R. B. Morin and M. Gorman, eds., "The Chemistry and Biology of β-Lactam Antibiotics," Vol. 3. Academic Press, New York, 1982.

[117] E. H. Flynn, ed., "Cephalosporins and Penicillins: Chemistry and Biology." Academic Press, New York, 1973.

[118] J. Elks, ed., "Recent Advances in the Chemistry of β-Lactam Antibiotics." Chemical Society, Burlington House, London, 1976.

ogy,[8,116,119–123,123a] pharmacology,[124] and toxicology[125] are available. Extremely comprehensive treatments of the β-lactam antibiotics are the recent volumes by Demain and Solomon.[125a]

The clinically useful natural and semisynthetic β-lactam antibiotics do not contain α-aminoadipate.[120] The natural β-lactams containing α-aminoadipate have only limited antibacterial activity but are important in the synthesis of semisynthetic β-lactam antibiotics. It is in the biosynthesis of the natural β-lactams that α-aminoadipate plays a role (Fig. 5). Because of space limitations and because an excellent review has recently been published providing experimental evidence and extensive references on the biosynthesis of the natural β-lactams,[8] only an overview will be presented here.

In 1960, Arnstien *et al.*[126] isolated the tripeptide aminoadipylcystienylvaline (ACV) (see Fig. 5) from *Penicilliium chrysogenum*. It was also later isolated from the mycelia of a number of organisms capable of producing penicillins, cephalosporins, or cephamycins. The LLD stereochemistry of ACV, isolated from *Cephalosporium acremonium*, was first reported by Loder and Abraham.[127] Subsequently, the ACV tripeptides produced by other organisms were shown to have the same stereochemistry. Studies of radioactive ACV incorporation into isopenicillin N and experiments with a variety of mutant organisms have unequivocally shown ACV to be a precursor in the biosynthesis of all the natural β-lactams.

Cell-free extracts of *C. acremonium* in the presence of an ATP generating system convert δ-(L-α-aminoadipyl)-L-cysteine plus DL-[^{14}C]valine to radioactive ACV. The extracts do not form ACV from α-[^{14}C]aminoadi-

[119] E. F. Gale, E. Cundliff, P. E. Reynolds, M. H. Richmond, and M. J. Waring, *in* "The Molecular Basis of Antibiotic Action," 2nd ed. Wiley, New York, 1980.

[120] H. Aoki and M. Okuhara, *Annu. Rev. Microbiol.* **34,** 159 (1980).

[121] J. F. Martin and A. L. Demain, *Microbiol. Rev.* **44,** 232 (1980).

[122] A. L. Demain, Y. M. Kennel, and Y. Aharonawitz, *Symp. Soc. Gen. Microbiol.* **29,** 163 (1979).

[123] D. J. Waxman and J. L. Strominger, *Annu. Rev. Biochem.* **52,** 825 (1983).

[123a] R. Hakenbeck, J. V. Hoeltje, and H. Labischinski, eds., *Proc. Int. FEMS Symp.* (*Berlin, West Germany*), March 13–18, 1983.

[124] L. D. Thrupp, *Annu. Rev. Pharmacol.* **14,** 435 (1974).

[125] W. E. Sanders, Jr. and C. C. Sanders, *Annu. Rev. Pharmacol. Toxicol.* **19,** 53 (1979).

[125a] A. L. Demain and N. A. Solomon, eds., "Antibiotics Containing the Beta-Lactam Structure," Parts I, II (Handbook of Experimental Pathology), Vol. 67. Springer-Verlag, Berlin and New York, 1983.

[126] H. R. V. Arnstein, M. Artman, D. Morris, and E. J. Toms, *Biochem. J.* **76,** 353 (1960).

[127] P. B. Loder and E. P. Abraham, *Biochem. J.* **123,** 471 (1971).

pate plus cysteinylvaline.[128] Furthermore, δ-(L-α-aminoadipyl)-L-cysteine has been isolated from *Streptomyces clavuligerus,* an organism that synthesizes cephamycin C. The valine substrate for the enzyme that converts δ-(L-α-aminoadipyl)-L-cysteine to ACV is L-valine not the D-isomer.[129] Experiments with L-[^{15}N]- and [U-^{14}C]valine[130] and L-[$^{18}O_2$]valine[131] have provided evidence consistent with an enzyme-bound form of valine during its incorporation into ACV rather than conversion to free D-valine prior to incorporation. These studies eliminate some mechanisms for the epimerization of L-valine but the actual one remains to be determined. These data are consistent with an ACV synthesis sequence of L-α-aminoadipate plus L-cysteine to give δ-(L-α-aminoadipyl)-L-cysteine followed by the addition of L-valine to yield LLD-ACV, a synthesis suggested to be analogous to the biosynthesis of glutathione.

While only the enzyme for conversion of δ-(L-α-aminoadipyl)-L-cysteine to ACV [δ-(α-aminoadipyl)cysteinylvaline synthetase] has been detected,[129] genetic evidence has been presented that indicate two enzymes are essential for ACV synthesis.[132] It has been proposed that activated carboxyl groups of the amino acids may be involved in the synthesis of ACV, but to date, there is no data to support this proposal.

Three ACV-type tripeptides containing α-aminoadipate with substitutions for the other amino acids have been isolated from the fermentation broth of *P. chrysogenum.*[133] In two of these peptides, serine or alanine replace cysteine. In the third, serine replaces cysteine and isodehydrovaline replaces valine. Since none of these peptides contains cysteine, and hence no sulfur atom, they cannot be precursors to the natural β-lactams. Their formation is considered to be the consequence of relaxed substrate specificity of the ACV synthesizing enzymes.

The cysteine and valine residues of ACV are cyclized into the fused lactam and thiozolidine ring systems of isopenicillin N (Fig. 6). The process occurs with stereospecific incorporation of the intact cysteine molecule and retention of configuration at carbon 3 of valine. The complete

[128] P. A. Fawcett, J. J. Usher, J. A. Huddleston, R. C. Bleany, J. J. Nisbet, and E. P. Abraham, *Biochem. J.* **157,** 651 (1976).

[129] P. Fawcett and E. P. Abraham, this series, Vol. 43, p. 471.

[130] H. Booth, B. W. Bycroft, C. M. Wels, K. Corbett, and A. P. Malony, *J. Chem. Soc., Chem. Commun.* p. 110 (1976).

[131] J. S. Delderfield, E. Mtetwa, R. Thomas, and T. E. Tyobeka, *J. Chem. Soc., Chem. Commun.* p. 650 (1981).

[132] C. H. Nash, N. De La Higuera, N. Neuss, and P. A. Lemke, *Dev. Ind. Microbiol.* **15,** 114 (1974).

[133] N. Neuss, R. D. Miller, C. A. Affolder, W. Nakatsukasa, W. Mabe, L. L. Huckstep, N. De La Higuera, J. L. Occolowitz, and J. H. Gilliam, *Helv. Chim. Acta* **63,** 1119 (1980).

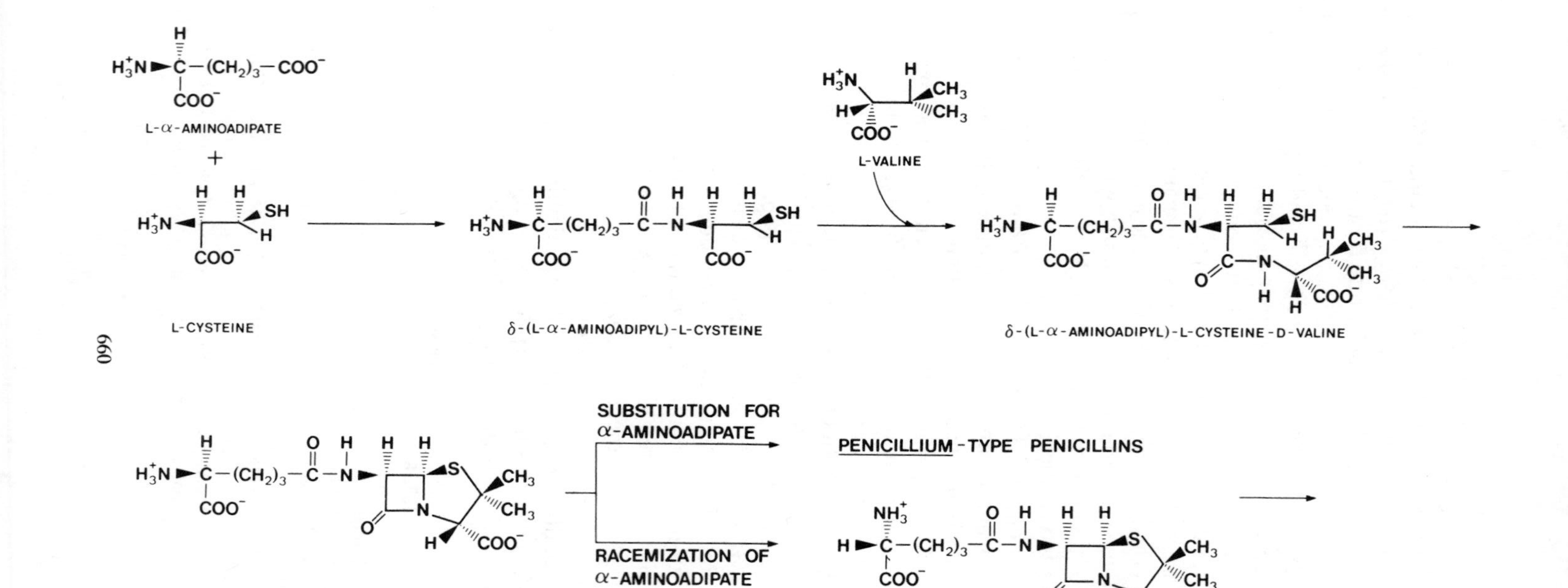
L-α-AMINOADIPATE
L-CYSTEINE
δ-(L-α-AMINOADIPYL)-L-CYSTEINE
L-VALINE
δ-(L-α-AMINOADIPYL)-L-CYSTEINE-D-VALINE
(L)
ISOPENICILLIN N
SUBSTITUTION FOR
α-AMINOADIPATE
RACEMIZATION OF
α-AMINOADIPATE
PENICILLIUM-TYPE PENICILLINS
(D)
PENICILLIN N

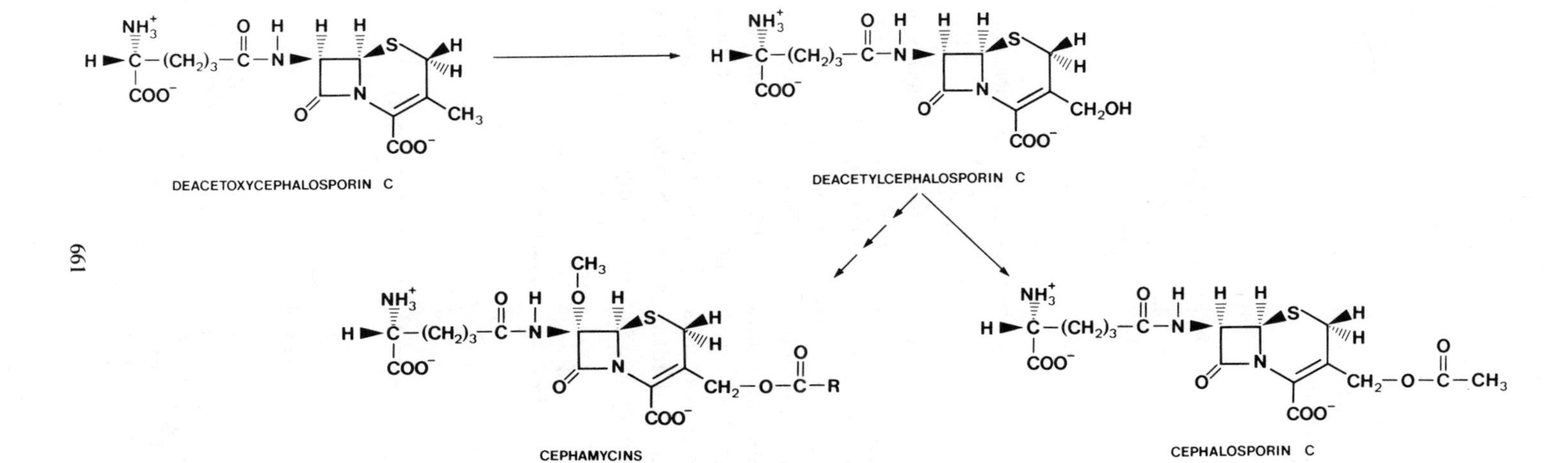

FIG. 5. Biosynthesis of the natural penicillins, cephalosporins, and cephamycins.

A β- lactam ring

A thiazolidine ring

The fused rings in the penicillins and cephalosporins

FIG. 6. The ring structures in β-lactam antibiotics.

cyclization may occur in two steps with the formation of the lactam ring (L-α-aminoadipyl-L-cyclocysteinyl-D-valine; a monocyclic azetidinone) followed by thiazolidine ring formation.[134] Unequivocal evidence supporting these reactions has not been established.[135,136]

Isopenicillin N is the precursor to the so-called *Penicillium*-type penicillins; for example, penicillin G or V (Fig. 7). Formation of the *Penicillium*-type penicillins necessitates replacement of the L-α-aminoadipate chain of isopenicillin N. Four enzymes have, thus far, been considered to participate in these conversions. They are (1) isopenicillin-*N*-amidolyase (6-aminopenicillanic acid forming) which hydrolyzes the peptide bond between α-aminoadipate and the lactam ring of isopenicillin N to produce

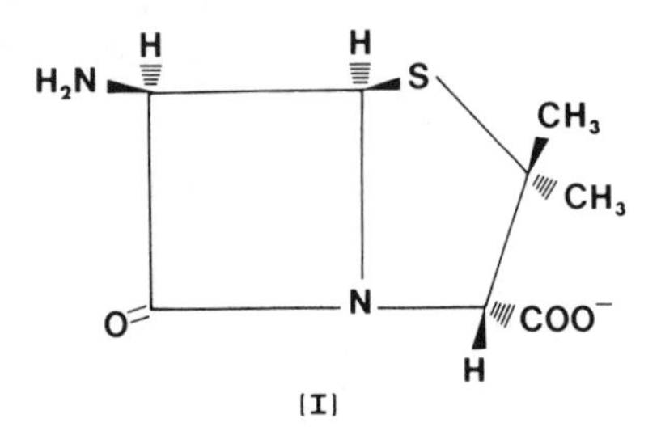

6 - Aminopenicillanic acid

[134] B. Meessechaert, P. Adriaens, and H. Eyssen, *J. Antibiot.* **31,** 1110 (1978).

[135] J. E. Baldwin and M. Jung, *J. Chem. Soc., Chem. Commun.* p. 609 (1978).

[136] J. E. Baldwin, M. Jung, P. Singh, T. Wan, S. Haber, S. Herchen, J. Kitchen, A. L. Demain, N. A. Hunt, M. Kohsaka, T. Konomi, and M. Yoshida, *Philos. Trans. R. Soc. London, Ser. B* **289,** 169 (1980).

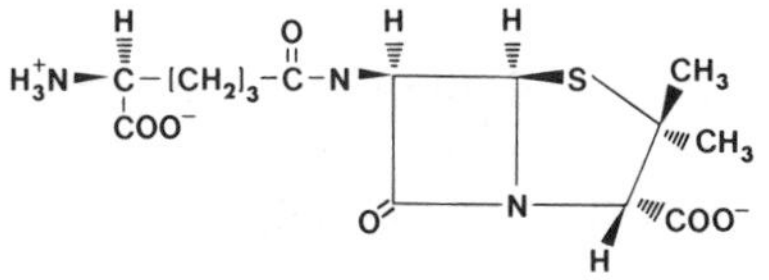

Isopenicillin N

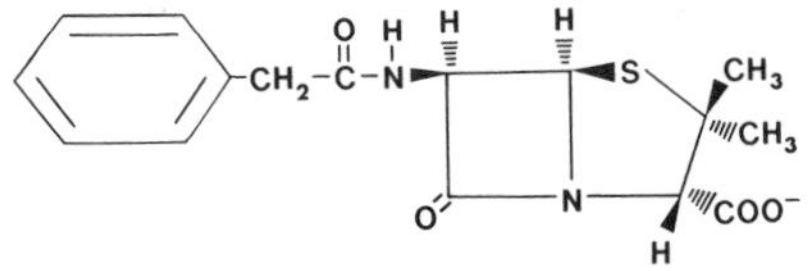

Penicillin G

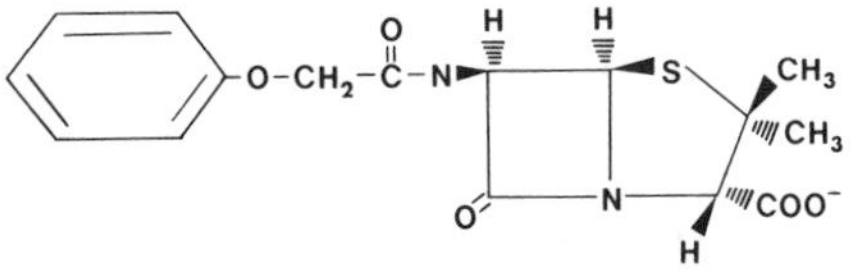

Penicillin V

FIG. 7. The structures of isopenicillin N and penicillin G and V.

6-aminopenicillanic acid **(I)**; (2) penicillin acylase which catalyzes a reversible acylation of 6-aminopenicillanic acid with monosubstituted acetic acids to form *Penicillium*-type penicillins; (3) acyl CoA : 6-aminopenicillanic acid acyl transferase which can form *Penicillium*-type penicillins by directly acylating 6-aminopenicillanic acid from a phenylacetyl-CoA or by exchanging the L-α-aminoadipate chain of isopenicillin N for the acyl group from a monsubstituted acetyl-CoA; (4) monosubstituted acetic acid : CoA ligase which forms acyl-CoA compounds at the expense of an ATP. This enzyme provides monosubstituted acetyl-CoA derivatives for the activity of the acyl CoA : 6-aminopenicillanic acid acyl transferase. The ligase is deemed important in the scheme of penicillin synthesis because of its appearance in extracts of *P. chrysogenum* just prior to and during rapid penicillin synthesis.[137] A model for the participation of these enzymes in the conversion of isopenicillin N to *Penicillium*-type penicillins can be found in the recent review by Queener and Neuss.[8]

Conversion of isopenicillin N to penicillin N, the precursor to the cephalosporins, requires only racemization of the L-α-aminoadipate chain of isopenicillin N to the D-configuration. This reaction has not been directly demonstrated in a cell-free system, but indirect evidence from preliminary experiments[138,139] suggests that a racemase responsible for the conversion may exist in *C. acremonium*. The racemase activity is easily destroyed during extraction from the cell and has a reaction equilibrium strongly favoring isopenicillin N.

[137] R. Brunner, M. Rohr, and M. Zinnrt, *Hoppe-Seyler's Z. Physiol. Chem.* **349,** 95 (1968).

[138] G. S. Jayatilake, J. A. Huddleston, and E. P. Abraham, *Biochem. J.* **194,** 645 (1981).

[139] A. L. Demain, "Conference on β-Lactam Antibiotic Biosynthesis." Madison, Wisconsin, 1978.

The steps in the formation of cephalosporin C from penicillin N have been deduced mainly through the use of blocked mutants.[120] Patterns of accumulation of intermediates are consistent with a biosynthetic sequential conversion of penicillin N to cephalosporin C through deacetoxycephalosporin C and deacetylcephalosporin C. Conversion of penicillin N to deactoxycephalosporin C is an oxidative expansion from a five-membered to a six-membered ring. The mechanism of this expansion is not known. The cephamycin antibiotics are formed by enzymatic transformations of deacetylcephalosporin C.[8]

Enzymes

The enzyme reported for which α-aminoadipate is a substrate or product of catalysis include α-ketoadipate : glutamate aminotransferase from bakers' yeast and *Neurospora*,[51] α-aminoadipate reductase from bakers' yeast,[51] α-aminoadipate-δ-semialdehyde : NAD oxidoreductase.[68,69] and α-aminoadipate transaminase from *P. putida*,[70] and α-aminoadipate : α-ketoglutarate aminotransferase from rat liver[106] and kidney.[106,107,140]

[140] See this volume [83].

[83] Kynurenine Aminotransferase from Kidney Supernatant

By RICHARD A. HARTLINE

$$^{-}OOC{-}(CH_2)_3{-}\overset{\overset{+}{N}H_3}{CH}{-}COO^{-} \xrightleftharpoons[\text{}]{\alpha\text{-Ketoglutarate} \quad \text{Glutamate}} {}^{-}OOC{-}(CH_2)_3{-}\overset{\overset{O}{\|}}{C}{-}COO^{-}$$

α-Aminoadipate α-Ketoadipate

The L-kynurenine : α-ketoglutarate aminotransferase (cyclizing) (EC 2.6.1.7) activity of rat kidney supernatant, an enzyme in tryptophan catabolism, copurifies with the activity of α-aminoadipate : α-ketoglutarate aminotransferase.[1] Evidence that the two activities are properties of a single protein are (1) copurification of the two activities with the ratio of their specific activities remaining constant, (2) similar chromatographic

[1] M. C. Tobes and M. Mason, *J. Biol. Chem.* **252**, 4591 (1977).

Copyright © 1985 by Academic Press, Inc.
All rights of reproduction in any form reserved.

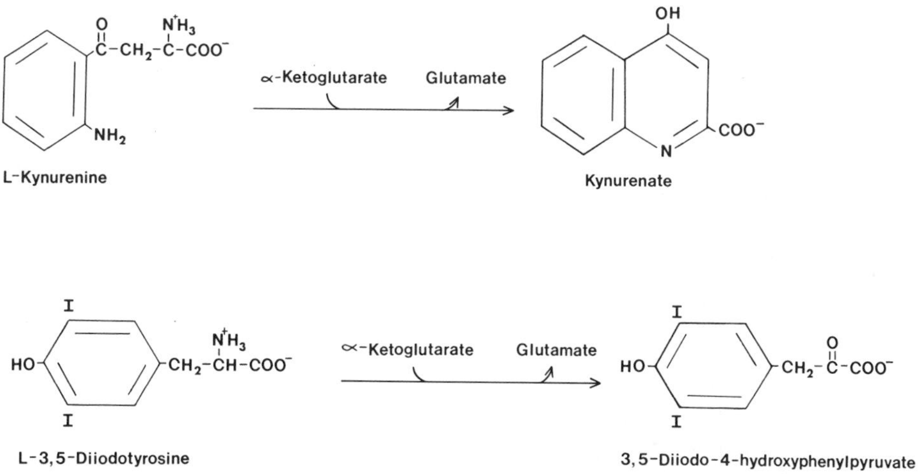

FIG. 1. Reactions catalyzed by the kynurenine and halogenated tyrosine aminotransferase activities of mammalian α-aminoadipate (kynurenine) : α-ketoglutarate aminotransferase.

behavior, (3) similarity in their dependence on added pyridoxal-5′-phosphate for activity, and (4) similar patterns of heat inactivation.

It has been suggested that the presence of aminoadipate (kynurenine) aminotransferase in kidney may represent an enzyme associated with lysine catabolism rather than with tryptophan degradation.[1,2] Consistent with this concept is the recent demonstration of lysine : α-ketoglutarate reductase and saccharopine dehydrogenase activities in human kidney.[3] A role in lysine catabolism could justify the presence of the relatively high level of kynurenine aminotransferase activity in kidney[4,5] which lacks a complete pathway for tryptophan catabolism. α-Aminoadipate (kynurenine) aminotransferase also catalyzes aminotransfer of 3,5-diiodotyrosine (halogenated tyrosine aminotransferase) suggesting a possible role in thyroid hormone metabolism.[6] The reactions catalyzed by the kynurenine and halogenated tyrosine aminotransferase activities are illustrated in Fig. 1.

α-Aminoadipate (kynurenine) : α-ketoglutarate aminotransferase activity is also present in rat kidney mitochondria[6] and the supernatant[7] and

[2] B. D. Manning and M. Mason, *Life Sci.* **17,** 225 (1975).

[3] J. Hutzler and J. Dancis, *Biochim. Biophys. Acta* **377,** 42 (1975).

[4] M. Mason, *J. Biol. Chem.* **211,** 839 (1954).

[5] M. Mason and C. P. Berg, *J. Biol. Chem.* **188,** 783 (1951).

[6] M. C. Tobes and M. Mason, *Life Sci.* **22,** 793 (1978).

[7] M. Tobes and M. Mason, *Biochem. Biophys. Res. Commun.* **62,** 390 (1975).

mitochondrial[8] fractions of rat liver. The kidney[6] and liver[8] mitochondrial enzymes have been partially purified.

L-α-Aminoadipate Aminotransferase Assay

$$\text{L-}\alpha\text{-Aminoadipate} + \alpha\text{-Ketoglutarate} \underset{b}{\overset{a}{\rightleftharpoons}} \alpha\text{-Ketoadipate} + \text{Glutamate}$$

Activity of this enzyme can be measured from direction a or b by the rate of disappearance (direction a) or appearance (direction b) of α-ketoglutarate with glutamate dehydrogenase.[8]

Reagents

Potassium phosphate, pH 7.5
Pyridoxal-5′-phosphate
α-Ketoglutarate, pH 7.5
Potassium L-α-aminoadipate, 0.1 *M*, pH 7.5
HCl, 1 *M*
KOH, 1 *M*
Ammonium chloride
NADH
Glutamate dehydrogenase

Procedure in the "a" Direction. The assay mixture contains 167 m*M* potassium phosphate buffer (pH 7.5), 0.26 m*M* pyridoxal-5′-phosphate, 1.67 m*M* α-ketoglutarate, and enzyme in a volume of 0.3 ml. After a 5 min incubation at 37°, the reaction is initiated by addition of 0.2 ml of 0.1 *M* potassium L-α-aminoadipate and incubated for 10 min at 37°. Incubation is terminated by addition of 0.1 ml 1 *M* HCl and the reaction mixture neutralized with 0.1 ml 1 *M* KOH. A 0.3-ml aliquot is analyzed for α-ketoglutarate. A unit of enzyme activity is defined as the amount of enzyme producing the disappearance of 1 μmol of α-ketoglutarate per min. The α-ketoglutarate that disappears is calculated by subtracting the α-ketoglutarate remaining from the α-ketoglutarate in a control containing no enzyme.

Procedure in the "b" Direction. Assay conditions are the same as for the "a" direction except that α-ketoglutarate and L-α-aminoadipate were replaced by α-ketoadipate and L-glutamate. With this assay the reaction is linear with respect to time and enzyme concentration until 0.1 μmol of α-ketoglutarate is produced. A unit of enzyme activity is defined as the amount of enzyme producing 1 μmol of α-ketoglutarate per min.

[8] Y. Nakatani, M. Fujioka, and K. Higashino, *Biochim. Biophys. Acta* **198,** 219 (1970).

Tobes and Mason[7] report the use of this assay with some modifications which include (1) potassium phosphate buffer at pH 7.0 rather than 7.5, (2) neutralization after addition of 0.1 ml 1 *M* HCl with 0.1 ml 1.1 *M* KOH to pH 7.5 rather than 1 *M* KOH to pH 7.0, and (3) centrifugation after neutralization at 5000 *g* for 5 min to remove precipitates. They also found that transferase activity in a mitochondrial suspension is increased by addition of $CaCl_2$ to 2 m*M*. The $CaCl_2$ has no effect on glutamate dehydrogenase activity. Addition of $CaCl_2$ does not increase transferase activity in the supernatant fraction of disrupted mitochondria.

α-Ketoglutarate Determination. α-Ketoglutarate is determined by the amount of NADH oxidized in the presence of NH_4^+ and glutamate dehydrogenase according to the following equation:

$$\alpha\text{-Ketoglutarate} + NH_4^+ \xrightarrow{NADH + H^+ \quad NAD^+} \text{Glutamate}$$

An incubation mixture contains 100 m*M* potassium phosphate buffer (pH 7.5), 50 m*M* NH_4Cl, 0.1 m*M* NADH, and 0.3 ml of neutralized aminotransferase reaction mixture in a final volume of 3.0 ml. The reaction is initiated by addition of glutamate dehydrogenase and the oxidation of NADH followed spectrophotometrically at 340 nm. The molar absorptivity of NADH at 340 nm is 6220. A unit of enzyme activity is defined as the amount of enzyme oxidizing 1 μmol of NADH per min.

Kynurenine Aminotransferase Assay

$$\text{L-Kynurenine} + \alpha\text{-Ketoglutarate} \longrightarrow \text{Kynurenate} + \text{Glutamate}$$

The activity of kynurenine aminotransferase is measured spectrophotometrically by following the appearance of kynurenate.[7]

Reagents

- Pyridoxal-5′-phosphate
- α-Ketoglutarate, pH 6.5
- Imidazole–HCl, pH 6.5
- L-Kynurenine sulfate, pH 6.5
- 1% Boric acid in 95% ethanol

Procedure. The incubation mixture contains 0.20 m*M* pyridoxal-5′-phosphate, 5.00 m*M* α-ketoglutarate, 3.26 m*M* L-kynurenine sulfate, 100 m*M* imidazole–HCl buffer (pH 6.5), and 0 to 0.1 ml of enzyme solution in a final volume of 0.60 ml. The mixture is incubated at room temperature for 10 min before the reaction is initiated by addition of L-kynurenine sulfate. After a 30 min incubation at 37° the reaction is terminated by addition of

10 ml of ethanolic boric acid solution. Precipitates are removed by centrifugation at 5000 *g* for 5 min. Kynurenate formed is determined spectrophotometrically at 333 nm with correction for any kynurenine present by its absorption at 365 nm. Kynurenate absorbance is calculated according to the equation[9]:

$$\text{Absorbance at 333 nm due to kynurenate} = (\text{absorbance at 333 nm}) - (0.39 \times \text{absorbance at 365 nm})$$

The molar absorptivity of kynurenate at 333 nm is 8850.[10] A unit of enzyme activity is defined as the amount of enzyme producing 1 μmol of kynurenate per min.

3,5-Diiodotyrosine Aminotransferase Assay

$$\text{L-3,5-Diiodotyrosine} + \alpha\text{-Ketoglutarate} \longrightarrow \text{L-3,5-Diiodo-4-hydroxyphenylpyruvate} + \text{Glutamate}$$

This transferase activity is determined by following the formation of the 3,5-diiodo-4-hydroxyphenylpyruvate product measured spectrophotometrically as the enol-borate complex.[11]

Reagents

Pyridoxal-5′-phosphate
L-3,5-Diiodotyrosine, pH 7.0 or 6.5
Sodium phosphate, pH 7.0 or 6.5
α-Ketoglutarate, pH 7.0 or 6.5
H_2SO_4, 6 *N*
Arsenate, 2 *M*, pH 6.5
Arsenate, 2 *M* which is 1 *M* in borate, pH 6.5

Procedure. The reaction mixture contains the 25 μ*M* pyridoxal-5′-phosphate, 2.5 m*M* L-3,5-diiodotyrosine, and 31.3 m*M* sodium phosphate buffer which is 0.31 m*M* in EDTA at pH 7.0[11] or 6.5,[6] and the enzyme in a final volume of 1.6 ml. After a 5 min incubation period at 37° the reaction is initiated by addition of α-ketoglutarate to 16 m*M*. The reaction is stopped, after incubation for 10 to 210 min at 37°, by addition of 0.1 ml of 6 *N* H_2SO_4. A 0.7-ml aliquot of the acidified incubation mixture is added to 2 ml of the arsenate solution (control). A second 0.7-ml aliquot is added to 2 ml of the arsenate-borate solution (enol-borate sample). Absorbance of the enol-borate sample is read at 330 nm. The molar absorptivity of the enol-borate complex of 3,5-diiodo-4-hydroxyphenylpyruvate at 330 nm is

[9] M. Tobes, personal communication.
[10] W. E. Knox, *Biochem. J.* **53,** 379 (1953).
[11] M. Nakano, *J. Biol. Chem.* **242,** 73 (1967).

8110.[11] A unit of activity is defined as the amount of enzyme required for the formation of 1 μmol of 3,5-diiodo-4-hydroxyphenylpyruvate per min.

Purification[1]

All procedures are carried out at 0 to 4°. Solutions are prepared with redistilled and deionized water. Buffers are all potassium phosphate containing 10 m*M* 2-mercaptoethanol.

Male adult albino rats (250 to 350 g) are sacrificed by decapitation and the kidneys immediately placed on ice. The kidneys are weighed, placed in a cold homogenizer in 4 ml of 0.25 *M* sucrose per g of tissue, homogenized, and centrifuged at 30,000 *g* for 20 min.[12,13] The supernatant is decanted and frozen immediately. The frozen supernatant is usually used within 3 weeks but loses little activity over a 3-month period at −15°. Kidneys from 30 rats yield approximately 200 ml of supernatant. Copurification of α-aminoadipate and kynurenine aminotransferase activities is summarized in the table.

Step 1. Acid Precipitation. Approximately 200 ml of frozen supernatant is thawed. The solution is made 10 μ*M* in pyridoxal-5′-phosphate, titrated to pH 5.25 with 1.0 *M* acetic acid, and allowed to stand 5 min with occasional stirring. After centrifugation at 30,000 *g* for 10 min the supernatant is titrated to pH 6.5 with 1.0 *M* NaOH.

Step 2. Ammonium Sulfate and Heat Fractionation. Supernatant from the acid precipitation step is made 35% saturated with solid ammonium sulfate[14] and, while maintaining the pH at 6.5 with addition of 1.0 *M* NaOH, the suspension is stirred for 1 hr. The solution is made 0.05 m*M* in pyridoxal-5′-phosphate, taken to 55° for 5 min with constant stirring, brought back to 4° in an ice bath, and centrifuged at 30,000 *g* for 20 min. The supernatant is taken to 65% saturation in ammonium sulfate, stirred for 1 to 2 hr while maintaining the pH at 6.5 with 1.0 *M* NaOH, and the resulting suspension centrifuged at 30,000 *g* for 20 min. The supernatant is discarded and the precipitate dissolved in a minimal volume (20 to 24 ml) of 8 m*M* buffer at pH 7.1 and dialyzed against 2.1 liters of buffer for 19 hr with one change of buffer.

Step 3. DEAE-Cellulose Chromatography. A 3.4 × 9.0 cm column is equilibrated with 8 m*M* buffer, pH 7.1, and the dialyzed solution applied to the column at approximately 45 ml/hr. Collecting 5 ml fractions, the enzyme is eluted off with the same buffer at the same flow rate. Aminotransferase activities which appear between fractions 40 to 60 are combined.

[12] M. Mason, J. Ford, and H. L. C. Wu, *Ann. N.Y. Acad. Sci.* **166**(1), 170 (1969).

[13] M. Mason, *Biochem. Biophys. Res. Commun.* **60,** 64 (1974).

[14] M. Dixon and E. C. Webb, "Enzymes," 2nd ed., p. 40. Academic Press, New York, 1964.

COPURIFICATION OF α-AMINOADIPATE AMINOTRANSFERASE AND KYNURENINE AMINOTRANSFERASE AND REMOVAL OF ASPARTATE AMINOTRANSFERASE FROM RAT KIDNEY SUPERNATANT[a]

Fraction and step	Total protein (mg)	Kynurenine aminotransferase				α-Aminoadipate aminotransferase				Aspartate aminotransferase		Specific activity ratio α-aminoadipate aminotransferase/Kynurenine aminotransferase
		Total units[b]	Recovery (overall)	Specific activity[c]	Purification (overall)	Total units[b]	Recovery (overall)	Specific activity[c]	Purification (overall)	Total units[b]	Recovery (overall)	
Supernatant	4150	39.2	100.0	0.009	1.0	378.0	100.0	0.092	1.0	2600.0	100.0	10.2
1. Acid precipitation	2220	32.3	82.0	0.014	1.6	316.0	83.7	0.142	1.5	2280.0	88.0	10.1
2. Ammonium sulfate-heat	670	29.5	75.2	0.044	4.9	277.0	73.3	0.414	4.5	770.0	29.7	9.4
3. DEAE-cellulose	51.3	21.1	53.7	0.410	45.6	207.0	54.6	4.02	43.7	604.0	23.3	9.8
4. Hydroxylapatite I	9.6	15.2	38.8	1.58	175.0	149.0	39.3	15.4	168.0	70.6	2.7	9.8
5. Hydroxylapatite II	3.1	8.94	22.8	2.90	323.0	89.4	23.7	29.0	315.0	7.2	0.3	10.0
6. Hydroxylapatite III	1.5	4.07	10.4	2.65	294.0	40.0	10.6	26.1	283.0	1.0	0.0	9.8

[a] Data from Tobes and Mason.[1]

[b] Units are defined as μmol/min at 37°.

[c] Specific activity is defined as units/mg protein.

Step 4. Hydroxylapatite I Chromatography. A 2.2 × 18.0 cm column is equilibrated with 8 m*M* buffer at pH 7.0 and the pooled DEAE eluate applied at 25 ml/hr. The column is eluted with 1.5 column volumes of 8 m*M* buffer and then eluted, at the same flow rate, with a linear gradient of 200 ml of 8 m*M* buffer, pH 7.0, and 200 ml of 210 m*M* buffer, pH 7.0. Fractions of 4 ml are collected and the active ones, from 117 to 128, are pooled and concentrated to 10 ml by means of a 50 ml Diaflo cell with a PM-10 membrane at an N_2 pressure of 10 to 15 psi. The concentrated enzyme is dialyzed overnight with one change of buffer against 1 liter of 70 m*M* buffer at pH 7.0.

Step 5. Hydroxylapatite II Chromatography. A 1.0 × 30.0 cm column is equilibrated with 70 m*M* buffer, pH 7.0, and the dialyzed sample applied at the rate of 7 ml/hr. Elution of the column is initiated with 1.5 column volumes of 70 m*M* buffer, pH 7.0, followed by a linear gradient elution of 94 ml of 70 m*M* buffer, pH 7.0, and 94 ml of 140 m*M* buffer at pH 7.0 at 8–9 ml/hr. Fractions of 2 ml are collected. The enzyme activity elutes off from fractions 65 to 75. With one change of buffer, the pooled active fractions are dialyzed overnight against 1 liter of 70 m*M* buffer, pH 7.0, containing 0.20 m*M* pyridoxal-5′-phosphate.

Step 6. Hydroxylapatite III Chromatography. A 1.0 × 33.0 cm column is equilibrated with 70 m*M* buffer, pH 7.0, and the dialyzed sample applied at 5 ml/hr. The initial elution is with 1.5 column volumes of 70 m*M* buffer at pH 7.0 followed by elution of enzyme activity with a linear gradient of 100 ml of 70 m*M* buffer, pH 7.0, and 100 ml of 140 m*M* buffer, pH 7.0, at a rate of 4 to 5 ml/hr. Collecting 2 ml fractions, the enzyme activity elutes from fractions 75 to 85. The pooled active fractions are made 0.20 m*M* in pyridoxal-5′-phosphate and concentrated to 2 ml by ultrafiltration in a 10 ml Diaflo cell with a PM-10 membrane at a N_2 pressure of 10 psi. After concentration, the enzyme is stored at 4° in 70 m*M* buffer, pH 7.0, that is 0.20 m*M* pyridoxal-5′-phosphate and 10 m*M* in 2-mercaptoethanol. The protein concentration of the stored enzyme is approximately 1.0 mg/ml.

After all chromatography steps the α-aminoadipate and kynurenine aminotransferase activities are dependent on added pyridoxal-5′-phosphate. Therefore, the enzyme purified as the apoenzyme.

The overall purification with respect to the crude supernatant is 283-fold for α-aminoadipate aminotransferase activity and 294-fold for kynurenine aminotransferase activity. Purification relative to the crude homogenate is 600-fold. The enzyme is nearly homogeneous according to analytical disc gel electrophoresis at pH 8.9 and 7.5, isoelectric focusing on polyacrylamide gels, and sodium dodecyl sulfate–polyacrylamide gel electrophoresis.

Stability. The apoenzyme is most stable under the purification conditions in the pH range of 6.5 to 7.5. Maximal activity is dependent on the presence of a reducing agent such as 2-mercaptoethanol. The apoenzyme is quite stable in phosphate buffer, but when dialyzed against Tris–HCl or imidazole–HCl, the enzyme activities are rapidly and irreversibly lost. In the Tris or imidazole buffers the apoenzyme is stable only in the presence of saturating levels of pyridoxal-5′-phosphate.

Physical Characteristics

The average molecular weight of the enzyme is 85,000. The average sedimentation coefficient is 5.8. No differences are observed for the sedimentation coefficient for sedimentation of apo- versus holoenzyme and crude versus purified enzyme. The enzyme is composed of two similar sized subunits (45,000 ± 850) that are not linked by intrachain disulfide bonds. The isoelectric point of the enzyme is 6.56 ± 0.06.

Specificity

The Purified Enzyme from Rat Kidney. α-Aminoadipate is more reactive with the enzyme than kynurenine (see the table). While no data are presented,[1] the specificity with other amino acids, using α-ketoglutarate as the amino-acceptor, is described to be similar to those observed with a partially purified preparation from rat liver mitochondria.[8] The exception to this similarity is activity with aspartate (see the table). While the available data are consistent with activity with aspartate reflecting a slight contamination by aspartate aminotransferase, the possibility remains that this activity is intrinsic. Additional specificities reported are strong activity with 3,5-diiodotyrosine[6] and weak activities with tryptophan, phenylalanine, and tyrosine.[15]

The Partially Purified Enzyme from Rat Liver. The specificity of a partially purified preparation of α-aminoadipate : α-ketoglutarate aminotransferase from rat liver mitochondria has been reported.[8] On a relative activity scale in which L-α-aminoadipate is assigned an arbitrary value of 100 and in which α-ketoglutarate is used as the amino-acceptor, DL-α-aminopimelate showed 14 and L-norleucine 15. Amino acids that are less than 2 include L-aspartate, L-alanine, L-α-aminobutyrate, glycine, L-valine, L-leucine, L-isoleucine, L-norvaline, L-ε-aminocaproate, L-threonine, L-serine, L-homoserine, L-cysteine, L-methionine, L-lysine, L-arginine, L-citrilline, L-ornithine, L-histidine, L-tyrosine, L-tryptophan, L-phenylalanine, L-proline, and L-glutamate.

[15] M. C. Tobes, Ph.D. Thesis, University of Michigan, Ann Arbor (1976).

Author Index

Numbers in parentheses are footnote reference numbers and indicate that an author's work is referred to although the name is not cited in the text.

A

C

D

E

G

H

I

J

L

M

N

O

P

T

U

V

W

Y

Z

Subject Index

A

D

E

F

L

R

S

T

V

X

Y

3 5282 00104 0792